page	symbol	definition
119	$\sqrt[q]{a^p} = a^{p/q}$	positive qth root of a^p
165	(a,b) or (a,b,c)	row vector
165	$\begin{pmatrix} a_1 & b_1 \\ a_2 & b_2 \end{pmatrix}$	matrix
165	$\begin{pmatrix} a_1 & b_1 & c_1 \\ a_2 & b_2 & c_2 \end{pmatrix}$	matrix
165	$\overrightarrow{PQ}$	vector
166	$\begin{pmatrix} a \\ b \end{pmatrix}$ or $\begin{pmatrix} a \\ b \\ c \end{pmatrix}$	column vector
166	$\mathbf{a}$	vector
168	$\mathbf{i}, \mathbf{j}, \mathbf{k}$	unit vectors along the axes
168	$(a_1,b_1,c_1) \cdot (a_2,b_2,c_2)$	inner (scalar) product
183	$\det A = \begin{vmatrix} a_1 & b_1 \\ a_2 & b_2 \end{vmatrix}$	determinant of $\begin{pmatrix} a_1 & b_1 \\ a_2 & b_2 \end{pmatrix}$
191	$(a_1,b_1,c_1) \wedge (a_2,b_2,c_2)$	outer (vector) product
242	f	function
242	$f(x)$	value of a function at x
248	$g \circ f$	composite of g and f
271	$P(x)$	polynomial
273	$P(x)/Q(x)$	rational function
298	$\approx$	approximately equal to
300	$e^x = \exp x$	exponential function
302	$\log x$	logarithmic function
324	(r,θ)	polar coordinates
361	$\theta^{(r)}$	radian measure of an angle
391	$\operatorname{Sin} x$	restricted sine function
392	$\operatorname{arc\,Sin} x = \operatorname{Sin}^{-1} x$	inverse sine function
462	$\lim_{x \to a} f(x)$	limit of the value of the function f as x approaches a
462	$\lim_{x \to \infty} f(x)$	limit of the value of the function f as x increases without bound
481	Σ	summation
482	$A_a^b = \lim_{n \to \infty} \sum_{i=1}^{n} f(x_i)\,\Delta x$	area under $y = f(x)$ from $x = a$ to $x = b$
482	$\int_a^b f(x)\,dx$	definite integral
501	Δx	change in x
501	Δf	change in f
504	$D_x f = f'(x)$	derivative of f
504	dy/dx	derivative
534	$\sinh \theta$	hyperbolic sine of θ

FUNDAMENTALS OF FRESHMAN MATHEMATICS

read 1.2 + 1.3
1.2 # 1,4,5,22,30
1.3 # 2,3,7,8

A) is there 1-1 corr. between {A, X, K} + {B, Q}

B) How many elements does {2,4,6,8,10} contain

C) How many elements does the set {2,4,6,8,10...}

D) What does 3 ∈ {3,6} mean in words

Read	2.5	2.6	1.1

problems 2.5
1,2,6,8,10,11,12
problems 2.6
1,2,4,12,14,25,26,32,31

linear programming
quadratic inequalities
TEST
6.5-6.12
all of 8

FUNDAMENTALS OF FRESHMAN MATHEMATICS

THIRD EDITION

CARL B. ALLENDOERFER
Professor of Mathematics
University of Washington

CLETUS O. OAKLEY
Professor and Department Head, Emeritus
Department of Mathematics
Haverford College

McGRAW-HILL BOOK COMPANY
New York / St. Louis / San Francisco / Düsseldorf / Johannesburg
Kuala Lumpur / London / Mexico / Montreal / New Delhi / Panama
Rio de Janeiro / Singapore / Sydney / Toronto

FUNDAMENTALS OF FRESHMAN MATHEMATICS

Library of Congress Catalog Card Number 74-37087
07-001366-7

1234567890DODO798765432

This book was set in Times Roman by York Graphic Services, Inc.,
and printed and bound by R. R. Donnelley & Sons Company.
The designer was J. Paul Kirouac;
new drawings were done by John Cordes, J. & R. Technical Services, Inc.
The editors were Howard S. Aksen, Lee W. Peterson, and Laura Warner.
Peter D. Guilmette supervised production.

CONTENTS

PREFACE xi

TO THE STUDENT xiii

CHAPTER 1 MATHEMATICAL METHOD 1

1.1 › Introduction 1
1.2 › Sets 3
1.3 › Open Sentences 8
1.4 › Set Operations 12
1.5 › Conditionals 16
1.6 › Derived Conditionals 20
1.7 › Alternative Expressions for Conditionals 23
1.8 › Direct Proof 27
1.9 › Indirect Proof 29
1.10 › Other Methods of Proof 32
1.11 › Methods of Disproof 34
1.12 › Mathematical Models 35

CHAPTER 2 THE NUMBER SYSTEM 37

2.1 › Introduction 37
2.2 › Binary Operations 39
2.3 › Properties of the Real Number System 44
2.4 › Theorems about Real Numbers 50
2.5 › The Natural Numbers 54
2.6 › The Integers—Mathematical Induction 56
2.7 › Rational Numbers 63
2.8 › Decimal Expansions 65
2.9 › Some Irrational Numbers 66
2.10 › Geometric Representation of Real Numbers 69
2.11 › The Use of Real Numbers in the Plane 70
2.12 › Lengths of Segments; Units on the Axes 72
2.13 › Complex Numbers 75
2.14 › Graphical Representation of Complex Numbers 81
2.15 › Solutions of Other Algebraic Equations 82
2.16 › Classification of Numbers 83

CHAPTER 3 POLYNOMIALS 84

3.1 › Algebraic Expressions 84
3.2 › Addition of Polynomials 85
3.3 › Multiplication of Polynomials 86
3.4 › Binomial Theorem 88

3.5 › Division of Polynomials 93
3.6 › Factoring 95

CHAPTER 4 ALGEBRAIC FRACTIONS 101

4.1 › Introduction 101
4.2 › Simplification of Fractions 101
4.3 › Addition 105
4.4 › Multiplication and Division 108
4.5 › Compound Fractions 111

CHAPTER 5 EXPONENTS AND RADICALS 113

5.1 › Positive Integral Exponents 113
5.2 › Zero and Negative Exponents 116
5.3 › Fractional Exponents 119
5.4 › Special Problems Concerning Square Roots 121
5.5 › Special Problems Concerning Odd Roots 123
5.6 › Unanswered Questions 123
5.7 › Rationalizing Denominators 125

CHAPTER 6 EQUATIONS 128

6.1 › Solutions of Equations 128
6.2 › Method of Solution 129
6.3 › Linear Equations in One Variable 130
6.4 › Quadratic Equations in One Variable 130
6.5 › Equations in Two Variables 134
6.6 › Equations Containing Fractions 139
6.7 › Equations Containing Radicals 140
6.8 › Simultaneous Linear Equations 143
6.9 › Simultaneous Linear Equations (Continued) 147
6.10 › Simultaneous Linear Equations in Three Unknowns 149
6.11 › Simultaneous Linear and Quadratic Equations 152
6.12 › Word Problems 157
6.13 › Transformation of Coordinates 160

CHAPTER 7 VECTORS AND MATRICES 165

7.1 › Introduction 165
7.2 › Vectors 165
7.3 › Products of Vectors 168
7.4 › Matrices 170
7.5 › Products of Matrices 172
7.6 › Inverse of a Square Matrix 180

7.7 › Determinants 182
7.8 › Applications of Matrices to Simultaneous Equations 188

CHAPTER 8 INEQUALITIES 195

8.1 › Fundamental Properties 195
8.2 › Theorems about Inequalities 196
8.3 › Linear Inequalities 199
8.4 › Quadratic Inequalities 205
8.5 › The Graph of a Linear Inequality 210
8.6 › Simultaneous Linear Inequalities 213
8.7 › The Graph of a Quadratic Inequality 219
8.8 › Applications 220
8.9 › Linear Programming 222

CHAPTER 9 FUNCTIONS AND RELATIONS 232

9.1 › Introduction 232
9.2 › Functions 241
9.3 › Notations 242
9.4 › Variables 244
9.5 › Algebra of Functions 246
9.6 › Graphs 250
9.7 › Graphs (Continued) 259
9.8 › Inverse Functions 260
9.9 › Functions Derived from Equations 267

CHAPTER 10 ALGEBRAIC FUNCTIONS 271

10.1 › Introduction 271
10.2 › Polynomial Functions 271
10.3 › Rational Functions 272
10.4 › Explicit Algebraic Functions 274
10.5 › Graphs and Continuity 276
10.6 › Properties of Polynomials 281
10.7 › Synthetic Division 285
10.8 › Roots of Polynomial Equations 289
10.9 › Rational Roots of Rational Polynomial Equations 291
10.10 › Real Roots of Real Polynomial Equations 293

CHAPTER 11 EXPONENTIAL AND LOGARITHMIC FUNCTIONS 297

11.1 › Exponential Functions 297
11.2 › The Number *e* 300

11.3 » Logarithmic Functions 302
11.4 » Graphs 305
11.5 » Applications 307
11.6 » The Logarithmic Scale 310

CHAPTER 12 TRIGONOMETRIC FUNCTIONS OF ANGLES 317

12.1 » Introduction 317
12.2 » Distance in the Plane 318
12.3 » Angles 320
12.4 » Polar Coordinates 324
12.5 » Sine and Cosine of a Directed Angle 325
12.6 » Sine and Cosine of Special Angles 327
12.7 » Other Trigonometric Functions 330
12.8 » Some Important Identities 332
12.9 » Trigonometric Tables 335
12.10 » Right Triangles 338
12.11 » Vectors 341
12.12 » Law of Sines 347
12.13 » Law of Cosines 351
12.14 » Law of Tangents 355

CHAPTER 13 TRIGONOMETRIC FUNCTIONS OF REAL NUMBERS 356

13.1 » Arc Length 356
13.2 » New Definitions of the Trigonometric Functions 359
13.3 » Computations 362
13.4 » Variation and Graphs of the Functions 364
13.5 » Amplitude, Period, Phase 370
13.6 » Addition Theorems 375
13.7 » Multiple- and Half-angle Formulas 380
13.8 » Identities 383
13.9 » Equations 387
13.10 » Inverse Trigonometric Functions 391
13.11 » Complex Numbers 395

CHAPTER 14 ANALYTIC GEOMETRY 402

14.1 » Introduction 402
14.2 » The Straight Line 403
14.3 » Other Forms of Equations of Lines 406
14.4 » General Equation of a Line 408
14.5 » Other Properties of Lines in the Plane 410

14.6 › Directed Lines and Vectors 414
14.7 › Applications to Plane Geometry 422
14.8 › Conic Sections 424
14.9 › Case I. The Circle 425
14.10 › Case II. The Parabola 427
14.11 › Case III. The Ellipse 430
14.12 › Case IV. The Hyperbola 434
14.13 › Applications 438
14.14 › Locus Problems 439
14.15 › Polar Coordinates 441
14.16 › Polar Coordinates (Continued) 445
14.17 › Parametric Equations 450

CHAPTER 15 INTUITIVE INTEGRATION 457

15.1 › Introduction 457
15.2 › Area of a Circle 458
15.3 › Some Limits 461
15.4 › Area under $y = x^2$ 469
15.5 › Area under $y = x^n$ 473
15.6 › Area under Graph of a Polynomial Function 477
15.7 › Area under $y = f(x)$ 480
15.8 › Integration 482
15.9 › Setting Up Problems; Applications 486

CHAPTER 16 INTUITIVE DIFFERENTIATION 492

16.1 › Introduction 492
16.2 › Notion of a Tangent 492
16.3 › Velocity and Acceleration 496
16.4 › Derivative 501
16.5 › Second Derivative 508
16.6 › The Chain Rule 509
16.7 › Maxima and Minima 512
16.8 › Related Rates 520
16.9 › Fundamental Theorem of Calculus 522
16.10 › Falling Bodies 528

CHAPTER 17 HYPERBOLIC FUNCTIONS 531

17.1 › Hyperbolic Functions 531
17.2 › Hyperbolic and Circular Trigonometric Functions 531
17.3 › Hyperbolic Trigonometry 534
17.4 › Euler's Formula 538

APPENDIX 541

Table I Values of e^x and e^{-x} 541
Table II Common Logarithms (Base 10) 542
Table III Natural Logarithms (Base e) 544
Table IV Trigonometric Functions of Real Numbers 547
Table V Trigonometric Functions of Angles 549
Table VI Some Important Constants 571
Table VII Greek Alphabet 571
Table VIII Squares, Cubes, Roots 572

ANSWERS TO ODD-NUMBERED PROBLEMS 573

INDEX 633

PREFACE

The purpose of this book is to give a modern treatment of those topics in mathematics which are needed to fill the gap between Intermediate Algebra and Analytic Geometry and Calculus. It is written primarily for college students whose high school mathematics has not prepared them to enter a regular course in Analytic Geometry and Calculus, but it has found use with accelerated students in the eleventh and twelfth grades of high school.

In writing this book, we have been influenced by the recommendations of many groups who have been working toward the reform of the mathematical curriculum, in particular the Committee on the Undergraduate Program in Mathematics of the Mathematical Association of America, and the Commission on Mathematics of the College Entrance Examination Board. The book, however, represents our own ideas for a modern course of instruction and is not written to follow any specific outline suggested by such official bodies.

The book is flexibly organized so that it can be used for a variety of courses, examples of which are given below. More detailed suggestions for its use are contained in the Teachers' Manual, published separately.

1 *Twelfth-grade Mathematics in High School.* As such it fully meets the syllabus of the College Entrance Examination Board in Elementary Functions.

2 *College Algebra.* Chapters 1 to 11 form a modern treatment of material usually covered in courses bearing this title.

3 *College Algebra and Trigonometry; or Elementary Functions.* Chapters 1 to 13 are suitable for a combined course in these subjects.

4 *Unified Freshman Course.* The whole book provides a year course in mathematics for students who enter college with some knowledge of Intermediate Algebra and who wish to proceed to Calculus in their sophomore year.

In preparing this third edition, we have benefited from our own teaching experience with the first two editions and from the many suggestions sent to us by users of the book. The chief modifications of the second edition are as follows:

1 Chapter 2 has been rewritten to emphasize the concept of a binary operation on a set. Since binary operations are automatically closed, the concept of closure does not appear.

2 The solution of equations and the proofs of trigonometric identities have been based upon the concept of equivalent open sentences so that a unified approach is followed.

3 The concept of the image of a function is introduced and is distinguished from that of the range. One-to-one functions and onto functions can then be discussed.
4 The problem sets have been rewritten and expanded.

This edition continues the following features which were well received in the second edition:

1 The more difficult problems have been marked with asterisks (*). Answers are given for odd-numbered problems. The even-numbered problems essentially duplicate the odd-numbered ones and can be used for review or alternative assignments. Answers to the even-numbered problems are given in the Teachers' Manual.
2 Color has been used to improve the clarity of the illustrations. Each figure is accompanied by a full legend so that the illustrations stand by themselves without reference to the text material.
3 Marginal notes are included to help the student locate essential ideas which might otherwise be missed.

The authors are grateful to James McEnerney, James Reill, and Justine Baker, who helped immeasurably with the problems and their solutions, and to Judy Perloe and Judy Rieben for their typing and preparation of the manuscript. Special appreciation is expressed to all those who have forwarded comments on the first two editions or have offered suggestions for the preparation of this new book.

"It is hoped that the book is relatively free of errors, but each author blames the other for those that may be discovered."

CARL B. ALLENDOERFER
CLETUS O. OAKLEY

TO THE STUDENT

As you approach the study of this book, many questions about it will surely occur to you. You may wonder how it differs from mathematics books you have previously studied, why we have chosen to include the topics which are given, what you are expected to gain from its study, or even what is in it for you. This foreword is written to answer questions of this type and to help you start your studies in the right frame of mind.

In the first place, we mean for you to read the book. Some students treat books of this kind as a collection of problems and never read the explanations; if you do this, you will miss many of our main objectives. We are trying to teach you the language of mathematics and to explain some of its most important ideas. Solving the problems is important, but understanding the ideas is far more essential.

Since mathematics is a technical subject, we shall need to introduce a new terminology, with unfamiliar words such as *set, union, intersection, distributive law, matrix, vector, function, derivative, integral,* and many others. Each of these involves an important idea and requires a definition. You must read these definitions and make sure that you understand them before you proceed. To help you with this, we frequently follow definitions or other important ideas by Exercise A, B, C, etc. You should answer these as you read; they are generally easy and will help to underscore subtle points which you might well have missed.

When you understand the language, you should then wish to use it for something which we hope will be meaningful to you. We shall begin by stating a general type of problem to be solved. Examples of such problems are: Can we solve a certain type of equation? If so, how do we do it? How can we tell what the graph of a certain equation will look like without plotting a large number of points? How do we use mathematics to express relationships between certain physical quantities? How do we define and measure the area bounded by a collection of curved lines? What do we mean by velocity and acceleration? Before proceeding to read the text, sit back and try to answer these questions yourself. If you can answer them without the help of the book, so much the better. Then read the book to see how well you did.

The book is organized into six main topics:

1 In Chaps. 1 and 2 we discuss the foundations of the subject, methods of reasoning, and the meaning of our number system.

2 In Chaps. 3 to 5 we present an algebra review. This may seem familiar to you, but it is more sophisticated than the usual beginning algebra book. We hope that after working through these chapters, you will have a deeper understanding of what you have been doing all these years.

3 Chapters 6 to 8 introduce you to various methods for solving equations and inequalities and to the ideas of vectors and matrices which arose from the problem of solving simultaneous linear equations.

4 In Chap. 9 we begin the study of relationships between two variables. Here we introduce the ideas of functions and relations which permeate the applications of mathematics to the physical and social sciences. You cannot manage to do your job, however, by just knowing about functions in general; you need to know a lot about the collections of "elementary functions" which play the most prominent role in college mathematics. These are the algebraic functions (Chap. 10), the exponential and logarithmic functions (Chap. 11), and the trigonometric functions (Chaps. 12 and 13). The list of elementary functions is completed by the hyperbolic functions. We cannot explain these until you have studied some calculus, so these are postponed until the end of the book (Chap. 17).

5 In Chap. 14 we approach the problem of the relationship of equations to their graphs. This subject is called "Analytic Geometry". You will find that it is much more powerful than the elementary geometry you studied earlier in your school career. Here we can deal with fancy curves and are no longer restricted to lines, triangles, parallelograms, and circles.

6 Finally (Chaps. 15 and 16) we introduce the basic ideas of calculus. This is a completely different subject from anything that you have studied before, for it involves the idea of a "limit". In your earlier studies you could solve any problem by a finite number of steps, and sometimes this portion of mathematics is called "Finite Mathematics". In calculus, however, we must take limits which require us to carry out an infinite number of steps. Of course, you will never have time to write down such an infinity of steps, and so we must show you the ways in which problems of this kind can be handled. The ideas here are sophisticated, and you will probably not understand them completely the first time through. We hope, however, that these two chapters will get you started so that you will be in a better position to understand these ideas when you study calculus over again in more detail later.

We hope that by studying this book you will acquire a mastery of the language and essential ideas of elementary mathematics. When you are finished with it, you should be prepared to go ahead with the more advanced ideas and techniques of calculus, differential equations, and modern algebra which are the keystones of our modern scientific and engineering developments.

FUNDAMENTALS OF FRESHMAN MATHEMATICS

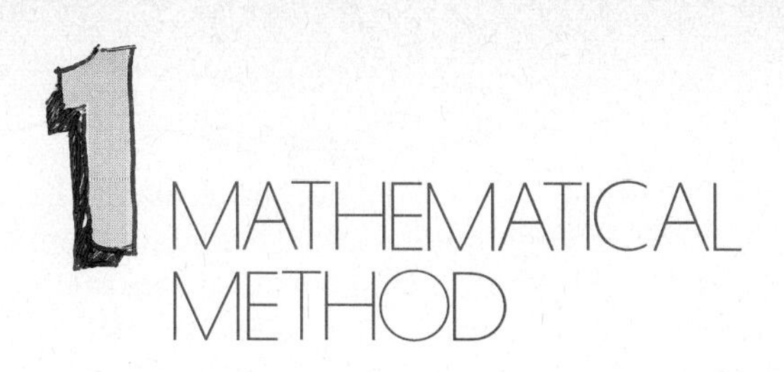

1 MATHEMATICAL METHOD

1.1 » INTRODUCTION

Most of you who are beginning this book are studying mathematics because of its importance in science, engineering, economics, psychology, biology, or the like. Although mathematics is an essential tool for applications to subjects like these, it is an entirely different kind of subject. Science is closely tied to the physical world, but mathematics is completely abstract. Many people shudder at the thought of anything abstract and consequently may have a mental block against mathematics. Actually, there is nothing so terrifying about the abstractness of mathematics once its true nature is understood. In order to assist this understanding, let us describe the essentials of a mathematical structure or theory.

As with any new subject, we begin mathematics by discussing the new, technical terms which we must introduce. Our intuition tells us that each of these should have a definition, but sooner or later we will find that our definitions are going in circles. To take a simple example, we may define:

Point: the common part of two intersecting *lines*
Line: the figure traced by a *point* which moves along the shortest path between two *points*

Here we have defined *point* in terms of *line* and *line* in terms of *point,* and so we have shed no real light on the nature of either *point* or *line.*

The situation is somewhat similar to that which we would encounter if we tried to learn a foreign language, say French, by using only an ordinary French dictionary—not a French-English dictionary. We look up a particular French word and find it described in more French words and we find ourselves no further ahead. Without a knowledge of a certain amount of French, a French dictionary is useless. In mathematics we have a similar difficulty.

Undefined words

The only way to avoid circular definitions in mathematics, or any other subject, is to take a small number of words as *undefined.* All other mathematical words will be defined in terms of these with the understanding that our definitions may also contain common English words ("is", "and", "the", etc.) which have no special mathematical meanings. It is not easy to decide which words should be left undefined and which should be defined in terms of the undefined words. Many choices can be made, and the final decision is largely based upon considerations of simplicity and elegance.

ILLUSTRATION 1 Let us suppose that *point, line,* and *between* are undefined. Then we may define:

Line segment: that portion of a line contained between two given points on a line.

The words in this definition other than *point, line,* and *between* are without special meanings and thus may be used freely.

Our use of undefined words is the first phase of our abstraction of mathematics from physical reality. The penciled line on our paper and the chalk line on our blackboard are physical realities, but *line,* the undefined mathematical concept, is something quite apart from them. In geometry we make statements about a *line* (which we shall call *axioms*) which correspond to observed properties of our physical lines, but if you insist on asking: "What is a *line?*" we must give you the somewhat disturbing answer: "We don't know; it isn't defined."

Once we have built up our vocabulary from undefined words and other words defined in terms of them, we are ready to make statements about these new terms. These statements will be ordinary declarative sentences which are so precisely stated that they are either true or false. We will exclude sentences which are ambiguous or which can be called true or false only after qualifications are imposed on them.

The following are acceptable statements:

All triangles are isosceles.
$6 = 2 + 5$
The number 4 is a perfect square.

Axioms

Our task, now, is to decide which of our statements are true and which are false. In order to give meaning to this task, we must first establish a frame of reference on which our later reasoning will be based. At the very beginning we must choose a few statements which we will call "true" by assumption; such statements are called "axioms". These axioms are statements about the technical words in our vocabulary and are completely abstract in character. They are not statements about the properties of the physical world. You must have heard that an "axiom is a self-evident truth", but axioms can be any statements at all, evident or not. Since mathematical theories can begin with any set of axioms at all, they are infinite in their variety; some of them are interesting and useful, others merely interesting, and still others only curiosities of little apparent value. The choice of a set of axioms which leads to an interesting and useful theory requires great skill and judgment, but for the most part such sets of axioms are obtained as models of the real world. We look about us, and from what we see we construct an abstract model in which our undefined words correspond to the most important objects that we have identified, and in which our axioms correspond to the basic properties of these objects. The mathematics which you will use as a scientist is entirely based on axioms which were derived in this fashion.

From our set of axioms (which we have assumed to be true) we now proceed to establish the truth or falsehood of other statements which arise. We must agree upon some rules of procedure, which we call the "Laws of Logic", and by means of these rules we seek to determine whether a given statement is true or false. We shall not dwell upon these Laws of Logic here, but if you are interested you can read about them in the References given at the end of the chapter. Except for a few tricky places which we will discuss below, you can rely upon your own good sense and previous experience in logical thinking. Whenever doubts arise, however, you must refer back to the full treatment of these logical principles.

Theorems

When we have shown that the truth of a given statement follows logically from the assumed truth of our axioms, we call this statement a "theorem" and say that "we have proved it". The truth of a theorem, therefore, is relative to a given set of axioms; absolute truth has no meaning when applied to mathematical statements. The main business of a mathematician is the invention of new theorems and the construction of proofs for them. The discovery of a new theorem depends upon deep intuition and intelligent guessing, and the process of making such a discovery is very much like that of creative effort in any field. After our intuition has led us to believe that a certain statement is true, we must still prove it; and this is where our use of logical deduction comes in.

Our abstract mathematical system, then, consists of four parts:

1 Undefined words
2 Defined words
3 Axioms, i.e., statements which are assumed to be true
4 Theorems, i.e., statements which are proved to be true

Since we shall need to have a good understanding of the nature of proof, we will devote the rest of this chapter to a discussion of various problems which you will meet in mathematical proofs.

1.2 › SETS

Before we can proceed we must introduce a concept which pervades all mathematics; namely that of a *set*.

We think of a set as a collection of objects: pencils, trees, numbers, points, etc. The individual components of the set are called its *elements*. As an example, consider the set consisting of the four boys named John, Joe, Jerry, Jim. This set has four elements. Sets may be of any size. We may think of the set of all particles of sand on a beach; this has a finite number of elements, but this number is certainly very large. A set, however, may have infinitely many elements. An example of an infinite set is the set of all positive integers: 1, 2, 3, 4, 5, ... Indeed a set may contain no elements, in which case we call it the *empty* set, or the *null* set.

We can describe sets in this way, but *set* is a primitive notion which cannot be defined. Hence we take *set* and *element* to be undefined. The statement: "a is an element of a set X" (notation: $a \in X$) is similarly an undefined relationship.

EXAMPLES AND NOTATION In the list below we give some typical examples of sets occurring in mathematics and indicate the notations appropriate for these. Note that we regularly use curly brackets { } to represent a set; but there are exceptions to this as we shall see in items 1, 4, and 5 below.

1 $\varnothing$, the empty, or null, set containing no elements. For example, $\varnothing$ is the set of all men living now who were born in 1600 A.D.

2 $\{3\}$, the finite set of which 3 is the only element. Note that this is quite different from the real number 3.

3 $\{2, 7, 15, 36\}$, a finite set of four elements.

4 $R = \{x \mid x \text{ is a real number}\}$, the set of all real numbers. This expression should be read: "The set R is the set of numbers x such that x is a real number", the vertical line standing for "such that".

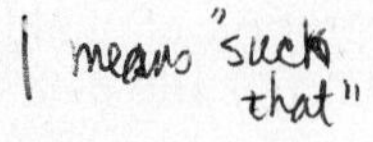

5 $R \times R = \{(x,y) \mid x \text{ and } y \text{ are real numbers}\}$, the set of all ordered pairs (x,y) of real numbers (see Sec. 2.11). This set is sometimes called the *Cartesian Product* of R with itself.

6 $\{x \mid x \text{ is a positive integer}\}$, the infinite set of all positive integers. We shall often write this as the set $\{1, 2, 3, 4, 5, \ldots\}$.

7 $\{x \mid x \text{ is an even positive integer}\}$, the infinite set of all even positive integers. We shall often write this as the set $\{2, 4, 6, 8, 10, \ldots\}$.

8 $\{R \mid R \text{ voted for Abraham Lincoln for President}\}$, the set of all people who voted for Lincoln.

9 $\{T \mid \text{Triangle } T \text{ is isosceles and lies in a given plane}\}$, the set of all isosceles triangles which lie in a given plane.

10 $\{L \mid L \text{ is a line parallel to line } M\}$, the set of all lines parallel to a given line M.

There are several types of relations between sets which we shall need in the future. One of these is the notion of *identity*.

DEFINITION Two sets are said to be *identical* if and only if every element of each is an element of the other. When A and B are identical, we write $A = B$.

ILLUSTRATION 1

a The sets $\{1, 7, 10, 12\}$ and $\{12, 7, 1, 10\}$ are identical. Note that the order in which the elements are written is not relevant.

b The sets $\{x \mid x \text{ is an even prime number}\}$ and $\{2\}$ are identical. These are just two different ways of defining the same set.

c The sets {Carl, Cletus} and {Dorothy, Louise} are not identical even though each contains two elements.

Subsets

When A and B are not identical, it is still possible that every element of A is an element of B. In such a case we say that A is a *proper subset* of B.

DEFINITION A set A is a proper subset of a set B if and only if (1) every element of A is an element of B and (2) A and B are not identical.

We write this relationship $A \subset B$, read "A is a proper subset of B".

ILLUSTRATION 2

a If $A = \{1, 4, 6\}$ and $B = \{1, 4, 6, 15\}$, then $A \subset B$.

b If $A = \{x \mid x \text{ is an even positive integer}\}$ and $B = \{x \mid x \text{ is a positive integer}\}$, then $A \subset B$.

If we are not certain whether or not A and B are identical, but if we know that every element of A is an element of B, we say that A is a *subset* of B.

DEFINITION A set A is a subset of B if and only if every element of A is an element of B.

We write this relationship $A \subseteq B$, read "A is a subset of B".

ILLUSTRATION 3 Let A be the set of rainy days in January of this year and B be the set of all days in January of this year. Then surely $A \subseteq B$. We cannot know the truth of $A \subset B$ unless we specify the place to which we are referring and investigate the records of the weather bureau.

REMARKS

1 If we know that $A \subset B$ we can conclude that $A \subseteq B$; but given $A \subseteq B$, we need more information before deciding whether or not $A \subset B$ is true.

2 By convention the empty set $\varnothing$ is a subset of every set, and is a proper subset of every set except itself.

One-to-one correspondence

The next notion is that of a one-to-one correspondence between two sets. Suppose that in your classroom there are 30 seats and 30 students (all present). When the students all sit down, one to a chair, they set up a correspondence between the set of students and the set of seats. Since there is one seat for each student and one student for each seat, this is called a "one-to-one correspondence".

DEFINITION Two sets $A = \{a_1, a_2, \ldots\}$ and $B = \{b_1, b_2, \ldots\}$ are said to be in *one-to-one correspondence* when there exists a pairing of the a's and the b's such that each a corresponds to one and only one b and each b corresponds to one and only one a.

ILLUSTRATION 4 Establish a one-to-one correspondence between the set of numbers $\{1, 2, \ldots, 26\}$ and the set of letters of the alphabet $\{a, b, \ldots, z\}$.

We make the pairing

$$\begin{array}{cccc} 1 & 2 & \cdots & 26 \\ \updownarrow & \updownarrow & & \updownarrow \\ a & b & \cdots & z \end{array}$$

However, there are many other possible pairings, such as

$$\begin{array}{ccccc} 2 & 3 & \cdots & 26 & 1 \\ \updownarrow & \updownarrow & & \updownarrow & \updownarrow \\ a & b & \cdots & y & z \end{array}$$

ILLUSTRATION 5 Establish a one-to-one correspondence between the set $\{1, 2, 3, 4, 5, \ldots\}$ and the set $\{2, 4, 6, 8, 10, \ldots\}$.

Let n represent an element of $\{1, 2, 3, 4, 5, \ldots\}$. Then the pairing $n \leftrightarrow 2n$ gives the required correspondence, examples of which are

$$\begin{array}{ccccccc} 1 & 2 & \cdots & 50 & \cdots & 100 & \cdots \\ \updownarrow & \updownarrow & & \updownarrow & & \updownarrow & \\ 2 & 4 & \cdots & 100 & \cdots & 200 & \cdots \end{array}$$

EXERCISE A Establish a one-to-one correspondence between the sets {John, Joe, Jerry, Jim} and {Mildred, Marcia, Ruth, Sandra}.

EXERCISE B Establish a one-to-one correspondence between the sets $\{1, 2, 3, 4, 5, \ldots\}$ and $\{3, 6, 9, 12, 15, \ldots\}$.

PROBLEMS 1.2

1 Which pairs of the following sets are identical?

a {James, John, Joe}	**b** $\{1, 2, 3\}$
c $\{c, b, a\}$	**d** $\{a, b, c\}$
e {John, Joe, James}	**f** $\{x, y, z\}$

2 Which pairs of the following sets are identical?

a {Seattle, London, New York, Chicago}
b $\{a, b, c, d\}$
c {English, French, Latin, Greek}
d {Seattle, Latin, New York, d}
e $\{d, c, a, b\}$
f {New York, Seattle, d, Latin}

In Probs. 3 to 8 list all subsets of the given set. Which of these are proper subsets?

3 $\{2, 9\}$	**4** $\{x, y\}$
5 $\{1, 2, 3\}$	**6** $\{x, y, z\}$
7 $\{1, 2, 3, 4\}$	**8** $\{a, b, c, d\}$

***9** Count the number of subsets in each of Probs. 3 to 8 that you have worked. Now guess a general formula for the number of subsets of a given, finite set. Now prove that your guess is correct.

10 Prove that if $A \subseteq B$ and $B \subseteq A$, then A and B are identical.

Problems 11 to 21 refer to the following situation: A four-sided die (a tetrahedron) has its faces marked with the numbers 1, 2, 3, 4, respectively. Two such dice are thrown and the numbers on the resulting bottom faces are recorded by symbols such as (2,1)—meaning that the number on the bottom face of the first die is 2 and that of the second die is 1. We call symbols like (2,1) the *outcomes* of throws of our two dice. What is the set of all possible outcomes of throws of our two dice such that the two numbers:

11 A: Are equal	**12** B: Are both even
13 C: Are both odd	**14** D: Have sum equal to 1
15 E: Have sum equal to 4	**16** F: Have product equal to 6
17 G: Have sum less than 11	**18** H: Have an odd sum
19 I: Have product equal to 15	**20** J: Are unrestricted

21 Write all relationships of the forms $A = B$, $A \subset B$, which are true for the above sets: $A, \ldots, J$.

In Probs. 22 to 30 establish a one-to-one correspondence between the given two sets whenever this is possible.

22 The set of positive integers; the set of all integers.
23 $\{2, 4, 6, 8, 10, \ldots\}$; $\{4, 8, 12, 16, 20, \ldots\}$
24 The set of states in the United States; the set of state capitals.
25 The set of human beings on Mars; the set of maidens on the moon.
26 The set of courses for which you are registered; the set of grades you expect to receive.
27 $\{1, 2, 3\}$; $\{1, 2, 3, \ldots\}$
28 $\{x \mid x \text{ is an even positive integer}\}$; $\{x \mid x \text{ is an odd positive integer}\}$
29 The set of first names of the members of your class; the set of their last names.
30 The set of letters in *deal;* the set of letters in *lead.*

1.3 › OPEN SENTENCES

We said above that a *statement* is a sentence which can meaningly be called true or false. In addition to statements, mathematicians frequently write assertions such as:

1 $x + 1 = 3$
2 $x^2 - 5x + 6 = 0$
3 $x^2 - 4 = (x - 2)(x + 2)$
4 $x^2 = 4$ and $x + 1 = 6$

It is not possible to say whether these are true or false, for we are not told the value of x. In the examples above, the assertion is:

1 True for $x = 2$, false for other values of x
2 True for $x = 2$ or 3, false for other values of x
3 True for all x, false for no x
4 True for no x, false for all x

Such assertions are called *open sentences.*

DEFINITION An open sentence is an assertion containing a variable x, which becomes a statement when x is given a specified value.

REMARKS
1 Open sentences may contain two or more variables, but for simplicity we begin by considering the case of a single variable.
2 This definition is faulty in that it refers to the "variable" x which so far has not been defined. We shall repair this deficiency in a moment.

EXERCISE A Write several open sentences containing two or more variables.

When we confront the problem of assigning values to x, we first must decide what values of x are permissible. That is, we must agree upon a set of numbers, geometric figures, people, etc., which will be the subject of our discourse. This set is called the *universal set* (or replacement set), U. The understanding is that the symbol x in our open sentence may be replaced by any element of U. If x appears more than once in a given open sentence, it must be replaced by the same element of U each time that it appears. When we make such a replacement, we say that "we have given x a value". These ideas permit us to define a *variable*.

Variable

DEFINITION A variable is a symbol in an assertion which may be replaced by any element of a given universal set U.

Variables are commonly, but not exclusively, represented by letters at the end of the alphabet: x, y, z.

In contrast to a variable, there is the notion of a *constant*.

DEFINITION A constant is a fixed element of a given set.

Frequently this given set is a set of numbers, and a constant is represented by the desired numeral such as $2, \frac{1}{2}, \pi$, etc. In other numerical cases the constant is represented by a letter in the first part of the alphabet such as a, b, c, d. Here we mean that a stands for a fixed, but arbitrary element of the given set.

When we are given an open sentence with variable x, we can consider the set of those elements of U (i.e., values of x) whose substitution for x converts the open sentence into a true statement. We call this the *truth set* of the given open sentence.

Truth set

DEFINITION The truth set of an open sentence is the set of elements of the universal set U whose substitution for x converts the open sentence into a true statement.

REMARKS

1 We shall write truth sets in the form

$$\{x \mid x^2 - 5x + 6 = 0\}$$

where $x^2 - 5x + 6 = 0$ is the given open sentence. Generally the universal set U is evident from the context, but in case of doubt we write expressions such as

$$\{x(\text{integers}) \mid x^2 - 5x + 6 = 0\}$$

which implies that U is the set of integers.

2 For more formal work, we shall use notations, such as p_x, q_x, r_x, etc.,

to represent open sentences with variable x. The truth set of p_x will be written as

$$P \qquad \text{or} \qquad \{x \mid p_x\}$$

3 If the open sentence is an equation which is true for *all* values of x, we call it an *identity*. In such a case we can write

$$P = U \qquad \text{or} \qquad \{x \mid p_x\} = U$$

For example,

$$\{x \mid x^2 - 4 = (x - 2)(x + 2)\} = U$$

4 The process commonly called "solving an equation" is the same as "finding the truth set of an open sentence which is an equation".

The steps in many mathematical processes amount to replacing an open sentence by one *equivalent* to it. For example we proceed from

$$2x + 1 = 5$$

to

$$2x = 4$$

to

$$x = 2$$

Equivalent open sentences

At each step we have replaced the previous open sentence by a simpler one which has the same truth set. Hence we define equivalent open sentences as follows:

DEFINITION Two open sentences p_x and q_x are equivalent if and only if their truth sets are identical, i.e., if $P = Q$. When p_x and q_x are equivalent, we write $p_x \leftrightarrow q_x$.

ILLUSTRATION 1 The open sentences

$$p_x\text{:} \quad x^2 + 5x = -6$$

and

$$q_x\text{:} \quad x^2 + 5x + 6 = 0$$

are equivalent since $P = Q = \{-2, -3\}$.

ILLUSTRATION 2 The open sentences

$$p_x\text{:} \quad x = 2$$

and

$$q_x\text{:} \quad x^2 = 4$$

are not equivalent, since $P = \{2\}$ and $Q = \{-2, 2\}$.

Negations

Later in this chapter we shall need to consider the *negation* of an open sentence, p_x. This is defined as follows.

DEFINITION The *negation* of an open sentence p_x is the open sentence "It is false that p_x" or any open sentence equivalent to this. We use the symbol $\sim p_x$ to represent the negation of p_x.

In simple cases the negation can be formed by inserting a *not* in an appropriate place, as suggested in Illustration 3.

ILLUSTRATION 3 Let U be the set of integers. Then negations are formed as follows:

Open sentence	*Negation*
$x^2 - 4 = 0$	$x^2 - 4 \neq 0$
x is an even integer	x is not an even integer or x is an odd integer

REMARKS

1 It is evident that the truth set of $\sim p_x$ consists of those elements of U which are not elements of the truth set of p_x. Thus the truth set of $\sim(x^2 = 4)$, where U is the set of integers, is the set of all integers different from 2 and -2.

2 The formation of the negation of more complicated open sentences can be quite tricky. For example, the negation of

$$x^2 = 4 \quad \text{and} \quad x + 1 = 6$$

is

$$x^2 \neq 4 \quad \text{or} \quad x + 1 \neq 6$$

Similarly the negation of

$$x^2 = 9 \quad \text{or} \quad x + 2 = 7$$

is

$$x^2 \neq 9 \quad \text{and} \quad x + 2 \neq 7$$

PROBLEMS 1.3

1 Which of the following are statements and which are open sentences?
   a Bob Hope is a mathematician.
   b 2 is a prime number.
   c $x^2 - 5 = 0$
   d $x^2 - 4 = (x + 2)(x - 2)$
   e For all integers x: $x^2 \geq 0$.

2 Find the truth sets of the following open sentences. The universal set U is the set of integers.
   a $x + 3 = 8$
   b $x^2 = 4$
   c $x^2 - 4 = (x + 2)(x - 2)$
   d $2x = 5$
   e x is a positive even number less than 8.

3 Which of the following pairs of open sentences are equivalent? The universal set is the set of integers.
   a $6x + 5 = 2x + 13 \qquad x + 1 = 3$

4x 8 x=2

b	$x^2 = 4$	$x = -2$
c	$x = 3$ or $x = -3$	$x^2 = 9$
d	$x^2 - 4x + 3 = 0$	$(x - 1)(x - 3) = 0$
e	$(x - 3)(x + 3) = x^2 - 9$	$(x + 3)(x + 3) = x^2 + 9$

4 Which of the following pairs of open sentences are equivalent? The universal set is the set of integers.

a	$2x + 3 = 2x - 3$	$3x = 2$
b	$(x + 2)(3x - 2) = 0$	$x + 2 = 0$
c	$x^2 - 7x + 6 = (x - 6)(x - 1)$	x is an integer
d	$3x + 5 = 0$	$2x - 7 = 0$
e	x is prime	x is odd

In Probs. 5 to 10 write the negation of the given open sentence, and find the truth set of this negation, assuming that $U = \{1, 2, 3, 4, 5\}$.

5 $3x = 6$ **6** $2x + 3 = 9$
7 $(x - 2)(x - 1)(x - 3) = 0$ **8** $(x - 1)(x - 5) = 0$
9 $x^2 = 4$ and $x + 2 = 5$ **10** $x^2 = 25$ or $x + 1 = 4$

1.4 › SET OPERATIONS

The examples just given of the way of finding the truth set of $\sim p_x$ suggest a more general situation. Let us suppose that we have decided on a universal set U, and have before us a subset of U, namely A. Then we can form the set consisting of those elements of U that are not in A. We call this the *complement* of A (relative to U) and write it as A'.

Complement

DEFINITION Given a universal set U and a subset $A \subseteq U$, then the *complement* of A (written A') is the set of elements of U which are not elements of A (Fig. 1.1).

ILLUSTRATION 1

a Let $U = \{1, 2, 3, 4, 5, 6, 7, 8, 9, 10\}$
and $A = \{1, 3, 5, 7, 9\}$
Then $A' = \{2, 4, 6, 8, 10\}$

A' is the complement of A relative to U.

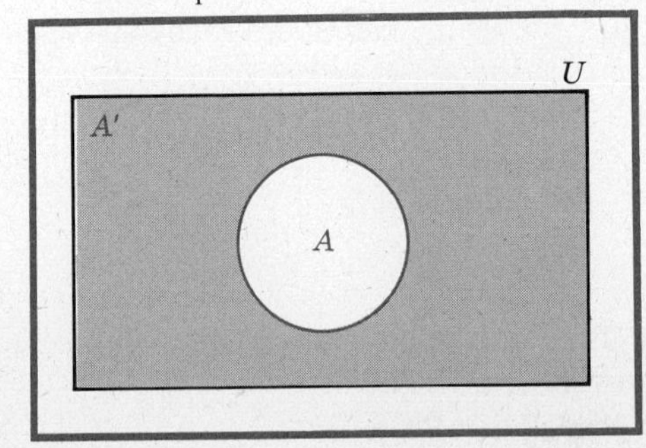

FIGURE 1.1 The complement of A relative to U consists of those elements of U which are not in A.

b Let U be the set of positive integers and A be the set of even positive integers. Then A' is the set of odd positive integers.
c If $A = U$, then $A' = \varnothing$.
d If $A = \varnothing$, then $A' = U$.
e $\{x \mid \sim p_x\} = \{x \mid p_x\}'$

EXERCISE A Show that $(A')' = A$.

Union

If we are given two sets, A and B, there are two important ways of operating with these to produce new sets. The first is called the *union* of A and B.

DEFINITION

The union of A and B, written $A \cup B$, is the set of elements which belong to either A or B or to both A and B (Fig. 1.2).

ILLUSTRATION 2

a Let $A = \{1, 3, 5\}$ and $B = \{2, 4, 6\}$
Then $A \cup B = \{1, 2, 3, 4, 5, 6\}$

b Let $A = \{1, 2, 6\}$ and $B = \{1, 2, 3\}$
Then $A \cup B = \{1, 2, 3, 6\}$

c Let A be the set of even positive integers and B be the set of odd positive integers. Then $A \cup B$ is the set of all positive integers.

d Let
$$A = \{x \mid x^2 = 9\} = \{-3, 3\}$$
$$B = \{x \mid (x^2 - 4x + 3 = 0)\} = \{1, 3\}$$
Then $A \cup B = \{x \mid (x^2 = 9) \text{ or } (x^2 - 4x + 3 = 0)\} = \{-3, 1, 3\}$

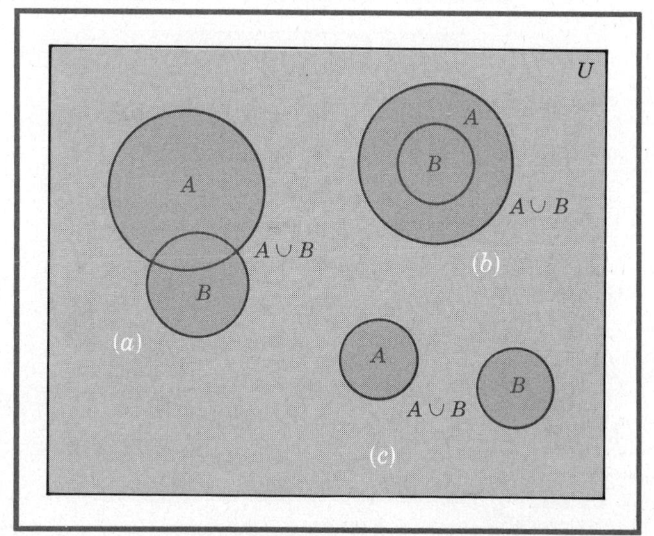

FIGURE 1.2
The union of A and B consists of those elements which are in A or in B or in both A and B.

When "or" is used to combine two open sentences as above, we interpret it to mean "either or both" in the same sense as the legal phrase "and/or".

e If P and Q are the truth sets of p_x and q_x, respectively, then $P \cup Q = \{x \mid p_x \text{ or } q_x\}$.

Intersection

The second operation of this kind is called *intersection*.

DEFINITION The intersection of A and B, written $A \cap B$, is the set of elements that belong to both A and B (Fig. 1.3).

ILLUSTRATION 3

a Let $A = \{1, 3, 5\}$ and $B = \{2, 4, 6\}$
Then $$A \cap B = \varnothing$$

b Let $A = \{1, 2, 6\}$ and $B = \{1, 2, 3\}$
Then $$A \cap B = \{1, 2\}$$

c Let $A = \{x \mid x^2 = 9\} = \{-3, 3\}$
$B = \{x \mid x^2 - 4x + 3 = 0\} = \{1, 3\}$
Then $A \cap B = \{x \mid (x^2 = 9) \quad \text{and} \quad (x^2 - 4x + 3 = 0)\} = \{3\}$

d If P and Q are the truth sets of p_x and q_x, respectively,

$$P \cap Q = \{x \mid p_x \text{ and } q_x\}$$

That is, $P \cap Q$ is the set of those elements in U (values of x) for which *both* p_x and q_x are true.

Intersection of A and B

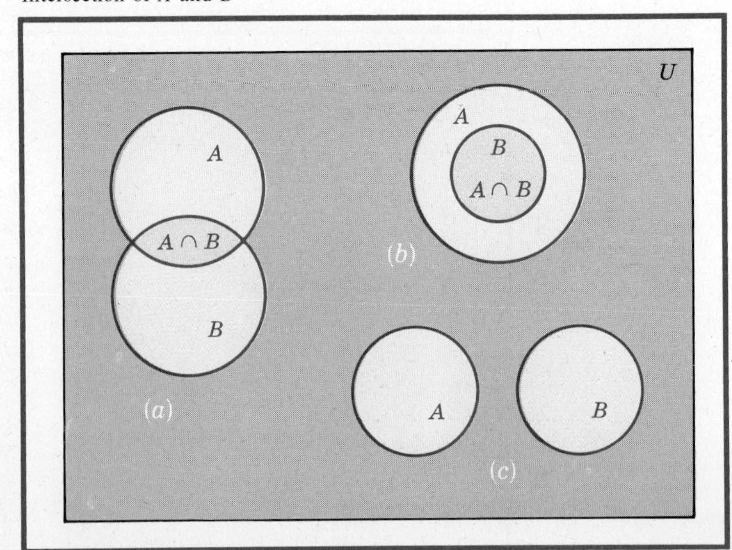

FIGURE 1.3 The intersection of A and B consists of those elements which are in both A and B. In (c), $A \cap B = \varnothing$.

PROBLEMS 1.4 In Probs. 1 to 6 let $U = \{1, 2, 3, 4, 5, 6, 7, 8, 9, 10\}$; $A = \{1, 3, 6, 8, 10\}$; $B = \{2, 4, 5, 6, 8\}$; and $C = \{1, 4, 6, 10\}$. Find simpler expressions for the following sets:

1 $A \cup B$
2 $A \cap B$
3 $(A \cap B)'$
4 $(A \cup B)'$
5 $(A \cap B) \cup C$
6 $(A \cup B) \cap C$

In Probs. 7 to 15 find simpler expressions for the given sets.

7 $\{x \mid x \text{ is an even integer}\} \cap \{x \mid x \text{ is an odd integer}\}$
8 $\{a, b, c, d\} \cup \varnothing$
9 $\{1, 2, 3, 4\} \cap \varnothing$
10 $\{x \mid x \text{ has blue eyes}\} \cap \{x \mid x \text{ has blonde hair}\}$
11 The set of cities in Illinois whose population is over 20 million.
12 $\{x(\text{people}) \mid x \text{ is a student}\} \cup \{x \mid x \text{ is over } 30\}$
13 $\{x(\text{people}) \mid x \text{ is a student}\} \cap \{x \mid x \text{ is over } 30\}$
14 $\{x(\text{airplanes}) \mid x \text{ is a Boeing 747}\} \cup \{x \mid x \text{ is owned by United Air Lines}\}$
15 $\{x(\text{airplanes}) \mid x \text{ is a Boeing 747}\} \cap \{x \mid x \text{ is owned by United Air Lines}\}$

16 If U is the set of undergraduate students in your university, and A is the set of freshmen, find A'.
17 For any set A, find $A \cap U$ and $A \cup U$.
18 For any set A, find $A \cap \varnothing$ and $A \cup \varnothing$.
19 For any set A, find $A \cap A'$ and $A \cup A'$.
20 **a** For any sets A and B, is $A \cup B = B \cup A$? Why?
b Answer the same questions for $A \cap B$.

In Probs. 21 to 30 one of the following relations is true: $A \subset B$, $A = B$, $B \subset A$. Write the correct relation in each case. The universal set in Probs. 21 to 28 is the set of all integers.

	A	B
21	$\{x \mid 2x + 3 = 11 - 2x\}$	$\{x \mid 5x + 4 = x + 12\}$
22	$\{x \mid x^2 + 4 = 6x - 5\}$	$\{x \mid 4 + 2x = 10\}$
23	$\{x \mid (x - 2)(x - 3) = 0\}$	$\{x \mid x = 2\}$
24	$\{x \mid (x + 4) = 0\}$	$\{x \mid x(x + 4) = 0\}$
25	$\{x \mid x - 1 = 0\} \cup \{x \mid x - 2 = 0\}$	$\{x \mid x^2 - 3x + 2 = 0\}$
26	$\{x \mid x = 3\}$	$\{x \mid x \text{ is an odd integer}\}$
27	$\{x \mid x + 3 = 4\}$	$\{x \mid (x + 3)^2 = 16\}$
28	$\{x \mid x^2 = 25\}$	$\{x \mid x + 2 = 7\}$
29	$\{T(\text{triangles in a plane}) \mid T \text{ is equilateral}\}$	$\{T(\text{triangles in a plane}) \mid T \text{ is isosceles}\}$
30	The set of squares in a plane	The set of rectangles in a plane

1.5 » CONDITIONALS

The great bulk of mathematical theorems are assertions of the form: "If . . . , then . . ." where the dots are to be filled with either statements or open sentences. Such assertions are called conditionals; a few typical examples are:

If x is an odd integer, then x^2 is an odd integer.
If $x^2 = 1$, then $(x + 1)(x - 1) = 0$.
If triangle x is equilateral, then x is equiangular.

In these examples, as in almost all interesting cases, the dots above are replaced by open sentences, and the conditional can be written in the standard form

If p_x, then q_x.

or, more briefly,

$$p_x \to q_x$$

where p_x and q_x are open sentences.

There is, moreover, a subtlety here which is often ignored. In the example above:

If x is an odd integer, then x^2 is an odd integer.

we really mean:

For *all* integers x: If x is an odd integer, then x^2 is an odd integer.

In other words, our conditionals are *general* statements which are to be true for *all* values of the variable involved. In order to be complete, we should write our conditionals in the form

$$\forall_x(p_x \to q_x)$$

which is read

For all x: If p_x, then q_x.

True conditionals

Notice that this is no longer an open sentence, but is a statement which is true or false. We are at liberty to define the circumstances under which it is true and do so as follows.

DEFINITION The statement $\forall_x(p_x \to q_x)$ is *true* if and only if $P \subseteq Q$, where P and Q are the truth sets of p_x and q_x, respectively.

This definition may seem arbitrary, but it is motivated by intuition obtained from special cases such as those given below. Observe that we can make any definition we please, and that no proof is required. Nevertheless, if our mathematics is to be useful in our daily lives, our definitions should not be in conflict with our intuition.

ILLUSTRATION 1 It seems reasonable to agree to the truth of

For all x: If $x = 2$, then $x^2 = 4$.

Let p_x be $x = 2$; q_x be $x^2 = 4$. Then $P = \{2\}$ and $Q = \{-2, 2\}$, and $P \subseteq Q$. In this case, therefore, the above definition agrees with common sense.

ILLUSTRATION 2 It does not seem reasonable to assert the truth of

For all x: If $x^2 = 4$, then $(x - 2)(x - 3) = 0$.

In particular, when $x = -2$ this becomes

If $4 = 4$, then $(-4)(-5) = 0$.

which is certainly false. Let us examine the truth sets. Let p_x be $x^2 = 4$ and q_x be $(x - 2)(x - 3)$. Then $P = \{-2, 2\}$ and $Q = \{2, 3\}$. In this case the statement $P \subseteq Q$ is false; and again our intuition gives the same result as the above definition.

This situation can be illustrated by the following diagrams. Let the interior of the rectangle represent the universal set U. Let P be represented by the points in the interior of the circle labeled P (see Fig. 1.4), and let Q be represented by the points in the interior of the circle labeled Q. Of the many relative positions in which circles P and Q may lie, there are just two that correspond to the statement $P \subseteq Q$. These are drawn in Fig. 1.5. This helps us formulate a test that will tell us when $\forall_x(p_x \to q_x)$ is true.

RULE

1. If q_x is true for each x for which p_x is true, then $\forall_x(p_x \to q_x)$ is true.
2. If there is at least one value of x for which p_x is true and q_x is false, then $\forall_x(p_x \to q_x)$ is false.

Graphical representation of sets

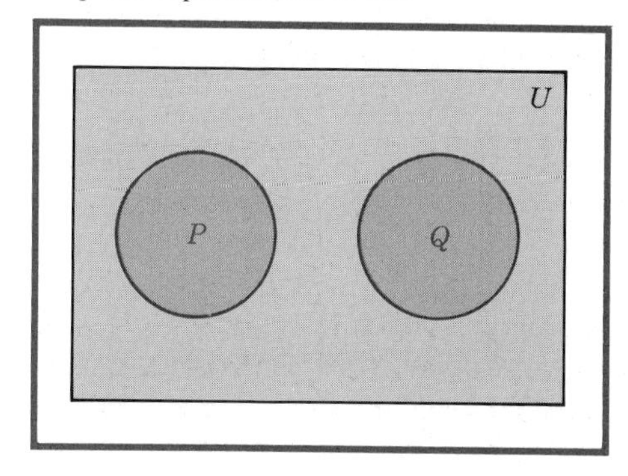

FIGURE 1.4
Sets are represented by the interiors of circles.

Possible positions of P and Q when $P \subseteq Q$

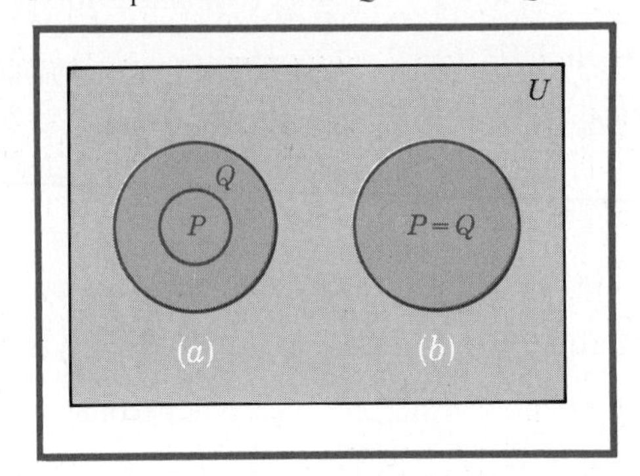

FIGURE 1.5
If $P \subseteq Q$, then P must either be inside Q or be identical with Q.

Let us apply this to the conditionals in Illustrations 1 and 2.

ILLUSTRATION 3 For all x: If $x = 2$, then $x^2 = 4$. There is only one value for which $x = 2$ is true, namely 2. When we substitute this for x in $x^2 = 4$, we obtain $4 = 4$, which is true. Hence the conditional is true (Fig. 1.6).

ILLUSTRATION 4 For all x: If $x^2 = 4$, then $(x - 2)(x - 3) = 0$. There are two values of x for which $x^2 = 4$ is true, namely 2 and -2. If we substitute 2 for x in $(x - 2)(x - 3) = 0$, we obtain $(2 - 2)(2 - 3) = 0(-1) = 0$, which is true. If, however, we substitute -2 for x in $(x - 2)(x - 3) = 0$, we obtain $(-2 - 2)(-2 - 3) = (-4)(-5) = 0$, which is false. Hence the conditional is false (Fig. 1.7).

$P = \{x \mid x = 2\}$; $Q = \{x \mid x^2 = 4\}$; $P \subseteq Q$

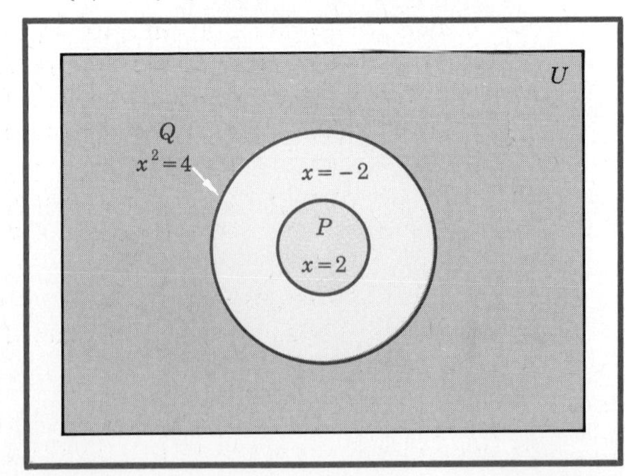

FIGURE 1.6
Since the set where $x = 2$ is included in the set where $x^2 = 4$, the conditional "For all x: If $x = 2$, then $x^2 = 4$" is true.

$P = \{x | x^2 = 4\}$; $Q = \{x | (x - 2)(x - 3) = 0\}$; $P \not\subseteq Q$

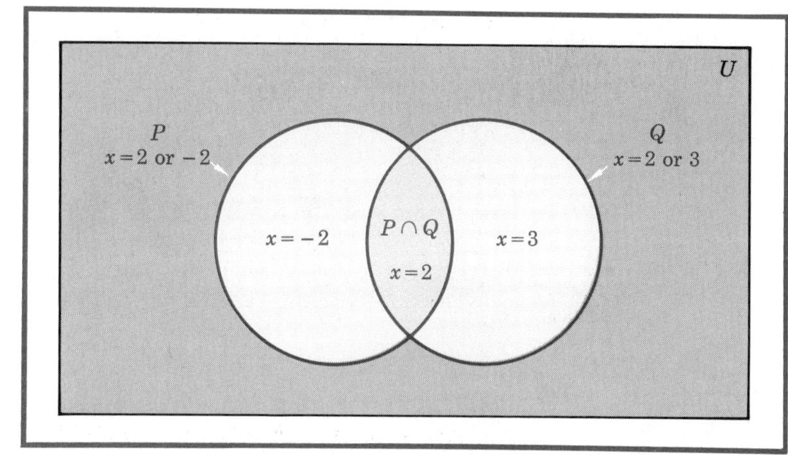

FIGURE 1.7 Since the set where $x^2 = 4$ is not contained in the set where $x = 2$ or 3, the conditional "For all x: If $x^2 = 4$, then $(x - 2)(x - 3) = 0$" is false.

In the conditional $p_x \to q_x$ it is entirely possible that there is no value for x for which p_x is true; i.e., $P = \varnothing$. Since $\varnothing$ is a subset of any set, it follows that $P \subseteq Q$ and that $p_x \to q_x$ is true. (See Illustration 5.)

ILLUSTRATION 5 Consider the following conditional:

For all odd integers x: If $x = 4$, then $x^2 = 6$.

Since no odd integer equals 4, $P = \varnothing$, and the conditional is true.

PROBLEMS 1.5

1. According to the definition, the statement $\forall_x(p_x \to q_x)$ is true if and only if what relationship is true for the corresponding truth sets P and Q?
2. If there are no values of x in U for which p_x is true, what conclusion can be drawn concerning the truth of $\forall_x(p_x \to q_x)$?
3. If there is at least one value of x in U for which p_x is true and q_x is false, what conclusion can be drawn concerning the truth of $\forall_x(p_x \to q_x)$?
4. If for all values of x in U both p_x and q_x are false, what conclusion can be drawn concerning the truth of $\forall_x(p_x \to q_x)$?

In Probs. 5 to 10 determine the truth or falsehood of the given conditional by finding P and Q and checking the truth of $P \subseteq Q$. The universal set is the set of integers.

5. For all x: If $x^2 = 4$, then $x = 2$.
6. For all x: If $(x + 2)(x - 4) = 0$, then $x = 4$.

7 For all x: If $4x + 3 = x - 6$, then $3x = -9$.
8 For all x: If $x + 3 = 7$, then $x^2 = 16$.
9 For all x: If $2x^2 = 7$, then $x = 4$.
10 For all x: If $3x + 2 = 9$, then $x = \frac{7}{2}$.

1.6 » DERIVED CONDITIONALS

Associated with the conditional $\forall_x(p_x \to q_x)$ there are three other conditionals with which it is often confused:

Converse
Inverse
Contrapositive

Converse: $\forall_x(q_x \to p_x)$
Inverse: $\forall_x(\sim p_x \to \sim q_x)$
Contrapositive: $\forall_x(\sim q_x \to \sim p_x)$

In terms of the truth sets, P and Q, of p_x and q_x, these become:

Given conditional: $P \subseteq Q$
Converse: $Q \subseteq P$
Inverse: $P' \subseteq Q'$
Contrapositive: $Q' \subseteq P'$

We must now examine how the truth of each of these is related to the truth of each of the others. The case when $P = Q$ is trivial, for in this case all four statements are true.

THEOREM 1 Let $\forall_x(p_x \to q_x)$ be a given conditional where p_x is equivalent to q_x (i.e., $P = Q$). Then the given conditional, its converse, its inverse, and its contrapositive are all true.

Let us now examine the case where $P \neq Q$. First let us assume that $P \subset Q$ is true and see what can be said about the truth of $Q \subset P$, $P' \subset Q'$, and $Q' \subset P'$. Our assumption tells us that Fig. 1.8*a* represents the true situation. Then clearly $Q \subset P$ is false. Moreover, P' (the outside of P) contains Q' (the outside of Q); so $P' \subset Q'$ is false and $Q' \subset P'$ is true.

By similar reasoning with Fig. 1.8*b*, *c*, and *d* we can draw the following conclusions.

TABLE 1
COMPARATIVE TRUTH OF A CONDITIONAL AND ITS DERIVED CONDITIONALS WHEN $P \neq Q$

Given conditional	*Converse*	*Inverse*	*Contrapositive*
Assumed true	Proved false	Proved false	Proved true
Proved false	*Assumed true*	Proved true	Proved false
Proved false	Proved true	*Assumed true*	Proved false
Proved true	Proved false	Proved false	*Assumed true*

$P \subseteq Q$ and its derived conditionals

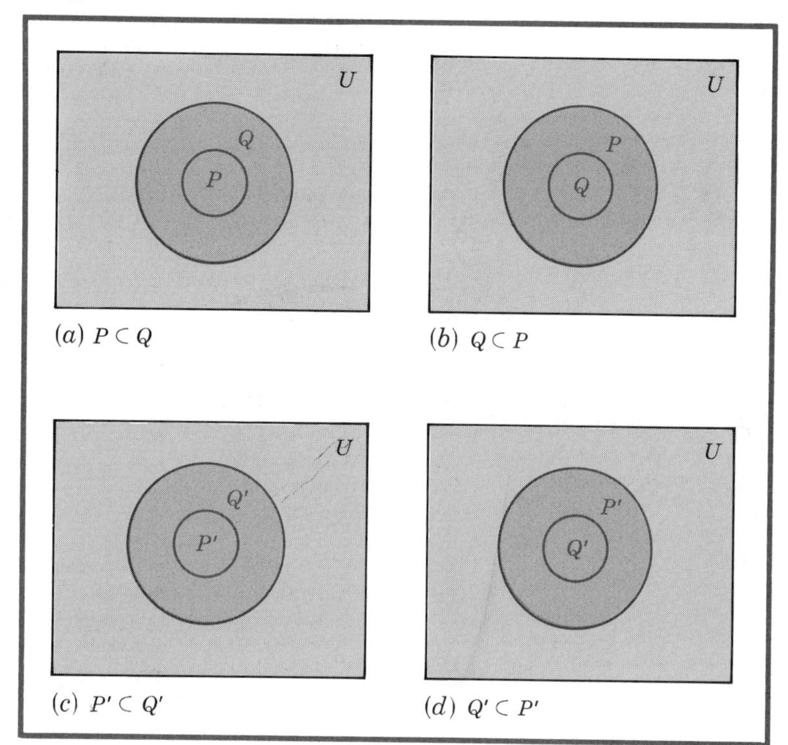

FIGURE 1.8 These diagrams, where $P \neq Q$, help us to discover the relative truth or falsehood of a conditional, its converse, its inverse, and its contrapositive.

We summarize the most important conclusions from this table in Theorem 2.

THEOREM 2 Let $\forall_x(p_x \rightarrow q_x)$ be a given conditional where p_x is not equivalent to q_x (i.e., $P \neq Q$). Then:

Conditional equivalent to its contrapositive

a If either the given conditional or its contrapositive is true, both are true.
b If either the converse or the inverse is true, both are true.
c The given conditional and its converse are not both true.
d The given conditional and its inverse are not both true.

EXERCISE A What can be said about the converse and the contrapositive?

EXERCISE B What can be said about the inverse and the contrapositive?

EXERCISE C By examining Fig. 1.8*b*, *c*, and *d*, complete the argument which establishes the last three lines of Table 1.

EXERCISE D Consider the situation where $P \subseteq Q$ and $Q \subseteq P$ are both false. Draw an appropriate figure, and show that in this case all four conditionals are false.

REMARKS

1 This theorem warns us against confusing the truth of a conditional with that of its converse or inverse.
2 It also tells us that the contrapositive is just another way of stating a given conditional. If we have trouble proving the truth of a conditional, we may shift to the contrapositive. If the contrapositive is true, so is the conditional.

ILLUSTRATION 1 The contrapositive of

For all x: If x is odd, then x^2 is odd.

is

For all x: If x^2 is not odd, then x is not odd.

or better

For all x: If x^2 is even, then x is even.

ILLUSTRATION 2 The contrapositive of

For all pairs of lines L and M: If L is parallel to M, then L and M are coplanar.

is

For all pairs of lines L and M: If L and M are not coplanar, then L is not parallel to M.

EXERCISE E Write the converse and inverse of the conditionals of Illustrations 1 and 2. Are these true or false?

The idea of a contrapositive can be extended to more complicated conditionals such as

$$\forall_x[(p_x \text{ and } q_x) \rightarrow r_x]$$

The contrapositive is $\forall_x[(\sim r_x) \rightarrow \sim(p_x \text{ and } q_x)]$, which can be rewritten

$$\forall_x[(\sim r_x) \rightarrow (\sim p_x \text{ or } \sim q_x)] \tag{1}$$

There are also two partial contrapositives, each of which is true if and only if the given conditional is true, namely:

$$\forall_x[(\sim r_x \text{ and } q_x) \rightarrow (\sim p_x)] \tag{2}$$
$$\forall_x[(p_x \text{ and } \sim r_x) \rightarrow (\sim q_x)] \tag{3}$$

ILLUSTRATION 3 The three contrapositives of

For all x: If x is even and x is a prime, then $x = 2$.

are

a For all x: If $x \neq 2$, then x is odd or x is not a prime.
b For all x: If $x \neq 2$ and x is a prime, then x is odd.
c For all x: If x is even and $x \neq 2$, then x is not a prime.

1.7 ALTERNATIVE EXPRESSIONS FOR CONDITIONALS

Mathematics frequently expresses conditionals in language different from that used above, and consequently you must learn to recognize conditionals even when they are disguised in a fashion which may seem confusing at first. In the expressions that follow we have omitted the symbol $\forall_x$ and the corresponding phrase "For all x"; for hereafter these will be understood even though they do not appear explicitly. This omission is the usual practice in mathematics.

DEFINITION The following six statements all carry the same meaning:

$p_x \rightarrow q_x$
$P \subseteq Q$
Sufficient — p_x is sufficient for q_x.
Necessary — q_x is necessary for p_x.
If — If p_x, then q_x.
Only if — Only if q_x, then p_x.

The last four statements can be understood by referring to Fig. 1.8a. From this we see that:

For x to be in Q, it is sufficient that it be in P.
For x to be in P, it is necessary that it be in Q.
If x is in P, then x is in Q.
Only if x is in Q is it in P.

ILLUSTRATION 1 The conditional "If a polygon is a square, then it is a rectangle", can be rewritten in the following ways:

The fact that a polygon is a square is a sufficient condition that it is a rectangle.
The fact that a polygon is a rectangle is a necessary condition that it is a square.
A polygon is a square only if it is a rectangle.

The first two of these may be rephrased:

A sufficient condition that a polygon is a rectangle is that it is a square.
A necessary condition that a polygon is a square is that it is a rectangle.

REMARK From the definition above it follows that the converse of a conditional can be obtained in any of the following ways, depending on how the given conditional is phrased:

Interchange p_x and q_x.
Interchange P and Q.
Replace "necessary" by "sufficient".
Replace "sufficient" by "necessary".
Replace "if" by "only if".
Replace "only if" by "if".

In a similar fashion, the equivalence $p_x \leftrightarrow q_x$ can be expressed in a number of alternative ways.

DEFINITION The following seven statements all carry the same meaning:

$p_x \leftrightarrow q_x$
$(p_x \rightarrow q_x)$ and $(q_x \rightarrow p_x)$
$P = Q$
p_x is necessary and sufficient for q_x.
q_x is necessary and sufficient for p_x.
If and only if — p_x if and only if q_x.
q_x if and only if p_x.

ILLUSTRATION 2 The equivalence, "A triangle is equilateral if and only if it is equiangular", can be rewritten in the following ways:

A triangle is equiangular if and only if it is equilateral.
A necessary and sufficient condition that a triangle is equilateral is that it is equiangular.
A necessary and sufficient condition that a triangle is equiangular is that it is equilateral.

PROBLEMS 1.7

In Probs. 1 to 4 state the given conditional in "if . . . , then . . ." language.

1. A necessary condition that a positive integer is divisible by 9 is that the sum of its digits is divisible by 3.
2. A sufficient condition that a triangle is isosceles is that it is equiangular.
3. Let x be a real number. Then $x^2 > 0$ only if $x \neq 0$.
4. $(x^2 = 9$ and x is positive$) \rightarrow (x = 3)$

In Probs. 5 and 6 express the given equivalence as the conjunction of two conditionals using "if . . . , then . . ." language.

5 Let x be a positive integer. Then x^2 is even if and only if x is even.
6 A necessary and sufficient condition that two triangles are similar is that their pairs of corresponding angles are equal.

In Probs. 7 to 12 state the converse, inverse, and contrapositive of the given conditional.

7 For all integers x: If x is even, then x^2 is even.
8 For all integers x: If $x \neq 0$, then x^2 is positive.
9 For all triangles T: If T is equiangular, then T is isosceles.
10 For all quadrilaterals Q: If Q is a square, then Q is a rectangle.
11 For all pairs of lines L and M: If L and M are parallel, then L and M do not intersect.
12 For all pairs of triangles T and S: If T is congruent to S, then T is similar to S.

13 Write the contrapositive of the inverse of $\forall_x(p_x \rightarrow q_x)$.
14 Write the converse of the contrapositive of $\forall_x(p_x \rightarrow q_x)$.
15 Write the inverse of the converse of $\forall_x(p_x \rightarrow q_x)$.
16 Write the inverse of the contrapositive of $\forall_x(p_x \rightarrow q_x)$.

In Probs. 17 to 20 write the contrapositive and the two (or more) partial contrapositives of the given conditional.

17 For all integers x: If x is negative and $x^2 = 16$, then $x = -4$.
18 For all integers x: If x is divisible by 2 and by 3, then x is divisible by 6.
19 For all quadrilaterals Q with vertices A, B, C, and D: If $AB = CD$ and $AB \| CD$, then Q is a parallelogram.
20 For all pairs of triangles ABC and $A'B'C'$:
If AB is congruent to $A'B'$, and
  AC is congruent to $A'C'$, and
  angle C is congruent to angle C',
then ABC is congruent to $A'B'C'$.

In Probs. 21 to 24 consider the conditional $\forall_x[(p_x \text{ and } q_x) \rightarrow r_x]$, and let P, Q, R be the truth sets of p_x, q_x, and r_x, respectively. By definition this conditional is true if and only if $(P \cap Q) \subseteq R$.

21 Draw a circle diagram illustrating $(P \cap Q) \subset R$, where $P \neq Q$ and $(P \cap Q) \neq R$.
22 Draw a diagram illustrating the partial contrapositive $(R' \cap Q) \subset P'$, and show that this is equivalent to that drawn for Prob. 21. Assume $R' \neq Q$ and $(R' \cap Q) \neq P$.

23 Draw a diagram illustrating the partial contrapositive $(R' \cap P) \subset Q'$, and show that this is equivalent to that drawn for Prob. 21. Assume $P \neq R'$ and $(R' \cap P) \neq Q$.

24 Draw a circle diagram illustrating the contrapositive $R' \subset (P' \cup Q')$, and show that this is equivalent to that drawn for Prob. 21. Assume $P \neq Q$ and $R' \neq (P' \cup Q')$.

In Probs. 25 to 30 write the given conditional using the "sufficient condition" language.

25 If the base angles of a triangle are equal, the triangle is isosceles.

26 If two triangles are congruent, their corresponding altitudes are equal.

27 If two lines are perpendicular to the same line, they are parallel.

28 If two spherical triangles have their corresponding angles equal, they are congruent.

29 If $5x - 3 = x + 5$, then $x = 2$.

30 If $x = 0$, then $x^2 = 0$.

In Probs. 31 to 36 write the given conditional using "necessary condition" language.

31 If two triangles are congruent, their pairs of corresponding angles are equal.

32 If $x = 2$, then $x^2 = 4$.

33 If x and y are both even, then $x + y$ is even.

34 If x and y are both odd, then $x + y$ is even.

35 If $a + b$ is divisible by 3, then $10a + b$ is divisible by 3 (a and b are positive integers).

36 If two triangles are congruent, they are similar.

In Probs. 37 to 42 write the given conditional using the phrase "only if".

37 The conditional of Prob. 31

38 The conditional of Prob. 32

39 The conditional of Prob. 33

40 The conditional of Prob. 34

41 The conditional of Prob. 35

42 The conditional of Prob. 36

In Probs. 43 to 48 write the converse of the given conditional using "necessary" and then "sufficient" language. Give two answers to each problem.

43 The conditional of Prob. 25

44 The conditional of Prob. 26

45 The conditional of Prob. 27

46 The conditional of Prob. 28

47 The conditional of Prob. 29

48 The conditional of Prob. 30

In Probs. 49 to 54 write the converse of the given conditional using the phrase "only if".

49 The conditional of Prob. 31
50 The conditional of Prob. 32
51 The conditional of Prob. 33
52 The conditional of Prob. 34
53 The conditional of Prob. 35
54 The conditional of Prob. 36

In Probs. 55 to 58 write the given equivalence in "necessary and sufficient" language.

55 Two lines are parallel if and only if they do not intersect.
56 Two triangles are equilateral if and only if their pairs of corresponding angles are equal.
57 x^2 is odd if and only if x is odd.
58 x^2 is even if and only if x is even.

59 A man promised his girl: "I will marry you only if I get a job." He got the job and refused to marry her. She sued for breach of promise. Can she logically win her suit? Why?

1.8 » DIRECT PROOF

We recall that the procedure for proving the truth of $p_x \to q_x$ [remember this means $\forall_x(p_x \to q_x)$] was to show that q_x is true for each x for which p_x is true, or that $P \subseteq Q$. In the examples of Sec. 1.5 this was easy, for there were only finitely many values of x for which p_x was true. Usually, however, there are infinitely many such values of x, and so we must turn to more sophisticated methods.

In the method known as direct proof, we are supposed to know the truth of a number of conditionals, say:

$$p_x \to q_x \qquad q_x \to r_x \qquad \text{and} \qquad r_x \to s_x$$

Then we wish to conclude the truth of

$$p_x \to s_x$$

The argument is immediate, for we are, in fact, given the diagrams shown in Fig. 1.9. From these we derive Fig. 1.10. Thus P is contained in S, and $p_x \to s_x$ is true.

Three given conditionals: $p_x \to q_x$; $q_x \to r_x$; $r_x \to s_x$

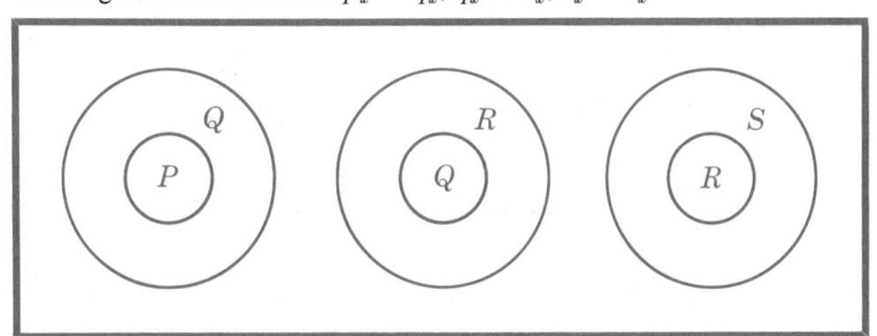

FIGURE 1.9
A conditional is represented by a pair of concentric circles.

The chain of conditionals leading to $p_x \to s_x$

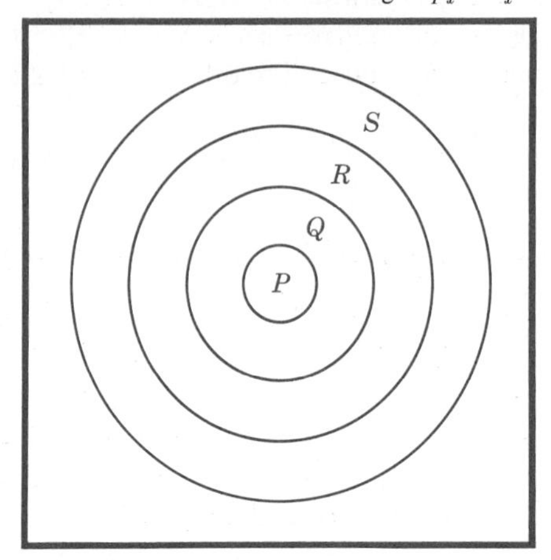

FIGURE 1.10
A chain of conditionals is represented by a collection of concentric circles.

There are endless variations of this pattern, and great skill is required in arranging the work so that the proof is valid. One common trick, for instance, is to replace a conditional by its contrapositive if this will simplify matters.

ILLUSTRATION 1

GIVEN: $(p_x \to \sim q_x)$
$(r_x \to q_x)$

PROVE: $(p_x \to \sim r_x)$

PROOF Since $(r_x \to q_x)$ is equivalent to $(\sim q_x \to \sim r_x)$ we have $(p_x \to \sim q_x)$; $(\sim q_x \to \sim r_x)$. Hence we conclude that $(p_x \to \sim r_x)$.

ILLUSTRATION 2 As a specific example, consider the following proof of

If x is odd, then x^2 is odd. [x is an integer.]

This amounts to the following:

GIVEN:

a If x is odd, then $x = 2a + 1$ where a is an integer.
b If $x = 2a + 1$ where a is an integer, then x is odd.
c The rules of algebra.

PROVE: If x is odd, then x^2 is odd.

PROOF

1	If x is odd, then $x = 2a + 1$.	Given
2	If $x = 2a + 1$, then $x^2 = 4a^2 + 4a + 1$.	Algebra
3	$4a^2 + 4a + 1 = 2(2a^2 + 2a) + 1 = 2b + 1$, where b is an integer.	Algebra

4	If $x = 2a + 1$, then $x^2 = 2b + 1$.	Substitution
5	If $x^2 = 2b + 1$, then x^2 is odd.	Given
6	If x is odd, then x^2 is odd.	From (1), (4), and (5)

PROBLEMS 1.8

In Probs. 1 to 4 find the error in the given direct "proofs".

1 GIVEN: 1 If $x = 2$, then $x^2 - 3x + 2 = 0$.
2 If $x = 1$, then $x^2 - 3x + 2 = 0$.
PROVE: 3 If $x = 2$, then $x = 1$.
PROOF: 4 If $x = 2$, then $x^2 - 3x + 2 = 0$. Given
5 If $x^2 - 3x + 2 = 0$, then $x = 1$. Given
6 If $x = 2$, then $x = 1$. From (4) and (5)

2 GIVEN: 1 If $x = 3$, then $x^2 - 5x + 6 = 0$.
2 If $x = 4$, then $x^2 - 5x + 6 \neq 0$.
PROVE: 3 If $x \neq 4$, then $x = 3$.
PROOF: 4 If $x \neq 4$, then $x^2 - 5x + 6 = 0$. Contrapositive of (2)
5 If $x^2 - 5x + 6 = 0$, then $x = 3$. From (1), given
6 If $x \neq 4$, then $x = 3$. From (4) and (5)

3 GIVEN: 1 If two triangles have all three pairs of corresponding angles equal, then they are similar.
2 If two triangles are congruent, then they are similar.
PROVE: 3 If two triangles have all three pairs of corresponding angles equal, then they are congruent.
PROOF: 4 Immediate from (1) and (2).

4 GIVEN: 1 If integer x is even, then x^2 is even.
PROVE: 2 If integer x is odd, then x^2 is odd.
PROOF: 3 (2) is the contrapositive of (1).

1.9 › INDIRECT PROOF

If you have difficulty in constructing a direct proof, you can sometimes make progress by using other tactics. The method of indirect proof is based upon the fact that $p_x \to q_x$ [remember that this means $\forall_x(p_x \to q_x)$] must be either true or false. If we can show that it is not false, then it must be true. We proceed by assuming that it is false, combining this assumption with other known facts, and (if we are successful) arriving at a contradiction. Since contradictions are impossible in correct thinking, we must have made a mistake somewhere. Our only dubious statement was the assumption that $p_x \to q_x$ is false. Hence this must be in error, and $p_x \to q_x$ must be true.

In practice, then, we must form the statement "$p_x \to q_x$ is false". It is at this point that most errors occur, for students forget what we said in Sec. 1.5. Our discussion there amounts to the following:

The conditional $p_x \rightarrow q_x$ is false if and only if there exists at least one value of x for which p_x is true *and* q_x is false (i.e., $\sim q_x$ is true). We proceed by making precisely this assumption.

ILLUSTRATION 1 Suppose that we are trying to construct indirect proofs of the following:

a If x^2 is even, then x is even. [x is an integer.]
b If x is not a perfect square, then x cannot be written in the form $x = a^2/b^2$, where a and b are integers. [x is an integer.]
c If x^2 is greater than zero, then $x \neq 0$. [x is an integer.]

Assume $p_x \rightarrow q_x$ is false

To start the indirect proof, we assume that for at least one value of x:

a x^2 is even *and* x is odd.
b x is not a perfect square, *and* x can be written in the form $x = a^2/b^2$, where a and b are integers.
c x^2 is greater than zero, *and* $x = 0$.

ILLUSTRATION 2 In other situations the variable x may represent something other than a number. For example, consider:

a If a *pair* of lines are cut by a transversal so that a pair of alternate interior angles are equal, then the lines are parallel. [x represents a *pair* of lines.]
b If *triangle ABC* has $a \neq b$, then triangle ABC has angle $A \neq$ angle B. [x represents triangle ABC.]

To start the indirect proof, we assume that:

a There exists at least one pair of lines, such that the lines are cut by a transversal so that alternate interior angles are equal, *and* the lines are not parallel.
b There exists at least one triangle ABC such that $a \neq b$ *and* angle $A =$ angle B.

Standard notation for a triangle

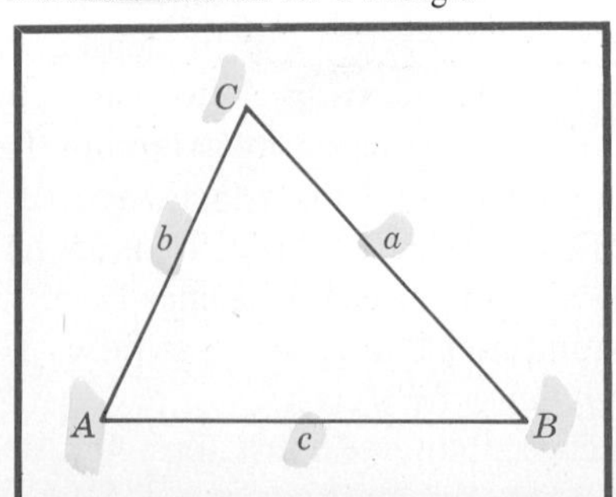

FIGURE 1.11

Indirect proof also can be applied to statements which are not conditionals. To begin the indirect proof, we assume the negation of the given statement.

ILLUSTRATION 3 Consider constructing indirect proofs of the following:

a There are infinitely many prime numbers.
b The integer 2 cannot be written in the form $2 = a^2/b^2$, where a and b are integers.

To start the indirect proof, we assume that:

a There are only finitely many prime numbers.
b The integer 2 can be written in the form $2 = a^2/b^2$, where a and b are integers.

Find a contradiction

The remainder of an indirect proof consists in establishing a contradiction based upon this assumption and other known facts. There is no standard way of doing this, but the following illustrations should give the idea.

ILLUSTRATION 4 Construct an indirect proof of

If x^2 is even, then x is even. [x is an integer.]

We assume that there is at least one integer x for which x^2 is even *and* x is odd.

From Sec. 1.8, Illustration 2, we know that if x is odd, then x^2 is odd. Hence it is impossible to have x odd and x^2 even. This is the desired contradiction.

ILLUSTRATION 5 Construct an indirect proof of the following: If a pair of lines are cut by a transversal so that a pair of alternate interior angles are equal, then the lines are parallel.

We assume that there is at least one pair of lines such that the lines are cut by a transversal so that alternate interior angles are equal, *and*

Assumption at beginning of indirect proof

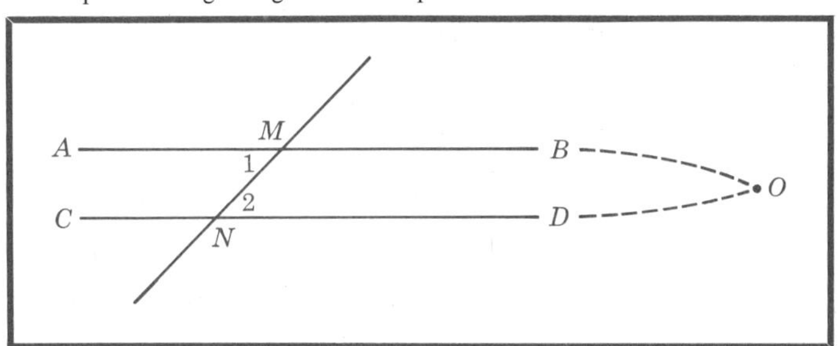

FIGURE 1.12
The figure assumes that angle 1 = angle 2 and that lines AB and CD meet at O.

the lines are not parallel. That is: assume that $\angle 1 = \angle 2$ and that AB and CD meet at O. Now in triangle MON, $\angle 1$ is an exterior angle, and hence it must be greater than the interior angle $\angle 2$. This contradicts the assumption that $\angle 1 = \angle 2$.

PROBLEMS 1.9 In Probs. 1 to 10 construct an indirect proof of the given conditional. You may use any theorems of geometry which you learned in high school.

1. If triangle ABC (see Fig. 1.11) has $a \neq b$, then triangle ABC has $\angle A \neq \angle B$.
2. If triangle ABC (see Fig. 1.11) has $\angle A \neq \angle B$, then triangle ABC has $a \neq b$.
3. If quadrilateral $ABCD$ has unequal diagonals, then quadrilateral $ABCD$ is not a rectangle.
4. If point P is not on line L, then there is no more than one line through P perpendicular to L.
5. If point P is on line L, then in a plane through L there is no more than one line through P perpendicular to L.
6. In triangles ABC and $A'B'C'$; if $b = b'$, $c = c'$, $\angle A \neq \angle A'$, then $a \neq a'$.
7. Two distinct circles can have at most two distinct points in common.
8. Two distinct planes can have at most a line in common.
9. Define a tangent to a circle at point P as the line perpendicular to the radius OP at P. Prove that P is the only point which is on both the circle and the tangent.
10. Define a tangent to a circle as a line which has one and only one point in common with the circle, namely point P. Prove that this tangent at P is perpendicular to the radius OP.

1.10 › OTHER METHODS OF PROOF

USE OF THE CONTRAPOSITIVE When we are trying to prove the truth of a conditional $p_x \to q_x$, we can just as well prove the contrapositive, $\sim q_x \to \sim p_x$. For we have seen that a conditional and its contrapositive are equivalent. Sometimes the contrapositive is easier to prove, and then we should choose this method. Often there are great similarities between indirect proof and the proof of the contrapositive.

ILLUSTRATION 1 Let us consider the theorem of Sec. 1.9, Illustration 5. The contrapositive of the stated conditional is: If two lines are not parallel, the alternate interior angles obtained by cutting these lines by a transversal are not equal.

We establish this by the precise argument used in Illustration 5 above.

PROOF BY ENUMERATION Suppose we wish to prove that for *all* x, p_x is true, under the assumption that the variable x refers to a *finite* universal set U. We can then give a complete proof by checking the truth of p_x for each x that is an element of U.

ILLUSTRATION 2 Let U be the set $\{0, 1\}$, and let "addition", $\oplus$, be defined by the following table.

$\oplus$	0	1
0	0	1
1	1	0

Prove that for all x in U, $x \oplus 1 = 1 \oplus x$. To do so we let x take the values 0 and 1 in turn.

$$0 \oplus 1 = 1 \qquad 1 \oplus 0 = 1$$

Hence

$$0 \oplus 1 = 1 \oplus 0$$

When $x = 1$, both sides of $x \oplus 1 = 1 \oplus x$ become $1 \oplus 1$; so they are equal. Therefore, for all x in U,

$$x \oplus 1 = 1 \oplus x$$

REMARK This method applies only when U is finite. If U is infinite, we could never finish the job of checking each value of x, and this method fails.

PROOF OF EXISTENCE Before you spend a lot of time and money (on a high-speed computer, say) trying to solve a problem, it is a good idea to determine in advance that the problem actually does have a solution. You have probably never seen problems that do not have solutions, for most textbooks and teachers consider it to be bad form to ask students to do something which is impossible. In actual practice, however, such problems may arise and it is a good idea to know how to recognize them. A very simple example of such a problem is the following:

Find all the integers x which satisfy the equation

$$7x + 5 = 2x + 9$$

In order to reassure you that you are working on problems that do have solutions, mathematicians have developed a number of "existence theorems". These are statements of the following form:

There exists a number x which has a given property.

An important example of such a theorem is this one:

If a and b are any real numbers such that $a \neq 0$, there exists a real number x which satisfies the equation $ax + b = 0$.

The best way of proving such a theorem is to exhibit a number x with the required property. The proof of the above theorem amounts to checking that $x = -(b/a)$ satisfies the given equation.

Although there are other forms of existence proofs, a constructive proof of this kind is considered to be of greater merit, and this method is used widely in establishing the existence of solutions of various types of equations.

1.11 › METHODS OF DISPROOF

If you have tried unsuccessfully to prove a conjectured theorem, you may well spend some time trying to disprove it. There are two standard methods for disproving such statements.

DISPROOF BY CONTRADICTION In this case we assume that the given statement is true and then proceed to derive consequences from it. If we succeed in arriving at a consequence which contradicts a known theorem, we have shown that the given statement is false.

ILLUSTRATION 1 Disprove the statement: "The square of every odd number is even."

Of course, this immediately contradicts our previous result (Sec. 1.8, Illustration 2) that the square of every odd number is odd. But let us disprove it from first principles. Since every odd number can be written in the form $2a + 1$, where a is an integer, and since every even number can be written in the form $2b$, where b is an integer, the given statement implies that

$$(2a + 1)^2 = 2b \qquad \text{for some } a \text{ and } b$$

or

$$4a^2 + 4a + 1 = 2b$$

Both sides are supposed to represent the same integer, but the left-hand sides *is not* divisible by 2, while the right-hand side *is* divisible by 2. This is surely a contradiction, and so the given statement is false.

DISPROOF BY COUNTEREXAMPLE This method is effective in disproving statements of the form

For *all* values of x, a certain statement involving x is true.

An example is the following:

For all values of x, $x^2 + 16 = (x + 4)(x - 4)$.

Counterexample

In order to disprove such an assertion, we proceed to find a "counterexample". In other words, we look for *one* value of x for which the statement is false; and since the statement was supposed to be true for *all* values of x, this single counterexample is the end of the matter. In the above example, $x = 0$ does the job.

ILLUSTRATION 2 Disprove the statement: "The square of every odd number is even."

All we have to do is to find a single odd number whose square is odd. Since $3^2 = 9$, we have established the disproof.

PROBLEMS 1.11

In Probs. 1 to 8 disprove the given statement.

1 The sum of two odd integers is odd.
2 The product of two odd integers is even.
3 $(x + 3)^2 = x^2 + 9$
4 If $x^2 + 2x = -1$, then $x + 2 = -1$ or $x = 1$.
5 If $x^2 = 4$, then $x = 2$.
6 $(2x + 1)/2 = x + 1$
7 $\sqrt{x^2} = x$ HINT: Let x be negative.
8 Given any three segments AB, CD, and EF, there is a triangle whose sides have lengths equal to the lengths of these given segments.

9 Give an indirect proof of the theorem: "There exist an infinite number of primes." If you are unable to do so, consult Courant and Robbins, "What Is Mathematics?", page 22. This theorem is due to Euclid.

1.12 » MATHEMATICAL MODELS

By this time you should have begun to understand what we mean by saying that mathematics is abstract. Mathematical proof is a process of reasoning by given rules from a set of axioms (which are assumed to be true) to a valid conclusion, which we call a "theorem". Because of the abstract character of mathematics, we cannot expect to prove anything about our physical world by purely mathematical means.

Scientists, however, spend their lives uncovering the secrets of nature, and engineers put these discoveries to work for the benefit of our society. You may quite properly wonder how an abstract subject like mathematics has become such an important tool for scientists and engineers. The key to this matter is the concept of a "mathematical model" of nature. The first step in the study of any branch of science is that of observing nature. When enough facts have been collected, the scientist begins to organize them into some pattern. In quantitative sciences like astronomy, chemistry, and physics this pattern is expressed in terms of mathematics. The undefined terms of the abstract mathematics (points, lines, etc.) represent physical objects; refined abstract concepts (velocity, acceleration, force, etc.) are then defined to correspond to intuitive ideas which seem important to the scientist. Then mathematical equations involving these concepts are used as axioms to describe the observed behavior of nature. All of these, taken together, constitute our mathematical model. This model, of course, is only a picture of nature; it differs from nature just as a model of an aircraft differs from the real plane itself. But just as a great deal can be learned about a plane from a model which is studied in a wind tunnel, we can use our mathematical model to help us understand nature. From our axioms, we can deduce theorems, which are true only in our abstract sense. Nevertheless, if our model is well constructed, these theorems will correspond to observable properties of nature which we may well not have suspected in advance. At the very worst, these theorems are intelligent guesses about nature and serve as guides for our experimental work. At their best, when the model is a good one as is the case in most physical sciences, our mathematical results can almost be identified with physical truth. In those portions of science which you are likely to be studying along with this book, this correspondence is so close that you may not realize the difference between mathematics and nature itself. It is our hope that the study of this chapter will have helped you to appreciate this important distinction.

REFERENCES

Carroll, Lewis: "Logical Nonsense", Putnam, New York (1934).

Courant, Richard, and Herbert Robbins: "What Is Mathematics?" Oxford, New York (1941).

Stabler, E. R.: "An Introduction to Mathematical Thought", Addison-Wesley, Reading, Mass. (1953).

Stoll, Robert R.: "Sets, Logic, and Axiomatic Theories", Freeman, San Francisco (1961).

Suppes, Patrick: "Introduction to Logic", Van Nostrand, Princeton, N.J. (1957).

Tarski, Alfred: "Introduction to Logic", Oxford, New York (1946).

2 THE NUMBER SYSTEM

2.1 INTRODUCTION

Since numbers are basic ideas in mathematics, we shall devote this chapter to a discussion of the most important properties of our number system. We do not give a complete account of this subject, and you are likely to study it in more detail when you take more advanced courses in mathematics. Numerous suggestions for further reading are given at the end of the chapter.

Let us retrace briefly the development of numbers as it is usually presented in schools. As a young child you first learned to count, and thus became acquainted with the *natural numbers* 1, 2, 3, ... In your early study of arithmetic you learned how to add, subtract, multiply, and divide pairs of natural numbers. Although some divisions such as $6 \div 3 = 2$ were possible, it soon developed that new numbers had to be invented so as to give meaning to expressions like $7 \div 2$ and $3 \div 5$. To handle such situations, fractions were introduced, and the arithmetic of fractions was developed.

It should be noted that the invention of fractions was a major step in the development of mathematics. In the early days many strange practices were followed. The Babylonians considered only fractions whose denominators were 60, the Romans only those whose denominators were 12. The Egyptians insisted that the numerators must be 1, and wrote $\frac{1}{3} + \frac{1}{15}$ instead of $\frac{2}{5}$. Our modern notation dates from Leonardo of Pisa (also called Fibonacci), whose great work *Liber Abaci* was published in A.D. 1202.

Later on you became acquainted with zero and negative numbers such as -7, -3, $-\frac{5}{3}$, $-4\frac{1}{5}$, etc., and you learned how to calculate with these. The entire collection consisting of the positive and negative integers, zero, and the positive and negative fractions is called the system of *rational numbers*. The advantage of using this system in contrast to the system of purely positive numbers is that it is possible to subtract any rational number from any rational number. With only positive numbers available, 3 − 5, for instance, is meaningless. It is interesting to note that it took many years before negative numbers were permanently established in mathematics. Although they were used to some extent by the early Chinese, Indians, and Arabs, it was not until the beginning of the seventeenth century that mathematicians accepted negative numbers on an even footing with positive numbers.

When you were introduced to *irrational* numbers such as $\sqrt{2}$ and π, you were told that these could not be expressed as ordinary fractions. Instead, they are written in the form of infinite decimal expansions such as 1.4142 . . . and 3.1415 . . . . The decimal expansions of the rational numbers are also infinite; for example,

$$\begin{aligned} \tfrac{1}{4} &= 0.25000\ldots \\ \tfrac{1}{3} &= 0.33333\ldots \\ 2 &= 2.00000\ldots \\ \tfrac{1}{7} &= 0.142857142857\ldots \end{aligned}$$

These, however, repeat after a certain point, whereas the irrationals do not have this property. The collection of all these, the rationals plus the irrationals, is called the system of *real* numbers. It is quite difficult to give a completely satisfactory definition of a real number, but for our present purposes the following will suffice.

Real number

DEFINITION A *real number* is a number which can be represented by an infinite decimal expansion.

If you wish a more subtle definition of a real number, read Courant and Robbins, "What Is Mathematics?", chap. 2.

Although a real number is a definite mathematical object, we can express such a number in a great variety of notations. For example, we can write 7 in the following ways:

$$\text{VII}, \quad 111_{\text{two}} \text{ (base two)}, \quad \tfrac{21}{3}, \quad 7.000\ldots, \quad 9-2$$

The rational number usually written $\frac{1}{2}$ can also be written

$$\frac{2}{4}, \quad \frac{8\pi}{16\pi}, \quad 0.5000\ldots, \quad \frac{\frac{1}{8}}{\frac{1}{4}}, \quad \frac{1}{4}+\frac{1}{4}, \quad \left(\frac{1}{\sqrt{2}}\right)^2, \quad \sqrt{\frac{1}{4}}$$

For each real number it is customary to adopt a "simplest" expression which is commonly used to represent it (7 and $\frac{1}{2}$ in the examples above), but we shall not hesitate to use other representations when they are more convenient.

Since $\frac{1}{2}$ and $\frac{2}{4}$ are merely names for the same number, we call them equal and write

$$\frac{1}{2} = \frac{2}{4}$$

Equality

More generally, we define equality for symbols representing real numbers as follows.

DEFINITION Two symbols, a and b, representing real numbers are equal if and only if they represent the same real number.

From this definition, Theorem 1 follows at once.

THEOREM 1 If a, b, and c represent real numbers and if $a = b$; then $a + c = b + c$, $a - c = b - c$, $ac = bc$, and $a/c = b/c$ (provided $c \neq 0$).

For, since a and b represent the same real number, $a + c$ and $b + c$ represent the same real number. Hence $a + c = b + c$. The other cases follow the same argument.

EXERCISE A Show that if $a = b$, then $b = a$.

EXERCISE B Show that if $a = b$ and $b = c$, then $a = c$.

2.2 › BINARY OPERATIONS

The usual operations of arithmetic such as $2 + 3 = 5$, $4 \times 5 = 20$, $8 - 6 = 2$, and $10 \div 5 = 2$ are called *binary* because, if we choose any *two* numbers, the operation generates a third number. That is, there are *two* inputs and one output. Set union and intersection, $A \cup B = C$ and $A \cap B = C$, are examples of binary operations on sets. On the other hand set complementation A' has a single input, A, and hence is called *unary*. Operations with three or more inputs can be defined, but since they do not occur in elementary mathematics we shall not introduce them here.

Binary operation

The general notion of a binary operation on a set is defined as follows:

DEFINITION Let $S = \{a, b, c, \ldots\}$ be any set. The operation $*$ is a *binary operation on S* if and only if to every ordered pair (a,b), where a and b are elements of S, there is assigned a unique element $a * b$ of S. We indicate this assignment in the notation

$$(a,b) \rightarrow (a * b)$$

There are several delicate points to observe in this definition:

1. The order of a and b may be important, for (a,b) is an *ordered* pair. Thus it may happen that $a * b \neq b * a$.
2. The operation must be defined for *every* pair (a,b) where a and b are elements of S.
3. The "output" $a * b$ must be an element of S.

ILLUSTRATION 1 Let S be the set of real numbers R. Then you know that

Addition: $(a,b) \rightarrow (a + b)$
Multiplication: $(a,b) \rightarrow (a \times b)$
Subtraction: $(a,b) \rightarrow (a - b)$

are binary operations on R.

ILLUSTRATION 2 By contrast,

Division: $(a,b) \rightarrow (a \div b)$

is *not* a binary operation on R, for $(a,0) \rightarrow (a \div 0)$ is not defined. Division, however, *is* a binary operation on the set of *nonzero* real numbers.

ILLUSTRATION 3 Let S be the set of subsets of a universal set U. Then

Union: $(A,B) \rightarrow (A \cup B)$
Intersection: $(A,B) \rightarrow (A \cap B)$

are binary operations on S.

EXERCISE A Are addition, multiplication, subtraction, and division binary operations on the set of positive integers?

COMMUTATIVE PROPERTY The *order* in which a binary operation is performed is important: switching the order sometimes gives different results.

DEFINITION
Commutative property

The binary operation $*$ on a set S is called *commutative* if and only if for every ordered pair (a,b) of elements in S

$$a * b = b * a$$

ILLUSTRATION 4

a. Let S be the set of real numbers R. Then addition and multiplication are commutative on R, since for every a and b,

$$a + b = b + a \qquad \text{and} \qquad a \times b = b \times a$$

b Subtraction on R is not commutative, since, for example, $8 - 2 \neq 2 - 8$.

EXERCISE B Are set union and intersection commutative?

EXERCISE C Is division a commutative binary operation on the set of nonzero real numbers?

ASSOCIATIVE PROPERTY A binary operation on a set S permits us to combine two members of S and thus obtain a third element of S. But it does not tell us how to combine *three* elements of S such as $a * b * c$. For this reason we shall introduce the associative property of a binary operation and show that if the operation satisfies this property, then we can give a meaning to the symbol $a * b * c$.

To introduce the concept of associativity let us take the addition of real numbers as an illustration. If we write the numbers 2, 5, and 8 in the order 2, 5, 8, we wish to find the value of

$$2 + 5 + 8$$

Since addition is a binary operation, we can add only two numbers at a time and so we seem to be in difficulty. We can, however, compute $2 + 5 + 8$ by introducing the grouping

$$(2 + 5) + 8$$

Now add $2 + 5 = 7$, and then add $7 + 8 = 15$. So it appears that

$$2 + 5 + 8 = 15$$

Another grouping is possible,

$$2 + (5 + 8) = 2 + 13 = 15$$

This suggests that the two groupings will always give the same result. As a matter of fact, it is true that for all real numbers a, b, and c

$$(a + b) + c = a + (b + c)$$

This is the statement of the associative property of the binary operation of addition.

The general definition of an associative binary operation is the following:

DEFINITION Associative property

The binary operation $*$ on a set S is associative if and only if for every triple a, b, and c of elements of S,

$$(a * b) * c = a * (b * c)$$

ILLUSTRATION 5

a Addition and multiplication are associative binary operations on R.

b Subtraction is *not* associative on R, since, for example,

$$(12 - 8) - 2 = 2 \qquad \text{and} \qquad 12 - (8 - 2) = 6$$

c Division is *not* associative on R, since, for example,

$$(24 \div 6) \div 2 = 2 \qquad \text{and} \qquad 24 \div (6 \div 2) = 8$$

Now that you understand the meaning of the associative property, let us return to our original problem, namely to give meaning to the expression $a * b * c$ where a, b, and c are elements of the set S on which $*$ is defined as a binary operation. If $*$ is associative, it is reasonable to define $a * b * c$ to be equal to either of the equal expressions $(a * b) * c$ and $a * (b * c)$. The formal definition is the following:

DEFINITION *$a * b * c$* Let $*$ be an associative binary operation on a set S, and let a, b, and c be any elements of S, written in the *order* a, b, c. Then by definition the expression $a * b * c$ is equal to either of the two equal expressions $(a * b) * c$ and $a * (b * c)$.

REMARK The definition of $a * b * c$ depends on the order in which a, b, and c are written. Thus we do not assert that $a * b * c = b * a * c$, although this may be true in special cases.

ILLUSTRATION 6

a Since addition on R is associative, we define $a + b + c$ to be equal to either of the equal expressions $(a + b) + c$ or $a + (b + c)$.

b Similarly we define $a \times b \times c$ to be equal to either of the equal expressions $(a \times b) \times c$ or $a \times (b \times c)$.

c Since subtraction and division are not associative, the expressions $12 - 6 - 3$ and $24 \div 6 \div 2$ are not defined without further discussion.

EXERCISE D Let the binary operation $\triangle$ be defined on R by $a \triangle b = a + 2b$ for all real numbers a and b. Is $\triangle$ commutative? Associative?

EXERCISE E Let the binary operation $\circ$ be defined on R by $a \circ b = a$ for all real numbers a and b. Is $\circ$ commutative? Associative?

GENERALIZED COMMUTATIVE PROPERTY The associative property of addition permits us to define $a + b + c$ for any three real numbers written in this order. But what if we alter the order and consider $b + c + a$? Is $a + b + c = b + c + a$? From your earlier experience you know that indeed they are equal, and that in general the sum of three real numbers is independent of the order in which they are written. This is a statement of the *generalized commutative property* of the addition of real numbers.

Generalized commutative property

The general definition of the generalized commutative property of a binary operation is as follows:

DEFINITION The binary operation $*$ on a set S has the generalized commutative property if and only if all the following six expressions are defined and are equal:

$$\begin{matrix} a * b * c & b * a * c & c * a * b \\ a * c * b & b * c * a & c * b * a \end{matrix}$$

THEOREM 2 If $*$ is a binary operation on a set S which is both commutative and associative, then $*$ satisfies the generalized commutative property.

PROOF Since $*$ is associative, all six of the expressions are defined. We must prove them all to be equal. Instead of carrying out the full proof which is rather tedious, we shall illustrate the method by proving that, for example,

$$a * b * c = c * b * a$$

1	$a * b * c = (a * b) * c$	Since $*$ is associative
2	$= (b * a) * c$	Since $*$ is commutative
3	$= c * (b * a)$	Since $*$ is commutative
4	$= c * b * a$	Since $*$ is associative

The rest of the theorem can be proved in a similar fashion.

EXERCISE F Which of the operations $+$, $-$, $\times$, and $\div$ on the real numbers satisfy the generalized commutative property?

PROBLEMS 2.2

1 A binary operation involves how many inputs?
2 A binary operation $*$ on a set S is an assignment $(a,b) \rightarrow a * b$. **(a)** What are a and b? **(b)** Is $a * b$ required to be an element of S? **(c)** Must $a * b$ be unique?
3 The binary operation $+$ on the set of integers assigns what integer to the pair (5,7)?
4 Why is subtraction not a binary operation on the set of positive integers? Give a counterexample.
5 Name a set on which subtraction *is* a binary operation.
6 Why is division not a binary operation on the set of real numbers.
7 Name a set on which division is a binary operation.
8 Is it always true that $a * b = b * a$?

9 If for all a and b in S, $a * b = b * a$, then the operation $*$ is said to have what property?

10 What binary operations of arithmetic on the set of real numbers are commutative?

11 Why is subtraction on the set of real numbers not commutative? Give a counterexample.

12 Why is division on the set of nonzero real numbers not commutative? Give a counterexample.

13 The binary operation $\otimes$ is defined on the set $\{1, 2, 3\}$ by the table. Is this operation commutative?

$\otimes$	1	2	3
1	1	2	3
2	3	1	2
3	2	3	1

14 The binary operation $\triangle$ is defined on the set of real numbers by the assignment $(a,b) \to a \triangle b = a + 2b$. Is the operation $\triangle$ commutative?

15 The binary operation $\square$ is defined on the set of real numbers by the assignment $(a,b) \to a \square b = a + b + ab$. Is the operation $\square$ commutative?

16 The binary operation $*$ on a set S is associative if for all a, b, and c in S what relationship is true?

17 What binary operations of arithmetic are associative on the set of real numbers?

18 What binary operations of arithmetic are not associative on the set of real numbers? Give counterexamples.

19 The binary operation $\circ$ is defined by the assignment $(a,b) \to a \circ b = a$ on the set of real numbers. Is the operation $\circ$ associative? Commutative?

20 A binary operation on a set S satisfies the generalized commutative property if what relations are true for all a, b, c in S?

21 Does the operation $\circ$ of Prob. 19 satisfy the generalized commutative property?

22 From what other properties of a binary operation can you conclude that the generalized commutative property is true?

23 What binary operations of arithmetic on the set of real numbers satisfy the generalized commutative property?

2.3 › PROPERTIES OF THE REAL NUMBER SYSTEM

In this section we shall discuss in a systematic fashion the arithmetic properties of the real numbers. We shall base this discussion on the properties of addition and multiplication, and shall derive the properties of subtraction and division from these. For the record let us first repeat the properties discussed in Sec. 2.2.

R1 Addition and multiplication are binary operations on the set R of real numbers.
R2 Addition and multiplication are commutative operations.
R3 Addition and multiplication are associative operations.

Identity elements

The next two properties involve the additive and multiplicative identity elements.

DEFINITION A binary operation on S has an identity element, called e, if and only if e is an element of S, and for all elements a of S,

$$a * e = a \qquad \text{and} \qquad e * a = a$$

and e is the only element of S that satisfies these equations.

Since we know that for all real numbers a,

$$a + 0 = a \qquad \text{and} \qquad 0 + a = a$$

it follows that *zero* is the identity element for addition.

Similarly, since for all real numbers a,

$$a \times 1 = a \qquad \text{and} \qquad 1 \times a = a$$

it follows that *one* is the identity element for multiplication.

Thus we have the properties:

R4 The real numbers have a unique additive identity, namely *zero*.
R5 The real numbers have a unique multiplicative identity, namely *one*.

Inverse elements

The next two properties involve inverse elements.

DEFINITION If $*$ is a binary operation on S which has an identity element e, and if a is a given element of S, then the element a^{-1} of S is called the inverse of a (relative to the operation $*$) if and only if

$$a * a^{-1} = e \qquad \text{and} \qquad a^{-1} * a = e$$

and if a^{-1} is the only element of S satisfying these equations.

REMARK The notation a^{-1} is the usual way of writing inverse elements. It is possibly confusing, however, because you may interpret the raised -1 as a negative exponent. Be careful not to do so, for a^{-1} is defined as above and the -1 is *not* an exponent in the usual sense.

For addition, it is easy to find the additive inverse of any real number since, for example:

The additive inverse of 3 is -3 since

$$3 + (-3) = -3 + 3 = 0$$

The additive inverse of -4 is 4 since

$$-4 + 4 = 4 + (-4) = 0$$

The additive inverse of 0 is 0 since

$$0 + 0 = 0$$

The additive inverse of a real number is often called its "opposite", and the additive inverse of a is written $-a$. The additive inverse of $(-a)$, is $-(-a)$; and since $a + (-a) = 0$, it follows that $-(-a) = a$.

Similarly, there is no difficulty with the multiplicative inverses of most real numbers, since, for example:

The multiplicative inverse of 5 is $\frac{1}{5}$ since

$$5 \times \tfrac{1}{5} = \tfrac{1}{5} \times 5 = 1$$

The multiplicative inverse of $\frac{2}{3}$ is $\frac{3}{2}$ since

$$\tfrac{2}{3} \times \tfrac{3}{2} = \tfrac{3}{2} \times \tfrac{2}{3} = 1$$

The real number zero, however, does not have such an inverse. Suppose there were a multiplicative inverse, a, for 0. Then we must have $0 \times a = 1$. But for all a, $0 \times a = 0$ (Theorem 3, page 50). So zero has no multiplicative inverse. The multiplicative inverse of a real number is often called its reciprocal, and the multiplicative inverse of $a(\neq 0)$ is written $1/a$.

Thus we have the two properties:

R6 Every real number a has a unique additive inverse: $-a$.

R7 Every real number a except zero has a unique multiplicative inverse: $1/a$.

The existence of these inverses permits us to define subtraction and division in terms of addition and multiplication as follows:

DEFINITIONS

Subtraction

1 The difference $a - b$ of two real numbers is defined by the equality

$$a - b = a + (-b)$$

Division

2 The quotient $a \div b$ (for $b \neq 0$) of two real numbers is defined by the equality

$$a \div b = a \times \left(\frac{1}{b}\right)$$

REMARKS

1 Since zero does not have a multiplicative inverse, division by zero is not defined.

2 As an extension of the definition of subtraction, it is customary to define $a - b - c$ to be equal to $a + (-b) + (-c)$, to which the properties of addition can be applied. For example,

$$12 - 6 - 3 = 12 + (-6) + (-3) = [12 + (-6)] + (-3)$$
$$= 6 - 3 = 3$$

Alternatively,

$$\begin{aligned} 12 - 6 - 3 &= 12 + [(-6) + (-3)] \\ &= 12 + (-9) \\ &= 12 - 9 \\ &= 3 \end{aligned}$$

A further extension of this is the following convention: Any sequence of additions and subtractions such as $a - b + c + e - f - g$ is defined to be equal to the sum

$$a + (-b) + c + e + (-f) + (-g) = (a + c + e) - (b + f + g)$$

Thus

$$\begin{aligned} 15 - 6 + 10 + 3 - 4 - 8 &= (15 + 10 + 3) - (6 + 4 + 8) \\ &= 28 - 18 \\ &= 10 \end{aligned}$$

Similarly, $a \div b \div c$ is defined to be $a \times (1/b) \times (1/c)$. So, for example, $24 \div 6 \div 2 = 24 \times \frac{1}{6} \times \frac{1}{2} = 2$. And $a \div b \times c \times e \div f \div g$ is defined to be

$$a \times \left(\frac{1}{b}\right) \times c \times e \times \left(\frac{1}{f}\right) \times \left(\frac{1}{g}\right) = \frac{a \times c \times e}{b \times f \times g}$$

Thus

$$12 \div 3 \times 5 \times 6 \div 4 \div 10 = \frac{12 \times 5 \times 6}{3 \times 4 \times 10} = \frac{360}{120} = 3$$

There is a relation which connects addition and multiplication and which is used over and over again in arithmetic and algebra. It is called the *distributive property of multiplication over addition.*

Distributive property

R8 *Distributive Property of Multiplication over Addition:* For all real numbers a, b, and c,

$$a \times (b + c) = (a \times b) + (a \times c)$$

ILLUSTRATION 1

a $4 \times (2 + 3) = (4 \times 2) + (4 \times 3)$

b $2 \times (x + y) = 2x + 2y$

c $(a + b) \times (c + d) = [(a + b) \times c] + [(a + b) \times d)]$
$$= ac + bc + ad + bd$$

REMARK The distributive property involves parentheses around $b + c$, and results in an unambiguous expression. But what meaning is to be given to the expression $a \times b + c$ without parentheses? Does this mean $(a \times b) + c$ or $a \times (b + c)$ or what? So that such expressions have a unique meaning, it is customary to adopt the convention: In a sequence of operations involving $+, -, \times, \div$, in which parentheses do not appear, all multiplications and divisions are to be completed first and then are to be followed by additions. Thus, $a \times b + c = (a \times b) + c$. Nevertheless, some people are confused by this convention, and to avoid this confusion most authors insert the parentheses as shown. On the other hand, the expression $ab + c$ is clear to most readers, and parentheses are usually omitted.

Properties R1 to R8 constitute the algebraic foundation of the real numbers, and from them we shall derive other important laws of algebra. In more advanced mathematics R1 to R8 are taken to be the axioms of an abstract system called a *field*. Hence we may say that the real numbers form a field.

Field

Within the set of real numbers R there is the important subset R^+ of *positive* real numbers. This subset has the important properties of which you should already be aware:

Positive numbers

R9 The sum of two positive real numbers is positive.
R10 The product of two positive real numbers is positive.
R11 *Law of Trichotomy:* For any real number a, one and only one of the following is true:
- **a** a is positive.
- **b** $-a$ is positive.
- **c** a is zero.

If $-a$ is positive, we say that a is negative. Thus R is subdivided into three disjoint subsets:

- **a** The positive reals
- **b** The negative reals
- **c** Zero

To any real number a we can associate a nonnegative real number called its absolute value $|a|$ according to the following definition:

Absolute value

DEFINITION
1. If a is positive, then $|a| = a$.
2. If $-a$ is positive (and hence a is negative), $|a| = -a$.
3. If $a = 0$, then $|a| = 0$.

ILLUSTRATION 2
- **a** $|4| = 4$
- **b** $|-2| = 2$
- **c** $|0| = 0$

Absolute value

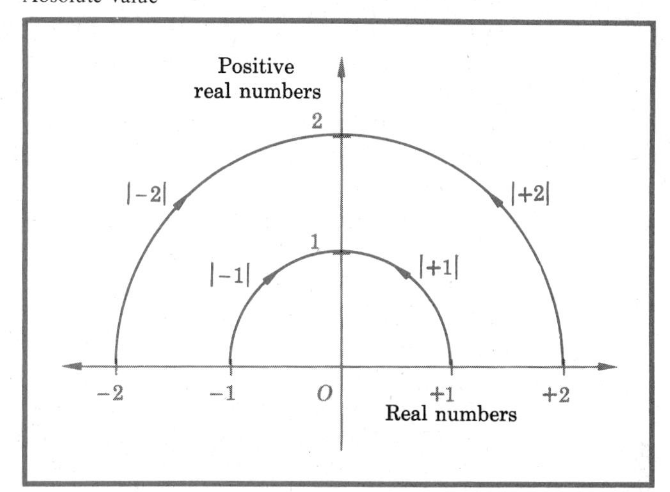

FIGURE 2.1
To each real number there is assigned a positive real number (or zero) called its absolute value. This assignment is indicated by the arrows in the figure.

This subdivision of the real numbers into the subsets of positive, negative, and zero real numbers permits us (in Sec. 2.10 and Chap. 8) to introduce the concepts of order such as *greater than* and *less than* into the real number system. If an abstract field has properties R9, R10, and R11 as well as R1 to R8, it is called an *ordered field.* Thus we may say that the real numbers form an ordered field.

Ordered field

SUMMARY: PROPERTIES OF THE REAL NUMBERS

R1 Addition and multiplication are binary operations on the set R of real numbers.

R2 Addition and multiplication are commutative operations.

R3 Addition and multiplication are associative operations.

R4 The real numbers have a unique additive identity, namely *zero.*

R5 The real numbers have a unique multiplicative identity, namely *one.*

R6 Every real number a has a unique additive inverse: $-a$.

R7 Every real number a except zero has a unique multiplicative inverse: $1/a$.

R8 *Distributive Property of Multiplication over Addition:* For all real numbers a, b, and c,

$$a \times (b + c) = (a \times b) + (a \times c)$$

R9 The sum of two positive real numbers is positive.

R10 The product of two positive real numbers is positive.

R11 *Law of Trichotomy:* For any real number a, one and only one of the following is true:

- **a** a is positive.
- **b** $-a$ is positive; that is, a is negative.
- **c** a is zero.

2.4 THEOREMS ABOUT REAL NUMBERS

There are many theorems about real numbers that can be proved from properties R1 to R11. The first of these is Theorem 3 to which we have already referred. It is a direct consequence of the distributive property.

THEOREM 3 For every real number a: $a \times 0 = 0$.

PROOF

$a \times 0 = 0$

1	$0 = 0 + 0$	Definition, Sec. 2.2
2	$a \times 0 = a \times (0 + 0)$	Theorem 1
3	$a \times 0 = (a \times 0) + (a \times 0)$	Distributive property
4	$a \times 0 = a \times 0$	An identity
5	$0 = a \times 0$	Subtracting (4) from (3)

From this theorem we conclude the following useful result.

THEOREM 4 If a and b are two real numbers such that $ab = 0$, then $a = 0$, or $b = 0$.

PROOF If $a = 0$, the theorem is immediately verified.
If $a \neq 0$, then $1/a$ is defined. Hence we may write

$$\left(\frac{1}{a}\right) \times ab = \left(\frac{1}{a}\right) \times 0$$

Using the associative property for multiplication and Theorem 3, we find that $b = 0$, which proves the theorem.

This theorem has very many applications, especially in the solution of equations.

ILLUSTRATION 1 Solve $x^2 - 5x + 6 = 0$.
By factoring we find that $(x - 2)(x - 3) = 0$.
From Theorem 4 we see that either

$$x - 2 = 0 \quad \text{and} \quad x = 2$$

or

$$x - 3 = 0 \quad \text{and} \quad x = 3$$

Hence 2 and 3 are solutions of the given equation.

THE PRODUCT OF TWO REAL NUMBERS IS ZERO IF AND ONLY IF AT LEAST ONE OF THE TWO FACTORS IS ZERO.

A second consequence of the distributive property is the set of rules for multiplying signed numbers. These are easily derived from the following theorem.

THEOREM 5 For any real number a, $(-1) \times a = -a$.

PROOF

1	$1 + (-1) = 0$	Definition of additive inverse
2	$[1 \times a] + [(-1) \times a] = 0 \times a$	Distributive property
3	$0 \times a = 0$	Theorem 3
4	$1 \times a = a$	Definition of 1
5	$a + [(-1) \times a] = 0$	(2), (3), and (4)
6	$a + (-a) = 0$	Definition of additive inverse
7	$a + (-1) \times a = a + (-a)$	(5) and (6)
8	$(-1) \times a = -a$	Subtraction of a from both sides

COROLLARY $(-1) \times (-1) = 1$

Put $a = -1$ in Theorem 5 and apply the fact that $-(-a) = a$. Now we can prove the usual rules as follows.

THEOREM 6 Let p and q be any positive real numbers. Then:

$p \times (-q) = -(pq)$
$(-p) \times (-q) = pq$

a $p \times (-q) = -(pq)$
b $(-p) \times (-q) = pq$

PROOF Write $(-p) = (-1) \times p$; $(-q) = (-1) \times q$; and

$$-(pq) = (-1) \times p \times q$$

Then the identities of the theorem follow from the associative and commutative properties of multiplication and the corollary to Theorem 5.

PROBLEMS 2.4

In Probs. 1 to 16 state whether the given equality is true or false. For those that are true, state what property of the real numbers justifies the equality; for those that are false, what property is violated.

1 $2 + (4 + 5) = 2 + (5 + 4)$
2 $(2 + 6) + 8 = 2 + (6 + 8)$
3 $12 - (8 - 1) = (12 - 8) - 1$
4 $2 + (5 - 3) = 2 + (3 - 5)$
5 $6 + (8 - 5) = (8 - 5) + 6$
6 $7 + 9 + 3 = 3 + 9 + 7$
7 $10 - (7 - 3) = (7 - 3) - 10$
8 $(15 - 10) - 2 = 15 - (2 - 10)$
9 $(7 \times 3) \times 8 = 7 \times (3 \times 8)$
10 $(16 \div 8) \div 2 = 16 \div (8 \div 2)$
11 $30 \div 6 = 6 \div 30$

12 $4 \times 5 \times 6 = 6 \times 5 \times 4$
13 $5 \times (6 + 3) = (5 \times 6) + (5 \times 3)$
14 $4(y + z) = 4y + 4z$
15 $3x + 6y = 3(x + 6y)$
16 $\frac{1}{2}(a + b) = (\frac{1}{2}a) + b$

17 Without using the generalized commutative property, prove that $a + b + c = b + c + a$.
18 As in Prob. 17, prove that $b \times a \times c = c \times b \times a$.
19 $[(a + b) + c] + d = (a + b) + (c + d)$
$= a + [b + (c + d)]$. Why?
20 Define $a + b + c + d$ to be equal to either of the equal (Prob. 19) expressions: $[a + b + c] + d$; $a + [b + c + d]$. Then prove that $a + b + c + d = d + c + b + a$.
21 Define $a \times b \times c \times d$.
22 Is there an identity element for the binary operation of subtraction on R? If so, what is it?
23 Is there an identity element for the binary operation of division on the set of nonzero reals? If so, what is it?
24 What is the additive inverse of each of the following: 3, -4, π, 0, $-\sqrt{7}$?
25 What is the multiplicative inverse of each of the following: 2, $\frac{3}{4}$, $-\frac{6}{5}$, 1, 0?
26 What is the absolute value of each of the following: 6, -8, $\frac{2}{3}$, 0, -12?
27 What meaning is to be attached to each of the following?

$$\frac{6}{0} \qquad \frac{0}{6} \qquad \frac{6}{6} \qquad \frac{0}{\frac{2}{3}} \qquad \frac{0}{0}$$

28 For what real values of x are the following expressions undefined?

$$\frac{5x + 2}{x + 1} \qquad \frac{x + 3}{2x + 6} \qquad \frac{x + 4}{x^2 - 16} \qquad \frac{7x + 2}{0} \qquad \frac{0}{x^2 + 10}$$

29 Why is the formula $(a + b) \times c = ac + bc$ true? This is another version of the distributive property of multiplication over addition.
30 Why is the distributive property of multiplication over subtraction true? Namely,

$$a \times (b - c) = ab - ac$$

HINT: Write $b - c = b + (-c)$, and apply the distributive property of multiplication over addition.
31 Which (if either) of the following distributive properties of division over addition is true?
a $a \div (b + c) = (a \div b) + (a \div c)$
b $(a + b) \div c = (a \div c) + (b \div c)$
32 Why is formula (*a*) in Prob. 31 false? HINT: Give a counterexample.

33 Why is formula (b) in Prob. 31 true? HINT: Write in the form $(a + b) \times (1/c)$.

34 Why does formula 31(a) not follow from 31(b) by writing

$$a \div (b + c) = (b + c) \div a = (b \div a) + (c \div a)$$
$$= (a \div b) + (a \div c)$$

35 Is the following relation true or false?

$$a \times \left(\frac{1}{b + c}\right) = \frac{a}{b} + \frac{a}{c}$$

In Probs. 36 to 42 evaluate the given expression.

36 $(-2)[3(4 - 2) + 6] + 5[-2(-3 + 8) + 9]$
37 $5[-9(5 - 2) + 4] - 4[3(4 + 8) - 45]$
38 $3[8(-2 + 4) - 5(7 - 9)] + 5[(4 - 8)6 - (9 - 4)3]$
39 $-6[3(7 + 6) - 4(3 - 8)] - 4[(7 - 3)9 - 16]$
40 $25 - 12 - 3 + 16 - 10$
41 $120 \div 4 \div 6 \div 5$
42 $26 - 3 \times 5 + 7 + 4 \times 6$

In Probs. 43 to 46 factor and solve for x.

43 $x^2 - 5x + 6 = 0$ **44** $x^2 + 8x + 12 = 0$
45 $x^2 - 8 = 1$ **46** $x^2 - 9 = 16$

In Probs. 47 to 52 prove or disprove the given statement. You may use R1 to R8 as given axioms.

47 $(a + b) \times c = (a \times c) + (b \times c)$
48 $a + (b \times c) = (a + b) \times (a + c)$
49 $a \div (b + c) = (a \div b) + (a \div c)$
50 $(a - b) + b = a$
51 If $a \neq 0$, $ax + b = 0$ has a unique solution.
52 To any real number a there corresponds a real number x such that $0x = a$.

53 Let "addiplication" be defined (with symbol $\odot$) as follows:

$$a \odot b = (a + b) + (a \times b)$$

Is addiplication a binary operation on the reals? Is addiplication commutative; associative? Is there an identity; an addiplicative inverse?

In Probs. 54 to 57 state the reason for each step in the given proof.

54 If $a + b = c$, then $b = (-a) + c$.
PROOF: **1** $a + b = c$
2 $(-a) + (a + b) = (-a) + c$

3 $[(-a) + a] + b = (-a) + c$
4 $0 + b = (-a) + c$
5 $b = (-a) + c$

55 If $ab = c$ and $a \neq 0$, then $b = \left(\frac{1}{a}\right) \times c$.

PROOF: 1 $ab = c$

2 $\left(\frac{1}{a}\right) \times ab = \left(\frac{1}{a}\right) \times c$

3 $\left(\frac{1}{a} \times a\right)b = \left(\frac{1}{a}\right) \times c$

4 $(1)b = \left(\frac{1}{a}\right) \times c$

5 $b = \left(\frac{1}{a}\right) \times c$

56 If $a + c = b + c$, then $a = b$.

PROOF: 1 $a + c = b + c$
2 $(a + c) + (-c) = (b + c) + (-c)$
3 $a + [c + (-c)] = b + [c + (-c)]$
4 $a + 0 = b + 0$
5 $a = b$

57 If $ac = bc$ and $c \neq 0$, then $a = b$.

PROOF: 1 $ac = bc$

2 $ac\left(\frac{1}{c}\right) = bc\left(\frac{1}{c}\right)$

3 $a\left(c \times \frac{1}{c}\right) = b\left(c \times \frac{1}{c}\right)$

4 $a \times 1 = b \times 1$
5 $a = b$

2.5 › THE NATURAL NUMBERS

The counting numbers, 1, 2, 3, . . . , are called the *natural numbers* or the *positive integers*. These are special cases of the real numbers, but they do not have all of the properties R1 to R11. We leave it to you to verify that the natural numbers do satisfy R1, R2, R3, R5, and R8.

EXERCISE A Choose $a = 2$, $b = 3$, $c = 5$, and for these natural numbers verify R1, R2, R3, R5, and R8.

Let us look at the other properties. The natural numbers cannot satisfy R4 or R6, since R4 involves zero and R6 involves negative numbers and neither zero nor the negative numbers are natural numbers. The natural numbers cannot satisfy R7 since fractions of the form $\frac{1}{2}$, $\frac{4}{3}$, etc., are not natural numbers. Similarly, R9 and R10 are true but R11 does not apply in this case.

EXERCISE B Prove or disprove the statement: "For every pair of natural numbers a and b, there is a natural number x such that $a + x = b$."

EXERCISE C Prove or disprove the statement: "For every pair of natural numbers a and b, where $b \neq 0$, there is a natural number x such that $bx = a$."

The natural numbers, however, do have properties which are not shared by some of the real numbers. Many of these have to do with their divisibility. We recall the following definition.

Divisibility

DEFINITION A natural number b is called a divisor (or factor) of a natural number a if and only if there is a natural number x such that $a = bx$. In this case a is said to be divisible by b.

Given a natural number, it is often important to make a list of its factors. Some numbers, called *primes,* have only two factors. We recall their definition.

Prime number

DEFINITION A natural number is called *prime* if and only if it has no natural numbers as factors except itself and 1. For special reasons 1 is usually not considered prime.

ILLUSTRATION 1 2, 3, 5, 7, 11, . . . are primes, whereas 4, 6, 8, 9, 10, . . . are not primes.

In factoring a natural number like 60, we may write

$$60 = 20 \times 3$$

and then factor these factors and continue factoring until only prime numbers are left as factors. Thus

$$60 = 20 \times 3 = 4 \times 5 \times 3 = 2 \times 2 \times 5 \times 3$$

This can be carried out in other ways, such as

$$60 = 15 \times 4 = 5 \times 3 \times 2 \times 2$$

Notice that these two sets of prime factors of 60 are the same except for their order. This illustrates a general property of the natural numbers which is stated as a theorem.

THEOREM 7 **UNIQUE FACTORIZATION THEOREM** A natural number $\neq 1$ can be expressed as a product of primes in a way which is unique except for the order of the factors.

We omit the proof of this theorem. You can find it, for instance, in "Principles of Mathematics", 3d ed., by Allendoerfer and Oakley, page 127.

The full collection of theorems about the natural numbers is called *Number Theory*. This is one of the most appealing branches of mathematics, and is generally the subject of an advanced college course. A few of the many reference books on the subject are listed at the end of this chapter.

PROBLEMS 2.5

1 Which of the operations $+$, $-$, $\times$, $\div$, are binary operations on the set of natural numbers?
2 Why is there no additive identity in the set of natural numbers?
3 Is there a multiplicative identity in the set of natural numbers? If so, what is it?
4 Are there additive and multiplicative inverses in the set of natural numbers?
5 In the natural numbers is multiplication distributive over addition?
6 Does the equation $5 + x = 3$ have a solution in the set of natural numbers?
7 Does the equation $2x = 7$ have a solution in the set of natural numbers?
8 Write all the divisors of 12.
9 Make a list of the first 10 primes.
10 Which of the following natural numbers are prime: 1, 6, 11, 31, 35, 47, 55, 69, 71, 83, 99?
11 Factor 24 into a product of primes.
12 Factor 162 into a product of primes.

2.6 › THE INTEGERS—MATHEMATICAL INDUCTION

The integers consist of the natural numbers, zero, and the negatives of the natural numbers: $\ldots -3, -2, -1, 0, 1, 2, 3, \ldots$. Often we call these respectively the "positive integers", "zero", and the "negative integers". The integers are thus special cases of the real numbers, but they fail to have all the properties R1 to R11. We leave it to you to verify that the integers do satisfy all of R1 to R11 except R7. From these properties we can infer the truth of a most important theorem for the integers, which was false for the natural numbers (Sec. 2.5, Exercise B).

THEOREM 8 For every pair of integers a and b, there exists a unique integer x such that $a + x = b$.

PROOF

1 *Existence:* $x = b - a$ is such an integer.
2 *Unicity:* Suppose that x_1 and x_2 are two solutions of $a + x = b$. Then $a + x_1 = a + x_2$, or $x_1 = x_2$.

EXERCISE A Prove or disprove the statement: "For every pair of integers a and b, where $b \neq 0$, there is an integer x such that $bx = a$."

The integers have another property which is essential for many parts of mathematics. This property permits us to use a method of proof called *Mathematical Induction* when we are concerned with theorems such as those given below. In the statements of these theorems we use the notation $a > b$, to mean "a is greater than b" and the notation $a \geq b$ to mean "a is greater than or equal to b". We defer the full treatment of such inequalities to Chap. 8.

1 For all integers $n \geq 2$, the product of n odd integers is odd; i.e., the product of any number of odd integers is odd.
2 For all positive integers n, and for any pair of integers x, y, where $x \neq y$, $x^n - y^n$ is divisible by $x - y$.
3 For all integers $n \geq 1$:

$$1 + 2 + \cdots + n = \frac{n(n+1)}{2}$$

4 For all integers $n \geq 2$, and all real numbers $a, b_1, \ldots, b_n$:

$$a(b_1 + \cdots + b_n) = ab_1 + \cdots + ab_n$$

5 For all integers $n \geq 1$, $n(n + 1)(n + 2)$ is divisible by 3; i.e., the product of every three consecutive positive integers is divisible by 3.

All these theorems have the common feature: their statement contains an open sentence p_n with variable n (Sec. 1.3), which is to be proved true for all n that are elements of a given infinite subset of the integers. Since we cannot prove such a theorem by verification in any finite number of cases, we must devise a general proof by some other means.

We shall illustrate the ideas behind such a proof in several informal examples.

ILLUSTRATION 1 PROVE: For all integers $n \geq 2$, the product of n odd integers is odd.

Let $\{a_1, a_2, \ldots, a_n\}$ be a set of n odd integers. We must show that $a_1 \times a_2 \times \cdots \times a_n$ is odd. First we note that we have proved earlier that the product of two odd integers is odd. Therefore $a_1 \times a_2$ is odd. Now $a_1 \times a_2 \times a_3 = (a_1 \times a_2) \times a_3$, which is the product of the odd integers $(a_1 \times a_2)$ and a_3. Since the product of two odd integers is odd, it follows that $a_1 \times a_2 \times a_3$ is odd. This argument can be repeated successively to show, in turn, that $a_1 \times a_2 \times a_3 \times a_4$ is odd, etc. Hence we are convinced of the truth of the theorem for all $n \geq 2$.

ILLUSTRATION 2 PROVE: For all integers $n \geq 2$ and all real numbers a, $b_1, \ldots, b_n$, it is true that $a(b_1 + \cdots + b_n) = ab_1 + \cdots + ab_n$.

First we recall (Sec. 2.3) that

$$a(b_1 + b_2) = ab_1 + ab_2 \qquad \text{Distributive property}$$
$$b_1 + b_2 + b_3 = (b_1 + b_2) + b_3 \qquad \text{Definition}$$

Hence

$$\begin{aligned} a(b_1 + b_2 + b_3) &= a[(b_1 + b_2) + b_3] && \text{Theorem 1} \\ &= a(b_1 + b_2) + ab_3 && \text{Distributive property} \\ &= (ab_1 + ab_2) + ab_3 && \text{Distributive property} \\ &= ab_1 + ab_2 + ab_3 && \text{Definition} \end{aligned}$$

Continuing in this fashion, we have

$$\begin{aligned} a(b_1 + b_2 + b_3 + b_4) &= a(b_1 + b_2 + b_3) + ab_4 \\ &= (ab_1 + ab_2 + ab_3) + ab_4 \\ &= ab_1 + ab_2 + ab_3 + ab_4 \end{aligned}$$

Since the process can continue indefinitely, we have again convinced ourselves of the truth of the theorem.

The trouble with these informal arguments is that in the end we must say: "this continues indefinitely", "etc.", "and so on", or something of the kind. How do we really know that we shall not be blocked at some stage far in the future? Since there is no logical means of giving such an assurance, we must formulate an axiom which effectively says that no such blocks will appear. In order to formulate this axiom, let us analyze the informal arguments above. In each case we began with a true fact for $n = 2$, namely, in Illustration 1, that the product of two odd integers is odd and, in Illustration 2, that $a(b_1 + b_2) = ab_1 + ab_2$. We used this fact to derive a true statement for $n = 3$, $n = 4$, $n = 5$, etc., in turn, and concluded that the process would never stop. By analogy we can think of the positive integers as the rungs of an infinitely tall ladder based on the ground and reaching to the sky. The bottom rung is 1, the next 2, and so on. We wish to climb this ladder to any desired rung. To do so, there are two essential steps:

I We must first get our foot on a low rung (the second rung in the above illustrations).

II We must be able to climb from any rung to the next rung. Clearly, if we can do these two things, we can climb up as far as we please.

To formalize this idea, we now state our axiom.

AXIOM OF MATHEMATICAL INDUCTION

Let a be an integer (positive, negative, or zero), and let A be the set of integers which are greater than or equal to a; that is, $A = \{n \mid n \geq a\}$.
If S is a subset of A with the two properties:

I S contains a.
II For all integers k in A: if k belongs to S, then $k + 1$ belongs to S.

then the set S is equal to the set A.

In many applications of this axiom we have $a = 1$, so that A is the set of all positive integers.

When we use this axiom to prove theorems of the type under consideration, the set A and the open sentence p_n are given to us in the statement of the theorem. We choose S to be the subset of A consisting of those integers for which p_n is true. With these interpretations, we can reformulate the axiom as an operational procedure, which we call the Principle of Mathematical Induction.

PRINCIPLE OF MATHEMATICAL INDUCTION

Let us be given a set of integers $A = \{n \mid n \geq a\}$ and a proposition of the form: For all n in A, p_n. We can prove the truth of this proposition by establishing the following:

I p_a is true. (We use the symbol p_a to denote the proposition obtained from the open sentence p_n by substituting a for n.)
II For all k in A, the conditional $p_k \rightarrow p_{k+1}$ is true.

Let us now illustrate this method.

ILLUSTRATION 3 (See Illustration 1.) PROVE: For all integers $n \geq 2$, the product of n odd integers is odd.

We have

$$a = 2 \qquad A = \{n \mid n \geq 2\}$$
$$p_n\text{: the product of } n \text{ odd integers is odd.}$$

Following the Principle of Mathematical Induction, we establish the two necessary facts:

I p_2 is true, for the product of two odd integers is odd.
II For all integers $k \geq 2$: if the product of k odd integers is odd, then the product of $k + 1$ odd integers is odd.

PROOF OF II Let $a_1, a_2, a_3, \ldots, a_k, a_{k+1}$ be odd integers. By hypothesis, $a_1 \times a_2 \times \cdots \times a_k$ is odd. Moreover, $a_1 \times \cdots \times a_k \times a_{k+1} = (a_1 \times \cdots \times a_k) \times a_{k+1}$ by definition. The expression in parentheses is odd by hypothesis, and so this is the product of two odd integers. Since such a product is odd, we conclude that $a_1 \times \cdots \times a_k \times a_{k+1}$ is odd.

From the Principle of Mathematical Induction we now conclude that the given proposition is true.

ILLUSTRATION 4 (See Illustration 2.) PROVE: For all integers $n \geq 2$ and all real numbers $a, b_1, \ldots, b_n$, it is true that

$$a(b_1 + \cdots + b_n) = ab_1 + \cdots + ab_n$$

We have $\quad a = 2 \quad A = \{n \mid n \geq 2\}$

$$p_n\colon\ a(b_1 + \cdots + b_n) = ab_1 + \cdots + ab_n$$

I p_2 is true, for $a(b_1 + b_2) = ab_1 + ab_2$ by the distributive property.
II For all integers $k \geq 2$: if $a(b_1 + \cdots + b_k) = ab_1 + \cdots + ab_k$, then $a(b_1 + \cdots + b_{k+1}) = ab_1 + \cdots + ab_{k+1}$.

PROOF OF II By definition, $b_1 + \cdots + b_{k+1} = (b_1 + \cdots + b_k) + b_{k+1}$. Hence

$$\begin{aligned} a(b_1 + \cdots + b_{k+1}) &= a[(b_1 + \cdots + b_k) + b_{k+1}] \\ &= a(b_1 + \cdots + b_k) + ab_{k+1} \end{aligned}$$

by the distributive property. By hypothesis, $a(b_1 + \cdots + b_k) = ab_1 + \cdots + ab_k$. Hence

$$\begin{aligned} a(b_1 + \cdots + b_{k+1}) &= (ab_1 + \cdots + ab_k) + ab_{k+1} \\ &= ab_1 + \cdots + ab_k + ab_{k+1} \end{aligned}$$

The stated theorem is now proved.

ILLUSTRATION 5 PROVE: For all integers ≥ 1,

$$1 + 2 + \cdots + n = \frac{n(n+1)}{2}$$

We have

$$a = 1 \quad A = \{n \mid n \geq 1\}$$

$$p_n\colon\ 1 + 2 + \cdots + n = \frac{n(n+1)}{2}$$

I p_1 is true, for $1 = \dfrac{1(2)}{2}$.

II For all integers $k \geq 1$, if

$$1 + 2 + \cdots + k = \frac{k(k+1)}{2}$$

then $$1 + 2 + \cdots + k + k + 1 = \frac{(k+1)(k+2)}{2}$$

PROOF OF II By hypothesis,

$$1 + 2 + \cdots + k = \frac{k(k+1)}{2}$$

Adding $k + 1$ to each side of the equality, we have

$$\begin{aligned} 1 + 2 + \cdots + k + k + 1 &= \frac{k(k+1)}{2} + k + 1 \\ &= \frac{k^2 + k + 2k + 2}{2} \\ &= \frac{k^2 + 3k + 2}{2} \\ &= \frac{(k+1)(k+2)}{2} \end{aligned}$$

ILLUSTRATION 6 PROVE: For all integers $n \geq 1$, $3^{2n} - 1$ is divisible by 8.

We have

$$a = 1 \qquad A = \{n \mid n \geq 1\}$$
$$p_n\colon\ 3^{2n} - 1 \text{ is divisible by } 8.$$

I p_1 is true, for $3^2 - 1 = 8$ is divisible by 8.
II For all integers $k \geq 1$, if $3^{2k} - 1$ is divisible by 8, then $3^{2k+2} - 1$ is divisible by 8.

PROOF OF II By hypothesis, $3^{2k} - 1$ is divisible by 8, so that there is an integer x such that $3^{2k} - 1 = 8x$, or $3^{2k} = 1 + 8x$. Multiplying both sides of this equality by $3^2 = 9$, we get

$$3^{2k+2} = 9 + 8(9x)$$

or

$$\begin{aligned} 3^{2k+2} &= 1 + [8 + 8(9x)] \\ &= 1 + 8(1 + 9x) \end{aligned}$$

So $3^{2k+2} - 1 = 8(1 + 9x)$, or $3^{2k+2} - 1$ is divisible by 8.

PROBLEMS 2.6

MATHEMATICAL INDUCTION In Probs. 1 to 28 use mathematical induction to show the truth of the stated proposition for all integers $n \geq 1$.

1 $1 + 3 + 5 + \cdots + (2n - 1) = n^2$

2 $1 + 4 + 7 + \cdots + (3n - 2) = \dfrac{n(3n - 1)}{2}$

3 $2 + 7 + 12 + \cdots + (5n - 3) = \frac{n}{2}(5n - 1)$

4 $2 + 6 + 10 + \cdots + (4n - 2) = 2n^2$

5 $n^3 + 2n$ is divisible by 3.

6 $n^2 + n$ is divisible by 2.

7 $n(n + 1)(n + 2)$ is divisible by 3.

8 $n^3 + 5n$ is divisible by 3.

9 $1 + 2 + 4 + 8 + \cdots + 2^{n-1} = 2^n - 1$

10 $2 + 6 + 18 + \cdots + 2 \cdot 3^{n-1} = 3^n - 1$

11 $\frac{1}{1 \cdot 2} + \frac{1}{2 \cdot 3} + \cdots + \frac{1}{n(n + 1)} = \frac{n}{n + 1}$

12 $1^2 + 2^2 + 3^2 + \cdots + n^2 = \frac{n(n + 1)(2n + 1)}{6}$

13 $1^3 + 2^3 + 3^3 + \cdots + n^3 = \frac{n^2(n + 1)^2}{4}$

14 $n^4 + 2n^3 + n^2$ is divisible by 4.

15 $a + (a + d) + (a + 2d) + \cdots + [a + (n - 1)d]$

$$= \frac{n}{2}[2a + (n - 1)d]$$

16 $a + ar + ar^2 + \cdots + ar^n = \frac{a(1 - r^{n+1})}{1 - r}, \; r \neq 1$

17 $4^n - 1$ is divisible by 3.

18 $5^n - 1$ is divisible by 4.

19 $3^{2n} - 1$ is divisible by 4.

20 $2^{2n} - 1$ is divisible by 3.

21 $x^n - 1$ is divisible by $x - 1$, where x is an integer $\neq 1$. HINT: Use the identity:

$$x^{n+1} - 1 = x(x^n - 1) + (x - 1)$$

22 $x^{2n} - 1$ is divisible by $x + 1$, where x is an integer $\neq -1$. HINT: First prove that $x^{2n} - 1$ is divisible by $x^2 - 1$ as in Prob. 21.

23 $x^n - y^n$ is divisible by $x - y$, where x and y are integers and $x \neq y$. HINT: Use the identity

$$x^{n+1} - y^{n+1} = x(x^n - y^n) + y^n(x - y)$$

24 $x^{2n} - y^{2n}$ is divisible by $x + y$, where x and y are integers and $x + y \neq 0$.

25 If a is a real number, we define $a^1 = a$ and $a^{n+1} = a^n \cdot a$. Prove that $1^n = 1$.

26 The product of $(2n + 1)$ negative real numbers $a_1 \times a_2 \times \cdots \times a_{2n+1}$ is negative.

27 The product of $2n$ negative real numbers $a_1 \times a_2 \times \cdots \times a_{2n}$ is positive.

28 $(ab)^n = a^n b^n$

29 For all integers $n \geq 2$, prove that the sum of n integers $a_1 + \cdots + a_n$ is an integer.

30 For all integers $n \geq 2$, prove that the product of n integers $a_1 \times a_2 \times \cdots \times a_n$ is an integer.

In Probs. 31 to 34 try to establish the indicated relation by mathematical induction. Point out why the method fails. (Each relation is false.)

31 $3 + 6 + 9 + \cdots + 3n = \dfrac{3n(n+1)}{2} + 1$

32 $3 + 5 + 7 + \cdots + (2n+1) = n^2 + 2$

33 $n^2 - 1$ is divisible by 3 for all $n \geq 2$.

34 $n^2 + n$ is divisible by 4 for all $n \geq 3$.

INTEGERS

35 Verify that R1 to R6 and R8 are satisfied when $a = 2, b = -3, c = 4$.

36 Verify that R1 to R6 and R8 are satisfied when $a = 4, b = 3, c = -2$.

37 Show by a counterexample that the integers do not satisfy R7.

38 Is subtraction a binary operation on the set of integers?

39 Is division a binary operation on the set of integers?

40 Do the integers form an ordered field?

2.7 RATIONAL NUMBERS

A rational number is really nothing but a fraction whose numerator and denominator are both integers. Let us give a formal definition.

DEFINITION A *rational number* is a real number which can be expressed in the form a/b where a and b are integers and $b \neq 0$.

As we know from the example $\frac{1}{2} = \frac{2}{4}$, there are many expressions of the form a/b which represent the same rational number. So that we can identify such cases easily, we need the following definition.

DEFINITION The expressions a/b and c/d where a, b, c, and d are integers and $b \neq 0$ and $d \neq 0$ represent the *same rational number* if and only if $ad = bc$.

Hence we may write

$$\frac{a}{b} = \frac{c}{d} \qquad \text{when} \qquad ad = bc$$

We can now state the following theorem, which expresses the most important and useful property of rational numbers.

Prove

THEOREM 9 Given any pair of integers a and b ($\neq 0$), there exists a rational number x such that $bx = a$. Moreover, any two rational numbers x_1 and x_2 with this property are equal.

PROOF The existence of a solution is immediate, for $x = a/b$ has the required property. In order to establish the second part of the theorem, we suppose that x_1 and x_2 both satisfy $bx = a$. Then

$$bx_1 = a$$
$$bx_2 = a$$

Subtracting, we have

$$b(x_1 - x_2) = 0$$

or

$$x_1 - x_2 = 0 \qquad \text{Theorem 4, since } b \neq 0$$

Finally we must remind you of the rules for adding and multiplying rational numbers. These are given by the theorems below.

THEOREM 10 $\dfrac{a}{b} + \dfrac{c}{d} = \dfrac{ad + bc}{bd}$

Addition

PROOF Let $x = a/b$; then $bx = a$.
Let $y = c/d$; then $dy = c$.

From these two equations we obtain

$$bdx = ad$$
$$bdy = bc$$

Adding and using the distributive property, we get

$$bd(x + y) = ad + bc$$

So

$$x + y = \frac{ad + bc}{bd}$$

THEOREM 11 $\dfrac{a}{b} \times \dfrac{c}{d} = \dfrac{ac}{bd}$

Multiplication

PROOF Using the notation in the proof of Theorem 10, we have again

$$bx = a$$
$$dy = c$$

Multiplying the left-hand sides and the right-hand sides separately, we have

$$(bd)(xy) = ac$$

Therefore $$xy = \frac{ac}{bd}$$

With these concepts of addition and multiplication we can now check to see how many of R1 to R11 are satisfied by the rational numbers. As a matter of fact we find that all of these are satisfied. This means that the arithmetic of rational numbers is just like that of the real numbers. This might lead us to believe that there is no difference between the real numbers and their special case, the rational numbers. However, we shall see that real numbers such as $\sqrt{2}$ are not rational and hence that a distinction must be made.

2.8 » DECIMAL EXPANSIONS

By carrying out the ordinary process of division, any rational number can be represented as a decimal. Some representations "terminate" after a finite number of steps; i.e., all later terms in the expansion are zero. For example,

$$\tfrac{1}{2} = 0.5000 \cdots$$
$$\tfrac{1}{4} = 0.2500 \cdots$$

But other expansions never terminate, such as

$$\tfrac{1}{3} = 0.3333 \cdots$$
$$1\tfrac{1}{7} = 1.142857142857 \cdots$$

By experimenting you may assure yourself that in each expansion the digits after a certain point repeat themselves in certain groups like (0), (3), and (142857) above. This is always true for rational numbers.

It is awkward to express numbers in this form since we cannot be sure what the $\cdots$ at the end really mean. To clear up this ambiguity, we place a bar over the set of numbers which is to be repeated indefinitely. In this notation we write

Repeating decimal

$$\tfrac{1}{2} = 0.5\overline{0}$$
$$\tfrac{1}{4} = 0.25\overline{0}$$
$$\tfrac{1}{3} = 0.\overline{3}$$
$$1\tfrac{1}{7} = 1.\overline{142857}$$

It is also true that any repeating decimal expansion of this type represents a rational number. We state this as Theorem 12.

THEOREM 12 Every repeating decimal expansion is a rational number.

Before giving the general proof, we give several illustrations.

ILLUSTRATION 1 Prove that $a = 3.\overline{3}$ is a rational number.

SOLUTION If we multiply by 10, we merely shift the decimal point; thus

$$10a = 33.\overline{3} = 30 + a$$

Hence

$$9a = 30$$

$$a = \tfrac{30}{9} = 3\tfrac{1}{3}$$

ILLUSTRATION 2 Now consider the harder case where $b = 25.\overline{12}$.

SOLUTION

$$100b = 2{,}512.\overline{12}$$

$$b = 25.\overline{12}$$

Subtracting, we find

$$99b = 2{,}487$$

$$b = \frac{2{,}487}{99} = \frac{829}{33} = 25\tfrac{4}{33}$$

ILLUSTRATION 3 Finally consider $c = 2.3\overline{12}$.

SOLUTION The change here is that the repeating part begins one place to the right of the decimal point. We can correct this easily by writing

$$10c = 23.\overline{12}$$

and then proceeding as in Illustration 2.

To prove the general theorem, suppose that

$$c = a_0 \,.\, a_1 \cdots a_k\overline{b_1 \cdots b_p}$$

where the a's and b's represent digits in the expansion of c. Using the idea of Illustration 3, we write

$$d = 10^k c = a_0 a_1 \cdots a_k \,.\, \overline{b_1 \cdots b_p}$$

Then

$$10^p d = a_0 a_1 \cdots a_k b_1 \cdots b_p \,.\, \overline{b_1 \cdots b_p}$$

$$(10^p d) - d = (a_0 a_1 \cdots a_k b_1 \cdots b_p - a_0 a_1 \cdots a_k)$$

Solving, we find that d is rational and that therefore c is rational.

2.9 › SOME IRRATIONAL NUMBERS

We have seen that the set of rational numbers is identical with the set of repeating decimals. However, we may perfectly well conceive of a nonrepeating decimal.

For example, consider the decimal expansion:

$$0.101001000100001\ldots$$

in which the number of zeros between successive ones increases by one at each step. Such a number is not rational but is included among the reals; and thus the reals include irrationals as well as rationals.

Perhaps this example is farfetched, and therefore we consider the very practical question of solving the equation $x^2 = 2$. The value of x is equal to the length of the hypotenuse of a right triangle whose legs are each 1. We now wish to show that $x = \sqrt{2}$ is not rational. We first prove a preliminary theorem, in which a is assumed to be an integer.

THEOREM 13 If a^2 is divisible by 2, then a is divisible by 2.

We have encountered this theorem before in Sec. 1.9, Illustration 4, but let us give another proof here.

PROOF Every integer a can be written in one of the two forms

$$a = \begin{cases} 2n \\ 2n + 1 \end{cases} \qquad \text{where } n \text{ is an integer}$$

Hence $$a^2 = \begin{cases} 4n^2 \\ 4n^2 + 4n + 1 \end{cases}$$

Since a^2 is divisible by 2, according to the hypothesis, a^2 must equal $4n^2$. Hence $a = 2n$, and a is divisible by 2.

THEOREM 14 $\sqrt{2}$ is not a rational number.

$\sqrt{2}$ is irrational

PROOF (INDIRECT) Suppose p/q is a rational number in lowest terms; that is, p and q have no common factor. Suppose also that $p^2/q^2 = 2$, or that $p^2 = 2q^2$.

Then p^2 is divisible by 2, and thus p is divisible by 2 (Theorem 13). Write $p = 2r$, where r is an integer. Then $4r^2 = 2q^2$, or $2r^2 = q^2$. Hence q^2 is divisible by 2, and thus q is divisible by 2 (Theorem 13). Hence p and q have a common factor contrary to our assumption. This proves the theorem.

PROBLEMS 2.9

RATIONAL NUMBERS

1 Verify that R1 to R8 are satisfied when $a = \frac{1}{2}$, $b = 3$, $c = -4$.
2 Verify that R1 and R8 are satisfied when $a = 2$, $b = \frac{1}{3}$, $c = 4$.
3 Show that the natural numbers are special cases of the rationals.
4 Show that the integers are special cases of the rationals.
5 PROVE: For any pair of rational numbers a and b, there exists a rational number x such that $a + x = b$. Moreover, any two rationals x_1 and x_2 with this property are equal.
6 PROVE: For any three rational numbers a, b, and c, where $a \neq 0$, there exists a rational number x such that $ax + b = c$. Moreover, any two rationals x_1 and x_2 with this property are equal.

7 PROVE: For any two rational numbers a/b and c/d, the quotient

$$\frac{a/b}{c/d} = \frac{ad}{bc}$$

8 PROVE: The two expressions $(-a)/b$ and $-(a/b)$ represent the same rational number. HINT: Show that each represents the additive inverse of (a/b).

9 PROVE: There is no rational number which is nearer to 0 than every other rational number.

10 PROVE: Given any two rational numbers a and b, with $a < b$, there is at least one rational number greater than a and less than b. How many rational numbers are there between a and b?

In Probs. 11 to 16 find decimal expansions for the given rational numbers.

11 $\frac{3}{7}$ **12** $\frac{1}{6}$ **13** $\frac{5}{9}$
14 $\frac{7}{8}$ **15** $2\frac{1}{11}$ **16** $3\frac{2}{9}$

In Probs. 17 to 22 find expressions of the form a/b for the given decimal expansions.

17 $0.\overline{5}$ **18** $3.1\overline{2}$ **19** $11.\overline{34}$
20 $4.3\overline{57}$ **21** $3.7\overline{428}$ **22** $1.\overline{9}$ (comment)

23 Prove that the decimal expansion of any rational number is repeating. HINT: Try dividing, and see what happens.

24 When a/b is expressed as a repeating decimal, what is the maximum length of the period? HINT: Try dividing, and see what happens.

25 Prove that $\sqrt{3}$ is irrational. HINT: First prove the analog of Theorem 13: "If a^2 is divisible by 3, then a is divisible by 3." To do so, note that every integer can be written in one of the forms:

$$a = \begin{cases} 3n \\ 3n+1 \\ 3n+2 \end{cases} \quad \text{where } n \text{ is an integer}$$

Hence
$$a^2 = \begin{cases} 9n^2 \\ 9n^2+6n+1 \\ 9n^2+12n+4 \end{cases}$$

Since a^2 is divisible by 3, according to the hypothesis, a^2 must equal $9n^2$, etc.

26 Prove that $\sqrt{5}$ is irrational.

27 Where does the method of Probs. 25 and 26 fail when we try to prove $\sqrt{4}$ to be irrational?

28 Prove that $1 + \sqrt{2}$ is irrational. HINT: Suppose that $1 + \sqrt{2} = a/b$. Then $\sqrt{2} = a/b - 1 = (a - b)/b$. Why is this impossible?

29 Prove that $2 - \sqrt{3}$ is irrational.

***30** Prove that $\sqrt{2} + \sqrt{3}$ is irrational.

2.10 » GEOMETRIC REPRESENTATION OF REAL NUMBERS

In the connection between arithmetic and geometry, the representation of real numbers as points on a line is most important. You are probably familiar with this idea, which is illustrated in Fig. 2.2. In order to obtain this representation, we start with the points 0 and 1 chosen at random, except that 0 is to the left of 1. The segment [0,1] is said to have length 1 by definition of "length". It is now assumed that this segment can be slid along the line without altering its length. Doing this step by step, we locate the other integers, so that the length of the segment between any two successive integers is still equal to 1. The location of $1/b$ (where b is a natural number) is found by dividing [0,1] into b equal parts by the usual geometric construction. Then by sliding the segment $[0,1/b]$ along the line we locate the points a/b. The location of the irrational numbers is more complicated, and we pass over this point. Their approximate positions, however, can be obtained from the first few decimals in their decimal expansions.

The most important fact about this representation is *that every point corresponds to one and only one real number and that every real number corresponds to one and only one point.* This property is often taken as one of the axioms of plane geometry.

This representation has another important property, namely, it preserves *order.* Before we can state this precisely, we must define a notion of order for real numbers and a similar notion for points on a line. Let us start with real numbers.

DEFINITION We say that a is greater than b (written $a > b$) if $a - b$ is *positive.* Similarly a is less than b (written $a < b$) if $a - b$ is *negative.*

Inequality

The symbols $\geq$ and $\leq$ mean, respectively, "greater than or equal to" and "less than or equal to". It is easy to see that, for any pair of real numbers, one and only one of the following relations is true:

$$a < b \qquad a = b \qquad a > b$$

We shall study the properties of these inequalities in some detail in Chap. 8.

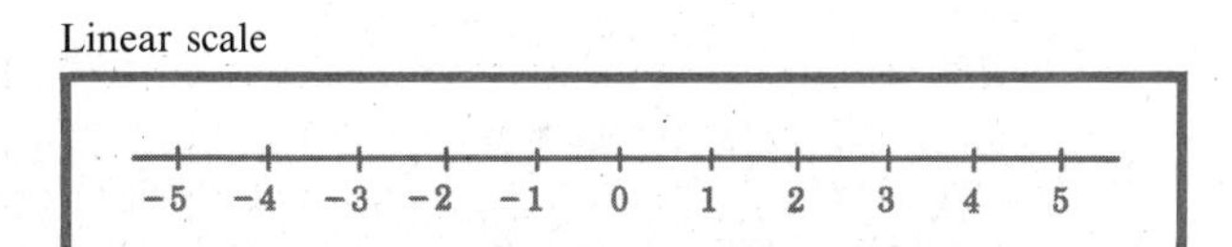

FIGURE 2.2

For a line, we introduce order by means of the notion "beyond". First of all we place an arrow on one end of the line and thus define a "positive direction" on the line as the direction toward the arrow. We now call this line a "directed line". It is customary to direct horizontal lines to the *right* and vertical lines *upward*. Then we define "beyond" as follows.

DEFINITION A point P is *beyond* a point Q on a directed line if the segment (or vector) from Q to P points in the given positive direction. If P is beyond Q, we write $P > Q$.

Let us now return to our assumption about real numbers and points on the line and describe how this preserves order. Let P_a be the point corresponding to the real number a and P_b to b. Then our correspondence is such that

$$P_a > P_b \textit{ if and only if } a > b$$

In summary, we have defined a correspondence between real numbers and points on a line which is one-to-one and which preserves *order*. The number associated with a point is called its "coordinate", and we can use coordinates to identify points. Thus, the point whose coordinate is the number a will henceforth be written as the point a and not as P_a, as was done above. This use of coordinates is the foundation of the application of real numbers to geometry and to geometrical representations of nature.

Length

By means of coordinates we can now define the length of an arbitrary segment whose endpoints are a and b. The notation for such a segment is $[a,b]$.

DEFINITION The *length* of the segment $[a,b]$ is the real number $|b - a|$.

2.11 » THE USE OF REAL NUMBERS IN THE PLANE

We shall now use the correspondence of the last section to set up a relationship between ordered pairs of real numbers and points in the plane. This is based upon an idea of René Descartes (1596–1650). Since ordered pairs will turn up in several other places in this book, let us say what they are.

DEFINITION

Ordered pair

An *ordered pair* (x,y) of real numbers is a pair in which x is the first element and in which y is the second element. Because of the ordering (x,y) is to be distinguished from (y,x) if $x \neq y$.

Rectangular coordinate system

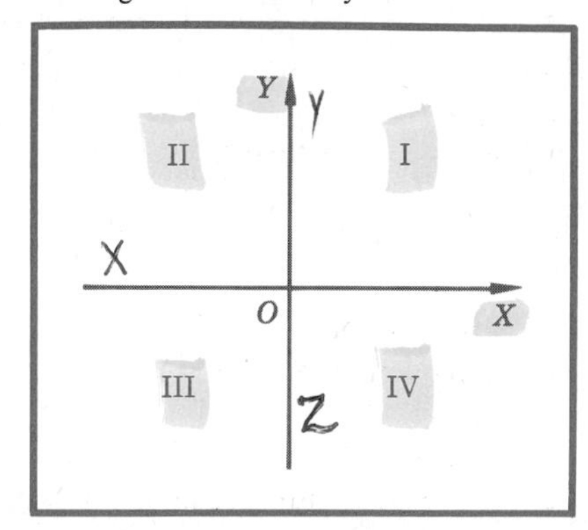

FIGURE 2.3
Roman numerals represent quadrants.

First, we construct two perpendicular lines in the plane (Fig. 2.3) which we call the X-axis and the Y-axis. Their point of intersection is called the origin O. We put the X-axis into an exact correspondence with the real numbers by placing zero at O, the positive reals to the right of O and the negative reals to its left. We do the same for the Y-axis, putting the positive reals above O and the negative reals below O. We remind ourselves of these conventions by putting arrows on the right end of the X-axis and the upper end of the Y-axis. These lines divide the plane into four regions called "quadrants" which are numbered I, II, III, and IV in Fig. 2.3.

Using this scheme, we can now associate an ordered pair of real numbers (x,y) with each point P of the plane. Let P be a point on the X-axis. It corresponds to a real number x on this axis. We associate with P the ordered pair $(x,0)$. Now let P be a point on the Y-axis. Similarly we associate the ordered pair $(0,y)$ with it. When P is not on either axis, draw PQ perpendicular to the X-axis and PR perpendicular to the Y-axis (Fig. 2.4). Suppose that Q corresponds to the real number x on the X-axis

Rectangular coordinate system

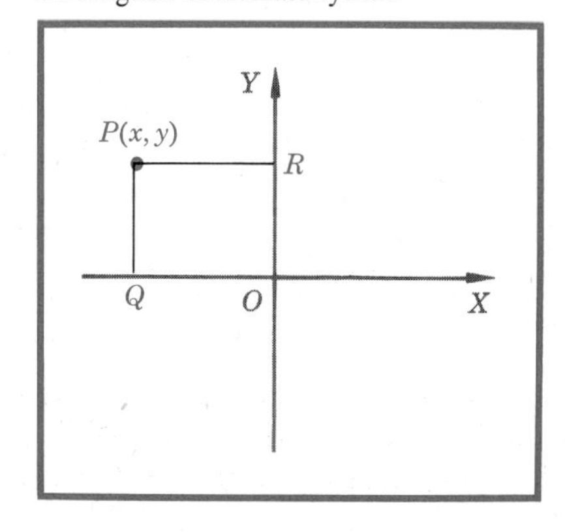

FIGURE 2.4
An ordered pair of numbers (x,y) determines a point in the plane.

and that R corresponds to the real number y on the Y-axis. Then we associate the ordered pair (x,y) with P.

By this process we find an ordered pair (x,y) which corresponds to each P in the plane. It is also evident that every pair (x,y) determines a point in the plane, for suppose (x,y) is given. These locate points Q and R (Fig. 2.4). Draw PQ and PR as perpendiculars to the X-axis and the Y-axis at Q and R, respectively. These lines intersect at P, which is the desired point.

Thus we have established a one-to-one correspondence between the points of the plane and the ordered pairs (x,y).

DEFINITION The real numbers x and y in the ordered pair (x,y) are called the *coordinates* of the point P. Sometimes x is called the *x-coordinate,* or the *abscissa;* and y is called the *y-coordinate,* or the *ordinate.*

We often identify the point P with its pair of coordinates and speak of the "point (x,y)". By using this identification, we can convert geometric statements about points into algebraic statements about numbers and can convert geometric reasoning into algebraic manipulation. The methods of algebra are usually simpler than those of geometry, and therefore the algebraic approach is now the common one. The detailed elaboration of this method is called "analytic geometry", which is discussed in Chap. 14.

2.12 › LENGTHS OF SEGMENTS; UNITS ON THE AXES

Suppose that P_1 and P_2 lie on a line parallel to the X-axis. Then we may write their coordinates as $P_1(x_1,a)$ and $P_2(x_2,a)$ (Fig. 2.5). We wish to have an expression for the length of P_1P_2. Draw P_1R and P_2S perpendicular to the X-axis. Then R has coordinates $(x_1,0)$ and S has coordinates $(x_2,0)$. Moreover, the lengths P_1P_2 and RS are equal, since opposite sides of a rectangle are equal.

Length of a horizontal segment

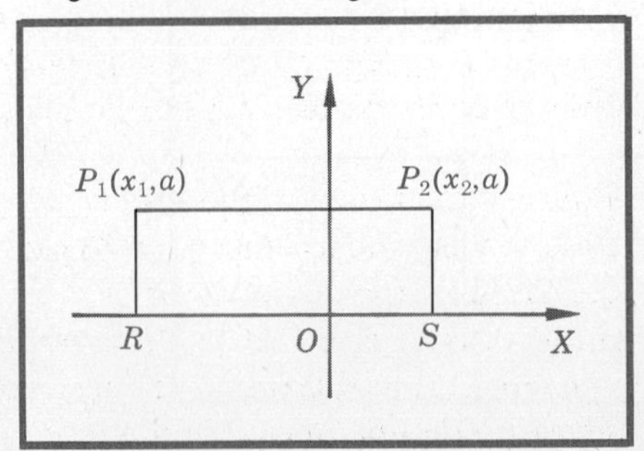

FIGURE 2.5
For a horizontal segment, the length of P_1P_2 is $|x_2 - x_1|$.

From Sec. 2.10 we know that the length $RS = |x_2 - x_1|$. Hence $P_1P_2 = |x_2 - x_1|$. This gives us Theorem 15.

THEOREM 15 The length of the segment between $P_1(x_1,a)$ and $P_2(x_2,a)$ is given by $P_1P_2 = |x_2 - x_1|$.

A similar proof gives Theorem 16.

THEOREM 16 The length of the segment between $Q_1(a,y_1)$ and $Q_2(a,y_2)$ is given by $Q_1Q_2 = |y_2 - y_1|$.

We have said nothing about the relation of distance on the X-axis to that on the Y-axis, and we prefer not to make any rigid requirements about this at present. Indeed, it is often useful to use different scales of measurement on the two axes. Unequal scales are used for a variety of reasons of which the following are the most common:

1 The range of values to be plotted on the Y-axis is much greater (or smaller) than the range to be plotted on the X-axis. In this case we must contract (or expand) the scale on the Y-axis in order to get a graph on a reasonably shaped piece of paper.

ILLUSTRATION 1 Suppose that we are plotting $y = x^{10}$ for x in the range 0 to 2. Then y lies in the range 0 to 1,024. In this case it would be extremely awkward to use equal scales on the two axes.

2 In applications to science the physical significance of the numbers on the two axes may be very different. In such cases the physical units of measurement (such as time, distance, velocity, etc.) are not comparable; and suitable scales on the two axes should be chosen independently.

ILLUSTRATION 2 In order to illustrate the motion of a particle, it is customary to plot the distance traveled on the vertical axis and the corresponding time on the horizontal axis. The units of measurement are *feet* and *seconds,* respectively, and it would be absurd to equate feet and seconds. Hence separate, convenient scales are used on the two axes.

In geometry, however, it is necessary to plot distance on each of the axes and to use the same scale on each. When we do this, it is meaningful to speak of the lengths of segments on slanting lines and we shall develop a formula for this in Sec. 12.2. The notion of slant distance, however, is quite meaningless in Illustration 2 above, and we shall avoid mention of it until we begin our study of geometry.

PROBLEMS 2.12

1 Use the symbol $>$ to represent the correct inequality between each of the following pairs of numbers: 1 and 4; -2 and 6; 3 and -3; -2 and -5; -4 and 0.

2 Use the symbol $<$ to represent the correct inequality between each of the following pairs of numbers: 7 and -8; -6 and -8; 5 and 2; -3 and 7; 0 and 6.

3 Write a set of inequalities expressing the fact that c lies inside the segment $[a,b]$ where $a < b$.

4 Write a set of inequalities expressing the fact that c lies outside the segment $[a,b]$ where $a < b$.

5 Find the lengths of the following segments: $[12,3]$, $[-8,6]$, $[5,-7]$, $[-6,-4]$, $[11,-12]$.

6 Find the lengths of the following segments: $[10,-4]$, $[-8,2]$, $[3,10]$, $[-16,0]$, $[-13,-20]$.

7 Plot the points where coordinates are $(2,5)$, $(-3,4)$, $(-2,-3)$, $(3,-2)$, $(0,0)$.

8 Plot the points where coordinates are $(1,6)$, $(-2,5)$, $(-4,-2)$, $(3,-4)$, $(0,0)$.

9 What signs do the coordinates of the points in quadrant I have; in quadrant III?

10 What signs do the coordinates of the points in quadrant II have; in quadrant IV?

11 State the quadrant in which each of the following points lies: $(4,-6)$, $(1,8)$, $(-3,-7)$, $(-4,5)$, $(-8,2)$.

12 State the quadrant in which each of the following points lies: $(-3,-6)$, $(15,-2)$, $(-3,6)$, $(3,3)$, $(7,-6)$.

13 Find the lengths of the segments joining the following pairs of points: $(2,4)$ and $(5,4)$; $(-3,-5)$ and $(-3,-8)$; $(-4,3)$ and $(6,3)$; $(5,5)$ and $(5,-7)$; $(0,0)$ and $(0,4)$.

14 Find the lengths of the segments joining the following pairs of points: $(4,3)$ and $(4,-6)$; $(3,-6)$ and $(6,-6)$; $(2,2)$ and $(13,2)$; $(4,0)$ and $(0,0)$; $(-5,8)$ and $(-5,-11)$.

15 If P_1P_2 is parallel to the X-axis, show that the length of P_1P_2 is equal to the length of P_2P_1.

16 If Q_1Q_2 is parallel to the Y-axis, show that the length of Q_1Q_2 is equal to the length of Q_2Q_1.

17 If P_1P_2 is parallel to the X-axis, show that the square of its length is $(x_2 - x_1)^2 = (x_1 - x_2)^2$.

18 If Q_1Q_2 is parallel to the Y-axis, show that the square of its length is $(y_2 - y_1)^2 = (y_1 - y_2)^2$.

TRANSFORMATION OF COORDINATES If we are given a coordinate system on a line in terms of numbers x, we can define a new coordinate system x' by giving a relationship between x and x'. This relabels the points with new numbers and is called a *transformation of coordinates.* The following problems give some important illustrations of these.

In Probs. 19 to 22 we take $x' = a + x$. This transformation is called a *translation*.

19 Prove that a translation leaves the lengths of segments unchanged. HINT: Prove that $|x_2' - x_1'| = |x_2 - x_1|$

20 PROVE: If the coordinate of any one point is left unchanged by a translation, then the coordinates of all points are unchanged. HINT: Prove that $a = 0$.

21 Express as a translation the relationship between absolute temperature K (degrees Kelvin) and centigrade temperature C (degrees centigrade or Celsius).

22 Express as a translation the relationship between the distance s of a rocket from the center of the earth and its height h above the surface of the earth.

In Probs. 23 to 26 we take $x' = ax$, where $a \neq 0$. This transformation is called a *dilatation*.

23 Prove that a dilatation multiplies the lengths of segments by $|a|$.

24 Prove that, if the coordinate of any *one* point (other than $x = x' = 0$) is left unchanged by a dilatation, then the coordinates of all points are unchanged.

25 Express the relationship between feet F and inches I as a dilatation.

26 Express the relationship between seconds S and hours H as a dilatation.

In Probs. 27 to 30 we take $x' = ax + b$, where $a \neq 0$. This transformation is called an *affine transformation*.

27 Prove that an affine transformation multiplies the lengths of segments by $|a|$.

28 Prove that an affine transformation with $a \neq 1$ leaves the coordinate of just one point unchanged. Find this point.

29 Express the relationship between degrees Fahrenheit F and degrees centigrade C as an affine transformation. What temperature is the same in both systems?

30 Express the relationship between degrees Fahrenheit F and degrees Kelvin K as an affine transformation. What temperature is the same in both systems?

2.13 › COMPLEX NUMBERS

There are, unfortunately, many problems that cannot be solved by the use of real numbers alone. For instance, we are unable to solve $x^2 = -1$.

In order to handle such situations the new symbol i is introduced, which, by definition, is to have the property that $i^2 = -1$. We then write expressions like $a + bi$ where a and b are real numbers and call these expressions *complex numbers*. The number a is the *real part of* $a + bi$, and b is its *imaginary part*. So that we can treat these like numbers, we must define the usual arithmetic operations on them.

DEFINITIONS The arithmetic operations on complex numbers are defined as follows:

Equality: $a + bi = c + di$ if and only if $a = c$ and $b = d$
Addition: $(a + bi) + (c + di) = (a + c) + (b + d)i$
Multiplication: $(a + bi) \times (c + di) = (ac - bd) + (bc + ad)i$

Note that the definition of multiplication is consistent with the property that $i^2 = -1$. For we can multiply $(a + bi)(c + di)$ by ordinary algebra and obtain $ac + i(bc + ad) + i^2(bd)$. When we replace i^2 with -1 and rearrange, we obtain the formula in the definition.

ILLUSTRATION 1

a $(3 + 6i) + (2 - 3i) = 5 + 3i$

b $(7 + 5i) - (1 + 2i) = 6 + 3i$

c $(5 + 7i)(3 + 4i) = 15 + 41i + 28i^2$
$= (15 - 28) + 41i = -13 + 41i$

d $(2 - 3i)(-1 + 4i) = -2 + 11i - 12i^2$
$= (-2 + 12) + 11i = 10 + 11i$

We also must consider division. We wish to express $1/(a + bi)$ as a complex number. This is best approached through the use of the conjugate complex number $a - bi$.

DEFINITION Conjugate

The complex numbers $a + bi$ and $a - bi$ are called *conjugates*.

We write

$$\frac{1}{a + bi} = \left(\frac{1}{a + bi}\right)\left(\frac{a - bi}{a - bi}\right) = \frac{a - bi}{a^2 + b^2} = \frac{a}{a^2 + b^2} + \frac{-b}{a^2 + b^2}i$$

which is the required complex number equal to $1/(a + bi)$. By extension of this method we can evaluate general quotients $(a + bi)/(c + di)$:

$$\frac{a + bi}{c + di} = \left(\frac{a + bi}{c + di}\right)\left(\frac{c - di}{c - di}\right) = \frac{(ac + bd) + (bc - ad)i}{c^2 + d^2}$$

Hence we have the rule for division.

Division

DIVISION In order to form the quotient $(a + bi)/(c + di)$, multiply numerator and denominator by the conjugate complex number $c - di$ and simplify the result.

ILLUSTRATION 2

$$\frac{4+i}{2-3i} = \frac{4+i}{2-3i} \times \frac{2+3i}{2+3i} = \frac{(4+i)(2+3i)}{(2-3i)(2+3i)} = \frac{5+14i}{13}$$

We could write this answer as $\frac{5}{13} + \frac{14}{13}i$, but this is an unnecessary refinement.

Finally, let us solve some equations involving complex numbers. The general method is suggested by the illustration below.

ILLUSTRATION 3 Solve $(x + yi)(2 - 3i) = 4 + i$. We could do this by writing $x + yi = (4 + i)/(2 - 3i)$ and evaluating the quotient on the right. But let us use another method. If we multiply out the left-hand side, we get

$$(2x + 3y) + (-3x + 2y)i = 4 + i$$

From our definition of the equality of two complex numbers, the *real parts* of both sides must be equal, and similarly the *imaginary* parts must be equal. Hence:

$$\begin{aligned} 2x + 3y &= 4 \\ -3x + 2y &= 1 \end{aligned}$$

We can solve these simultaneous equations and obtain $x = \frac{5}{13}$ and $y = \frac{14}{13}$.

This method of equating real and imaginary parts is of great importance in the application of complex numbers to engineering, and you should be certain that you understand it.

There are a number of other important properties of complex numbers that need to be discussed. Since these depend upon a knowledge of trigonometry, we defer their treatment to Chap. 13.

Another approach

The above definition of a complex number is somewhat lacking in intuitive appeal. We have said that there is no real number x such that $x^2 = -1$, but immediately we introduce i with this property. What, then, is i? Mathematicians were sufficiently disturbed about this to call i an *imaginary* number and $a + bi$ *complex* numbers; by contrast, other numbers in our system are *real*. Our purpose now is to give an alternative development of the complex numbers in a logical and nonimaginary fashion.

DEFINITIONS *Complex number: A complex number* is an ordered pair of real numbers (a,b).

Real part of a complex number: The complex number $(a,0)$ is called the *real part* of the complex number (a,b). We shall see that the pairs $(a,0)$ can be identified with the real numbers a in a natural fashion.

Pure imaginary number: A complex number of the form $(0,b)$ is called a pure imaginary number.

The arithmetic of complex numbers is given by the following basic definitions:

DEFINITIONS *Equality:* Two complex numbers (a,b) and (c,d) are said to be *equal* if and only if $a = c$ and $b = d$.
Addition: $(a,b) + (c,d) = (a + c,\ b + d)$
Multiplication: $(a,b) \times (c,d) = (ac - bd,\ bc + ad)$

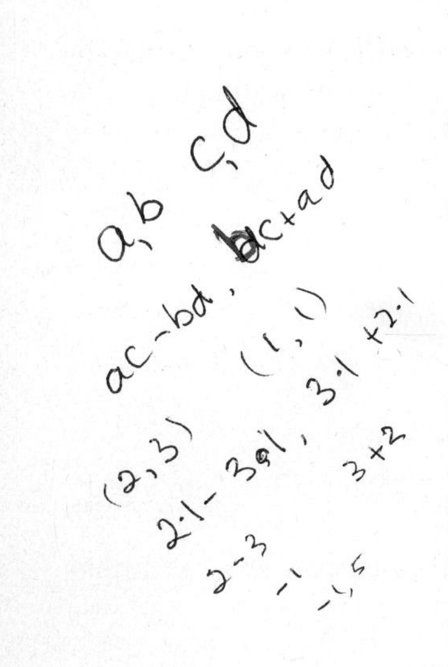

It is evident that there is a one-to-one correspondence between the complex numbers $(a,0)$ and the real numbers a which is defined by $(a,0) \leftrightarrow a$. This is a particularly useful correspondence, for under it sums correspond to sums and products to products. That is:

$$\begin{array}{ccccc} (a,0) & + & (c,0) & = & (a + c, 0) \\ \updownarrow & & \updownarrow & & \updownarrow \\ a & + & c & = & a + c \end{array} \qquad \begin{array}{ccccc} (a,0) & \times & (c,0) & = & (ac,0) \\ \updownarrow & & \updownarrow & & \updownarrow \\ a & \times & c & = & ac \end{array}$$

Such a correspondence is called an *isomorphism,* and we say that the set of complex numbers $(a,0)$ is isomorphic to the set of real numbers a relative to addition and multiplication. We are therefore justified in identifying these two systems and in calling $(a,0)$ a real number when there is no source of confusion.

Although the complex numbers $(a,0)$ are really nothing new, the pure imaginaries $(0,b)$ *are* something new. Their arithmetic, as derived from the definitions, is given by the following rules.

ADDITION $(0,b) + (0,d) = (0,\ b + d)$

MULTIPLICATION $(0,b) \times (0,d) = (-bd,0)$

It is important to note that the product of two pure imaginaries is a real number. In particular,

$$(0,1) \times (0,1) = (-1,0)$$

We now recall that our motivation for introducing the complex numbers was our inability to solve the equation $x^2 = -1$ in terms of real numbers. Let us see how the introduction of complex numbers enables us to provide such a solution. By means of the isomorphism above, we see that the equation $x^2 = -1$ corresponds to the equation

$$(x,y)^2 = (x,y) \times (x,y) = (-1,0)$$

As we have noted, $(x,y) = (0,1)$ is a solution of this equation, and we also see that $(x,y) = (0,-1)$ is another solution. Therefore our introduc-

tion of complex numbers permits us to solve equations of this type, which had no solution in terms of real numbers.

In order to complete our discussion we need to show the correspondence between our two definitions of complex numbers. In preparation for this we note the following identities:

$$(0,b) = (b,0) \times (0,1)$$
$$(a,b) = (a,0) + [(b,0) \times (0,1)]$$

EXERCISE A Verify the above identities.

We then set up the following relationship between the two notations:

	(a,b) *notation*	$a + bi$ *notation*
Real numbers	$(a,0)$	a
Unit imaginary	$(0,1)$	i

Using the identities above, we then derive the correspondences:

	(a,b) *notation*	$a + bi$ *notation*
Pure imaginaries	$(0,b)$	bi
Complex numbers	(a,b)	$a + bi$

From these we show that the rules for the equality, addition, and multiplication of complex numbers in the $a + bi$ notation, which were stated as definitions at the beginning of this section, are in agreement with the corresponding definitions in the (a,b) notation.

Finally, we observe that with these definitions the complex numbers form a field, but not an ordered field. The details of the proof of this are included in the problems.

PROBLEMS 2.13

In Probs. 1 to 14 find the sum or difference of the given complex numbers.

1 $(3 + 6i) + (8 - i)$
2 $(-14 + 2i) + (9 + i)$
3 $(13 - 4i) - (25 + 6i)$
4 $(-11 + 6i) - (19 + i)$
5 $(8 + 2i) + (-6 - 20i)$
6 $(4 - 7i) + (-9 + 5i)$
7 $-(6 - 5i) + (7 + 3i)$
8 $-(41 + 10i) + (14 - 5i)$
9 $-(4 + 3i) - (-3 + 7i)$
10 $-(-5 + 2i) - (4 - 6i)$
11 $18 - (13 - 2i)$
12 $(17 - 3i) + 12$
13 $(14 + 11i) - 13i$
14 $(19 - 14i) + 24i$

In Probs. 15 to 30 find the product of the given complex numbers.

15 $(2 + 5i)(-3 + i)$
16 $(4 - 7i)(4 + 2i)$
17 $(4 + 11i)(6 - i)$
18 $(8 + 2i)(3 + 10i)$
19 $(\sqrt{7} + i)(\sqrt{7} - i)$
20 $(\sqrt{3} - 2i)(\sqrt{3} + 2i)$
21 $(4 + 2i)(4 - 2i)$
22 $(1 - 10i)(1 + 10i)$
23 $9(4 - 6i)$
24 $6(-4 + 3i)$
25 $(7i)(3 + 6i)$
26 $(-8i)(-4 + 3i)$
27 $(4i)(9i)$
28 $(3i)(-8i)$
29 i^6
30 i^5

In Probs. 31 to 40 find the quotient of the given complex numbers.

— **31** $(3 + 2i)/(4 + 3i)$
32 $(2 + 5i)/(2 + i)$
33 $(14 + 3i)/(2 - i)$
34 $(4 + 7i)/(5 - i)$
35 $(-6 + 2i)/(-3 + 2i)$
36 $(-4 + 2i)/(-6 - 5i)$
37 $3i/(5 + 4i)$
38 $6i/(5 - 3i)$
39 $(4 + 2i)/7i$
40 $(6 - i)/2i$

In Probs. 41 to 46 solve for x and y by equating real and imaginary parts.

41 $(x + iy)(3 - 2i) = 4 + i$
42 $(x + iy)(6 - 3i) = 5 + 6i$
43 $(x + iy)(-4 + 7i) = -16 + 3i$
44 $(x + iy)(1 + i) = 3 - i$
45 $(x + iy)(4i) = 8$
46 $(x + iy)(-i) = 14i$

In Probs. 47 to 52 show that the given complex number satisfies the given equation.

47 $1 + 2i;\ x^2 - 2x + 5 = 0$
48 $1 - 2i;\ x^2 - 2x + 5 = 0$
49 $3 + 2i;\ x^2 - (7 + 3i)x + (10 + 11i) = 0$
50 $4 + i;\ x^2 - (7 + 3i)x + (10 + 11i) = 0$
51 $5 - 3i;\ x^2 + (-11 + 8i)x + (15 - 43i) = 0$
52 $6 - 5i;\ x^2 + (-11 + 8i)x + (15 - 43i) = 0$

53 Verify that R1 to R8 are satisfied when

$$a = 2 + i \qquad b = 3 - 2i \qquad c = 1 - 3i$$

54 What is the additive inverse of $a + bi$? The multiplicative inverse?

In Probs. 55 to 60 carry out the indicated operations on complex numbers in the (a,b) notation.

55 $(6,2) + (1,4)$
56 $(3,1) - (-2,7)$
57 $(6,1) \times (3,2)$
58 $(-1,3) \times (4,-1)$
59 $(3,2)/(1,1)$
60 $(5,10)/(2,3)$

61 To show that the complex numbers cannot satisfy R9, R10, and R11, let us suppose that we call some subset of the complex numbers "positive". Then i is positive or negative; let us say positive. Hence:

i^2 is positive.	Why?
-1 is positive.	Since $i^2 = -1$
$(-1)(i)$ is positive.	Why?
$-i$ is positive.	

Why is this impossible?

62 Repeat Prob. 61 assuming that i is negative, and arrive at a similar contradiction.

2.14 » GRAPHICAL REPRESENTATION OF COMPLEX NUMBERS

In contrast to the real numbers, the complex numbers cannot be represented in a useful way by the points on a line. By the same token we shall not define the notion of inequality between two complex numbers.

If we do wish to have a graphical representation of complex numbers, the most convenient procedure is to plot them as points in the plane (Fig. 2.6). To do so we measure the real part a of $a + bi$ along the horizontal (or *real*) axis and the imaginary part b along the vertical (or *imaginary*) axis. The details are the same as those for plotting the ordered pair (a,b) in Sec. 2.11. In this way we can establish a one-to-one correspondence between the complex numbers and the points of the plane.

Addition

Addition of complex numbers has a convenient graphical interpretation within this framework (Fig. 2.7). Let P represent $a + bi$, and Q represent $c + di$. Then, as in the figure, complete the parallelogram $POQR$. The point R now represents $(a + bi) + (c + di)$ or $(a + c) + (b + d)i$.

To prove this is a simple exercise in geometry. In Fig. 2.8, $OT = a$ and $OS = c$. Since triangle PRU is congruent to triangle OQS, $PU = OS = c$. Hence, $OV = OT + TV = OT + PU = a + c$. A similar argument shows that $RV = b + d$.

Complex plane

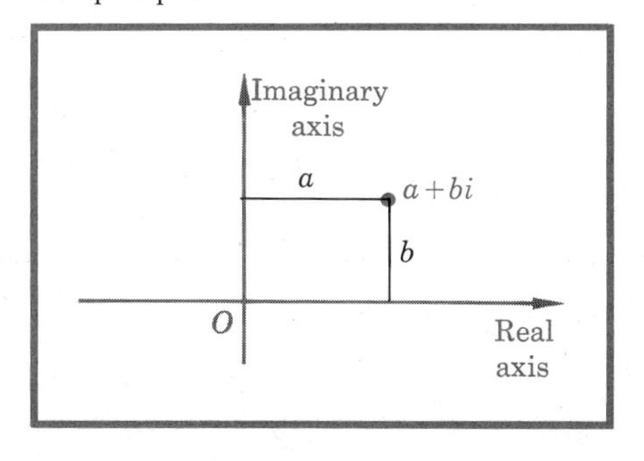

FIGURE 2.6
The real part a of $a + bi$ is measured horizontally, the imaginary part b vertically.

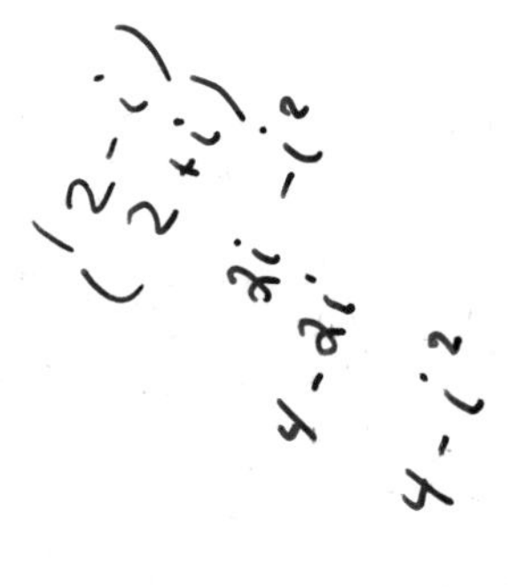

Sum of two complex numbers

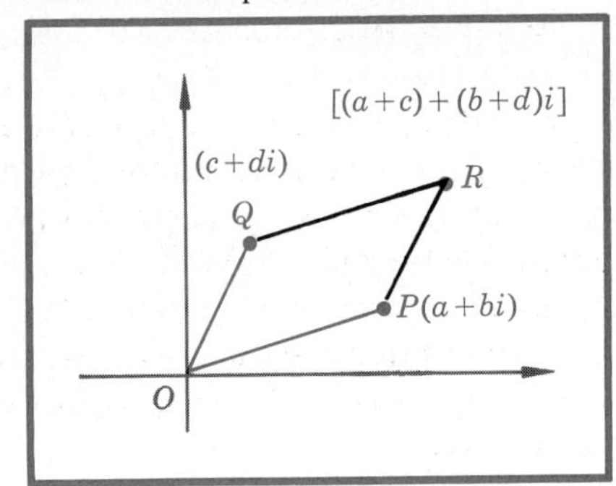

FIGURE 2.7
To add two complex numbers, complete the indicated parallelogram.

Sum of two complex numbers

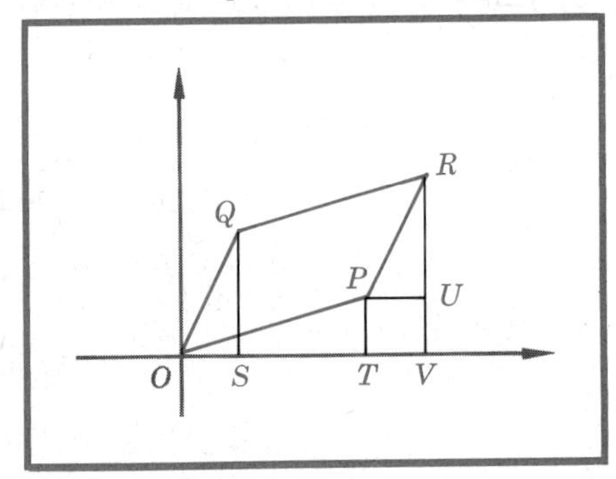

FIGURE 2.8
This diagram is used to prove that the construction employed in Fig. 2.7 is correct.

EXERCISE A Add $2 + 3i$ and $4 + i$ algebraically and graphically.

EXERCISE B Devise a graphical construction for the difference $(a + bi) - (c + di)$. Check this by computing $(4 + 2i) - (3 + 5i)$ both algebraically and graphically.

Unfortunately the multiplication of complex numbers by graphical means is more complicated. We shall discuss this in Sec. 13.11.

2.15 › SOLUTIONS OF OTHER ALGEBRAIC EQUATIONS

Since we have extended our number system by considering progressively more and more complicated equations, it is reasonable to suppose that we may be led to "supercomplex" numbers in an effort to solve equations of the form

$$ax^n + bx^{n-1} + \cdots + cx + d = 0 \qquad a \neq 0$$

where $a, b, \ldots, c, d$ are complex numbers. As a matter of fact no new types of numbers need to be introduced for this purpose. This is a consequence of the so-called "Fundamental Theorem of Algebra":

THEOREM 17 The equation above with complex coefficients always has a complex number $x = u + iv$ as a solution.

The proof of this theorem is beyond the scope of this book.

Since we do not have to invent any new numbers to solve this equation, we say that the set of complex numbers is *algebraically closed.* As a consequence we end our development of the number system at this point. It should be remarked, however, that for other purposes mathematicians have developed systems of "hypercomplex" numbers, two of which are called "quaternions" and "Cayley Numbers". We do not discuss these here.

2.16 » CLASSIFICATION OF NUMBERS

The following chart shows the various types of numbers and their relationships.

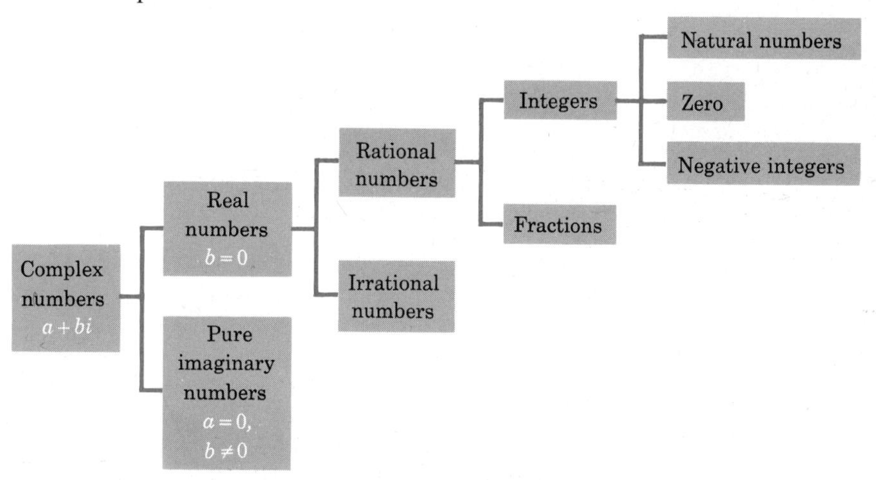

REFERENCES

Birkhoff, Garrett, and Saunders MacLane: "A Survey of Modern Algebra", chaps. 1 to 3, Macmillan, New York (1965).
Courant, Richard, and Herbert Robbins: "What Is Mathematics?" chaps. 1 and 2, Oxford, New York (1941).
Dantzig, Tobias: "Number, the Language of Science", Macmillan, New York (1930).
Henkin, Leon: "Mathematical Induction", Mathematical Association of America, Washington, D.C. (1961).
Niven, Ivan, and Herbert S. Zuckerman: "An Introduction to the Theory of Numbers", Wiley, New York (1960).
Sorinskii, I. S.: "The Method of Mathematical Induction", Blaisdell, New York (1961).
Youse, B. K.: "Mathematical Induction", Prentice-Hall, Englewood Cliffs, N.J. (1964).

3 POLYNOMIALS

3.1 » ALGEBRAIC EXPRESSIONS

In this chapter and in the following two chapters we present a review of some of the most important notions in elementary algebra. These deal with the properties of *algebraic expressions.* In order to form such an expression, we first fix upon a particular field of numbers, which is generally the real or complex numbers. We then select a finite set of numbers belonging to this field and a finite collection of variables whose universal set is this field. Then we combine these in a finite number of steps through the processes of addition, subtraction, multiplication, division, and extraction of roots. The end result is an algebraic expression. The following are examples of such expressions:

$$2a + \frac{3}{b^2} \qquad \frac{6x + 5x^2}{2 - 3x} \qquad \sqrt{3r + 5s^2}$$

Polynomials

For the present we shall restrict ourselves to the operations of addition, subtraction, and multiplication; and we call the resulting algebraic expressions *polynomials.* A polynomial is the sum of a finite number of *terms,* each of which is the product of a finite collection of numbers and variables. A term may also include a negative sign, corresponding to a -1 contained in this product. In writing the expressions for these terms we write, as usual, a^2 for $a \times a$; b^3 for $b \times b \times b$; etc. You should note that the only exponents which occur on variables are *positive integers;* fractional and negative exponents are excluded. The numbers involved, however, may be any elements of the underlying field. Examples of polynomials are

$$4x^3 - 2x^2 + x - 1 \qquad \tfrac{1}{2}a + 3bc$$
$$5xy^2 - \tfrac{1}{3}yz + \sqrt{2} \qquad (3 + 2i)x^2 - 6i$$

Polynomials, such as $8x^2y^6$, which contain only one term are called *monomials;* those with two terms, such as $4a^2 - 7bc$, are called *binomials;* and those with three terms, such as $7a^2 + 4ab + 3b^2$, are called *trinomials.*

In writing a term it is customary to multiply together all the numbers involved and to put this product in front. This number is called the *numerical coefficient* of the term. Similarly, all factors involving the same variable are brought together and are written as this variable with a suitable exponent. Thus we write

$$2 \times a \times 4 \times b \times a \times b \times b \qquad \text{as} \qquad 8a^2b^3$$

Like terms

Two terms which differ only in their numerical coefficient are called *like terms*. For instance, $4x^2y^3$ and $6x^2y^3$ are like terms, but $3xy^2$ and $7x^2y$ are *not* like terms; similarly $4x^2$ and $3a^2$ are *not* like terms. In writing a polynomial it is customary to combine like terms by the use of the distributive law; for example:

$$4x^2y^3 + 6x^2y^3 = 10x^2y^3$$

As a result a polynomial will be expressed as a finite sum of unlike terms, each of which either is a number or consists of a numerical coefficient multiplied by the product of a finite number of distinct variables, each of which carries an exponent which is a positive integer.

EXERCISE A Which of the following expressions are polynomials?

1 $3x^2 - x + \frac{1}{2}$

2 $x^3 + \dfrac{1}{x} + 3$

3 $\sqrt{2}x + 4x^2$

4 $2\sqrt{x} + 5x^2 - 2$

5 $\dfrac{3}{2x^2 - x + 6}$

3.2 ADDITION OF POLYNOMIALS

The procedure for adding polynomials is a direct consequence of the commutative and associative laws for addition and of the distributive law (R2, R3, R8, Sec. 2.3). Let us illustrate by adding:

$$(5x^2y + x - 3xy^2 + 2) + (-2x + 3y + 7xy^2 - 5)$$

The first step is to use the commutative and associative laws to group like terms together. In the above example we obtain

$$(5x^2y) + (x - 2x) + (-3xy^2 + 7xy^2) + (3y) + (2 - 5)$$

Then we use the distributive law to combine the like terms. In our example we have the final result

$$5x^2y - x + 4xy^2 + 3y - 3$$

The process of addition can be conveniently carried out by arranging the work in columns, where each column contains like terms. In the above example we write

$$\begin{array}{r} 5x^2y + x - 3xy^2 + 2 \phantom{{}+3y} \\ -\,2x + 7xy^2 - 5 + 3y \\ \hline 5x^2y - x + 4xy^2 - 3 + 3y \end{array}$$

This arrangement of the work is particularly helpful when three or more polynomials are to be added.

In order to subtract two polynomials, we convert the problem to addition (see the definition of subtraction, Sec. 2.3) and then proceed as above. For example:

$$(4x^2 - 3xy + 2) - (5x^2 + x - 3)$$

is written as the sum

$$(4x^2 - 3xy + 2) + (-5x^2 - x + 3)$$

and equals $$-x^2 - 3xy - x + 5$$

PROBLEMS 3.2

Perform the indicated operations.

1 $(3a^2 + 2ab + c) + (3c - 4a^2 - ab)$
2 $(2x^2 - 3y^2 + 4x) + (3x + 7y^2 - 4x^2)$
3 $(pq + 2q^3 + 3p^2) - (2p^2 - 4q^3 + 5pq)$
4 $(5r + 6s - 2r^2s^3) - (2s + 4r + 2r^2s^3)$
5 $(10x^2y + 5xy^2 - 3xy + 2) + (4xy + x^2 - 2y^2)$
6 $(4abc + 3ab^2 - 6ab + 4) + (5ab + 3ab^2 + 6b)$
7 $(7x^3 - 6y^3 + 3xy) - (5x^3 + 6y + 2xy)$
8 $(-3pq + 7p^2 - 5q) - (3q^2 + q - 6p^2)$
9 $(x^2 - 2y^2 + 3x) + (5a^2 + 4b^2 + 3y)$
10 $(5r^2 + 3rs + 2s^2) + (2x^2 + y^2 + s^3)$
11 $(2x^2y - 4xy^2 + 6xy) + (4x^3 + 8x^2y + 2xy^2) + (-xy^2 + 4xy + 2)$
12 $(-8a^3 + 15b^2 - 5a^2b) + (9a^3 + 9b^3 - a^2b) + (3a^2 + 8ab + b^2)$
13 $(2x^2 - 16y^2) + (3x^2 + 8y^2) + (-3r^2 + 5s^2)$
14 $(-pq + 3q^2) + (4pq - q^2) + (2a^2 + b^2)$
15 $(6x^4 - 3a^2 + 4xy) - (2x^4 + 5a^2 - xy) + (3x^4 + 2a^2 + 3xy)$
16 $(3a^2 - 3b^2 + 8c^3) + (2a^2 + 4b^2 - 6c^3) - (a^2 + b^2 + c^3)$
17 $(4xy + x^2 - 4y^4) + (x^2 + y^2 - 2xy) - (4x^2 - 3y^2 + 7xy)$
18 $(6r^2s + r^3 + 4s^5) - (3r^3 - 4s^5 + 6rs^2) + (10r^2s + 15rs^2)$
19 $[(4a^2 - 3b^2) - (7ab + b^2)] - (5a^2 + 6ab + 10b^2)$
20 $(20x^2 - 12xy + 15y^2) - [4(x^2 + 2y^2) - (10xy - 5y^2)]$

3.3 › MULTIPLICATION OF POLYNOMIALS

The procedure for the multiplication of polynomials is based upon the method of multiplying monomials, together with the repeated use of the distributive law (R8). We recall that a monomial is either a number or the product of a numerical coefficient and a collection of variables carrying positive integral exponents. The product of two monomials is, therefore, the product of all the factors of the two given monomials taken together.

ILLUSTRATION 1

a $(2x^2y)(5x^3y^4) = 10x^5y^5$
b $(-5a^3bc)(4ab^2g^3) = -20a^4b^3cg^3$

In carrying out this product we recall that

$$a^p \cdot a^q = a^{p+q}$$

for a^p is the product of p a's, and a^q is the product of q a's. There are, therefore, $(p + q)$ a's in the combined product.

We now use the distributive law to reduce the problem of multiplying polynomials to that of multiplying monomials.

ILLUSTRATION 2 Multiply $(a^2 + 2b)(3a^2 + 4b + 1)$. One use of the distributive law permits us to write this as

$$a^2(3a^2 + 4b + 1) + 2b(3a^2 + 4b + 1)$$

A second application of the distributive law gives

$$(3a^4 + 4a^2b + a^2) + (6a^2b + 8b^2 + 2b) = 3a^4 + 10a^2b + a^2 + 8b^2 + 2b$$

The work can be conveniently arranged as shown below:

$$\begin{array}{l} 3a^2 + \ \ 4b + 1 \\ \ \ a^2 + \ \ 2b \\ \hline 3a^4 + \ \ 4a^2b + a^2 \\ \quad\ + \ \ 6a^2b \qquad\quad + 8b^2 + 2b \\ \hline 3a^4 + 10a^2b + a^2 + 8b^2 + 2b \end{array}$$

A particularly important special case is that of the product of two polynomials which involve powers of a single variable only. In this case it is convenient to arrange the order of the terms so that the exponents decrease from term to term, i.e., "according to decreasing powers". Thus we rearrange

$$7x^2 + 21x^5 - x^3 + 2x - 1 + 5x^4$$

to read $\qquad 21x^5 + 5x^4 - x^3 + 7x^2 + 2x - 1$

This will help us to keep things straight in our multiplications and later in our divisions.

ILLUSTRATION 3 Multiply:

$$
\begin{array}{rrrrrrr}
5x^4 - & 8x^3 + & x^2 + & 5x & - 3 & & \\
x^2 + & 2x & - 1 & & & & \\
\hline
5x^6 - & 8x^5 + & x^4 + & 5x^3 - & 3x^2 & & \\
& + 10x^5 - & 16x^4 + & 2x^3 + & 10x^2 - & 6x & \\
& & - 5x^4 + & 8x^3 - & x^2 - & 5x + & 3 \\
\hline
5x^6 + & 2x^5 - & 20x^4 + & 15x^3 + & 6x^2 - & 11x + & 3
\end{array}
$$

PROBLEMS 3.3 Perform the indicated operations.

1	$(3a^5b^2)(-2a^4b^3)$	**2**	$(-4p^3q)(3p^5q^7)$
3	$(8x^2y^3z^5)(3x^5wy^2)$	**4**	$(-5r^4s^2t)(9s^3t^7u^3)$
5	$(3a + 6b)(a - 5b)$	**6**	$(3x - 5y)(2x - 7y)$
7	$(3x + 2y)^2$	**8**	$(4r - 3s)^2$
9	$(2p + q)^3$	**10**	$(2r - s)^3$
11	$(2x + 5y)(2x - 5y)$	**12**	$(4a - 9b)(4a + 9b)$

13 $(4x^2 - 3x + 5)(x^2 - 4x + 3)$
14 $(2x^3 - 4x + 7)(x^3 + 2x^2 + 3)$
15 $(3x^2 - x^3 - 4x^4 + 2)(3x - 4 + x^2)$
16 $(2x + x^4 - 3x^3 + 4)(3x + 6 - 5x^2)$
17 $(4a^4 - 6ab^2 + b^3)(3a^3 - b^2 + 2)$
18 $(r^3 + 3r^2s^2 + s^4)(s^2 - 2s + 6)$
19 $(5x^4 - 3x^3y + 6x^2y^2 + xy^3 - y^4)(2x^2 + xy - y^2)$
20 $(4a^4 - 2a^3b + 5a^2b^2 - 3ab^3 + b^4)(a^2 - ab - b^2)$

3.4 ▸ BINOMIAL THEOREM

By direct multiplication, as in the last section, we can easily establish the following formulas:

$$
\begin{aligned}
(a + b)^2 &= a^2 + 2ab + b^2 \\
(a + b)^3 &= a^3 + 3a^2b + 3ab^2 + b^3 \\
(a + b)^4 &= a^4 + 4a^3b + 6a^2b^2 + 4ab^3 + b^4 \\
(a + b)^5 &= a^5 + 5a^4b + 10a^3b^2 + 10a^2b^3 + 5ab^4 + b^5
\end{aligned}
$$

Pascal's Triangle

The coefficients in these products have a pattern which is illustrated by the following scheme, known as *Pascal's Triangle.*

$$
\begin{array}{lccccccccccc}
(a + b)^0 & & & & & & 1 & & & & & \\
(a + b)^1 & & & & & 1 & & 1 & & & & \\
(a + b)^2 & & & & 1 & & 2 & & 1 & & & \\
(a + b)^3 & & & 1 & & 3 & & 3 & & 1 & & \\
(a + b)^4 & & 1 & & 4 & & 6 & & 4 & & 1 & \\
(a + b)^5 & 1 & & 5 & & 10 & & 10 & & 5 & & 1 \\
\cdots & & & & & & & & & & &
\end{array}
$$

In this array each horizontal line begins and ends with a 1, and each other entry is the sum of the two numbers to its left and right in the horizontal row above.

EXERCISE A By direct multiplication verify the above formulas for $(a + b)^2$, $(a + b)^3$, $(a + b)^4$, $(a + b)^5$.

EXERCISE B From Pascal's Triangle determine the coefficients of the terms in the expansion of $(a + b)^6$, and verify your result by direct multiplication. (*Ans.*:1, 6, 15, 20, 15, 6, 1.)

Whenever we discover a pattern like this, we suspect that there must be some general way of describing it and so we are led to ask whether there is some general formula for $(a + b)^n$, where n is any positive integer. There is indeed such a formula, known as the *Binomial Formula,* and we shall now proceed to develop it. By way of preparation we must introduce some new notations.

DEFINITION
Factorial

The symbol $n!$ (for n a positive integer), read "*n factorial*", stands for the product

$$n! = 1 \times 2 \times 3 \times \cdots \times n$$

Further, $0!$ is defined to be 1. Factorials are not defined for negative integers or for other real numbers.

EXERCISE C Compute $2!$, $3!$, $4!$, $5!$, $6!$.

EXERCISE D Show that $n!/n = (n - 1)!$.

EXERCISE E Compute $7!/3!$; $n!/(n - 1)!$; $n!(n - 2)$; $n!/r!$, where $r < n$.

DEFINITION

The symbol $\binom{n}{r}$, where n and r are integers ≥ 0 and $n \geq r$, is defined to be

$$\binom{n}{r} = \frac{n!}{(n - r)!r!}$$

These symbols are called *binomial coefficients.*

EXERCISE F Show that

$$\binom{4}{2} = 6 \qquad \binom{6}{4} = 15 \qquad \binom{5}{3} = 10 \qquad \binom{4}{4} = 1 \qquad \binom{5}{1} = 5$$

EXERCISE G From the definition show that

$$\binom{n}{r} = \binom{n}{n-r}$$

The connection between these symbols and the expression for $(a+b)^n$ is easily seen from the fact that Pascal's Triangle now can be written in the form

$$
\begin{array}{ll}
(a+b)^0 & \binom{0}{0} \\
(a+b)^1 & \binom{1}{0} \quad \binom{1}{1} \\
(a+b)^2 & \binom{2}{0} \quad \binom{2}{1} \quad \binom{2}{2} \\
(a+b)^3 & \binom{3}{0} \quad \binom{3}{1} \quad \binom{3}{2} \quad \binom{3}{3} \\
(a+b)^4 & \binom{4}{0} \quad \binom{4}{1} \quad \binom{4}{2} \quad \binom{4}{3} \quad \binom{4}{4} \\
(a+b)^5 & \binom{5}{0} \quad \binom{5}{1} \quad \binom{5}{2} \quad \binom{5}{3} \quad \binom{5}{4} \quad \binom{5}{5} \\
\cdots & \cdots
\end{array}
$$

EXERCISE H Verify that this representation of Pascal's Triangle agrees with the one given earlier in this section.

We have now given you a broad hint regarding the nature of the Binomial Formula. Can you guess the correct expression for it? Cover up the next few lines of this page, and write down your guess. Please do not peek until you have written it down, for intelligent guessing is a most important part of the process of mathematical discovery, and you need practice in doing it. Now you can look, and we hope that you have written something like the following.

THEOREM 1 **BINOMIAL THEOREM** Let n be a positive integer. Then

$$(a+b)^n = a^n + \binom{n}{1}a^{n-1}b + \binom{n}{2}a^{n-2}b^2 + \cdots$$
$$+ \binom{n}{n-2}a^2b^{n-2} + \binom{n}{n-1}ab^{n-1} + b^n$$

Or, in the expansion of $(a+b)^n$, the coefficient of $a^{n-r}b^r$ is $\binom{n}{r}$.

PROOF Since this theorem is to be proved for all values of n, a reasonable approach is to try mathematical induction. The formula is trivially verified for $n = 1$, and indeed we have verified it for $n = 1, 2, 3, 4, 5$. We must now establish the truth of step II of the Principle of Mathematical Induction (Sec. 2.6). In this case we must show that:

For all $k \geq 1$: If the coefficient of $a^{k-r}b^r$ in the expansion of $(a + b)^k$ is $\binom{k}{r}$, then the coefficient of $a^{k+1-r}b^r$ in the expansion of $(a + b)^{k+1}$ is $\binom{k+1}{r}$. We write

$$(a + b)^k = a^k + \cdots + \binom{k}{r-1}a^{k+1-r}b^{r-1} + \binom{k}{r}a^{k-r}b^r + \cdots + b^k$$

Then $(a + b)^{k+1}$ is given by the product

$$\begin{array}{lll}
a^k + \cdots & + \dbinom{k}{r-1}a^{k+1-r}b^{r-1} + \dbinom{k}{r}a^{k-r}b^r + \cdots + b^k & \\
a + b & & \\
\hline
a^{k+1} + \cdots & + \dbinom{k}{r-1}a^{k+2-r}b^{r-1} + \dbinom{k}{r}a^{k+1-r}b^r + \cdots + ab^k & \\
\quad a^kb + \cdots & \qquad + \dbinom{k}{r-1}a^{k+1-r}b^r + \dbinom{k}{r}a^{k-r}b^{r+1} & \\
 & \qquad\qquad + \cdots + b^{k+1} & \\
\hline
a^{k+1} + \cdots & + \left[\dbinom{k}{r} + \dbinom{k}{r-1}\right]a^{k+1-r}b^r + \cdots + b^{k+1} &
\end{array}$$

Therefore the coefficient of $a^{k+1-r}b^r$ is $\binom{k}{r} + \binom{k}{r-1}$. We must now simplify this.

$$\begin{aligned}
\binom{k}{r} + \binom{k}{r-1} &= \frac{k!}{r!(k-r)!} + \frac{k!}{(r-1)!(k-r+1)!} \\
&= \frac{k!(k-r+1) + k!r}{r!(k-r+1)!} \\
&= \frac{k!(k+1)}{r!(k+1-r)!} = \frac{(k+1)!}{r!(k+1-r)!} = \binom{k+1}{r}
\end{aligned}$$

EXERCISE I Relate the last computation in the proof above to the method of constructing Pascal's Triangle.

The Binomial Theorem permits us to write down rather quickly the expansions of powers of binomials which are tedious to compute by repeated multiplication.

ILLUSTRATION 1 Expand $(2x + 5y)^3$.

$$\begin{aligned}
(2x + 5y)^3 &= (2x)^3 + 3(2x)^2(5y) + 3(2x)(5y)^2 + (5y)^3 \\
&= 8x^3 + 60x^2y + 150xy^2 + 125y^3
\end{aligned}$$

ILLUSTRATION 2 Expand $(3x - 2y)^4$.

$$(3x - 2y)^4 = (3x)^4 + 4(3x)^3(-2y) + 6(3x)^2(-2y)^2 + 4(3x)(-2y)^3 + (-2y)^4$$
$$= 81x^4 - 216x^3y + 216x^2y^2 - 96xy^3 + 16y^4$$

ILLUSTRATION 3 Compute the term involving x^4y^3 in the expansion of $(3x - 5y)^7$. This term will involve $(3x)^4(-5y)^3$ with an appropriate coefficient. The theorem tells us that this coefficient is $\binom{7}{3} = 35$. So the term is

$$35(3x)^4(-5y)^3 = -354{,}375x^4y^3$$

PROBLEMS 3.4

1 Compute $\binom{5}{2}$; $\binom{6}{3}$; $\binom{7}{4}$; $\binom{4}{0}$; $\binom{3}{1}$.

2 Compute $\binom{5}{3}$; $\binom{6}{0}$; $\binom{7}{2}$; $\binom{8}{4}$; $\binom{7}{3}$.

In Probs. 3 to 6 verify the given formulas by direct computation.

3 $\binom{5}{2} + \binom{5}{3} = \binom{6}{3}$

4 $\binom{7}{3} + \binom{7}{4} = \binom{8}{4}$

5 $1 + \binom{3}{1} + \binom{3}{2} + \binom{3}{3} = 2^3$

6 $\binom{5}{0} + \binom{5}{1} + \binom{5}{2} + \binom{5}{3} + \binom{5}{4} + \binom{5}{5} = 2^5$

In Probs. 7 to 16 expand by the Binomial Theorem.

7 $(x + 2y)^5$

8 $(3a + b)^4$

9 $(r - s)^6$

10 $(3p - 4q)^4$

11 $(x + \frac{1}{3}y)^6$

12 $(\frac{1}{2}x + y)^6$

13 $\left(\frac{3}{x} + x^2\right)^3$

14 $\left(x^2 - \frac{2}{x}\right)^4$

15 $(1.01)^5 = (1 + 0.01)^5$

16 $(1.98)^4 = (2 - 0.02)^4$

In Probs. 17 to 22 find the coefficient of the given term in the given expansion.

17 a^7b^6 in $(a + b)^{13}$

18 r^2s^7 in $(2r + 3s)^9$

19 x^2y^5 in $(2x - 3y)^7$

20 p^4q^2 in $(4p - 5q)^6$

21 x^2y^9 in $(x^2 + y^3)^4$

22 x^4 in $\left(x^2 + \frac{1}{x}\right)^5$

23 Write $(x + y + z)^3 = [x + (y + z)]^3$, and expand through repeated use of the Binomial Theorem.

24 Write $(x + y - z)^3 = [(x + y) - z]^3$, and expand through repeated use of the Binomial Theorem.

25 In the Binomial Formula put $a = 1$ and $b = 1$, and then prove that

$$2^n = 1 + \binom{n}{1} + \binom{n}{2} + \cdots + \binom{n}{n-2} + \binom{n}{n-1} + \binom{n}{n}$$

3.5 › DIVISION OF POLYNOMIALS

In this section we shall restrict ourselves to polynomials which involve only a single variable, and we shall assume them to be arranged in descending powers of that variable. Examples of such polynomials are

$$15x^5 + 8x^3 - 4x^2 + x + 7 \qquad 8a^4 + \tfrac{1}{2}a^3 - a^2 + 2a + 5$$

The highest exponent which appears is called the *degree* of the polynomial. The degrees of the two examples above are 5 and 4, respectively. We shall denote such polynomials by the symbols $P(x)$, $Q(x)$, $D(x)$, etc., where the variable in the parentheses indicates the variable of the polynomial and P, Q, R, etc., are names to represent different polynomials of this type.

Suppose that we have two polynomials $P(x)$ of degree n and $D(x)$ of degree r, where $n \geq r$. We wish to consider what happens when we divide $P(x)$ by $D(x)$. This process of division needs a definition, but we shall postpone giving this until the method is clear. We illustrate by an example.

ILLUSTRATION 1 Divide $6x^4 + 7x^3 + 12x^2 + 10x + 1$ by $2x^2 + x + 4$.

$$\begin{array}{r|l}
 & 3x^2 + 2x - 1 \\ \hline
2x^2 + x + 4 & 6x^4 + 7x^3 + 12x^2 + 10x + 1 \\
 & 6x^4 + 3x^3 + 12x^2 \\ \cline{2-2}
 & \qquad 4x^3 + 0x^2 + 10x + 1 \\
 & \qquad 4x^3 + 2x^2 + 8x \\ \cline{2-2}
 & \qquad\qquad - 2x^2 + 2x + 1 \\
 & \qquad\qquad - 2x^2 - x - 4 \\ \cline{2-2}
 & \qquad\qquad\qquad 3x + 5
\end{array}$$

We have a quotient of $3x^2 + 2x - 1$ and a remainder of $3x + 5$.

The process of division as illustrated above is straightforward and should already be familiar to you. Let us suppose that we have carried through the division

$$D(x)\overline{)P(x)}$$

and obtained a *quotient* $Q(x)$ and a *remainder* $R(x)$. $D(x)$ is called the *divisor*, and $P(x)$ is called *the dividend*. There are several points worthy of mention:

1. The division continues step by step until a remainder is reached whose *degree is less than the degree of the divisor*.
2. If the remainder is zero, the division is said to be *exact*.
3. By reversing the steps of the computation we can show that

$$P(x) = D(x) \cdot Q(x) + R(x)$$

   for all values of x.

This leads us to the following statement.

THEOREM 2
Division algorithm

THE DIVISION ALGORITHM Let $P(x)$ and $D(x)$ be polynomials of degrees n and r, respectively, where $n \geq r$. Then there exist polynomials $Q(x)$, called the quotient, and $R(x)$, called the remainder, such that:

a. $P(x) = D(x) \cdot Q(x) + R(x)$ for all x.
b. The degree of $R(x)$ is less than the degree of $D(x)$.

The proof of this algorithm can be obtained by a generalization of the process shown in Illustration 1 or by more sophisticated means. Since either of these is too complicated to explain in detail here, we shall omit the proof.

In the problems below, follow the method of the illustration. You may run into trouble if a term is missing from the dividend or the divisor as in

$$4x^3 + 2x^2 - 3$$

which has no term in x. It will help you to keep matters straight if you will supply the missing term (or terms) with zero coefficients and write the above as

$$4x^3 + 2x^2 + 0x - 3$$

PROBLEMS 3.5

In Probs. 1 to 20 obtain the quotient and remainder, and check your result by substituting back into the equation $P(x) = D(x) \cdot Q(x) + R(x)$.

	Dividend	*Divisor*
1	$2x^3 + 5x^2 - 22x + 15$	$2x - 3$
2	$6x^3 - x^2 + 11x + 4$	$3x + 1$
3	$3x^3 + 14x^2 + 17x + 11$	$x + 3$
4	$8x^3 - 10x^2 + 5x + 4$	$2x - 1$
5	$2x^4 + x^3 + 6x^2 + 3x + 6$	$x^2 - x + 2$
6	$3x^4 - 7x^3 + x^2 - 7x + 9$	$3x^2 + 2x + 4$
7	$8x^5 - 18x^3 - 6x^2 - 6x + 22$	$2x^2 - 5$

8	$x^5 + 6x^3 + 3x^2 + 8x + 10$	$x^2 + 4$
9	$4x^5 - x^4 + 12x^3 + 2x^2 + x + 5$	$4x^3 - x^2 + 1$
10	$3x^5 + 4x^4 + 7x^3 + 7x^2 + 8x + 4$	$x^3 + 2$
11	$13x^3 + 3x + 10x^5 - 16 - 6x^2 + 4x^4$	$3 + 2x^2$
12	$30x^2 - 11x^3 + 10x^5 - 7 - x$	$-3 + 5x^2$
13	$x^4 - 1$	$x^2 - 1$
14	$x^6 - 1$	$x^3 - 1$
15	$x^6 - 1$	$x^2 + 1$
16	$x^5 - 1$	$x^2 + 1$
17	$x^6 - y^6$	$x - y$
18	$x^8 - y^8$	$x + y$
19	$x^3 + \frac{20}{3}x^2 + 16x + 10$	$3x + 2$
20	$(3 + i)x^3 + 8x^2 + (5 - 6i)x - (1 + 12i)$	$x + (2 - i)$

21 If $P(x)$ is of degree n and $D(x)$ of degree r and if $r > n$, show that the division algorithm is satisfied trivially with $Q(x) = 0$ and $R(x) = P(x)$.

3.6 › FACTORING

In factoring we seek to undo the process of multiplication. We are given a polynomial and are asked to discover how it can be expressed as a product of other polynomials, called its factors. You will find factoring important when you are asked to simplify algebraic fractions or to solve certain types of equations.

DEFINITION If a polynomial $P(x)$ is the product of r polynomials $Q_1(x), \ldots, Q_r(x)$, that is, if

$$P(x) = Q_1(x) \cdot Q_2(x) \cdots Q_r(x)$$

then $Q_1(x), \ldots, Q_r(x)$ are called *factors* of $P(x)$.

EXERCISE A If $Q(x)$ is a factor of $P(x)$, show that the quotient of $P(x)$ by $Q(x)$ involves no remainder, i.e., that the division is exact.

In general, the factors $Q_1(x), \ldots, Q_r(x)$ may have any coefficients in the field over which $P(x)$ is defined. In special cases, however, we may specify that the coefficients of the factors are to be integers and hence ask for the "integral factors" of $P(x)$. Similarly we may require the coefficients of the factors to be rational numbers or real numbers, etc. Unless some such restriction is stated, it is to be understood that the coefficients of the factors are unrestricted, i.e., that they may be of the general form $a + bi$.

The problem of factoring a general polynomial can be a quite complicated affair. As we shall see in Chap. 10, it is equivalent to the problem of finding all the roots of a polynomial equation, and this is by no means an elementary topic. We have a more limited objective here, namely, to factor certain simple polynomials by methods which are elementary and straightforward. Although these methods are not adequate to factor a general polynomial, they are very useful when they do apply. Hence they are worth mastering.

Common factor

REMOVAL OF A COMMON FACTOR When all the terms of the given polynomial have a factor in common, this expression may be factored out by use of the distributive law in reverse. These common factors should be removed before other methods of factoring are employed.

ILLUSTRATION 1

a $2x^3 + 6x^2 - 10 = 2(x^3 + 3x^2 - 5)$

b $x^4 - x^2 = x^2(x^2 - 1)$

c $ac + bc + ad + bd = (ac + bc) + (ad + bd)$
$= c(a + b) + d(a + b) = (c + d)(a + b)$

Trinomials

TRINOMIALS WITH INTEGRAL COEFFICIENTS We consider here trinomials of the form

$$ax^2 + bx + c$$

where a, b, and c are *integers*. We seek to write this as the product of two linear polynomials with *integral* coefficients, i.e.,

$$ax^2 + bx + c = (px + q)(rx + s)$$

where p, q, r, and s are integers. This factorization is not always possible, but our method produces the required factors whenever they exist.

Let us first consider the simpler situation where $a = 1$. Then p and r must also equal 1, and we write

$$x^2 + bx + c = (x + q)(x + s)$$

Multiplying out the right-hand side, we find

$$x^2 + bx + c = x^2 + (q + s)x + qs$$

Thus, we are looking for two integers, q and s, such that

$$q + s = b \qquad qs = c$$

To find these, we factor c into all possible pairs (q,s) such that $qs = c$. Then we examine these pairs (q,s) to determine whether or not in any of them $q + s = b$. If we find such a pair, we have solved the problem; otherwise there are no factors of the prescribed form. The details of the method are best shown by illustrations.

ILLUSTRATION 2 Factor $x^2 + 5x + 6$.

The integral pairs of factors of 6 are

$$(1,6) \qquad (2,3) \qquad (-1,-6) \qquad (-2,-3)$$

We exclude the last two immediately since the sum of two negative integers cannot be $+5$. Examining the other two in turn, we find that $(2,3)$ is a solution since $2 + 3 = 5$. Therefore,

$$x^2 + 5x + 6 = (x + 2)(x + 3)$$

ILLUSTRATION 3 Factor $x^2 - 6x + 8$.

The integral pairs of factors of 8 are

$$(1,8) \qquad (2,4) \qquad (-1,-8) \qquad (-2,-4)$$

We exclude the first two immediately since the middle coefficient, -6, is negative. We find that $(-2,-4)$ is a solution since $(-2) + (-4) = -6$. Therefore

$$x^2 - 6x + 8 = (x - 2)(x - 4)$$

ILLUSTRATION 4 Factor $x^2 + 3x - 10$.

The integral pairs of factors of -10 are

$$(1,-10) \qquad (-1,10) \qquad (2,-5) \qquad (5,-2)$$

Examining these in turn, we find that $(5,-2)$ is a solution since $5 + (-2) = 3$. Therefore

$$x^2 + 3x - 10 = (x + 5)(x - 2)$$

ILLUSTRATION 5 Factor $x^2 + 3x + 4$.

The integral pairs of factors of 4 are

$$(1,4) \qquad (2,2) \qquad (-1,-4) \qquad (-2,-2)$$

None of these is a solution, and hence there are no factors of the prescribed form.

In the case of the general trinomial, we write

$$\begin{aligned} ax^2 + bx + c &= (px + q)(rx + s) \\ &= prx^2 + (ps + qr)x + qs \end{aligned}$$

Therefore, $a = pr$, $b = ps + qr$, $c = qs$. The method is similar to the special case above, but here we have more possibilities. We find the pairs (p,r) which factor a and the pairs (q,s) which factor c. Then we examine each pair (p,r) in connection with each pair (q,s) to see whether or not for any of these combinations $ps + qr = b$. If so, we have a solution; otherwise, there are no factors of this form.

The number of possibilities will be reduced if we always take $a > 0$. This can always be arranged by removing the common factor, -1, if

a is initially negative. Moreover, we can assume that p and r are both positive without losing any possible solutions; we must, however, allow q and s to take all appropriate positive and negative signs.

ILLUSTRATION 6 Factor $8x^2 + 2x - 15$.

The pairs of factors of 8 are (1,8) and (2,4). The pairs of factors of -15 are $(1,-15)$, $(-1,15)$, $(3,-5)$, $(5,-3)$, $(-15,1)$, $(15,-1)$, $(-5,3)$, $(-3,5)$. In this case it is necessary to write the numbers in each pair in the two possible orders in order to cover all cases.

Now write one of the first pairs and one of the second pairs as shown below:

$$(1,8),\ (1,-15)$$

Multiply the two outside numbers, 1 and -15, and add to this the product of the two inside numbers, 8 and 1. This gives $(1)(-15) + (8)(1) = -7$. This should equal the coefficient of x, namely 2. Since it does not, these two pairs do not give a solution. Try each combination of a first pair and a second pair. Among these we find that the solution is

$$(2,4),\ (3,-5)$$

for $(2)(-5) + (4)(3) = 2$. Hence

$$8x^2 + 2x - 15 = (2x + 3)(4x - 5)$$

Difference of two squares

DIFFERENCE OF TWO SQUARES We consider expressions of the form

$$x^2 - a^2$$

An elementary calculation shows that

$$x^2 - a^2 = (x + a)(x - a)$$

Hence, the factors may be written down at sight.

ILLUSTRATION 7

a $x^2 - 9 = (x + 3)(x - 3)$
b $16x^2 - 25 = (4x + 5)(4x - 5)$
c $25x^4 - 36y^6 = (5x^2 + 6y^3)(5x^2 - 6y^3)$
d $(3x + 5)^2 - (2x - 1)^2 = [(3x + 5) + (2x - 1)] \times [(3x + 5) - (2x - 1)] = (5x + 4)(x + 6)$
e $x^2 - 2 = (x + \sqrt{2})(x - \sqrt{2})$

This method may be extended to cover expressions of the form $x^2 + a^2$. For

$$\begin{aligned} x^2 + a^2 &= x^2 - (ia)^2 \\ &= (x + ia)(x - ia) \end{aligned}$$

ILLUSTRATION 8

a $4x^2 + 9 = (2x + 3i)(2x - 3i)$
b $x^2 + 3 = (x + i\sqrt{3})(x - i\sqrt{3})$

SUM AND DIFFERENCE OF TWO CUBES We rely upon the two formulas

$$x^3 + a^3 = (x + a)(x^2 - ax + a^2)$$
$$x^3 - a^3 = (x - a)(x^2 + ax + a^2)$$

which require no further explanation.

COMBINATIONS OF THE ABOVE METHODS It is often possible to factor complicated looking expressions by using two or more of these methods in turn.

ILLUSTRATION 9 Factor $4x^2 + 24x + 32 - 16y^2 + 16y$.

$$\begin{aligned} 4x^2 + 24x + 32 - 16y^2 + 16y \\ &= 4x^2 + 24x + 36 - 16y^2 + 16y - 4 \\ &= 4[(x^2 + 6x + 9) - (4y^2 - 4y + 1)] \\ &= 4[(x + 3)^2 - (2y - 1)^2] \\ &= 4[(x + 3 + 2y - 1)(x + 3 - 2y + 1)] \\ &= 4(x + 2y + 2)(x - 2y + 4) \end{aligned}$$

PROBLEMS 3.6

In Probs. 1 to 6 factor the given expression.

1 $xy + 4x + 3y + 12$
2 $xy + 5x - 6y - 30$
3 $x^3 - x^2 + x - 1$
4 $ay - by + 4xy$
5 $5x + 5y - vx - vy$
6 $x^2y - 3y + x^2 - 3$

In Probs. 7 to 32 find the factors with integral coefficients, if any.

7 $x^2 + 2x - 8$
8 $x^2 - 3x - 4$
9 $x^2 - 13x + 36$
10 $x^2 + 12x + 35$
11 $32 + 12x + x^2$
12 $20 - 9x + x^2$
13 $y^2 - 14xy + 45x^2$
14 $a^2 - 13ab + 30b^2$
15 $x^2 + 7x + 11$
16 $x^2 - 4x - 6$
17 $3x^2 + 5x + 2$
18 $6x^2 - 13x - 5$
19 $3x^2 - 8x + 4$
20 $2x^2 + 11x + 15$
21 $14x^2 + 22x - 12$
22 $5x^2 - x - 4$
23 $10x^2 - xy - 2y^2$
24 $x^2 - xy - 2y^2$
25 $10x^2 + 2x - 12$
26 $6x^2 - 2x - 28$
27 $12x^2 - 27$
28 $24x^2 + 17x - 20$
29 $12x^2 + 11x - 5$
30 $45x^2 + 53x - 14$
31 $3x^2 - 2x + 8$
32 $x^2 + x + 1$

In Probs. 33 to 56 factor the given expression.

33 $4x^2 - 25$

34 $36x^2 - 121$

35 $x^2 - 5$

36 $x^2 - 3$

37 $4x^2 + 9$

38 $9x^2 + 25$

39 $(x + 5)^2 - (x + 2)^2$

40 $(x + 3)^2 - (x + 7)^2$

41 $x^4 - y^2$

42 $x^6 - y^6$

43 $(x^2 + 4x + 4) - (y^2 - 6y + 9)$

44 $(x^2 + 2x + 1) - (y^2 + 8y + 16)$

45 $8x^3 - y^3$

46 $27x^3 - 1$

47 $x^3 + 8$

48 $x^3 + 125$

49 $a^2 - 9b^2 + 4a + 4$

50 $k^2 + 9 + 6k - x^2$

51 $x^4 - 7x^2 + 12$

52 $x^4 - 20x^2 + 64$

53 $2x^3 + 7x^2y - 4xy^2$

54 $2x^2y - 5xy^2 - 3y^3$

55 $3x^2 - 7xy + 2y^2 + 19x - 13y + 20$

56 $6x^2 + xy - 2y^2 - x + 11y - 15$

4 ALGEBRAIC FRACTIONS

4.1 » INTRODUCTION

An algebraic fraction is the quotient of two algebraic expressions. Examples of these are

$$(1)\ \frac{3x^2 + 4a}{x + 1} \qquad (2)\ \frac{\sqrt{x^3 - 1} + 4x}{\sqrt[3]{x + 7}} \qquad (3)\ \frac{1/x^2 + \sqrt{3}\,x}{2/(x + 5)}$$

We recall that the variables in these expressions stand for arbitrary numbers. In the most general circumstances these numbers are complex, but in particular situations it may be specified that they are real or rational. This leads to two important remarks:

1. The algebra of fractions can be derived from properties R1 to R8 of Sec. 2.3.
2. It is understood that we cannot assign values to any variable which makes any denominator equal to zero. Thus in example (1) above we exclude $x = -1$; in example (2) we exclude $x = -7$; in example (3) we exclude $x = 0$ and $x = -5$. It is tedious to state these exclusions each time that we write a fraction, and so you will have to supply this information yourself and take necessary precautions.

Throughout this chapter we shall restrict ourselves to fractions whose numerators and denominators are either polynomials or quotients of polynomials. We do this for simplicity of exposition, and not because the theory is restricted to such cases. Fractions containing radicals will be treated in Chap. 5.

4.2 » SIMPLIFICATION OF FRACTIONS

Since fractions are troublesome enough in any case, we wish to be able to simplify any fraction that turns up as much as possible before putting it back into some further calculation. The most important way of doing this is nothing more than the familiar "reducing to lowest terms". In spite of the simplicity of this method, its misuse is a source of frequent errors on the part of careless students—so read this section carefully.

BASIC PRINCIPLE The method depends upon the familiar relation

$$\frac{ka}{kb} = \frac{a}{b} \qquad \text{for } k \neq 0$$

In other words: If we divide the numerator and denominator of a given fraction by the same quantity (not zero) the result is a fraction equal to the given one.

In order to apply this principle to algebraic fractions, we factor the numerator and denominator, look for common factors, and divide top and bottom by any factor which is common to both.

ILLUSTRATION 1

a $\dfrac{x^2 - 5x + 6}{x^2 - 4x + 3} = \dfrac{(x-3)(x-2)}{(x-3)(x-1)} = \dfrac{x-2}{x-1}$

b $\dfrac{4x^2 + 7x}{x^2} = \dfrac{x(4x+7)}{x(x)} = \dfrac{4x+7}{x}$

c $\dfrac{x^2 + 4x + 4}{x^2 + 4x + 3} = \dfrac{(x+2)(x+2)}{(x+1)(x+3)}$

This does not simplify since the numerator and denominator have no common factor.

The matter of excluded values of x raises the question of what we mean by the equality of two fractions. In Illustration 1*a*, the fraction on the left is defined for all values of x except $x = 1$ and $x = 3$; on the other hand, the fraction on the right is defined for all values of x except $x = 1$. A strict use of equality between these two would lead us to the following nonsensical relation where we put $x = 3$:

$$\frac{0}{0} = \text{nonsense} = \frac{3-2}{3-1} = \frac{1}{2}$$

Equality

In order to avoid such difficulties, let us define equality of algebraic fractions as follows.

DEFINITION *Two algebraic fractions* involving the variable x are *equal* if and only if they have the same numerical values when x is put equal to any number for which both fractions are defined.

COMMON ERRORS Some of the mistakes which students make are based upon the following erroneous relation:

$$\frac{k+a}{k+b} = \frac{a}{b} \qquad \textit{FALSE}$$

In other words, it is incorrect to simplify a fraction by subtracting the same quantity from numerator and denominator!

ILLUSTRATION 2

a $\dfrac{x^2+4x+4}{x^2+4x+3}$ does not equal $\dfrac{4}{3}$. See Illustration 1*c*.

b $\dfrac{2x+3}{2x+1}$ does not equal $\dfrac{3}{1} = 3$.

EXERCISE A Disprove the statements

$$\frac{x^2+4x+4}{x^2+4x+3} = \frac{4}{3} \qquad \text{and} \qquad \frac{2x+3}{2x+1} = 3$$

by finding suitable counterexamples.

Other errors are caused by failure to remember the distributive law. These are based upon the following erroneous relation:

$$\frac{k+a}{k} = a \qquad \textit{FALSE}$$

The trouble here is that k has been treated as a factor of the numerator. Proceeding correctly, we can write, however,

$$\frac{k+a}{k} = \frac{k\left(1+\frac{a}{k}\right)}{k} = 1 + \frac{a}{k} \qquad \textit{TRUE}$$

Another approach to this is the following:

$$\frac{k+a}{k} = \frac{1}{k}(k+a)$$

Now apply the distributive law, which gives

$$\frac{1}{k}(k+a) = \frac{1}{k}(k) + \frac{1}{k}(a) = 1 + \frac{a}{k} \qquad \textit{TRUE}$$

ILLUSTRATION 3

a $\dfrac{5x+7}{5x}$ is not equal to 7 but does equal $1 + \dfrac{7}{5x}$.

b $\dfrac{(x+3)^2+x-2}{x+3}$ is not equal to $(x+3)+x-2$ but does equal $x+3+\dfrac{x-2}{x+3}$.

EXERCISE B Disprove the statements

$$\frac{5x+7}{5x} = 7 \quad \text{and} \quad \frac{(x+3)^2 + x - 2}{x+3} = (x+3) + x - 2$$

by finding suitable counterexamples.

In the problems which follow, some fractions will simplify, and others will not. In your zeal to effect a simplification do not commit either of the common errors illustrated above.

PROBLEMS 4.2 In Probs. 1 to 20 simplify where possible.

1 $\dfrac{x^2 - 16}{x^2 + 8x + 16}$

2 $\dfrac{x^2 + 6x + 9}{x^2 - 9}$

3 $\dfrac{x^2 - 8x + 15}{x^2 - 7x + 12}$

4 $\dfrac{x^2 - x - 6}{x^2 + 7x + 10}$

5 $\dfrac{x^2 + 2x}{x^2 + 4x + 3}$

6 $\dfrac{x^2 + 7}{7x}$

7 $\dfrac{2x^2 + 2x - 4}{x^2 + 4x - 5}$

8 $\dfrac{x^2 + 5x + 6}{x + 5}$

9 $\dfrac{4x^2 - 9}{2x^2 - 5x - 12}$

10 $\dfrac{6x^2 + 13x - 5}{3x^2 - 22x + 7}$

11 $\dfrac{ac + ad - bc - bd}{4a - 4b - ay + by}$

12 $\dfrac{y^2 - 2yz - 3z^2}{ay + by + az + bz}$

13 $\dfrac{x^2 + y^2}{x^3 + y^3}$

14 $\dfrac{x^2 - y^2}{x^3 - y^3}$

15 $\dfrac{a^4 - b^4}{a^2 + b^2}$

16 $\dfrac{9x^4 - y^2}{9x^4 + y^2}$

17 $\dfrac{a^3x^3 - a^2x^3 + 5ax^3 - 5x^3}{a^4x - 25x}$

18 $\dfrac{a^6 - b^6}{a^2 - b^2}$

19 $\dfrac{x^3 + 7x^2 + 19x}{x + 7x^2 + 19}$

20 $\dfrac{x^2 + 5x - 6}{y + 6}$

In Probs. 21 to 26 find counterexamples which disprove the given statements.

21 $\dfrac{4x + 9}{4x} = 9$

22 $\dfrac{5x + 3}{5x} = 3$

23 $\dfrac{8x + x^3}{8x + x^2} = x$

24 $\dfrac{7x + 6}{7x - 2} = 3$

25 $\frac{7x}{4} + \frac{9x}{5} = \frac{16x}{9}$

26 $\frac{6x}{5} - \frac{2x}{3} = \frac{4x}{2} = 2x$

4.3 ADDITION

The addition of fractions is a straightforward application of the formula below, which we derived in Sec. 2.7, Theorem 10:

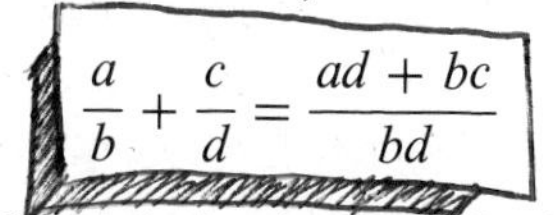

$$\frac{a}{b} + \frac{c}{d} = \frac{ad + bc}{bd}$$

When the fraction on the right is obtained, it should then be simplified as much as possible.

ILLUSTRATION 1

$$\frac{2x - 1}{x + 3} + \frac{x^2}{3x - 1} = \frac{(2x - 1)(3x - 1) + (x + 3)x^2}{(x + 3)(3x - 1)}$$
$$= \frac{x^3 + 9x^2 - 5x + 1}{3x^2 + 8x - 3}$$

ILLUSTRATION 2

$$\frac{x}{x + 3} + \frac{5x^2}{x^2 - 9} = \frac{x(x^2 - 9) + (x + 3)(5x^2)}{(x + 3)(x^2 - 9)}$$
$$= \frac{6x^3 + 15x^2 - 9x}{(x + 3)(x^2 - 9)}$$
$$= \frac{3x(2x - 1)(x + 3)}{(x + 3)(x^2 - 9)}$$
$$= \frac{3x(2x - 1)}{x^2 - 9}$$

Least common denominator

Although the above process always gives the correct result, it may involve unnecessary complexities. These occur because the process leads to a denominator which is not necessarily the *least common denominator* (LCD). You have doubtless met the notion of a least common denominator in arithmetic. When you added

$$\frac{4}{9} + \frac{7}{12}$$

you learned to write

$$\frac{4}{9} + \frac{7}{12} = \frac{16}{36} + \frac{21}{36} = \frac{37}{36}$$

and you avoided the use of $9 \times 12 = 108$ as a denominator. In arithmetic the least common denominator is the smallest number which contains the given denominators as factors. We found it above by first factoring $9 = 3^2$ and $12 = 2^2 \times 3$ into prime factors. Then we formed a number (the LCD) by multiplying together the several distinct factors we had found (namely, 2 and 3), each raised to the larger of the two powers to which it was raised in the given two numbers. Thus the LCD of $\frac{4}{9}$ and $\frac{7}{12}$ is $3^2 \times 2^2 = 36$.

In algebra we would like to follow the same procedure, but practical difficulties may occur. For example, one of the denominators may be $x^2 + 3x + 3$, and you may not know how to factor this into factors of the first degree. So you may be unable to find the LCD. But you should factor the denominators as far as you can and come as near as you can to the LCD. The point to remember is that the use of the LCD is a great convenience when it can be found easily but that the method of the LCD cannot be applied universally. The use of the LCD is not, therefore, a fixed requirement of the addition process, but it should be used wherever possible. In most of the problems below its use is recommended.

ILLUSTRATION 3 $\dfrac{3x + 4}{x^2 - 16} + \dfrac{x - 3}{x^2 + 8x + 16}$

We write $x^2 - 16 = (x + 4)(x - 4)$ and $x^2 + 8x + 16 = (x + 4)(x + 4)$. Hence we form the LCD, which is $(x + 4)^2(x - 4)$. Then

$$\begin{aligned}\frac{3x + 4}{x^2 - 16} + \frac{x - 3}{x^2 + 8x + 16} &= \frac{3x + 4}{(x + 4)(x - 4)} + \frac{x - 3}{(x + 4)^2} \\ &= \frac{(3x + 4)(x + 4)}{(x + 4)^2(x - 4)} + \frac{(x - 3)(x - 4)}{(x + 4)^2(x - 4)} \\ &= \frac{4x^2 + 9x + 28}{(x + 4)^2(x - 4)}\end{aligned}$$

Since this fraction does not simplify, it is the final answer.

ILLUSTRATION 4 $\dfrac{x}{x + 3} + \dfrac{5x^2}{x^2 - 9}$ (see Illustration 2)

The LCD is $x^2 - 9$. Hence we write

$$\frac{x(x - 3)}{x^2 - 9} + \frac{5x^2}{x^2 - 9} = \frac{6x^2 - 3x}{x^2 - 9} = \frac{3x(2x - 1)}{x^2 - 9}$$

ILLUSTRATION 5 $\dfrac{x}{x^2 - 2x + 5} + \dfrac{3}{x - 1}$

In this case our procedure breaks down, for we do not know how to factor $x^2 - 2x + 5$. So we cannot find the LCD. Instead we use $(x^2 - 2x + 5)(x - 1)$ as our common denominator, and write

$$\frac{x}{x^2 - 2x + 5} + \frac{3}{x - 1} = \frac{(x^2 - x) + 3(x^2 - 2x + 5)}{(x^2 - 2x + 5)(x - 1)}$$
$$= \frac{4x^2 - 7x + 15}{(x^2 - 2x + 5)(x - 1)}$$

ILLUSTRATION 6 $\dfrac{x}{x^2 - 2x + 5} + \dfrac{3}{x - 1 + 2i}$

As in Illustration 5, the only thing you can do is to write

$$\frac{x}{x^2 - 2x + 5} + \frac{3}{x - 1 + 2i} = \frac{(x^2 - x + 2ix) + (3x^2 - 6x + 15)}{(x^2 - 2x + 5)(x - 1 + 2i)}$$
$$= \frac{4x^2 - 7x + 2ix + 15}{(x^2 - 2x + 5)(x - 1 + 2i)}$$

Actually, however, a little more knowledge will give you a better result. If you had been clever, you might have noted that

$$x^2 - 2x + 5 = (x - 1 + 2i)(x - 1 - 2i)$$

So you could have written

$$\frac{x}{x^2 - 2x + 5} + \frac{3}{x - 1 + 2i} = \frac{x + 3(x - 1 - 2i)}{x^2 - 2x + 5}$$
$$= \frac{4x - 6i - 3}{x^2 - 2x + 5}$$

This is a better answer than that given above, but it was obtained by methods which you are not likely to have thought of. We shall not discuss these methods here, but this illustration should be a sufficient hint for a good student.

PROBLEMS 4.3

Carry out the indicated operations.

1 $\dfrac{3}{x + 2} + \dfrac{2}{x - 2}$

2 $\dfrac{3}{2 + x} + \dfrac{5}{6 - x}$

3 $\dfrac{4}{2a + 3b} + \dfrac{7}{3b - 2a}$

4 $\dfrac{3}{2x + 4y} - \dfrac{5}{x + 2y}$

5 $\dfrac{5}{a} - \dfrac{2}{a - b}$

6 $\dfrac{6}{x + 2} - \dfrac{9}{x + 1}$

7 $\dfrac{4}{x^2 + xy} + \dfrac{7}{xy + y^2}$

8 $\dfrac{x}{x + y} + \dfrac{3}{x^2 - y^2}$

9 $\dfrac{2x}{x^2 + 3x + 2} - \dfrac{x}{x^2 - 4}$

10 $\dfrac{x+5}{x^2+5x+6}+\dfrac{3x}{x^2+x-2}$

11 $\dfrac{3x}{x^2+2x+1}-\dfrac{x^2}{x^2+4x+4}$

12 $\dfrac{x^2}{x^2-4x+4}+\dfrac{2x}{x^2+6x+9}$

13 $\dfrac{2x+1}{2x+1}+\dfrac{x^2-9}{x^2-4}$

14 $\dfrac{x^2+6}{x^2-3}-\dfrac{2x+2}{x+1}$

15 $\dfrac{3x}{x+1}-\dfrac{2}{x}+\dfrac{2}{x-1}$

16 $3-\dfrac{2}{a+1}+\dfrac{4}{a^2-1}$

17 $\dfrac{3}{x^2+3x+2}-\dfrac{2}{x^2+4x+3}+\dfrac{5}{x^2+5x+6}$

18 $\dfrac{4}{x^2-3x-4}+\dfrac{3}{x^2-16}-\dfrac{7}{x^2+5x+4}$

19 $\dfrac{2x}{x^2-4x+13}+\dfrac{5}{x-2+3i}$

20 $\dfrac{3x}{x^2+8x+25}-\dfrac{2}{x+4-3i}$

4.4 › MULTIPLICATION AND DIVISION

We saw in Sec. 2.7, Theorem 11, that the multiplication of fractions follows the rule

$$\frac{a}{b}\times\frac{c}{d}=\frac{ac}{bd}$$

Also from Prob. 7, Sec. 2.9, we know that division follows the rule

$$\frac{a}{b}\div\frac{c}{d}=\frac{a}{b}\times\frac{d}{c}=\frac{ad}{bc}$$

These rules apply equally well to algebraic fractions. When a, b, c, and d are polynomials, it is desirable to factor them if possible so that simplifications in the final answer can be made easily. As shown in the illustrations below, it is usually convenient to make these simplifications at an early stage of the work rather than to wait to carry them out after the product has been found.

ILLUSTRATION 1

$$\frac{x^2-y^2}{4y}\times\frac{2y}{x+y}=\frac{(x-y)(x+y)}{4y}\times\frac{2y}{x+y}$$

$$=\frac{(x-y)(x+y)(2y)}{4y(x+y)}=\frac{x-y}{2}$$

In the next to the last fraction we have divided numerator and denominator by $(x + y)(2y)$ in order to obtain the final result. You might just as well, however, have carried out this division at the previous stage and thus have shortened the calculation as follows:

$$\frac{x^2 - y^2}{4y} \times \frac{2y}{x + y} = \frac{(x - y)\overset{\checkmark}{(x + y)}}{\underset{\underset{2}{\checkmark\checkmark}}{4y}} \times \frac{\overset{\checkmark\checkmark}{2y}}{\underset{\checkmark}{x + y}} = \frac{x - y}{2}$$

In order to keep track of our divisions, we have placed check marks ($\checkmark$) above those factors that have been divided out. We have also written a 2 as the quotient of 4 by 2. Instead of using check marks many people cross out these factors and say that they have been "canceled". There is no harm in canceling if it is done with understanding, but too often it is used blindly without an appreciation of the fact that *division* is the true operation involved.

Cancellation

ILLUSTRATION 2

$$\frac{x^2 + 3x + 2}{x^2 - x - 2} \times \frac{x^2 + x - 2}{x^2 + 4x + 4} \div \frac{x^2 - x}{x^2 + x - 6}$$

$$= \frac{\overset{\checkmark}{(x + 2)}\overset{\checkmark}{(x + 1)}}{\underset{\checkmark}{(x - 2)}\underset{\checkmark}{(x + 1)}} \times \frac{\overset{\checkmark}{(x + 2)}\overset{\checkmark}{(x - 1)}}{\underset{\checkmark}{(x + 2)}\underset{\checkmark}{(x + 2)}} \times \frac{\overset{\checkmark}{(x - 2)}(x + 3)}{x\underset{\checkmark}{(x - 1)}} = \frac{x + 3}{x}$$

ILLUSTRATION 3

$$\frac{x^2 + 5x + 6}{x^2 - 1} \div \frac{x^2 + 1}{x + 4} = \frac{(x + 2)(x + 3)}{(x + 1)(x - 1)} \times \frac{x + 4}{x^2 + 1}$$

$$= \frac{(x + 2)(x + 3)(x + 4)}{(x + 1)(x - 1)(x^2 + 1)}$$

There are no common factors to divide out, and so no simplification is possible.

Do not expect that every problem will simplify. You will make errors if you force yourself to simplify *every* problem just because simplifications do occur in *many* problems proposed in books. Actual problems derived from nature rarely simplify, but you must know the process just in case you are lucky enough to find a problem in your work which does simplify.

PROBLEMS 4.4 Carry out the indicated operations, and simplify where possible.

1 $\frac{x^2 + 6x + 9}{x^2 - 9} \times \frac{x - 3}{4}$

2 $\frac{x^2 - 16}{x^2 + 3x} \times \frac{x + 3}{x - 4}$

3 $\frac{x^2 - 2x - 3}{x^2 + 5x + 6} \div \frac{x^2 - 2x - 3}{x^2 + 6x + 8}$

4 $\frac{x^2 + 3x - 4}{x^2 - 2x - 3} \div \frac{x^2 + x - 2}{x^2 + 2x - 15}$

5 $\frac{x^2 + 4x + 4}{x^2 + 2x + 1} \times \frac{x^2 - 4}{(x + 1)^2}$

6 $\frac{x^2 - 9}{x^2 - 16} \div \frac{x^2 - 3}{9x^2 - 1}$

7 $\frac{x^2 - x - 6}{x^2 + x - 6} \times \frac{x^2 - x - 12}{x - 1}$

8 $\frac{x^2 + 4x}{x^2 + 2x - 8} \times \frac{x^2 + 3x - 10}{x^2 + 2x - 15}$

9 $\frac{5r + 10s}{4rs} \div \frac{2(r + 2s)}{6rs}$

10 $\frac{xy + 2x}{5y} \div \frac{4x^2y + 8x^2}{3y}$

11 $\frac{2x^2 + x - 3}{x^2 - 1} \times \frac{x^2 - x}{4x + 6}$

12 $\frac{8x^2 - 2x - 3}{5x^2 + 7x + 2} \times \frac{3x^2 - 4x - 7}{8x^2 + 14x - 15}$

13 $\frac{6x^2 - 7x - 20}{2x^2 + x - 21} \div \frac{6x^2 - x - 12}{4x^2 - 7x - 15}$

14 $\frac{x^2 + 5x + 6}{x^2 + 5x + 4} \div \frac{x^2 + 15x + 56}{x^2 + 11x + 30}$

15 $\frac{xy + xz}{xy - xz} \times \frac{y}{y + z} \times \frac{y - z}{y}$

16 $\frac{xy - x}{y^2 - 1} \times \frac{y + 1}{x + 2} \times \frac{2x + 4}{5x}$

17 $\frac{x^2 - y^2}{x} \times \frac{y^2}{x^3 - y^3} \div \frac{x^2 + 2xy + y^2}{x^2 + xy + y^2}$

18 $\frac{r^3 + 8}{r^2 + 4r + 4} \times \frac{r^2 - 2r}{8 - 2r - r^2} \div \frac{r^3 - 2r^2 + 4r}{r + 4}$

19 $\dfrac{p}{p+q-r} \times \dfrac{(p+q)^2 - r^2}{rp + rq + r^2}$

20 $\dfrac{(a+b)^2 - 16}{a+b+4} \times \dfrac{a^2 + 4a + ab}{(a+4)^2 - b^2}$

4.5 › COMPOUND FRACTIONS

The operations of adding, subtracting, multiplying, and dividing fractions can be combined in various ways. The only situation which calls for special comment is that of simplifying a fraction whose numerator and denominator are themselves sums of fractions. In this case the numerator and denominator should be simplified separately, and finally the division should be performed.

ILLUSTRATION 1

$$\frac{\dfrac{1}{x+1} - \dfrac{1}{x-1}}{\dfrac{1}{x+1} + \dfrac{1}{x-1}} = \frac{\dfrac{(x-1)-(x+1)}{(x+1)(x-1)}}{\dfrac{(x-1)+(x+1)}{(x+1)(x-1)}} = \frac{\dfrac{-2}{x^2-1}}{\dfrac{2x}{x^2-1}}$$

$$= \frac{-2}{x^2-1} \times \frac{x^2-1}{2x} = -\frac{1}{x}$$

PROBLEMS 4.5

GENERAL REVIEW OF FRACTIONS Carry out the indicated operations, and simplify where possible.

1 $\left(\dfrac{2}{x-3} - \dfrac{3}{x-2}\right) \times \dfrac{3x}{x-5}$

2 $\left(\dfrac{2}{x+1} + \dfrac{4}{x+3}\right) \times \dfrac{3x-1}{3x+5}$

3 $\left(\dfrac{x}{x-3} + \dfrac{3x}{-2x+4}\right) \div \dfrac{x-5}{2x}$

4 $\left(\dfrac{x}{x+1} - \dfrac{7}{x+7}\right) \div \dfrac{4x-5}{x+1}$

5 $\dfrac{\dfrac{x}{x+y} - \dfrac{y}{x-y}}{\dfrac{x}{x-y} + \dfrac{y}{x+y}}$

6 $\dfrac{2 + \dfrac{3a}{4a+b}}{\dfrac{5a}{4a+b}}$

7 $2x - \dfrac{3}{5x - \dfrac{6}{7x}}$

8 $3x + \dfrac{5}{2x - \dfrac{3}{x}}$

9 $\left(\dfrac{x^2 - x - 2}{x^3 - 4x} \times \dfrac{x^2 - x - 6}{x^2 + 5x + 4}\right) + \left(\dfrac{x + 3}{x} \times \dfrac{x + 5}{x^2 + 4x + 3}\right)$

10 $\left(\dfrac{x^2 - 4}{4x^2 - 1} \times \dfrac{1 - 2x}{x + 2}\right) - \left(\dfrac{x + 4}{x^2 + 2x - 15} \times \dfrac{x + 5}{x + 4}\right)$

11 $\left(\dfrac{3x - 1}{x + 1} \times \dfrac{3}{x + 2}\right) + \left(\dfrac{4}{x + 2} \div \dfrac{x - 3}{x - 5}\right) - \left(\dfrac{2}{x + 1} \times \dfrac{2x + 5}{x - 3}\right)$

12 $\left(\dfrac{2}{x} \div \dfrac{x + 1}{3x + 2}\right) - \left(\dfrac{4}{x + 1} \times \dfrac{2x + 1}{x - 4}\right) + \left(\dfrac{x + 2}{x - 4} \div \dfrac{x}{3}\right)$

13 $\dfrac{\dfrac{x + 3}{4x^2 - 1}}{\dfrac{x + 2}{6x^2 - 5x - 4}}$

14 $\dfrac{\dfrac{1 - 2x}{4x^2 + 6x + 2}}{\dfrac{x + 4}{2x^2 - 5x - 3}}$

15 $\left(\dfrac{2x + 1}{x} - \dfrac{x}{2x + 1}\right) \times \left(\dfrac{4x - 3}{x} + \dfrac{x}{4x - 3}\right)$

16 $\left(\dfrac{x^2}{x + 3} + \dfrac{x + 3}{x^2}\right) \times \left(\dfrac{x}{3x + 2} - \dfrac{3x + 2}{x}\right)$

17 $\dfrac{\dfrac{x^2}{x + 3} + \dfrac{x - 1}{x + 2}}{\dfrac{x^2}{x - 3} - \dfrac{2}{x - 2}}$

18 $\dfrac{\dfrac{x}{x + 1} + \dfrac{2x}{2x - 3}}{\dfrac{x^2}{x + 3} - \dfrac{x + 1}{x + 2}}$

19 $\dfrac{x - 2}{x^2 - 2x + 5} - \dfrac{3x}{x - 1 + 2i}$

20 $\dfrac{2 - 3x}{x^2 + 4x + 13} + \dfrac{x + 3}{x + 2 + 3i}$

5 EXPONENTS AND RADICALS

5.1 » POSITIVE INTEGRAL EXPONENTS

By this time you should be well acquainted with the notation a^n, where a is any real or complex number and n is a positive integer. The simplest approach to its meaning is the statement that a^n stands for the product of n factors, each equal to a. We call a^n "the nth power of a", and n the exponent.

Inductive definition of a^n

For our purposes in this chapter we prefer to use instead the following *inductive* definition of a^n.

DEFINITION The symbol a^n (where a is any number and n is a positive integer) is defined inductively by the formulas $a^1 = a$, $a^{n+1} = a^n \times a$.

According to this definition

$$\begin{aligned} a^1 &= a \\ a^2 &= a^1 \times a = a \times a \\ a^3 &= a^2 \times a = (a \times a) \times a = a \times a \times a \\ a^4 &= a^3 \times a = (a \times a \times a) \times a = a \times a \times a \times a \\ &\text{etc.} \end{aligned}$$

We must now examine the rules for handling these symbols. All of them are derived from the definition and from properties R1 to R8 (Sec. 2.3).

THEOREM 1 Let m and n be positive integers, and let a be any number. Then

$$a^m \times a^n = a^{m+n}$$

?→ PROOF Let us fix m and construct an induction on n. As step I we must verify that $a^m \times a^1 = a^{m+1}$. But this is immediate, for by definition $a^{m+1} = a^m \times a = a^m \times a^1$.

In step II of the induction we must prove that for all $k \geq 1$: If $a^m \times a^k = a^{m+k}$, then $a^m \times a^{k+1} = a^{m+k+1}$. To show this we consider the product $a^m \times a^k \times a$. Now

$$\begin{aligned} a^m \times a^k \times a &= (a^m \times a^k) \times a \\ &= (a^{m+k}) \times a && \text{Hypothesis} \\ &= a^{m+k+1} && \text{Definition} \end{aligned}$$

Also

$$\begin{aligned} a^m \times a^k \times a &= a^m \times (a^k \times a) \\ &= a^m \times a^{k+1} && \text{Definition} \end{aligned}$$

Equating these two expressions for $a^m \times a^k \times a$, we have

$$a^m \times a^{k+1} = a^{m+k+1}$$

which we were to prove.

EXERCISE A Where have we used the associative law of multiplication in the above proof?

COROLLARY I If $m > n$ and $a \neq 0$, then $a^m/a^n = a^{m-n}$.

PROOF First, $$\frac{a^m}{a^n} \times a^n = a^m$$

Moreover, $$a^{m-n} \times a^n = a^m \qquad \text{Theorem 1}$$

Therefore $$\frac{a^m}{a^n} \times a^n = a^{m-n} \times a^n$$

Dividing both sides by a^n, we complete the proof.

COROLLARY II If $m < n$ and $a \neq 0$, then $a^m/a^n = 1/a^{n-m}$.

PROOF First, $$\frac{a^m}{a^n} \times \frac{1}{a^m} = \frac{1}{a^n}$$

Moreover,

$$\frac{1}{a^{n-m}} \times \frac{1}{a^m} = \frac{1}{a^{n-m} \times a^m} = \frac{1}{a^n} \qquad \text{Theorem 1}$$

Hence $$\frac{a^m}{a^n} \times \frac{1}{a^m} = \frac{1}{a^{n-m}} \times \frac{1}{a^m}$$

and $$\frac{a^m}{a^n} = \frac{1}{a^{n-m}}$$

ILLUSTRATION 1

a $2^5 \times 2^3 = 2^{5+3} = 2^8 = 256$
b $r^8 \times r^{15} = r^{23}$
c $3^6/3^2 = 3^4$
d $4^3/4^7 = 1/4^4$
e But note that the theorem does not apply to $2^4 \times 3^6$, which cannot be written more simply in terms of exponents.

THEOREM 2 Let m and n be positive integers, and let a be any number. Then

$$(a^m)^n = a^{mn}$$

PROOF Let us fix m and construct an induction on n. As step I we must verify that $(a^m)^1 = a^{m\times 1}$, a result which is immediately evident.

In step II of the induction we must prove that for all $k \geq 1$: If $(a^m)^k = a^{mk}$, then $(a^m)^{k+1} = a^{m(k+1)}$. To do so we consider the product $(a^m)^k \times a^m$. From the definition

$$(a^m)^k \times a^m = (a^m)^{k+1}$$

On the other hand,

$$\begin{aligned} (a^m)^k \times a^m &= a^{mk} \times a^m && \text{Hypothesis} \\ &= a^{mk+m} && \text{Theorem 1} \\ &= a^{m(k+1)} \end{aligned}$$

Equating these two expressions for $(a^m)^k \times a^m$, we have

$$(a^m)^{k+1} = a^{m(k+1)}$$

which we were to prove.

EXERCISE B Which of R1 to R8 have we used in the above proof?

EXERCISE C Find a counterexample to the following false relation, which is sometimes confused with Theorem 2:

$$(a^m)^n = a^{(m^n)}$$

ILLUSTRATION 2

a $(4^2)^3 = 4^6$
b $(x^4)^2 = x^8$
c $2^3 \times 4^5 = 2^3 \times (2^2)^5 = 2^3 \times 2^{10} = 2^{13}$

THEOREM 3 Let n be a positive integer, and let a and b be any numbers. Then

$$(ab)^n = a^n \times b^n$$

PROOF Again we use mathematical induction. In step I, we immediately verify that $(ab)^1 = a^1 \times b^1$. As step II we must prove that for all $k \geq 1$:

If $(ab)^k = a^k \times b^k$, then $(ab)^{k+1} = a^{k+1} \times b^{k+1}$. To do so we consider the product $(ab)^k \times (ab)$. On the one hand we have

$$(ab)^k \times (ab) = (ab)^{k+1} \qquad \text{Definition}$$

On the other hand

$$\begin{aligned} (ab)^k \times (ab) &= a^k \times b^k \times (ab) && \text{Hypothesis} \\ &= (a^k \times a) \times (b^k \times b) \\ &= a^{k+1} \times b^{k+1} && \text{Definition} \end{aligned}$$

Equating these two expressions for $(ab)^k \times (ab)$, we have

$$(ab)^{k+1} = a^{k+1} \times b^{k+1}$$

which we were to prove.

EXERCISE D Which of R1 to R8 have we used in this proof?

ILLUSTRATION 3
a $(3 \times 5)^4 = 3^4 \times 5^4$
b $(xy)^7 = x^7 \times y^7$

5.2 ZERO AND NEGATIVE EXPONENTS

So far in our discussion of a^n we have required that n be a positive integer. We now wish to extend the definition of a^n to permit n to be zero or a negative integer. How shall we define a^0 and a^{-p} where p is a positive integer? It would seem reasonable to seek definitions which would permit us to extend Theorem 1 to this more general situation. Thus, for instance, we should choose a definition for a^0 which makes the following equation true:

$a^0 = 1$

$$a^0 \times a^n = a^{0+n} = a^n$$

Thus, provided that $a \neq 0$,

$$a^0 = \frac{a^n}{a^n} = 1$$

We use this argument to motivate the following definition.

DEFINITION The symbol a^0 (where a is any number $\neq 0$) is equal to 1. We give no meaning to the symbol 0^0.

In a similar fashion we wish to define a^{-p} so that

$$a^{-p} \times a^p = a^{-p+p} = a^0 = 1$$

Provided that $a \neq 0$, this leads to the result

$$a^{-p} = \frac{1}{a^p}$$

DEFINITION $a^{-p} = 1/a^p$ The symbol a^{-p} (where p is a positive integer and a is any number $\neq 0$) stands for the quotient $1/a^p$. We give no meaning to the symbol 0^{-p}.

We can now reexamine Theorems 1, 2, and 3 of Sec. 5.1 in order to see how they generalize when the exponents are arbitrary integers (positive, negative, or zero).

THEOREM 1′ Let m and n be any integers, and let a be any number $\neq 0$. Then

$$a^m \times a^n = a^{m+n}$$

PROOF To prove this, we must treat various cases separately.

1 $m > 0$, $n > 0$. This is then Theorem 1 of Sec. 5.1.
2 m arbitrary, $n = 0$. Then

$$a^m \times a^n = a^m \times a^0 = a^m \times 1 = a^m = a^{m+0} = a^{m+n}$$

3 $m > 0$, $n < 0$. Let $n = -p$, and suppose $m > p$. Then

$$a^m \times a^n = a^m \times a^{-p} = \frac{a^m}{a^p} = a^{m-p} = a^{m+n}$$

Now suppose $m < p$. Then

$$a^m \times a^n = a^m \times a^{-p} = \frac{a^m}{a^p} = \frac{1}{a^{p-m}} = a^{m-p} = a^{m+n}$$

Finally suppose $m = p$. Then

$$a^m \times a^n = a^p \times a^{-p} = \frac{a^p}{a^p} = 1 = a^0 = a^{p-p} = a^{m+n}$$

4 $m < 0$, $n < 0$. Let $m = -p$; $n = -q$. Then

$$a^m \times a^n = a^{-p} \times a^{-q} = \frac{1}{a^p} \times \frac{1}{a^q}$$

$$= \frac{1}{a^p \times a^q} = \frac{1}{a^{p+q}} = a^{-(p+q)} = a^{-p-q} = a^{m+n}$$

THEOREM 2′ Let m and n be any integers, and let a be any number $\neq 0$. Then

$$(a^m)^n = a^{mn}$$

The proof is by cases as above and is included in the Problems.

THEOREM 3′ Let n be any integer, and let a and b be any numbers $\neq 0$. Then

$$(ab)^n = a^n \times b^n$$

The proof is again by cases and is included in the Problems.

EXERCISE A Why must we exclude $a = 0$ in the statements of Theorems 1′, 2′, and 3′?

PROBLEMS 5.2 In Probs. 1 to 20 perform the indicated operations. Write your answers in a form which uses *positive* exponents only.

1 $5^3 \times 5^6$

2 $3^5 \times 3^8$

3 $4^{-2} \times 4^4 \times 4^0$

4 $8^5 \times 8^{-4} \times 8^0$

5 $3^4 \times 6^{-3} \times 6^4 \times 3^{-7}$

6 $8^5 \times 15^6 \times 8^{-7} \times 15^{-4}$

7 $(-3)^5 \times 9^2 \times (-3)^3 \times 9^3$

8 $4^4 \times (-2)^5 \times 4^{-3} \times (-2)^3$

9 $\dfrac{(4^{-3})6 + 4 + 4(6^{-2})}{(4^{-2})(6^{-4})}$

10 $\dfrac{3(5^{-2}) + 4 + (3^{-2})6}{(5^{-2})(3^{-1})}$

11 $\dfrac{x^2y^{-3} + 2 + x^{-6}y^2}{x^{-4}y^{-5}}$

12 $\dfrac{p^{-4}q + 4 + p^3q^{-5}}{p^{-5}q^{-3}}$

13 $\dfrac{ab^2c^{-3} - a^{-2}b^{-1}c^3}{a^2b^{-1}c^4}$

14 $(2z^{-1} + z^{-3})(3z^{-2} - z^2)$

15 $(y^2 + 3y^{-2})(2y - y^{-3})$

16 $(1 + x^{-1} + x^{-2})(1 + x)$

17 $\dfrac{5x^{-2} + 2 - 4x}{4x^{-3} + 2}$

18 $\dfrac{6x^2 + 3x - 9x^{-4}}{3x^{-3} + 2x}$

19 $\dfrac{4x^{-2} + 5x^{-3} + 3x^{-4}}{6x^2 + x^0}$

20 $\dfrac{3x^4 - 2x^{-3} + 5x^0 - x^{-2}}{x^3 - x^{-3}}$

21 Prove by induction: For all integers $n \geq 1$, and for any numbers a and b ($\neq 0$),

$$\left(\frac{a}{b}\right)^n = \frac{a^n}{b^n}$$

22 Prove Theorem 2′. Consider the cases: ($m > 0$, $n > 0$); ($m > 0$, $n < 0$); ($m < 0$, $n > 0$); ($m < 0$, $n < 0$); ($m = 0$, n arbitrary); (m arbitrary, $n = 0$).

23 Prove Theorem 3′. Consider the cases: $n > 0$, $n = 0$, $n < 0$.

5.3 › FRACTIONAL EXPONENTS

We now come to the matter of taking the square root, the cube root, and fourth root, etc., of a number. In order to avoid some troublesome difficulties, in this section we shall assume that we are dealing with the roots of *positive real* numbers only.

a is positive

Let a be a positive real number, and suppose that $b^n = a$. Then we say that "b is *an* nth root of a"; we must not say "*the* nth root", for indeed there may be several of these. For instance,

$$2^2 = 4 \quad \text{and} \quad (-2)^2 = 4$$

so that both 2 and (-2) are square roots of 4. As a general principle a mathematical symbol should stand for just one mathematical object rather than for several such objects. For this reason we use $\sqrt[n]{a}$ in the carefully defined sense given below.

DEFINITION

$\sqrt[n]{a}$ is positive nth root of a

Let a be a positive real number and n a positive integer. Then the symbols $\sqrt[n]{a}$ and $a^{1/n}$ will be used interchangeably to mean that particular one of the nth roots of a which is a positive real number.

ILLUSTRATION 1 $\sqrt{4} = 4^{1/2} = +2$; $\sqrt{25} = 25^{1/2} = +5$; $-\sqrt{36} = -36^{1/2} = -6$. We never write $\sqrt{25} = \pm 5$.

EXERCISE A Why is $a^{1/n}$ a desirable notation for $\sqrt[n]{a}$? HINT: Consider a reasonable generalization of Theorem 2.

$a^{p/q} = (a^{1/q})^p$

Now that we know what we mean by $a^{1/n}$, we must extend our definition to symbols of the form $a^{p/q}$.

DEFINITION

Let a be a positive real number, and let p and q be positive integers. Then the symbols $(\sqrt[q]{a})^p$ and $a^{p/q}$ are used interchangeably to mean the pth power of $a^{1/q}$. That is, $a^{p/q} = (a^{1/q})^p$.

Another meaning for $a^{p/q}$ is derived from Theorem 4.

THEOREM 4

$a^{p/q} = \sqrt[q]{a^p}$, where $\sqrt[q]{a^p}$ denotes the positive qth root of a^p. That is, $(a^{p/q})^q = a^p$.

PROOF We must show that $(a^{p/q})^q = a^p$. From the above definition we have

$$\begin{aligned} (a^{p/q})^q &= [(a^{1/q})^p]^q \\ &= [a^{1/q}]^{pq} && \text{Theorem 2} \\ &= [(a^{1/q})^q]^p && \text{Theorem 2} \\ &= a^p && \text{Definition of } a^{1/q} \end{aligned}$$

Finally we extend our definition of negative exponents to our fractional exponents.

DEFINITION Let a be a positive real number and p and q be positive integers. Then

$$a^{-p/q} = \frac{1}{a^{p/q}}$$

We have now completely defined the symbol a^r where a is positive and r is any rational number. Let us see how Theorems 1, 2, and 3 generalize to this situation.

THEOREM 1″ Let a be a positive real number and r and s any rational numbers. Then

$$a^r \times a^s = a^{r+s}$$

PROOF Let $r = p/q$ and $s = u/v$, where p, q, u, and v are integers. Then we must prove that

$$a^{p/q} \times a^{u/v} = a^{(pv+qu)/qv}$$

Because of Theorem 4 this is equivalent to proving that

$$(a^{p/q} \times a^{u/v})^{qv} = a^{pv+qu}$$

However:

$$\begin{aligned}
(a^{p/q} \times a^{u/v})^{qv} &= (a^{p/q})^{qv} \times (a^{u/v})^{qv} && \text{Theorem 3}' \\
&= [(a^{p/q})^q]^v \times [(a^{u/v})^v]^q && \text{Theorem 2}' \\
&= (a^p)^v \times (a^u)^q && \text{Theorem 4} \\
&= a^{pv} \times a^{qu} && \text{Theorem 2}' \\
&= a^{pv+qu} && \text{Theorem 1}'
\end{aligned}$$

THEOREM 2″ Let a be a positive real number and r and s any rational numbers. Then

$$(a^r)^s = a^{rs}$$

THEOREM 3″ Let a and b be any positive real numbers and r any rational number. Then

$$(ab)^r = a^r \times b^r$$

The proofs of these theorems are similar to that of Theorem 1″ and are included in the Problems.

EXERCISE B Define $0^r = 0$ where r is a *positive* rational number. Show that Theorems 1″, 2″, and 3″ can be extended to include the case where $a = 0$ and r and s are *positive* rational numbers.

PROBLEMS 5.3

In Probs. 1 to 5 find a simpler expression for the given quantity.

1 a $\sqrt{9}$
b $-\sqrt{9}$
c $27^{1/3}$
d $32^{1/5}$

2 a $16^{3/2}$
b $25^{5/2}$
c $\sqrt[3]{8^2}$
d $\sqrt{2^4}$

3 a $25^{-1/2}$
b $8^{-2/3}$
c $-(4^{-1/2})$
d $64^{1/6}$

4 a $2^{1/2} \times 2^{3/2}$
b $5^{1/4} \times 5^{3/4}$
c $3^{1/2} \times 3^{-1/2}$
d $7^{2/3} \times 7^{-8/3}$

5 a $(5^{1/2})^2$
b $(7^2)^{1/2}$
c $(6^{1/2})^{1/3}$
d $(3^{1/2})^{-2/3}$

In Probs. 6 to 13 perform the given operations. The universal set for each variable mentioned is the set of positive real numbers.

6 $(4x^{5/2} - 2x^{-3/2})x^{3/2}$

7 $(3x^{2/3} - 2x^{-5/3})x^{-1/3}$

8 $(x^{1/3} + y^{1/3})^3$

9 $(a^{1/2} + b^{1/2})^2$

10 $(p^{1/2} + q^{1/2})(p^{1/2} - q^{1/2})$

11 $(x^{-5/2} + y^{3/2})(x^{1/2} - y^{5/2})$

12 $\dfrac{x^2 + 3x^{5/2}}{2x^{1/2}}$

13 $\dfrac{x^3 + 4x^{-3} - x^{-5/2}}{x^{3/2}}$

In Probs. 14 to 17 change the given expression to an equivalent form in which all exponents are positive and no compound fractions appear. All variables are positive.

14 $\dfrac{3x^{2/3} + x^{4/3} - 3x^{-2/3}}{3x^{2/3} + x^{-5/3}}$

HINT: Multiply numerator and denominator by $x^{5/3}$.

15 $\dfrac{3y^{1/5} - 4y^{-3/5} + 5y^{4/5}}{4y^2 - y^{-3/5}}$

16 $\dfrac{5a^{2/3}b^{-1/3} + 4a^{-1/3}b^{7/3}}{5a^{-2}b^{1/3} + 3a^{2/3}b^{4/3}}$

17 $\dfrac{4r^{-2/5}s^{3/5} - 3r^{2/5}s^{-3/5}}{2rs + 7r^2s^{-4}}$

5.4 › SPECIAL PROBLEMS CONCERNING SQUARE ROOTS

In the last section we required that a be a positive real number. Here we relax that restriction and consider two special difficulties that occur when we take square roots.

THE SQUARE ROOT OF a^2, OR $\sqrt{a^2} = (a^2)^{1/2}$ When a is positive, Theorem 2″ tells us that $(a^2)^{1/2} = a$. For example, $(3^2)^{1/2} = 3$. This result is correct since $(3^2)^{1/2} = 9^{1/2} = 3$. If, however, a is negative, a difficulty arises.

For example, let $a = -3$. Then if Theorem 2″ were true for negative a, we would conclude that $[(-3)^2]^{1/2} = -3$. On the other hand, consider the direct computation:

$$[(-3)^2]^{1/2} = 9^{1/2} = +3$$

Hence there is a contradiction, and so Theorem 2″ is false for negative values of a.

A correct result that includes both positive and negative values of a is given by the following theorem.

THEOREM 5 $\sqrt{a^2} = |a|$ For any real number a, $(a^2)^{1/2} = |a|$, where $|a|$ denotes the absolute value of a.

PROOF When $a \geq 0$, Theorem 5 is a consequence of Theorem 2″. When $a < 0$, let $a = -b$, where $b > 0$. Then

$$(a^2)^{1/2} = [(-b)^2]^{1/2} = (b^2)^{1/2} = b = |a|$$

ILLUSTRATION 1

a $(-5^2)^{1/2} = \sqrt{(-5)^2} = +5$

b $(x^2)^{1/2} = \sqrt{x^2} = |x|$

c $(x^2 + 2x + 1)^{1/2} = \sqrt{x^2 + 2x + 1} = |x + 1|$

The result of Theorem 5 can be extended at once to any even root.

THEOREM 6 For any real number a and any positive integer n, $(a^{2n})^{1/(2n)} = |a|$.

PROOF $(a^{2n})^{1/(2n)} = [(a^2)^n]^{1/(2n)} = (a^2)^{1/2} = |a|$

EXERCISE A State the justification for each step in the proof of Theorem 6.

$\sqrt{-a} = i\sqrt{a}$ **THE SQUARE ROOT OF A NEGATIVE NUMBER** Let a be a positive real number, and consider $(-a)^{1/2}$. This needs to be defined.

DEFINITION For any positive real number a,

$$(-a)^{1/2} = i(a^{1/2})$$

ILLUSTRATION 2 $\sqrt{-3} = (-3)^{1/2} = i\sqrt{3} = i3^{1/2}$

Expressions of this kind need special care, for Theorem 3 does not hold in this case. Let us consider $(-a)^{1/2} \times (-b)^{1/2}$, where a and b are both positive.

$$(-a)^{1/2} \times (-b)^{1/2} = i(a^{1/2}) \times i(b^{1/2}) = i^2 a^{1/2} b^{1/2} = -(ab)^{1/2}$$

The application of Theorem 3 would have given the incorrect result:

$$(-a)^{1/2} \times (-b)^{1/2} = +(ab)^{1/2}$$

ILLUSTRATION 3

a $(-3)^{1/2} \times (-5)^{1/2} = i(3)^{1/2} \times i(5)^{1/2} = -15^{1/2}$

b $\sqrt{-10} \times \sqrt{-7} = i\sqrt{10} \times i\sqrt{7} = -\sqrt{70}$

c $\sqrt{13} \times \sqrt{-3} = \sqrt{13} \times i\sqrt{3} = i\sqrt{39}$

5.5 SPECIAL PROBLEMS CONCERNING ODD ROOTS

Again let a be a positive real number, and consider $\sqrt[3]{-a}$. There is always a negative real number $-b$ such that $(-b)^3 = -a$, and we shall write $-b = \sqrt[3]{-a} = (-a)^{1/3}$. In general we proceed as follows.

DEFINITION Let a be a positive real number and p be any odd positive integer. Then the symbols $\sqrt[p]{-a}$ and $(-a)^{1/p}$ will be used interchangeably to denote the negative real number, $-b$ such that $(-b)^p = -a$.

ILLUSTRATION 1

a $\sqrt[3]{-27} = (-27)^{1/3} = -3$

b $\sqrt[5]{-32} = (-32)^{1/5} = -2$

c $\sqrt[11]{-x^{11}} = -x^1 = -x$

REMARKS

1. The notation $(-a)^{1/p}$ is used by many authors in a fashion different from that defined above. In their usage this symbol denotes a certain complex number called the "principal pth root of $-a$". We shall discuss this notation in Sec. 13.11.
2. The symbol $(-a)^{p/q}$ for a positive and q odd can now be defined in a fashion analogous to the definition of $a^{p/q}$ in Sec. 5.3. These symbols then obey Theorems 1″, 2″, and 3″. The proofs are included in the Problems.

5.6 UNANSWERED QUESTIONS

Although we have discussed the meaning of the symbol a^b, where a and b are certain types of real numbers, there are still several cases which we have omitted. We call your attention to the following situations, which we shall treat later in this book:

1 $(-a)^{p/q}$, where a is positive, and q is even. This is included as a special case in Sec. 13.11.
2 a^b, where a is positive and b is irrational. This is discussed in Sec. 11.2.
3 We must omit entirely the complicated, but fascinating story of a^b, where a and b are any complex numbers. For this consult G. H. Hardy, "Pure Mathematics", 10th edition, pages 409 to 410, 457 to 459, Cambridge, New York, 1963.

PROBLEMS 5.6 In Probs. 1 to 8 simplify the given expressions. The universal set for each variable mentioned is the set of real numbers.

1 $(x^2 + 4x + 4)^{1/2}$
2 $(16x^2 - 8x + 1)^{1/2}$
3 $(x^2 + 2x + 1)^{1/2} + (x^2 - 2x + 1)^{1/2}$. Give a counterexample to show that $2x$ is an incorrect answer.
4 $(x^2 + 4x + 4)^{1/2} - (x^2 - 4x + 4)^{1/2}$. Give a counterexample to show that 4 is an incorrect answer.
5 $\dfrac{(9x^2 - 12x + 4)^{1/2}}{3x - 2}$
6 $\dfrac{(25x^2 + 40x + 16)^{1/2}}{5x + 4}$
7 $x\sqrt{1 + x^{-2}}$
8 $(x + 3)\sqrt{(x + 3)^{-2} + 1}$

In Probs. 9 to 18 simplify the given expression, assuming that the variables are restricted as indicated.

9 $(x^2 + 8x + 16)^{1/2}$ where $x + 4 \geq 0$
10 $(x^2 - 10x + 25)^{1/2}$ where $x - 5 \geq 0$
11 $(x^2 + 8x + 16)^{1/2}$ where $x + 4 \leq 0$
12 $(x^2 - 10x + 25)^{1/2}$ where $x - 5 \leq 0$
13 $(x^2 + 4x + 4)^{1/2} + (x^2 + 6x + 9)^{1/2}$ where $x + 2 \geq 0$
14 $(x^2 - 6x + 9)^{1/2} + (x^2 - 2x + 1)^{1/2}$ where $x - 3 \geq 0$
15 $(x^2 + 4x + 4)^{1/2} + (x^2 + 6x + 9)^{1/2}$ where $x + 3 \leq 0$
16 $(x^2 - 6x + 9)^{1/2} + (x^2 - 2x + 1)^{1/2}$ where $x - 1 < 0$
17 $(x^2 + 4x + 4)^{1/2} + (x^2 + 6x + 9)^{1/2}$ where $-3 \leq x \leq -2$
18 $(x^2 - 6x + 9)^{1/2} + (x^2 - 2x + 1)^{1/2}$ where $1 \leq x \leq 3$

In Probs. 19 to 34 perform the indicated operations.

19 $\sqrt{-6} \times \sqrt{-5}$
20 $\sqrt{-16} \times \sqrt{-9}$
21 $\sqrt{16} \times \sqrt{-9}$
22 $\sqrt{25} \times \sqrt{-4}$
23 $\sqrt[3]{-8} \times \sqrt{36}$
24 $\sqrt{49} \times \sqrt[3]{-64}$
25 $\sqrt[3]{-8} \times \sqrt[5]{32}$
26 $\sqrt[3]{-125} \times \sqrt[3]{-27}$

27 $\sqrt{-9} \times \sqrt[3]{-343}$

28 $\sqrt[4]{16} \times \sqrt[5]{-32}$

29 $\sqrt{2} + \sqrt{8}$

HINT: $\sqrt{8} = 2\sqrt{2}$

30 $\sqrt{5} + \sqrt{20}$

31 $\sqrt{44} - \sqrt{11}$

32 $\sqrt{11} + 3\sqrt{13}$

33 $3\sqrt{2} - \sqrt{50} + 4\sqrt{-32}$

34 $2\sqrt{3} + \sqrt{27} + \sqrt{-3}$

35 Prove Theorem 2″.

36 Prove Theorem 3″.

37 Define $(-a)^{p/q}$, where a is a positive real number, and p and q are integers with q odd.

38 Review the proof of Theorems 4, 1″, 2″, and 3″. What changes are needed in these proofs so that these theorems apply to $(-a)^{p/q}$ where a is positive, p any integer, and q an odd positive integer?

In Probs. 39 to 44, we consider the symbol a^b where a and b are positive real numbers. We assume that these symbols are defined (Sec. 11.2) and that for these the analogs of Theorems 1, 2, and 3 are true.

39 Is a^b a binary operation on the set of positive real numbers?

40 Is this operation commutative?

41 Is this operation associative?

42 Is there an identity?

43 Is there an inverse?

44 What is the distributive property of exponentiation over multiplication? Is it true?

5.7 RATIONALIZING DENOMINATORS

From time to time you will meet fractions containing square roots in the denominator, such as

$$\frac{1}{\sqrt{2}} \qquad \frac{2}{\sqrt{3} - \sqrt{5}} \qquad \frac{x + 5}{\sqrt{x + 1} + \sqrt{2x - 3}}$$

Let us consider problems which these present. If we wish to express $1/\sqrt{2}$ as a decimal, it is awkward to divide 1/1.414. A simpler procedure is to write

$$\frac{1}{\sqrt{2}} = \frac{1}{\sqrt{2}} \cdot \frac{\sqrt{2}}{\sqrt{2}} = \frac{\sqrt{2}}{2} \approx \frac{1.414}{2} = 0.707$$

In some textbooks it is required that all answers be written with rational denominators; thus $1/\sqrt{2}$ is incorrect, and $\sqrt{2}/2$ is correct. This is an absurd requirement, and we shall accept either answer as correct. The choice between them depends on how we are to use them. Consider the examples below.

ILLUSTRATION 1 Find $1/\sqrt{2} + 1/\sqrt{3}$.
Here we find it wise to write

$$\frac{1}{\sqrt{2}} + \frac{1}{\sqrt{3}} = \frac{\sqrt{2}}{2} + \frac{\sqrt{3}}{3} = \frac{3\sqrt{2} + 2\sqrt{3}}{6}$$

ILLUSTRATION 2 Find $1/\sqrt{2} \times 1/\sqrt{2}$.
Here we can write

$$\frac{1}{\sqrt{2}} \times \frac{1}{\sqrt{2}} = \frac{1}{\sqrt{2} \times \sqrt{2}} = \frac{1}{2}$$

It would be silly to write

$$\frac{1}{\sqrt{2}} \times \frac{1}{\sqrt{2}} = \frac{\sqrt{2}}{2} \times \frac{\sqrt{2}}{2} = \frac{2}{4} = \frac{1}{2}$$

Hence, leave your answer in whatever form is most convenient for later use.

When we are faced with $2/(\sqrt{3} - \sqrt{5})$, another technique is needed if we wish a rational denominator. We can rationalize this one as follows:

$$\frac{2}{\sqrt{3} - \sqrt{5}} = \frac{2}{\sqrt{3} - \sqrt{5}} \times \frac{\sqrt{3} + \sqrt{5}}{\sqrt{3} + \sqrt{5}} = \frac{2\sqrt{3} + 2\sqrt{5}}{3 - 5}$$
$$= \frac{2\sqrt{3} + 2\sqrt{5}}{-2} = -\sqrt{3} - \sqrt{5}$$

We can apply this method to various cases as shown in the Illustrations below.

ILLUSTRATION 3

$$\frac{3}{1 + \sqrt{3}} = \frac{3}{1 + \sqrt{3}} \times \frac{1 - \sqrt{3}}{1 - \sqrt{3}} = \frac{3 - 3\sqrt{3}}{1 - 3} = -\frac{3 - 3\sqrt{3}}{2}$$

ILLUSTRATION 4

$$\frac{x + 5}{\sqrt{x + 1} + \sqrt{2x - 3}} = \frac{x + 5}{\sqrt{x + 1} + \sqrt{2x - 3}} \times \frac{\sqrt{x + 1} - \sqrt{2x - 3}}{\sqrt{x + 1} - \sqrt{2x - 3}}$$
$$= \frac{(x + 5)\sqrt{x + 1} - (x + 5)\sqrt{2x - 3}}{(x + 1) - (2x - 3)}$$
$$= \frac{(x + 5)\sqrt{x + 1} - (x + 5)\sqrt{2x - 3}}{-x + 4}$$

ILLUSTRATION 5

$$\frac{x}{1+\sqrt{x}} + \frac{2}{\sqrt{x+1}} = \frac{x}{1+\sqrt{x}} \times \frac{1-\sqrt{x}}{1-\sqrt{x}} + \frac{2\sqrt{x+1}}{x+1}$$

$$= \frac{x - x\sqrt{x}}{1-x} + \frac{2\sqrt{x+1}}{x+1}$$

$$= \frac{(x - x\sqrt{x})(x+1) + 2(1-x)\sqrt{x+1}}{1-x^2}$$

PROBLEMS 5.7 In Probs. 1 to 10 rationalize the denominator in the given expressions.

1 $\frac{2}{\sqrt{5}}$

2 $\frac{-3}{\sqrt{7}}$

3 $\frac{-4}{\sqrt{2}+\sqrt{3}}$

4 $\frac{1}{\sqrt{7}-\sqrt{5}}$

5 $\frac{4}{\sqrt{x-2}}$

6 $\frac{3x^2}{\sqrt{1-2x^2}}$

7 $\frac{2}{\sqrt{x-2}-\sqrt{x+1}}$

8 $\frac{x}{\sqrt{x^2-3}+\sqrt{x-4}}$

9 $\frac{3}{2\sqrt{2}-\sqrt{3}+\sqrt{8}}$

10 $\frac{7}{2\sqrt{5}+\sqrt{20}-\sqrt{3}}$

In Probs. 11 to 18 perform the stated operation. State your answers with rationalized denominators.

11 $\frac{1}{\sqrt{2}} \times \frac{1}{\sqrt{8}}$

12 $\frac{1}{\sqrt{5}} \times \frac{1}{\sqrt{125}}$

13 $\frac{1}{\sqrt{3}} + \frac{1}{\sqrt{7}}$

14 $\frac{1}{\sqrt{11}} - \frac{1}{\sqrt{7}}$

15 $\frac{1}{2-\sqrt{x}} - \frac{2}{x}$

16 $\frac{4}{\sqrt{x+2}-3} - \frac{1}{x^2}$

17 $\frac{x^2}{1+\sqrt{x}} \times \frac{4}{1-\sqrt{x}}$

18 $\frac{2}{3+\sqrt{1+2x}} \times \frac{3}{3-\sqrt{1+2x}}$

In Probs. 19 to 22 use Table VIII of the Appendix to compute the best approximation of the following to two decimal places.

19 $\frac{5}{\sqrt{3}}$

20 $\frac{3}{\sqrt{6}}$

21 $\frac{1}{\sqrt{7}-\sqrt{3}}$

22 $\frac{3}{\sqrt{5}-\sqrt{7}}$

128

6 EQUATIONS

6.1 SOLUTIONS OF EQUATIONS

One of the most common problems in mathematics is *solving an equation*. In this chapter we shall tackle a number of problems of this kind, but first we must know what is meant by the *solution of an equation*.

In order to avoid unnecessary complications, we shall begin by considering the simple cases of linear and quadratic equations in one variable. These are equations of the forms

$$ax + b = 0 \qquad \text{and} \qquad ax^2 + bx + c = 0$$

where a, b, and c are elements of a chosen field F. We shall generally choose F to be the field of complex numbers, but on occasion may restrict it to the real numbers. In any case, the field involved must be specified in advance.

Before we can say what is meant by a solution of equations like these, we must first choose a universal set U from which the values of x are to be selected. U is frequently the given field, but in some cases we may wish to take U to be a subset of this field. For example, if F is the field of real numbers, we may wish to chose U to be the set of integers.

DEFINITION The solution of an equation in one variable is the subset S of U consisting of those elements of U which satisfy the given equation. S, which is often called the *solution set*, is therefore the truth set of the open sentence defined by the given equation.

Solution set

ILLUSTRATION 1

a Let U be the set of real numbers. Then the solution set of $2x + 5 = 11$ is $S = \{3\}$.

b Let U be the set of integers. Then the solution set of $2x = 5$ is $S = \varnothing$.

6.2 » METHOD OF SOLUTION

In order to lay the groundwork for more complicated situations which come later, let us review the familiar procedure for solving a linear equation in one variable, such as

(1) $$4x + 2 = 10$$

First we subtract 2 from both sides and obtain the simpler equation

(2) $$4x = 8$$

Then we divide both sides by 4 and obtain the still simpler equation,

(3) $$x = 2$$

Therefore, $$S = \{2\}$$

Equivalence

This process is a special case of the following more general method. To solve our equation (1), we first convert it into a simpler equation (2), which is equivalent to it. We recall that two equations are equivalent if and only if they have the same solution set (Sec. 1.3). In the second step we convert Eq. (2) into a still simpler equation, (3), which is equivalent to it. Then (3) is equivalent to the given equation (1). Now the solution S of (3) is obvious. Hence S is the solution of (1).

In order to apply this method, we must have a collection of operations which convert an equation of this type into an equivalent equation. Two such operations are given in Theorems 1 and 2.

THEOREM 1 Let $P(x)$ represent either $ax + b$ or $ax^2 + bx + c$ with coefficients in a field F. Then the equation $P(x) = 0$ is equivalent to the equation $P(x) + d = d$, where d is any element of F.

Addition to both sides

In other words, the addition of the same element of F to both sides of the equation results in an equivalent equation.

PROOF Let $A = \{x \mid P(x) = 0\}$ and $B = \{x \mid P(x) + d = d\}$. We must show that $A = B$. The best way of proving this is to show that $A \subseteq B$ and that $B \subseteq A$.

1. $A \subseteq B$. Let r be an element of A. Then $P(r) = 0$. Therefore, from Theorem 1, Chap. 2, $P(r) + d = d$, which means that r satisfies $P(x) + d = d$. Hence every element of A is an element of B.
2. $B \subseteq A$. Let s be an element of B. Then $P(s) + d = d$. Subtracting d from each side (Theorem 1, Chap. 2), we obtain $P(s) = 0$. This means that s satisfies $P(x) = 0$. Hence every element of B is an element of A.

THEOREM 2 Multiplication on both sides

The equation $P(x) = 0$ (as defined in Theorem 1) is equivalent to the equation $P(x) \cdot d = 0$, where d is any nonzero element of F.

The proof is very similar to that of Theorem 1.

EXERCISE A Prove Theorem 2.

EXERCISE B Why must we exclude $d = 0$ in Theorem 2?

EXERCISE C Show that Theorems 1 and 2 also apply to equations of the form $P(x) = Q(x)$ where $P(x)$ and $Q(x)$ are expressions of the form $ax + b$ or $ax^2 + bx + c$.

This method of solution is basic to the solutions of much more complicated types of problems. Be sure that you understand it here so that you can apply it later.

6.3 LINEAR EQUATIONS IN ONE VARIABLE

The process just outlined in Sec. 6.2 leads to the proof of the following theorem:

THEOREM 3 If a and b are elements of a field F and $a \neq 0$, and if the universal set $U = F$, then the equation $ax + b = 0$ has a unique solution.

PROOF From Theorem 1, $ax + b = 0$ is equivalent to $ax = -b$. From Theorem 2, this is equivalent to $x = -b/a$. Since $-(b/a)$ is an element of F, it is the one and only solution of $ax + b = 0$.

6.4 QUADRATIC EQUATIONS IN ONE VARIABLE

Earlier we saw that the quadratic equation $x^2 = 2$ could not be solved in the field of rational numbers and that $x^2 = -1$ could not be solved in the field of real numbers. Thus it is clear that the axioms of a field are not strong enough to assure the solution of every quadratic equation. It was for this reason that we invented the field of complex numbers, in which every quadratic equation has a solution.

In order to prove this, let us first consider the equation

(1)
$$ax^2 + bx + c = 0$$

where a, b, and c are complex and $a \neq 0$. We proceed to solve this by a method known as completing the square. This depends upon the fact that

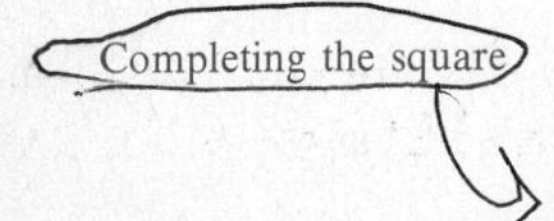

(2)
$$x^2 + 2dx + d^2 = (x + d)^2$$

Since $a \neq 0$, let us write Eq. (3), which is equivalent to (1):

(3) $$x^2 + \frac{b}{a}x + \frac{c}{a} = 0 \qquad \text{Theorem 2}$$

If we put $d = b/2a$, the first two terms on the left side of (2) are equal to the corresponding terms in (3). In general, however, $d^2 \neq c/a$. Therefore we write (4), which is equivalent to (3):

(4) $$x^2 + \frac{b}{a}x + \left(\frac{b}{2a}\right)^2 = \left(\frac{b}{2a}\right)^2 - \frac{c}{a}$$

Now the left-hand side of (4) is of the same form as the left-hand side of (2). Thus:

(5) $$\left(x + \frac{b}{2a}\right)^2 = \frac{b^2 - 4ac}{4a^2}$$

Using the formula for factoring the difference of two squares, we can write (5) in the form

$$\left(x + \frac{b}{2a} - \frac{\sqrt{b^2 - 4ac}}{2a}\right)\left(x + \frac{b}{2a} + \frac{\sqrt{b^2 - 4ac}}{2a}\right) = 0$$

Then from Theorem 4, Chap. 2,

$$x = \frac{-b + \sqrt{b^2 - 4ac}}{2a} \qquad \text{or} \qquad x = \frac{-b - \sqrt{b^2 - 4ac}}{2a}$$

This proves the following theorem:

THEOREM 4 The quadratic equation

$$ax^2 + bx + c = 0 \qquad a \neq 0$$

Quadratic formula

where a, b, and c are complex numbers, and U is the set of complex numbers, has two (and only two) solutions, namely,

$$x = \frac{-b \pm \sqrt{b^2 - 4ac}}{2a}$$

REMARKS

1. Theorem 4 would have been false, even for real a, b, and c, if we had restricted x to belong to the set of real numbers, for $x^2 + 1 = 0$ has no real solutions.
2. If a, b, and c are real, we may easily deduce the following properties of the solutions:
   When $b^2 - 4ac$ is positive, the two solutions are real and unequal.
   When $b^2 - 4ac$ is zero, the two solutions are real and equal; i.e., there is only one solution.
   When $b^2 - 4ac$ is negative, the two solutions are unequal and neither of them is real.

3 Let r_1 and r_2 be the two roots of $ax^2 + bx + c = 0$; that is,

(6) $$r_1 = \frac{-b + \sqrt{b^2 - 4ac}}{2a} \qquad r_2 = \frac{-b - \sqrt{b^2 - 4ac}}{2a}$$

Then

(7) $$r_1 + r_2 = -\frac{b}{a} \qquad r_1 r_2 = \frac{c}{a}$$

EXERCISE A By direct calculation verify the above statements.

4 The above formulas permit us to give a general expression for the factors of $ax^2 + bx + c$, namely:

(8) $$ax^2 + bx + c = a(x - r_1)(x - r_2)$$

where r_1 and r_2 are defined in (6) above. To prove this correct, we note that

$$\begin{aligned} a(x - r_1)(x - r_2) &= ax^2 - a(r_1 + r_2)x + ar_1r_2 \\ &= ax^2 - a\left(-\frac{b}{a}\right)x + a\left(\frac{c}{a}\right) \\ &= ax^2 + bx + c \end{aligned}$$

The use of formula (8) permits us to factor trinomials which could not be factored by the methods of Chap. 3. It also provides a direct approach to certain trinomials which can be factored by the methods of Chap. 3 but whose factorization by those methods is likely to be long and tedious.

ILLUSTRATION 1 Factor $x^2 + 2x + 5$.

$$r_1 = \frac{-2 + \sqrt{-16}}{2} = -1 + 2i$$

$$r_2 = \frac{-2 - \sqrt{-16}}{2} = -1 - 2i$$

So $x^2 + 2x + 5 = (x + 1 - 2i)(x + 1 + 2i)$.

ILLUSTRATION 2 Factor $35x^2 - 11x - 72$.

$$r_1 = \frac{11 + \sqrt{10{,}201}}{70} = \frac{11 + 101}{70} = \frac{112}{70} = \frac{8}{5}$$

$$r_2 = \frac{11 - \sqrt{10{,}201}}{70} = \frac{11 - 101}{70} = -\frac{90}{70} = -\frac{9}{7}$$

So $35x^2 - 11x - 72 = 35(x - \frac{8}{5})(x + \frac{9}{7}) = (5x - 8)(7x + 9)$.

PROBLEMS 6.4

In Probs. 1 to 10 find the solutions of the quadratic equations.

1 $x^2 + 5x + 5 = 0$
2 $x^2 + 4x + 2 = 0$
3 $16x^2 - 8x + 1 = 0$
4 $9x^2 + 12x + 4 = 0$
5 $6x^2 + 13x - 5 = 0$
6 $6x^2 - 5x - 21 = 0$
7 $x^2 + 2x + 2 = 0$
8 $4x^2 + 4x + 5 = 0$
9 $x^2 - 6(1 + i)x - 1 + 18i = 0$
10 $x^2 - 2(2 - i)x - 6 - 4i = 0$

In Probs. 11 to 16 add terms to both sides so that the left-hand side becomes a perfect square.

11 $x^2 + 5x + 3 = 0$
12 $x^2 - 2x - 1 = 0$
13 $x^2 - 8x + 3 = 0$
14 $x^2 + 7x - 5 = 0$
15 $4x^2 - 10x + 7 = 0$
16 $3x^2 + 12x + 5 = 0$

In Probs. 17 to 22 find the sum and product of the solutions without solving the equation.

17 $x^2 + 5x + 9 = 0$
18 $x^2 - 6x - 11 = 0$
19 $3x^2 + 5x + 4 = 0$
20 $5x^2 - 4x + 7 = 0$
21 $(1 - i)x^2 + (2 - 3i)x + (1 - i) = 0$
22 $(2 + 5i)x^2 + (1 - 4i)x + 5 = 0$

In Probs. 23 to 28 find the value of k for which the solutions of the given equation are equal.

23 $x^2 + 7x + k = 0$
24 $x^2 - 9x + k = 0$
25 $2x^2 + 5x + k = 0$
26 $3x^2 - 4x + k = 0$
27 $x^2 + 5kx + 2 = 0$
28 $x^2 - 4kx + 7 = 0$

In Probs. 29 to 34 find a quadratic equation the sum and product of whose solutions have the given values.

29 Sum 3, product 5
30 Sum 0, product -6
31 Sum $\frac{1}{2}$, product $\frac{5}{2}$
32 Sum $\frac{3}{5}$, product $-\frac{4}{5}$
33 Sum $2 + i$, product $3 + 4i$
34 Sum $1 + 2i$, product $3 - i$

In Probs. 35 to 40 factor the given trinomial.

35 $x^2 - 5x + 3$
36 $x^2 - 3x + 1$
37 $3x^2 + 2x + 5$
38 $2x^2 + x + 3$
39 $54x^2 - 33x - 35$
40 $32x^2 + 4x - 45$

41 Find the value of k for which the sum of the solutions of the following equation is twice their product:

$$4x^2 + 5x + k = 0$$

42 Find a quadratic equation whose solutions are the reciprocals of those of $2x^2 + 7x + 6 = 0$.

HINT: $\frac{1}{p} + \frac{1}{q} = \frac{p+q}{pq} \qquad \left(\frac{1}{p}\right)\left(\frac{1}{q}\right) = \frac{1}{pq}$

43 Find a quadratic equation whose solutions are the squares of those of $x^2 + 7x + 5 = 0$.

HINT: $p^2 + q^2 = (p+q)^2 - 2pq \qquad (p^2)(q^2) = (pq)^2$

44 Find a quadratic equation whose solutions are respectively the sum and product of the solutions of $x^2 + 7x + 11 = 0$.

45 Find a quadratic equation whose solutions are 3 smaller, respectively, than the solutions of $x^2 + 5x - 9 = 0$.

46 Each solution of $x^2 + x - 6 = 0$ differs from the square of the other by the same number c. Without solving the equation, find the value of c.

6.5 › EQUATIONS IN TWO VARIABLES

As a generalization of the previous section, let us consider equations in two variables of the forms

$$y = ax + b \qquad y = ax^2 + bx + c$$

or

$$y - (ax + b) = 0 \qquad y - (ax^2 + bx + c) = 0$$

where a, b, and c are elements of a field F. For simplicity we write any equation like these as $P(x,y) = 0$. The question now is: What is a solution of $P(x,y) = 0$? As in the situation for equations in one variable, we must choose a universal set U from which the values of the variables x and y are to be taken. A solution of $P(x,y) = 0$ will be an ordered pair (x,y) where $x \in U$ and $y \in U$. Hence a solution is an element of $U \times U$. Thus we have the definition:

DEFINITION The solution of an equation $P(x,y) = 0$ is the subset S of $U \times U$ consisting of those ordered pairs (x,y) that satisfy the equation. Again, S is called the solution set of $P(x,y) = 0$.

Solution set

ILLUSTRATION 1 Let U be the set of real numbers. Then a few of the elements of the solution set of $y = 3x + 2$ are: $(-2,-4)$, $(-1,-1)$, $(0,2)$, $(\frac{1}{3},3)$. The solution set S, indeed, contains infinitely many elements.

ILLUSTRATION 2 Let U be the set of integers. Then a few of the elements of the solution set of $y = x^2 - 6x + 5$ are $(-1,12)$, $(0,5)$, $(1,0)$, $(2,-3)$, $(4,-3)$, $(5,0)$, $(6,5)$, and $(7,12)$.

The method of finding elements of such a solution set is undoubtedly familiar to you. To do so, choose any value of x in U, say x_0, and from the given equation calculate the corresponding value of y, say y_0. If y_0 is an element of U, the pair (x_0,y_0) is an element of the corresponding solution set. When $U = F$, the solution set S of $P(x,y) = 0$ is thus an *infinite* set. Thus we can never list all the elements of S, and to describe S we must use another technique, namely graphing. This method requires that we restrict the underlying field F to the set of real numbers; and, hereafter, in this discussion we shall assume that this has been done.

Earlier (Sec. 2.11), we have seen that there is a one-to-one correspondence between ordered pairs of real numbers and points in the plane. Using this correspondence, we can define the notion of the graph of an equation of the form $P(x,y) = 0$.

DEFINITION The graph of an equation $P(x,y) = 0$ (over the reals) is the set of points in the plane whose ordered pairs of coordinates are elements of the solution set of $P(x,y) = 0$.

Graphs

The usual procedure for plotting the graph of such an equation is to compute a small number of points on it and then to connect these with a smooth curve.

ILLUSTRATION 3 Plot the graph of

$$y = 3x + 2$$

We write the following table of pairs (see Illustration 1).

x	-2	-1	0	1
y	-4	-1	2	5

We then draw Fig. 6.1.

REMARKS

1 The graph in Fig. 6.1 is a straight line. In Chap. 14 we shall prove that the graph of any equation of the form $y = ax + b$ is a straight line. Hence you need compute only two points to determine this graph. To catch errors in computation, however, it is wise to calculate at least three points.

2 The point $(-\frac{2}{3},0)$ where this line cuts the X-axis corresponds to the solution $x = -\frac{2}{3}$ of $3x + 2 = 0$. In general, the graph of $y = ax + b$ will cut the X-axis at the point $(-b/a,0)$ which corresponds to the solution $x = -b/a$ of the equation $ax + b = 0$. This suggests a

Straight line

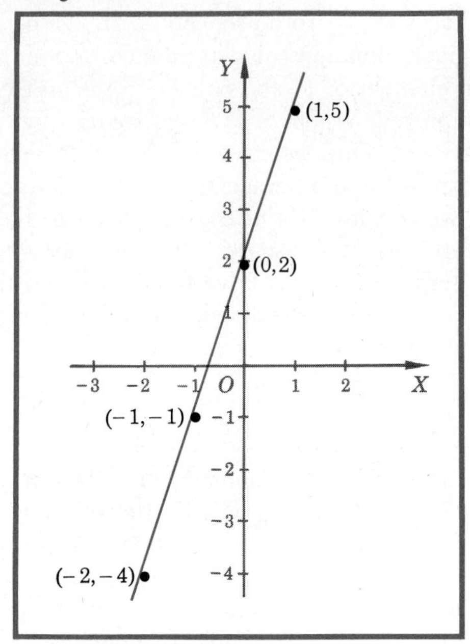

FIGURE 6.1
The graph of $y = ax + b$ is a straight line. Here $y = 3x + 2$.

graphical method for solving $ax + b = 0$, namely: Plot the graph of $y = ax + b$, and find the x-coordinate of the point in which this line intersects the X-axis. This coordinate is the desired solution.

ILLUSTRATION 4 Plot the graph of

$$y = x^2 - 6x + 5$$

We write the following table of pairs (see Illustration 2) and draw Fig. 6.2.

x	−1	0	1	2	3	4	5	6	7
y	12	5	0	−3	−4	−3	0	5	12

ILLUSTRATION 5 Plot the graph of

$$y = -2x^2 + x - 2$$

The table of pairs is as follows, and we draw Fig. 6.3.

x	−2	−1	0	1	2	3
y	−12	−5	−2	−3	−8	−17

Parabola

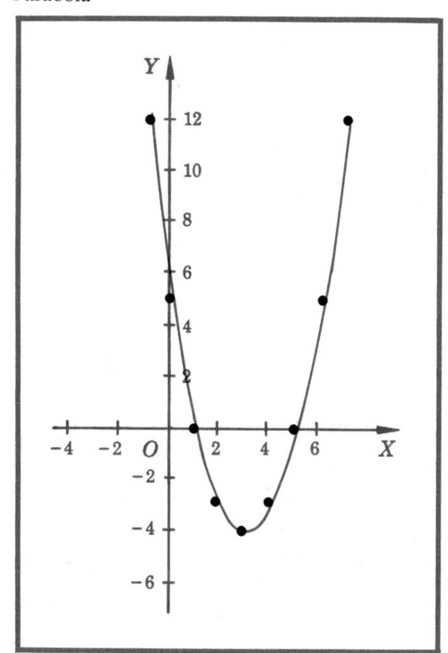

FIGURE 6.2
The graph of $y = ax^2 + bx + c$ where $a > 0$ is a parabola opening upward. Here $y = x^2 - 6x + 5$.

Parabola

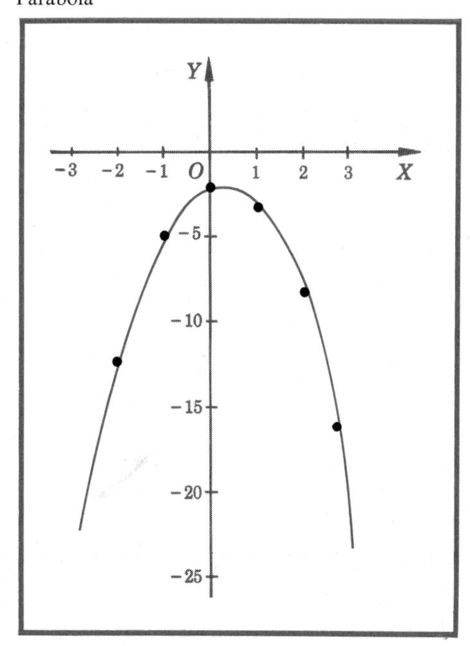

FIGURE 6.3
When $a < 0$, the graph of $ax^2 + bx + c$ opens downward. Here $y = -2x^2 + x - 2$.

DEFINITION Parabola

The graph of $y = ax^2 + bx + c$ $(a \neq 0)$ is called a *parabola*. Its highest (or lowest) point is called its *vertex*.

REMARKS The graphs of these equations have the following important properties:

1. If $a > 0$, the graph opens upward.
2. If $a < 0$, the graph opens downward.
3. If $ax^2 + bx + c = 0$ has the real solutions r_1 and r_2, where $r_1 \neq r_2$, the graph crosses the X-axis at $x = r_1$ and $x = r_2$.
4. If $ax^2 + bx + c = 0$ has two equal real solutions r, the graph is tangent to the X-axis at $x = r$.
5. If $ax^2 + bx + c = 0$ has nonreal solutions, the graph does not cross the X-axis.
6. The graph is symmetric about the line

$$x = -\frac{b}{2a}$$

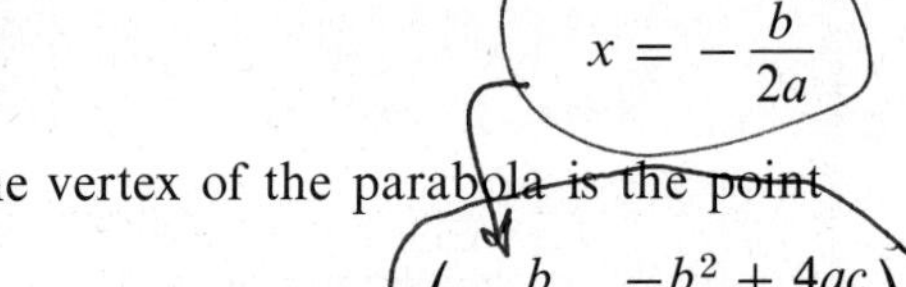

7. The vertex of the parabola is the point

$$\left(-\frac{b}{2a}, \frac{-b^2 + 4ac}{4a}\right)$$

EXERCISE A Verify as many as possible of these properties by examining the graphs in Illustrations 4 and 5 above.

PROBLEMS 6.5

In Probs. 1 to 16 plot the graphs of the given equations, and from these find the solutions of the corresponding linear or quadratic equations.

1 $y = 3x + 6$
2 $y = 2x - 10$
3 $y = -3x + 12$
4 $y = -2x + 6$
5 $y = -x - 4$
6 $y = -3x - 9$
7 $y = x^2 + 7x + 12$
8 $y = x^2 + 8x + 12$
9 $y = x^2 + x + 1$
10 $y = x^2 - x + 1$
11 $y = 4x^2 - 4x + 1$
12 $y = 9x^2 - 6x + 1$
13 $y = 4 - x^2$
14 $y = -3x^2 - 2x + 1$
15 $y = -x^2 - 6x - 9$
16 $y = -4x^2 + 4x - 1$

Problems 17 to 23 refer to properties 1 to 7 of parabolas, which are stated at the top of this page.

17 Why is property 1 true?
18 Why is property 2 true?
19 Why is property 3 true?
20 Why is property 4 true?
21 Why is property 5 true?

22 Why is property 6 true? HINT: Write $y = ax^2 + bx + c$ in the form

$$y + \frac{b^2 - 4ac}{4a} = a\left(x + \frac{b}{2a}\right)^2$$

23 Why is property 7 true? HINT: Write $y = ax^2 + bx + c$ in the form

$$y = \frac{-b^2 + 4ac}{4a} + a\left(x + \frac{b}{2a}\right)^2$$

6.6 » EQUATIONS CONTAINING FRACTIONS

It is not uncommon for you to encounter equations like

(1)
$$\frac{1}{x} + \frac{2}{x+1} = 3$$

or

(2)
$$\frac{x-1}{x+4} - \frac{x+2}{x-3} = \frac{5}{x+2}$$

in which algebraic fractions appear. In order to discuss the solution of these, we must reexamine the matters of the meaning of such equations and the collection of operations which convert such an equation into an equivalent equation.

Let us suppose that the coefficients are real and that we are seeking real solutions. Then what is the universal set for the variable x? It cannot be the set of real numbers R, for neither 0 nor -1 can be substituted for x in Eq. (1). Similarly, -4, 3, and -2 are excluded from R in (2).

Universal set

Thus the set U is R with the exception of those values of x for which any denominator is zero.

Theorems 1 and 2 of Sec. 6.2 apply equally well to equations of this type, but we shall need to use an additional operation, which is given in Theorem 5.

THEOREM 5 Let $R(x) = P(x)/Q(x)$, where $P(x)$ and $Q(x)$ are polynomials with real coefficients. Further let U be the set of real numbers excluding all x for which $Q(x) = 0$. Let $S(x)$ be any polynomial which is nonzero for all x in U. Then the equation $R(x) = 0$ is equivalent to the equation $R(x) \cdot S(x) = 0$.

Rx = 0
Rx × Sx = 0

PROOF Let $A = \{x \mid R(x) = 0\}$ and $B = \{x \mid R(x) \cdot S(x) = 0\}$. If a is an element of U such that $R(a) = 0$, then a is an element of B. Hence $A \subseteq B$. Moreover if b is an element of U such that $R(b) \cdot S(b) = 0$, then since $S(b) \neq 0$, we have $R(b) = 0$, and b is an element of A. Hence $B \subseteq A$. Therefore, $A = B$.

We shall apply Theorem 5 as in the following illustrations.

ILLUSTRATION 1 Solve $\dfrac{1}{x} + \dfrac{6}{x+4} = 1.$

Then U is R except for 0 and -4. Hence we may multiply both sides by $x(x+4)$, since this is nonzero for all x in U. The result is

$$x + 4 + 6x = x^2 + 4x \qquad \text{or} \qquad x^2 - 3x - 4 = 0$$

The solutions are $x = -1$ and 4, which are both in U. Hence the solution of the original equation is the set $\{-1, 4\}$.

ILLUSTRATION 2 Solve $\dfrac{7}{x-1} - \dfrac{6}{x^2-1} = 5.$

Then U is R except for -1 and 1. So we multiply both sides by $x^2 - 1$ and obtain

$$7(x+1) - 6 = 5(x^2 - 1)$$

or

$$5x^2 - 7x - 6 = (x-2)(5x+3) = 0$$

The solution is $\{2, -\frac{3}{5}\}$, since both of the elements of this set are in U.

ILLUSTRATION 3 Solve $\dfrac{x}{x+2} - \dfrac{4}{x+1} = \dfrac{-2}{x+2}.$

Then U is R except for -1 and -2. So we multiply both sides by $(x+2)(x+1)$ and obtain

$$x(x+1) - 4(x+2) = -2(x+1)$$

or

$$x^2 - x - 6 = (x-3)(x+2) = 0$$

whose solution in U is $x = 3$. (We must discard $x = -2$, since -2 is not in U.)

ILLUSTRATION 4 Solve $\dfrac{1}{x-1} + x = 1 + \dfrac{1}{x-1}.$

Then U is R except for 1. We multiply both sides by $x - 1$ and obtain

$$1 + x(x-1) = (x-1) + 1$$

or

$$x^2 - 2x + 1 = (x-1)^2 = 0$$

This equation has no solution in U, and so there is no solution for the given equation.

6.7 » EQUATIONS CONTAINING RADICALS

In this section we are interested in equations like

$$\sqrt{x+13} - \sqrt{7-x} = 2 \qquad \text{or} \qquad 2\sqrt{x+4} - x = 1$$

in which x appears under a radical. For simplicity we shall consider square roots only. The only reasonable procedure is to square both sides in order to eliminate a radical. But before using such an operation we must ask whether it converts the given equation into an equivalent equation. The trouble is that the answer is "No"! If we square both sides of an equation, we are not at all sure that the new equation is equivalent to the given one. Consider the following equation:

$$x - 3 = 1 \qquad \text{for which } x = 4$$

If we square both sides, we obtain

$$(x - 3)^2 = 1 \qquad \text{or} \qquad x^2 - 6x + 8 = 0$$

whose solutions are $x = 4$ and $x = 2$. The two equations are not equivalent.

Nevertheless, we shall go ahead and square both sides, but with the realization that this process may well introduce unwanted roots. Therefore, all solutions are tentative and must be checked in the given equation.

Solutions must be tested

ILLUSTRATION 1 Solve $2\sqrt{x + 4} - x = 1$.

$$\begin{aligned} 2\sqrt{x+4} &= x + 1 \\ 4(x + 4) &= x^2 + 2x + 1 \\ x^2 - 2x - 15 &= 0 \\ (x - 5)(x + 3) &= 0 \\ x &= 5, \ -3 \end{aligned}$$

Checking $x = 5$, we have $2\sqrt{9} - 5 = 1$, or $6 - 5 = 1$, which is true.

Checking $x = -3$, we have $2\sqrt{1} + 3 = 1$, or $2 + 3 = 1$, which is false.

The solution set of the given equation is therefore $\{5\}$.

When there are two radicals, the method is similar, but two squarings are required. Proceed as in the illustration below.

ILLUSTRATION 2 Solve $\sqrt{x + 13} - \sqrt{7 - x} = 2$.

$$\begin{aligned} \sqrt{x + 13} &= 2 + \sqrt{7 - x} \\ x + 13 &= 4 + 4\sqrt{7 - x} + 7 - x \\ 2x + 2 &= 4\sqrt{7 - x} \\ x + 1 &= 2\sqrt{7 - x} \\ x^2 + 2x + 1 &= 28 - 4x \\ x^2 + 6x - 27 &= 0 \\ (x - 3)(x + 9) &= 0 \\ x &= 3, \ -9 \end{aligned}$$

Checking $x = 3$, we have $\sqrt{16} - \sqrt{4} = 2$, or $4 - 2 = 2$, which is true.

Checking $x = -9$, we have $\sqrt{4} - \sqrt{16} = 2$, or $2 - 4 = 2$, which is false.

Hence the solution set is $\{3\}$.

ILLUSTRATION 3 Solve $\sqrt{x+1} - \sqrt{x+6} = 1$.

$$\sqrt{x+1} = 1 + \sqrt{x+6}$$
$$x + 1 = 1 + 2\sqrt{x+6} + x + 6$$
$$-6 = 2\sqrt{x+6}$$
$$36 = 4(x+6)$$
$$4x = 12$$
$$x = 3$$

Testing, we find

$$\sqrt{3+1} - \sqrt{3+6} \neq 1$$

Therefore the equation has no solution, and the solution set is the null set $\varnothing$.

There is another way of observing that this equation has no solution. Since $x + 6 > x + 1$ for all x,

$$\sqrt{x+6} > \sqrt{x+1} \qquad \text{and} \qquad \sqrt{x+1} - \sqrt{x+6}$$

must be negative. Hence it cannot equal 1 for any x.

PROBLEMS 6.7

In Probs. 1 to 10 describe the universal set U as a subset of R in each of the following equations and solve:

1 $\dfrac{5}{x+3} + \dfrac{4}{x} = 3$

2 $\dfrac{8}{x-4} - \dfrac{6}{x} = 3$

3 $\dfrac{4}{x+1} + \dfrac{8}{x^2-1} = 2$

4 $\dfrac{6}{x-2} - \dfrac{24}{x^2-4} = 1$

5 $\dfrac{5}{x} - \dfrac{1}{x-1} + \dfrac{9}{x+3} = 0$

6 $\dfrac{9}{x} + \dfrac{8}{x+1} + \dfrac{5}{x-4} = 0$

7 $\dfrac{-5}{x+1} + \dfrac{4x+1}{x^2+1} = 0$

8 $\dfrac{2}{x+5} + \dfrac{3x-5}{x^2+5} = 0$

9 $\dfrac{2x}{x+3} - \dfrac{1}{x-3} = \dfrac{-6}{x+3}$

10 $\dfrac{-2x}{x+2} + \dfrac{1+2x}{x} = \dfrac{4}{x+2}$

In Probs. 11 to 20 solve the given equation.

11 $\sqrt{x+6} + x - 6 = 0$

12 $\sqrt{x-2} - x + 4 = 0$

13 $\sqrt{x+2} + \sqrt{x+7} = 0$

14 $\sqrt{x-10} + \sqrt{x+11} = 0$

15 $\sqrt{x+3} - \sqrt{x+8} + 1 = 0$

16 $\sqrt{x+5} + \sqrt{x-3} - 4 = 0$

17 $-\sqrt{2x+9}+\sqrt{3x+16}=1$ 18 $\sqrt{3x+16}-\sqrt{2x+9}=-1$

19 $\dfrac{1}{\sqrt{x}}-\dfrac{2}{\sqrt{x+12}}=0$ 20 $\dfrac{x}{\sqrt{2x+3}}-\dfrac{2x}{3\sqrt{x+1}}=0$

6.8 » SIMULTANEOUS LINEAR EQUATIONS

Here we complicate the situation a little by considering a pair of simultaneous linear equations in two variables. The general expression for such a system is

$$\text{(1)}\qquad \begin{cases} a_1x + b_1y + c_1 = 0 & a_1 \neq 0 \text{ or } b_1 \neq 0 \\ a_2x + b_2y + c_2 = 0 & a_2 \neq 0 \text{ or } b_2 \neq 0 \end{cases}$$

where the coefficients are real numbers, and the universal set of the variables is also the set of real numbers. By a solution of (1) we mean an ordered pair (x,y) which satisfies both equations.

In set language, the solution set of (1) is the set

$$\{(x,y) \mid a_1x + b_1y + c_1 = 0\} \cap \{(x,y) \mid a_2x + b_2y + c_2 = 0\}$$

Since the graph of each equation in (1) is a straight line, we are looking for the points which lie on both these lines.

Method of addition and subtraction

The method of solution is to transform (1) into an equivalent system whose solution is obvious. You are undoubtedly familiar with this procedure as the method of elimination by *addition and subtraction.* Let us state it formally.

THEOREM 6 The linear systems

$$\text{(1)}\qquad \begin{cases} a_1x + b_1y + c_1 = 0 & a_1 \neq 0 \text{ or } b_1 \neq 0 \\ a_2x + b_2y + c_2 = 0 & a_2 \neq 0 \text{ or } b_2 \neq 0 \end{cases}$$

and

$$\text{(2)}\qquad \begin{cases} a_1x + b_1y + c_1 = 0 \\ k_1(a_1x + b_1y + c_1) + k_2(a_2x + b_2y + c_2) = 0 \end{cases}$$

with $k_2 \neq 0$ are equivalent.

PROOF It is clear that every solution of (1) satisfies (2). Conversely, if $(\bar{x},\bar{y})$ satisfies (2), it follows that $a_1\bar{x} + b_1\bar{y} + c_1 = 0$, and

$$k_2(a_2\bar{x} + b_2\bar{y} + c_2) = 0$$

Since $k_2 \neq 0$, $(\bar{x},\bar{y})$ satisfies (1).

Using this theorem, we can eliminate x and y in turn by choosing suitable values of k_1 and k_2 and thus solve the system. Let us do this systematically. By hypothesis, either $a_1 \neq 0$ or $b_1 \neq 0$. Let us suppose that $a_1 \neq 0$. Then, choosing $k_1 = -a_2$ and $k_2 = a_1$ in the system (2), we replace (1) by

(3)
$$\begin{cases} a_1x_1 \qquad +b_1y \qquad\qquad +c_1 = 0 \\ \qquad (a_1b_2 - a_2b_1)y + (a_1c_2 - a_2c_1) = 0 \end{cases}$$

If $a_1b_2 - a_2b_1 \neq 0$, we solve the second equation of (3) for y and obtain

$$y = -\frac{a_1c_2 - a_2c_1}{a_1b_2 - a_2b_1}$$

We may substitute this value of y in the first equation of (3) and solve for x:

$$x = -\frac{b_2c_1 - b_1c_2}{a_1b_2 - a_2b_1}$$

If we suppose that $b_1 \neq 0$, we choose $k_1 = b_2$ and $k_2 = -b_1$ and replace (1) by

(4)
$$\begin{cases} a_1x \qquad\qquad +b_1y \qquad\qquad +c_1 = 0 \\ (a_1b_2 - a_2b_1)x \qquad +(b_2c_1 - b_1c_2) = 0 \end{cases}$$

Solving as before, we obtain the values for x and y derived above.

If, on the other hand, $a_1b_2 - a_2b_1 = 0$, there are two possibilities:

1 $a_1c_2 - a_2c_1 = 0$, and $b_2c_1 - b_1c_2 = 0$. Then the system is equivalent to the single equation $a_1x + b_1y + c_1 = 0$, which has infinitely many solutions.

2 At least one of $(a_1c_2 - a_2c_1)$ and $(b_2c_1 - b_1c_2)$ is not zero. Then (3) or (4) or both contains a contradiction and there is no solution.

EXERCISE A Show that if $a_1b_2 - a_2b_1 = 0$ and $a_1c_2 - a_2c_1 = 0$, then

$$b_2c_1 - b_1c_2 = 0$$

We summarize these results in the following theorem.

THEOREM 7 The simultaneous equations

$$\begin{matrix} a_1x + b_1y + c_1 = 0 & \quad a_1 \neq 0 \text{ or } b_1 \neq 0 \\ a_2x + b_2y + c_2 = 0 & \quad a_2 \neq 0 \text{ or } b_2 \neq 0 \end{matrix}$$

a Have a unique solution if $a_1b_2 - a_2b_1 \neq 0$.

b Have no solution if $a_1b_2 - a_2b_1 = 0$ and at least one of $(a_1c_2 - a_2c_1)$ and $(b_2c_1 - b_1c_2)$ is not zero.

c Have infinitely many solutions if $a_1b_2 - a_2b_1 = 0$, $a_1c_2 - a_2c_1 = 0$, and $b_2c_1 - b_1c_2 = 0$.

REMARK Since the coefficients are real, case *a* of Theorem 7 corresponds to two intersecting lines, case *b* to two parallel lines, and case *c* to a single line.

Intersecting lines; Theorem 7, case *a*

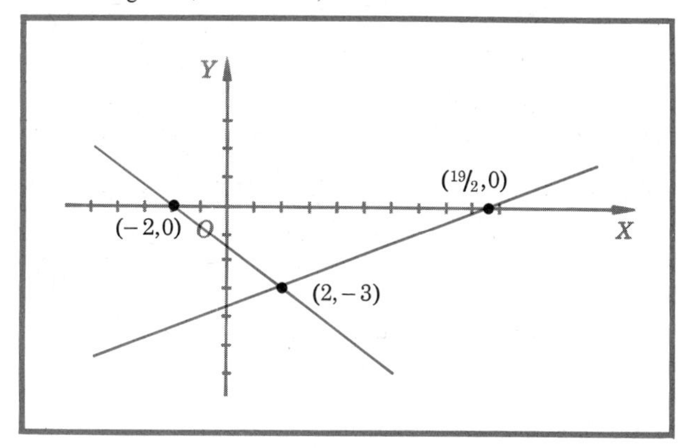

FIGURE 6.4
$2x - 5y - 19 = 0$
$3x + 4y + 6 = 0$

ILLUSTRATION 1 Solve

$$\begin{aligned} 2x - 5y - 19 &= 0 \\ 3x + 4y + \ \ 6 &= 0 \end{aligned} \qquad \text{(Fig. 6.4)}$$

To eliminate x, we multiply the first equation by 3 and the second by 2. This gives us

$$\begin{aligned} 6x - 15y - 57 &= 0 \\ 6x + \ \ 8y + 12 &= 0 \end{aligned}$$

Subtracting, we have

$$-23y - 69 = 0 \qquad \text{or} \qquad y + 3 = 0$$

This equation, combined with the first equation of the stated system, gives us the equivalent system

$$\begin{aligned} 2x - 5y - 19 &= 0 \\ y + \ \ 3 &= 0 \end{aligned}$$

We solve the second equation for y and get $y = -3$. Putting $y = -3$ in the first equation and solving for x, we have $x = 2$. Hence the solution is the pair $(2,-3)$.

ILLUSTRATION 2 Solve

$$\begin{aligned} 3x + 2y + 5 &= 0 \\ 6x + 4y - 4 &= 0 \end{aligned} \qquad \text{(Fig. 6.5)}$$

Elimination of x as in Illustration 1 gives us the equivalent system

$$\begin{aligned} 3x + 2y + 5 &= 0 \\ 14 &= 0 \end{aligned}$$

Since the last equation is, in fact, not an equality, there can be no solution. The lines are parallel.

Parallel lines; Theorem 7, case b

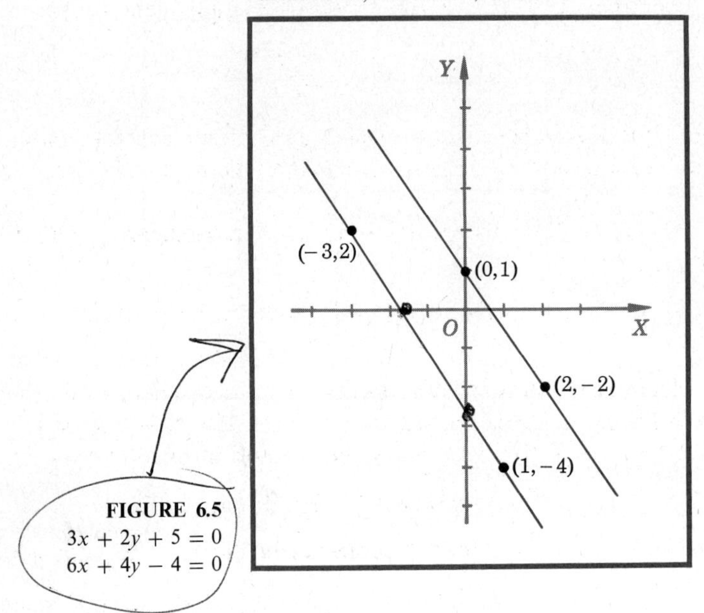

FIGURE 6.5
$3x + 2y + 5 = 0$
$6x + 4y - 4 = 0$

ILLUSTRATION 3 Solve

$$\begin{aligned} 4x - y + 3 &= 0 \\ 8x - 2y + 6 &= 0 \end{aligned} \qquad \text{(Fig. 6.6)}$$

Single line; Theorem 7, case c

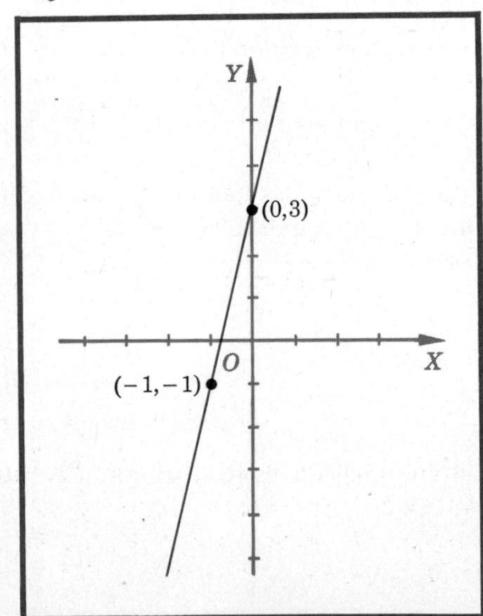

FIGURE 6.6
$4x - y + 3 = 0$
$8x - 2y + 6 = 0$

Elimination of x as above gives us the equivalent system

$$\begin{aligned} 4x - y + 3 &= 0 \\ 0 &= 0 \end{aligned}$$

Hence the system reduces to a single equation: i.e., the lines are coincident, and there are infinitely many solutions.

ILLUSTRATION 4 Consider the following homogeneous system of equations:

Homogeneous system

$$\begin{aligned} a_1x + b_1y &= 0 \\ a_2x + b_2y &= 0 \end{aligned}$$

The pair (0,0) is obviously a solution and is called the trivial solution. From Theorem 7, this is the only solution if $a_1b_2 - a_2b_1 \neq 0$. But if $a_1b_2 - a_2b_1 = 0$, this system has the infinite set of solutions,

$$x = kb_1 \qquad y = -ka_1$$

where k is any real number. Geometrically, the solution is a line through the origin.

PROBLEMS 6.8

Solve the given pair of equations algebraically. Then plot the graph of the two lines, and check your solution graphically.

1. $2x - 5y + 8 = 0$, $x + 4y - 9 = 0$
2. $3x + 2y - 3 = 0$, $4x - y + 7 = 0$
3. $5x + 2y - 4 = 0$, $3x - y - 9 = 0$
4. $6x + y - 19 = 0$, $-2x + 3y + 3 = 0$
5. $4x - y + 5 = 0$, $3x + 2y - 10 = 0$
6. $3x - 2y - 8 = 0$, $-x + 3y - 2 = 0$
7. $x + y + 9 = 0$, $2x + 2y - 1 = 0$
8. $3x - 2y + 5 = 0$, $-6x + 4y + 3 = 0$
9. $-3x + 5y + 11 = 0$, $6x - 10y - 22 = 0$
10. $2x + 6y = 0$, $-x - 3y = 0$

6.9 › SIMULTANEOUS LINEAR EQUATIONS (Continued)

In this section we consider a further idea which will help you to understand what we have just been discussing. Let us multiply the left members of the system (1) of Sec. 6.8 by k_1 and k_2, respectively, where k_1 and k_2 are not both zero, and then add. The result is

$$k_1(a_1x + b_1y + c_1) + k_2(a_2x + b_2y + c_2) = 0 \tag{5}$$

For all values of k_1 and k_2 (not both zero) this is the equation of some line. What line is it?

To simplify matters let us assume that the two lines intersect (case *a* of Sec. 6.8) and call the point of intersection $P(x_0,y_0)$. Then it follows that line (5) passes through P. To see this substitute (x_0,y_0) in (5); the result is zero since each parenthesis is zero by hypothesis. As k_1 and k_2 take different values, we get a family of lines all passing through P.

Family of lines through P

THEOREM 8 If the lines $a_1x + b_1y + c_1 = 0$ and $a_2x + b_2y + c_2 = 0$ intersect at a point P, Eq. (5) represents a family of lines, each of which passes through P.

Let us not forget our original problem—to solve the simultaneous system (1) of Sec. 6.8. The point P of intersection of the two given lines can be found equally well by solving the equations of any two other lines through P. In other words, we will get the same point P if we solve any pair of equations chosen from the family (5). In the terminology of Sec. 6.2, the given system of two equations is equivalent to any system of two distinct equations obtained from (5). So let us pick the simplest possible pair of equations from (5). These will correspond to the horizontal and vertical lines through P. To find the horizontal line, choose k_1 and k_2 so that the coefficient of x in (5) is zero; i.e., eliminate x. This gives the solution for y. Similarly, to find the vertical line through P, choose k_1 and k_2 so that y disappears, and solve for x. This solves the problem.

Incidentally, Eq. (5) permits us to obtain easy solutions to a number of other problems. The illustration below will give you the idea.

Equivalent linear systems

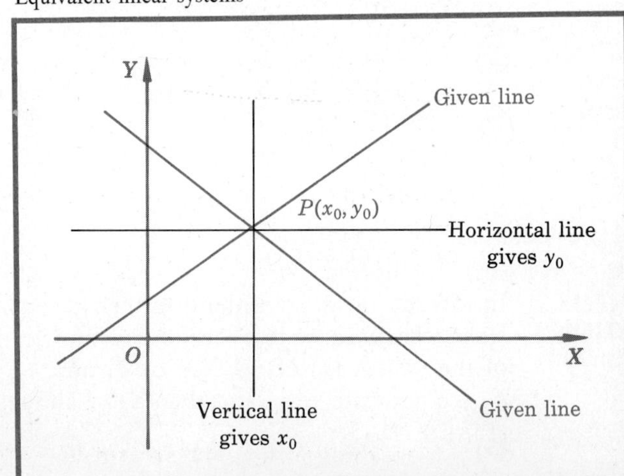

FIGURE 6.7
If the lines corresponding to a system of linear equations intersect at P, then the lines corresponding to any equivalent system also intersect at P.

ILLUSTRATION 1 Find the equation of the line passing through the point $(2,-1)$ and through the point of intersection of the lines $2x + y - 5 = 0$ and $x + 2y - 7 = 0$.

Using (5), we obtain the equation of the family of lines through this point of intersection:

$$k_1(2x + y - 5) + k_2(x + 2y - 7) = 0$$

We want to pick out the one passing through $(2,-1)$. So substitute $(2,-1)$ for (x,y) in the above equation. This gives

$$k_1(-2) + k_2(-7) = 0$$

Choose any k_1 and k_2 for which this is true, say $k_1 = 7$, $k_2 = -2$. This gives the required equation, namely:

$$7(2x + y - 5) + (-2)(x + 2y - 7) = 0$$

or

$$12x + 3y - 21 = 0$$

which is the answer. Observe that we never found the point (1,3), which is the point of intersection of the two given lines.

PROBLEMS 6.9

In Probs. 1 to 5 find an equation of the line passing through the intersection of the two given lines and the given point.

1. $x + 2y - 5 = 0$; $2x - 3y + 4 = 0$ and (2,3)
2. $2x - 3y - 7 = 0$; $3x + y - 5 = 0$ and (1,3)
3. $4x - 5y - 2 = 0$; $x + 3y - 9 = 0$ and (1,1)
4. $x = 2$, $y = 4$; $(-3,-1)$
5. $x = -2$, $y = 5$; (1,4)
6. Find an equation of the line through $(2,-1)$ and (1,5). HINT: $x - 2 = 0$ and $y + 1 = 0$ pass through $(2,-1)$.
7. Find an equation of the line through $(-1,3)$ and (2,5). See hint for Prob. 6.
8. Find an equation of the line through (x_1,y_1) and (x_2,y_2).
9. What does the equation

$$k_1(a_1x + b_1y + c_1) + k_2(a_2x + b_2y + c_2) = 0$$

represent when the two given lines are parallel?

10. What does the equation

$$k_1(a_1x + b_1y + c_1) + k_2(a_2x + b_2y + c_2) = 0$$

represent when the two given lines are coincident?

6.10 » SIMULTANEOUS LINEAR EQUATIONS IN THREE UNKNOWNS

The method of Sec. 6.8 can be applied without substantial change to simultaneous systems of three equations. The general expression for such a system is

$$
\text{(1)} \quad \begin{aligned} a_1x + b_1y + c_1z + d_1 &= 0 \qquad \text{Not all } a_1, b_1, c_1 = 0 \\ a_2x + b_2y + c_2z + d_2 &= 0 \qquad \text{Not all } a_2, b_2, c_2 = 0 \\ a_3x + b_3y + c_3z + d_3 &= 0 \qquad \text{Not all } a_3, b_3, c_3 = 0 \end{aligned}
$$

A solution is an ordered triple (x,y,z) which satisfies all three equations. If we wish to plot ordered triples, we need three dimensions and so we shall not draw the graphs of these equations. Their geometric interpretation, however, is helpful. It can be proved that the equation $ax + by + cz + d = 0$ corresponds to a plane in 3-space; so the system (1) represents three planes. The number of possible configurations for three planes is a little large, but here they are:

The graph of each equation is a plane.

1 The three planes intersect in a point; hence (1) has a unique solution.
2 The three planes are mutually parallel; hence (1) has no solutions.
3 Two planes coincide, and the third plane is parallel to this common plane; hence (1) has no solutions.
4 All three planes coincide; hence (1) has a plane of solutions.
5 The three planes intersect in three parallel lines; hence (1) has no solutions.
6 Two planes are parallel, and the third intersects them in two parallel lines; hence (1) has no solutions.
7 The three planes intersect in a common line; hence (1) has a line of solutions.

In summary, (1) may have a unique solution, a line of solutions, a plane of solutions, or no solution.

The method of solution, which handles all these cases, is best explained by the illustrations below.

ILLUSTRATION 1 Solve

$$
\begin{aligned} 2x - y + 3z + 9 &= 0 \\ x + 3y - z - 10 &= 0 \\ 3x + y - z - 8 &= 0 \end{aligned}
$$

Unique solution

First we eliminate x between the first and second and between the first and third equations. The result, together with the unchanged first equation, is the equivalent system

$$
\begin{aligned} 2x - y + 3z + 9 &= 0 \\ 7y - 5z - 29 &= 0 \\ 5y - 11z - 43 &= 0 \end{aligned}
$$

Next eliminate y between the second and third equations. Leaving the first two equations unchanged, we have the equivalent system

$$
\begin{aligned} 2x - y + 3z + 9 &= 0 \\ 7y - 5z - 29 &= 0 \\ z + 3 &= 0 \end{aligned}
$$

From the last equation, $z = -3$. Putting $z = -3$ in the second equation enables us to find $y = 2$; and putting $y = 2$ and $z = -3$ in the first equation gives us $x = 1$.

Hence the solution is $(1,2,-3)$.

ILLUSTRATION 2 Solve

$$\begin{aligned} x + 2y - z + 3 &= 0 \\ 2x + y + z - 1 &= 0 \\ 3x + 3y \qquad + 2 &= 0 \end{aligned}$$

The first elimination (of x) gives us the equivalent system

Line of solutions

$$\begin{aligned} x + 2y - z + 3 &= 0 \\ 3y - 3z + 7 &= 0 \\ 3y - 3z + 7 &= 0 \end{aligned}$$

The final elimination (of y) gives the equivalent system

$$\begin{aligned} x + 2y - z + 3 &= 0 \\ 3y - 3z + 7 &= 0 \\ 0 &= 0 \end{aligned}$$

Thus the system really reduces to two equations, i.e., to two planes which meet in a line. Hence there is a line of solutions.

ILLUSTRATION 3 Solve

Plane of solutions

$$\begin{aligned} x - y + 2z + 1 &= 0 \\ 2x - 2y + 4z + 2 &= 0 \\ 3x - 3y + 6z + 3 &= 0 \end{aligned}$$

The first elimination (of x) gives

$$\begin{aligned} x - y + 2z + 1 &= 0 \\ 0 &= 0 \\ 0 &= 0 \end{aligned}$$

Thus the system reduces to a single equation, i.e., to a single plane all of whose points are solutions. The three planes are coincident, and there is a plane of solutions.

ILLUSTRATION 4 Solve

No solution

$$\begin{aligned} x + y + z - 1 &= 0 \\ 2x - 3y - 2z + 4 &= 0 \\ 3x - 2y - z + 2 &= 0 \end{aligned}$$

The first elimination (of x) gives the system

$$\begin{aligned} x + y + z - 1 &= 0 \\ 5y + 4z - 6 &= 0 \\ 5y + 4z - 5 &= 0 \end{aligned}$$

The final elimination (of y) gives the system

$$\begin{aligned} x + y + z - 1 &= 0 \\ 5y + 4z - 6 &= 0 \\ -1 &= 0 \end{aligned}$$

Since the last equation is not, in fact, an equality, the system has no solution. There is no need to look further into the geometry of the case.

PROBLEMS 6.10 Solve for x, y, and z.

1 $$\begin{aligned} 3x + y - 2z - 2 &= 0 \\ x + 3y - z - 3 &= 0 \\ 2x - y + 4z - 5 &= 0 \end{aligned}$$

2 $$\begin{aligned} 2x - y + 3z - 14 &= 0 \\ -3x + y - z + 10 &= 0 \\ x + y + z - 4 &= 0 \end{aligned}$$

3 $$\begin{aligned} 3x + 2z + 5 &= 0 \\ 2y - 3z - 12 &= 0 \\ x + y - 1 &= 0 \end{aligned}$$

4 $$\begin{aligned} 5x - y + 3z - 3 &= 0 \\ -2x + y - 2z + 2 &= 0 \\ 3x + 2y + z - 1 &= 0 \end{aligned}$$

5 $$\begin{aligned} 2x - y + 4z - 1 &= 0 \\ 3x + 2y - 5z + 4 &= 0 \\ 5x + y - z + 3 &= 0 \end{aligned}$$

6 $$\begin{aligned} 3x - 2y + 4z - 5 &= 0 \\ -6x + 4y - 8z + 10 &= 0 \\ 9x - 6y + 12z - 15 &= 0 \end{aligned}$$

7 $$\begin{aligned} x - 2y + z - 3 &= 0 \\ 4x - 8y + 4z - 12 &= 0 \\ -2x + 4y - 2z + 6 &= 0 \end{aligned}$$

8 $$\begin{aligned} 3x + y - 5z + 4 &= 0 \\ -x - y + z + 2 &= 0 \\ x - y - 3z + 8 &= 0 \end{aligned}$$

9 $$\begin{aligned} x + 2y - z + 3 &= 0 \\ 2x - y + 3z + 1 &= 0 \\ 3x + y + 2z - 2 &= 0 \end{aligned}$$

10 $$\begin{aligned} 3x - 5y + z + 1 &= 0 \\ x + y - z + 3 &= 0 \\ x - 7y + 3z + 4 &= 0 \end{aligned}$$

6.11 › SIMULTANEOUS LINEAR AND QUADRATIC EQUATIONS

Suppose that we are given a mixed system consisting of one linear equation and one quadratic equation, each involving two variables, such as

(1) $$\begin{cases} y = ax + b \\ y = cx^2 + dx + e \end{cases}$$

We are looking for those ordered pairs (x,y) which satisfy both equations. When x and y are real, this amounts to finding the points of intersection (if any) of the line and the parabola, which are the graphs of the given equations.

The method of solution is quite straightforward: Set the two right-hand sides equal and solve the resulting quadratic equation for x. Then substitute the solution (or solutions) into the linear equation and find the corresponding values for y. The various possibilities are suggested by the following illustrations.

ILLUSTRATION 1 Solve

$$\begin{cases} y = x + 2 \\ y = x^2 - 6x + 8 \end{cases}$$

Putting the right-hand sides equal, we obtain

$$x + 2 = x^2 - 6x + 8 \quad \text{or} \quad x^2 - 7x + 6 = 0$$
$$(x - 6)(x - 1) = 0 \quad \text{or} \quad x = 1, 6$$

When $x = 1$, $y = 3$ and when $x = 6$, $y = 8$. So the solutions are (1,3) and (6,8). The line intersects the parabola in two points (Fig. 6.8).

ILLUSTRATION 2 Solve

$$\begin{cases} y = 2x - 4 \\ y = x^2 - 4x + 5 \end{cases}$$

We obtain

$$2x - 4 = x^2 - 4x + 5$$

or

$$x^2 - 6x + 9 = 0 \quad \text{or} \quad x = 3$$

Thus the only solution is (3,2). The line is tangent to the parabola (Fig. 6.9).

ILLUSTRATION 3 Solve

$$\begin{cases} y = -4x + 2 \\ y = x^2 - 4x + 3 \end{cases}$$

Straight line and parabola

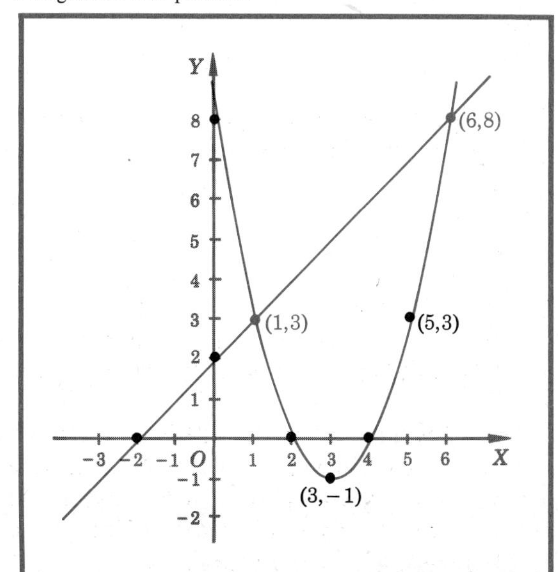

FIGURE 6.8 There are two points of intersection.
$y = x + 2$
$y = x^2 - 6x + 8$

Straight line and parabola

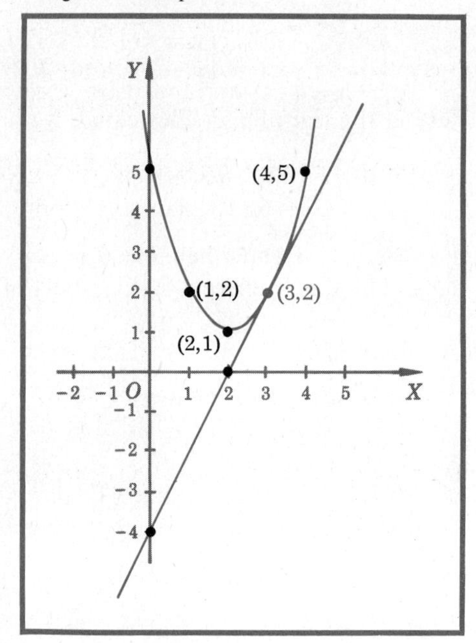

FIGURE 6.9
The line is tangent to the parabola.
$y = 2x - 4$
$y = x^2 - 4x + 5$

Straight line and parabola

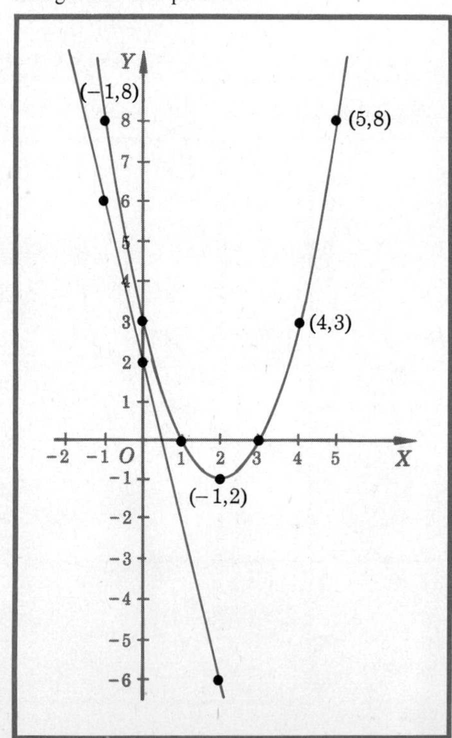

FIGURE 6.10
The line and the parabola do not intersect.
$y = -4x + 2$
$y = x^2 - 4x + 3$

We obtain

$$-4x + 2 = x^2 - 4x + 3 \qquad \text{or} \qquad x^2 + 1 = 0$$

Hence $x = +i, -i$, and $y = 2 - 4i, 2 + 4i$. The solutions are $(i, 2 - 4i)$, $(-i, 2 + 4i)$. The line and the parabola do not intersect (Fig. 6.10).

The same methods apply to simultaneous systems such as

$$\text{(2)} \qquad \begin{cases} y = ax^2 + bx + c \\ y = px^2 + qx + r \end{cases}$$

and

$$\text{(3)} \qquad \begin{cases} y = ax + b \\ y = \dfrac{k}{cx + d} \end{cases}$$

Simple transformations will often convert other simultaneous systems to one of these types.

ILLUSTRATION 4 Solve

$$\begin{cases} 2x - y - 2 = 0 \\ xy = 4 \end{cases}$$

Straight line and hyperbola

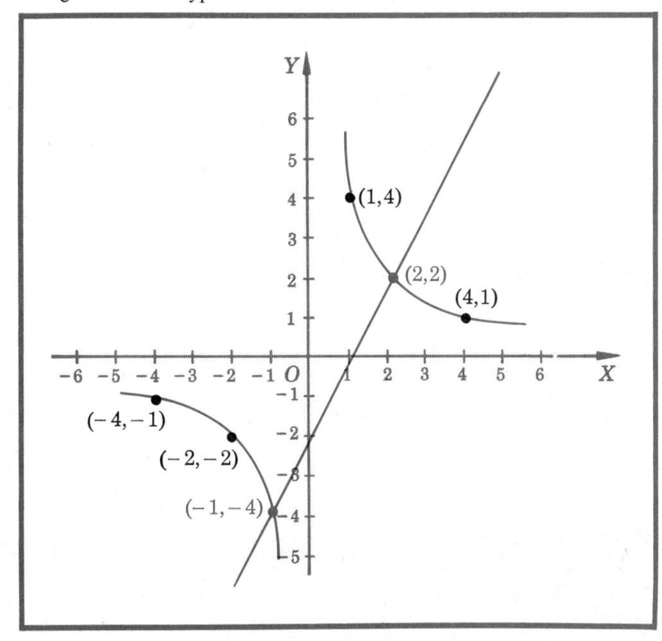

FIGURE 6.11
There are two points of intersection.
$2x - y - 2 = 0$
$xy = 4$

This is equivalent to

$$\begin{cases} y = 2x - 2 \\ y = \dfrac{4}{x} \end{cases}$$

So we solve

$$2x - 2 = \frac{4}{x}$$

or

$$2x^2 - 2x - 4 = 0$$

or

$$x^2 - x - 2 = 0$$

or

$$(x - 2)(x + 1) = 0$$

Thus $x = 2$, or -1. The corresponding values of y are 2 and -4. Hence the solutions are (2,2) and $(-1, -4)$ (Fig. 6.11).

PROBLEMS 6.11 In Probs. 1 to 12 solve algebraically and graphically.

1 $y = 1 - x$
$y = x^2 + 2x - 3$

2 $y = 6 - 2x$
$y = x^2 - 2x + 2$

3 $y = x - 4$
$y = -x^2 + 8$

4 $y = 2x - 2$
$y = -x^2 + x + 4$

5 $y = 2x - 1$
$y = x^2 - 2x + 3$

6 $y = 2x + 1$
$y = -x^2 - 4x - 8$

7 $y = -x - 1$
$y = x^2 + x + 3$

8 $y = 4x + 1$
$y = -x^2 + 2x - 1$

9 $y = x^2 - 2x + 2$
$y = -x^2 + 2x + 2$

10 $y = x^2 + 4x + 4$
$y = x^2 + 8x + 8$

11 $y = \dfrac{x + 1}{4}$
$y = \dfrac{4}{x + 1}$

12 $y = \dfrac{3x - 6}{2}$
$y = \dfrac{6}{x - 2}$

In Probs. 13 to 20 solve algebraically. To solve these, you will need to invent your own methods.

13 $x^2 + y^2 = 25$
$x^2 + y^2 - 5x = 0$

14 $8x + 3y = 0$
$xy + 24 = 0$

15 $x^2 + y^2 = 25$
$x + y = 1$

16 $4x^2 - y^2 = 7$
$2x^2 + 3y^2 = 35$

17 $x^2 + 3xy + y^2 - 25 = 0$
$x = y$

18 $x^2 + 2xy + y^2 = 25$
$x^2 - 2xy + y^2 = 9$

19 $x - y = 1$
$y + z = 4$
$y^2 + xz = 7$

20 $xy = -12$
$yz = -4$
$xz = 3$

6.12 WORD PROBLEMS

As we pointed out in Chap. 1, the ability to solve a given set of equations is not the only mathematical skill which a person requires. He must also be able to translate his practical problems into mathematical terms. In order to develop this ability of translation from nature to mathematics, we include a set of "word problems" here. Since the real problems which you are likely to meet in practice are too complicated for you to handle at this stage, these word problems represent situations which have been greatly simplified. They are worth your attention, however, for by solving them you will be preparing yourself to handle less artificial problems.

The method of solution involves the following steps:

1. Let variables x, y, z, etc., represent the unknown quantities. If the problem permits, draw a sketch labeling the known and unknown parts.
2. Translate the statement of the problem into a set of equations involving the chosen variables.
3. Solve these equations for the required quantities.
4. Check the solution to make sure that it fits the conditions as stated in the original problem. In particular be sure that the solution makes sense practically; i.e., distances cannot be negative, numbers of people must be positive integers, etc.

ILLUSTRATION 1 Find the dimensions of a rectangle whose perimeter is 24 ft and whose area is 35 ft².

1. Let x be the length and y the width in feet. Draw Fig. 6.12.
2. Then
$$2x + 2y = 24$$
$$xy = 35$$
3. Solving, we obtain $y = 12 - x$ and $y = 35/x$. Hence
$$(12 - x)x = 35$$
or
$$x^2 - 12x + 35 = 0$$
$$(x - 5)(x - 7) = 0$$
$$x = 5 \text{ or } 7$$
Thus the solutions are $x = 5$, $y = 7$ and $x = 7$, $y = 5$. *Ans.*: 5 by 7 ft.
4. These check in the original problem.

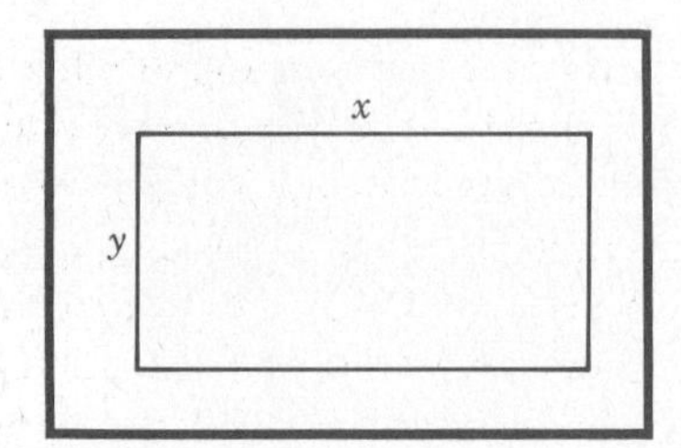

FIGURE 6.12

PROBLEMS 6.12

1. The sum of the digits of a certain two-digit number is 9. If the order of the digits is reversed, the number is increased by 45. What is the number?
2. One side of a rectangular field is 5 yd less than the other side, and the area of the field is 150 yd^2. Find the dimensions of the field.
3. Of two positive numbers, the first is three more than the second, and their product is 54. Find the numbers.
4. When 2 ft is subtracted from each side of a certain square, its area is decreased by 100 ft^2. What was the original area of the square?
5. A farmer wishes to fence a rectangular field with 260 yd of wire fence. If a stone wall runs along one side of the field, and no fence is required there, and if the area of the field is to be 8,400 yd^2, what are its dimensions?
6. A legislative assembly contains 135 members. If there are five more women than men, what are the numbers of men and women in the assembly.
7. In an election for the mayor of Oxbridge, the Conservative candidate received 5,819 more votes than the Labour candidate. A total of 18,653 votes was cast. How many people voted for the winner?
8. An eastbound, nonstop flight of 7,000 miles requires 10 hr. A similar westbound flight requires 14 hr. Assuming a constant westerly wind throughout, find the speed of the wind and the airspeed of the plane.
9. When two bricklayers, A and B, are working separately, A lays 4 more bricks/min than B. When they work together, each of their rates of laying drops to three-fourths of what it was when they worked alone, and together they lay 15 bricks/min. What are their rates of laying when they work separately?
10. A jet plane flies three times as fast as a propeller plane. Over a route of 3,000 miles, the propeller plane takes 10 hr longer than the jet plane. Find the speeds of the two planes.
11. The price of 4 milk shakes and 2 coffees is \$2.70. The cost of 1 milk shake and 3 coffees is \$1.05. Find the price of a milk shake and of a coffee.
12. The annual cost C of operating a new car is $C = f + cm$, where f is the fixed cost (depreciation, insurance, license, etc.), c is the oper-

ating cost per mile, and m is the number of miles driven. The total cost for 10,000 miles is \$2,000, and the cost for 15,000 miles is \$2,600. Find the fixed cost and the cost per mile.

13 At supermarkets in Suburbia the price of a pack of cigarettes includes a tax of 25 cents, which is the same for all brands. In Suburbia 5 packs of Notar cigarettes cost the same as 4 packs of Coffin Nail cigarettes. In the free port of Utopia there are no taxes, and the price of a pack of cigarettes is 25 cents lower than that in Suburbia. In Utopia 5 packs of Notar cigarettes cost the same as 3 packs of Coffin Nail cigarettes. Find the prices of the cigarettes in Suburbia.

14 A citizen of the nation of Nancago has an annual taxable income of \$11,880. The income tax rate in Nancago is 20 percent. Moreover, the province of Camford also imposes an income tax of 5 percent. The arrangement is that the national tax is based upon the annual taxable income less the provincial tax paid, and the provincial tax is based upon the annual taxable income less the national tax paid. Find the tax payable to Nancago and also that payable to its province, Camford.

15 In order to test the relative effectiveness of newspaper versus TV advertising, a detergent manufacturer advertised a new product, Nodirt, in city A by using 4 full-page newspaper ads and 15 TV spots. The result was the sale of 6,250 boxes of Nodirt. In city B of the same size, the use of 5 full-page ads and 10 TV spots produced the sale of 6,500 boxes. How many sales can be attributed to each newspaper ad and to each TV spot, assuming that there is no interaction between these.

16 At a time when the economy is balanced between boom and depression, an investor wishes to place \$14,000 in a suitable combination of bonds and stocks. He estimates that his expected gains (or losses) per \$100 invested in the next year are as given in the table below:

	Bonds	*Stocks*
Boom	$-\$10$	$+\$25$
Depression	$+\$5$	$-\$30$

Since he does not know whether a boom or a depression is coming, he decides to distribute his funds in such a way that he may expect to obtain the same result in either eventuality. How should he distribute his funds? Would he be better off to put his money in a savings bank?

17 A tourist has a collection of 33 coins consisting of Belgian francs (worth 2 cents each), British shillings (14 cents each), and German marks (25 cents each). The total value of the collection is \$6.33. He has three times as many marks as shillings. How many coins of each kind has he?

18 A man has two *major medical* insurance policies. Policy A will pay him 80 percent of the difference between the cost of an illness and the portion of this cost paid by policy B. Policy B will pay him 90 percent of the difference between the cost of an illness and the portion of this cost paid by policy A. If the total cost of an illness is \$2,520, how much of this cost is paid by each policy? How much must the man pay himself?

19 Neverstart automobiles use regular gasoline at 30 cents/gal and get 20 miles/gal. Everknock automobiles use premium gasoline at 34 cents/gal and get 17 miles/gal. In one day, the sum of the distances traveled by a Neverstart and by an Everknock was 91 miles, and the total cost for gasoline was \$1.62. How far did each go?

20 In Redtape College, all courses carry one *unit* credit, and *grade points* for each course are assigned as follows: A, 4 grade points; B, 3 points; C, 2 points; D, 1 point; E, 0 points. A student's *grade point average* is the quotient of the sum of his grade points divided by the number of units completed. After spending some time at Redtape, a student is enrolled for the current term in algebra and economics (only). If he receives a C in algebra and a D in economics, his grade point average at the end of this term will be 2.75. If he receives an A in algebra and a C in economics, it will be 3.00. Find his grade point average at the beginning of the current term.

6.13 TRANSFORMATION OF COORDINATES

If we have a coordinate system on a line with coordinates x, we can obtain a new coordinate system by means of the affine transformation

(1) $$x' = ax + b \qquad \text{where } a \neq 0$$

The effect of this is to give new names, x', to the points in place of their old names, x. In Probs. 2.12 we developed certain properties of such affine transformations and discussed the following two special cases:

$$x' = x + a \qquad\qquad \text{(Translation)}$$
$$x' = ax \qquad a \neq 0 \qquad \text{(Dilatation)}$$

Affine transformation

We observed that a translation merely shifts the origin, and that a dilatation multiplies lengths of segments by $|a|$. A general affine transformation combines these two operations into one.

Now we turn to the similar problem in the plane. If we have a coordinate system (x,y), we can define a new coordinate system (x',y') by means of the *affine transformation*.

$$x' = a_1x + b_1y + c_1$$
$$y' = a_2x + b_2y + c_2$$

where we assume $a_1b_2 - a_2b_1 \neq 0$. The new pairs (x',y') serve as new labels for the points in the plane. The point O', where $x' = 0$, $y' = 0$, is the new origin, the X'-axis is the line $y' = 0$, and the Y'-axis is the line $x' = 0$. These axes, however, need not be at right angles (Fig. 6.13).

A point P is called a *fixed point* if its coordinates in both systems are equal, i.e., if $x' = x$, $y' = y$ at P. If every point is a fixed point, the transformation has the equations $x' = x$, $y' = y$ and is called the *identity transformation*.

We shall develop the properties of these affine transformations in Probs. 6.13. In working some of these, you will need to anticipate the following result, which will be proved in Sec. 12.2. Let d be the distance between $P_1(x_1,y_1)$ and $P_2(x_2,y_2)$. Then

$$d^2 = (x_2 - x_1)^2 + (y_2 - y_1)^2 \tag{2}$$

This is really nothing but the Pythagorean Theorem.

Before entering into these details, we need to say a few words about the terminology to be used. We have defined affine transformations as transformations of the forms

$$x' = ax + b \qquad \text{and} \qquad \begin{cases} x' = a_1x + b_1y + c_1 \\ y' = a_2x + b_2y + c_2 \end{cases}$$

The term *affine* is used to express the fact that every finite point is transformed into a finite point. By way of contrast, the transformation $x' = 1/x$ is not affine, for the finite point $x = 0$ is transformed into infinity.

Since the equations of an affine transformation are linear, it might be thought reasonable to call such a transformation a *linear transformation*. The current practice, however, is to reserve the name *linear transformation* for the special cases of affine transformations whose equations are of the forms

$$x' = ax \qquad \text{and} \qquad \begin{cases} x' = a_1x + b_1y \\ y' = a_2x + b_2y \end{cases}$$

Oblique axes

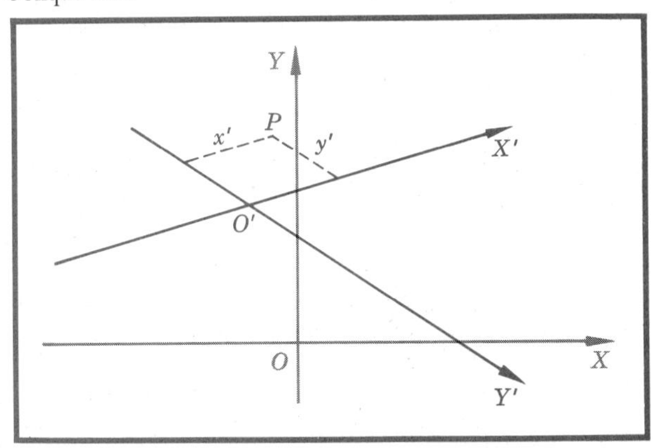

FIGURE 6.13
An affine transformation results in a new set of axes.

Thus all the transformations considered below are affine, and those in Probs. 9 to 26 are also linear.

PROBLEMS 6.13

Problems 1 to 4 refer to affine transformations on a line.

1. Find the affine transformation which relabels $x = 1$ with $x' = -1$, and $x = 4$ with $x' = 5$. HINT: Solve for a and b in Eq. (1).
2. Given that 0°C corresponds to 32°F and 100°C corresponds to 212°F, find the affine transformation that expresses F in terms of C.
3. In one grading system 50 is passing and 80 is perfect. In a second grading system 60 is passing and 100 is perfect. Find an affine transformation which converts the first grading system into the second and which takes passing into passing and perfect into perfect. What grade remains unchanged?
4. A faulty speedometer reads 15 miles/hr for an actual speed of 10 miles/hr, and 75 miles/hr for an actual speed of 80 miles/hr. Assuming an affine relationship, find a formula by means of which the correct speed can be calculated from the speedometer reading. For what speed is the speedometer correct?

Translation

Problems 5 to 26 refer to affine transformations in the plane. In Probs. 5 to 8 we take $x' = x + a$, $y' = y + b$. This transformation is called a *translation.*

5. Prove that a translation leaves the lengths of segments unchanged.
6. Prove: if a translation has a fixed point, then it is the identity transformation.
7. Show that the correspondence $(x,y) \leftrightarrow (x',y')$ defined by a translation is one-to-one.
8. For the translation $x' = x - 2$, $y' = y - 5$, find the new origin and sketch the new axes.

Dilatation

In Probs. 9 to 12 we take $x' = ax$, $y' = ay$, where $a \neq 0$. This transformation is called a *dilatation.*

9. Prove that a dilatation multiplies lengths of segments by $|a|$ and areas of rectangles by a^2. Hence show that a triangle is transformed into a similar triangle.
10. Prove that the origin is a fixed point under a dilatation.
11. Prove that, if a dilatation has a fixed point in addition to the origin, then it is the identity transformation.
12. Show that the correspondence $(x,y) \leftrightarrow (x',y')$ defined by a dilatation is one-to-one.

Reflection

In Probs. 13 to 15 we take $x' = -x, y' = y$. This transformation is called a *reflection* in the Y-axis.

13 Prove that a reflection leaves the lengths of segments unchanged.
14 Find the fixed points for the above reflection.
15 Prove that the correspondence $(x,y) \leftrightarrow (x',y')$ defined by a reflection is one-to-one.

In Probs. 16 to 20 we take

Rotation

$$\left.\begin{aligned} x' &= ax + by \\ y' &= -bx + ay \end{aligned}\right\} \quad \text{where } a^2 + b^2 = 1$$

This transformation is called a *rotation*.

16 Sketch the new axes when

$$x' = \frac{1}{\sqrt{2}}x + \frac{1}{\sqrt{2}}y$$

$$y' = -\frac{1}{\sqrt{2}}x + \frac{1}{\sqrt{2}}y$$

17 Prove that a rotation leaves the lengths of segments unchanged.
18 Prove that the origin is a fixed point under a rotation.
19 Prove that, if a rotation has a fixed point other than the origin, then it is the identity transformation. HINT: Solve

$$\left.\begin{aligned} x &= ax + by \\ y &= -bx + ay \end{aligned}\right\} \quad \text{for } a \text{ and } b$$

assuming $(x,y) \neq (0,0)$.
20 Prove that the correspondence $(x,y) \leftrightarrow (x',y')$ defined by a rotation is one-to-one.

Linear transformation

In Probs. 21 to 26 we consider the linear transformation

$$\left.\begin{aligned} x' &= a_1x + b_1y \\ y' &= a_2x + b_2y \end{aligned}\right\} \quad \text{where } a_1b_2 - a_2b_1 \neq 0$$

21 Prove that every point on the line $x + y = 0$ is a fixed point for the transformation

$$\begin{aligned} x' &= 2x + y \\ y' &= x + 2y \end{aligned}$$

HINT: Solve

$$\left.\begin{aligned} x &= 2x + y \\ y &= x + 2y \end{aligned}\right\} \quad \text{for } x \text{ and } y$$

22 Find the fixed points of the transformation

$$\begin{aligned} x' &= 3x - y \\ y' &= 2x \end{aligned}$$

23 Prove that the origin is the only fixed point of the general linear transformation unless $a_1b_2 - a_2b_1 - b_2 - a_1 + 1 = 0$.

24 Solve the equations of the general linear transformation for (x,y) in terms of (x',y'). This is the *inverse* transformation.

25 Prove that the correspondence $(x,y) \leftrightarrow (x',y')$ defined by such a linear transformation is one-to-one.

26 Consider the pair of transformations

$$x' = ax + by \qquad x'' = px' + qy'$$
$$y' = cx + dy \qquad y'' = rx' + sy'$$

Find (x'',y'') in terms of (x,y). This formula will be used in Sec. 7.5 to motivate our definition for the product of two matrices.

REFERENCE

Allendoerfer, C. B.: The Method of Equivalence, *The Mathematics Teacher,* vol. 59, pp. 531–535 (1966).

7 VECTORS AND MATRICES

7.1 » INTRODUCTION

In this chapter we shall study a new kind of algebra in which the elements are as follows:

Vector

1 Ordered pairs like (x,y), ordered triples like (x,y,z). We call these vectors.
2 Rectangular arrays like those in the coefficients of our simultaneous equations, such as

$$\begin{pmatrix} a_1 & b_1 \\ a_2 & b_2 \end{pmatrix} \quad \begin{pmatrix} a_1 & b_1 & c_1 \\ a_2 & b_2 & c_2 \end{pmatrix}$$

$$\begin{pmatrix} a_1 & b_1 & c_1 \\ a_2 & b_2 & c_2 \\ a_3 & b_3 & c_3 \end{pmatrix} \begin{pmatrix} a_1 & b_1 & c_1 & d_1 \\ a_2 & b_2 & c_2 & d_2 \\ a_3 & b_3 & c_3 & d_3 \end{pmatrix}$$

We call these matrices.

DEFINITION
Matrix

A *matrix* is any rectangular (or square) array of numbers, and a *vector* is a special case of a matrix which has only one row or one column.

7.2 » VECTORS

Let us begin by restricting ourselves to vectors. You have probably met vectors before in your study of physics, and may wonder about the connection between the vectors of physics and those defined above. In physics a vector is represented in the plane as a directed distance $\overrightarrow{PQ}$ and is said to have magnitude and direction. The magnitude is represented by the length of the line segment in the plane, and the direction is given by the angle which this line makes with the horizontal and by the sense in which the arrow points. Common examples of vectors in physics are velocity, acceleration, and force.

Corresponding to the vector $\overrightarrow{PQ}$, we may draw a right triangle PQR (Fig. 7.1) with PR horizontal and QR vertical. The length of PR is the "x-component a" of $\overrightarrow{PQ}$; a is positive if $\overrightarrow{PQ}$ points to the right and is negative if $\overrightarrow{PQ}$ points to the left. Similarly RQ is the "y-component b" of $\overrightarrow{PQ}$; b is positive if $\overrightarrow{PQ}$ points up and negative if $\overrightarrow{PQ}$ points down. Clearly these components are known if the vector is known, and, conversely, a pair of components determines a vector. To simplify the discussion, we shall suppose that all of our vectors have the origin O as their initial point, so that the coordinates of their endpoints are equal to the components of the vectors. Then any vector is determined by the ordered pair of numbers (a,b). In the same way vectors in space have three components and are determined by a triple (a,b,c).

This gives us the connection between our vector and the vectors of physics. You should note, however, that every physical vector can be represented by a pair or a triple, but that vectors as we have defined them do not necessarily have physical interpretations. This is a good example of a mathematical concept which has arisen as a generalization of a concrete physical object.

NOTATION FOR VECTORS We shall write our vectors as "row vectors": (a,b) or (a,b,c) or as "column vectors": $\begin{pmatrix} a \\ b \end{pmatrix}$ or $\begin{pmatrix} a \\ b \\ c \end{pmatrix}$. There is no real distinction between row vectors and column vectors, but it will be convenient to use both notations in the applications which follow. Sometimes we use a single boldface letter such as $\mathbf{a}$ or $\mathbf{i}$ to stand for a vector. An equation of the form $\mathbf{a} = (a,b,c)$ means that $\mathbf{a}$ and (a,b,c) are two different names for the same vector.

Components of a vector

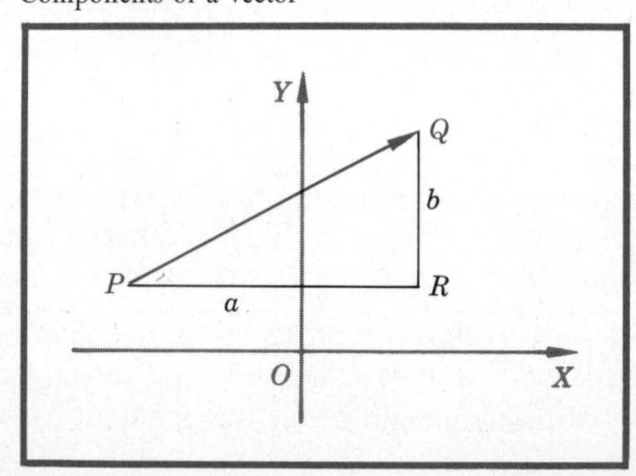

FIGURE 7.1
A vector can be resolved into horizontal and vertical components.

Addition

ADDITION OF VECTORS Since vectors are not numbers, the sum of two vectors is a new idea and must be defined.

DEFINITION The *sum of two vectors* is defined by the formulas

$$(a,b) + (c,d) = (a + c, b + d)$$
$$(a,b,c) + (d,e,f) = (a + d, b + e, c + f)$$

Thus, to add two vectors of the same dimension we add their corresponding components. This has an important geometric interpretation, which we illustrate in the plane (Fig. 7.2). In order to add $\overrightarrow{OP}$ to $\overrightarrow{OQ}$, we find point R, which is the fourth vertex of the parallelogram having O, P, and Q as its other vertices. Then triangle PRS is congruent to triangle OQT so that $PS = c$ and $RS = d$. Now $OU = a$, and OV is the x-component of $\overrightarrow{OR}$. From the figure

$$\begin{aligned} OV &= OU + UV \\ &= OU + PS \\ &= a + c \end{aligned}$$

Similarly $RV = b + d$. Thus $\overrightarrow{OR} = \overrightarrow{OP} + \overrightarrow{OQ}$. This interpretation is the source of the graphical method for adding vectors which is used widely in physics and navigation.

You will observe that this process is identical with that which we described in Sec. 2.14 for the graphical addition of two complex numbers. Although this shows that complex numbers and two-dimensional vectors have the same rules for addition, we must warn you that their rules for multiplication are completely different. Do not be misled into the common error of confusing vectors with complex numbers.

Addition of vectors

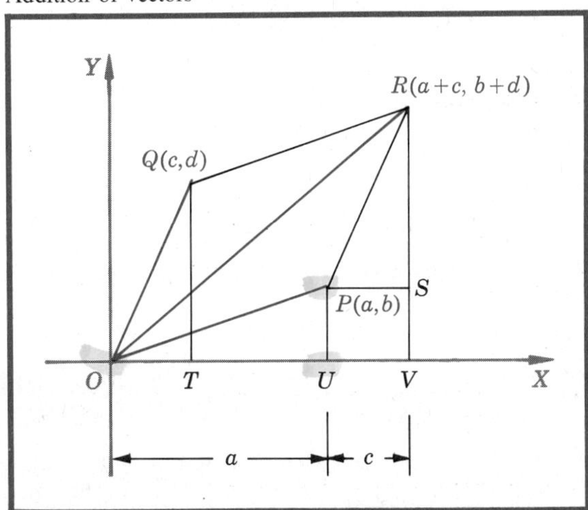

FIGURE 7.2
The sum of two vectors is the vector obtained by completing the parallelogram.

12 24
3, 6

ILLUSTRATION 1

a $(1,-3,2) + (3,4,-1) = (4,1,1)$
b $(5,2,-1) + (-5,-2,1) = (0,0,0)$, the "zero vector"
c $(-4,7,3) - (2,-1,4) = (-6,8,-1)$
d $(6,4,3) - (6,4,3) = (0,0,0)$

7.3 PRODUCTS OF VECTORS

When we are speaking of vectors, we shall refer to an ordinary real number as a *scalar*. We now define the product of a scalar times a vector.

DEFINITION If (a,b,c) is a vector and k is a scalar, the *product* $k(a,b,c)$ is defined to be the vector (ka,kb,kc).

ILLUSTRATION 1

a $2(1,3,-4) = (2,6,-8)$
b $-1(2,1,3) = (-2,-1,-3)$
c $0(a,b,c) = (0,0,0)$

It is often useful to define three *base vectors* **i**, **j**, and **k** as follows. These are vectors of length 1 drawn along the positive directions of the three coordinate axes.

DEFINITION $\mathbf{i} = (1,0,0)$; $\mathbf{j} = (0,1,0)$; $\mathbf{k} = (0,0,1)$

In terms of these, any vector (a,b,c) can be written

$$(a,b,c) = a\mathbf{i} + b\mathbf{j} + c\mathbf{k}$$

Although this notation is quite common in physics and engineering, we shall not use it regularly in this book.

EXERCISE A Using the definitions of **i**, **j**, and **k**, show that (a,b,c) is correctly expressed as $a\mathbf{i} + b\mathbf{j} + c\mathbf{k}$.

Inner product

The product of a vector by a vector is another concept which needs definition. There are, in fact, three kinds of products in common use; but here we shall discuss only the inner (or "scalar", or "dot") product.

DEFINITION The *inner product* of two vectors (a_1,b_1,c_1) and (a_2,b_2,c_2) is defined to be the scalar $a_1a_2 + b_1b_2 + c_1c_2$. This product is denoted by a dot, so that

$$(a_1,b_1,c_1) \cdot (a_2,b_2,c_2) = a_1a_2 + b_1b_2 + c_1c_2$$

ILLUSTRATION 2

a $(3,1,-2) \cdot (1,3,4) = (3)(1) + (1)(3) + (-2)(4) = -2$
b $(5,2,6) \cdot (1,1,1) = 5 + 2 + 6 = 13$
c $(-4,1,7) \cdot (0,0,0) = 0 + 0 + 0 = 0$
d $\mathbf{i} \cdot \mathbf{j} = (1,0,0) \cdot (0,1,0) = 0 + 0 + 0 = 0$

Length

In terms of inner products we can define the *length* of a vector.

DEFINITION The *length* of a vector (a,b,c) is the positive square root of the inner product $(a,b,c) \cdot (a,b,c)$. That is:

$$\text{Length of } (a,b,c) = \sqrt{(a,b,c) \cdot (a,b,c)} = \sqrt{a^2 + b^2 + c^2}$$

ILLUSTRATION 3

a The length of $(3,2,4) = \sqrt{9 + 4 + 16} = \sqrt{29}$
b The length of $\mathbf{i} = \sqrt{1 + 0 + 0} = 1$
c The length of $(0,0,0) = \sqrt{0 + 0 + 0} = 0$

The importance of the inner product in physics lies in the following geometrical interpretation, which you can understand if you have an elementary knowledge of trigonometry:

The inner product $(a_1,b_1,c_1) \cdot (a_2,b_2,c_2)$ is equal to the length of (a_1,b_1,c_1) times the length of (a_2,b_2,c_2) times the cosine of the angle between these two vectors. This statement is equivalent to the Law of Cosines for a triangle (see Sec. 12.13). Since $\cos 90° = 0$, it follows that (a_1,b_1,c_1) is perpendicular to (a_2,b_2,c_2) if and only if their inner product is zero.

Perpendicular vectors

Physical concepts are frequently defined in terms of the inner product. For example, if a force $\mathbf{F} = (f_1,f_2,f_3)$ in pounds acts during a displacement $\mathbf{s} = (s_1,s_2,s_3)$ in feet, the work W which is done is defined to be

$$W = \mathbf{F} \cdot \mathbf{s} = (f_1,f_2,f_3) \cdot (s_1,s_2,s_3) \qquad \text{ft-lb}$$

PROBLEMS 7.3

In Probs. 1 to 6 add the given vectors algebraically, and check your result graphically.

1 $(2,3) + (-1,4)$ **2** $(4,-2) + (-2,4)$
3 $(-4,7) + (5,-9)$ **4** $(2,6) + (-6,9)$
5 $(7,2) - (3,5)$ **6** $(8,-2) - (4,-6)$

In Probs. 7 to 12 write a vector equal to the given expression.

7 $(2,1,4) + (-3,6,5) - (4,-1,3)$
8 $(0,2,-1) - (7,-3,6) + (2,-4,3)$

9 $2(2,1,1) + 3(0,4,6) + 4(2,-4,1)$
10 $-3(1,6,2) + 4(3,1,-2) - 2(2,-1,3)$
11 $4(2\mathbf{i} - 3\mathbf{j} + 5\mathbf{k}) + 3(4\mathbf{i} + 2\mathbf{j} - \mathbf{k})$
12 $-2(\mathbf{i} - 2\mathbf{j} + \mathbf{k}) + 5(2\mathbf{i} + \mathbf{j} - \mathbf{k})$

In Probs. 13 to 16 prove the given statement.

13 Addition of vectors is commutative.
14 Addition of vectors is associative.
15 The zero vector, $(0,0,0)$, is the additive identity for vectors.
16 The vector $(-a,-b,-c)$ is the additive inverse of (a,b,c).

In Probs. 17 to 22 compute the indicated inner products.

17 $(1,3,2) \cdot (2,-1,4)$
18 $(5,-2,1) \cdot (2,3,5)$
19 $(1,3,4) \cdot (6,2,-1)$
20 $(2,-5,4) \cdot (0,0,3)$
21 $\mathbf{i} \cdot \mathbf{j}$
22 $\mathbf{i} \cdot \mathbf{k}$

In Probs. 23 to 30 find the length of the given vector.

23 $(2,1,4)$
24 $(1,-3,5)$
25 $(3,0,1)$
26 $(4,-2,5)$
27 $\mathbf{i}$
28 $\mathbf{k}$
29 $4\mathbf{i} + 2\mathbf{j} - \mathbf{k}$
30 $2\mathbf{i} - \mathbf{j} + 3\mathbf{k}$

In Probs. 31 to 33 prove the given statement.

31 In the multiplication of a scalar times a vector the following distributive laws hold:

$$k[(a_1,b_1,c_1) + (a_2,b_2,c_2)] = k(a_1,b_1,c_1) + k(a_2,b_2,c_2)$$

and

$$(k_1 + k_2)(a,b,c) = k_1(a,b,c) + k_2(a,b,c)$$

32 The inner product is commutative.
33 For the inner product the following distributive law holds:

$$[(a_1,b_1,c_1) + (a_2,b_2,c_2)] \cdot (a_3,b_3,c_3) = (a_1,b_1,c_1) \cdot (a_3,b_3,c_3) + (a_2,b_2,c_2) \cdot (a_3,b_3,c_3)$$

7.4 › MATRICES

As we have said above, a matrix is a square or rectangular array of numbers. The numbers of which a matrix is composed are called its elements. You are already familiar with many examples of them, such as the statistical tables which compose the bulk of the "World Almanac". In mathematics, matrices first appeared as the arrays of coefficients in simultaneous linear equations. In physics they are widely used in quantum theory and appear in elementary physics as (1) the set of moments and products of inertia of a rigid body or (2) the set of pressures at a point in a viscous fluid.

Although matrices may be of any dimensions, in this book we shall deal only with those of dimensions 2×2 (that is, two rows and two columns), 2×3, 3×2, and 3×3. As special cases we have already discussed vectors, which are matrices of dimensions $2 \times 1, 3 \times 1, 1 \times 2$, and 1×3. We shall now develop the elementary algebra of matrices.

DEFINITION Two matrices are *equal* if and only if they have the same dimensions and are *identical.*

That is, for example,

$$\begin{pmatrix} a & b \\ c & d \end{pmatrix} = \begin{pmatrix} x & y \\ z & w \end{pmatrix}$$

if and only if

$$a = x \qquad b = y \qquad c = z \qquad d = w$$

Addition

The sum of two matrices is analogous to the sum of two vectors.

DEFINITION The *sum* of two matrices of the same dimensions is a matrix whose elements are the sums of the corresponding elements of the given matrices.

For example,

$$\begin{pmatrix} a & b \\ c & d \end{pmatrix} + \begin{pmatrix} x & y \\ z & w \end{pmatrix} = \begin{pmatrix} a + x & b + y \\ c + z & d + w \end{pmatrix}$$

DEFINITION Zero matrix

A matrix (of any pair of dimensions) each of whose elements is zero is called the *zero matrix* (for that pair of dimensions).

Examples of zero matrices are

$$(0) \qquad \begin{pmatrix} 0 \\ 0 \\ 0 \end{pmatrix} \qquad \begin{pmatrix} 0 & 0 \\ 0 & 0 \\ 0 & 0 \end{pmatrix} \qquad \begin{pmatrix} 0 & 0 \\ 0 & 0 \end{pmatrix}$$

Again, as for vectors, we can define the product of a scalar times a matrix.

DEFINITION The *product* of a scalar k times a matrix is a matrix whose elements are k times the corresponding elements of the given matrix.

For example,

$$k\begin{pmatrix} a & b \\ c & d \end{pmatrix} = \begin{pmatrix} ka & kb \\ kc & kd \end{pmatrix}$$

7.5 › PRODUCTS OF MATRICES

We now turn our attention to the product of two matrices, when this can be defined. This concept is a generalization of the inner product of two vectors.

In order to motivate this definition, let us consider the pair of 2×2 matrices which are formed from the coefficients of the linear transformations (Sec. 6.13):

$$\begin{aligned} x' &= ax + by \qquad & x'' &= px' + qy' \\ y' &= cx + dy \qquad & y'' &= rx' + sy' \end{aligned}$$

namely:

$$\begin{pmatrix} a & b \\ c & d \end{pmatrix} \quad \text{and} \quad \begin{pmatrix} p & q \\ r & s \end{pmatrix}$$

In Sec. 6.13, Prob. 26, we saw that the combination of these two linear transformations is

$$\begin{aligned} x'' &= (ap + cq)x + (bp + dq)y \\ y'' &= (ar + cs)x + (br + ds)y \end{aligned}$$

If you did not work this problem previously, do so now, to verify this computation. The matrix of the coefficients of the combined linear transformation is then

$$\begin{pmatrix} ap + cq & bp + dq \\ ar + cs & br + ds \end{pmatrix}$$

The question, now, is: "How are these three matrices related?" We shall say that the third matrix is the *product* of the first two and shall write

$$\begin{pmatrix} p & q \\ r & s \end{pmatrix}\begin{pmatrix} a & b \\ c & d \end{pmatrix} = \begin{pmatrix} ap + cq & bp + dq \\ ar + cs & br + ds \end{pmatrix}$$

From this equation we observe that:

1 The element $ap + cq$ in the first row and first column of the product is equal to the inner product of the first row vector (p,q) of the first factor with the first column vector $\begin{pmatrix} a \\ c \end{pmatrix}$ of the second factor.

2 The element $bp + dq$ in the first row and second column of the product is equal to the inner product of the first row vector (p,q) of the first factor with the second column vector $\begin{pmatrix} b \\ d \end{pmatrix}$ of the second factor.

3 Similarly, $$ar + cs = (r,s) \cdot \begin{pmatrix} a \\ c \end{pmatrix}$$

and $$br + ds = (r,s) \cdot \begin{pmatrix} b \\ d \end{pmatrix}$$

As another piece of motivation, let us see how we can combine (by multiplication) the matrices

$$\begin{pmatrix} a & b \\ c & d \end{pmatrix} \quad \text{and} \quad \begin{pmatrix} x \\ y \end{pmatrix}$$

to obtain the column vector $\begin{pmatrix} ax + by \\ cx + dy \end{pmatrix}$.

We shall write

$$\begin{pmatrix} a & b \\ c & d \end{pmatrix}\begin{pmatrix} x \\ y \end{pmatrix} = \begin{pmatrix} ax + by \\ cx + dy \end{pmatrix}$$

Notice that

$$ax + by = (a,b) \cdot \begin{pmatrix} x \\ y \end{pmatrix}$$

$$cx + dy = (c,d) \cdot \begin{pmatrix} x \\ y \end{pmatrix}$$

Perhaps the pattern is becoming clear, but before giving a precise definition of this product, let us consider some more examples.

ILLUSTRATION 1

a $$\begin{pmatrix} a_1 & b_1 & c_1 \\ a_2 & b_2 & c_2 \\ a_3 & b_3 & c_3 \end{pmatrix}\begin{pmatrix} x \\ y \\ z \end{pmatrix} = \begin{pmatrix} a_1x + b_1y + c_1z \\ a_2x + b_2y + c_2z \\ a_3x + b_3y + c_3z \end{pmatrix}$$

b $$\begin{pmatrix} 1 & -2 & 3 \\ 2 & 1 & 4 \end{pmatrix}\begin{pmatrix} 2 \\ 1 \\ 5 \end{pmatrix} = \begin{pmatrix} 15 \\ 25 \end{pmatrix}$$

c $$\begin{pmatrix} 2 & -3 \\ 1 & 4 \end{pmatrix}\begin{pmatrix} 3 \\ -2 \end{pmatrix} = \begin{pmatrix} 12 \\ -5 \end{pmatrix}$$

d $$\begin{pmatrix} 4 & 1 & -2 \\ 3 & 2 & 5 \\ -1 & 2 & 1 \end{pmatrix}\begin{pmatrix} 1 \\ 3 \\ -1 \end{pmatrix} = \begin{pmatrix} 9 \\ 4 \\ 4 \end{pmatrix}$$

ILLUSTRATION 2 In view of the above,

$$\begin{pmatrix} a_1 & b_1 \\ a_2 & b_2 \end{pmatrix}\begin{pmatrix} x \\ y \end{pmatrix} = \begin{pmatrix} a_1x + b_1y \\ a_2x + b_2y \end{pmatrix}$$

and $$\begin{pmatrix} a_1 & b_1 \\ a_2 & b_2 \end{pmatrix}\begin{pmatrix} x \\ y \end{pmatrix} + \begin{pmatrix} c_1 \\ c_2 \end{pmatrix} = \begin{pmatrix} a_1x + b_1y + c_1 \\ a_2x + b_2y + c_2 \end{pmatrix}$$

We can, therefore, write the system of simultaneous equations

$$\begin{aligned} a_1x + b_1y &= c_1 \\ a_2x + b_2y &= c_2 \end{aligned}$$

in the compact form

$$AX = C$$

where $\quad A = \begin{pmatrix} a_1 & b_1 \\ a_2 & b_2 \end{pmatrix} \quad X = \begin{pmatrix} x \\ y \end{pmatrix} \quad C = \begin{pmatrix} c_1 \\ c_2 \end{pmatrix}$

EXERCISE A Show that the simultaneous system

$$\begin{aligned} a_1x + b_1y + c_1z &= d_1 \\ a_2x + b_2y + c_2z &= d_2 \\ a_3x + b_3y + c_3z &= d_3 \end{aligned}$$

can be written in the form

$$AX = D$$

where $\quad A = \begin{pmatrix} a_1 & b_1 & c_1 \\ a_2 & b_2 & c_2 \\ a_3 & b_3 & c_3 \end{pmatrix} \quad X = \begin{pmatrix} x \\ y \\ z \end{pmatrix} \quad D = \begin{pmatrix} d_1 \\ d_2 \\ d_3 \end{pmatrix}$

We are now ready to define the product $AB = C$ of a $p \times q$-dimensional matrix A and a $q \times r$-dimensional matrix B. A consists of p q-dimensional row vectors, and B consists of r q-dimensional column vectors. The elements of C are the inner products of the row vectors of A times the column vectors of B.

DEFINITION Product

Let A be a $p \times q$-dimensional matrix and B a $q \times r$-dimensional matrix. Their *product* $AB = C$ is a $p \times r$-dimensional matrix whose elements are as follows: The element in the ith row and the jth column of C is the inner product of the ith row vector of A with the jth column vector of B.

ILLUSTRATION 3

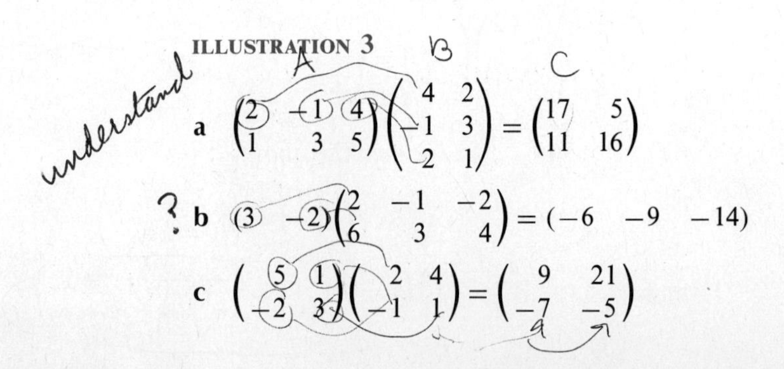

a $\begin{pmatrix} 2 & -1 & 4 \\ 1 & 3 & 5 \end{pmatrix} \begin{pmatrix} 4 & 2 \\ -1 & 3 \\ 2 & 1 \end{pmatrix} = \begin{pmatrix} 17 & 5 \\ 11 & 16 \end{pmatrix}$

b $(3 \quad -2) \begin{pmatrix} 2 & -1 & -2 \\ 6 & 3 & 4 \end{pmatrix} = (-6 \quad -9 \quad -14)$

c $\begin{pmatrix} 5 & 1 \\ -2 & 3 \end{pmatrix} \begin{pmatrix} 2 & 4 \\ -1 & 1 \end{pmatrix} = \begin{pmatrix} 9 & 21 \\ -7 & -5 \end{pmatrix}$

REMARKS

1 The product AB is in this order: A first, B second. The necessity for this follows from the definition in which A and B are treated differently; we multiply the *rows* of A by the *columns* of B.
2 The product AB is defined only when the dimension of the row vectors of A equals that of the column vectors of B. That is, the number of columns in A must equal the number of rows in B.
3 When A and B are square and of the same dimension, both AB and BA are defined. However, in general, AB does not equal BA; that is, multiplication of square matrices is *not* commutative.

ILLUSTRATION 4

$$\begin{pmatrix} 2 & 1 \\ -3 & 4 \end{pmatrix}\begin{pmatrix} 1 & 3 \\ 5 & -1 \end{pmatrix} = \begin{pmatrix} 7 & 5 \\ 17 & -13 \end{pmatrix}$$

but
$$\begin{pmatrix} 1 & 3 \\ 5 & -1 \end{pmatrix}\begin{pmatrix} 2 & 1 \\ -3 & 4 \end{pmatrix} = \begin{pmatrix} -7 & 13 \\ 13 & 1 \end{pmatrix}$$

On the other hand, it can be shown that matrix multiplication *is* associative; that is, $A(BC) = (AB)C$.

There is a multiplicative identity for matrices which is described in the following definition.

DEFINITION The square matrices

$$I_2 = \begin{pmatrix} 1 & 0 \\ 0 & 1 \end{pmatrix} \quad \text{and} \quad I_3 = \begin{pmatrix} 1 & 0 & 0 \\ 0 & 1 & 0 \\ 0 & 0 & 1 \end{pmatrix}$$

Identity matrix — are called *identity matrices.*

The name *identity matrix* follows from the following theorems.

THEOREM 1 If A is a matrix having two columns and any number of rows, then $AI_2 = A$. Similarly, if A has two rows and any number of columns, then $I_2A = A$.

PROOF For example, let

$$A = \begin{pmatrix} a & b \\ c & d \\ e & f \end{pmatrix}$$

Then
$$AI_2 = \begin{pmatrix} a & b \\ c & d \\ e & f \end{pmatrix}\begin{pmatrix} 1 & 0 \\ 0 & 1 \end{pmatrix} = \begin{pmatrix} a & b \\ c & d \\ e & f \end{pmatrix} = A$$

Similarly, let

$$A = \begin{pmatrix} a & c & e \\ b & d & f \end{pmatrix}$$

Then

$$I_2A = \begin{pmatrix} 1 & 0 \\ 0 & 1 \end{pmatrix}\begin{pmatrix} a & c & e \\ b & d & f \end{pmatrix} = \begin{pmatrix} a & c & e \\ b & d & f \end{pmatrix} = A$$

EXERCISE B If $A = (a \quad b)$, show that $AI_2 = A$.

EXERCISE C If $A = \begin{pmatrix} a & b \\ c & d \end{pmatrix}$, show that $AI_2 = A$.

EXERCISE D If $A = \begin{pmatrix} a \\ b \end{pmatrix}$, show that $I_2A = A$.

EXERCISE E If $A = \begin{pmatrix} a & b \\ c & d \end{pmatrix}$, show that $I_2A = A$.

There is an analogous theorem for I_3.

THEOREM 2 If A is a matrix having three columns and any number of rows, then $AI_3 = A$. Similarly, if A has three rows and any number of columns, then $I_3A = A$.

The proof is the same as that for Theorem 1.

REMARK Whenever the dimension of the identity matrix is clear from the context, we shall omit the subscripts on I_2, I_3, etc., and merely write I.

PROBLEMS 7.5

In Probs. 1 to 6 carry out the indicated operations where possible.

1 $\begin{pmatrix} 3 & -2 \\ 1 & 4 \end{pmatrix} + \begin{pmatrix} 2 & 5 \\ -3 & 6 \end{pmatrix}$ **2** $\begin{pmatrix} 4 & -1 \\ -3 & 6 \end{pmatrix} + \begin{pmatrix} 3 & 7 \\ 0 & -4 \end{pmatrix}$

3 $\begin{pmatrix} 5 & -2 \\ -4 & 3 \end{pmatrix} - \begin{pmatrix} 6 & 8 \\ -5 & -2 \end{pmatrix}$ **4** $\begin{pmatrix} -2 & 5 \\ 3 & 4 \end{pmatrix} - \begin{pmatrix} 7 & 6 \\ 2 & -3 \end{pmatrix}$

5 $2\begin{pmatrix} 4 & 2 \\ 3 & 1 \end{pmatrix} + 3\begin{pmatrix} 2 & 1 \\ 0 & 3 \end{pmatrix}$ **6** $4\begin{pmatrix} 1 & 4 \\ 2 & -1 \end{pmatrix} - 5\begin{pmatrix} 3 & -1 \\ -4 & 5 \end{pmatrix}$

In Probs. 7 to 10 prove the given statement.

7 Addition of matrices is commutative.
8 Addition of matrices is associative.

9 The zero matrix, $\begin{pmatrix} 0 & 0 \\ 0 & 0 \end{pmatrix}$, is the additive identity for 2×2 matrices.

10 The matrix $\begin{pmatrix} -a & -b \\ -c & -d \end{pmatrix}$ is the additive inverse of the matrix $\begin{pmatrix} a & b \\ c & d \end{pmatrix}$.

In Probs. 11 to 34 find the products of the given matrices.

11 $(1 \quad -2 \quad 3)\begin{pmatrix} 2 \\ 1 \\ 5 \end{pmatrix}$

12 $(3 \quad 1 \quad -4)\begin{pmatrix} 1 \\ 3 \\ -6 \end{pmatrix}$

13 $\begin{pmatrix} -1 & 2 & 5 \\ 3 & 4 & -1 \end{pmatrix}\begin{pmatrix} 2 \\ -3 \\ 1 \end{pmatrix}$

14 $\begin{pmatrix} 3 & 8 & -2 \\ 4 & 0 & 6 \end{pmatrix}\begin{pmatrix} 2 \\ 5 \\ -3 \end{pmatrix}$

15 $\begin{pmatrix} 2 & 8 & -3 \\ 6 & 0 & 2 \end{pmatrix}\begin{pmatrix} -3 & 2 \\ 5 & 2 \\ 1 & 1 \end{pmatrix}$

16 $\begin{pmatrix} 4 & -1 & 1 \\ 2 & 3 & 0 \end{pmatrix}\begin{pmatrix} 7 & -3 \\ 5 & 4 \\ 1 & 2 \end{pmatrix}$

17 $\begin{pmatrix} 4 & -1 \\ 2 & 5 \end{pmatrix}\begin{pmatrix} 2 & 4 \\ 1 & 6 \end{pmatrix}$

18 $\begin{pmatrix} 3 & -2 \\ 1 & 4 \end{pmatrix}\begin{pmatrix} 4 & 5 \\ -6 & 2 \end{pmatrix}$

19 $\begin{pmatrix} 4 & 1 \\ -2 & 3 \end{pmatrix}\begin{pmatrix} 2 & 4 \\ 1 & -3 \end{pmatrix}$

20 $\begin{pmatrix} -5 & 2 \\ 1 & 3 \end{pmatrix}\begin{pmatrix} 3 & 6 \\ 1 & -2 \end{pmatrix}$

21 $\begin{pmatrix} 5 & 3 \\ 2 & 4 \\ 1 & 7 \end{pmatrix}\begin{pmatrix} 2 & -1 \\ 4 & 0 \end{pmatrix}$

22 $\begin{pmatrix} 4 & 1 \\ 0 & -2 \\ -7 & 3 \end{pmatrix}\begin{pmatrix} 2 & -3 \\ 1 & 4 \end{pmatrix}$

23 $\begin{pmatrix} 1 \\ -5 \\ 2 \end{pmatrix}(2 \quad 1 \quad 4)$ (The resulting 3×3 matrix is the *tensor product* of the two vectors.)

24 $\begin{pmatrix} -1 \\ 0 \\ 5 \end{pmatrix}(2 \quad 3 \quad -1)$

25 $\begin{pmatrix} 5 & 2 & 1 \\ -3 & 1 & 7 \\ 0 & 1 & 2 \end{pmatrix}\begin{pmatrix} 2 & -1 & 4 \\ 4 & -3 & 1 \\ 1 & 2 & 1 \end{pmatrix}$

26 $\begin{pmatrix} 3 & -1 & 5 \\ 2 & 0 & 2 \\ 1 & 1 & 1 \end{pmatrix}\begin{pmatrix} 1 & 2 & -3 \\ 1 & -2 & 4 \\ 1 & -3 & 1 \end{pmatrix}$

27 $\begin{pmatrix} -2 & 1 & 3 \\ 2 & 0 & 4 \\ 1 & 2 & -3 \end{pmatrix}\begin{pmatrix} -8 & 9 & 4 \\ 10 & 3 & 14 \\ 4 & 5 & -2 \end{pmatrix}$

28 $\begin{pmatrix} 2 & 3 & -5 \\ -1 & 1 & -4 \\ 4 & 3 & -2 \end{pmatrix}\begin{pmatrix} 10 & -9 & -7 \\ -18 & 16 & 13 \\ -7 & 6 & 5 \end{pmatrix}$

29 $\begin{pmatrix} 2 & 2 \\ -1 & 3 \\ 4 & 1 \end{pmatrix}\begin{pmatrix} 1 & 0 \\ 0 & 1 \end{pmatrix}$

30 $\begin{pmatrix} 1 & 0 & 0 \\ 0 & 1 & 0 \\ 0 & 0 & 1 \end{pmatrix}\begin{pmatrix} 3 & 2 \\ 1 & 6 \\ -4 & 5 \end{pmatrix}$

31 $\begin{pmatrix} 1 & 2 \\ -2 & -4 \end{pmatrix}\begin{pmatrix} 4 & 6 \\ -2 & -3 \end{pmatrix}$ Could this happen for the product of real numbers?

32 $\begin{pmatrix} 1 & 4 \\ -2 & -8 \end{pmatrix}\begin{pmatrix} 4 & 4 \\ -1 & -1 \end{pmatrix}$ Why is this result surprising?

33 $\begin{pmatrix} 1 & 0 & 2 \\ 3 & 4 & -2 \end{pmatrix}\begin{pmatrix} 1 & 2 \\ 4 & 0 \\ -3 & 1 \end{pmatrix}$ and $\begin{pmatrix} 1 & 2 \\ 4 & 0 \\ -3 & 1 \end{pmatrix}\begin{pmatrix} 1 & 0 & 2 \\ 3 & 4 & -2 \end{pmatrix}$

34 $\begin{pmatrix} 2 & -1 & 1 \\ 1 & 3 & -2 \end{pmatrix}\begin{pmatrix} 3 & 4 \\ 0 & 5 \\ 2 & 0 \end{pmatrix}$ and $\begin{pmatrix} 3 & 4 \\ 0 & 5 \\ 2 & 0 \end{pmatrix}\begin{pmatrix} 2 & -1 & 1 \\ 1 & 3 & -2 \end{pmatrix}$

***35** Why do $A = \begin{pmatrix} a & b & c \\ d & e & f \end{pmatrix}$ and $B = \begin{pmatrix} r & u \\ s & v \\ t & w \end{pmatrix}$ not commute under multiplication? Do not perform the multiplication.

***36** Why do $(a \quad b \quad c)$ and $\begin{pmatrix} d \\ e \\ f \end{pmatrix}$ not commute under multiplication?

37 Let $A = \begin{pmatrix} 2 & 1 \\ 3 & 5 \end{pmatrix}$. Compute $A^2 - 7A + 7I_2$.

38 Let $A = \begin{pmatrix} 1 & 3 \\ -1 & 5 \end{pmatrix}$. Compute $A^2 - 6A + 8I_2$.

39 Let $A = \begin{pmatrix} 1 & 0 \\ 0 & -1 \end{pmatrix}$. Compute $A^4 - I_2$.

***40** Let $A = \begin{pmatrix} a & -b \\ b & a \end{pmatrix}$ and $B = \begin{pmatrix} c & -d \\ d & c \end{pmatrix}$. Compute $A + B$ and AB. Compare your results with the sum and product of the complex numbers $a + bi$ and $c + di$.

In Probs. 41 to 44 write the system of linear equations in matrix form.

41 $\begin{aligned} 2x - 5y &= -8 \\ x + 3y &= 7 \end{aligned}$

42 $\begin{aligned} 3x + y &= 8 \\ 2x - y &= 7 \end{aligned}$

43 $\begin{aligned} 2x - 3y + 4z &= 3 \\ x - 2y + z &= 0 \\ 4x + 5y - 5z &= 4 \end{aligned}$

44 $\begin{aligned} x \qquad - z &= 0 \\ y + z &= 1 \\ 2x - y \qquad &= 5 \end{aligned}$

45 Write the systems

$$\begin{aligned} x' &= ax + by \\ y' &= cx + dy \end{aligned} \qquad \text{and} \qquad \begin{aligned} x'' &= px' + qy' \\ y'' &= rx' + sy' \end{aligned}$$

in the matrix forms $X' = AX$; $X'' = BX'$. Hence find X'' in terms of X.

46 Compute the product

$$(x \quad y)\begin{pmatrix} 2 & 1 \\ 1 & 3 \end{pmatrix}\begin{pmatrix} x \\ y \end{pmatrix}$$

47 Compute the product

$$(x \quad y)\begin{pmatrix} 4 & 2 \\ 2 & 5 \end{pmatrix}\begin{pmatrix} x \\ y \end{pmatrix}$$

48 Compute the product

$$(x \quad y \quad z)\begin{pmatrix} 4 & 0 & 0 \\ 0 & -1 & 0 \\ 0 & 0 & 2 \end{pmatrix}\begin{pmatrix} x \\ y \\ z \end{pmatrix}$$

49 Consider the matrix $\begin{pmatrix} a & b \\ -b & a \end{pmatrix}$ where $a^2 + b^2 = 1$. This is the matrix associated with a rotation (see Probs. 16 to 20 of Sec. 6.13). Show that

$$\begin{pmatrix} a & -b \\ b & a \end{pmatrix}\begin{pmatrix} a & b \\ -b & a \end{pmatrix} = \begin{pmatrix} 1 & 0 \\ 0 & 1 \end{pmatrix}$$

50 For the rotation (see Prob. 49)

$$\begin{aligned} x' &= ax + by \\ y' &= -bx + ay \end{aligned} \qquad \text{where } a^2 + b^2 = 1$$

show that

$$\begin{pmatrix} x' \\ y' \end{pmatrix} = \begin{pmatrix} a & b \\ -b & a \end{pmatrix}\begin{pmatrix} x \\ y \end{pmatrix}$$

and

$$(x', y') = (x, y)\begin{pmatrix} a & -b \\ b & a \end{pmatrix}$$

51 Show that

$$\begin{aligned} d^2 &= (x_1 - x_2)^2 + (y_1 - y_2)^2 \\ &= (x_1 - x_2, y_1 - y_2)\begin{pmatrix} x_1 - x_2 \\ y_1 - y_2 \end{pmatrix} \end{aligned}$$

52 Using the results of Probs. 49, 50, and 51, show that

$$\begin{aligned} d'^2 &= (x_1' - x_2')^2 + (y_1' - y_2')^2 \\ &= (x_1' - x_2', y_1' - y_2')\begin{pmatrix} x_1' - x_2' \\ y_1' - y_2' \end{pmatrix} \\ &= (x_1 - x_2, y_1 - y_2)\begin{pmatrix} a & -b \\ b & a \end{pmatrix}\begin{pmatrix} a & b \\ -b & a \end{pmatrix}\begin{pmatrix} x_1 - x_2 \\ y_1 - y_2 \end{pmatrix} \\ &= (x_1 - x_2)^2 + (y_1 - y_2)^2 \\ &= d^2 \end{aligned}$$

7.6 » INVERSE OF A SQUARE MATRIX

The multiplication of square matrices has many of the multiplicative properties of real numbers; it is commutative, associative, and has an identity. In this section we investigate the existence of a multiplicative inverse.

DEFINITION Inverse

If A is a square matrix, an *inverse* is a square matrix A^{-1} (read "A inverse") which satisfies the equations

$$AA^{-1} = I \qquad \text{and} \qquad A^{-1}A = I$$

REMARKS

1 In the notation A^{-1} the -1 is not an exponent; it is merely a symbol indicating the inverse. Do *not* write $A^{-1} = 1/A$. We shall not define the quotient of two matrices.

2 Some square matrices do not have inverses. See Theorems 3 and 4 below.

In the 2×2 case we are given a matrix A and are looking for an A^{-1} which satisfies the definition. First let us require that

$$AA^{-1} = I_2$$

If A is $\begin{pmatrix} a & b \\ c & d \end{pmatrix}$ and $A^{-1} = \begin{pmatrix} w & x \\ y & z \end{pmatrix}$, we are asked to solve the matrix equation

$$\begin{pmatrix} a & b \\ c & d \end{pmatrix}\begin{pmatrix} w & x \\ y & z \end{pmatrix} = \begin{pmatrix} 1 & 0 \\ 0 & 1 \end{pmatrix}$$

Taking the product on the left, we have

$$\begin{pmatrix} aw + by & ax + bz \\ cw + dy & cx + dz \end{pmatrix} = \begin{pmatrix} 1 & 0 \\ 0 & 1 \end{pmatrix}$$

From the definition of the equality of two matrices, this gives us the two simultaneous systems:

(1)
$$\begin{cases} aw + by = 1 \\ cw + dy = 0 \end{cases}$$

(2)
$$\begin{cases} ax + bz = 0 \\ cx + dz = 1 \end{cases}$$

Writing $\Delta = ad - bc$, and supposing this not to be zero, we find that the unique solution is

$$w = \frac{d}{\Delta} \qquad y = -\frac{c}{\Delta} \qquad x = -\frac{b}{\Delta} \qquad z = \frac{a}{\Delta}$$

Therefore $$A^{-1} = \frac{1}{\Delta}\begin{pmatrix} d & -b \\ -c & a \end{pmatrix}$$

As a bonus we find that

$$A^{-1}A = \frac{1}{\Delta}\begin{pmatrix} d & -b \\ -c & a \end{pmatrix}\begin{pmatrix} a & b \\ c & d \end{pmatrix} = \begin{pmatrix} 1 & 0 \\ 0 & 1 \end{pmatrix} = I$$

This gives us Theorem 3, in which $A = \begin{pmatrix} a & b \\ c & d \end{pmatrix}$.

THEOREM 3 If $ad - bc \neq 0$, the 2×2 matrix A has a unique inverse A^{-1} such that

$$AA^{-1} = A^{-1}A = I$$

THEOREM 4 If $ad - bc = 0$, the inverse of A does not exist.

PROOF If an inverse exists, the simultaneous systems (1) and (2) must have solutions. But when $ad - bc = 0$, Theorem 7 of Chap. 6 tells us that no such solutions exist.

To find the inverse of a particular matrix, you may either use the formula just given or you may solve Eqs. (1) and (2).

ILLUSTRATION 1 If $A = \begin{pmatrix} 2 & 5 \\ -1 & 4 \end{pmatrix}$, then by the formula

$$A^{-1} = \frac{1}{13}\begin{pmatrix} 4 & -5 \\ 1 & 2 \end{pmatrix} = \begin{pmatrix} \frac{4}{13} & -\frac{5}{13} \\ \frac{1}{13} & \frac{2}{13} \end{pmatrix}$$

ILLUSTRATION 2 To find the inverse of $\begin{pmatrix} 2 & 5 \\ -1 & 4 \end{pmatrix}$ from first principles we write

$$\begin{pmatrix} 2 & 5 \\ -1 & 4 \end{pmatrix}\begin{pmatrix} w & x \\ y & z \end{pmatrix} = \begin{pmatrix} 1 & 0 \\ 0 & 1 \end{pmatrix}$$

Thus we derive

$$\begin{cases} 2w + 5y = 1 \\ -w + 4y = 0 \end{cases} \qquad \begin{cases} 2x + 5z = 0 \\ -x + 4z = 1 \end{cases}$$

These are equivalent to

$$\begin{cases} 2w + 5y = 1 \\ -2w + 8y = 0 \end{cases} \qquad \begin{cases} 2x + 5z = 0 \\ -2x + 8z = 2 \end{cases}$$

Hence $13y = 1$, $y = \frac{1}{13}$, $w = \frac{4}{13}$ and $13z = 2$, $z = \frac{2}{13}$, $x = -\frac{5}{13}$. Therefore

$$A^{-1} = \begin{pmatrix} \frac{4}{13} & -\frac{5}{13} \\ \frac{1}{13} & \frac{2}{13} \end{pmatrix}$$

ILLUSTRATION 3 If $A = \begin{pmatrix} 2 & -3 \\ 4 & -6 \end{pmatrix}$, A^{-1} does not exist.

Nonsingular matrix

A 2×2 matrix for which $ad - bc \neq 0$ is called *nonsingular;* if $ad - bc = 0$, it is *singular*. Hence a *matrix has an inverse if and only if it is nonsingular.*

In the case of 3×3 matrices, inverses can be computed by the method just described, but the computations are very tedious. We shall approach this problem from a simpler point of view in the following section.

PROBLEMS 7.6 Find the inverses of the given matrices when they exist. Check your inverses in the formulas $AA^{-1} = I$, $A^{-1}A = I$.

1 $\begin{pmatrix} 2 & 3 \\ -1 & 4 \end{pmatrix}$

2 $\begin{pmatrix} 1 & 3 \\ 5 & -2 \end{pmatrix}$

3 $\begin{pmatrix} -3 & 4 \\ 2 & 0 \end{pmatrix}$

4 $\begin{pmatrix} 5 & 2 \\ -3 & 4 \end{pmatrix}$

5 $\begin{pmatrix} 2 & 4 \\ -1 & -2 \end{pmatrix}$

6 $\begin{pmatrix} 5 & -2 \\ 10 & -4 \end{pmatrix}$

7 $\begin{pmatrix} 1 & 0 \\ 0 & 1 \end{pmatrix}$

8 $\begin{pmatrix} 0 & 0 \\ 0 & 0 \end{pmatrix}$

9 $\begin{pmatrix} \frac{1}{\sqrt{2}} & -\frac{1}{\sqrt{2}} \\ \frac{1}{\sqrt{2}} & \frac{1}{\sqrt{2}} \end{pmatrix}$

10 $\begin{pmatrix} 0 & 1 \\ 1 & 0 \end{pmatrix}$

7.7 › DETERMINANTS

In finding the inverse of a 2×2 matrix and in solving a system of two simultaneous equations, we have run across expressions like $ad - bc$ and $a_1b_2 - a_2b_1$ in critical places. It is time we gave these a formal discussion. This brings us to determinants, which is the name given to expressions of this kind.

DEFINITION Let A be the 2×2 matrix $\begin{pmatrix} a_1 & b_1 \\ a_2 & b_2 \end{pmatrix}$. Then we define the expression $a_1b_2 - a_2b_1$ to be the *determinant* of A and write

$$\det A = \begin{vmatrix} a_1 & b_1 \\ a_2 & b_2 \end{vmatrix} = a_1b_2 - a_2b_1$$

EXERCISE A Prove that the determinant of a 2×2 matrix is zero if the two columns (rows) are proportional or equal.

REMARKS

1 We use parentheses for matrices and parallel lines for the corresponding determinants.
2 A determinant is a single number associated with a square matrix. The determinant is *not* the array; the array is the matrix.

For 3×3 matrices we define the determinant in the following fashion. Let

$$A = \begin{pmatrix} a_1 & b_1 & c_1 \\ a_2 & b_2 & c_2 \\ a_3 & b_3 & c_3 \end{pmatrix}$$

be a given 3×3 matrix. If we strike out the row and column containing any element, we are left with a 2×2 matrix whose determinant has already been defined. This determinant is called the *minor* of the corresponding element. We list a few examples of these:

Minor

Element	*Minor*
a_1	$b_2c_3 - b_3c_2$
b_1	$a_2c_3 - a_3c_2$
c_2	$a_1b_3 - a_3b_1$
c_3	$a_1b_2 - a_2b_1$

We now attach an algebraic sign to each minor in the following way: Consider the corresponding element, and move it by a series of horizontal and/or vertical steps to the upper left-hand corner. The sign is + if the number of steps required is even, − if this number is odd. The product of the minor times this sign is called the *cofactor* of the corresponding element. The cofactor of any element will be denoted by the corresponding capital letter; for instance, the cofactor of a_1 is A_1. We list a few examples:

Cofactor

Element	*Cofactor*
a_1	$A_1 = b_2c_3 - b_3c_2$
b_1	$B_1 = -(a_2c_3 - a_3c_2)$
c_2	$C_2 = -(a_1b_3 - a_3b_1)$
c_3	$C_3 = a_1b_2 - a_2b_1$

To define the determinant, we now consider the first row and define

$$\begin{aligned}\det A &= a_1A_1 + b_1B_1 + c_1C_1\\ &= a_1(b_2c_3 - b_3c_2) - b_1(a_2c_3 - a_3c_2) + c_1(a_2b_3 - a_3b_2)\\ &= a_1b_2c_3 + a_2b_3c_1 + a_3b_1c_2 - a_1b_3c_2 - a_2b_1c_3 - a_3b_2c_1\end{aligned}$$

We might equally well have done this for any row or column, and at first sight you would expect the results to be six different numbers. They are, in fact, all equal.

EXERCISE B By direct computation show that

$$\det A = b_1B_1 + b_2B_2 + b_3B_3$$

DEFINITION The *determinant* of a 3×3 matrix is equal to the inner product of any row vector (or column vector) with the vector of its corresponding cofactors.

ILLUSTRATION 1 Find

$$\det \begin{pmatrix} 1 & -2 & 3 \\ 4 & 1 & -1 \\ 1 & 2 & 1 \end{pmatrix} = \begin{vmatrix} 1 & -2 & 3 \\ 4 & 1 & -1 \\ 1 & 2 & 1 \end{vmatrix}$$

Choosing the first row, we find that the cofactors are

$$\begin{vmatrix} 1 & -1 \\ 2 & 1 \end{vmatrix} = 3 \qquad -\begin{vmatrix} 4 & -1 \\ 1 & 1 \end{vmatrix} = -5 \qquad \begin{vmatrix} 4 & 1 \\ 1 & 2 \end{vmatrix} = 7$$

So the answer is $(1)(3) + (-2)(-5) + (3)(7) = 34$.

As an alternative solution, choose the first column. The cofactors are respectively

$$\begin{vmatrix} 1 & -1 \\ 2 & 1 \end{vmatrix} = 3 \qquad -\begin{vmatrix} -2 & 3 \\ 2 & 1 \end{vmatrix} = 8 \qquad \begin{vmatrix} -2 & 3 \\ 1 & -1 \end{vmatrix} = -1$$

So the answer is $(1)(3) + (4)(8) + (1)(-1) = 34$.

The following two theorems about determinants are of great utility.

THEOREM 5 If two rows (columns) of a matrix are proportional (or equal), its determinant is zero.

PROOF (FOR 3×3 MATRICES) First choose the third row (column) not involved in the proportionality. Then all the corresponding cofactors are zero (see Exercise A for the case of 2×2 matrices).

THEOREM 6 The inner product of any row (column) vector and the vector of cofactors of a different row (column) is zero.

PROOF (FOR 3×3 MATRICES) Consider for example $a_1B_1 + a_2B_2 + a_3B_3$. This, however, is the determinant of the matrix

$$\begin{pmatrix} a_1 & c_1 & a_1 \\ a_2 & c_2 & a_2 \\ a_3 & c_3 & a_3 \end{pmatrix}$$

which is zero since two columns are equal.

Finally we can calculate the inverse of a 3×3 matrix.

THEOREM 7 Let $A = \begin{pmatrix} a_1 & b_1 & c_1 \\ a_2 & b_2 & c_2 \\ a_3 & b_3 & c_3 \end{pmatrix}$, and let $\det A = \Delta$. Then, if $\Delta \neq 0$,

Formula for A^{-1}

$$A^{-1} = \frac{1}{\Delta}\begin{pmatrix} A_1 & A_2 & A_3 \\ B_1 & B_2 & B_3 \\ C_1 & C_2 & C_3 \end{pmatrix}$$

PROOF We must show that $AA^{-1} = A^{-1}A = I$.

$$AA^{-1} = \frac{1}{\Delta}\begin{pmatrix} a_1 & b_1 & c_1 \\ a_2 & b_2 & c_2 \\ a_3 & b_3 & c_3 \end{pmatrix}\begin{pmatrix} A_1 & A_2 & A_3 \\ B_1 & B_2 & B_3 \\ C_1 & C_2 & C_3 \end{pmatrix} = \frac{1}{\Delta}\begin{pmatrix} \Delta & 0 & 0 \\ 0 & \Delta & 0 \\ 0 & 0 & \Delta \end{pmatrix}$$

$$= \begin{pmatrix} 1 & 0 & 0 \\ 0 & 1 & 0 \\ 0 & 0 & 1 \end{pmatrix} = I$$

because of the definition of the determinant and Theorem 6. A similar proof gives $A^{-1}A = I$.

Determinants of square matrices of higher dimension are defined inductively in a similar fashion. For a 4×4 matrix, for example, the cofactors are $\pm$ determinants of 3×3 matrices. The determinant, in an obvious notation, is defined to be $a_1A_1 + a_2A_2 + a_3A_3 + a_4A_4$. We can continue step by step to define determinants of square matrices of any size.

We conclude by giving without proof the following theorem for any square matrices.

THEOREM 8 If $AB = C$, then $(\det A) \times (\det B) = \det C$.

PROBLEMS 7.7 In Probs. 1 to 10 find the determinants of the given matrices.

1 $\begin{pmatrix} 1 & -2 & 4 \\ 3 & 1 & 1 \\ 2 & -1 & 3 \end{pmatrix}$ **2** $\begin{pmatrix} 1 & 0 & 2 \\ 3 & -1 & 4 \\ 2 & 1 & 0 \end{pmatrix}$

3 $\begin{pmatrix} 3 & 1 & 5 \\ 0 & 2 & -1 \\ 4 & -3 & 2 \end{pmatrix}$ **4** $\begin{pmatrix} 4 & 1 & 3 \\ 2 & 0 & -2 \\ 5 & 1 & 1 \end{pmatrix}$

5 $\begin{pmatrix} 2 & -1 & 4 \\ 3 & 5 & 0 \\ 1 & 0 & 3 \end{pmatrix}$ **6** $\begin{pmatrix} -1 & 2 & 0 \\ 3 & 1 & 5 \\ 2 & 2 & 4 \end{pmatrix}$

7 $\begin{pmatrix} 6 & 2 & 4 \\ 0 & 3 & 5 \\ -1 & 2 & -1 \end{pmatrix}$ **8** $\begin{pmatrix} 3 & 1 & 7 \\ -1 & 0 & 2 \\ 4 & 1 & 6 \end{pmatrix}$

9 $\begin{pmatrix} x & y & 1 \\ 1 & 2 & 1 \\ 3 & 1 & 1 \end{pmatrix}$ **10** $\begin{pmatrix} x & y & 1 \\ -1 & 4 & 1 \\ 3 & -2 & 1 \end{pmatrix}$

11 Prove that

$$\det\begin{pmatrix} a_1 & b_1 & c_1 \\ a_2 & b_2 & c_2 \\ a_3 & b_3 & c_3 \end{pmatrix} = \det\begin{pmatrix} a_1 + kb_1 & b_1 & c_1 \\ a_2 + kb_2 & b_2 & c_2 \\ a_3 + kb_3 & b_3 & c_3 \end{pmatrix}$$

HINT: Expand the second determinant in terms of its first column.

***12** Prove that

$$\det\begin{pmatrix} a_1 & b_1 & c_1 \\ a_2 & b_2 & c_2 \\ a_3 & b_3 & c_3 \end{pmatrix} = -\det\begin{pmatrix} b_1 & a_1 & c_1 \\ b_2 & a_2 & c_2 \\ b_3 & a_3 & c_3 \end{pmatrix}$$

HINT: Expand both determinants in terms of their third columns.

In Probs. 13 to 20 find the inverse of the given matrix, if it exists, and check your inverses in the formula $AA^{-1} = I$.

13 $\begin{pmatrix} 1 & 2 & -1 \\ 0 & -1 & 1 \\ 3 & 0 & -2 \end{pmatrix}$ **14** $\begin{pmatrix} 2 & -3 & 1 \\ 4 & 0 & 2 \\ -1 & 2 & 0 \end{pmatrix}$

15 $\begin{pmatrix} 1 & 2 & 3 \\ 3 & 1 & 0 \\ 2 & 2 & 1 \end{pmatrix}$ **16** $\begin{pmatrix} 4 & -1 & 2 \\ -1 & 2 & 1 \\ 3 & 1 & 1 \end{pmatrix}$

17 $\begin{pmatrix} 2 & 4 & 3 \\ -1 & -2 & 1 \\ 0 & 0 & 5 \end{pmatrix}$ **18** $\begin{pmatrix} -3 & 0 & 2 \\ 1 & 4 & 3 \\ 2 & -1 & 0 \end{pmatrix}$

19 $\begin{pmatrix} 3 & 2 & -1 \\ -4 & 3 & 0 \\ 2 & 5 & 1 \end{pmatrix}$ **20** $\begin{pmatrix} 2 & -4 & 10 \\ -1 & 2 & -5 \\ 3 & -6 & 15 \end{pmatrix}$

21 Prove that $\det A = 1/(\det A^{-1})$. HINT: Use Theorem 8, and the fact that $AA^{-1} = I$.

22 Verify the steps in the following derivation: Let the vertices of a triangle be labeled as in Fig. 7.3. The subscripts are numbered in the counterclockwise direction around the triangle. Then the areas of the triangle and the three trapezoids satisfy

$$\begin{aligned} P_1P_2P_3 &= P_1P_3Q_1Q_3 - P_1P_2Q_1Q_2 - P_2P_3Q_2Q_3 \\ &= \tfrac{1}{2}[(x_3 - x_1)(y_1 + y_3) - (x_2 - x_1)(y_1 + y_2) \\ &\qquad - (x_3 - x_2)(y_2 + y_3)] \\ &= \tfrac{1}{2}[(x_2y_3 - x_3y_2) - (x_1y_3 - x_3y_1) + (x_1y_2 - x_2y_1)] \end{aligned}$$

Therefore,

$$\text{Area of } P_1P_2P_3 = \frac{1}{2}\begin{vmatrix} x_1 & y_1 & 1 \\ x_2 & y_2 & 1 \\ x_3 & y_3 & 1 \end{vmatrix}$$

How is this result altered if we number the vertices in the clockwise direction?

23 Use the method of this section to find the inverse of $\begin{pmatrix} a & b \\ c & d \end{pmatrix}$. Show that the result agrees with the formula of Sec. 7.6.

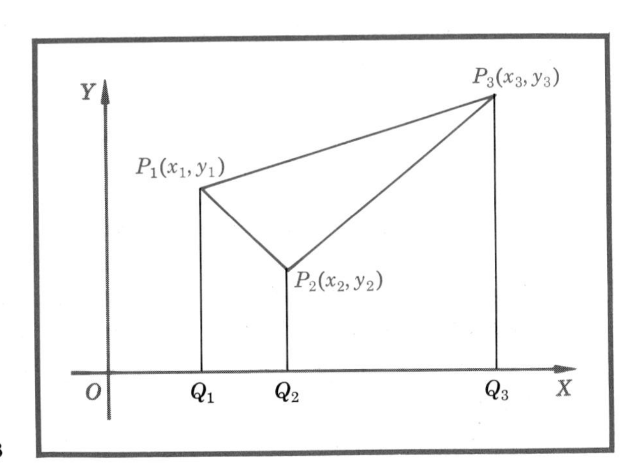

FIGURE 7.3

24 Prove that $\det\begin{pmatrix} x & y & 1 \\ x_1 & y_1 & 1 \\ x_2 & y_2 & 1 \end{pmatrix} = 0$ is an equation of the line passing through (x_1,y_1) and (x_2,y_2). HINT: **(a)** Expand by the first row to show that the equation is linear. **(b)** Use Theorem 5 to show that (x_1,y_1) and (x_2,y_2) satisfy this equation.

25 Verify Theorem 8 for the product

$$\begin{pmatrix} 2 & 5 \\ 1 & 4 \end{pmatrix}\begin{pmatrix} 1 & -3 \\ 2 & 5 \end{pmatrix}$$

26 Verify Theorem 8 for the product

$$\begin{pmatrix} 1 & 5 \\ -3 & 2 \end{pmatrix}\begin{pmatrix} 2 & -4 \\ 3 & 7 \end{pmatrix}$$

7.8 » APPLICATIONS OF MATRICES TO SIMULTANEOUS EQUATIONS

We have seen above that the simultaneous system

$$\begin{aligned} a_1x + b_1y &= d_1 \\ a_2x + b_2y &= d_2 \end{aligned}$$

can be written in the compact form

$$AX = D$$

where $\qquad A = \begin{pmatrix} a_1 & b_1 \\ a_2 & b_2 \end{pmatrix} \qquad X = \begin{pmatrix} x \\ y \end{pmatrix} \qquad D = \begin{pmatrix} d_1 \\ d_2 \end{pmatrix}$

Similarly,

$$\begin{aligned} a_1x + b_1y + c_1z &= d_1 \\ a_2x + b_2y + c_2z &= d_2 \\ a_3x + b_3y + c_3z &= d_3 \end{aligned}$$

can be written in the form $AX = D$, where

$$A = \begin{pmatrix} a_1 & b_1 & c_1 \\ a_2 & b_2 & c_2 \\ a_3 & b_3 & c_3 \end{pmatrix} \qquad X = \begin{pmatrix} x \\ y \\ z \end{pmatrix} \qquad D = \begin{pmatrix} d_1 \\ d_2 \\ d_3 \end{pmatrix}$$

This suggests that the problem of solving simultaneous equations is really that of solving the matrix equation

$$AX = D$$

But this is now an easy problem for us. Multiplying both sides, on the left, by A^{-1}, we have

$$A^{-1}AX = A^{-1}D$$

or

$$IX = A^{-1}D$$

since $A^{-1}A = I$, or

$$X = A^{-1}D \qquad \text{the solution}$$

Cramer's Rule

A possible method of solution, therefore, is to compute A^{-1} by the methods of Secs. 7.6 and 7.7 and then to find $A^{-1}D$. Although this is not the best method in practice, the above formula is of considerable theoretical value. It is known as "Cramer's Rule".

ILLUSTRATION 1 Solve

$$\begin{aligned} 2x + 5y &= 6 \\ x - 2y &= -5 \end{aligned}$$

$$A = \begin{pmatrix} 2 & 5 \\ 1 & -2 \end{pmatrix} \qquad \det A = -9$$

$$A^{-1} = -\frac{1}{9}\begin{pmatrix} -2 & -5 \\ -1 & 2 \end{pmatrix} \qquad D = \begin{pmatrix} 6 \\ -5 \end{pmatrix}$$

$$X = A^{-1}D = -\frac{1}{9}\begin{pmatrix} -2 & -5 \\ -1 & 2 \end{pmatrix}\begin{pmatrix} 6 \\ -5 \end{pmatrix} = -\frac{1}{9}\begin{pmatrix} 13 \\ -16 \end{pmatrix} = \begin{pmatrix} -\frac{13}{9} \\ \frac{16}{9} \end{pmatrix}$$

So $x = -\frac{13}{9}$; $y = \frac{16}{9}$.

Cramer's Rule is sometimes written in a different form. To illustrate this, let us consider the system

$$\begin{aligned} a_1x + b_1y &= d_1 \\ a_2x + b_2y &= d_2 \end{aligned}$$

Then $A^{-1} = \frac{1}{\Delta}\begin{pmatrix} b_2 & -b_1 \\ -a_2 & a_1 \end{pmatrix}$, where $\Delta = a_1b_2 - a_2b_1$. Hence

$$X = A^{-1}D = \frac{1}{\Delta}\begin{pmatrix} b_2 & -b_1 \\ -a_2 & a_1 \end{pmatrix}\begin{pmatrix} d_1 \\ d_2 \end{pmatrix} = \frac{1}{\Delta}\begin{pmatrix} b_2d_1 - b_1d_2 \\ -a_2d_1 + a_1d_2 \end{pmatrix}$$

Therefore

$$x = \frac{b_2d_1 - b_1d_2}{\Delta} = \frac{\begin{vmatrix} d_1 & b_1 \\ d_2 & b_2 \end{vmatrix}}{\begin{vmatrix} a_1 & b_1 \\ a_2 & b_2 \end{vmatrix}}$$

and

$$y = \frac{a_1d_2 - a_2d_1}{\Delta} = \frac{\begin{vmatrix} a_1 & d_1 \\ a_2 & d_2 \end{vmatrix}}{\begin{vmatrix} a_1 & b_1 \\ a_2 & b_2 \end{vmatrix}}$$

You will observe that the numerator for x is obtained from the denominator by replacing the column of a's (the coefficients of x) by the column of d's. Similarly to obtain the numerator of y we replace the column of b's in the denominator by the column of d's. For the corresponding formula for three equations in three variables see Sec. 7.8, Prob. 1

ILLUSTRATION 2 Applying this method to the system in Illustration 1, we obtain

$$x = \frac{\begin{vmatrix} 6 & 5 \\ -5 & -2 \end{vmatrix}}{\begin{vmatrix} 2 & 5 \\ 1 & -2 \end{vmatrix}} = \frac{13}{-9} = -\frac{13}{9}$$

$$y = \frac{\begin{vmatrix} 2 & 6 \\ 1 & -5 \end{vmatrix}}{\begin{vmatrix} 2 & 5 \\ 1 & -2 \end{vmatrix}} = \frac{-16}{-9} = \frac{16}{9}$$

The problem of solving systems of linear equations is theoretically handled by Cramer's Rule, and we have a practical means for their solution in simple cases. In applied mathematics, however, one meets systems of linear equations containing 100 or more unknowns with coefficients which are 5- to 10-place decimals. The practical problem of solution is quite formidable, even on a high-speed machine. Modern research has developed elaborate techniques for tackling this problem, but improvements in these are currently under study.

Two homogeneous equations

Finally let us consider the following "homogeneous" system of two equations in three unknowns:

$$\begin{aligned} a_1x + b_1y + c_1z &= 0 \\ a_2x + b_2y + c_2z &= 0 \end{aligned}$$

Geometrically these equations represent two planes through the origin, and so we expect to find a *line of solutions.* By the use of determinants we can express this solution in a very elegant form.

THEOREM 9 The solutions of

$$\begin{aligned} a_1x + b_1y + c_1z &= 0 \\ a_2x + b_2y + c_2z &= 0 \end{aligned}$$

are

$$x = k\begin{vmatrix} b_1 & c_1 \\ b_2 & c_2 \end{vmatrix} \qquad y = -k\begin{vmatrix} a_1 & c_1 \\ a_2 & c_2 \end{vmatrix} \qquad z = k\begin{vmatrix} a_1 & b_1 \\ a_2 & b_2 \end{vmatrix}$$

where k is an arbitrary scalar (provided that at least one of these is different from zero).

PROOF If we substitute these values of x, y, and z into the left-hand side of the first equation, we get

$$k\left\{a_1\begin{vmatrix} b_1 & c_1 \\ b_2 & c_2 \end{vmatrix} - b_1\begin{vmatrix} a_1 & c_1 \\ a_2 & c_2 \end{vmatrix} + c_1\begin{vmatrix} a_1 & b_1 \\ a_2 & b_2 \end{vmatrix}\right\}$$

The expression in braces, however, is precisely the determinant of the matrix

$$\begin{pmatrix} a_1 & b_1 & c_1 \\ a_1 & b_1 & c_1 \\ a_2 & b_2 & c_2 \end{pmatrix}$$

which is zero by Theorem 5. Therefore the first equation is satisfied. A similar argument shows that the second equation is satisfied.

ILLUSTRATION 3 Solve

$$\begin{aligned} 3x - 2y + z &= 0 \\ x + 4y + 2z &= 0 \end{aligned}$$

By Theorem 9:

$$x = k\begin{vmatrix} -2 & 1 \\ 4 & 2 \end{vmatrix} \qquad y = -k\begin{vmatrix} 3 & 1 \\ 1 & 2 \end{vmatrix} \qquad z = k\begin{vmatrix} 3 & -2 \\ 1 & 4 \end{vmatrix}$$

or

$$x = -8k \qquad y = -5k \qquad z = 14k$$

ILLUSTRATION 4 Find the vector (x,y,z) whose inner products with each of the vectors $(4,1,-2)$ and $(2,1,3)$ are zero.

The required conditions are

$$\begin{aligned} 4x + y - 2z &= 0 \\ 2x + y + 3z &= 0 \end{aligned}$$

Hence $x = 5k$, $y = -16k$, $z = 2k$. The required vector is $k(5,-16,2)$.

Illustration 4 motivates the following definition of the outer (or *vector* or *cross*) product of two vectors.

DEFINITION Outer product

The *outer product* of the two vectors (a_1,b_1,c_1) and (a_2,b_2,c_2) is the vector

$$\left(\begin{vmatrix} b_1 & c_1 \\ b_2 & c_2 \end{vmatrix}, -\begin{vmatrix} a_1 & c_1 \\ a_2 & c_2 \end{vmatrix}, \begin{vmatrix} a_1 & b_1 \\ a_2 & b_2 \end{vmatrix}\right)$$

The notation for this product is $(a_1,b_1,c_1) \wedge (a_2,b_2,c_2)$.†

† The $\wedge$ symbol used here has no connection with the similar symbol which sometimes denotes *conjunction* in symbolic logic. Physicists and engineers commonly use a cross, $\times$, to denote this product and write $(a_1,b_1,c_1) \times (a_2,b_2,c_2)$.

ILLUSTRATION 5 $(2,4,-3) \wedge (1,-2,6) = (18,-15,-8)$

REMARKS

1. The outer product is defined for three-dimensional vectors only. Thus an expression such as $(a_1,b_1) \wedge (a_2,b_2)$ has no meaning.
2. The inner product of $(a_1,b_1,c_1) \wedge (a_2,b_2,c_2)$ with each of its factors is zero. Hence $(a_1,b_1,c_1) \wedge (a_2,b_2,c_2)$ is perpendicular to (a_1,b_1,c_1) and to (a_2,b_2,c_2).

3 $(a_1,b_1,c_1) \wedge (a_2,b_2,c_2) = -(a_2,b_2,c_2) \wedge (a_1,b_1,c_1)$. The "wedge" symbol $\wedge$ is commonly used in higher mathematics to denote "skew-commutative" multiplication, that is to say, multiplication for which

$$a \wedge b = -b \wedge a$$

4 A convenient way of expressing this product is the expansion of the following symbolic determinant by means of its first row:

$$\begin{vmatrix} \mathbf{i} & \mathbf{j} & \mathbf{k} \\ a_1 & b_1 & c_1 \\ a_2 & b_2 & c_2 \end{vmatrix}$$

ILLUSTRATION 6

$$(1,-3,2) \wedge (4,1,3) = \begin{vmatrix} \mathbf{i} & \mathbf{j} & \mathbf{k} \\ 1 & -3 & 2 \\ 4 & 1 & 3 \end{vmatrix}$$
$$= -11\mathbf{i} + 5\mathbf{j} + 13\mathbf{k}$$
$$= (-11,5,13)$$

5 Let $\mathbf{A} = (a_1,b_1,c_1)$ and $\mathbf{B} = (a_2,b_2,c_2)$. Then the length of $\mathbf{A} \wedge \mathbf{B}$ is equal to the length of **A** times the length of **B** times the sine of the angle between **A** and **B.**

6 Let **F** be a force acting on a body at point P; let O be a reference point and $OP = \mathbf{r}$. Then the vector moment **M** is defined to be

$$\mathbf{M} = \mathbf{r} \wedge \mathbf{F}$$

Vector moment

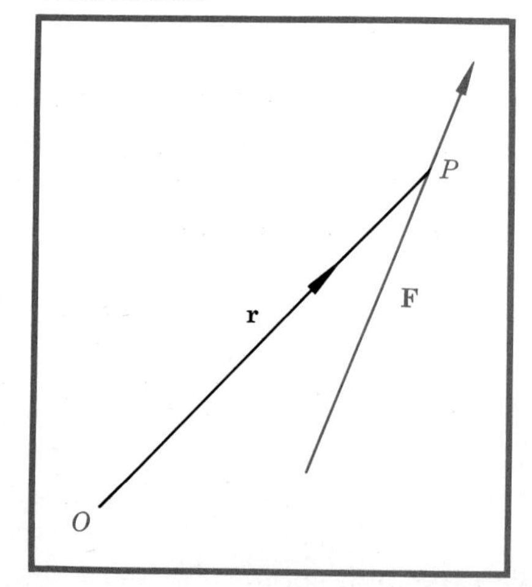

FIGURE 7.4
To find the vector moment of **F** about O, choose P on **F**, draw $\mathbf{r} = \overrightarrow{OP}$, and compute $\mathbf{M} = \mathbf{r} \wedge \mathbf{F}$. The result does not depend upon the choice of P.

PROBLEMS 7.8

In Probs. 1 to 14 use Cramer's Rule to solve the given system of equations (if possible).

1 $2x + 3y = 8$
$-x + 4y = 7$
(See Prob. 1, Sec. 7.6)

2 $x + 3y = 0$
$5x - 2y = 17$
(See Prob. 2, Sec. 7.6)

3 $-3x + 4y = -2$
$2x = 4$
(See Prob. 3, Sec. 7.6)

4 $5x + 2y = 1$
$-3x + 4y = 15$
(See Prob. 4, Sec. 7.6)

5 $2x + 4y = 6$
$-x - 2y = 5$
(See Prob. 5, Sec. 7.6)

6 $5x - 2y = 1$
$10x - 4y = 2$
(See Prob. 6, Sec. 7.6)

7 $x + 2y - z = 4$
$-y + z = -1$
$3x - 2z = 1$
(See Prob. 13, Sec. 7.7)

8 $2x - 3y + z = -3$
$4x + 2z = 0$
$-x + 2y = 2$
(See Prob. 14, Sec. 7.7)

9 $x + 2y + 3z = 6$
$3x + y = 5$
$2x + 2y + z = 4$
(See Prob. 15, Sec. 7.7)

10 $4x - y + 2z = 7$
$-x + 2y + z = 0$
$3x + y + z = 7$
(See Prob. 16, Sec. 7.7)

11 $2x + 4y + 3z = 1$
$-x - 2y + z = 2$
$5z = 5$
(See Prob. 17, Sec. 7.7)

12 $-3x + 2z = 8$
$x + 4y + 3z = 13$
$2x - y = -7$
(See Prob. 18, Sec. 7.7)

13 $3x + 2y - z = -6$
$-4x + 3y = 18$
$2x + 5y + z = 5$
(See Prob. 19, Sec. 7.7)

14 $2x - 4y + 10z = 12$
$-x + 2y - 5z = -6$
$3x - 6y + 15z = 18$
(See Prob. 20, Sec. 7.7)

***15** Given the system

$$\begin{aligned} a_1x + b_1y + c_1z &= d_1 \\ a_2x + b_2y + c_2z &= d_2 \\ a_3x + b_3y + c_3z &= d_3 \end{aligned}$$

Use Cramer's Rule to show that

$$x = \frac{d_1A_1 + d_2A_2 + d_3A_3}{\Delta}$$

$$y = \frac{d_1B_1 + d_2B_2 + d_3B_3}{\Delta}$$

$$z = \frac{d_1C_1 + d_2C_2 + d_3C_3}{\Delta}$$

Hence write formulas for x, y, and z as quotients of determinants.

16 Solve Prob. 7 above by using the formulas derived in Prob. 15.

In Probs. 17 to 20 solve the given system of equations.

17 $2x + 3y - 4z = 0$, $x - 2y + 3z = 0$

18 $3x - y + 2z = 0$, $4x + y - z = 0$

19 $x - 3y = 0$, $4x + y + 5z = 0$

20 $2x - y + z = 0$, $x - z = 0$

In Probs. 21 to 24 find a vector whose inner products with each of the two given vectors is zero.

21 $(2,1,2)$, $(3,-1,4)$

22 $(4,-1,2)$, $(1,0,3)$

23 $(3,3,1)$, $(-2,1,4)$

24 $(5,0,3)$, $(-3,4,2)$

25 Why does the method of Theorem 9 fail for the system

$$\begin{aligned} x - 2y + 5z &= 0 \\ 2x - 4y + 10z &= 0 \end{aligned}$$

26 Why can Theorem 9 not be applied to the system

$$\begin{aligned} x - 2y + z &= 5 \\ 2x - y + 3z &= 4 \end{aligned}$$

In Probs. 27 to 36 find the given outer products.

27 $(1,3,-1) \wedge (2,1,2)$

28 $(-1,-3,2) \wedge (2,-2,2)$

29 $(5,-1,0) \wedge (1,-5,0)$

30 $(4,3,2) \wedge (2,3,4)$

31 $\mathbf{i} \wedge \mathbf{j}$

32 $\mathbf{j} \wedge \mathbf{k}$

33 $\mathbf{k} \wedge \mathbf{i}$

34 $(-1,3,2) \wedge (-2,6,4)$

35 $(2,5,1) \wedge (4,10,2)$

36 $\mathbf{i} \wedge \mathbf{i}$

REFERENCES

Hohn, F. E.: "Elementary Matrix Algebra", Macmillan, New York (1964).
Murdoch, D. C.: "Linear Algebra", John Wiley, New York (1970).

8 INEQUALITIES

8.1 FUNDAMENTAL PROPERTIES

When we discussed the real numbers in Chap. 2 (Sec. 2.3) we introduced the subset of *positive* real numbers with the important properties:

Positive numbers

R9 The sum of two positive real numbers is positive.
R10 The product of two positive real numbers is positive.
R11 (*Law of Trichotomy*). For any real number a, one and only one of the following is true:
- **a** a is positive.
- **b** $-a$ is positive.
- **c** a is zero.

Then we said that a is negative if and only if $-a$ is positive.

In Sec. 2.10 we introduced the notions of inequality for real numbers, and stated:

DEFINITION By definition a is greater than b (written $a > b$) if $a - b$ is positive. Similarly a is less than b (written $a < b$) if $a - b$ is negative.

In the language of inequalities, we can, therefore, rewrite R11 in the form:

R11′ For each pair of real numbers a and b, one and only one of the following relations is true:

$$a < b \qquad a = b \qquad a > b$$

Properties of inequalities

We are now in position to prove some elementary properties of inequalities. First consider the inequalities

$$a > b \qquad b > c$$

From these we can conclude that $a > c$. To see this we note that $a - b$ is positive and $b - c$ is positive. Hence the sum $(a - b) + (b - c) = a - c$ is positive, which shows that $a > c$. We call this the transitive property for inequalities and write it as Theorem 1.

THEOREM 1 **TRANSITIVE PROPERTY** If a, b, and c are real numbers, and if $a > b$ and $b > c$, then $a > c$.

ILLUSTRATION 1 Since $2 > -3$ and $8 > 2$, it follows that $8 > -3$.

Next let us suppose that $a > b$, so that $a - b$ is positive. Then it follows that $a + c > b + c$, for $(a + c) - (b + c) = a - b$, which is positive. We write this as Theorem 2.

THEOREM 2 If a, b, and c are real numbers and if $a > b$, then

Addition

$$a + c > b + c$$

ILLUSTRATION 2 Since $5 > 3$, it follows that $5 + 2 > 3 + 2$, or $7 > 5$.

Finally, we again suppose that $a > b$ and that c is positive. Then $ac - bc = (a - b)c$ is positive, for each factor is positive. Therefore $ac > bc$. We write this as Theorem 3.

THEOREM 3 If a, b, and c are real numbers, and if $a > b$, and $c > 0$, then $ac > bc$.

Multiplication by positive number

ILLUSTRATION 3 Since $6 > 2$, it follows that $6 \times 3 > 2 \times 3$, or that $18 > 6$.

EXERCISE A The conclusion of Theorem 3 is false if c is negative. Find a counterexample which illustrates this fact.

EXERCISE B Do the rational numbers have properties R9 to R11?

EXERCISE C Show that $a > 0$ if and only if a is positive.

EXERCISE D Show that $a < b$ if and only if $b > a$.

8.2 › THEOREMS ABOUT INEQUALITIES

From properties R1 to R11 and Theorems 1, 2, and 3 we shall now derive the chief theorems on inequalities. These will enable us to manipulate our inequalities and to solve problems involving them. All letters refer to real numbers.

THEOREM 4 If $a > b$ and $c > d$, then $(a + c) > (b + d)$.

PROOF

1	$a + c > b + c$	Theorem 2
2	$b + c > b + d$	Theorem 2
3	$a + c > b + d$	Theorem 1

ILLUSTRATION 1 From $6 > -3$ and $8 > 4$ we conclude from Theorem 4 that $6 + 8 > -3 + 4$, or $14 > 1$. Note that Theorem 4 says nothing about adding the corresponding sides of two inequalities, one of which contains a *less than* ($<$) and the other a *greater than* ($>$).

EXERCISE A Prove that, if $a < b$ and $c < d$, then $(a + c) < (b + d)$.

THEOREM 5 $a > 0$ if and only if $-a < 0$.

PROOF (LEFT TO RIGHT)

1	$a + (-a) > 0 + (-a)$	Theorem 2
2	$0 > -a$	R4 and R6
3	$-a < 0$	Sec. 8.1, Exercise D

EXERCISE B Complete the proof (right to left) of Theorem 5.

EXERCISE C PROVE: $a < 0$ if and only if $-a > 0$.

THEOREM 6 $a > b$ if and only if $-a < -b$.

PROOF (LEFT TO RIGHT)

1	$[(-a) + (-b)] + a > [(-a) + (-b)] + b$	Theorem 2
2	$-b > -a$	Theorem 2, Chap. 2, R4 and R6
3	$-a < -b$	Sec. 8.1, Exercise D

EXERCISE D Complete the proof (right to left) of Theorem 6.

EXERCISE E PROVE: $a < b$ if and only if $-a > -b$.

ILLUSTRATION 2
From $7 > 2$, we conclude that $-7 < -2$.
From $10 > -3$, we conclude that $-10 < 3$.
From $-4 > -8$, we conclude that $4 < 8$.

This theorem is sometimes stated in the form:

If we change the signs of both sides of an inequality, we change its sense. By changing the sense of an inequality we mean that we have replaced $>$ by $<$ or $<$ by $>$.

THEOREM 7 If $a > b$ and $c < 0$, then $ac < bc$.

Multiplication by negative number reverses sense of inequality

PROOF

1	Let $c = -d$, where $d > 0$.	
2	$ad > bd$	Theorem 3
3	$-ad < -bd$	Theorem 6
4	$a(-d) < b(-d)$	Sec. 2.4, Theorem 6
5	$ac < bc$	Substitution from (1)

This means that if we multiply both sides of an inequality by a negative number, we change its sense.

THEOREM 8 If $a \neq 0$, then $a^2 > 0$.

PROOF If a is positive, then a^2 is positive; for the product of two positive numbers is positive. If a is negative, then $a < 0$. Hence, multiplying both sides by a and using Theorem 7, we find that $a^2 > 0$.

THEOREM 9 $a > 0$ if and only if $(1/a) > 0$.

PROOF (LEFT TO RIGHT)

1	Since $(1/a) \neq 0$, $(1/a)^2 > 0$	Theorem 8
2	$a(1/a)^2 > 0 \cdot (1/a)^2$	Theorem 3
3	$(1/a) > 0$	Sec. 2.3

EXERCISE F Complete the proof (right to left) of Theorem 9.

EXERCISE G PROVE: $a < 0$ if and only if $(1/a) < 0$.

THEOREM 10 If $a > b$ and $c > 0$, then $(a/c) > (b/c)$.

PROOF Multiply both sides by $(1/c)$, which is positive because of Theorem 9.

EXERCISE H PROVE: If $a < b$ and $c > 0$, then $(a/c) < (b/c)$.

THEOREM 11 If $a > b$ and $c < 0$, then $(a/c) < (b/c)$.
If $a < b$ and $c < 0$, then $(a/c) > (b/c)$.

Proof is left to the reader.

EXERCISE I Prove Theorem 11.

The net result of these theorems is that inequalities behave *almost* like equalities. We can add the corresponding sides of two inequalities having the same sense and obtain a true inequality. We can add (or subtract) equal quantities to (or from) both sides of an inequality. We can multiply or divide both sides of an inequality by a *positive* number. The only difference in the behavior of inequalities as compared to that of equations is that when we multiply or divide by a *negative* number, we must *change the sense of the inequality.*

8.3 LINEAR INEQUALITIES

In many practical situations we encounter inequalities such as the following:

$$3x + 5 < x - 7 \qquad \text{or} \qquad 2x^2 - 4x + 9 < 0$$

We wish to "solve" each of these inequalities; i.e., we seek to identify those values of x which satisfy the inequalities. That is, we are interested in the sets $\{x \mid 3x + 5 < x - 7\}$ and $\{x \mid 2x^2 - 4x + 9 < 0\}$, and wish to rewrite the definitions of these sets in a simpler form from which we can read off their elements at once. These sets will be subsets of the real numbers and will generally consist of intervals or unions of intervals.

Let us first examine linear inequalities, i.e., those like $ax + b < 0$, $ax + b > 0$, $ax + b \leq 0$, $ax + b \geq 0$. We can solve these at once by using the theorems introduced in Sec. 8.2. We proceed as in the illustrations below.

ILLUSTRATION 1 Solve $3x + 5 < x - 7$.

By Theorem 2 we may subtract $x + 5$ from each side. Doing this, we obtain $2x < -12$. By Theorem 10 we may divide both sides by 2, and we thus conclude that $x < -6$.

The graph of this solution set is given in Fig. 8.1*a*, where we use an open circle above -6 to indicate that this point is *not* part of the set.

ILLUSTRATION 2 Solve $x + 8 \geq 5x - 12$.

$$-4x + 8 \geq -12$$
$$-4x \geq -20$$
$$x \leq 5$$

Solution sets of linear inequalities

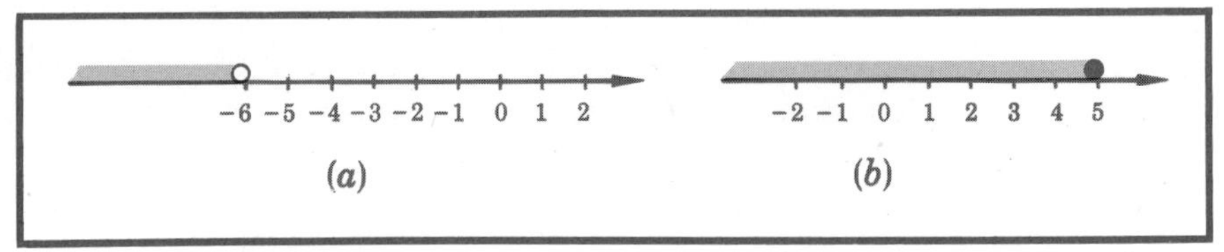

FIGURE 8.1
(*a*) $3x + 5 < x - 7$
(*b*) $x + 8 \geq 5x - 12$
The solution set is indicated by the colored strip. The open circle indicates that the endpoint is *not* part of the solution set. A filled circle indicates that the endpoint *is* included in the solution set.

The graph of this solution set is given in Fig. 8.1*b*, where we use a solid circle above 5 to indicate that this point *is* part of the set.

Thus we see that the method of solving inequalities is very similar to that for solving equations. By using the operations which we justified in Sec. 8.2 we convert the given inequality into a series of equivalent inequalities, i.e., inequalities which define the same set as the given inequality. From the last of these we can read off the answer.

As a variation on this type of problem, let us consider linear inequalities involving absolute values.

Solution of $|x| > a$

INEQUALITIES OF THE FORM $|x| > a$ If a is negative, this inequality is true for all x. If a is zero, it is true for all nonzero x. So the important case is that of a positive. Such an inequality says, geometrically, that the unsigned distance of x from zero is greater than a. So we must have either $x > a$ or $x < -a$.

ILLUSTRATION 3 The inequality $|x| > 4$ is equivalent to the statement that $x > 4$ or $x < -4$. The graph is given in Fig. 8.2*a*.

Solution of $|x - b| > a$

INEQUALITIES OF THE FORM $|x - b| > a$ Again we assume that a is positive. Then such an equality says, geometrically, that the unsigned distance of x from b is greater than a. So we must have $x - b > a$ or $x - b < -a$. Or in other words, $x > b + a$ or $x < b - a$.

ILLUSTRATION 4 In the inequality $|x + 1| > 3$ it is apparent that $b = -1$ and $a = 3$. So this is equivalent to the statement that $x > -1 + 3$ or $x < -1 - 3$. Or, in other words, $x > 2$ or $x < -4$. The graph is given in Fig. 8.2*b*.

ILLUSTRATION 5 Write an inequality which states that the distance from x to 2 is greater than 6.

SOLUTION $|x - 2| > 6$. This can be rewritten as $x > 2 + 6$ or $x < 2 - 6$, or more simply as $x > 8$ or $x < -4$.

Inequalities containing absolute values

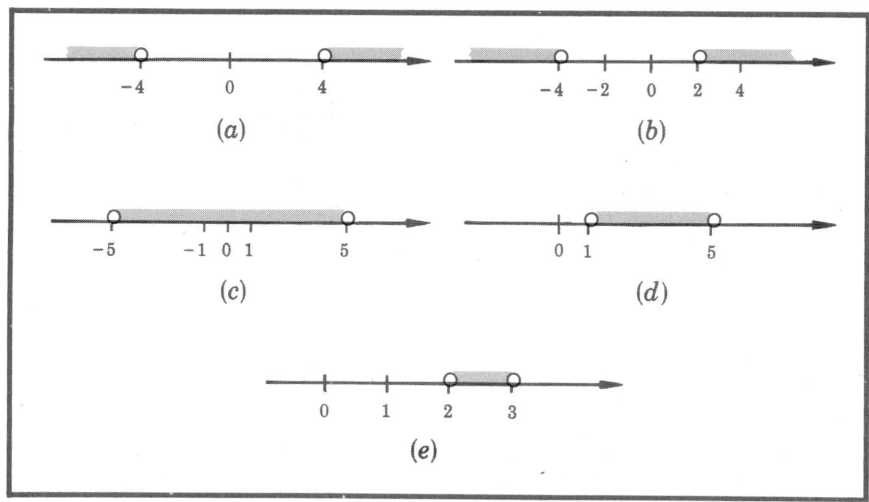

FIGURE 8.2
(*a*) $|x| > 4$
(*b*) $|x + 1| > 3$
(*c*) $|x| < 5$
(*d*) $|x - 3| < 2$
(*e*) $|2x - 5| < 1$
The graphs of inequalities can be finite intervals, infinite intervals, or unions of these.

INEQUALITIES OF THE FORM $|x| < a$ Since $|x|$ is nonnegative, there are no solutions of this inequality if a is zero or negative; so hereafter we assume that a is positive. Then the inequality $|x| < a$ says, geometrically, that the unsigned distance of x from zero is less than a. In other words, x lies in the interval between $-a$ and $+a$ (not including the endpoints). We can express this fact by writing the double inequality $-a < x < a$.

Solution of $|x| < a$

ILLUSTRATION 6 The inequality $|x| < 5$ is equivalent to the double inequality $-5 < x < 5$. The graph is given in Fig. 8.2*c*.

Solution of $|x - b| < a$

INEQUALITIES OF THE FORM $|x - b| < a$ Again there are solutions only if a is positive. Such an inequality says, geometrically, that the unsigned distance of x from b is less than a. In other words, x lies in the interval between $b - a$ and $b + a$ (not including the endpoints). We can express this by means of the double inequality $b - a < x < b + a$.

ILLUSTRATION 7 The inequality $|x - 3| < 2$ is equivalent to the double inequality $3 - 2 < x < 3 + 2$ or $1 < x < 5$. The graph is given in Fig. 8.2*d*.

ILLUSTRATION 8 Write an inequality which states that the distance from x to 5 is less than 8.

SOLUTION $|x - 5| < 8$, or $5 - 8 < x < 5 + 8$, or $-3 < x < 13$.

Solution of $|cx - b| < a$

INEQUALITIES OF THE FORM $|cx - b| < a$ Again we assume that a is positive. This is similar to the preceding case, and is equivalent to $b - a < cx < b + a$. If c is positive, we can finally write

$$\frac{b - a}{c} < x < \frac{b + a}{c}$$

ILLUSTRATION 9 The inequality $|2x - 5| < 1$ is equivalent to $5 - 1 < 2x < 5 + 1$, or to $4 < 2x < 6$, or to $2 < x < 3$. The graph is given in Fig. 8.2*e*.

The above theorems also help us to prove certain relationships by means of mathematical induction (Sec. 2.6).

Use of mathematical induction

ILLUSTRATION 10 PROVE: For all integers $n \geq 1, 2^n \geq 1 + n$. Using the notation of Sec. 2.6, we have

$$a = 1 \qquad A = \{n \geq 1\}$$
$$p_n\colon \quad 2^n \geq 1 + n$$

I p_1 is true, for $2 \geq 1 + 1$.
II For all integers $k \geq 1$: If $2^k \geq 1 + k$, then $2^{k+1} \geq 1 + (k + 1) = 2 + k$.

PROOF OF II By hypothesis $2^k \geq 1 + k$. Multiplying both sides of this inequality by 2, we obtain

$$2^{k+1} \geq 2 + 2k$$

Since $2 > 1$ and k is positive, it follows that $2k > k$. Hence

$$2^{k+1} \geq 2 + 2k > 2 + k \qquad \text{or} \qquad 2^{k+1} > 2 + k$$

which is the desired result.

ILLUSTRATION 11 PROVE: For all integers $n \geq 4$, $n! > 2^n$. We have

$$a = 4 \qquad A = \{n \mid n \geq 4\}$$
$$p_n\colon \quad n! > 2^n$$

I p_4 is true, for $4! = 24$, $2^4 = 16$, and $24 > 16$.
II For all integers $k \geq 4$: If $k! > 2^k$, then $(k + 1)! > 2^{k+1}$.

PROOF OF II By hypothesis $k! > 2^k$. Hence

$$(k + 1)! = k!(k + 1) > 2^k(k + 1)$$

Since $k \geq 4$,

$$k + 1 > 4 \qquad \text{and} \qquad 2^k(k + 1) > 2^k \cdot 4 > 2^k \cdot 2 = 2^{k+1}$$

Therefore $$(k+1)! > 2^{k+1}$$

which is the desired result.

ILLUSTRATION 12 PROVE: For all integers $n \geq -2$, $2n^3 + 3n^2 + n + 6 \geq 0$. We have

$$a = -2 \qquad A = \{n \mid n \geq -2\}$$
$$p_n\colon\ 2n^3 + 3n^2 + n + 6 \geq 0$$

I p_{-2} is true, since $2(-2)^3 + 3(-2)^2 + (-2) + 6 \geq 0$.
II For all integers $k \geq -2$: If $2k^3 + 3k^2 + k + 6 \geq 0$, then

$$2(k+1)^3 + 3(k+1)^2 + (k+1) + 6 \geq 0$$

PROOF OF II We write

$$\begin{aligned} 2(k+1)^3 + 3(k+1)^2 + (k+1) + 6 \\ &= 2k^3 + 9k^2 + 13k + 12 \\ &= (2k^3 + 3k^2 + k + 6) + (6k^2 + 12k + 6) \\ &= (2k^3 + 3k^2 + k + 6) + 6(k+1)^2 \end{aligned}$$

In the last expression, $(2k^3 + 3k^2 + k + 6) \geq 0$ by hypothesis, and $(k+1)^2 \geq 0$. Hence the sum of these is greater than or equal to zero.

EXERCISE A Show that $2n^3 + 3n^2 + n + 6 = (n+1)(2n^2 + n) + 6$. Hence construct a proof of the inequality in Illustration 12 without using induction.

PROBLEMS 8.3

In Probs. 1 to 20 solve the stated inequalities, and plot their solution sets on the line.

1 $x + 5 > 0$
2 $3x + 9 > 0$
3 $-2x - 6 \geq 0$
4 $-4x + 8 \leq 0$
5 $3x + 5 > x + 7$
6 $-3x + 4 < 2x - 6$
7 $|x| > 3$
8 $|x + 2| \geq 6$
9 $|x - 1| < 5$
10 $|x - 2| < 4$
11 $|2x - 5| \leq 9$
12 $|4x + 3| > -5$
13 $|x + 1| < -1$
14 $|x + 3| < 7$
15 $|2x + 4| < 10$
16 $|4x - 13| \leq 5$
17 $|2x + 7| \geq 9$
18 $|-3x + 6| < 9$
19 $|x - 3| < 0.1$
20 $|x - 3| < 0.01$

In Probs. 21 to 28 write an inequality which states that:

21 x is within 4 units of 3.
22 x is within 5 units of -2.
23 $2x$ is within 6 units of 4.
24 $3x$ is within 2 units of 4.

25 The distance from x to 3 is greater than 2.
26 The distance from x to -4 is greater than 6.
27 The distance from $3x$ to -4 is less than 5.
28 The distance from $-2x$ to 6 is less than 3.

29 Prove that the affine transformation $x' = ax + b$ with $a > 0$ preserves the order relationship; i.e., if $x_1 > x_2$, then $x_1' > x_2'$.
30 Prove that the affine transformation $x' = ax + b$ with $a < 0$ reverses the order relationship; i.e., if $x_1 > x_2$, then $x_1' < x_2'$.

In Probs. 31 to 36 prove the given statement by mathematical induction.

31 For all integers $n \geq 1$, $5^n \geq 1 + 4n$.
32 For all integers $n \geq 1$, $3^n \geq 1 + 2n$.
33 For all integers $n \geq 1$ and for $h \geq -1$, $(1 + h)^n \geq 1 + nh$.
34 For all integers $n \geq 1$ and for $h \geq 0$,

$$(1 + h)^n \geq 1 + nh + \frac{n(n-1)}{2}h^2$$

35 For all integers $n \geq -1$, $2n^3 - 9n^2 + 13n + 25 > 0$.
36 For all integers $n \geq -2$, $2n^3 + 9n^2 + 13n + 7 > 0$.

37 PROVE: For all integers $n \geq 1$ and for $-1 < h \leq \sqrt{2}$, $(1 - h)^n \leq 1/(1 + h)^n$. HINT: $(1 - h)^n(1 + h)^n = (1 - h^2)^n \leq 1$. Now divide this inequality by $(1 + h)^n$.
38 PROVE: If $a \geq b$ and $ab > 0$, then $(1/a) \leq (1/b)$.
39 PROVE: For all integers $n \geq 1$ and for $0 \leq h \leq \sqrt{2}$, $(1 - h)^n \leq 1/(1 + nh)$. HINT: Use the results of Probs. 37, 33, and 38.
40 PROVE: For all integers $n \geq 1$ and for $0 \leq h < \sqrt{2}$,

$$(1 - h)^n \leq 1\Big/\left(1 + nh + \frac{n(n-1)}{2}h^2\right)$$

41 Show that $2x^3 - 3x^2 - 5x + 3 \leq 2|x|^3 + 3|x|^2 + 5|x| + 3$. Hence show that when $-3 < x < 1$, $2x^3 - 3x^2 - 5x + 3 < 99$. HINT: From $-3 < x < 1$, we conclude that $|x| < 3$.
42 Using the method of Prob. 41, show that when $-1 < x < 2$, $x^3 - 4x^2 - 3x + 5 < 35$.

***43** Show that if $|x| < 1$, then $\left|\dfrac{x+1}{x-2}\right| < 2$.

HINT: Use the following steps:
a $-1 < x < 1$; so $0 < x + 1 < 2$ and $-3 < x - 2 < -1$.
b $|x + 1| < 2$ and $|x - 2| > 1$

c $\left|\dfrac{x+1}{x-2}\right| = \left|\dfrac{x+1}{x-2}\right| < \dfrac{2}{1} = 2$

In this step we obtain an upper bound for the given expression by making its numerator as large as it can be and its denominator as small as it can be.

***44** Show that if $|x| < 2$, then $\left|\dfrac{x-3}{x+4}\right| < \dfrac{5}{2}$.

***45** Show that if $|x-3| < 1$, then $\left|\dfrac{x+5}{x+1}\right| < 3$.

***46** Show that if $|x| < 1$, then $\left|\dfrac{x-2}{x+3}\right| < \dfrac{3}{2}$.

***47** Show that if $|x-2| < 1$, then $|x^2-4| < 5$.
***48** Show that if $|x-2| < 0.1$, then $|x^2-4| < 0.41$.
***49** Show that if $|x-2| < 0.01$, then $|x^2-4| < 0.0401$.
***50** Show that if $|x-2| < a$, then $|x^2-4| < 4a + a^2$.

8.4 › QUADRATIC INEQUALITIES

The properties of quadratic inequalities follow easily from those of quadratic equations. We are concerned with the inequalities

$$ax^2 + bx + c > 0$$
$$ax^2 + bx + c < 0$$
$$ax^2 + bx + c \geq 0$$
$$ax^2 + bx + c \leq 0$$

where $a > 0$, and a, b, and c are real. Only real values of x will be permitted.

Let us suppose, first, that $b^2 - 4ac > 0$, so that the solutions of $ax^2 + bx + c = 0$ are real and unequal. Then we can write

$$ax^2 + bx + c = a(x - r_1)(x - r_2)$$

where r_1, r_2 are real and $r_1 \neq r_2$. Suppose $r_1 < r_2$. Let us imagine x moving from left to right along the real line, and consider how the sign of $ax^2 + bx + c$ varies in the process. At the extreme left (where $x < r_1 < r_2$), $(x - r_1)$ is negative, and $(x - r_2)$ is negative. Hence, in this region, $a(x - r_1)(x - r_2)$ is positive. When x reaches r_1, $a(x - r_1)(x - r_2)$ becomes zero. Between r_1 and r_2, $(x - r_1)$ is positive and $(x - r_2)$ is negative. So in this region, $a(x - r_1)(x - r_2)$ is negative. At r_2, $a(x - r_1) \times (x - r_2)$ is zero. To the right of r_2, $(x - r_1)$ and $(x - r_2)$ are both positive, so that $a(x - r_1)(x - r_2)$ is positive. These results are summarized in Fig. 8.3. We state them formally in the theorem.

Sign pattern for $ax^2 + bx + c$

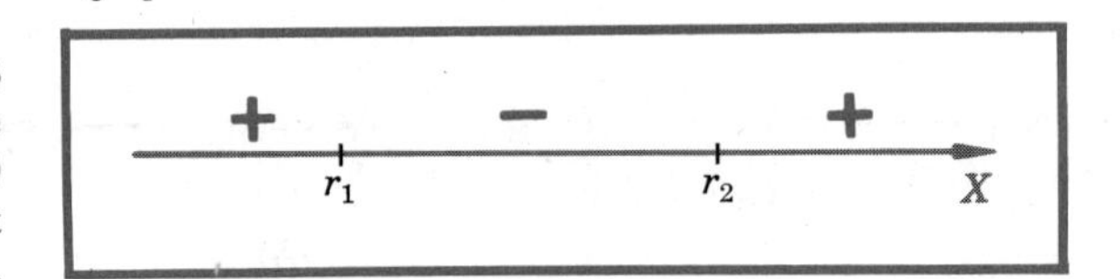

FIGURE 8.3 This figure assumes that the roots of $ax^2 + bx + c = 0$ are real and distinct and that $a > 0$.

THEOREM 12 — Quadratic has real solutions

If $ax^2 + bx + c = 0$ $(a > 0)$ has distinct real solutions r_1 and r_2 with $r_1 < r_2$, then

$$\begin{array}{ll} ax^2 + bx + c > 0 & x < r_1 \\ ax^2 + bx + c = 0 & x = r_1 \\ ax^2 + bx + c < 0 & r_1 < x < r_2 \\ ax^2 + bx + c = 0 & x = r_2 \\ ax^2 + bx + c > 0 & x > r_2 \end{array}$$

A similar method gives the next result.

THEOREM 13

If $ax^2 + bx + c = 0$ $(a > 0)$ has equal real solutions $r_1 = r_2 = r$, then

$$\begin{array}{ll} ax^2 + bx + c > 0 & x < r \\ ax^2 + bx + c = 0 & x = r \\ ax^2 + bx + c > 0 & x > r \end{array}$$

Quadratic has nonreal solutions

The final situation is that in which $ax^2 + bx + c = 0$ has nonreal solutions; that is, $b^2 - 4ac < 0$. As we have just seen, the only way for $ax^2 + bx + c$ to change sign as x moves from left to right is for x to pass through a point at which $ax^2 + bx + c = 0$. In this case, however, this equation has no real solutions; hence there is no such point. Therefore, $ax^2 + bx + c$ must have a constant sign. At the extreme right (or left), $ax^2 + bx + c$ is certainly positive. Hence it is positive for all values of x. This proves the theorem.

THEOREM 14

If $ax^2 + bx + c = 0$ $(a > 0)$ has nonreal solutions, then for all x, $ax^2 + bx + c > 0$.

EXERCISE A Prove the converse of Theorem 14.

ILLUSTRATION 1

a Solve $3(x - 1)(x - 4) > 0$. *Ans.*: $\{x \mid (x < 1) \text{ or } (x > 4)\}$. (See Fig. 8.4*a*.)

b Solve $2(x + 1)(x - 2) < 0$. *Ans.*: $\{x \mid -1 < x < 2\}$. (See Fig. 8.4*b*.)

c Solve $4(x - 3)^2 \geq 0$. *Ans.*: All real x. (See Fig. 8.4*c*.)

Solution sets of quadratic inequalities

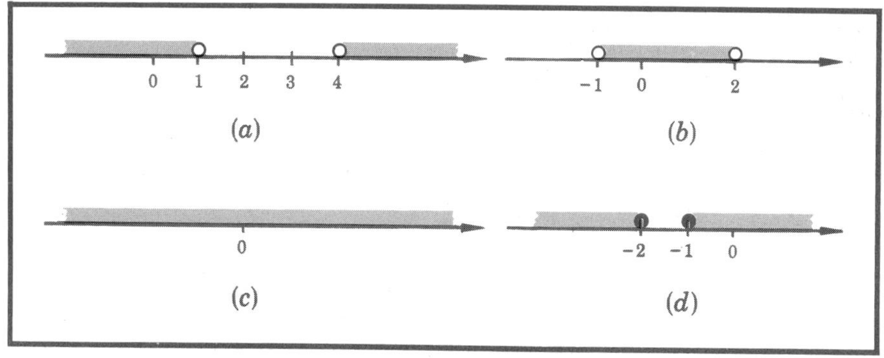

FIGURE 8.4
(a) $3(x-1)(x-4)>0$
(b) $2(x+1)(x-2)<0$
(c) $4(x-3)^2 \geq 0$ and $x^2+x+1>0$
(d) $(x+1)(x+2) \geq 0$
To solve a quadratic inequality, first find the corresponding roots and then test each of the resulting intervals.

d Solve $(x+1)(x+2) \geq 0$. *Ans.*: $\{x \mid (x \leq -2) \text{ or } (x \geq -1)\}$. (See Fig. 8.4*d*.)

e Solve $x^2+x+1>0$. *Ans.*: All real x. (See Fig. 8.4*c*.)

The method of this section is easily extended to inequalities of the form

$$a(x-r_1)(x-r_2)(x-r_3)\cdots(x-r_n)>0, <0$$

etc., where $a>0$ and $r_1, \ldots, r_n$ are real.

ILLUSTRATION 2 Solve $2(x-1)(x-2)(x-3)>0$.

When $x<1$, each factor is negative; so the product is negative.

When $1<x<2$, $(x-1)$ is positive, and $(x-2)$ and $(x-3)$ are negative. Hence the product is positive.

Continuing in this way we find the solution set: $\{x \mid (1<x<2) \text{ or } (3<x)\}$, whose graph is given in Fig. 8.5.

Inequalities involving fractions

A further extension of this method permits us to solve inequalities involving algebraic fractions. As in Sec. 6.6, we must first exclude from the universal set U (of permissible real values of x) exactly those values of x for which any denominator is zero. In the case of equations, we then obtained an equivalent equation by multiplying by any polynomial which was nonzero for all x in U. Here, however, we need an extra condition if we are to obtain an equivalent inequality: namely, the

Solution set of a cubic inequality

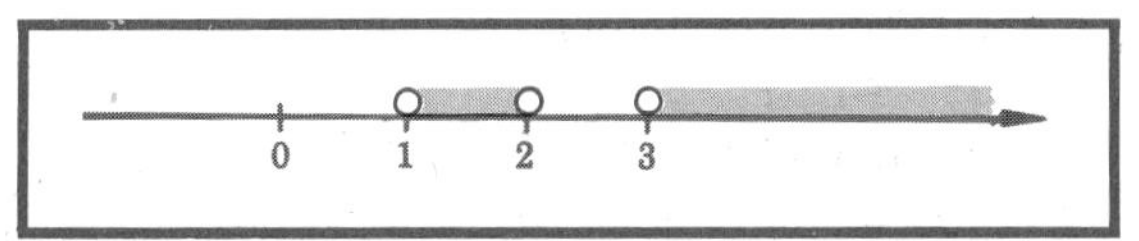

FIGURE 8.5
$2(x-1)(x-2)(x-3)>0$
The same method applies to products of three or more factors.

Solution set of a fractional inequality

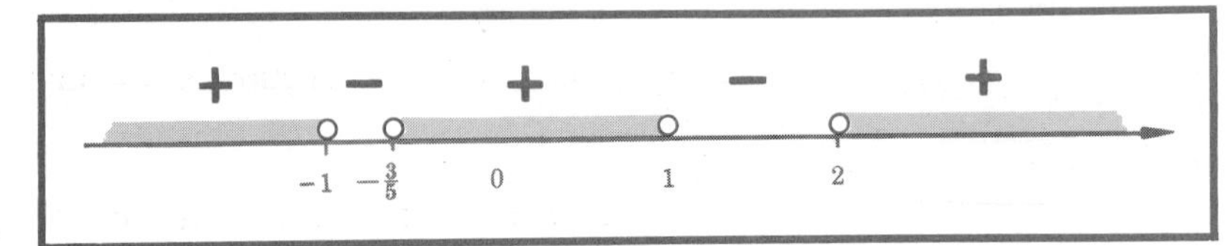

FIGURE 8.6
$\dfrac{7}{x-1} - \dfrac{6}{x^2-1} < 5$ or
$(x+1)(5x+3) \times (x-1)(x-2) > 0$
The method also applies to an inequality involving a fraction whose numerator and denominator are products of linear factors.

polynomial by which we multiply must be *positive* for all x in U. We then solve the resulting inequality by the method of Illustration 2.

ILLUSTRATION 3 Solve $\dfrac{7}{x-1} - \dfrac{6}{x^2-1} < 5$.

Then U is R except for -1 and 1. It would now be incorrect to multiply both sides by $x^2 - 1$, for $x^2 - 1$ is positive for some x in U and negative for other x in U. So instead we multiply by $(x^2 - 1)^2$ which is positive for all x in U. The result is

Multiplier must be positive

$$7(x^2-1)(x+1) - 6(x^2-1) < 5(x^2-1)^2$$

After simplification this becomes

$$(x+1)(5x+3)(x-1)(x-2) > 0$$

The sign of the left-hand side changes at -1, $-\frac{3}{5}$, 1, and 2. We then plot the graph (Fig. 8.6) and examine each interval. The answer is

$$\{x \mid x < -1, \text{ or } -\tfrac{3}{5} < x < 1, \text{ or } x > 2\}$$

PROBLEMS 8.4

In Probs. 1 to 20 solve the given inequality, and plot the graph of its solution set.

1 $(x-1)(x+3) > 0$
2 $(x+2)(x-4) > 0$
3 $5(x-1)^2 \geq 0$
4 $-2(x-4)^2 \leq 0$
5 $2x^2 - 5x - 3 \geq 0$
6 $3x^2 + 2x - 8 \leq 0$
7 $4x^2 + 20x + 24 < 0$
8 $2x^2 + 5x - 3 < 0$
9 $x^2 + x + 1 \geq 0$
10 $x^2 + 2x + 2 < 0$
11 $-x^2 - 3x - 3 > 0$
12 $2x^2 + x + 5 > 0$
13 $3(x+3)(x+1)(x-4) > 0$
14 $4(x+2)(x+1)(x-3) < 0$
15 $-4(x+1)^2(x-5) \geq 0$
16 $3(x+5)(x)(x-3) > 0$
17 $\dfrac{4}{x+1} - \dfrac{3}{x+2} > 1$
18 $\dfrac{5}{x+3} + \dfrac{1}{x-1} > 2$
19 $\dfrac{5}{x+7} + \dfrac{3}{x+5} < 2$
20 $-\dfrac{4}{x-1} + \dfrac{7}{x+4} < -1$

***21** By mathematical induction show that: For all integers $n \geq 5$, $2^n > n^2$.

***22** By mathematical induction show that: For all integers $n \geq 5$, $4^n > n^4$.

In Probs. 23 to 30 try to establish the indicated theorem by mathematical induction. Point out why the method fails.

23 For $n \geq 1$, $2^n > n^2$

24 For $n \geq 1$, $4^n > n^4$

25 For $n \geq 2$, $2^n > n^2$

26 For $n \geq 2$, $4^n > n^4$

27 For $n \geq 3$, $2^n > n^2$

28 For $n \geq 3$, $4^n > n^4$

29 For $n \geq 4$, $2^n > n^2$

30 For $n \geq 4$, $4^n > n^4$

APPLICATIONS

31 PROVE: Let $a > 0$ and $b > 0$. Then, if $a^2 > b^2$, it follows that $a > b$. HINT: Consider $a^2 - b^2 > 0$ or $(a + b)(a - b) > 0$.

32 From the Pythagorean relation $a^2 + b^2 = c^2$, prove that the hypotenuse of a right triangle is longer than either leg.

33 Apply the result of Prob. 32 to the general triangle given in Fig. 8.7 to show that $a + b > c + d$. Hence the sum of two sides of a triangle is greater than the third side.

34 In Fig. 8.8 assume that $c > a$. Use the result of Prob. 33 to show that $c - a < b$. Hence any side of a triangle is greater than the absolute value of the difference of the other two sides.

35 Prove that $|a + b| \leq |a| + |b|$. HINT: Since $|a|^2 = a^2$, etc., let us consider

$$\begin{aligned}(a + b)^2 &= a^2 + 2ab + b^2 \\ &\leq a^2 + 2|a||b| + b^2 \\ &\leq [|a| + |b|]^2\end{aligned}$$

When does equality hold in the above relation?

36 By mathematical induction prove that: For all integers $n \geq 2$, $|x_1 + x_2 + \cdots + x_n| \leq |x_1| + |x_2| + \cdots + |x_n|$.

37 Using the result of Prob. 35, prove that

$$|a - b| \geq |a| - |b|$$

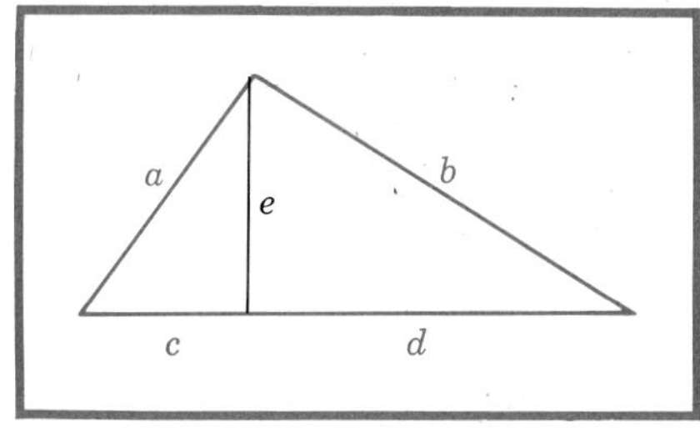

FIGURE 8.7

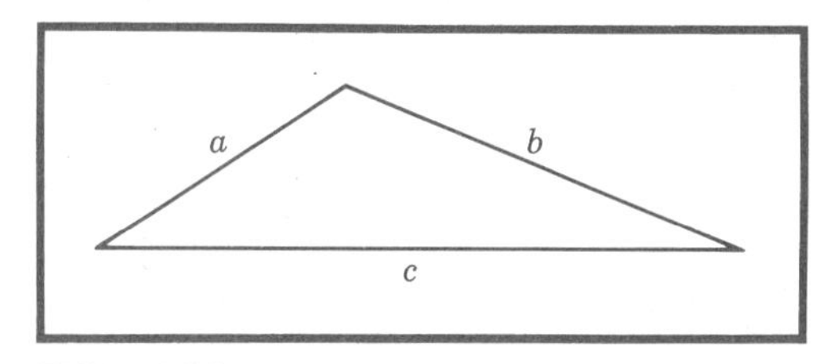

FIGURE 8.8

38 Prove that $\sqrt{ab} \le \dfrac{a+b}{2}$ if a and b are positive. HINT: Show that this is equivalent to $0 \le a^2 - 2ab + b^2$. $\sqrt{ab}$ is called the "geometric mean" of a and b; $(a+b)/2$ is called the "arithmetic mean".

The above result is a special case of the general theorem which states that, for a set of positive quantities $a_1, \ldots, a_n$, the geometric mean $\sqrt[n]{a_1 a_2 \cdots a_n}$ is less than or equal to the arithmetic mean $(1/n)(a_1 + \cdots + a_n)$. When does equality hold in the above relation?

39 Prove that for any real numbers a_1, a_2, b_1, and b_2

$$(a_1 b_1 + a_2 b_2)^2 \le (a_1{}^2 + a_2{}^2)(b_1{}^2 + b_2{}^2)$$

This is known as "Cauchy's Inequality". HINT: Consider

$$(a_1 x + b_1)^2 = a_1{}^2 x^2 + 2a_1 b_1 x + b_1{}^2 \ge 0$$
$$(a_2 x + b_2)^2 = a_2{}^2 x^2 + 2a_2 b_2 x + b_2{}^2 \ge 0$$

Adding, we find that

$$(a_1{}^2 + a_2{}^2)x^2 + 2(a_1 b_1 + a_2 b_2)x + (b_1{}^2 + b_2{}^2) \ge 0 \qquad \text{for all } x$$

Now apply the result of Exercise A, Sec. 8.4. When does equality hold in the above relation?

40 Generalize the result of Prob. 39 to two sets of numbers: $a_1, \ldots, a_n$ and $b_1, \ldots, b_n$. The formula so obtained is of great importance in statistics and higher geometry.

41 Let **A** be the vector $(a_1, \ldots, a_n)$; **B** be the vector $(b_1, \ldots, b_n)$; $|\mathbf{A}|$ be the length of **A**; and $|\mathbf{B}|$ be the length of **B**. Then from Cauchy's Inequality show that

$$\left(\frac{\mathbf{A} \cdot \mathbf{B}}{|\mathbf{A}||\mathbf{B}|}\right)^2 \le 1$$

42 Show that of all rectangles with a given perimeter, the square has the largest area. HINT: Let the rectangle's dimensions be x and y. The perimeter $P = 2x + 2y$. The area A is xy. From Prob. 38,

$$xy \le \frac{(x+y)^2}{4} \qquad \text{or} \qquad A \le \frac{P^2}{16}$$

Since P is fixed, A will be greatest when equality holds. When is this?

8.5 THE GRAPH OF A LINEAR INEQUALITY

Here we are concerned with the graph of the linear inequality $ax + by + c > 0$, or more properly with the graph of the set

$$\{(x,y) \mid ax + by + c > 0\}$$

We assume that at least one of the coefficients a and b is not zero.

First we plot the graph of the equation

$$ax + by + c = 0$$

which we know is a line.

This line divides the plane into two open half-planes A and B, neither of which contains the line. These half-planes are each *connected,* in the sense that, if points P_1 and P_2 lie in one of them, then there is a smooth curve joining P_1 and P_2 which lies entirely in that half-plane (Fig. 8.9). In particular, the curve does not have a point in common with the given line. Our procedure for graphing the solution set of $ax + by + c > 0$ depends strongly on the next theorem.

THEOREM 15 Let P lie in one of the two half-planes into which the graph of $ax + by + c = 0$ divides the plane. If $ax + by + c > 0$ at P, then $ax + by + c > 0$ at every point of the half-plane in which P lies.

PROOF Assume that P_1 and P_2 lie in the same half-plane, that $ax + by + c > 0$ at P_1, and that $ax + by + c \leq 0$ at P_2.

We shall prove that this is absurd (indirect proof).

Let P_1 have coordinates (x_1,y_1) and P_2 have coordinates (x_2,y_2).

In the first place, we cannot have $ax_2 + by_2 + c = 0$, for in this case P_2 is on the given line, which is contrary to its assumed location. Hence our hypotheses reduce to: $ax_1 + by_1 + c > 0$; $ax_2 + by_2 + c < 0$; P_1 and P_2 in the same half-plane.

Draw the curve P_1P_2, which lies entirely in the given open half-plane. Let $\bar{P}(\bar{x},\bar{y})$ move along this curve from P_1 to P_2. At P_1, $a\bar{x} + b\bar{y} + c > 0$, and at P_2, $a\bar{x} + b\bar{y} + c < 0$. Hence there must be a

Separation of the plane by a line

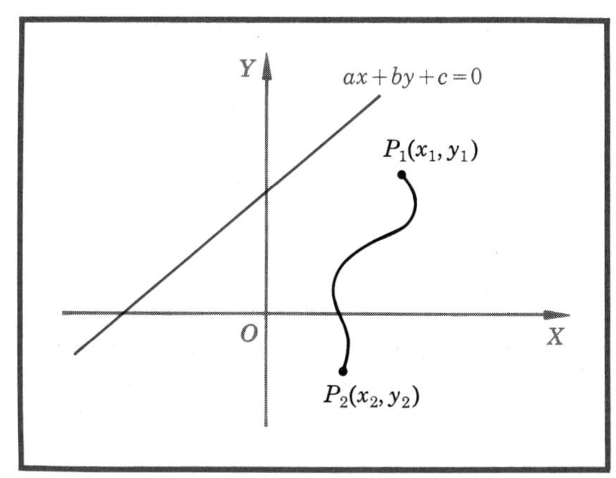

FIGURE 8.9
A line divides the plane into three disjoint subsets: (1) the line, (2) a half-plane on one side of it, (3) a half-plane on the other side of it.

point on the curve P_1P_2 at which $a\bar{x} + b\bar{y} + c = 0$. This, however, is impossible, for such a point would have to be on the given line, and P_1P_2 has no points in common with this line.†

† This proof assumes properties of continuity which you have not studied so far. Although precise details must be omitted here, the idea should be clear.

EXERCISE A State similar theorems for the inequalities: $ax + by + c < 0$, $ax + by + c \geq 0$, and $ax + by + c \leq 0$.

Test point in each half-plane

As a result of this theorem, we have the following procedure for plotting the graph of the solution set of a linear inequality. (1) Plot the graph of the line $ax + by + c = 0$. (2) For a single point in each half-plane determined by this line, test the truth of the given inequality. (3) Shade the half-plane or half-planes corresponding to points at which the given inequality is true.

REMARK You will observe that if $ax + by + c > 0$ in one half-plane, then $ax + by + c < 0$ in the other half-plane. Nevertheless, it is wise to check both half-planes. There are situations in which an inequality is true in both half-planes. For instance, this is the case for the inequality

$$(ax + by + c)^2 > 0$$

ILLUSTRATION 1 Plot the graph of the solution set of $3x + 2y - 6 > 0$.

The points (0,3) and (2,0) satisfy the equation $3x + 2y - 6 = 0$, so we plot the line through them (Fig. 8.10). The point (0,0) is in the lower half-plane, and $0 + 0 - 6 > 0$ is false. On the other hand, (3,0) is in the upper half-plane and $9 + 0 - 6 > 0$ is true. Therefore we color the

Linear inequality in two variables

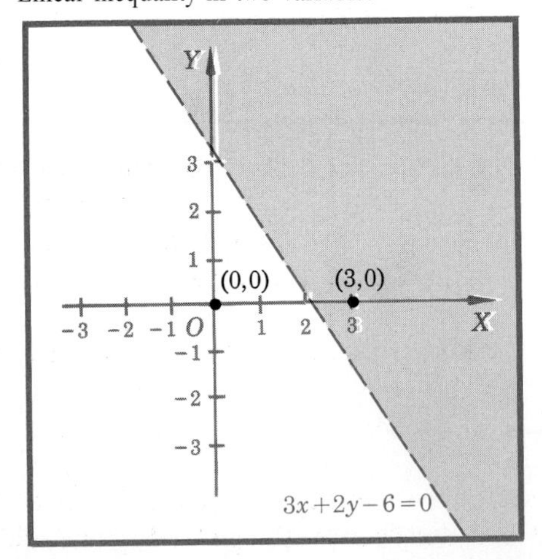

FIGURE 8.10
$3x + 2y - 6 > 0$
The solution set of a linear inequality is the colored half-plane and may or may not also include the line. A dashed line is not part of the solution set; a solid line is included in the solution set (see Fig. 8.11).

Linear inequality in two variables

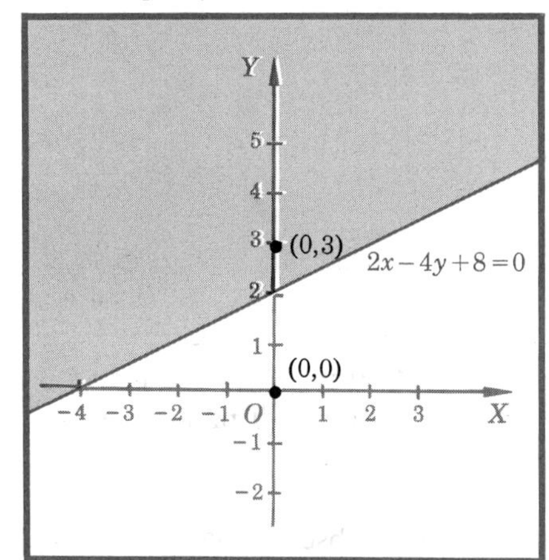

FIGURE 8.11
$2x - 4y + 8 \leq 0$
The solution set of a linear inequality is the colored half-plane and may or may not also include the line. A dashed line is not part of the solution set (see Fig. 8.10); a solid line is included in the solution set.

upper half-plane. The line is dashed, since it is not part of the desired graph.

ILLUSTRATION 2 Plot the graph of the solution set of $2x - 4y + 8 \leq 0$.

We draw the line through $(-4,0)$ and $(0,2)$ (Fig. 8.11). The point $(0,0)$ does not satisfy the inequality. The point $(0,3)$ does satisfy the inequality, so we color the upper half-plane, drawing the line solid, since it is part of the graph.

8.6 SIMULTANEOUS LINEAR INEQUALITIES

We now extend our treatment of linear inequalities (Sec. 8.5) to the case of systems of inequalities. A typical system of two linear inequalities is

$$\begin{aligned} a_1x + b_1y + c_1 &> 0 \qquad a_1 \neq 0 \text{ or } b_1 \neq 0 \\ a_2x + b_2y + c_2 &> 0 \qquad a_2 \neq 0 \text{ or } b_2 \neq 0 \end{aligned}$$

where we assume that the coefficients are real numbers. The solution set of each inequality is a half-plane, and the solution set of the pair of inequalities is the intersection of these two half-planes. The best method of procedure is graphical.

ILLUSTRATION 1 Graph the set determined by

$$\begin{aligned} 2x + y - 3 &> 0 \\ x - 2y + 1 &< 0 \end{aligned}$$

First draw the two lines, which intersect at $(1,1)$. We find that $2x + y - 3 > 0$ determines its right half-plane and that $x - 2y + 1 < 0$ de-

Two simultaneous linear inequalities

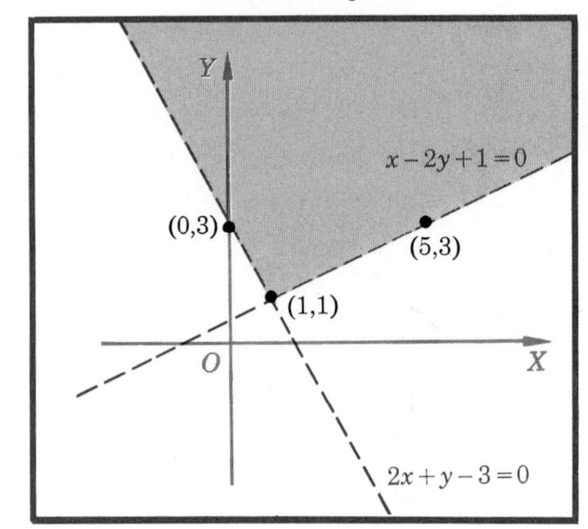

FIGURE 8.12
$2x + y - 3 > 0$
$x - 2y + 1 < 0$
The solution set of a system of inequalities is the intersection of their individual solution sets.

termines its left half-plane. The region common to the two is colored in Fig. 8.12. It is the interior of an angle whose vertex is (1,1).

This illustration is typical for two inequalities. But we can consider three or more simultaneous inequalities. The ideas and procedure are the same.

ILLUSTRATION 2 Graph the set determined by

$$\begin{aligned} 2x + y - 3 &> 0 \\ x - 2y + 1 &< 0 \\ y - 3 &< 0 \end{aligned}$$

The first two inequalities are the same as in Illustration 1; so we merely add the third line to Fig. 8.12. This gives a triangle with vertices (1,1), (0,3), and (5,3). This line divides the colored region of Fig. 8.12 into two parts. We see that $y - 3 < 0$ determines its lower half-plane. Hence the desired set is the interior of the colored triangle in Fig. 8.13.

ILLUSTRATION 3 Let us add one more inequality to our picture, and consider the system

$$\begin{aligned} 2x + y - 3 &> 0 \\ x - 2y + 1 &< 0 \\ y - 3 &< 0 \\ x + y - 5 &< 0 \end{aligned}$$

Since $x + y - 5 < 0$ determines its left half-plane, the desired set is the interior of the colored quadrilateral in Fig. 8.14.

Three simultaneous linear inequalities

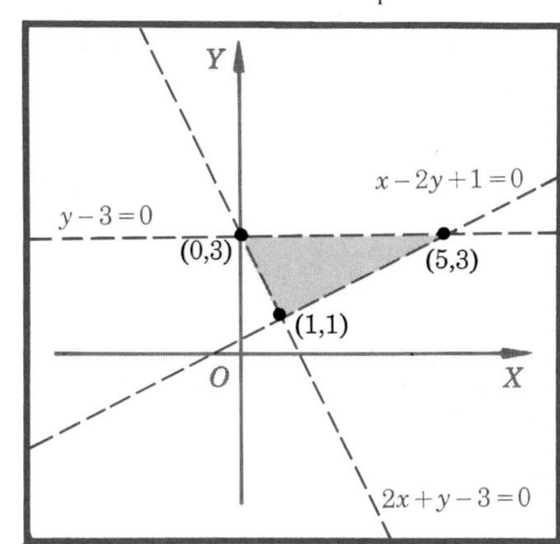

FIGURE 8.13
$$\begin{aligned} 2x + y - 3 &> 0 \\ x - 2y + 1 &< 0 \\ y - 3 &< 0 \end{aligned}$$
The solution set of a system of inequalities is the intersection of their individual solution sets.

Four simultaneous linear inequalities

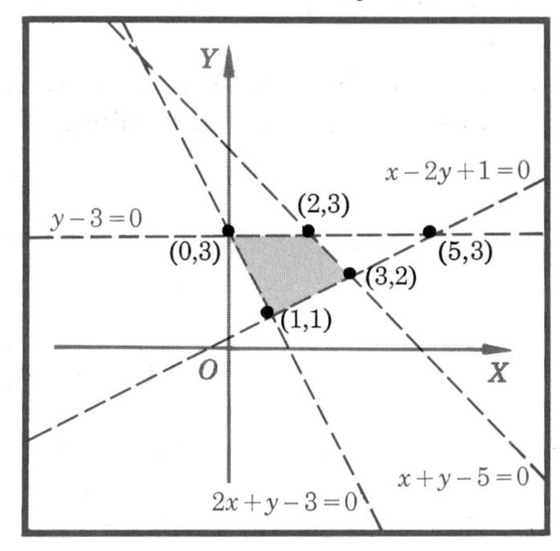

FIGURE 8.14
$$\begin{aligned} 2x + y - 3 &> 0 \\ x - 2y + 1 &< 0 \\ y - 3 &< 0 \\ x + y - 5 &< 0 \end{aligned}$$
The solution set of a system of inequalities may be the interior of a polygon.

ILLUSTRATION 4 Instead of the system of Illustration 3, consider

$$\begin{aligned} 2x + y - 3 &> 0 \\ x - 2y + 1 &< 0 \\ y - 3 &< 0 \\ x + y + 2 &< 0 \end{aligned}$$

The inequality $x + y + 2 < 0$ determines its left half-plane, which has no points in common with the triangle of Fig. 8.13. Hence the desired set is empty, and nothing is colored in Fig. 8.15.

Four simultaneous linear inequalities

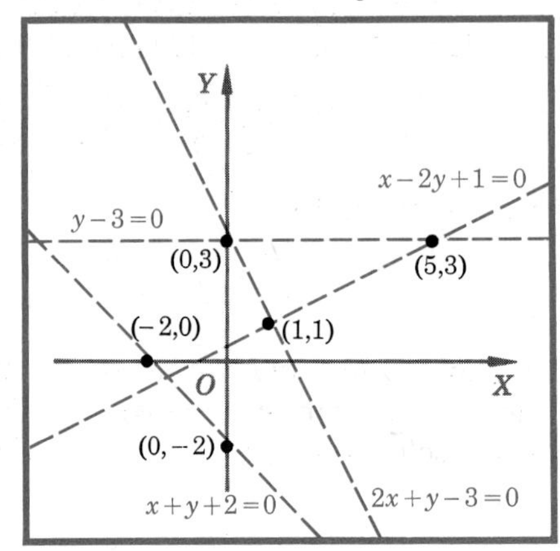

FIGURE 8.15
$2x + y - 3 > 0$
$x - 2y + 1 < 0$
$y - 3 < 0$
$x + y + 2 < 0$
The solution set of a system of inequalities may be the empty set.

Simultaneous inequalities determine a polygon

We see in this way that the graph of a system of linear inequalities can take many forms. In the usual cases it is the interior of a convex polygon, which may be bounded as in Figs. 8.13 and 8.14 or unbounded as in Fig. 8.12. In other cases it may be the empty set. When the inequalities include $\geq$ or $\leq$, the possibilities of variation in form are even more numerous.

Simultaneous inequalities can arise in unexpected places in practical situations. Let us consider two of these.

Football problem

ILLUSTRATION 5 The situation in a simplified version of a football game is as follows. There are just two plays, a running play and a pass play. We assume these facts to be true:

Play	*Distance gained, yd*	*Time required, sec*
Running	3	30
Pass	6	10

Also suppose that there are 60 yd to go for a touchdown and that 150 sec remain in the game. Ignore the requirement of having to make 10 yd in four downs and other considerations of score and strategy.

PROBLEM What combinations of running and pass plays will secure a touchdown in the allotted time?

SOLUTION Let r and p, respectively, represent the numbers of running and pass plays. Then the conditions of the problem may be written

$$3r + 6p \geq 60 \qquad r \geq 0$$
$$30r + 10p \leq 150 \qquad p \geq 0$$

These may be simplified to

$$r + 2p \geq 20 \qquad r \geq 0$$
$$3r + p \leq 15 \qquad p \geq 0$$

Plot the graph of this simultaneous system as in Fig. 8.16. The solution set is the colored triangle including its sides. Since r and p must be integers, the solution consists of the following combinations:

r	0	0	···	0	1	1	···	1	2
p	10	11	···	15	10	11	···	12	9

The quarterback can then select his pattern of plays from this array according to his best judgment regarding strategy or other matters.

Lumber problem

ILLUSTRATION 6 The Minneapolis and Seattle Lumber Company can convert logs into either lumber or plywood. In a given week the mill

Solution of the football problem

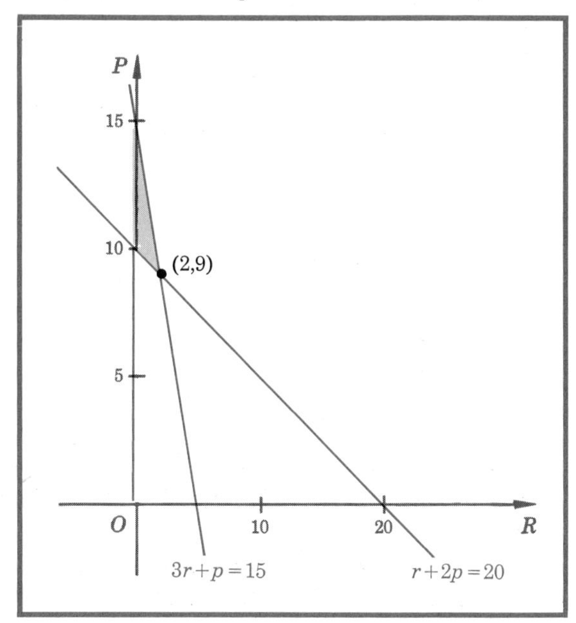

FIGURE 8.16
$r + 2p \geq 20 \qquad r \geq 0$
$3r + p \leq 15 \qquad p \geq 0$
The quarterback should use a combination of plays corresponding to a point in the interior or on the boundary of the colored triangle.

can turn out 400 units of production, of which 100 units of lumber and 150 units of plywood are required by regular customers. The profit on a unit of lumber is \$20 and on a unit of plywood is \$30.

PROBLEM How many units of lumber and plywood should the mill produce (totaling 400) in order to maximize the total profit?

SOLUTION Let L and P, respectively, represent the number of units of lumber and plywood. Then the conditions of the problem are

$$L + P = 400 \qquad L \geq 100 \qquad P \geq 150$$
$$\text{Total profit} = 20L + 30P$$

Let us graph these in Fig. 8.17.

The possible solutions must lie on the line $L + P = 400$ and in the colored region of the plane. Hence they lie on the segment AB. The total profit at A is \$11,000 and at B is \$9,500. At other points of AB the profit lies between these values. Hence the mill should produce 100 units of lumber and 300 units of plywood.

Perhaps this solution is obvious to you without all this analysis, and it should be. The point of the illustration is that methods of this kind are extremely useful in more complicated problems where the answers are far from obvious. The mathematical subject which deals with such problems is called *Linear Programming* which we treat in more detail in Sec. 8.9.

Solution of the lumber problem

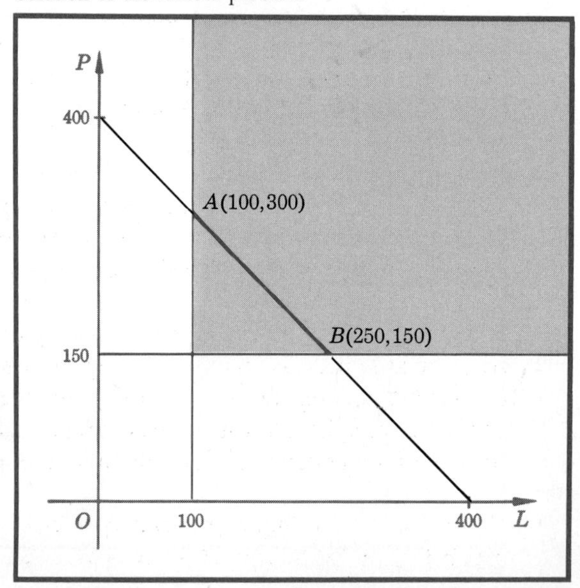

FIGURE 8.17
$L + P = 400$
$L \geq 100$
$P \geq 150$
Production must correspond to some point on the closed segment AB.

PROBLEMS 8.6

In Probs. 1 to 6 plot the graph of the solution set of the given inequality.

1 $2x - 3y + 6 > 0$
2 $2x + 6y - 12 > 0$
3 $x + y + 2 \leq 0$
4 $x - y - 3 \geq 0$
5 $(3x - 4y + 12)^2 \geq 0$
6 $(4x + 3y - 12)^2 < 0$

In Probs. 7 to 12 plot the graph of the solution set of the simultaneous inequalities.

7 $x + 3y - 7 > 0$
$3x - 2y + 1 > 0$

8 $x + 2y - 3 > 0$
$2x - y - 1 < 0$

9 $x + 3y - 7 > 0$
$3x - 2y + 1 > 0$
$4x + y - 17 < 0$

10 $x + 2y - 3 > 0$
$2x - y - 1 < 0$
$x - 3y + 12 < 0$

11 $x + 3y - 7 > 0$
$3x - 2y + 1 > 0$
$4x + y - 17 > 0$

12 $x + 2y - 3 > 0$
$2x - y - 1 < 0$
$x - 3y + 12 > 0$

8.7 THE GRAPH OF A QUADRATIC INEQUALITY

In connection with the quadratic equation in two variables (Sec. 6.5) we can consider related inequalities of the form

$$y > ax^2 + bx + c \qquad y < ax^2 + bx + c$$

The parabola whose equation is

$$y = ax^2 + bx + c$$

divides the plane into two regions, in one of which $y > ax^2 + bx + c$, and in the other $y < ax^2 + bx + c$. Neither region contains the parabola itself. By an argument similar to that in the proof of Theorem 15, we can establish the following result.

THEOREM 16 Let P lie in one of the regions into which a parabola divides the plane. If $y > ax^2 + bx + c$ at P, or if $y < ax^2 + bx + c$ at P, then the corresponding inequality is satisfied at every point of this region.

As a result of this theorem we can plot the graph of the solution set of inequalities of this form by (1) drawing the graph of $y = ax^2 + bx + c$, (2) testing the truth of the given inequality at a point in each of the two regions into which this parabola divides the plane, (3) coloring the region or regions in which the above test is affirmative.

ILLUSTRATION 1 Plot the graph of $y > x^2 - 6x + 5$. First draw the graph (Fig. 8.18) of $y = x^2 - 6x + 5$. Then (0,0) lies below the graph,

Quadratic inequality in two variables

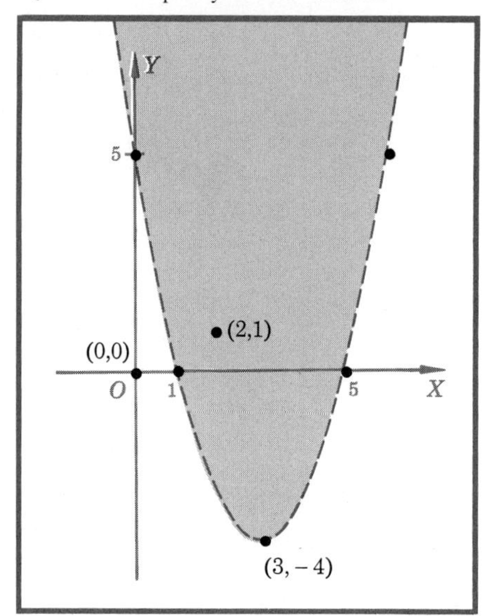

FIGURE 8.18
$y > x^2 - 6x + 5$
The dashed parabola is not included in the colored region which represents the solution set.

and $0 > 0 - 0 + 5$ is false. On the other hand, (2,1) lies above the graph, and $1 > 4 - 12 + 5$ is true. Hence we color the region above the graph.

This technique also permits us to solve simultaneous inequalities as suggested by the next illustration.

ILLUSTRATION 2 Plot the graph of

$$y > x^2 - 6x + 5$$
$$4x + 2y - 10 < 0$$

The graph of the first is given by Illustration 1. Now draw the line (Fig. 8.19) $4x + 2y - 10 = 0$. Since (0,0) lies below the line and satisfies the inequality, the graph of the second inequality is the half-plane below this line. The solution of the pair of inequalities is the intersection of these two regions.

8.8 › APPLICATIONS

Since you have probably not met inequalities and systems of inequalities in your previous mathematical education, you may wonder how they arise in practice and what they are good for. In the first place the acceptable solution of a problem may be limited by practical considerations such as unavoidable restrictions on space, cost, use of materials, manpower,

Simultaneous linear and quadratic inequalities

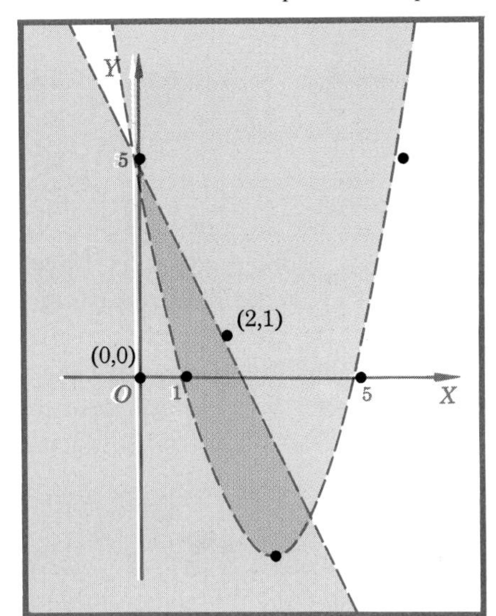

FIGURE 8.19
$y > x^2 - 6x + 5$
$4x + y - 10 < 0$
The darker-colored region is the intersection of the individual solution sets of the two inequalities.

or time. Each of these would be expressed analytically as an inequality. On the other hand, the solution may have to satisfy certain minimum requirements such as load-carrying capacity, food intake necessary to maintain health, serviceable life of a product, or volume of sales necessary to justify the production of a certain item.

Our objectives are often stated in terms of inequalities. We do not try to manufacture a bolt whose diameter is exactly $\frac{1}{2}$ in.; we are quite content if it is $\frac{1}{2}$ in. plus or minus certain tolerances, which are expressed as inequalities. We want unemployment to be less than 4 percent and the minimum family income to be greater than $4,000. To handle the analysis of such situations, we need inequalities.

Inequalities are of importance in modern mathematical economics and also in recent theories such as game theory and linear programming.

PROBLEMS 8.8

In Probs. 1 to 10 plot the graph of the solution set of the given inequality.

1	$y > x^2 + 3x + 2$	**2**	$y > x^2 - 6x + 5$
3	$y \leq -x^2 + 8x - 15$	**4**	$y \geq -x^2 - 7x - 10$
5	$y < 9x^2 + 6x + 1$	**6**	$y \leq 4x^2 + 4x + 1$
7	$y - 2x^2 + 3x - 1 \geq 0$	**8**	$y - 3x^2 - 5x + 2 < 0$
9	$(y + x^2 - 2x + 1)^2 \geq 0$	**10**	$(y + 16x^2 - 8x + 1)^2 < 0$

In Probs. 11 to 14 plot the graph of the solution set of the simultaneous inequalities.

11 $y \geq x^2 + 6x + 5$
$x + y - 5 < 0$

12 $y > x^2 - 8x + 12$
$x - y - 2 > 0$

13 $y > x^2$
$y + 1 < 0$

14 $y < -x^2 - 2x + 8$
$x - y - 2 > 0$

15 In the football problem (Sec. 8.6, Illustration 5), what combinations of plays will meet the required conditions if there are 180 sec left to play? If there are 90 sec left to play?

16 How should the Minneapolis and Seattle Lumber Company (Sec. 8.6, Illustration 6) arrange its production if the profit on a unit of plywood is \$10 and the profit on a unit of lumber is \$15?

8.9 › LINEAR PROGRAMMING

This subject was invented during World War II by George Dantzig as a method of solving certain logistic problems for the U.S. Air Force. In the subsequent years it has developed as an important tool in business and economic analysis. The basic problem in two dimensions is the following:

BASIC PROBLEM We are given a convex polygonal set R in the plane (defined below) and also a linear polynomial $f(x,y) = px + qy + r$ (p, q, and r are real and either $p \neq 0$ or $q \neq 0$). We are required to find those points of R at which $f(x,y)$ has its greatest (or least) value.

In this section we shall show you how to solve this problem by graphical means. The analogous problem occurs with more variables in higher dimensions; but since graphing is impractical in such cases, more sophisticated methods are needed for their analysis. You can find a discussion of these methods in the many books devoted to this subject.

A typical two-dimensional problem is the following simplified form of the Diet Problem. This was one of the original problems which led to the development of this subject.

Diet problem

ILLUSTRATION 1 Let us suppose that we have two types of synthetic foods, which we call A and B. You may think of K rations and C rations if you like, but the numbers given below are purely artificial and apply to no foods on the market. Let us suppose that the two foods contain the following nutritional components:

Food	*Calories per ounce*	*Protein per ounce*	*Fat per ounce*
A	100	50	0
B	200	10	30

The units for protein and fat are arbitrary and need not be specified. Let us also suppose that the minimum daily requirements for an active man are calories, 2,500; protein, 350; fat, 150.

PROBLEM Which food or combination of foods should be employed in order to (1) fulfill the minimum daily nutritional requirements and (2) minimize the total weight?

Such a problem, for example, would be of importance to a mountain climber. In order to attack the problem, let us introduce some notation.

Let a represent the number of ounces of food A that are required.
Let b represent the number of ounces of food B that are required.
Then the minimum daily requirements will be met if:

$$\begin{array}{lrl} \textit{Calories:} & 100a + 200b & \geq 2{,}500 \\ \textit{Protein:} & 50a + 10b & \geq 350 \\ \textit{Fat:} & 30b & \geq 150 \end{array}$$

These three inequalities can be simplified to the following simultaneous system:

$$\begin{aligned} a + 2b &\geq 25 \\ 5a + b &\geq 35 \\ b &\geq 5 \end{aligned}$$

To these we should add the practical requirements that $a \geq 0$ and $b \geq 0$, for one cannot eat a negative amount of food. The graph of this simultaneous system is given in Fig. 8.20. The colored region is the region R referred to above.

The quantity to be minimized is the total weight $W = a + b$. So the problem is to find the point (or points) of R at which W is a minimum. Theorem 17 (below) tells us that this minimum will occur at one or more of the *vertices* labeled P, Q, R in the figure. In order to determine which vertex to use, let us do some arithmetic.

At P: $a = 0,\ b = 35,\ W = a + b = 35$
At Q: $a = 5,\ b = 10,\ W = a + b = 15$
At R: $a = 15,\ b = 5,\ W = a + b = 20$

Therefore, W is a minimum at Q, and the diet should consist of 5 oz of food A and 10 oz of food B.

The solution of problems of this type depends on the following definitions and theorems.

DEFINITION A convex polygonal set R is the set of points in the plane whose coordinates satisfy the finite system of inequalities

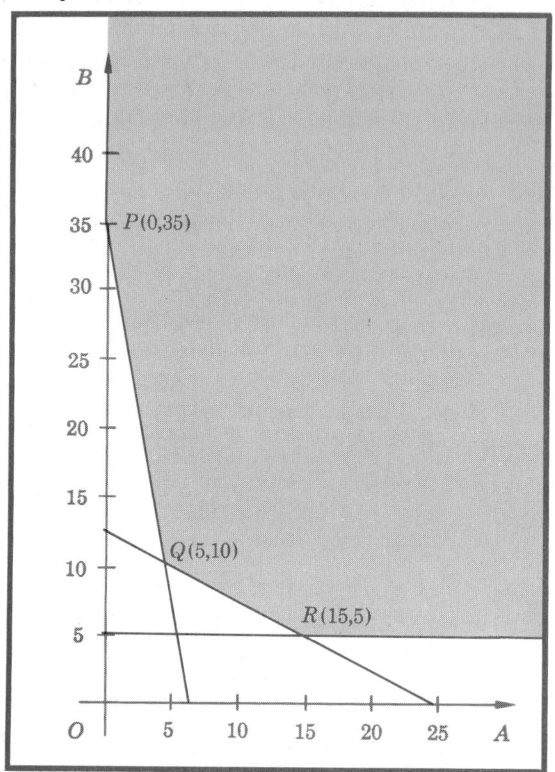

FIGURE 8.20
The colored region represents the points (a,b) which meet the minimum daily nutritional requirements.

$$\begin{aligned} a_1x + b_1y &\geq c_1 \\ a_2x + b_2y &\geq c_2 \\ \cdots&\cdots\cdots \\ a_nx + b_ny &\geq c_n \end{aligned}$$

where the a's, b's, and c's are real.

When R is two-dimensional, it is called a *convex polygonal region.*

Examples of such sets are given in Figs. 8.12 to 8.14, except that in the present situation the bounding segments are to be included in the set. In Fig. 8.16 the colored triangle and in Fig. 8.17 the segment AB are sets of this type.

EXERCISE A Show that the segment AB of Fig. 8.17 is defined by

$$\begin{aligned} L + P &\geq 400 \qquad & L &\geq 100 \\ -L - P &\geq -400 \qquad & P &\geq 150 \end{aligned}$$

It is possible that R is empty as in Fig. 8.15, in which case our problem has no solution. It is also possible that R consists of a single point, so that the solution to our problem is trivial. So we assume that R contains at least two points; this implies that R is a line, a ray, a line segment (Fig. 8.17) or a nonempty convex polygonal region in the plane (Figs. 8.12 to 8.14 and 8.16). A point of R is called a *feasible* solution of our problem.

DEFINITION The point $P(\bar{x},\bar{y})$ of R is called a *maximum* for the given polynomial $f(x,y) = px + qy + r$ if and only if $f(x,y) \leq f(\bar{x},\bar{y})$ for all (x,y) in R. Similarly, P is a *minimum* if and only if $f(x,y) \geq f(\bar{x},\bar{y})$ for all (x,y) in R. A maximum or minimum point P is called an *optimal* solution of the problem.

max, min

THEOREM 17 If $P(\bar{x},\bar{y})$ is a maximum point of $f(x,y) = px + qy + r$ (where $p \neq 0$ or $q \neq 0$) in R, then:

1 If R is a convex polygonal region in the plane,
- **a** P is a vertex of R and, at all other points of R, $f(x,y) < f(\bar{x},\bar{y})$, or
- **b** P is a point of a side S of R such that, on S, $f(x,y) = f(\bar{x},\bar{y})$ and, at all other points of R, $f(x,y) < f(\bar{x},\bar{y})$.

2 If R is a line segment or a ray,
- **a** P is an endpoint of R and, at all other points of R, $f(x,y) < f(\bar{x},\bar{y})$, or
- **b** $f(x,y) = f(\bar{x},\bar{y})$ at all points of R.

PROOF Consider the line

$$L\colon f(x,y) - f(\bar{x},\bar{y}) = 0$$

or

$$(px + qy + r) - (p\bar{x} + q\bar{y} + r) = 0$$

As discussed in Sec. 8.5, this line divides the plane into two half-planes in one of which, say I,

$$f(x,y) - f(\bar{x},\bar{y}) < 0 \qquad \text{or} \qquad f(x,y) < f(\bar{x},\bar{y})$$

Moreover, since $P(x,y)$ is a maximum,

$$f(x,y) \leq f(\bar{x},\bar{y})$$

for every point (x,y) of R. It follows, therefore, that the points of R are either on L or on half-plane I. This means that:

1 If R is a convex polygonal region in the plane, the intersection of L and R (which contains P) is either:
- **a** A vertex of R which is the only point of R where $f(x,y) = f(\bar{x},\bar{y})$ and, at all other points of R, $f(x,y) < f(\bar{x},\bar{y})$ (Fig. 8.21*a*), or
- **b** A side S of R along which $f(x,y) = f(\bar{x},\bar{y})$ and, at all other points of R, $f(x,y) < f(\bar{x},\bar{y})$ (Fig. 8.21*b*).

Possible locations of maximum point P when R is a region

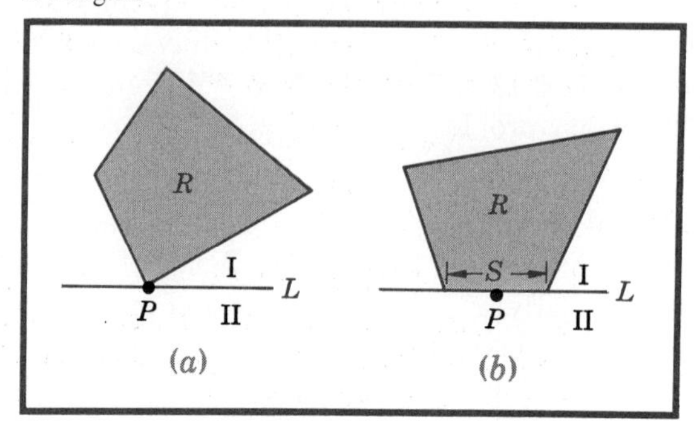

FIGURE 8.21
If P is a maximum point of $f(x,y)$ for (x,y) in R, it must lie on the boundary of R. It is, therefore, a vertex of R or a point of a side of R along which $f(x,y)$ is constant.

2 If R is a line segment or a ray, the intersection of L and R (which contains P) is either:

  a An endpoint of R which is the only point of R where $f(x,y) = f(\bar{x},\bar{y})$ and, at all other points of R, $f(x,y) < f(\bar{x},\bar{y})$ (Fig. 8.22*a*), or

  b The whole of R, so that at all points of R $f(x,y) = f(\bar{x},\bar{y})$ (Fig. 8.22*b*).

A similar result is true if P is a minimum.

From this theorem our working procedure for finding a maximum point involves the following steps:

1 Check the situation to see whether a maximum exists. It always exists if R is bounded.
2 Find all vertices (or endpoints) of R and the value of $f(x,y)$ at each of these.
3 If a maximum exists and if there is one vertex (or endpoint) at which $f(x,y)$ is greater than at all the other vertices, this is the maximum point.

Possible locations of maximum point P when R is a segment

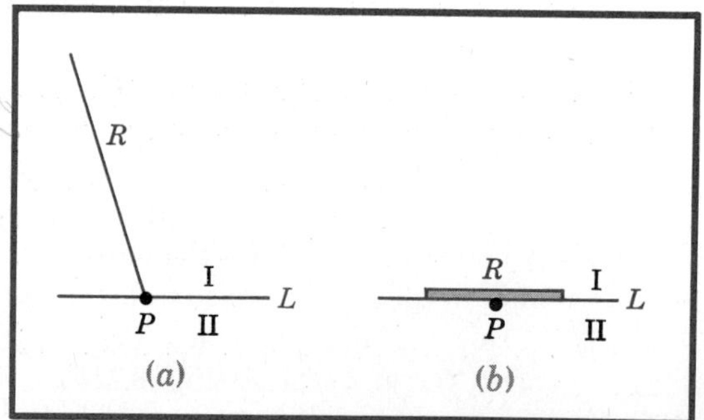

FIGURE 8.22
In this case P must be an endpoint of R, or R lies on L and $f(x,y)$ is constant on R.

4 If a maximum exists and if there are two adjacent vertices (or end-points) at which $f(x,y)$ has equal values which are greater than its values at the other vertices, then all points of the side joining these vertices are maximum points.

No other possibilities can occur. A similar procedure holds for finding a minimum point.

PROBLEMS 8.9

1 In the Diet Problem (Illustration 1 above), assume all the information given there plus the fact that food A costs 10 cents/oz and food B costs 20 cents/oz. What combination of A and B should be used to provide an adequate diet at least cost?

2 In the Diet Problem (Illustration 1 above), suppose that 1 oz of food A occupies 2 in.^{3} and that 1 oz of food B occupies 3 in.^{3}. Then what combination of A and B will provide an adequate diet and minimize the total volume?

3 In the Football Problem (Sec. 8.6, Illustration 5), suppose that on the average there is one injury in each five running plays and one injury in each ten pass plays. Then what combination of plays should the quarterback call to secure the touchdown in the allotted time with the minimum risk of injuries?

4 To provide sufficient nitrogen (N), phosphorus (P), and lime (L) for his 400-acre farm, Mr. Hill buys x bags of A and y bags of B per acre. A and B are two kinds of commercial fertilizer. The necessary data are shown in tabular form:

	N	*P*	*L*	*First year cost/bag*	*Second year cost/bag*
A	6	2	8	\$ 5	\$ 6
B	6	9	15	\$10	\$10
Minimum need/acre	108	85	235		

Find the values of x and y which will meet at least the minimum need for fertilizer and for which the cost per acre is a minimum, each year.

5 A diet contains carbohydrates (H), protein (P), and fats (F). Two standard rations are available, with the data in tabular form:

	H	*P*	*F*	*Cost*	*Time*	*Weight*
Ration *A*	100	50	30	\$10	k	1
Ration *B*	200	10	0	\$30	$2k$	1
ANPM	2,500	350	150			

ANPM means "minimum annual need per man".

x is the number of Ration A used and y the number of Ration B. Find the value of x and y which will provide at least the ANPM, where **(a)** cost is a minimum; **(b)** time (of preparation) is a minimum; **(c)** weight is a minimum.

6 A ship arrives in port with 2,800 tons of grain. It must be unloaded in at most 35 hr. Available machinery is a conveyor belt, which runs with a crew of three men, costs \$80 per hr, and unloads 120 tons/hr and also a power shovel, which costs \$10 per hr, runs with a crew of ten men, and unloads 50 tons/hr. In each case, \$4 per hr per man of crew must be added. The conveyor belt and the shovel cannot operate at the same time.
The union contract stipulates that at least 1 hr of work must be provided some union man for each 20 tons of cargo on each ship. That means that at least 140 hr of labor must be provided. The conveyor belt runs for x hr and the power shovel for y hr. Find the minimum cost of unloading the ship. Also find the values of x and y which produce the minimum.

7 Accommodations must be provided for at least 435 persons at a remote mine. Buildings must be of two types, A and B. The pertinent data are:

House	*Units of wood*	*Units of concrete*	*Capacity (persons)*	*Hours to build*	*Cost*	*No. of houses*
A	10	7	12	40	\$ 5,000	x
B	5	25	25	50	\$12,000	y

At least 125 units of wood and 285 units of concrete must be used. Find the values of x and y for which **(a)** the cost of construction is a minimum; **(b)** the time of construction is a minimum.

8 A certain machine tool company has found that it must have a core of at least 700 reliable customers to remain competitive in the local market. Furthermore, to earn the desired rate of profit, total sales of at least \$160,000 must be made yearly for a total selling cost of no more than \$4,600.
To meet these requirements, the sales manager has suggested contacting some customers directly by traveling salesmen and contacting others indirectly by written correspondence. The problem is determining how many to contact each way. When the traveling-salesman method of contact is used, it has been found that the company makes \$400 of sales per customer for a selling cost of \$7 per customer. When the written correspondence method of contact is used, it has been found that the company makes \$100 of sales per customer for a selling cost of \$4 per customer.

The company has a shortage of experienced employees and must attempt to minimize the use of labor hours per customer. Salesman contact requires 6 hr per customer, correspondence contact requires only 0.5 hr per customer. How many customers should be contacted by each method?

9 The personnel director of a corporation must find the most suitable method of testing the qualifications of job applicants. At least 35 men must be tested; the total cost of the test administration cannot exceed \$6.25; and the total time for test administration cannot exceed 8.75 hr. Two types of qualification tests are available: a written aptitude test which indicates the applicant's potential in various job situations and a performance test which indicates the applicant's actual capabilities. The administration cost is \$20 per 100 aptitude tests and \$5 per 100 performance tests. The administration time is 10 hr per 100 aptitude tests and 25 hr per 100 performance tests. The personnel director must stay within the three restrictions given above when deciding how many of each type of test to give, but he also must minimize the total number of inaccurate appraisals. If he has found that the appraisals of 18 percent of the aptitude tests and 22 percent of the performance tests are inaccurate, how many of each should he administer? Let x be the number of aptitude tests and y be the number of performance tests.

10 A certain large winery is considering automating the wine-testing process. Below are the data comparing the old manual testers (a man with a glass) with the new electronic testers (a computer with a tube):

Electronic testers Initial cost, \$50 per tester; hourly cost of operation, \$15 per tester; number of vats tested per hour, 23 per tester; daily man-hours of labor needed, 5 per tester; cups of wine wastage per day, 3 per tester.

Manual testers Initial costs, \$0 per tester; hourly cost of operation, \$20 per tester; number of vats tested per hour, 4 per tester; daily man-hours of labor needed, 20 per tester; cups of wine wastage per day, 12 per tester.

In an effort to fight unemployment, the local union representing the winery workers objects to the introduction of electronic testers. The following compromise is finally agreed to by the union and the winery:

Demands by the union (1) No more than \$2,500 will be spent on electronic wine testers. (2) A total of at least 850 man-hours of labor will be employed by the winery daily.

Demands by the winery (1) The total hourly costs of operating the testers cannot be greater than \$1,550. (2) A total of at least 830 vats must be tested per hour.

Find a number of electronic testers which will combine with a number of manual testers to satisfy the above compromise and will minimize the total cups of wine wastage per vat.

11 In the Diet Problem (Illustration 1 above) does $W = a + b$ have a maximum in R?

12 In the Diet Problem (Illustration 1 above) does the cost $C = 10a + 20b$ have a maximum in R?

13 Given that R is defined by the inequalities (see Sec. 8.6, Prob. 9)

$$\begin{aligned} x + 3y - 7 &\geq 0 \\ 3x - 2y + 1 &\geq 0 \\ 4x + y - 17 &\leq 0 \end{aligned}$$

Find the maximum and minimum values in R of $f(x,y) = 2x + y + 4$.

14 Given that R is defined by the inequalities (Sec 8.6, Prob. 12)

$$\begin{aligned} x + 2y - 3 &\geq 0 \\ 2x - y - 1 &\leq 0 \\ x - 3y + 12 &\geq 0 \end{aligned}$$

Find the maximum and minimum values in R of $3x - 4y + 5$.

15 Given that R is defined by

$$\begin{aligned} x + 2y - 12 &= 0 \\ 5x - 2y &\geq 0 \\ x - 2y &\leq 0 \end{aligned}$$

Find the maximum and minimum values in R of $f(x,y) = 3x - y + 5$.

16 Given that R is defined by

$$\begin{aligned} 3x - 2y &= 0 \\ 3x + y - 9 &\geq 0 \\ x + 2y - 8 &\leq 0 \end{aligned}$$

Find the maximum and minimum values in R of

$$f(x,y) = x + y + 5$$

17 What is the situation if both $p = 0$ and $q = 0$ in

$$f(x,y) = px + qy + r?$$

18 What is the situation if R is the whole of a line?

19 What conclusion can you draw if $f(x,y)$ is a maximum at an interior point of R?

20 What is the situation if R is the entire plane?

REFERENCES

Dorfman, R., P. A. Samuelson, and R. M. Solow: "Linear Programming and Economic Analysis", McGraw-Hill, New York (1958).

Ficken, F. A.: "The Simplex Method of Linear Programming", Holt, Rinehart, and Winston, New York (1961).

Freund, John E.: "College Mathematics with Business Applications", chap. 8, Prentice-Hall, Englewood Cliffs, N.J. (1969).

Kemeny, John G., A. Schleifer, J. Laurie Snell, and Gerald L. Thompson: "Finite Mathematics with Business Applications", chap. 7, Prentice-Hall, Englewood Cliffs, N.J. (1962).

9 FUNCTIONS AND RELATIONS

9.1 » INTRODUCTION

In earlier chapters we have discussed equations and inequalities in the two variables x and y and have seen how the values of x and y are related by means of a given condition. Our previous treatment was limited to quite simple situations such as

$$y = 3x + 5 \qquad y > 4x + 9$$
$$y = 2x^2 + 5x + 7 \qquad y \geq x^2 + 4x + 6$$
$$\begin{cases} x + 2y = 4 \\ 3x - y = 5 \end{cases} \qquad \begin{cases} y = 2x^2 - 3x + 4 \\ x + y \geq 0 \end{cases}$$

Relation

in which we described the relation between x and y by plotting a suitable graph in the plane. In this chapter we shall discuss this concept of a relation from a more general point of view. We begin with some examples.

ILLUSTRATION 1 We take 0, 1, and 4 as the elements of set A and their square roots, namely, 0, 1, -1, 2, -2, as the elements of set B. Thus $A = \{0, 1, 4\}$ and $B = \{0, 1, -1, 2, -2\}$. There are many ways of describing the relation between A and B. Here are some of them.

a The relation involved suggests a natural pairing which can be exhibited nicely by a diagram.

b The relation is a correspondence which makes 0 correspond to 0, 1 correspond to 1, 1 also correspond to -1, and 4 correspond both to 2 and to -2.

c The relation is a set of ordered pairs, namely, (0,0), (1,1), (1,-1), (4,2), and (4,-2).

Relation in Illustration 1

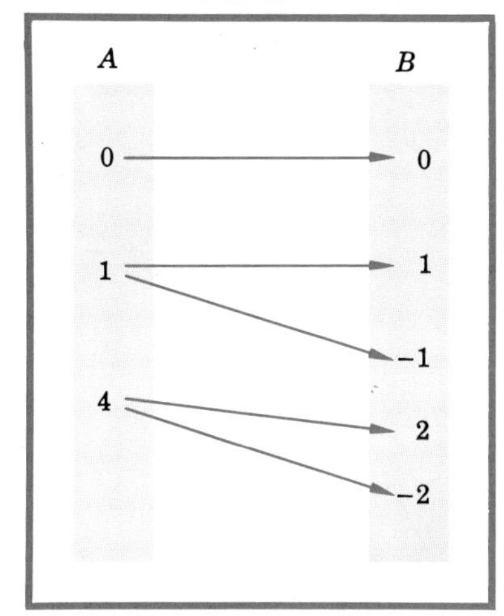

FIGURE 9.1
$A = \{0, 1, 4\}$ is related to $B = \{0, 1, -1, 2, -2\}$. The relation is defined by $y = \pm\sqrt{x}$.

d For x an element of A, x corresponds to $\pm\sqrt{x}$ *under* the relation r (or *by* the relation r). The letter r just stands for the relation. This is written $x \xrightarrow{r} \pm\sqrt{x}$.

e The relation is defined by the equation $y = \pm\sqrt{x}$.

ILLUSTRATION 2 Sets A and B do not have to be sets of numbers. Let $A = \{$Oscar, Amos, Kevin$\}$ and $B = \{$Laura, Elsa, Yvonne$\}$.

a The relation is described by "girl is younger than boy" and is drawn in Fig. 9.2.

b The relation is a correspondence which makes Oscar correspond to Laura and to Elsa; Amos to Elsa; and Kevin to Laura, Elsa, and Yvonne.

c The relation is a set of ordered pairs, namely (Oscar,Laura), (Oscar,Elsa), (Amos,Elsa), (Kevin,Laura), (Kevin,Elsa), and (Kevin, Yvonne).

ILLUSTRATION 3 In a certain school the year's overall performance marks are just four in number, 1 (best), 2, 3, and 4 (poorest). Let $A = \{$Patrick, Eileen, Ronald, Thomas, Hannah, William, Alice$\}$, and $B = \{1, 2, 3, 4\}$. In this class of students A, the relation is a mapping from A to B, that is, it is a mapping from *people* onto *grades* (Fig. 9.3). We might also describe this relation as "grade of student" and also by a set of ordered pairs.

ILLUSTRATION 4 A new building has four flats in which four young married couples live. Let A be the set of husbands and B the set of wives.

Relation in Illustration 2

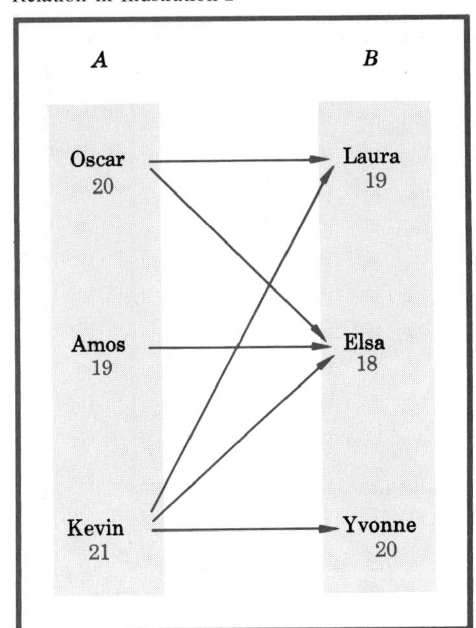

FIGURE 9.2
$A = \{$Oscar, Amos, Kevin$\}$ is related to $B = \{$Laura, Elsa, Yvonne$\}$. The relation is defined by "girl is younger than boy".

Relation in Illustration 3

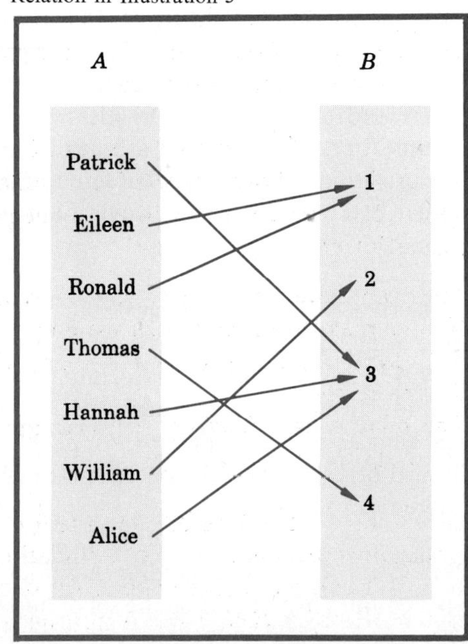

FIGURE 9.3
The set of students is mapped onto the set of grades.

Relation in Illustration 4

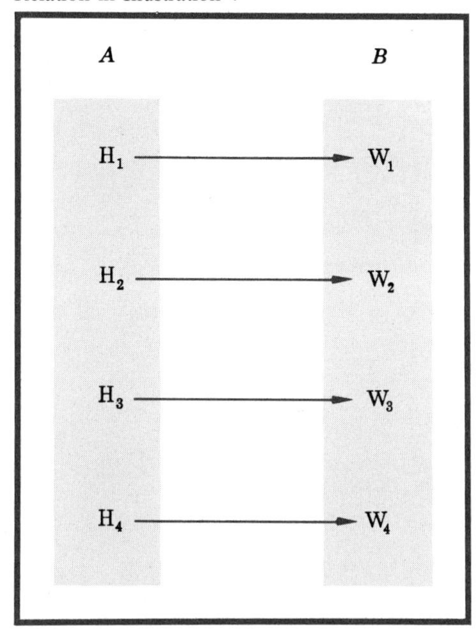

FIGURE 9.4
The set of husbands is in one-to-one correspondence with the set of wives.

One-to-one

The map is shown in Fig. 9.4. In this case we say the correspondence is one-to-one, that is, to a given element x in A there corresponds one and only one element y in B, and conversely to a given element y in B there corresponds one and only one element x in A. Such relations play an important role in mathematics.

ILLUSTRATION 5 Sets do not have to be finite. Indeed our most important examples are those sets which are infinite and whose members are numbers. Let $A = \{n \mid n \text{ is a positive integer}\}$ and $B = \{m \mid m = n^3, n \text{ a positive integer}\}$.

a The map is shown in Fig. 9.5. Clearly it is one-to-one.
b If x is a member of set A, then x is mapped onto x^3 under the relation f. In symbols, if $x \in A$, then $x \xrightarrow{f} x^3$, where the letter f just stands for the relation.
c The relation is defined by the equation (also called a *rule*) $y = x^3$ or by $f(x) = x^3$.

ILLUSTRATION 6 Let X be the set of real numbers, and let Y be the set of nonnegative real numbers. Let the relation f be "the square of". That is, for $x \in X$ and $y \in Y$, the relation f is defined in symbols by $x \xrightarrow{f} x^2$ or by $y = x^2$ or by $f(x) = x^2$. A portion of this simple map is shown in Fig. 9.6. To each element in X there corresponds one and only one

Relation in Illustration 5

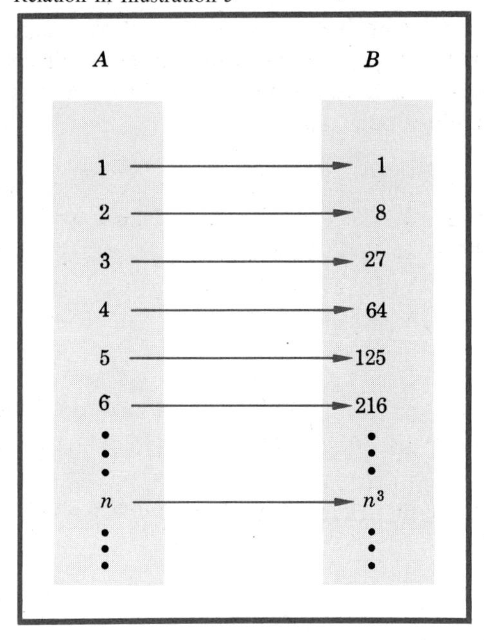

FIGURE 9.5
The map is defined by $y = x^3$.

Relation in Illustration 6

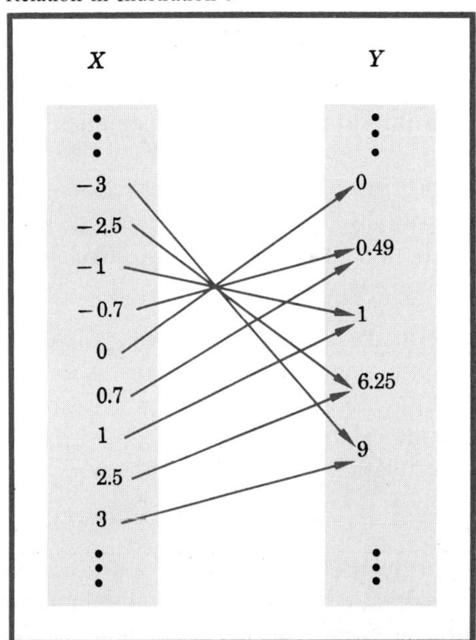

FIGURE 9.6
The map is defined by $y = x^2$.

element in Y, but the mapping is not one-to-one. You can read off the ordered pairs indicated in the simple map (Fig. 9.6). These we have plotted on rectangular coordinates in Fig. 9.7. Other points could be added to exhibit the complete graph, which is indeed a continuous curve, namely a parabola.

We have observed that each of the illustrated relations can be described as a set of ordered pairs, and this leads us to the general definition of a relation. Since we are given two sets A and B, we can form their Cartesian Product $A \times B$, namely the set of ordered pairs (a,b) where $a \in A$ and $b \in B$. Then a relation is defined as follows.

DEFINITION Domain and image

A relation from A to B is a subset of the Cartesian Product $A \times B$. The set A is called the *domain* of the relation, and the set B is its *image*.

As has been illustrated, there are many ways of defining a specific relation, the most important of which are given below.

Relation in Illustration 6

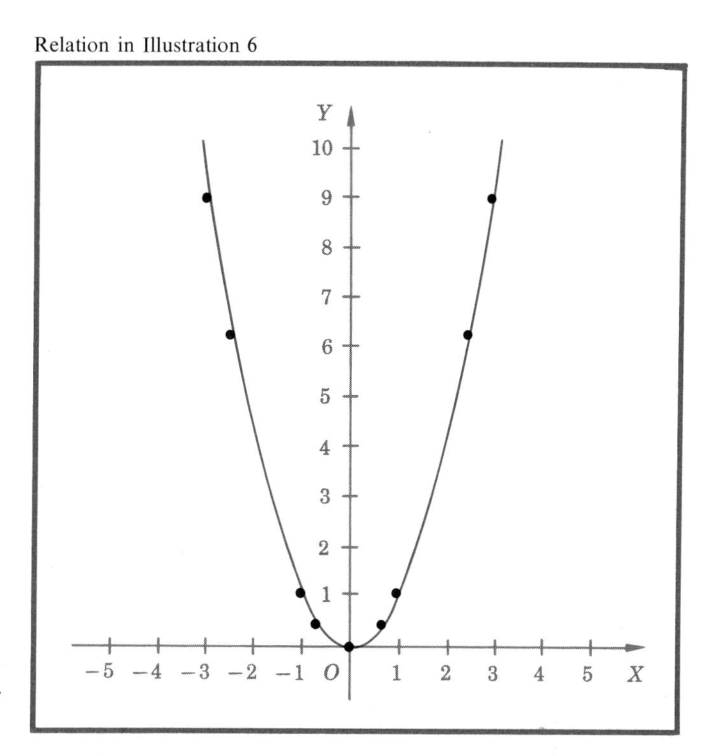

FIGURE 9.7 The curve is the graph of the relation $y = x^2$.

LIST OF ORDERED PAIRS When the subset of $A \times B$ consists of a reasonably small number of ordered pairs, we just write them down. See Illustrations 1 and 2.

ILLUSTRATION 7 Relations of this type frequently appear as tables. You have probably encountered tables of logarithms which are good examples of this. In these the ordered pairs are of the form $(x, \log x)$. The set of all such pairs of entries is the relation so defined.

EQUATIONS The subset is often defined to consist of those ordered pairs of real numbers that satisfy a certain equation. Then the relation is

$$\{(x,y) \mid (x,y) \text{ satisfies the given equation}\}$$

See Illustrations 1, 5, and 6.

INEQUALITIES Here the subset is defined to consist of those ordered pairs of real numbers which satisfy a certain inequality. An example is the relation

$$\{(x,y) \mid x + 3y > 2\}$$

SIMULTANEOUS SYSTEMS Relations may also be defined (as in Secs. 6.8, 6.10, 6.11, and 8.6) by a simultaneous system of equalities, or inequalities, or mixtures of these.

ILLUSTRATION 8 The following simultaneous systems define relations:

$$\begin{cases} 2x + y = 3 \\ \quad\;\; x > 1 \end{cases} \qquad \begin{cases} 3x + 2y = 9 \\ \;\; x + \;\; y = 1 \end{cases} \qquad \begin{cases} 2x - y < 0 \\ \;\; x + y \geq 1 \end{cases}$$

EXERCISE A Graph the relations of Illustration 8.

GRAPHS In science relations sometimes are defined by a graph. This amounts to a curve in the plane, to a shaded region, or to a discrete set of points. The coordinates (x,y) of all points in the graph give us the desired subset of $X \times Y$ and define the relation.

Relations can be classified according to the character of the correspondences involved in the following way.

Multivalued
1 *Multivalued.* At least one element of the domain corresponds to more than one element of the image. See Illustrations 1 and 2.

Single-valued
2 *Single-valued.* Each element of the domain corresponds to one and only one element of the image. Such relations are called *functions.* See Illustrations 3, 4, 5, and 6.

One-to-one
3 *One-to-one.* The correspondence is one-to-one as defined in Sec. 1.2.

These are special cases of functions called *one-to-one functions*. See Illustrations 4 and 5.

When the domain and image of a relation are subsets of the reals, the relation is often defined by an equation, an inequality, or a system of these. It is up to you to find the exact domain and image. A direct method of doing this is to plot the graph of the defining equation, inequality, or system of these, as has been done in preceding chapters. Then:

1 The domain is that subset of the X-axis consisting of those points x such that a vertical line through x intersects the graph (Fig. 9.8).
2 The image is that subset of the Y-axis consisting of the points y such that a horizontal line through y intersects the graph (Fig. 9.8).

The classification into multivalued, single-valued, and one-to-one can also be obtained from the graph.

1 The relation is multivalued if at least one vertical line through a point of the domain intersects the graph in more than one point (Fig. 9.9).
2 The relation is single-valued if every vertical line through a point of the domain intersects the graph in one and only one point (Fig. 9.10).
3 The relation is one-to-one if it is single-valued *and* if every horizontal line through a point of the image intersects the graph in one and only one point (Fig. 9.11).

Domain and image

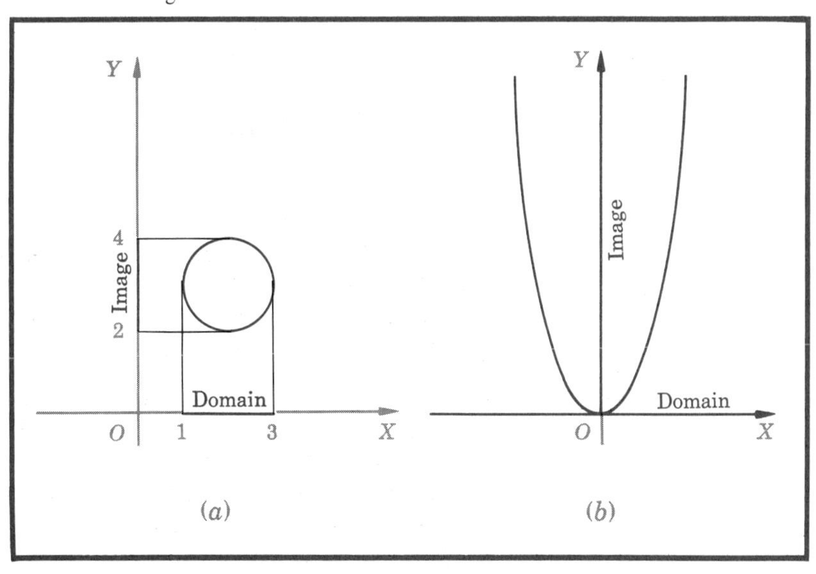

FIGURE 9.8
To find domain and image, first plot the graph of the relation. From this read off domain and image. For (*a*) domain is $\{x|1 \leq x \leq 3\}$; image is $\{y|2 \leq y \leq 4\}$. For (*b*) domain is all x; image is $\{y|y \geq 0\}$.

Multivalued relation

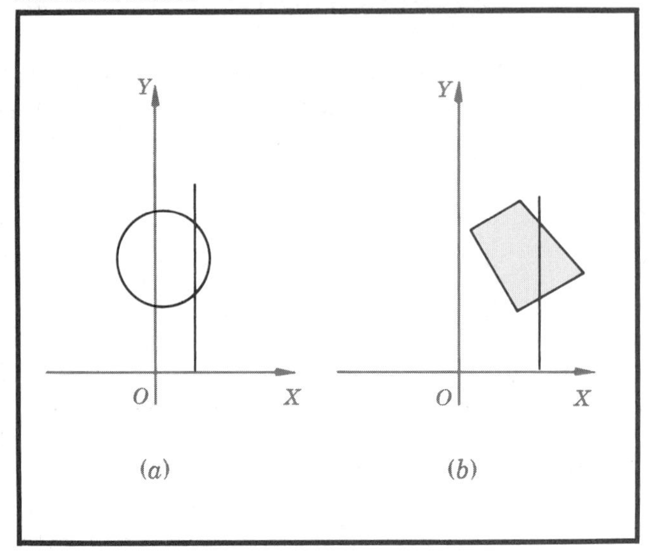

FIGURE 9.9
Many vertical lines intersect the graph in more than one point.

Single-valued relation

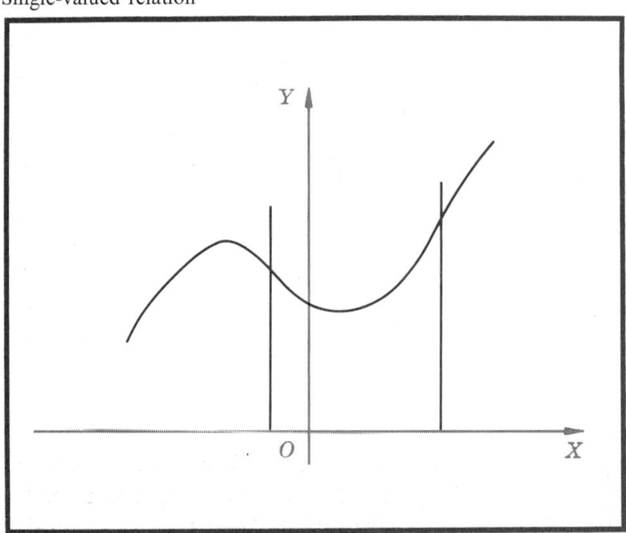

FIGURE 9.10
No vertical line intersects the graph in more than one point. The relation is not one-to-one.

One-to-one relation

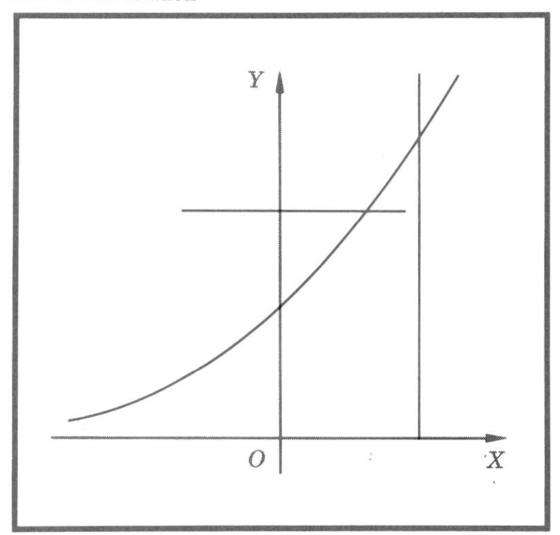

FIGURE 9.11
No vertical or horizontal line intersects the graph in more than one point.

PROBLEMS 9.1

In Probs. 1 to 20 discuss the domain and image of each relation whose defining rule is given below. State whether multivalued, single-valued (not one-to-one), or single-valued (one-to-one).

1 $y = 3x - 4$

2 $y = 10x - 9$

3 $y = 6x + 5$

4 $y = \frac{x}{6} + x^2$

5 $y = x^2 - 1$

6 $y = x^3$

7 $x^2 + y^2 = 81$

8 $x^2 \leq y$

9 $x - y \leq 0$

10 $x^2 + y^2 \leq 1$

11 $x + y > 0$

12 $x^2 + y > 3$

13 $\begin{cases} x - y = 1 \\ x \leq 2 \end{cases}$

14 $\begin{cases} 2x + y = 3 \\ y \geq 5 \end{cases}$

15 $\begin{cases} 2x + 3y = 8 \\ 4x - y = 2 \end{cases}$

16 $\begin{cases} x \geq 1 \\ x + y \geq 2 \end{cases}$

17 $\{(3,7), (7,8)\}$

18 $\{(3,1), (3,8)\}$

19 $\{(1,2), (2,3)\}$

20 $\{(0,3), (1,3), (6,3)\}$

9.2 » FUNCTIONS

As has been noted before, single-valued relations have a special name and are called functions. The concept of a function is so important that we shall start at the beginning and define it from scratch.

DEFINITIONS
Function

Let there be given two sets, the first X, the second Y. Then a function f from X to Y is a correspondence (or a rule or a mapping) that assigns to each $x \in X$ one and only one $y \in Y$.

Domain and range

1. The set X is called the *domain* of f.
2. The set Y is called the *range* of f.
3. The element $y \in Y$ into which a point $x \in X$ is mapped is called the *image* of x. It is written $f(x)$, a symbol that is read "f of x". The image of x is also called the value of the function f at x.
4. The set $\{y \mid y = f(x) \text{ for } x \in X\}$ is the set of the images of points in X. It is called the *image* of f and is written $f(X)$. The image $f(X)$ is necessarily a subset of Y. We may have $f(X) = Y$ or $f(X)$ a proper subset of Y.

Image

EXERCISE A Show that the above definitions of the domain and image of a function agree with those given in Sec. 9.1 for the domain and image of a single-valued relation.

Real-valued

The purpose of introducing the range of a function is to enable us to say what kind of values the function may assume without explicitly giving its image. Thus we may speak of a "real-valued function", that is, a function whose range is R. Or we may consider a "complex-valued function", that is, a function whose range is the set of all complex numbers; or an "integral-valued function", that is, one whose range is the set of integers; and so forth. Unless an explicit statement is made to the contrary, it is to be understood hereafter in this book that all functions are *real-valued,* namely that their range is the set of reals.

This distinction between image and range leads us to classify functions into two types.

DEFINITIONS
Onto and into

A function f is *onto* if its image equals its range.
A function f is *into* if its image is a proper subset of its range.

We also have the final definition:

DEFINITION
One-to-one

A function is one-to-one if the correspondence between its domain and image is one-to-one.

9.3 » NOTATIONS

In the functions to be considered hereafter, it is to be assumed that the domain X is some subset of R and that the range Y is R, that is, that the function is real-valued. There are several common notations for such a function.

1 $f{:}X \to Y$	This is read "f is the function from X to Y". This is a generic and not a specific notation.
2 $x \xrightarrow{f} f(x)$	This is read "under the function f, x is mapped onto $f(x)$".
3 $\{(x,y) \mid y = f(x)\}$	This is read "f is the function whose ordered pairs are (x,y) where the rule is $y = f(x)$".
4 $f{:}y = f(x)$	This is read "f is the function determined by the rule $y = f(x)$". It is a shortened version of (3).
5 $f{:}(x,y)$	This is read "f is the function which is the set of ordered pairs (x,y)". The ordered pairs may be determined by a given rule, or in simple cases may just be listed.

In notations 2, 3, and 4 the symbol $f(x)$ occurs. This is read "f of x" and stands for "the value of the function at x" and, as such, is an element of the image $f(X)$. The ordered pair (x,y) could just as well be written $(x,f(x))$. Do not confuse f, the function, with $f(x)$, the value of the function at x.

EXERCISE A The set of ordered pairs $\{(1,2), (2,2), (3,7)\}$ is a function. The domain is $\{1,2,3\}$, and the image is $\{2,7\}$. The function is not one-to-one. Why?

EXERCISE B The equation $y = 12x + 3$ defines a function. The equation $f(x) = 12x + 3$ defines the same function. To each y there is a unique x and to each x there is a unique y. Why? The domain X is the set of reals and so is the image Y. The function is one-to-one. Why?

When a function is given in this notation, its domain and image are not usually specified. Unless the domain is specifically restricted in some fashion (by requiring $-2 < x < 5$ for example), it is to be understood that the domain is the largest subset of the reals for which the rule $y = f(x)$ makes sense. The range is the set of reals, and it is your job to find the image. In easy cases you can find the image graphically, as described in Sec. 9.1. More difficult cases will be treated later.

Two important functions are:

Constant function

1 The constant function defined by $x \xrightarrow{f} k$, k a constant. Let X be the set of reals. Each element $x \in X$ is mapped onto one and the same number k. The function is also defined by the equation $y = k$ (Fig. 9.12). It is into and not one-to-one.

Graph of the constant function

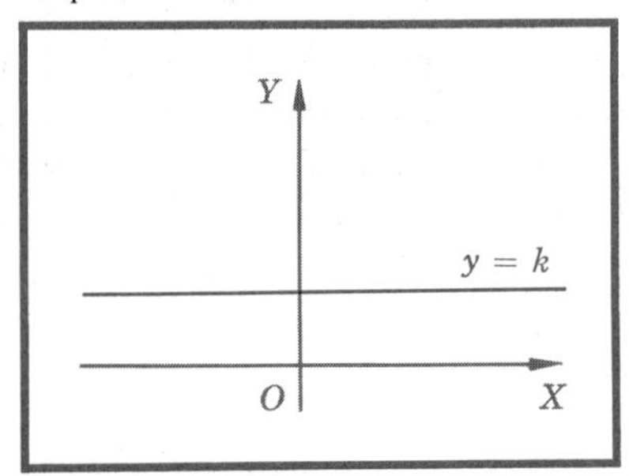

FIGURE 9.12

Absolute-value function

2 The absolute-value function, which is a mapping from R onto the set of nonnegative reals, is given by $y = |x|$ where

$$|x| = \begin{cases} x & \text{if } x > 0 \\ 0 & \text{if } x = 0 \\ -x & \text{if } x < 0 \end{cases}$$

The image is $\{y | y \geq 0\}$ and so this function is into. It is not one-to-one.

EXERCISE C Sketch the graph of $y = |x|$.

EXERCISE D Show that this function is the same as that given by $y = \sqrt{x^2}$.

9.4 › VARIABLES

We have already defined a variable as a symbol for which any element of the universal set may be substituted (Sec. 1.3). Where the function is a mapping from X to Y, we call $x \in X$ the *independent variable* and $y \in Y$ the *dependent variable.*

Independent, dependent variables

ILLUSTRATION 1 Let the function f be given by $f(x) = 2x$. Then:

a The ordered pairs are of the form (x,y), where $y = 2x$.
b The independent variable is x.
c The dependent variable is y.
d The value of the function at:
  i $x = 4 + \sqrt{3}$ is $f(4 + \sqrt{3}) = 2 \times (4 + \sqrt{3})$.
  ii $x = 1/a$ (where $a \neq 0$) is $f(1/a) = 2 \times (1/a)$.
e The domain is the set of reals.
f The image is the set of reals.
g The mapping is onto and is one-to-one.

ILLUSTRATION 2 Let the function f be given by $f(x) = 1/x$. Then:

a The ordered pairs are of the form $(x,1/x)$.
b The independent variable is x.
c The dependent variable is y where it is understood that the function is also given by $y = 1/x$.
d The value of the function at:

i $x = \frac{1}{2}$ is $f(\frac{1}{2}) = \frac{1}{\frac{1}{2}} = 2$.

ii $x = \dfrac{4}{1+\pi}$ is $f\left(\dfrac{1}{4/(1+\pi)}\right) = \dfrac{1+\pi}{4}$.

e Since $f(0)$ does not exist, the domain is the real numbers, omitting 0.
f Since $1/x$ is never zero, the image is the real numbers, omitting 0.
g The mapping is into and is one-to-one. Why?

PROBLEMS 9.4

In Probs. 1 to 4 which sets of ordered pairs are functions?

1 $\{(1,4), (3,4), (7,3)\}$
2 $\{(1,2), (1,3), (2,3)\}$
3 $\{(4,3), (4,7), (3,4)\}$
4 $\{(1,2), (2,3), (3,4)\}$

In Probs. 5 to 24 find the domain and the image of the indicated real-valued functions. Which are one-to-one, into, onto?

5 $f(x) = x^3$
6 $f(x) = x^2 + 1$
7 $f(x) = x/2$
8 $f(x) = x^4$

9

x	1	2
$f(x)$	7	7

10

x	1	3
$f(x)$	5	6

11

x	5	8	7
$f(x)$	1	2	1

12

x	0	3	5
$f(x)$	1	3	6

13

x	2	3
$f(x)$	7	7

14

x	2	3
$f(x)$	1	2

15 $f(x) = \begin{cases} 0 & x \text{ rational} \\ 1 & x \text{ irrational} \end{cases}$

16 $f(x) = \begin{cases} 0 & x \text{ an integer} \\ \text{undefined} & \text{otherwise} \end{cases}$

17 $f(x) = \begin{cases} 0 & 3 \le x \le 4 \\ 1 & 5 \le x \le 6 \\ 2 & \text{otherwise} \end{cases}$ **18** $f(x) = \begin{cases} \text{the smallest integer} \\ \text{greater than } x \end{cases}$

19 $y = \dfrac{1}{x^2 + 4}$ **20** $y = \dfrac{1}{x^2 - 4}$

21 $y = \dfrac{1}{x(x + 3)}$ **22** $y = \dfrac{x - 2}{x^2 - 4}$

23 $y = \dfrac{x + 2}{x^2 - 4}$ **24** $y = \dfrac{x - 1}{x(x - 2)}$

25 For $y = (x + 3)(x - 6)$ state the image if the domain is considered to be:
a $\{x \mid -3 \le x \le 3\}$ **b** $\{x \mid x \le -3\} \cap \{x \mid x \ge -3\}$

26 The sides of a rectangle are 10 ft and L ft. Show that the area A is a function of L, and state the domain and image.

27 Given $f(x) = 2x^2 + 1$; compute $f(x)$ for $x = 0, 1, 2, a + h, h$.

28 Given $f(z) = 6z + 3$; compute $f(z)$ for $z = 0, 1, a + 2$. Also compute $f(z) - f(z + h)$.

29 The equation $x^2 - y^2 = 81$ defines a relation. Graph the relation. Indicate three functions which are special cases (i.e., subsets) of this relation. State the domain, image, and defining rule of each.

30 The equation $x^2 + y^2 = 81$ defines a relation. Graph the relation. Indicate three functions which are special cases (i.e., subsets) of this relation. State the domain, image, and defining rule for each.

9.5 › ALGEBRA OF FUNCTIONS

We have studied the four elementary operations of arithmetic $+, -, \times, \div$ in connection with numbers (Chap. 2). These ideas can also be applied to functions according to the following definitions.

DEFINITIONS Consider the functions $f{:}(x,y)$ and $g{:}(x,z)$ whose domains are respectively indicated by d_f and d_g. The *sum* $f + g$, the *difference* $f - g$, the *product* fg, and the *quotient* f/g are defined as follows:

1 $f{:}(x,y) + g{:}(x,z) = (f + g){:}(x, y + z)$
2 $f{:}(x,y) - g{:}(x,z) = (f - g){:}(x, y - z)$
3 $f{:}(x,y) \times g{:}(x,z) = (fg){:}(x,yz)$
4 $\dfrac{f{:}(x,y)}{g{:}(x,z)} = \left(\dfrac{f}{g}\right){:}\left(x, \dfrac{y}{z}\right)$

1′ In the addition $f + g$ of two functions the functional values are added.

2′ In the subtraction $f - g$ the functional values are subtracted (in the proper order).
3′ In the multiplication fg the functional values are multiplied.
4′ In the division f/g the functional values are divided (in the proper order).

The functional values of f and g are y and z, respectively. Thus, when f and g are defined by $y = f(x)$ and $z = g(x)$, the above definitions become:

1″ $y + z = f(x) + g(x)$
2″ $y - z = f(x) - g(x)$
3″ $yz = f(x) \times g(x)$
4″ $\dfrac{y}{z} = \dfrac{f(x)}{g(x)}$

The domain of each of $f + g, f - g$, and fg is the set of all elements x common to the domains of f and g; that is, it is the intersection of the sets d_f and d_g. Thus, in symbols $d_{f+g} = d_f \cap d_g$, $d_{f-g} = d_f \cap d_g$, and $d_{fg} = d_f \cap d_g$. The domain $d_{f/g} = d_f \cap d_g$ except for those x's for which $g(x) = 0$. (Division by zero is impossible.)

ILLUSTRATION 1 Given the two functions f and g whose values are $y = x^2$ and $z = x^3$, the domain of each is a set of real numbers. Then

$$y + z = x^2 + x^3 \qquad d_{f+g} = X$$
$$y - z = x^2 - x^3 \qquad d_{f-g} = X$$
$$yz = x^2 \times x^3 = x^5 \qquad d_{fg} = X$$
$$\frac{y}{z} = \frac{x^2}{x^3} = \frac{1}{x} \qquad d_{f/g} = X \text{ except } 0$$

Note that $x = 0$ is not in the domain of f/g since $g(0) = 0$.

ILLUSTRATION 2 Let f and g have the values $y = 1 + 1/x$, $z = \sqrt{1 - x^2}$. The domain d_f is the set of all real numbers excluding 0; the domain d_g is the set of all real numbers between -1 and 1 inclusive. Then

$$y + z = 1 + \frac{1}{x} + \sqrt{1 - x^2} \quad d_{f+g} = \{x \mid (-1 \le x < 0) \text{ or } (0 < x \le 1)\}$$
$$y - z = 1 + \frac{1}{x} - \sqrt{1 - x^2} \quad d_{f-g} = d_{f+g}$$
$$yz = \left(1 + \frac{1}{x}\right)\sqrt{1 - x^2} \quad d_{fg} = d_{f+g}$$
$$\frac{y}{z} = \frac{1 + (1/x)}{\sqrt{1 - x^2}} \quad d_{f/g} = \{x \mid (-1 < x < 0) \text{ or } (0 < x < 1)\}$$

In f/g we must exclude $x = \pm 1$, since $g(-1) = g(1) = 0$.

Composite of two functions

One further operation in the algebra of functions is of great importance; it is the operation of forming the *composite* of two functions.

We illustrate with a special example. Consider first the two functions $f:(x,y)$ and $g:(y,z)$, whose values are given respectively by

$$y = 2x \qquad \begin{cases} \text{domain } X \\ \text{image } Y \end{cases}$$

$$z = +\sqrt{y^3 - 1} \qquad \begin{cases} \text{domain } \{y \mid y^3 - 1 \geq 0\}; \text{ that is, } \{y \mid y \geq 1\} \\ \text{image } \{z \mid z \geq 0\} \end{cases}$$

If there is no connection between these two functions f and g, then there is nothing but a notational difference intended in using ordered pairs of the form (x,y) for f and ordered pairs of the form (y,z) for g.

However, mathematics is filled with situations in which there *is* a connection—situations in which image values of one function f must serve as the domain values of another function g. In the above example, only those values $y \geq 1$ of the image of f can be used in the domain of g. This process leads to a *third* function whose ordered pairs are (x,z), where $z = +\sqrt{y^3 - 1} = +\sqrt{(2x)^3 - 1}$. This third and new function is called the *composite of g and f*, and we choose the symbol $g \circ f$ to represent it. The symbol $g(f)$ is also used.

Now let us describe the general situation where we are given two functions $f:(x,y)$ and $g:(y,z)$. Choose an x such that the y which f assigns to it is in the domain of g. Then g assigns a z to this y. This gives us the pair (x,z). The set of all pairs (x,z) which can be constructed in this fashion is the composite function $g \circ f$. The domain of $g \circ f$ is the set of all x's for which this process can be defined; if there are no such x's, the function $g \circ f$ is not defined.

DEFINITION For two given functions $f:(x,y)$ and $g:(y,z)$, the set of ordered pairs (x,z) described above defines a function called the *composite* of g and f and written $g \circ f:(x,z)$. The value of the composite of g and f is written $(g \circ f)(x)$ or $g(f(x))$.

ILLUSTRATION 3 Let $z = g(y) = 3y^2 - 2y + 1$ and $y = f(x) = 4x + 7$. The composite $g \circ f$ is given by

$$z = g(y) = 3y^2 - 2y + 1 = 3(4x + 7)^2 - 2(4x + 7) + 1$$

This can be simplified to yield

$$z = 48x^2 + 160x + 134$$

which defines $g \circ f:(x, 48x^2 + 160x + 134)$.

ILLUSTRATION 4 A stone is dropped into a liquid, forming circles which increase in radius with time according to the formula $r = 4t$. How does the area of a given circle depend upon time?

SOLUTION The area A of a circle is $A = \pi r^2$, and we are given that $r = 4t$. These define two functions $g:(r,A)$ and $f:(t,r)$; and we seek the composite $g \circ f:(t,A)$.

$$A = \pi r^2 = \pi(4t)^2$$

or, reduced,

$$A = 16\pi t^2$$

Hence $g \circ f:(t,16\pi t^2)$ is the composite of g and f. Here we are not interested in negative values of t, although, mathematically, the maximal domain is the set of all real numbers. For the physical problem, a subset such as $0 \leq t \leq t_1$, where t_1 is sufficiently large, would suffice.

Since the letters used to represent the independent and dependent variables of a function can be replaced by other letters without change of meaning, we can speak of the composite $g \circ f$, where the functions f and g are given in the usual form $f:(x,y)$ and $g:(x,y)$. For we can rewrite g in the form $g:(y,z)$ if we wish. If we were interested in the composite $f \circ g$ we would rewrite f in the form $f:(y,z)$.

ILLUSTRATION 5 Given f and g whose values are $f(x) = x^2 + 2$ and $g(x) = 1 - (1/x)$, form the composite functions $g \circ f$ and $f \circ g$.

SOLUTION

For $g \circ f$

Rewrite the defining equations in the form

$$y = f(x) = x^2 + 2$$
$$z = g(y) = 1 - (1/y)$$

Then

$$z = 1 - (1/y) = 1 - [1/(x^2 + 2)]$$

and the composite $g \circ f$ is

$$g \circ f:(x,z)$$

or $\quad g \circ f:[x, 1 - [1/(x^2 + 2)]]$

For $f \circ g$

Rewrite the defining equations in the form

$$z = f(y) = y^2 + 2$$
$$y = g(x) = 1 - (1/x)$$

Then

$$z = y^2 + 2 = [1 - (1/x)]^2 + 2$$

and the composite $f \circ g$ is

$$f \circ g:(x,z)$$

or $\quad f \circ g:[x, [1 - (1/x)]^2 + 2]$

ILLUSTRATION 6 Find $g \circ f$ when g and f have the values $g(x) = |x|$ and $f(x) = x^2 - 3x + 1$.

SOLUTION Write $z = g(y) = |y|$ and $y = f(x) = x^2 - 3x + 1$. Then $z = |y| = |x^2 - 3x + 1|$. Thus $g \circ f:(x, |x^2 - 3x + 1|)$. To evaluate

$|x^2 - 3x + 1|$ for a given x, say $x = 1$, we first find $x^2 - 3x + 1$, which equals -1. Then we take its absolute value, which is $+1$.

PROBLEMS 9.5 In Probs. 1 to 7 find the values $(f+g)(x)$, $(f-g)(x)$, $(fg)(x)$, and $(f/g)(x)$. In each case state the domain.

1 $f(x) = \dfrac{1}{x+1}$, $g(x) = \dfrac{1}{x+2}$ **2** $f(x) = \dfrac{1}{\sqrt{x}}$, $g(x) = \dfrac{1}{\sqrt{x+1}}$

3 $f(x) = \dfrac{1}{x+2}$, $g(x) = x + 1$ **4** $f(x) = x^2 - 1$, $g(x) = x + 1$

5 $f(x) = x + 1$, $g(x) = (x+2)^2$ **6** $f(x) = 0$, $g(x) = \dfrac{1}{x^2+1}$

7 $f(x) = 1 + \dfrac{1}{x}$, $g(x) = \dfrac{1}{x^2} - 1$

In Probs. 8 to 10 form the composite $(g \circ f)(x)$ and state the domain.

8 $z = \sqrt{y}$, $y = x^2 + 6x + 9$
9 $z = 3y^2 - 2y$, $y = 1/x$
10 $z = y^3$, $y = 2/x$

In Probs. 11 to 16 form $(g \circ f)(x)$ and $(f \circ g)(x)$. State the domains.

11 $f(x) = x$, $g(x) = 1/x$ **12** $f(x) = |x|$, $g(x) = \sqrt{x}$
13 $f(x) = x - 4$, $g(x) = x^2$ **14** $f(x) = x^2$, $g(x) = 1/x^3$
15 $f(x) = x - 1$, $g(x) = 3(x+1)^2$ **16** $f(x) = 1/x^2$, $g(x) = |x - 1|$

17 Simplify:

a $\sqrt{(x^2 - 6x + 9)(x - 3)}$ **b** $\dfrac{1}{\sqrt{|-x^4|}}$

9.6 ▸ GRAPHS

In Chaps. 6 and 8 we plotted the graphs of certain special relations and in Sec. 6.5 we discussed the general concept of the graph of a relation. Here we shall study general methods for graphing relations and functions. By way of review, we define graph as follows:

DEFINITION The *graph* of a relation whose ordered pairs are (x,y) is the set of points in the XY-plane whose coordinates are the given pairs.

When the relation is defined by an equation, the basic method of plotting its graph is to find a reasonable number of points (x,y) whose coordinates satisfy the equation. Then we join these points by a smooth curve. There are two disadvantages to this method: (1) We may need to compute a rather large number of points in order to see just how the graph should look. (2) Even then, we may overlook some abnormal features of the graph which occur in the gaps between the plotted points. It is, therefore, desirable for us to develop some general aids to graphing which will cut down on your work and improve your accuracy. These apply only to relations (or functions) which are defined by equations.

Intercepts

INTERCEPTS The x-intercepts are the x-coordinates of the points at which the graph crosses (or meets) the X-axis, and the y-intercepts are the y-coordinates of the corresponding points on the Y-axis. To find the x-intercepts, put $y = 0$ in the equation which defines the relation, and solve for x. To find the y-intercepts, put $x = 0$, and solve for y.

When the relation is a function, the x-intercepts correspond to those x's for which the value of the function is zero. These x's are called the *zeros of the function*. We shall devote considerable attention in Chap. 10 to methods for finding the zeros of polynomial functions.

ILLUSTRATION 1 Find the intercepts of the graph of $y = x^2 - 3x + 2$.

SOLUTION Setting $y = 0$, we find that the solutions of $x^2 - 3x + 2 = 0$ are $x = 1$ and $x = 2$. The x-intercepts are 1 and 2. By setting $x = 0$ we find $y = 2$, which is the y-intercept (Fig. 9.13).

Domain and image

DOMAIN AND IMAGE It is very useful to know the domain and image of a relation, for this knowledge tells us about regions of the plane to which the graph is confined or from which it is excluded. It is useless to try to plot points in excluded regions. There are two common situations in which the domain or image is restricted to a subset of the whole axis.

Intercepts

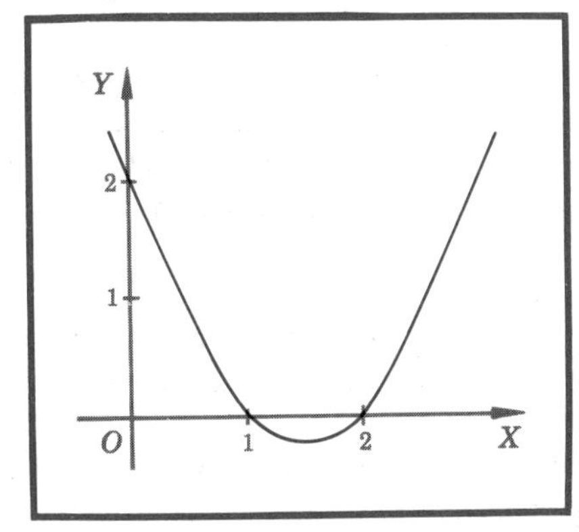

FIGURE 9.13
The zeros of a function are the x-intercepts of its graph. Here $y = x^2 - 3x + 2$.

The first of these is based upon the principle that y cannot take on values which require x to be complex and similarly that x cannot take on values which require y to be complex.

ILLUSTRATION 2 Discuss the domain and image of $x^2 + y^2 = 4$.

SOLUTION First solve for x, and obtain

$$x = \pm\sqrt{4 - y^2}$$

The right-hand side is real if and only if $4 - y^2 \geq 0$, or $4 \geq y^2$. Hence y must lie in the interval $-2 \leq y \leq 2$. Solving for y, we obtain $y = \pm\sqrt{4 - x^2}$ and arrive at the similar conclusion that x is in the interval $-2 \leq x \leq 2$.

ILLUSTRATION 3 Discuss the domain and image of $y = x^2 - 3x + 2$.

SOLUTION Since all values of x give real values of y, there is no restriction on x and the domain is the real line X. To find any possible restrictions on the image, solve for x. We have

$$x^2 - 3x + (2 - y) = 0$$

which yields
$$x = \frac{3 \pm \sqrt{9 - 4(2 - y)}}{2}$$
$$= \tfrac{3}{2} \pm \tfrac{1}{2}\sqrt{1 + 4y}$$

Since x must be real, this requires that y satisfy the inequality $1 + 4y \geq 0$, or $y \geq -\frac{1}{4}$. No part of the graph can therefore lie below the horizontal line $y = -\frac{1}{4}$ (Fig. 9.13).

The second principle is that expressions equal to a perfect square can never be negative. The application of this may give us inequalities which x or y must satisfy.

ILLUSTRATION 4 Discuss the domain and image of $y^2 = (x - 1)(x + 3)$.

SOLUTION Since $y^2 \geq 0$, we must have $(x - 1)(x + 3) \geq 0$. This is a quadratic inequality of the type discussed in Sec. 8.4. Using the methods developed there, we find that x cannot lie in the interval $-3 < x < 1$.

Solving for x, we find

$$x = -1 \pm \sqrt{4 + y^2}$$

Since $4 + y^2$ can never be negative, there are no restrictions on y (Fig. 9.14).

Symmetry

SYMMETRY The points (x,y) and $(x,-y)$ are symmetric with respect to the X-axis, the one being the mirror likeness of the other. Either point

Restrictions on domain

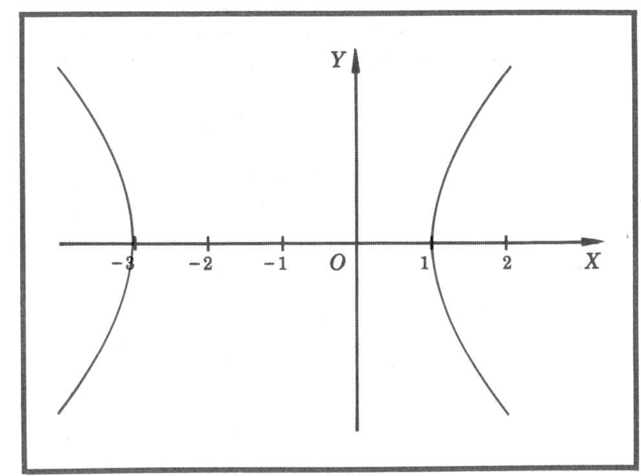

FIGURE 9.14
The domain of a real-valued function cannot include values of x which correspond to nonreal values of y. Here $y^2 = (x - 1)(x + 3)$.

is called a *reflection* of the other about the X-axis. The graph will be symmetric about the X-axis if for every point (x,y) on the graph the corresponding point $(x,-y)$ also lies on it. To test for symmetry, we therefore replace y in the equation of our relation by $-y$. If the resulting equation is the same as the given one, the graph is symmetric about the X-axis. In particular the graph is symmetric about the X-axis when y appears in the given equation to an *even* power only, for $y^{2k} = (-y)^{2k}$.

In a similar manner, a graph is symmetric about the Y-axis when replacement of x by $-x$ leaves the equation unchanged, e.g., when x occurs to an even power only.

Further, since a line joining (x,y) and $(-x,-y)$ passes through the origin and the distance from (x,y) to the origin is the same as the distance from $(-x,-y)$ to the origin, the graph will be symmetric about the origin if replacement of (x,y) with $(-x,-y)$ leaves the given equation unchanged.

EXERCISE A Examine $|y| - x = 0$, $y - |x| = 0$, $|x| + |y| - 1 = 0$ for symmetry.

EXERCISE B Show that if there is symmetry with respect to both axes there is, necessarily, symmetry with respect to the origin, but not conversely.

ILLUSTRATION 5

a The graph of $x^2 - x + y^4 - 2y^2 - 6 = 0$ is symmetric about the X-axis, but not about the Y-axis or the origin.

b The graph of $x^2 - x^4 + y - 5 = 0$ is symmetric about the Y-axis, but not about the X-axis or the origin.

c The graph of $x^4 + 2x^2y^2 + y^4 - 10 = 0$ is symmetric about both axes and the origin.

d The graph of $xy = 1$ is symmetric about the origin, but not about either axis.

Asymptotes

ASYMPTOTES When we solve the given equation for x or y, we may get an expression which contains a variable in the denominator. For example, we may have

$$y = \frac{x}{x - 1}$$

We have seen before that we cannot substitute $x = 1$ on the right, for this would make the denominator zero. We can, however, let x take values nearer and nearer to 1 and see how the graph behaves. Construct the table of values:

x	1.1	1.01	1.001	1.0001
y	11	101	1,001	10,001

It is clear that, as x approaches 1 from the right, y is becoming very large in the positive direction. Similarly, as x approaches 1 from the left, y becomes very large in the negative direction (Fig. 9.15).† The line $x = 1$ is now called a *vertical asymptote*.

If we solve the above equation for x, we obtain

$$x = \frac{y}{y - 1}$$

The same argument can now be applied to show that $y = 1$ is a *horizontal asymptote*.

To find asymptotes, the procedure is as follows: Solve for y and x if possible. Values of x or y which make the corresponding denominator zero correspond to vertical or horizontal asymptotes, provided that for these values the numerator is not also zero. The behavior of the graph near an asymptote must be determined by examining points near it, as was done above.

There is a more general definition of asymptote which applies to lines in other directions, but we shall not give it here.

†The language here is very imprecise, but is the best that can be presented to you at this stage. Later (Chap. 15) we shall write

$$\lim_{x \to 1^+} \frac{x}{x - 1} = +\infty$$

and

$$\lim_{x \to 1^-} \frac{x}{x - 1} = -\infty$$

and will define these terms more precisely.

ILLUSTRATION 6 Find the horizontal and vertical asymptotes, if any, of

$$y = \frac{x(x - 1)}{x + 2}$$

Vertical asymptote

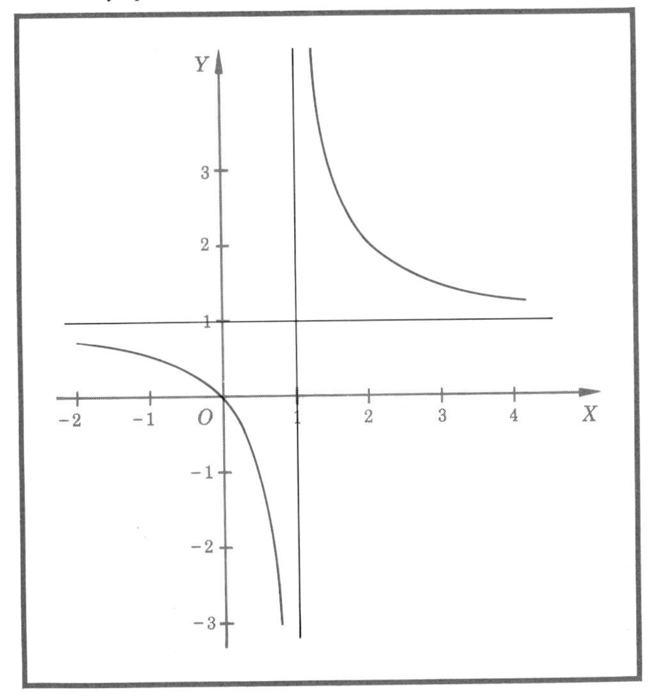

FIGURE 9.15 Values of x for which a denominator is zero but for which the corresponding numerator is not zero usually give vertical asymptotes. Similar values of y give horizontal asymptotes. Here $y = x/(x - 1)$.

SOLUTION Since the denominator is zero for $x = -2$, there is a vertical asymptote at $x = -2$. Solving for x, we find

$$x = \frac{1 + y \pm \sqrt{y^2 + 10y + 1}}{2}$$

Since y does not occur in the denominator, there are no horizontal asymptotes.

In the illustrations below we shall use these methods as needed to plot a number of graphs.

ILLUSTRATION 7 Plot the graph of the function whose values are $y = 4x^2 - 3$.

SOLUTION To determine the x-intercepts, or the zeros of the function, we set $4x^2 - 3 = 0$ and compute $x = \pm\sqrt{3}/2$. The y-intercept is $y = -3$.

Since any x gives a real value of y, the domain is the entire X-axis. We solve for x to determine the image. We find that $x = \pm\frac{1}{2}\sqrt{y + 3}$.

Graph of $y = \pm\sqrt{x^3 - x^2}$

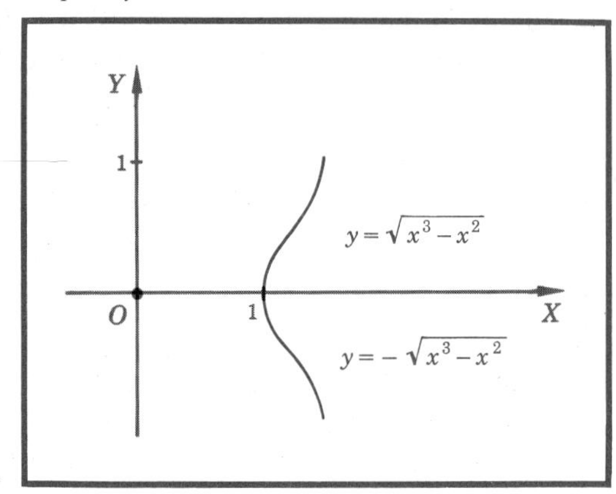

FIGURE 9.18
Watch for isolated points of graphs, such as (0,0) in this figure.

Each graph has an isolated point (0,0). Each graph is shown in Fig. 9.18.

PROBLEMS 9.6 In Probs. 1 to 30 plot the graphs of the relations.

1 $y = 3 + x$
2 $y = 1 - 2x$
3 $y = 3x^2 + 1$
4 $y = 3$
5 $y = 4x^2 - 1$
6 $y = -1/x$
7 $y = 2\sqrt{x}$
8 $y = (x - 3)(x + 3)$
9 $y = \sqrt{4 - x}$
10 $y = -4x^2 + 8x - 4$
11 $y = x^2 + x - 2$
12 $y = \sqrt{1 + x^2}$
13 $y = \sqrt{x^2 - 1}$
14 $y^3 = x$
15 $y = x(x^2 + 4)$
16 $y^2 = x^2(x - 4)$
17 $(x + y)^2 = 4$
18 $-x^2 + y^2 = x^3$
19 $y^3 = x^2$
20 $-y^3 = x^2$
21 $(-x)^3 = y^4$
22 $y^3 = x^4$
23 $x^2 + y^2 = 4$
24 $y = \dfrac{2\sqrt{x}}{3}$
25 $y = \dfrac{2}{\sqrt{x - 1}}$
26 $y = \dfrac{(x + 1)(x - 1)}{x - 1}$
27 $y = 1/x^2$
28 $y = x(x - 2)(x + 2)$
29 $y = \dfrac{x(x + 1)}{x - 1}$
30 $y = \sqrt{9 - x^2}$

In Probs. 31 to 43 analyze and sketch the graphs of the relations.

31 $4x^2 + 4y^2 = 4$
32 $4x^2 + 4y^2 = 1$
33 $4x^2 - y^2 = 1$
34 $x^2 - y^2 = 1$

35 $x^2 + y^2 = 0$

36 $x^2 + 4y^2 = 1$

37 $y^2 = -9x$

38 $y^2 = 9x$

39 $y^2 = x(x + 1)(x + 2)$

40 $y = (1 - x)(x + 1)(2 - x)$

41 $y^2 = -\dfrac{1}{x^3}$

42 $y^2 = \dfrac{x(x - 2)}{x + 3}$

***43** $y =$ the greatest integer smaller than x.

9.7 ▸ GRAPHS (Continued)

When the given relation is defined by an inequality or in some other way, additional ideas may be needed to plot its graph. We have seen (Secs. 8.5 and 8.7) how to plot the graphs of linear and quadratic inequalities. The following illustration extends this idea.

ILLUSTRATION 1 Plot the graph of the relation defined by the inequality $x^2 + y^2 < 9$.

We first plot the graph of $x^2 + y^2 = 9$, which is a circle (Fig. 9.19). This divides the plane into two regions, in one of which $x^2 + y^2 > 9$ and in the other $x^2 + y^2 < 9$. To see which region we want, test a typical point, say (0,1). At this point the inequality becomes $0^2 + 1^2 < 9$, which is true. Hence this is the desired region, and we color the graph as shown.

When the relation is defined in a different way, we must use methods suitable to the case in hand.

ILLUSTRATION 2 Plot the graph of the function f defined to be the set $\{(x,y)|x$ is a positive integer > 1, and y is the least prime not less than $x\}$.

The domain is the set of positive integers > 1, and the image is

Graph of $x^2 + y^2 < 9$

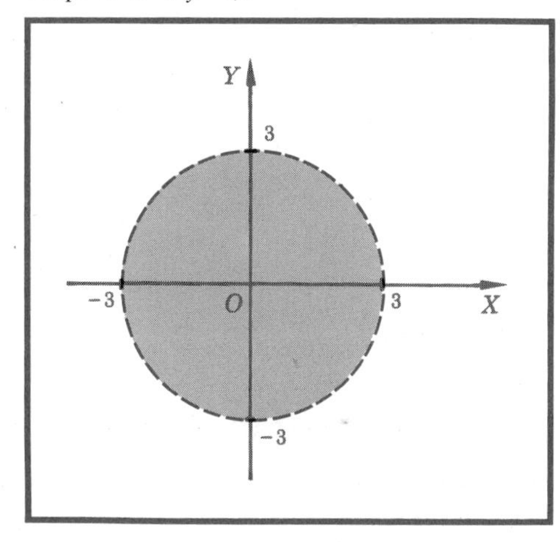

FIGURE 9.19
The graph of a relation may be a region in the plane.

Discrete set of points

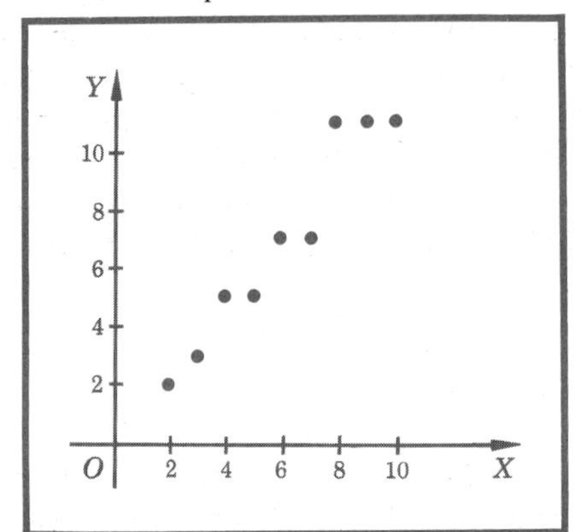

FIGURE 9.20 The graph of a relation may be a discrete set of points. This is the graph of the set $\{(x,y)\,|\,x$ is a positive integer >1 and y is the least prime not less than $x\}$.

the set of primes. The graph is a discrete set of points which must not be joined by a continuous curve (Fig. 9.20).

PROBLEMS 9.7

In Probs. 1 to 26 plot the graphs of the relations.

1 $x + y^2 = 0$

2 $x + y^2 > 0$

3 $x - y^2 = 0$

4 $x - y^2 \geq 0$

5 $x^2 - y^2 > 0$

6 $x^2 + y^2 > 0$

7 $x - y^4 \leq 0$

8 $x + y^4 > 0$

9 $y \geq x$

10 $y + x \geq 0$

11 $4x + 3y - 12 > 0$

12 $4x + 3y - 12 < 0$

13 $|y| = -x$

14 $|y| = |x|$

15 $x^2 + y^2 < 4$

16 $x^2 + y^2 \geq 4$

17 $|x| + |y| = 1$

18 $|x| + |y| > 1$

***19** $y = \begin{cases} 1 - x & x > 1 \\ x & x \leq 1 \end{cases}$

***20** $y = \begin{cases} -1 & x \text{ is an integer} \\ 2 & \text{otherwise} \end{cases}$

***21** $y = \begin{cases} 2 & x \text{ is a positive integer} \\ 0 & \text{otherwise} \end{cases}$

***22** $y = \begin{cases} 1/x & \text{if } x \text{ is a positive or negative integer} \\ 0 & \text{if } x = 0 \\ & \text{otherwise undefined} \end{cases}$

23 $y = -|x|$

24 $y = |x| - x$

***25** Let $[x]$ stand for the greatest integer not exceeding x. Plot $y = [x]$.

***26** Suppose that the postage on a certain class of domestic mail is 6 cents/oz or fraction thereof. This defines a function. Plot it.

9.8 › INVERSE FUNCTIONS

Let us return for the moment to a general relation with domain A and image B (Sec. 9.1). The given relation is *from A to B*. If we reverse the roles of A and B, we get a new relation *from* B to A with domain B and image A. This new relation is called the *inverse* of the given relation.

Inverse

Now suppose that the given relation is a function, which means that every element of A maps into a unique element of B. Is the inverse of this relation again a function? If so, every element of B must map into a unique element of A. But this happens if and only if the given function is one-to-one. Hence we have the theorem:

THEOREM 1 The inverse of a function f is again a function if and only if f is one-to-one.

NOTATION The function inverse to a one-to-one function f is denoted by the symbol f^{-1} (read "f inverse"). The "exponent" -1 indicates *inverse* and not *reciprocal. Do not confuse* f^{-1} *with* $1/f$. A similar notation is used in Sec. 2.3. In this notation we say that domain $f =$ image f^{-1} and image $f =$ domain f^{-1}.

When f is given as $y = f(x)$, we have called x the independent variable and y the dependent variable, and the ordered pairs are (x,y). The inverse function f^{-1} can then be written $x = f^{-1}(y)$, where now y is the independent variable, x is the dependent variable, and the ordered pairs are (y,x).

In graphing functions it is customary to plot the independent variable on the horizontal axis (regardless of its name x, y, z, t, etc.) and the dependent variable on the vertical axis. So if the graph of f is as in Fig. 9.21a, then the graph of f^{-1} is given by Fig. 9.21b. Since the names of the variables are unimportant (only the functional relations really matter), we may relabel the axes in Fig. 9.21b and plot both f and f^{-1}

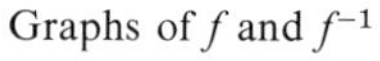
Graphs of f and f^{-1}

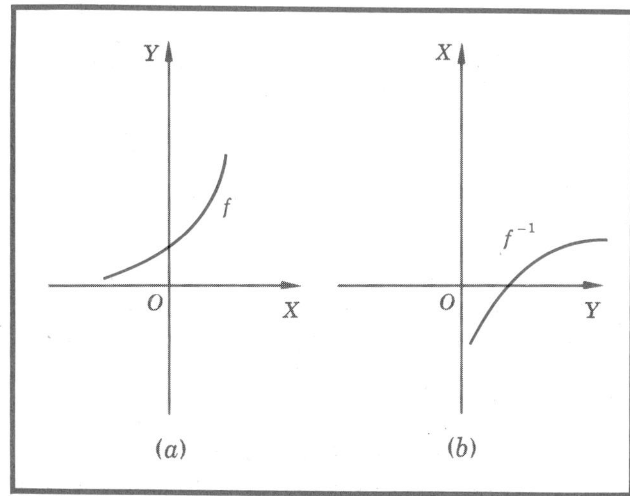

FIGURE 9.21
Note that in (b) the axes are interchanged.

Graphs of f and f^{-1}

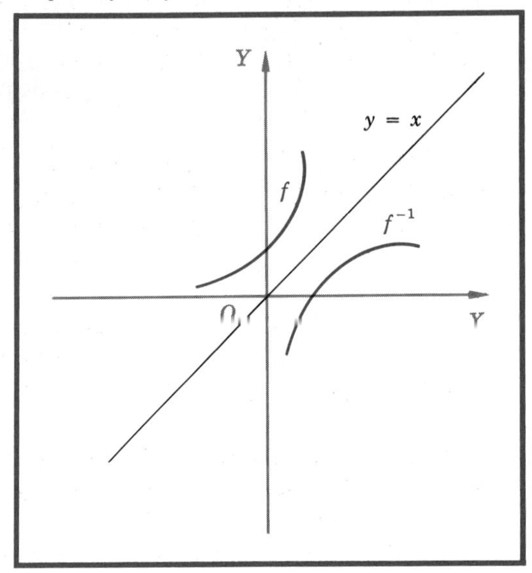

FIGURE 9.22
By interchanging the variables in f^{-1} we can plot f and f^{-1} on the same axes. Their graphs are symmetric about $y = x$.

on the same axes as in Fig. 9.22. In this figure it is apparent that the graphs of f and f^{-1} are symmetric about the line $y = x$.

If f is given by a simple formula $y = f(x)$, we can often obtain f^{-1} by solving this for x so that $x = f^{-1}(y)$. There are a number of difficulties with this procedure, which will be clarified by the following illustrations.

ILLUSTRATION 1 Let f be given by $y = 3x + 1$ and have the domain $\{x \mid 0 \leq x \leq 1\}$. The image of f is then $\{y \mid 1 \leq y \leq 4\}$. Find f^{-1} and its domain and image.

SOLUTION We solve the given equation for x and find

$$x = \frac{y - 1}{3}$$

This is the inverse function $f^{-1}:(y,x)$. Its domain is $\{y \mid 1 \leq y \leq 4\}$; its image is $\{x \mid 0 \leq x \leq 1\}$.

ILLUSTRATION 2 Let f be defined by $y = \frac{1}{2}\sqrt{4 - x^2}$ with the given domain $\{x \mid -2 \leq x \leq 0\}$ and image $\{y \mid 0 \leq y \leq 1\}$. Find f^{-1}, its domain and image.

SOLUTION In order to solve for x we square both sides and obtain

$$4y^2 = 4 - x^2$$

This process is risky, for $4y^2 = 4 - x^2$ is also obtained by squaring $y = -\frac{1}{2}\sqrt{4 - x^2}$, which is not the given function. With our fingers crossed, we now solve for x. The result is

$$x = \pm 2\sqrt{1 - y^2}$$

This is not exactly what we want, for we can use only one sign. We recall that the domain of f is $\{x \mid -2 \leq x \leq 0\}$, and this tells us to use the minus sign. The inverse function f^{-1} is therefore given by

$$x = -2\sqrt{1 - y^2}$$

Its domain is $\{y \mid 0 \leq y \leq 1\}$, and its image is $\{x \mid -2 \leq x \leq 0\}$.

To plot the graphs of f and f^{-1} on the same axes, we rewrite f^{-1} in the form $y = -2\sqrt{1 - x^2}$ and obtain Fig. 9.23.

If, however, we had chosen the maximum domain for f in Illustration 2, namely, $\{x \mid -2 \leq x \leq 2\}$, then f would not have been one-to-one, and f^{-1} would not have existed.

Restriction of domain

This illustration emphasizes an important point which will be discussed later: *If f is not one-to-one and hence does not have an inverse, we may be able to restrict its domain so that it is one-to-one in the restricted domain.* The restricted function then has an inverse (Sec. 13.10).

EXERCISE A The function $y = x^2(-\infty < x < \infty)$ does not have an

Graphs of function and its inverse

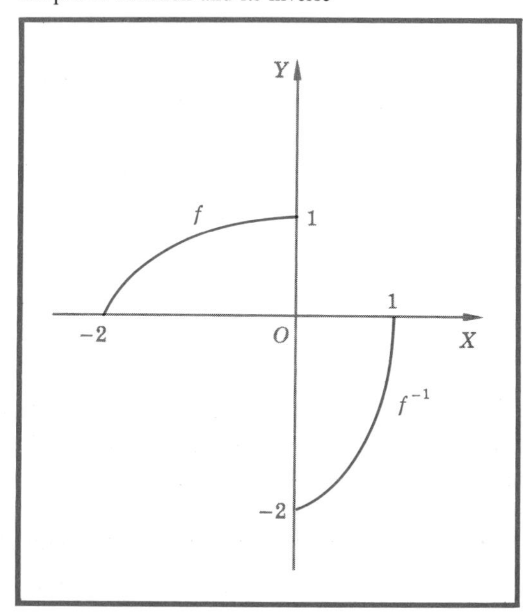

FIGURE 9.23
f is given by $y = \frac{1}{2}\sqrt{4 - x^2}$; f^{-1} is given by $y = -2\sqrt{1 - x^2}$.

inverse. Find a domain for x such that the restricted function does have an inverse. How many domains can you find?

ILLUSTRATION 3 Let f be defined by $y = 2^x$, where the domain is $\{x \mid -\infty < x < \infty\}$ and the corresponding image is $\{y \mid 0 < y < \infty\}$. Find f^{-1}.

SOLUTION Since the graph of $y = 2^x$ rises steadily as x increases (Fig. 9.24), no two values of x give the same value of y. So f is one-to-one, and the inverse f^{-1} clearly exists. The trouble is that we do not know how to solve for x in terms of y by means of any formula. The function f^{-1} is therefore a new function unnamed at present. If we have frequent occasion to refer to this function, we shall find it convenient to give it a name and to investigate its properties. This is indeed a common method of obtaining new functions from known functions. You may already know the name of f^{-1} when $f(x) = 2^x$. It is called the *logarithm of y to the base* 2, written $\log_2 y$. We shall study this function in detail in Chap. 11.

This illustration emphasizes another point of importance: *If a function f has an inverse which cannot be calculated by elementary means, we shall often give this inverse a name and add it to our list of useful functions.*

EXERCISE B Give a definition of $\sqrt[3]{y}\,(y > 0)$ as the inverse of some function.

EXERCISE C If you have studied trigonometry, show that

$$y = \sin x \qquad -\frac{\pi}{2} \leq x \leq \frac{\pi}{2}$$

has an inverse. What is the name of this inverse?

The example of $y = 2^x$ in Illustration 3 suggests the following new term.

Graph of $y = 2^x$

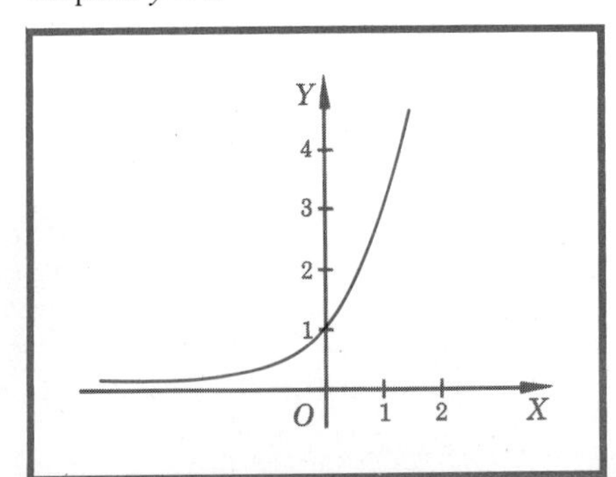

FIGURE 9.24
The inverse of a familiar function may be a new, unfamiliar function.

DEFINITION
Monotone function

A function f whose domain and image are subsets of the reals is *strictly monotone increasing* if and only if, for every pair x_1 and x_2 such that $x_2 > x_1$, we have $f(x_2) > f(x_1)$. Similarly, f is *strictly monotone decreasing* if and only if, for every pair x_1 and x_2 such that $x_2 > x_1$, we have $f(x_2) < f(x_1)$.

From the definition of a strictly monotone function f it follows that if $x_1 \neq x_2$ then $f(x_1) \neq f(x_2)$, and so f is one-to-one. This leads to the following theorem.

THEOREM 2 A strictly monotone function (either increasing or decreasing) is one-to-one and hence has an inverse.

REMARK The converse of this theorem is false, as is illustrated for the one-to-one function sketched in Fig. 9.25. If, however, f is defined in an interval in which it is not only one-to-one but is also continuous (in a sense to be defined in Sec. 16.9), then f is strictly monotone.

Identity function

In order to introduce a final important property of inverse functions, let us define the identity function whose domain and image are the same set X.

Graph of one-to-one function

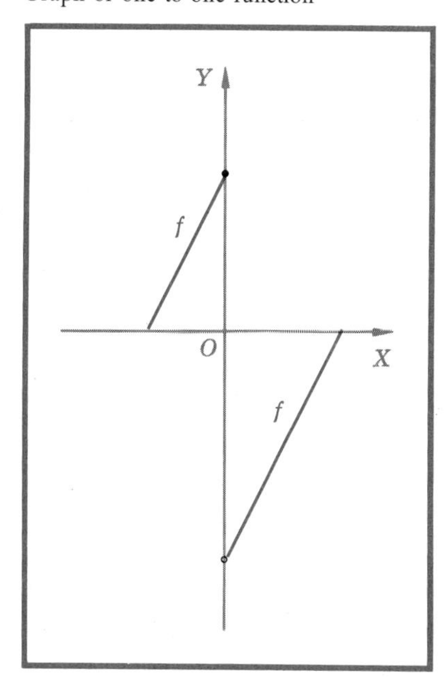

FIGURE 9.25
This function is one-to-one and has an inverse, but it is not monotone over its entire domain.

DEFINITION The function $E:X \to X$ whose elements are the ordered pairs (x,x) is called the identity function. This function maps each x onto itself. (See Fig. 9.26.)

The identity function

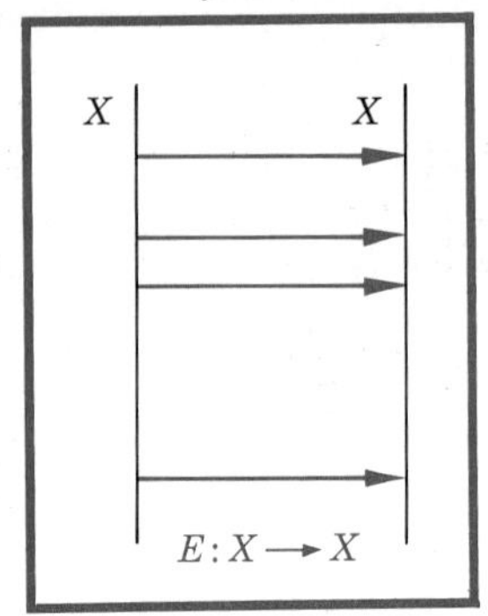

FIGURE 9.26 The identity function $E: X \to X$ consists of the ordered pairs (x,x).

Now let us consider a function $f:X \to Y$ and its inverse $f^{-1}:Y \to X$. The composite function $f^{-1}(f)$ maps each x onto some y (under f) and then back onto itself (under f^{-1}). We see then (Fig. 9.27) that

$$f^{-1}(f) = E \qquad \text{and similarly} \qquad f(f^{-1}) = E$$

ILLUSTRATION 4 In Illustration 1, $f(x) = 3x + 1$ and $f^{-1}(y) = (y - 1)/3$. Hence

$$f^{-1}(f(x)) = \frac{(3x + 1) - 1}{3} = x$$

Also

$$f(f^{-1}(x)) = 3\left(\frac{x - 1}{3}\right) + 1 = x$$

EXERCISE D In Illustration 2, verify that $f(f^{-1}) = f^{-1}(f) = E$.

PROBLEMS 9.8

In Probs. 1 to 12 find f^{-1} if it exists, its domain, and image, and sketch $y = f(x)$, $y = f^{-1}(x)$.

1 $y = x$

2 $y = x + 1$

3 $y = 3x + 6$

4 $y = ax + b$

5 $y = ax - b$

6 $y = x^2$, $x =$ all reals

7 $y = x^2 - 9$

8 $y = x^3$

9 $y = x^3 - 8$

10 $y = x^4$, $x =$ all reals

11 $y = |x|$

12 $y = \sqrt{1 - x^2}$, $0 \le x \le 1$

13 Find two functions $f(x)$ and $g(x)$ such that $f(g(x)) \neq g(f(x))$. Can $f(x) = x$ ever be one of the two?

The composite of f and f^{-1}

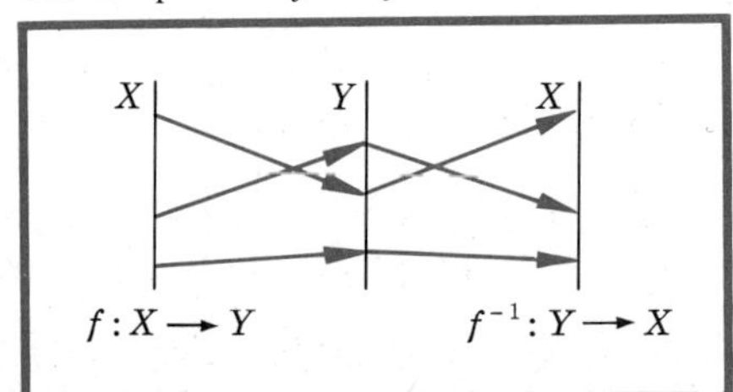

FIGURE 9.27 $f^{-1}(f) = E$

9.9 › FUNCTIONS DERIVED FROM EQUATIONS

At first sight it may seem that any equation can be used to define a function by solving it for one of the variables. This process, however, often has a number of difficulties which are suggested by the illustrations below.

ILLUSTRATION 1 The equation $2x - 3y + 1 = 0$ is called a "linear equation" because the pairs (x,y) which satisfy it lie on a straight line (see Sec. 14.8). From this equation we can derive two functions:

$$y = f(x) = \frac{2x + 1}{3} \qquad x = g(y) = \frac{-1 + 3y}{2}$$

EXERCISE A Show that $f^{-1} = g$.

ILLUSTRATION 2 The equation $s = 16t^2$ gives the distance s in feet through which a body falls from rest under the influence of gravity in t sec. As such it defines a function. We may ask, however: "How long does it take for the body to fall 64 ft?" To answer this, we solve for t:

$$t^2 = \frac{s}{16} \qquad t = \pm\tfrac{1}{4}\sqrt{s}$$

This gives two functions defined by: $t = \frac{1}{4}\sqrt{s}$; $t = -\frac{1}{4}\sqrt{s}$. In terms of the physical situation only the first makes *practical* sense, but *both* make *mathematical* sense. Therefore we choose $t = \frac{1}{4}\sqrt{s}$, put $s = 64$, and find $t = 2$.

EXERCISE B Are there any physical situations in which $t = -\frac{1}{4}\sqrt{s}$ makes *practical* sense?

This illustration makes the point that, although an equation may lead to several functions, not all of these necessarily have meaning in a practical situation. You will have to use your head and discard those which are nonsense.

ILLUSTRATION 3 Consider the equation $x^2 + y^2 = 4$, which represents a circle of radius 2 (Fig. 9.28). If we solve for y, we obtain $y = \pm\sqrt{4 - x^2}$. Of the many functions which can be obtained from this, two have outstanding importance:

$$y = f(x) = \sqrt{4 - x^2} \qquad -2 \le x \le 2$$
$$y = g(x) = -\sqrt{4 - x^2} \qquad -2 \le x \le 2$$

The graph of f is the upper semicircle, and the graph of g is the lower semicircle.

Graph of $x^2 + y^2 = 4$

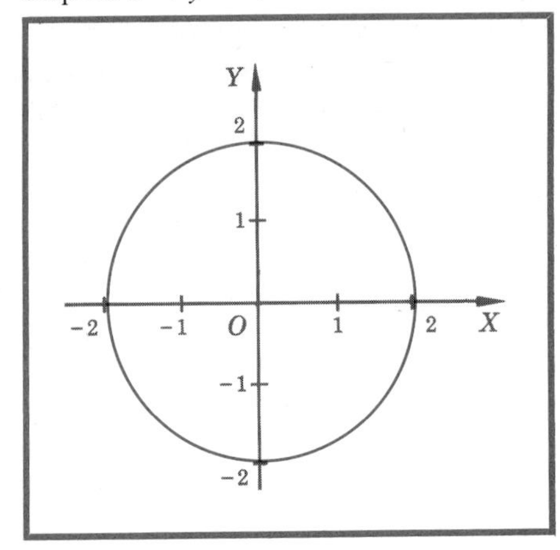

FIGURE 9.28
Two or more functions may be derived from an equation.

EXERCISE C Solve $x^2 + y^2 = 4$ in Illustration 3 for x, and describe the graphs of the two functions so obtained. Call these F and G.

EXERCISE D Does F^{-1} equal f or g?

ILLUSTRATION 4 The equation $x^2 + xy + 4 = 0$ is quadratic in x and thus has the solution

$$x = \frac{-y \pm \sqrt{y^2 - 16}}{2}$$

This yields two functions defined by

$$x = f(y) = \frac{-y + \sqrt{y^2 - 16}}{2} \qquad |y| \geq 4$$

$$x = g(y) = \frac{-y - \sqrt{y^2 - 16}}{2} \qquad |y| \geq 4$$

When we solve for y, we obtain

$$y = -\frac{x^2 + 4}{x}$$

which gives the function h, where

$$y = h(x) = -\frac{x^2 + 4}{x} \qquad x \neq 0$$

EXERCISE E What is the domain of $f \circ h$? Show that in this domain $f \circ h = E$. Answer the same question for $g \circ h$ and find its domain.

Graph of $x^5y + xy^5 = 2$

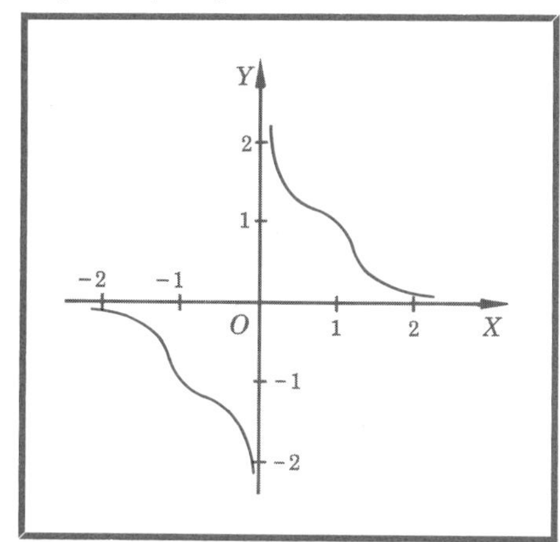

FIGURE 9.29 Functions may be derived from equations even though we cannot solve directly for x or y.

All these functions derived from equations have a common property: their elements (x,y) satisfy the equation. This suggests the more general definition.

DEFINITION If the elements (x,y) of a function f satisfy an equation in x and y, the function f is said to be *derived* from this equation.

Implicit function

In many textbooks and older works a function thus derived from an equation is said to be given "implicitly" by the equation. The functions themselves are called "implicit functions".

In the examples given above the functions derived from an equation were obtained by solving for one of the variables. It is important to note that derived functions may exist even when we are unable to carry through such a solution. We shall sometimes wish to consider such functions, an example of which is given below.

ILLUSTRATION 5 The equation $x^5y + xy^5 = 2$ has a graph given by Fig. 9.29. There is no simple way of solving this equation for x or y, but the graph indicates that functions $y = f(x)$ and $x = g(y)$ exist which are derived from this equation.

PROBLEMS 9.9

In Probs. 1 to 13 find some function derived from the given equation, and state its domain and image.

1 $4y + 3x = 8$

2 $8x - 3y + 7 = 0$

3 $2x^2 + y^2 = 2$

4 $x^2 - y^2 = 1$

5 $x^2y + x = 1$

6 $x^2y + y - 1 = 0$

7 $|x| - |y - 1| = 0$

8 $|x - y| = 0$

9 $2^x = 2^y$

10 $x < y$

11 $|x| < y$

12 $x + y \geq 1,\ x \geq 0,\ y \geq 0$

13 $v^2 - 2gs = 0$, where v is the velocity of a body falling from rest, s is the distance fallen, and g is a given positive constant.

10 ALGEBRAIC FUNCTIONS

10.1 › INTRODUCTION

In Chap. 2 we discussed some of the properties of real and complex numbers. The four operations of addition, subtraction (inverse of addition), multiplication, and division (inverse of multiplication) are called the *rational operations* of arithmetic. The *algebraic operations* of arithmetic include these and also those of taking roots and raising powers. In this chapter the universal set will be the field of complex numbers which is closed under each of these processes.

10.2 › POLYNOMIAL FUNCTIONS

We now wish to apply three of the rational operations to a *variable* x in order to generate a very important special type of function. We restrict ourselves to addition, subtraction, and multiplication, and with these we build up such functions as those given by

1 2 **2** $(3x - 5)(x + 6)$

3 $\sqrt{3}x^5 - \pi x^2 - 1$ **4** $(2^{1/3}x - x^2)(x + 2)^3 + (7 - 3i)x - i$

But we do not obtain such functions as those defined by

5 $(1/x) - 3i$ **6** $\sqrt{x}$

7 2^x **8** $|x|$

Polynomial function

The functions illustrated by 1 through 4 are special cases of what are called polynomial functions according to the following definition.

DEFINITION A function P is called a *polynomial function* if it is given by

(1) $$y = P(x) = a_0x^n + a_1x^{n-1} + \cdots + a_{n-1}x + a_n$$

where n is a positive integer or zero and the coefficients $a_0, a_1, \ldots, a_n$ are complex numbers. Its domain and range are each the set of complex numbers. Its image is a constant, if $n = 0$; otherwise it is the set of complex numbers.

EXERCISE A Show that function 4 above can be written in the form (1).

We say that P is of *degree n,* provided that $a_0 \neq 0$. In the theory of polynomial functions it is customary to call the right-hand side of (1), namely, the expression

(2) $$a_0x^n + a_1x^{n-1} + \cdots + a_{n-1}x + a_n$$

a *polynomial,* which we designate by the symbol $P(x)$, read "polynomial in the variable x". $P(x)$ also stands for the value of P at x, but this should not lead to any confusion. Note that, since n may be zero, a constant is to be considered a polynomial (and a polynomial function). If $a_0 = n = 0$, then $P(x) = 0$ is called the zero polynomial.

We have stated that the domain of P is the set of complex numbers and that its range is the set of complex numbers. It is possible to discuss polynomial functions with other sets as domain and range and with other types of coefficients as well. For example, we may consider real polynomials in which x, y, and the a's are all real numbers. Or we may require that the a's be rational numbers or even integers and then let x and y be real or complex. All of these are special cases of the general definition given above. In discussing these special cases, we shall have to take great pains in stating the types of coefficients and variables which are under consideration.

EXERCISE B Prove that the sum of two polynomials is a polynomial.

EXERCISE C Prove that the product of the two polynomials $(ax^2 + bx + c)$ and $(Ax^2 + Bx + C)$ is a polynomial. (As a matter of fact, the product of any two polynomials is a polynomial.)

EXERCISE D Show by an example that the quotient of two polynomials may be a polynomial. Find another example in which the quotient is not a polynomial.

EXERCISE E Prove that the composite $g \circ f$ of the two polynomial functions given by $f(x) = ax^2 + bx + c$, $g(x) = Ax^2 + Bx + C$ is a polynomial function. (As a matter of fact, the composites of any two polynomial functions are polynomial functions.)

10.3 › RATIONAL FUNCTIONS

The next simplest type of function is a *rational function* which is so called because we now permit the use of division along with the other rational operations.

DEFINITION
$R(x) = P(x)/Q(x)$

A function R defined by $y = R(x) = P(x)/Q(x)$, where $P(x)$ and $Q(x)$ are polynomials, is called a *rational function.*

The remarks made above about the domain and range of a polynomial function apply equally well to a rational function but here we must be a little careful: The function R is not defined at points where $Q(x) = 0$. This is made clear by the following illustrations.

ILLUSTRATION 1

a $y = 1/x$ is not defined at $x = 0$.
b $y = (x - 1)/(x + 2)$ is not defined at $x = -2$.
c $y = 3x^3/[(x - 1)(x^2 + 1)]$ is not defined at either $x = 1$ or at $x = \pm i$.

ILLUSTRATION 2

a $y = x/x$ is not defined at $x = 0$. For other values of x, however, $x/x = 1$. The two functions x/x and 1 are consequently not identical. This illustration brings up an important point: The cancellation of a common factor in the numerator and denominator may change the function involved.
b As a similar example, consider the two rational functions

$$y = \frac{x(x - 1)}{x - 1} \qquad \text{and} \qquad y = x$$

These have the same values for $x \neq 1$, but at $x = 1$ the first is undefined, whereas the second has the value 1. Hence they are different functions.

ILLUSTRATION 3 Some functions which are not written in the explicit form $y = P(x)/Q(x)$ are nevertheless equivalent to rational functions. Consider

$$\begin{aligned} y &= \left(\frac{1}{x} - \frac{4}{x + 1}\right)\left(1 + \frac{1}{x - 1}\right) \\ &= \frac{x + 1 - 4x}{x(x + 1)} \cdot \frac{x - 1 + 1}{x - 1} \\ &= \frac{(1 - 3x)(x)}{x(x + 1)(x - 1)} \\ &= \frac{-3x^2 + x}{x^3 - x} \end{aligned}$$

In this bit of algebra we did not cancel out the x in numerator and denominator. Note that the function

$$y = \frac{-3x + 1}{x^2 - 1}$$

is *not* equivalent to the given function. Why?

In Exercises A to D below assume that the sum, product, and composites of two polynomials are each polynomials.

EXERCISE A Prove that the product of two rational functions is a rational function.

EXERCISE B Prove that the sum of two rational functions is a rational function.

EXERCISE C Prove that the quotient of two rational functions is a rational function.

EXERCISE D Prove that the composites of two rational functions are rational functions.

10.4 › EXPLICIT ALGEBRAIC FUNCTIONS

Explicit algebraic functions constitute the next important class of functions. These include the polynomial and rational functions as special cases, and they are generated by a finite number of the algebraic operations. Thus the function whose values are given by

$$\frac{\sqrt{1+x} - \sqrt[3]{x^5}}{\sqrt[6]{(2+x-x^2)^3 - 8}}$$

is an example of an explicit algebraic function. Because of the possible appearance of (even) roots in the equation defining such a function, it may happen that the value y of the function is real only when x is restricted to a very limited subset of the real numbers. For the example above, it is seen first of all that x (real) must be greater than or equal to -1 if $\sqrt{1+x}$ in the numerator is to be real. Similarly in $\sqrt[6]{(2+x-x^2)^3 - 8}$ it must be true that $(2+x-x^2)^3 > 8$; that is, $2+x-x^2 > 2$ or $x(1-x) > 0$. This says that x must lie between 0 and 1 exclusively. The domain of definition is therefore $0 < x < 1$.

Of course, if x and y are not required to be real, then the only values of x for which the above function is not defined are $x = 0$ and $x = 1$.

Our interest in the most general explicit algebraic functions will be confined mainly to their graphs.

PROBLEMS 10.4

In Probs. 1 to 4 state which of the following define polynomial functions.

1 **a** $y = x^2$
b $y = 1/x^2$
c $y = (x+2)^x$

2 **a** $y = 2x + 4 - 2i$
b $y = 2x^2 + \sqrt{3}x - i$
c $y = ix^3 - \sqrt{4x} - 2$

3 **a** $y = \dfrac{x^6}{6!} - \dfrac{x^4}{4!} + \dfrac{x^2}{2!}$
b $y = (1 + i)(x + 3i)$
c $y = (9 - x)^3$

4 **a** $f:(x,3)$
b $g:(x,x^{-3})$
c $h:(x, x^{-3} + x)$

In Probs. 5 to 8 state which of the following define rational functions.

5 **a** $y = cx + d$
b $y = \dfrac{(x - 2)^{3/2}}{x - 2}$
c $y = |x|$

6 **a** $y = (x + 2)^x$
b $y = \dfrac{x - 1}{x}$
c $y = \dfrac{(x + 2)^2}{(x - 3)^2}$

7 **a** $y = \dfrac{1}{\pi x}$
b $y = 3^{x^2} + 2$
c $y = \dfrac{x^2 + 3x - 7}{x^2 + 5x - 6}$

8 **a** $y = \sqrt{x - 2}$
b $y = \dfrac{(x + 1)^2}{2x + i}$
c $y = \dfrac{x^n}{3^x}$

In Probs. 9 to 12 state which of the following define explicit algebraic functions. (Assume x real.)

9 **a** $y = x^{2/3} + x^{3/2}$
b $y = |x|$
c $y = \dfrac{x^5}{5} - \dfrac{2ix^3}{3}$

10 **a** $y = x^2 + 2\sqrt{x}$
b $y = \sqrt{\sqrt{x} + 2}$
c $y = \left(\dfrac{x^2}{12!} + \dfrac{x^6}{16!}\right)\sqrt{x}$

11 **a** $y = \begin{cases} x & x \text{ rational} \\ -x & x \text{ irrational} \end{cases}$
b $y = 4^x$
c $y = |x^2|$

12 **a** $y = \sqrt{x^4}$
b $y = |x^3|$
c $y = x^{1/3}$

In Probs. 13 to 18 state kind of function and domain (we assume x, y, etc., real, and n a positive integer).

13 **a** $y = x^2 + \sqrt{3}x$
b $y = x^3 + \sqrt{3x}$
c $y = \dfrac{\sqrt{x^3 + 3x}}{\sqrt{x + 1}}$

14 **a** $y = x^{1/5}$
b $y = \sqrt{x^2 + 3x}$
c $y = \sqrt{x - 2} - \sqrt{x - 4}$

15 **a** $y = \dfrac{x^2}{(x + 1)(x - 6)}$
b $y = \sqrt[3]{x^3}$
c $y = \sqrt{4 - x^2}$

16 **a** $y = \left(\dfrac{2}{x}\right)^{-n}$
b $y = \sqrt{x} + \dfrac{3}{x + 2}$
c $y = \dfrac{\sqrt{x} + x + 1}{\sqrt{x + 1}}$

17 a $f:(x, 2 - \sqrt{x - 2})$
b $g:\left(y, \left(1 - \frac{y}{y^2}\right)^n\right)$
c $h:\left(\theta, 1 + \frac{\theta}{2} + \frac{\theta^2}{3}\right)$

18 a $f:(y,n)$
b $g:\left(z, \frac{\sqrt{z - z^2}}{3z}\right)$
c $h:(x, \sqrt{x^3} + 2)$

10.5 » GRAPHS AND CONTINUITY

Continuity

We have already considered methods of plotting the graphs of functions and relations, or, what amounts to the same thing, of plotting the graph of an equation (Chap. 9). We must still rely upon intuition when we speak about the *continuity* of a function or about a *continuous* graph, but we wish at this time to make some pertinent remarks on the continuity of an algebraic function. For this discussion we restrict ourselves to the field of real numbers, since we plot only the real elements of a function. An element (x,y) is real when and only when x and y are both real. We shall refer indifferently to a continuous function or a continuous graph, the latter being merely descriptive geometric language.

POLYNOMIAL FUNCTIONS A polynomial function, defined by $y = P(x)$, where $P(x)$ is a polynomial, is continuous everywhere. The graph of a polynomial function is a continuous curve. The domain of definition is the set of real numbers; the image is a subset of the real numbers (which could be the whole set). As an example, see Sec. 9.6, Illustration 7.

DEFINITION The *zeros* of P are the values of x for which $P(x) = 0$.

Zeros of a polynomial

ILLUSTRATION 1 Sketch the graph of the polynomial function given by $y = x^4 - 2x^2$.

SOLUTION For purposes of graphing we now consider the domain as the set of real numbers. Since

$$x^4 - 2x^2 = x^2(x^2 - 2)$$

the zeros are seen to be $x = 0, 0, \pm\sqrt{2}$. (For the factor x^2, we write $x = 0, 0$; see Theorem 4, and following remark.)

The graph is continuous everywhere. It is symmetric with respect to the Y-axis since x appears to even powers only. Some values of the function defined by this equation are given in the following table.

x	-2	$-\sqrt{2}$	-1	0	1	$\sqrt{2}$	2
y	8	0	-1	0	-1	0	8

Graph of polynomial function

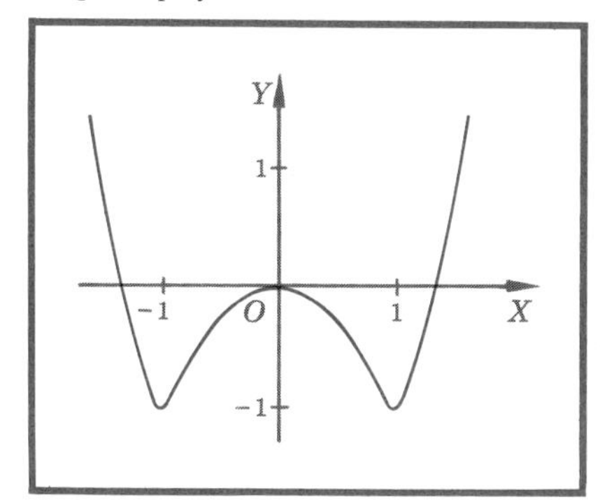

FIGURE 10.1
The graph of a polynomial function is continuous. Here $y = x^4 - 2x^2$.

The graph is shown in Fig. 10.1. The least value of y occurs when $x = \pm 1$, although we cannot prove this. (See Sec. 16.7 where maxima and minima are treated by methods of the calculus.) The image is $-1 \leq y < \infty$.

RATIONAL FUNCTIONS A rational function, defined by

$$y = \frac{P(x)}{Q(x)}$$

where $P(x)$ and $Q(x)$ are polynomials and $Q(x)$ is not the zero polynomial, is continuous everywhere with the exception of, at most, a finite number of isolated values of x, namely, those for which $Q(x) = 0$. These values must be excluded from the domain. The image is a subset of the reals. The graph of a rational function is a continuous curve with the exception of at most a finite number of points. It is not continuous at a point x_1 where $Q(x_1) = 0$. At such a point the function is undefined, as in $(x - 5)/(x - 5)$ at $x = 5$ or as in $1/x$ at $x = 0$.

ILLUSTRATION 2 Sketch the graph of the rational function defined by $y = f(x) = (x + 2)/(x - 1)^3$.

SOLUTION The domain is the set of real numbers excluding $x = 1$. The following intuitive argument will tell us something about the image: If x is just a little larger than 1, y is positive and very large; if x is a very large number, y is positive and very small. For $x > 1$, the image is, therefore, the set of positive real numbers. There is no value of y for $x = 1$, and the function is not continuous there. Further, at $x = -2$, $y = 0$ and $x = -2$ is the only zero of the function. If x is negative and a little larger than -2, y is negative and in absolute value very small; if x is positive and a little less than 1, y is negative and very large in absolute value. Therefore, for $-2 \leq x < 1$, y ranges over all nonpositive real numbers. Hence the image is the set of real numbers.

We compute the following table of values.

x	-4	-3.5	-3	-2	-1	0	$\frac{1}{2}$	1	1.5	2	3	11
y	$\frac{1}{62.5}$	$\frac{1}{60.75}$	$\frac{1}{64}$	0	$-\frac{1}{8}$	-2	-20	—	28	4	$\frac{5}{8}$	$\frac{13}{1{,}000}$

We have included the value of y at $x = -3.5$ because it is the largest value of y for $x < 1$. (Methods of the calculus are needed to prove this.) The graph is plotted in Fig. 10.2.

EXPLICIT ALGEBRAIC FUNCTIONS Explicit algebraic functions as defined include the rational functions as special cases. An explicit algebraic function may, therefore, fail to be continuous for the reasons given above. In addition it may have isolated points as indicated in the examples of Illustration 9, Sec. 9.6. It may also have endpoints as indicated in Illustration 3 below. The domain and image are subsets of the reals.

ILLUSTRATION 3 Sketch the graph of the explicit algebraic function defined by

$$y = \sqrt{x^2 - 1}$$

Graph of rational function

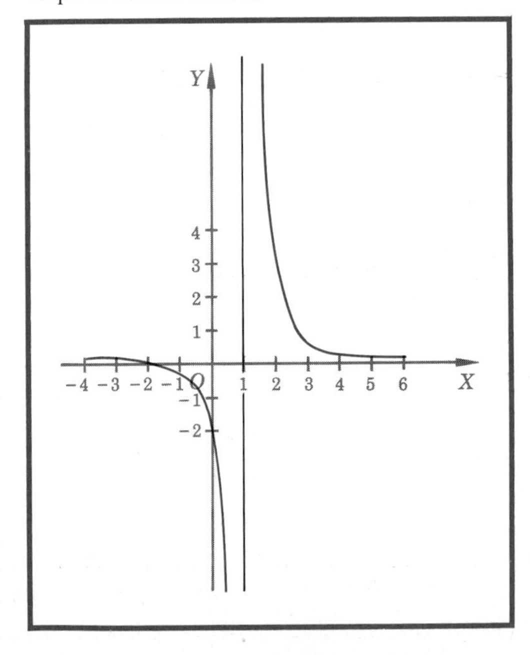

FIGURE 10.2
The graph of a rational function may not be continuous. Here $y = (x + 2)/(x - 1)^3$.

Since y must be real, we must have $x^2 - 1 \geq 0$, or $x^2 \geq 1$. Hence the domain of the function is $-\infty < x \leq -1;\ 1 \leq x < +\infty$. The image is $0 \leq y < \infty$. The graph is symmetric with respect to the Y-axis, since substitution of $-x$ for x in $\sqrt{x^2 - 1}$ does not change this expression. Thus we compute a table of values.

x	± 1	± 2	± 3	± 4
y	0	1.73	2.83	3.87

The graph is shown in Fig. 10.3. It has endpoints at $x = \pm 1$.

It is interesting to compare Fig. 10.3 with the graph of $y = x^2 - 1$ (Fig. 10.4) plotted from the table of values.

x	0	± 1	± 2	± 3
y	-1	0	3	8

The ordinates of Fig. 10.3 are the square roots of those of Fig. 10.4 except in the interval $-1 < x < 1$ in which these square roots become imaginary. In cases like this considerable information about graphs such as Fig. 10.3 can be obtained by first plotting the related graph such as Fig. 10.4.

ILLUSTRATION 4 Sketch the graph of

$$y = \frac{x}{\sqrt{x - 1}}$$

Graph of explicit algebraic function

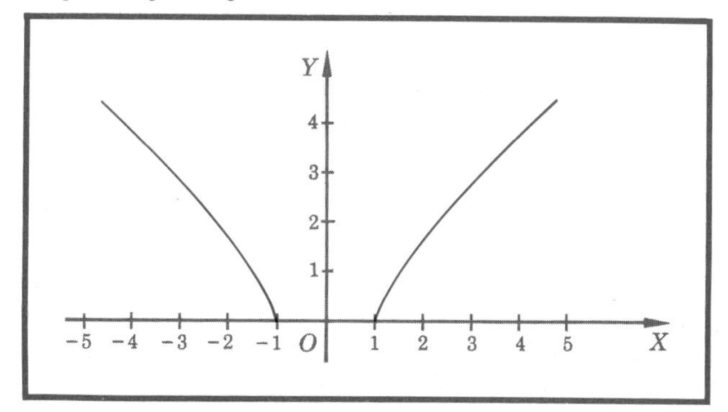

FIGURE 10.3
The graph of an explicit algebraic function may have endpoints. Here $y = \sqrt{x^2 - 1}$.

Graph related to Fig. 10.3

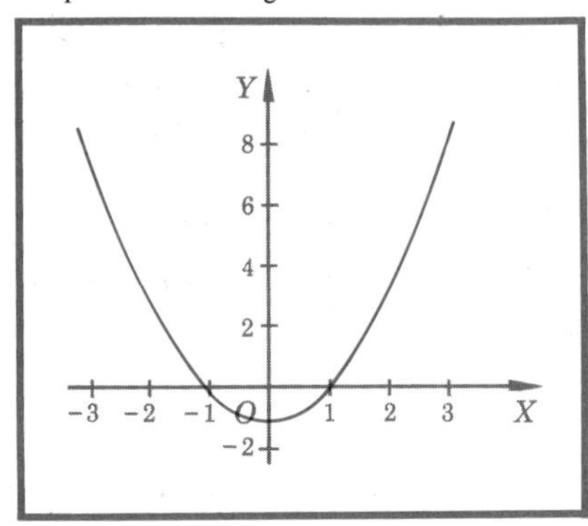

FIGURE 10.4
The graph of $y = \sqrt{f(x)}$ can be derived from the graph of $y = f(x)$. Here $y = x^2 - 1$. In this way we may obtain Fig. 10.3 from Fig. 10.4.

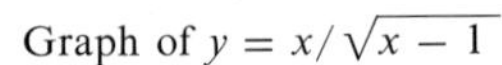

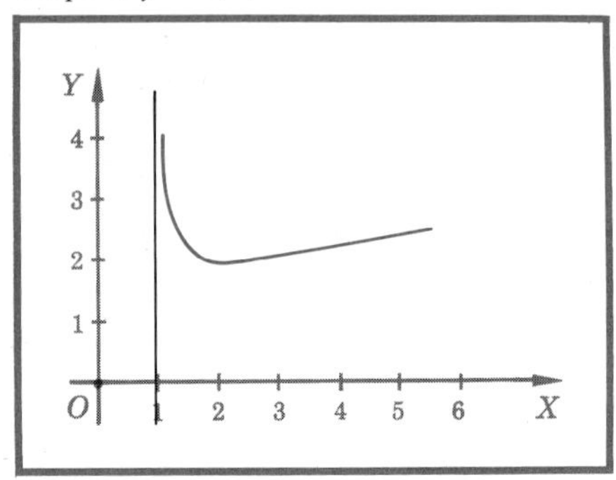

FIGURE 10.5
Be on the lookout for isolated points.

Since $\sqrt{x-1}$ must be real and nonzero, we must have $x > 1$. The domain is therefore $1 < x < +\infty$, and $x = 0$. At $x = 1$ there is a vertical asymptote. The value of y diminishes as x increases; but later, as x gets very large, y increases again. (By the methods of Sec. 16.7 it can be shown that, for $x > 1$, y is least when $x = 2$.)

We plot the table of values.

x	0	1	1.5	2	3	4	5
y	0	...	2.12	2	2.12	2.31	2.5

The graph is shown in Fig. 10.5. The point (0,0) is an isolated point, and the graph is not continuous at $x = 1$.

PROBLEMS 10.5

In Probs. 1 to 21 discuss type of algebraic function, domain, image (if possible), zeros, isolated points, endpoints, and asymptotes. Sketch.

1 $y = x^3 + x$

2 $y = 4x^3 + x$

3 $y = \sqrt{x^3 + x}$

4 $y = \sqrt{4x^3 + x}$

5 $y = 4x^2 - 4x^4$

6 $y = 4x^4 - 4x^2$

7 $y = \sqrt{4x^2 - 4x^4}$

8 $y = \sqrt{4x^4 - 4x^2}$

9 $y = x^3 + 2x^2$

10 $y = 1 - x^3$

11 $y = \sqrt{x^3 + 2x^2}$ **12** $y = \sqrt{1 - x^3}$
13 $y = 1/(x^2 + 1)$ **14** $y = \sqrt[3]{x^3 + 1}$
15 $y = 3x^4 - x^3 + 2$ (Minimum value occurs at $x = \frac{1}{4}$ and is $\frac{511}{256}$.)
16 $y = (x^{1/2} - 1)^2$ (Portion of a parabola)
17 $y = 27/(x^2 + 9)$
18 $y = \sqrt{x^3/(2 - x)}$ (Portion of the cissoid of Diocles)
19 $y = \sqrt{(1 - x^{2/3})^3}$ (Portion of a hypocycloid)
20 $y = \sqrt{(x - 1)(x + 2)(x - 3)}$
21 $y = -\sqrt{(x - 1)(x + 2)(x - 3)}$

In Probs. 22 to 27 sketch the graph of each equation, and discuss.

22 $2x^{1/2} + 2y^{1/2} = 1$ **23** $2x^{2/3} + 2y^{2/3} = 1$
24 $x^{2/3} + y^{2/3} = 1$ (See Prob. 19.) **25** $y^2 - 16x^2 = 0$
26 $y^2 + x^3 + x^2 = 0$ **27** $(x - y)(xy) = 0$

28 Sketch the graph of the function defined jointly by $y = 1/(x^2 - x)$ and the condition $y > 1$.
29 Discuss the graph $y = x^n$, n an even integer. Do the same for n odd.

10.6 › PROPERTIES OF POLYNOMIALS

We have defined a polynomial as an expression of the form

$$P(x) = a_0x^n + a_1x^{n-1} + \cdots + a_{n-1}x + a_n$$

We now wish to discuss some of the more important properties of polynomials and polynomial functions.

DEFINITIONS

Complex, real, rational polynomials

A polynomial with complex coefficients we shall call a *complex* polynomial. A polynomial with real coefficients we shall call a *real* polynomial. A polynomial with rational coefficients we shall call a *rational* polynomial. The domain of definition in any case is the set of complex numbers. Associated with every polynomial $P(x)$ is the polynomial function P defined by $y = P(x)$, whose zeros are also called the *roots* of the polynomial equation $P(x) = 0$. A polynomial equation is called complex, real, or rational according as the coefficients are complex, real, or rational, respectively.

We have previously referred to the following theorem.

THEOREM 1 **THE FUNDAMENTAL THEOREM OF ALGEBRA** Every complex polynomial equation $P(x) = 0$ of degree ≥ 1 has at least one root.

The proof of this theorem is beyond the scope of this text.

We pause here to tell you of a remarkable theorem first proved by the Norwegian mathematician Abel (1802–1829). The nature of the problem is easily described. You know that the general equations of the first and second degree

$$ax + b = 0 \qquad a \neq 0$$
$$ax^2 + bx + c = 0 \qquad a \neq 0$$

can be solved explicitly for x, giving respectively

$$x = -\frac{b}{a}$$

and

$$x = \frac{-b \pm \sqrt{b^2 - 4ac}}{2a}$$

So can the general equations of the third and fourth degree, namely,

$$ax^3 + bx^2 + cx + d = 0 \qquad a \neq 0$$

and

$$ax^4 + bx^3 + cx^2 + dx + e = 0 \qquad a \neq 0$$

be solved explicitly for x, using only a finite number of the algebraic operations. Abel proved that it is impossible to solve the general fifth-degree equation, using only a finite number of these operations. The proof is difficult. It is now known that no general equation of degree greater than four can be solved algebraically, i.e., by a finite number of the algebraic operations.

We seek properties of the roots of polynomial equations. First, we derive Theorem 2.

THEOREM 2 **REMAINDER THEOREM** If a complex polynomial $P(x)$ is divided by $x - b$ (where b is a complex number) until a remainder R free of x is obtained, then $P(b) = R$.

PROOF Let us divide $P(x)$ by $x - b$ until we obtain a remainder not containing x. In accordance with the Division Algorithm (Sec. 3.5), this division may be written

$$P(x) = (x - b)Q(x) + R \tag{1}$$

where $Q(x)$ is a complex polynomial of degree $n - 1$ called the quotient and R is a complex number.

Since Eq. (1) is true for all complex values of x, we may substitute b for x and obtain

$$P(b) = 0 + R \qquad \text{or} \qquad P(b) = R$$

ILLUSTRATION 1 Let $P(x) = x^3 + 2x^2 - 3$. Find $P(2)$ by the Remainder Theorem. By division, we have

$$\frac{x^3 + 2x^2 - 3}{x - 2} = x^2 + 4x + 8 + \frac{13}{x - 2}$$

Hence $R = 13$ and $P(2) = 13$. This can be checked by noting that

$$P(2) = 8 + 8 - 3 = 13$$

THEOREM 3 FACTOR THEOREM If r is a root of a complex polynomial equation $P(x) = 0$, then $x - r$ is a factor of $P(x)$.

PROOF The statement "r is a root of $P(x) = 0$" is equivalent to the statement "$P(r) = 0$". Divide $P(x)$ by $x - r$ as in Eq. (1). By the Remainder Theorem, $R = P(r) = 0$. Hence $x - r$ is a factor of $P(x)$.

EXERCISE A Prove the converse of the Factor Theorem: If $(x - r)$ is a factor of $P(x)$, then r is a root of $P(x) = 0$.

ILLUSTRATION 2 We may use the Factor Theorem to find a polynomial equation with given roots. Suppose we are given $r_1 = 1, r_2 = -2, r_3 = 0$ and are asked to find an equation with these roots. From the Factor Theorem $x - 1$, $x + 2$, and x are factors. Hence an equation with the desired property is

$$(x - 1)(x + 2)x = x^3 + x^2 - 2x = 0$$

ILLUSTRATION 3 We use the converse of the Factor Theorem to help us solve polynomial equations which we can factor. Consider the problem: Solve

$$(x + 2)(x - 1)(x^2 + x + 1) = 0$$

SOLUTION Since $x + 2$ and $x - 1$ are factors, we know that two roots are $r_1 = -2$, $r_2 = 1$. The other roots are solutions of

$$x^2 + x + 1 = 0$$

or

$$x = \frac{-1 \pm i\sqrt{3}}{2}$$

THEOREM 4 NUMBER-OF-ROOTS THEOREM A complex polynomial equation $P(x) = 0$ of degree n $(n \geq 1)$ has exactly n roots.

Multiple roots

It is possible, of course, for two or more of these roots to be equal. If k roots are all equal to r, say, the common language used is "r is said to be a root of multiplicity k". For Theorem 4 to be true, it is necessary to count a root of multiplicity k as k roots.

PROOF We have one root r_1 of $P(x) = 0$ from the Fundamental Theorem of Algebra. Therefore, from the Factor Theorem,

$$P(x) = (x - r_1)Q(x)$$

where $Q(x)$ is a polynomial of degree $n - 1$. Unless $Q(x)$ is a constant, the equation $Q(x) = 0$ also has a root r_2; thus $Q(x) = (x - r_2)S(x)$, where $S(x)$ is a polynomial of degree $n - 2$. Thus

$$P(x) = (x - r_1)(x - r_2)S(x)$$

Continue this process as long as possible. It must stop when n factors have been obtained; for the product of more factors would have a degree higher than n. Hence we have

$$P(x) = (x - r_1)(x - r_2)\cdots(x - r_n)a_0$$

This theorem tells us how many roots to look for. If we have an equation of fifth degree and have found three roots, we still have two more to find.

PROBLEMS 10.6 In Probs. 1 to 10 find a polynomial equation of lowest degree which has the given roots.

1 $0, 3$ **2** $-1, 1, 0$ **3** $-i$
4 $2i, i$ **5** $-1, i$ **6** $1, -1, i$
7 $-1, i, -i$ **8** $i\sqrt{2}, -i\sqrt{2}$ **9** $1 + i, -1 - i$
10 $1, -1, i, -i$

In Probs. 11 to 18 find a polynomial of third degree which has the following zeros.

11 $0, 1$ **12** $i, 2i, -i$
13 $1 + \sqrt{3}, 1 - \sqrt{3}, -1 - \sqrt{3}$ **14** $i, -i, 1$
15 a, b **16** $a + ib, a - ib$
17 0 **18** $-a - ib, -a + ib$

In Probs. 19 to 23 use the Remainder Theorem to find

19 $P(0)$ when $P(x) = x^4 + 3x^3 - 9x^2 + 2x + 1$
20 $P(2)$ when $P(x) = x^4 - 3x^3 + x^2 - 10x + 3$
21 $P(i)$ when $P(x) = x^3 + (3 - i)x^2 + (1 - 3i)x - i$
22 $P(4)$ when $P(x) = 4x^3 - 3x^2 + 6x + 2$
23 $P(-1)$ when $P(x) = x^3 + 3x^2 + 3x + 1$

In Probs. 24 to 27 by using the converse of the Factor Theorem find all the roots of

24 $x^3 + 3x^2 + 3x + 1 = 0$ **25** $x^4 - 16 = 0$
26 $x^3 - 27 = 0$ **27** $(x^2 + 1)(x^2 - 2x + 1) = 0$

28 How many roots does $x^5 - 3x^2 + 1 = 0$ have?
29 How many roots does $3x^4 - 3 = 0$ have?
30 How many roots does $(2i)x^2 - \sqrt{2} = 0$ have?
31 How many roots does $(i - 1)x - 1 = 0$ have?
32 How many zeros does the polynomial $x(x - r_1)(x - r_2) \cdots (x - r_k)$ have?
***33** As a consequence of Theorem 4 show that, if

$$P(x) = a_0x^n + a_1x^{n-1} + \cdots + a_{n-1}x + a_n = 0$$

has $n + 1$ roots, then each coefficient a_i is zero.
34 Show that a polynomial of degree n cannot have more than $n/2$ double roots. (A double root is a root of multiplicity 2.)
***35** Prove that $P(x) = 0$ and $Q(x) = 0$ have all their roots equal if and only if there is a constant $c \neq 0$ such that, for all x, $P(x) - cQ(x) = 0$.

10.7 » SYNTHETIC DIVISION

The Remainder Theorem gives us a convenient shortcut for finding the value $P(b)$, say, for it tells us that $P(b) = R$ and R is easy to compute. To perform the division called for in the Remainder Theorem, we use a short method, called synthetic division. To illustrate the method, we consider the case of the general cubic (complex) polynomial

$$P(x) = a_0x^3 + a_1x^2 + a_2x + a_3$$

which is to be divided by $x - b$. The work is exhibited in all detail below, where R is the remainder.

$$
\begin{array}{l}
x - b \,\big|\, a_0x^3 + a_1x^2 + a_2x + a_3 \,\big|\, a_0x^2 + (a_0b + a_1)x + (a_0b^2 + a_1b + a_2) \\
\quad a_0x^3 - a_0bx^2 \\
\qquad (a_0b + a_1)x^2 + a_2x \\
\qquad (a_0b + a_1)x^2 - (a_0b^2 + a_1b)x \\
\qquad\quad (a_0b^2 + a_1b + a_2)x + a_3 \\
\qquad\quad (a_0b^2 + a_1b + a_2)x - (a_0b^3 + a_1b^2 + a_2b) \\
\qquad\qquad\qquad\qquad a_0b^3 + a_1b^2 + a_2b + a_3 = R
\end{array}
$$

But, surely, we have written down more detail than we actually need; the following, where we have suppressed every x, is quite clear:

$$
\begin{array}{l}
-b \,|\, a_0 + a_1 + a_2 + a_3 \,\lfloor a_0 + (a_0 b + a_1) + (a_0 b^2 + a_1 b + a_2) \\
\qquad - a_0 b \\
\qquad (a_0 b + a_1) + a_2 \\
\qquad\qquad - (a_0 b^2 + a_1 b) \\
\qquad\qquad (a_0 b^2 + a_1 b + a_2) + a_3 \\
\qquad\qquad\qquad - (a_0 b^3 + a_1 b^2 + a_2 b) \\
\qquad\qquad\qquad a_0 b^3 + a_1 b^2 + a_2 b + a_3 = R
\end{array}
$$

We have also omitted the second writing of a_0, $a_0 b + a_1$, and $a_0 b^2 + a_1 b + a_2$ inasmuch as they are going to cancel by subtraction anyway. We will further simplify the process by changing the sign of $-b$ to $+b$ in the divisor, and hence the subtractive process to an additive one. Also there is no need of writing the quotient Q in the little box to the right since every term there is to be found in the work below, which is finally written on just three lines:

$$
\begin{array}{l|lllll}
b & a_0 & a_1 & a_2 & a_3 & \\
 & & a_0 b & a_0 b^2 + a_1 b & a_0 b^3 + a_1 b^2 + a_2 b & \\
\hline
 & a_0 & a_0 b + a_1 & a_0 b^2 + a_1 b + a_2 & a_0 b^3 + a_1 b^2 + a_2 b + a_3 = R &
\end{array}
$$

Although we have skeletonized the work, the details can still be extracted: We are dividing $a_0 x^3 + a_1 x^2 + a_2 x + a_3$ by $x - b$, and we get a quotient of $a_0 x^2 + (a_0 b + a_1)x + (a_0 b^2 + a_1 b + a_2)$ and a remainder of $a_0 b^3 + a_1 b^2 + a_2 b + a_3$. Note that the remainder is $P(b)$ as it should be by the Remainder Theorem.

ILLUSTRATION 1 Divide $x^4 - 3x^3 + x + 3$ by $x - 2$ synthetically.

SOLUTION Form the array, noting that the coefficient of x^2 is zero. (We normally place the "2" associated with the divisor $x - 2$ on the right.)

$$
\begin{array}{rrrrr|l}
1 & -3 & 0 & 1 & 3 & 2 \\
 & 2 & -2 & -4 & -6 & \\
\hline
1 & -1 & -2 & -3 & -3 = R &
\end{array}
$$

The quotient $Q(x) = x^3 - x^2 - 2x - 3$, and the remainder $R = -3$. By direct computation we also find that $P(2) = -3$.

ILLUSTRATION 2 Given $P(x) = 3x^4 - 4x^3 - 2x^2 + 1$, compute $P(-1)$, $P(0)$, $P(1)$, $P(2)$, $P(3)$, $P(-0.3)$, and sketch $y = P(x)$.

SOLUTION Directly from $P(x)$ we compute $P(0) = 1$, $P(1) = -2$, and $P(-1) = 6$. In the slightly more complicated cases of $P(2)$, $P(3)$, and $P(-0.3)$ we use synthetic division:

3	-4	-2	0	1	2
	6	4	4	8	
3	2	2	4	$9 = P(2)$	

3	-4	-2	0	1	3
	9	15	39	117	
3	5	13	39	$118 = P(3)$	

3	-4	-2	0	1	-0.3
	-0.9	1.47	0.159	-0.0477	
3	-4.9	-0.53	0.159	$0.9523 = P(-0.3)$	

The preceding table is self-explanatory. Note especially the value $P(-0.3) = 0.9523$ and the corresponding dip in the graph (Fig. 10.6). This kind of variation cannot be discovered in general without the methods of the calculus (Chap. 16).

EXERCISE A Compute $P(-1)$ by synthetic division, and note the alternating signs in the last line of your work. Explain why this tells us there are no real zeros of P to the left of $x = -1$. Generalize for the case where $P(x)$ is a real polynomial.

EXERCISE B Examine the line where $P(2) = 9$ and state what follows about real zeros of P to the right of $x = 2$. Generalize for the case where $P(x)$ is a real polynomial.

Graph of $y = 3x^4 - 4x^3 - 2x^2 + 1$

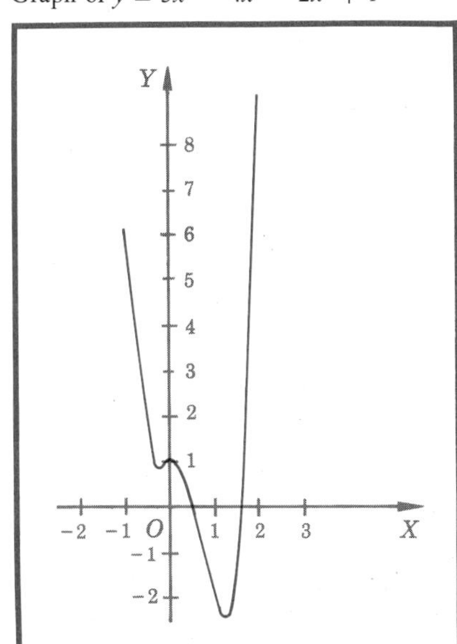

FIGURE 10.6
Some of the points are located by means of synthetic division.

PROBLEMS 10.7 In Probs. 1 to 6 use synthetic division to find:

1 $P(0)$ and $P(1)$ when $P(x) = 2x^3 - 3x^2 + 2x$
2 $P(-2)$ and $P(2)$ when $P(x) = 4x^4 + 2x^2 - 1$
3 $P(1)$ and $P(-1)$ when $P(x) = x^3 + x^2 + x + 1$
4 $P(-1)$ and $P(2)$ when $P(x) = x^5 + x^3 + x + 2$
5 $P(\frac{1}{2})$ and $P(-\frac{1}{2})$ when $P(x) = x^2 - 3x + 4$
6 $P(\frac{3}{2})$ and $P(\frac{1}{2})$ when $P(x) = 3x^3 + 2x^2 - 3$

In Probs. 7 to 16 use synthetic division to find quotient and remainder.

7 $\dfrac{3x^3 + 2x^2 + x + 1}{x - 1}$

8 $\dfrac{8x^4 + 2x^2 - 1}{x + 1}$

9 $\dfrac{x^3 - 2x^2 + 6x - 8}{x + 3}$

10 $\dfrac{3x^3 - 6x^2 + 2x - 9}{x - 1}$

11 $\dfrac{6x^4 - x^2 + 1}{x - 1}$

12 $\dfrac{2x^4 + 4x^3 - x^2 + 3}{x + 2}$

13 $\dfrac{(x^2/2) - 3x + 2}{x - \frac{1}{2}}$

14 $\dfrac{6x^2 - 2x + 10}{x - 3}$

15 $\dfrac{2x^3 - 3x + 9}{x + \frac{1}{3}}$

16 $\dfrac{x^3 - 2x^2 + 2x - 1}{x + \frac{7}{3}}$

In Probs. 17 to 22 use synthetic division to show that the first polynomial is a factor of the second.

17 $x - 3,\ x^3 - 18x + 27$
18 $x - 4,\ 2x^3 - 5x^2 - 11x - 4$
19 $x + 1,\ x^4 + 2x^3 + 2x^2 + 2x + 1$
20 $2x - 3,\ 2x^5 - 3x^4 - 6x^3 + 9x^2 + 2x - 3$
21 $x - \dfrac{1}{2},\ x^5 - \dfrac{7x^4}{2} - \dfrac{3x^3}{2} + x^2 - \dfrac{x}{2} + \dfrac{3}{8}$
22 $x + \dfrac{1}{2},\ x^4 - \dfrac{3x^2}{4} + \dfrac{x}{2} + \dfrac{3}{8}$

23 Divide $ax^2 + bx + c$ by $x + r$ by long division and also by synthetic division. Compare the results.
24 Divide $ax^2 + bx + c$ by $x - r$ by long division and also by synthetic division. Compare the results.
25 For what values of k does $kx^2 + x - 4$ yield the same remainder when divided by either $x + 1$ or $x - 1$?
26 For what values of k is $x^2 + kx - 2$ divisible by $x - k$?
27 If the polynomial $P(x) = Ax^4 - Ax^2 + x - 1$ is such that $P(3) = 0$, what is $P(2)$?
28 If the polynomial $P(x) = 3x^3 - kx^2 + 1$ is such that $P(1) = 2$, what is $P(3)$?

29 When $x^2 + 4x - 4$ is divided by $x + r$, the remainder is -8. Find r.

30 When $6x^2 + x - 9$ is divided by $x - r$, the remainder is -9. Find r.

31 Use the Factor Theorem to prove that $x^n + a^n$ is divisible by $x + a$ when n is an odd positive integer.

32 Use the Factor Theorem to prove that $x^n - a^n$ is divisible by $x - a$ when n is a positive integer.

10.8 ROOTS OF POLYNOMIAL EQUATIONS

Because of its practical importance, much effort has been spent on the question of how to calculate the roots of a polynomial equation. We have mentioned that formulas for the roots exist for $n = 1, 2, 3$, and 4; but there is no simple method of handling equations of higher degree. The general procedure consists of two steps:

I Find all roots which can be obtained by elementary means; then use the Factor Theorem or other methods to factor the polynomial into polynomials of lower degree.

II Find the zeros of the factors by known formulas or by approximate methods.

When the coefficients of $P(x)$ are general complex numbers, there is little that can be said here which will help you in these steps, for the known methods are too complicated to be treated in this book. We can make progress, however, if we consider only polynomials whose coefficients are real numbers. In this case we can prove the following theorem.

THEOREM 5 A real polynomial $P(x)$, with real coefficients, can always be represented as a product of factors each of which is either of the form $ax + b$ or $cx^2 + dx + e$, where a, b, c, d, and e are real numbers.

PROOF We know that the roots of $P(x) = 0$ are complex numbers, but some of them may actually be real. Corresponding to each real root r, the Factor Theorem tells us that there is a factor $(x - r)$. Therefore we can write

$$P(x) = (x - r_1)(x - r_2) \cdots (x - r_s)Q(x) = 0$$

where $r_1, r_2, \ldots, r_s$ are its real roots and $Q(x)$ is a polynomial of degree $n - s$ which has no real zeros. We must show that $Q(x)$ can be factored into quadratic factors of the form $cx^2 + dx + e$.

Suppose that $\alpha + i\beta$ with $\beta \neq 0$ is a root of $Q(x) = 0$. Construct the quadratic polynomial

$$(x - \alpha - i\beta)(x - \alpha + i\beta) = (x - \alpha)^2 + \beta^2 = S(x)$$

Note that $S(\alpha + i\beta) = 0$ and $S(\alpha - i\beta) = 0$. Now divide $Q(x)$ by $S(x)$, and obtain

$$Q(x) = S(x) \cdot R(x) + px + q$$

Substitute $x = \alpha + i\beta$ into this equation. Since $Q(\alpha + i\beta) = 0$ and $S(\alpha + i\beta) = 0$, we get

$$p(\alpha + i\beta) + q = 0$$

or

$$p\alpha + q = 0 \qquad p\beta = 0$$

Since $\beta \neq 0$, this shows that $p = 0$ and $q = 0$. Therefore $S(x)$ is a factor of $Q(x)$ and hence of $P(x)$. The same process can now be applied to $R(x)$, and we continue until we get

$$P(x) = (x - r_1) \cdots (x - r_s)S_1(x) \cdots S_t(x)a_0$$

where $s + 2t = n$. This is of the required form.

COROLLARY
Complex roots come in pairs

If $\alpha + i\beta$ is a root of a real polynomial equation, then $\alpha - i\beta$ is also a root of this equation.

EXERCISE A Construct an example which shows this corollary false when the coefficients of $P(x)$ are no longer real.

EXERCISE B Show that the degree of $Q(x)$ must be even.

This theorem tells us a lot about the nature of the roots of $P(x) = 0$, but it does not help us to find them. Special methods for finding the roots of certain simple types of equations are given in the next two sections.

PROBLEMS 10.8 Solve the following equations.

1 $2x^2 - 3x - 2 = 0$

2 $6x^2 + 4x - 16 = 0$

3 $x^2 + (k - 3)x - 3k = 0$ where k is real

4 $6x^2 - 13x + 6 = 0$

5 $2x^3 - 3x^2 - 3x + 2 = 0$

6 $4x^3 - 4x^2 + x = 0$

7 $(x + 1)^2 - 4(x + 1) + 4 = 0$

8 $\dfrac{1}{(x - 2)^4} - 1 = 0$

9 $\dfrac{1}{x^4} - 16 = 0$

10 $\dfrac{1}{(x + 1)^2} = 1$

11 $2x^3 + 3x^2 - 11x - 6 = 0$

12 $3x^3 + 7x^2 - 14x - 24 = 0$

13 $x^4 - x^2 = 0$

14 $y^4 + y^2 = 0$

15 $1 - \frac{3}{x} - \frac{54}{x^2} = 0$ **16** $2 + \frac{1}{x} - \frac{1}{x^2} = 0$

17 $2t^3 - 16t^2 + 30t = 0$ **18** $12v^4 - 41v^3 + 35v^2 = 0$

19 $z^6 - z^3 = 0$ **20** $x^4 + 8x^2 + 16 = 0$

21 $(x^2 - x - 3)^2 + (x^2 - x - 3) - 2 = 0$

22 $(z^2 + z - 1)^2 - 5(z^2 + z - 1) + 6 = 0$

10.9 › RATIONAL ROOTS OF RATIONAL POLYNOMIAL EQUATIONS

We now restrict ourselves to rational polynomial equations, i.e., to polynomial equations of the form $P(x) = 0$, where the coefficients in $P(x)$ are rational numbers.

EXERCISE A Show that a rational polynomial can be written in the form $A \cdot P(x)$, where A is a rational number and where $P(x)$ has integer coefficients. Hence show that a given rational polynomial equation has the same roots as a certain polynomial equation in which the coefficients are integers. [Multiplying both sides of an equation by a constant ($\neq 0$) does not change the roots.]

There is a simple method in this case for obtaining quickly all those roots of $P(x) = 0$ which happen to be rational numbers. Of course, there is no necessity that any of these roots be rational; therefore this method may not produce any of the roots at all since it exhibits only the rational roots.

THEOREM 6 **RATIONAL-ROOT THEOREM** If

$$P(x) = a_0x^n + a_1x^{n-1} + \cdots + a_{n-1}x + a_n$$

has integers for coefficients, and if $r = p/q$ is a rational root (in lowest terms) of $P(x) = 0$, then p is a factor of a_n and q is a factor of a_0.

PROOF We are given that

$$a_0\frac{p^n}{q^n} + a_1\frac{p^{n-1}}{q^{n-1}} + \cdots + a_{n-1}\frac{p}{q} + a_n = 0$$

Multiply through by q^n; the result is

(1) $$a_0p^n + a_1p^{n-1}q + \cdots + a_{n-1}pq^{n-1} + a_nq^n = 0$$

This may be written as

$$p(a_0p^{n-1} + a_1p^{n-2}q + \cdots + a_{n-1}q^{n-1}) = -a_nq^n$$

Now p is a factor of the left-hand side of this equation, and therefore p is a factor of the right-hand side, $-a_nq^n$. Since p/q is in lowest terms,

p and q^n are relatively prime; and since p is a factor of a_nq^n, it follows (from Theorem 7, Chap. 2) that *p is a factor of a_n*.

Equation (1) can also be written

$$a_0p^n = -q(a_1p^{n-1} + \cdots + a_{n-1}pq^{n-2} + a_nq^{n-1})$$

By a similar argument *q is a factor of a_0*.

ILLUSTRATION 1 Solve the equation $2x^4 + 5x^3 - x^2 + 5x - 3 = 0$.

SOLUTION The possible rational roots are ± 1, ± 3, $\pm\frac{1}{2}$, $\pm\frac{3}{2}$. Using synthetic division, we find that -3 is a root, for

$$\begin{array}{rrrrr|l} 2 & 5 & -1 & 5 & -3 & -3 \\ & -6 & 3 & -6 & 3 & \\ \hline 2 & -1 & 2 & -1 & 0 & \end{array}$$

Therefore

$$2x^4 + 5x^3 - x^2 + 5x - 3 = (x + 3)(2x^3 - x^2 + 2x - 1)$$

The remaining roots of the given equation are thus roots of the "reduced" equation

$$2x^3 - x^2 + 2x - 1 = 0$$

Its possible rational roots are ± 1, $\pm\frac{1}{2}$. Using synthetic division, we find that $\frac{1}{2}$ is a root, for

$$\begin{array}{rrrr|l} 2 & -1 & 2 & -1 & \frac{1}{2} \\ & 1 & 0 & 1 & \\ \hline 2 & 0 & 2 & 0 & \end{array}$$

The new reduced equation is

$$2x^2 + 2 = 0 \qquad \text{or} \qquad x^2 + 1 = 0$$

This is solved by the usual methods for quadratic equations and yields $x = \pm i$.

The roots of the original equation are therefore $\frac{1}{2}$, -3, i, $-i$. In this case each real root is a rational number.

PROBLEMS 10.9 In the following equations find the rational roots, and, where possible, solve completely.

1 $2x^3 + 3x^2 - 11x - 6 = 0$

2 $6x^3 + 7x^2 - 1 = 0$

3 $15x^3 + 19x^2 - 70x + 24 = 0$

4 $x^3 + x^2 - 8x + 6 = 0$

5 $2x^3 + 5x^2 - x - 1 = 0$

6 $2x^3 - 7x^2 - 27x - 18 = 0$

7 $2x^4 + x^3 + x^2 + x - 1 = 0$

8 $12x^4 - 67x^3 - 15x^2 + 484x - 480 = 0$
9 $4x^4 - 9x^2 + 8x - 12 = 0$
10 $32x^4 - 80x^3 + 70x^2 - 25x + 3 = 0$
11 $9x^4 + 18x^3 + 11x^2 - 30x + 8 = 0$

10.10 » REAL ROOTS OF REAL POLYNOMIAL EQUATIONS

In Sec. 10.9 we discussed the general method of obtaining the roots of rational polynomial equations when those roots are rational numbers. There is no simple general way in which a root can be determined exactly when it is not rational and when the degree of the polynomial exceeds 4. Indeed, about the only method available to us is an approximation method which is best described as a graphical one. This method will yield those roots of $P(x) = 0$ which are real but gives no information concerning other roots. The method applies equally to other types of equations as well, provided that the graphs of these equations are continuous.

A real root of $f(x) = 0$ or a zero of $f:(x,y)$ corresponds to a value of x at which the graph of $y = f(x)$ crosses or touches the X-axis. Hence the procedure is to construct an accurate graph from which the zeros may be read off (approximately).

Most graphs will only be accurate enough to locate the desired zero between successive integers, and a refined technique is needed to obtain more decimal places. To be definite, suppose that we have located a single root between 2 and 3, so that $f(2)$ and $f(3)$ have opposite signs. We may calculate $f(2), f(2.1), f(2.2), \ldots, f(2.9), f(3)$ in turn and thus locate the root between the adjacent pair of these which have opposite signs. Since this process is tedious, we try to speed it up graphically by a procedure which suggests which of these tenths to try first.

Suppose the situation is as in Fig. 10.7. Draw a straight line between the points $(2,f(2))$ and $(3,f(3))$, and observe where this crosses the axis. Now try tenths in the neighborhood of this crossing. When the root is located between successive tenths, the process may be repeated for hundredths, etc., as far as desired. Usually, however, the graphic method is abandoned after the tenths have been obtained, and refined numerical techniques (beyond the scope of this book) are employed.

Linear approximation to a root

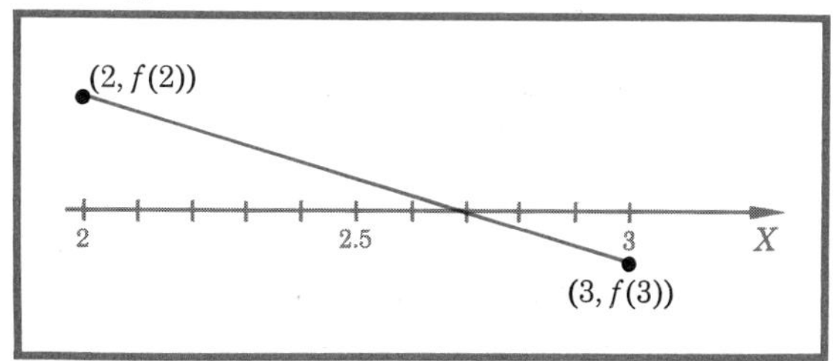

FIGURE 10.7 This graph helps us to estimate the first decimal place of the root.

We should say a final word about the use of a straight line with which to approximate a curve whose equation is $y = f(x)$ within some interval $[a,b]$. We shall assume that within $[a,b]$ the curve is continuous and strictly monotone (increasing or decreasing). Then we draw the line segment joining $(a,f(a))$ and $(b,f(b))$ and treat it as an approximation to the graph of $y = f(x)$ within $[a,b]$. The accuracy of the approximation depends upon the length of $[a,b]$ and the behavior of $y = f(x)$ within it; but for the functions used in elementary mathematics and for short intervals, the approximation is generally quite satisfactory for practical purposes. This is the basis on which we use (linear) interpolation in various tables (such as a table of logarithms).

ILLUSTRATION 1 Find the real zeros of the function defined by

$$y = x^3 - 2x^2 + x - 3$$

SOLUTION We find the table of values

x	-1	1	0	2	3
$f(x)$	-7	-3	-3	-1	9

and plot the graph as in Fig. 10.8. We see that there is a zero between 2 and 3. We cannot prove it with our present knowledge, but this is the

Graph of $y = x^3 - 2x^2 + x - 3$

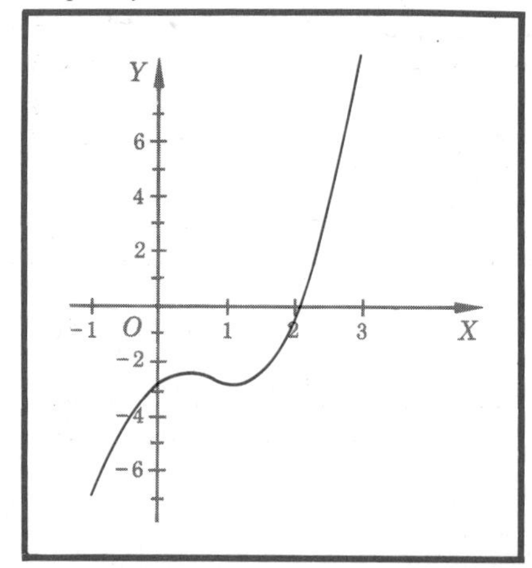

FIGURE 10.8
The graph suggests a root between 2 and 3.

Linear approximation in interval $2 \leq x \leq 3$

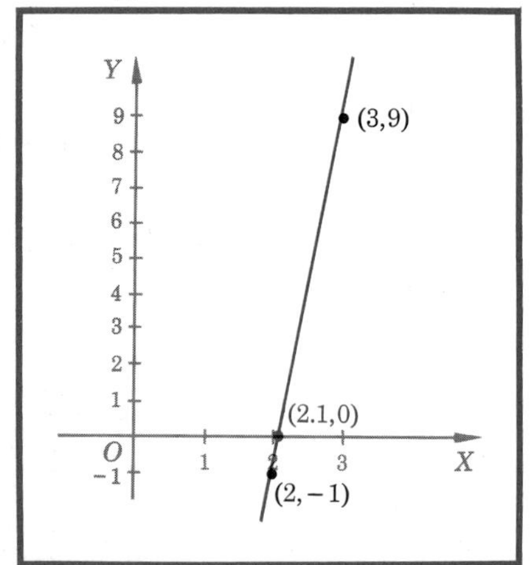

FIGURE 10.9
A linear approximation suggests a root near 2.1.

only real zero of this function. We plot Fig. 10.9. The line crosses the axis at exactly 2.1; therefore we calculate the following table.

x	2.0	2.1	2.2
$f(x)$	-1	-0.459	$+0.168$

Thus the zero is between 2.1 and 2.2.

To obtain the next decimal place, we plot Fig. 10.10. The line crosses the axis between 2.17 and 2.18; therefore we calculate the following table.

x	2.17	2.18
$f(x)$	-0.03	0.04

Hence the zero is 2.17+.

Repeated, this process will determine the decimal expansion of the root in question. But note that to obtain the best approximation to, say, two decimal places we should compute the expansion to three places and then round off to two places.

PROBLEMS 10.10

Find the first two decimal places (and round to one decimal place) of the real zero of least absolute value of the function defined as follows:

1. $2x^3 + x^2 - x + 5$
2. $x^3 - x^2 - 2x + 1$
3. $x^3 - 2x^2 - x + 1$
4. $2x^3 + 4x^2 - x + 2$
5. $x^3 + x^2 + x + 1$
6. $x^3 + \frac{x^2}{2} + x - 3$

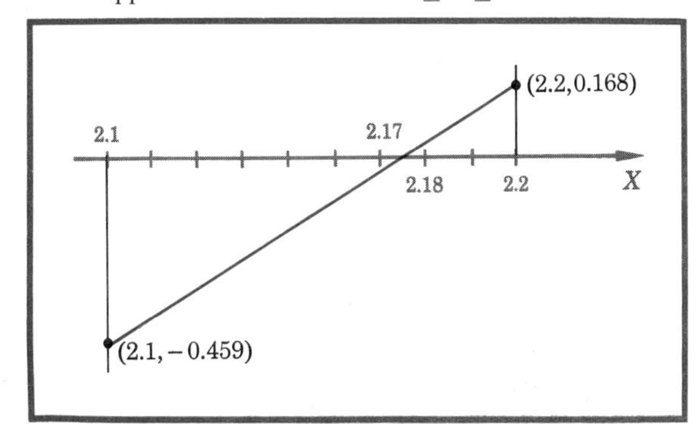

FIGURE 10.10 A further linear approximation suggests a root between 2.17 and 2.18.

7 $4x^3 - 3x^2 + 1$

8 $7x^3 + 19x^2 - 62x + 16$

9 $7x^4 + 2x^3 - 4x^2 + 3x - 1$

10 $2x^4 - 7x^3 + 2x + 1$

11 $2 - \dfrac{1}{x^2}$

12 $1 - \dfrac{2}{x^2}$

13 $(x + 5)(x - 0.16)(x + 2.5)$

14 $(x - 0.63)(x + 1.3)(2x - 0.63)$

15 $(x^2 - 4)(x + 3)^3$

16 $(x^2 + 2x + 4)(x - \frac{1}{2})(x + 2)$

11 EXPONENTIAL AND LOGARITHMIC FUNCTIONS

11.1 » EXPONENTIAL FUNCTIONS

In your earlier studies you have become acquainted with powers such as 2^3, $(-3)^4$, π^5, and the like. You have also encountered

$$7^{-2} = \frac{1}{7^2} \qquad 4^0 = 1 \qquad \pi^{-3} = \frac{1}{\pi^3}$$

The general expression for symbols like these is a^n, where a is any real number and n is an integer. You will also recall the use of fractional exponents to represent roots, such as

$$3^{1/2} = \sqrt{3} \qquad 5^{1/3} = \sqrt[3]{5} \qquad 2^{-1/7} = \frac{1}{\sqrt[7]{2}}$$

For our present purposes we are interested in the roots of positive real numbers only, and we know that every positive real number a has a unique real nth root which we write $a^{1/n}$, where n is a positive integer.

Also, we recall that $a^{p/q}$ is defined to be $a^{(1/q)p} = (a^p)^{1/q}$, where p and q are integers and a is positive. Hence we know the meaning of the function defined by

$$y = a^x \qquad a \text{ positive, } x \text{ rational}$$

We should like to extend the domain of definition of this function to the entire set of real numbers and thus give sense to numbers such as 2^π, $\pi^{-\sqrt{3}}$, and $3^{\sqrt{2}}$. The most natural way to obtain $3^{\sqrt{2}}$, for example, is to consider the successive decimal approximations to $\sqrt{2}$, such as 1.4, 1.41, 1.414, 1.4142, etc. Then $3^{1.4}$, $3^{1.41}$, $3^{1.414}$, $3^{1.4142}$, etc., are successive approximations to $3^{\sqrt{2}}$. A complete discussion of this matter is not feasible here, for it would require a study of the real numbers in more detail than we have treated them in Chap. 2. We shall content ourselves with the remarks that this extension is possible and that the value of a number like those above can be obtained to any desired approximation by choosing an expansion of each irrational to a sufficient number of decimal places. Thus

$$2^{\pi} \approx 2^{3.1416} \approx 8.825 \qquad 3^{\sqrt{2}} \approx 3^{1.414} \approx 4.728$$

where the symbol $\approx$ means "approximately equal to".

In summary we define a new function as follows.

DEFINITION $y = a^x$ $(a > 0)$ The function f defined by $y = a^x (a > 0)$ is called the exponential function with base a. Its domain of definition is the set of real numbers. We observe that its image is $0 < y < \infty$, if $a \neq 1$.

We now wish to develop some of its properties.

THEOREM 1 For $a > 0$ and $b > 0$ and x and y real:

a $a^x \times a^y = a^{x+y}$
b $(a^x)^y = a^{xy}$
c $(ab)^x = a^x \times b^x$

These theorems are proved in Chap. 5 for rational values of x. We do not give the proof for irrational values of x.

THEOREM 2 **a** $a^x > 1$ for $a > 1$, x real and > 0
b $a^x = 1$ for $a = 1$, x real and > 0
c $a^x < 1$ for $0 < a < 1$, x real and > 0

PROOF Theorem 2*a* is immediate when x is a positive integer, for the product of two numbers each of which is greater than 1 must itself exceed 1. When $x = 1/n$ (n a positive integer), Theorem 2*a* also is true. For if $a^{1/n}$ were to be less than 1 in this case, its nth power $(a^{1/n})^n = a$ would be less than 1. This follows from the fact that the product of two numbers each between zero and 1 must itself be less than 1. Finally, Theorem 2*a* is true for rational x by combining the above cases. We omit the proof for irrational values of x. Theorem 2*b* is immediate since all powers and roots of 1 are themselves 1. Theorem 2*c* is proved similarly to Theorem 2*a*.

EXERCISE A Write out the details of the proof of Theorem 2*c* for rational x.

THEOREM 3 Let x and y be real numbers such that $x < y$. Then

a $a^x < a^y$ for $a > 1$
b $a^x = a^y$ for $a = 1$
c $a^x > a^y$ for $0 < a < 1$

Graph of $y = a^x$, $a > 1$

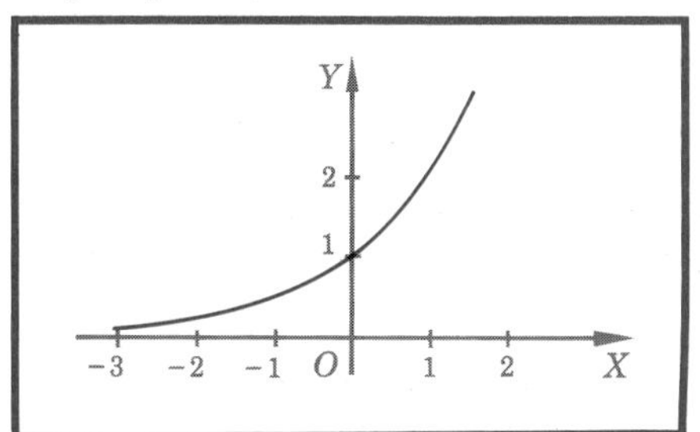

FIGURE 11.1
a^x is strictly monotone increasing when $a > 1$.

Graph of $y = a^x$, $a < 1$

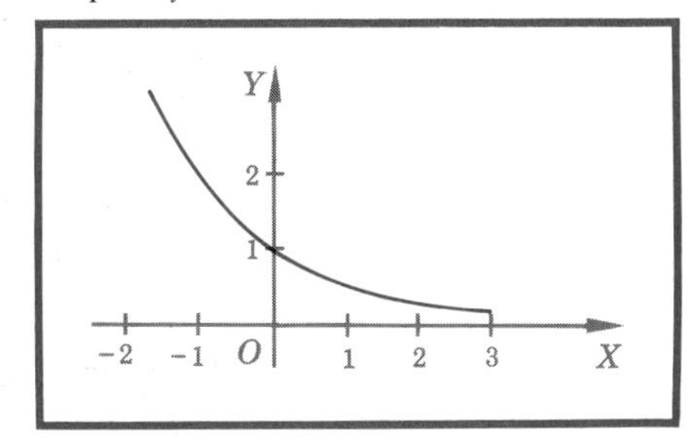

FIGURE 11.2
a^x is strictly monotone decreasing when $a < 1$.

The proof depends on Theorem 2. In all cases we know, from Theorem 1, that $a^y = a^{y-x} \cdot a^x$. By hypothesis, $y - x$ is positive. Thus, if $a > 1$, $a^{y-x} > 1$ and $a^y > a^x$, and similarly for the other cases.

EXERCISE B Complete the proof of Theorem 3.

a^x is monotone

We state without proof that the exponential functions are continuous for all values of x. Theorem 3 shows that when $a > 1$, the function a^x is strictly monotone increasing (Sec. 9.8) and that, when $a < 1$, it is strictly monotone decreasing. Typical graphs are given in Figs. 11.1 and 11.2.

There is an interesting symmetry between the graphs of $y = a^x$ and $y = (1/a)^x$, for $(1/a)^x = a^{-x}$, and the graph of $y = a^{-x}$ is just like the graph of $y = a^x$, with the X-axis reversed in direction.

EXERCISE C Draw the graph of $y = 1^x$.

PROBLEMS 11.1

1 Show that $a^x \times a^y$ defines an exponential function.
2 Show that $a^x \div a^y$ defines an exponential function.

In Probs. 3 to 14, simplify.

3 $2^{-1}6^{-2}$

4 $10^{-2}2^3$

5 $\dfrac{2^{-4}}{4^{-2}}$

6 $\dfrac{10^{-2}}{6^2}$

7 $\dfrac{2(\sqrt{3})^4}{2^{-1}3^2}$

8 $(36^{-1/2})8^2$

9 $4^{3/2}9^{-5/2}$

10 $\sqrt{3^2\sqrt{81}}$

11 $3^2(16)^{-1/2}$ **12** $\sqrt[3]{64}/\sqrt{64}$

13 $\sqrt{(1.69) \times 10^2}$ **14** $\sqrt[4]{16 \times (-3)^6}$

In Probs. 15 to 20 simplify, but leave the answer in exponential form.

15 $\dfrac{a^{4x}a^{-2x}}{a^x}$ **16** $\dfrac{a^{3x}a^{-2x}}{a^{5x}}$

17 $\dfrac{a^m a^{2m}}{a^{-m+2}}$ **18** $\dfrac{a^{nx}a^{2nx}}{a^{-nx}}$

19 $6\sqrt{60}\ \sqrt[3]{600}\ \sqrt[4]{6{,}000}$ **20** $10^{1/2} \cdot 10^{-3/2} \cdot 10^x \cdot 10^5$

In Probs. 21 to 24 plot the graphs ($x \geq 0$) on the same axes.

21 $y = x^3, y = 3^x$ **22** $y = x^2, y = 2^x$

23 $y = x^{1/3}, y = (\frac{1}{3})^x$ **24** $y = x^{1/2}, y = (\frac{1}{2})^x$

25 Plot the graph of $y = 2^x$. Now change the scale on the Y-axis so that the graph you have drawn is that of $y = 2 \cdot 2^x$.

26 Plot the graph of $y = 2 \cdot 1^x$.

11.2 » THE NUMBER *e*

We have defined the function $f:(x,y)$ whose values are given by $y = a^x$. The graphs of all these functions pass through the point (0,1); but as a varies, their tangents at this point make a variety of different angles with the Y-axis. There is just one of these graphs whose tangent at (0,1) makes an angle of 45° with the Y-axis; the corresponding function is called e^x, where e is an irrational number approximately equal to 2.71828. Thus

$$e \approx 2.71828$$

For applications of this particular function see Sec. 11.5.

This exponential function is in fact so important that we speak of it as *the* exponential function and neglect to mention its base. Sometimes this function is written

$$y = \exp x$$

where no base appears at all; the base is assumed to be e unless otherwise specified. Its values are tabulated in many convenient tables. Its graph is plotted in Fig. 11.3. In this figure we have also plotted $y = 2^x$.

A convenient approximation to e^x is given by the polynomial

$$1 + x + \frac{x^2}{2!} + \frac{x^3}{3!} + \cdots + \frac{x^n}{n!}$$

(By definition $n! = 1 \times 2 \times \cdots \times n$ for n a positive integer, and $0! = 1$.) As n increases, the approximation becomes closer and closer; as a matter of fact, e^x is exactly equal to the infinite sum

Two special cases of $y = a^x$

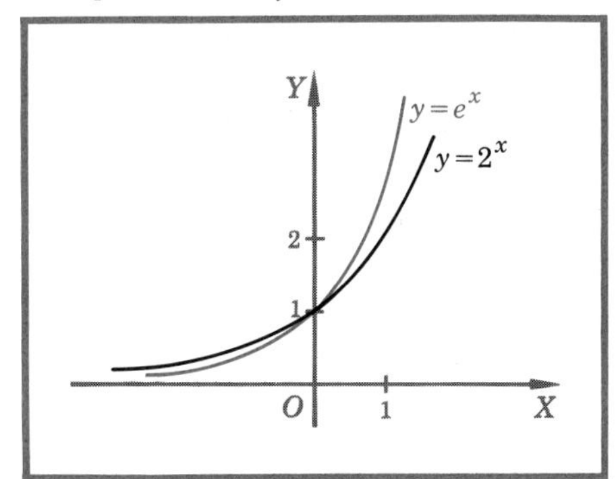

FIGURE 11.3
The graph of $y = e^x$ cuts the Y-axis at (0,1) at an angle of 45°.

$$e^x = 1 + x + \frac{x^2}{2!} + \cdots + \frac{x^n}{n!} + \cdots$$

Of course, we have not defined such an infinite sum. This can be done, however, and then one can give a proper definition of e and of the function e^x. Table I in the Appendix gives values of e^x and e^{-x}.

Hyperbolic functions

The exponential function is also used in the definitions of some other functions which occur in elementary mathematics. These are called the *hyperbolic functions* (Chap. 17). Their definitions are

$$\sinh x = \frac{e^x - e^{-x}}{2}$$

$$\cosh x = \frac{e^x + e^{-x}}{2}$$

$$\tanh x = \frac{e^x - e^{-x}}{e^x + e^{-x}}$$

The function $\sinh x$ is read "the hyperbolic sine of x"; $\cosh x$, "the hyperbolic cosine of x"; and $\tanh x$, "the hyperbolic tangent of x". These are somewhat related to the trigonometric functions with similar names which we shall study in Chap. 13.

PROBLEMS 11.2

In Probs. 1 to 10 obtain the values from Table I in the Appendix, interpolating where necessary.

1 $e^{0.5}$

2 $e^{1.2}$

3 $e^{-0.06}$

4 $e^{4.1}$

5 $2e^3$

6 $4e^{2.4}$

7 e^{-1}

8 $e^{-0.5}$

9 $e^{-4.65}$

10 $e^{-0.37}$

In Probs. 11 to 20 plot the graph, using the same scale on the two axes.

11 $y = 2e^{0.2x}$ **12** $y = 2e^{-0.1x}$
13 $y = -3e^{0.8x}$ **14** $y = -4e^{-0.8x}$
15 $y = 4e^{-0.05x}$ **16** $y = 6e^{0.05x}$
17 $y = e^{2.1x}$ **18** $y = 10^{-1}e^{-2.1x}$
19 $2y = e^{x} - e^{-x}$ **20** $2y = e^{x} + e^{-x}$

11.3 › LOGARITHMIC FUNCTIONS

Since $y = a^x$ defines a strictly monotone function, for $a \neq 1$, each value of y is obtained from a single x. Therefore the inverse function exists, and this is called the logarithmic function.

DEFINITION The function inverse to that given by $y = a^x$, $a > 0$, $a \neq 1$, is written $y = \log_a x$ and is called the *logarithm* of x to the base a.

For computational purposes, the base is usually taken to be 10, so that properties of our decimal system may be used to simplify the needed tables. For theoretical purposes, the base is always taken to be e. In advanced books, this base is omitted, and $\log x$ is to be understood to mean $\log_e x$. Frequently, the notation $\ln x$ is used for $\log_e x$. Logarithms to the base 10 are called "common" logarithms; those to the base e are called "natural" logarithms. Tables of both kinds are available in most collections of elementary tables. (See Appendix, Tables II and III.)

Natural logarithms

EXERCISE A Prove that $\log_a a^x = x$ and that $a^{\log_a x} = x$.

The graph of $y = \log_a x$ is obtained from that of $y = a^x$ by reflecting it in the line $y = x$. It is given in Figs. 11.4 and 11.5.

From the graphs we see that the domain and image are as follows.

DOMAIN AND IMAGE The domain of definition of $\log_a x$ is the set of positive real numbers. Its image is the set of all real numbers.

Note that the logarithms of negative numbers are not defined here. In advanced books, you will learn how to extend the definition of $\log_a x$ so that x can be negative. Its value turns out to be nonreal. We do not consider this case.

PROPERTIES The logarithmic function to the base a defined by $y = \log_a x$ is strictly monotone increasing for $a > 1$, strictly monotone decreasing for $0 < a < 1$, and not defined for $a = 1$. The following theorems have useful applications.

Properties of logarithms

THEOREM 4 $\log_a xy = \log_a x + \log_a y$

Graph of $y = \log_3 x$

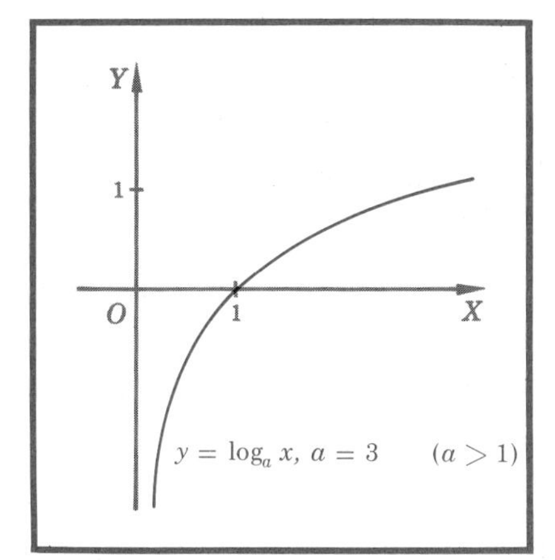

FIGURE 11.4
The graph of $y = \log_a x$ is strictly monotone increasing when $a > 1$.

Graph of $y = \log_{1/2} x$

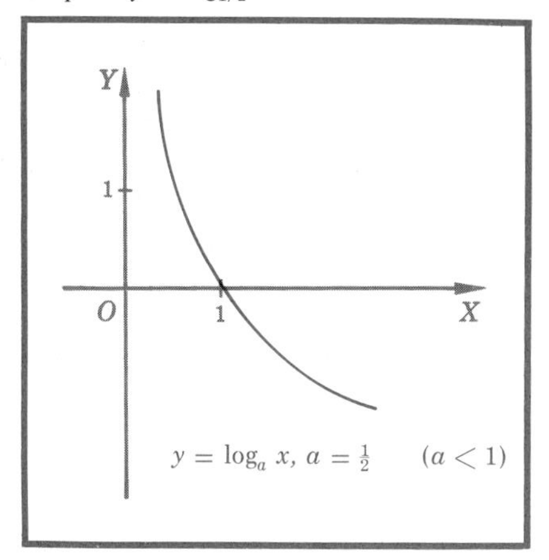

FIGURE 11.5
The graph of $y = \log_a x$ is strictly monotone decreasing when $a < 1$.

PROOF Let

$$z = \log_a xy \qquad \text{then } a^z = xy$$
$$u = \log_a x \qquad \text{then } a^u = x$$
$$v = \log_a y \qquad \text{then } a^v = y$$

Therefore $\quad a^z = xy = a^u \cdot a^v = a^{u+v}$

or $\quad z = u + v$

from which the theorem follows.

THEOREM 5 $\log_a \dfrac{1}{x} = -\log_a x$

PROOF Let $z = \log_a \dfrac{1}{x}$; then $a^z = \dfrac{1}{x}$, and $a^{-z} = x$. Therefore $-z = \log_a x$. Hence the theorem follows.

THEOREM 6 $\log_a \dfrac{y}{x} = \log_a y - \log_a x$

PROOF Combine Theorems 4 and 5.

THEOREM 7 $\log_a (x^y) = y \log_a x$

PROOF Let

$$z = \log_a (x^y) \qquad \text{then } a^z = x^y$$
$$u = \log_a x \qquad \text{then } a^u = x$$

Therefore $\quad (a^u)^y = a^z \quad$ or $\quad uy = z$

from which the theorem follows.

EXERCISE B Prove $\log_b a = \dfrac{1}{\log_a b}$.

EXERCISE C Prove $\log_a x = \dfrac{\log_b x}{\log_b a}$. A special case of this is

$$\log_a x = \frac{\log_e x}{\log_e a}$$

Theorems 4 to 6 are useful for numerical computations involving only products and quotients. Logarithms to the base 10 are generally employed.

ILLUSTRATION 1 Find $\dfrac{(33.0)(27.2)}{15.8}$

SOLUTION We compute

$$\log_{10} \frac{(33.0)(27.2)}{15.8} = \log_{10} 33.0 + \log_{10} 27.2 - \log_{10} 15.8$$

To find these logarithms, we use the table of common logarithms (Table II) in the Appendix. We find

$$\begin{aligned} \log_{10} 33.0 &= 1.5185 \\ \log_{10} 27.2 &= \underline{1.4346} \\ & 2.9531 \\ -\log_{10} 15.8 &= \underline{-1.1987} \\ \log_{10} \frac{(33.0)(27.2)}{15.8} &= 1.7544 \end{aligned}$$

Working backward from the table, we obtain

$$\frac{(33.0)(27.2)}{15.8} = 56.81$$

The importance of logarithms in problems of this sort is not as great as it was in former years. Calculations such as that above can be performed more rapidly on a slide rule, provided that the numbers involved do not contain more than three essential digits. When the numbers are more complicated, or when greater accuracy is desired, rapid results can be obtained from a desk computing machine. For this reason most students do not need to develop great skill in this use of logarithms.

On the other hand, logarithms must be used to compute exponentials such as $2^{1.42}$ by the use of Theorem 7.

ILLUSTRATION 2 Compute $2^{1.42}$.

SOLUTION From Theorem 7,

$$\log_{10} 2^{1.42} = 1.42 \log_{10} 2 = (1.42)(0.3010) = 0.4274$$

Therefore $$2^{1.42} = 2.675$$

PROBLEMS 11.3

In Probs. 1 to 6 compute, using Table II, Appendix:

1 $3^{1.2}$ **2** $5^{2.5}$ **3** $e^{2.1}$
4 $e^{0.4}$ **5** $10^{2.3}$ **6** $6^{1.8}$

In Probs. 7 to 12 compute, using Table III, Appendix:

7 $3^{1.2}$ **8** $5^{2.5}$ **9** $e^{2.1}$
10 $e^{0.4}$ **11** $10^{2.3}$ **12** $6^{1.8}$

In Probs. 13 to 18 evaluate or simplify:

13 $6^{\log_6 6}$ **14** $4^{\log_2 16}$ **15** $36^{\log_{36} 6}$
16 $4^{\log_{49} 7}$ **17** $16^{\log_8 2}$ **18** $6^{\log_6 x}$

19 Show that $x^x = a^{x \log_a x}$, $x > 0$.
20 Show that $(f(x))^{g(x)} = a^{g(x) \log_a f(x)}$, $f(x) > 0$.
21 Write the function f defined by $y = a^x b^x$ as an exponential function with base a.
22 By using seven-place common logarithms, we find that $(1 + \frac{1}{10})^{10} \approx 2.594$, $(1 + \frac{1}{100})^{100} \approx 2.705$, $(1 + \frac{1}{1,000})^{1,000} \approx 2.717$. Compare with the value of e.
23 Pick out the pairs of inverse functions, and state the domain and the image: 5^{6x}, 6^{5x}, $-\log_2 x$, $6 + \log_5 x$, $\frac{1}{5} \log_6 x$, 2^{-x}, $\log_6 (5x)$, $\frac{1}{6} \log_5 x$, $\log_2 (-x)$.

11.4 › GRAPHS

With a set of the standard mathematical tables at our disposal, we can now make light work of graphing various exponential, logarithmic, and related functions.

ILLUSTRATION 1 Plot the graph of the function given by $y = xe^{-x}$.

SOLUTION Again we prepare a table of x's and corresponding values of y and sketch the graph in Fig. 11.6.

x	-3	-2	-1	0	1	2	3
e^{-x}	20.09	7.39	2.72	1	0.37	0.14	0.05
$y = xe^{-x}$	-60.27	-14.78	-2.72	0	0.37	0.28	0.15

[By methods of the calculus (Chap. 16) we find that a maximum value of the function occurs at $x = 1$.]

ILLUSTRATION 2 On the same axes and to the same scale plot the graphs of $y = e^x$, $y = \log_e x$.

SOLUTION The functions defined by these equations are inverses of each other. We use a table of natural logarithms to prepare the following entries.

x	-3	-2	-1	0	0.2	0.5	1	2	3
$y = \log_e x$	...	...	...	$-\infty$	-1.61	-0.69	0	0.69	1.10
$y = e^x$	0.05	0.14	0.37	1	1.22	1.65	2.72	7.39	20.09

The graphs are plotted in Fig. 11.7.

Graph of $y = xe^{-x}$

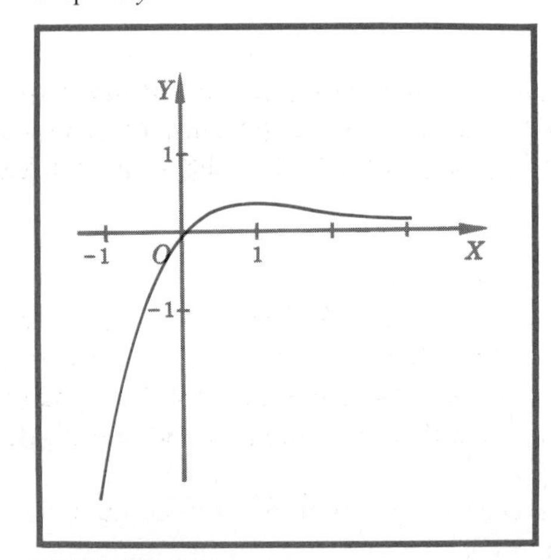

FIGURE 11.6
To find points on the graph of $y = xe^{-x}$, multiply the ordinates of $y = x$ and $y = e^{-x}$.

Comparison of $y = e^x$ and $y = \log_e x$

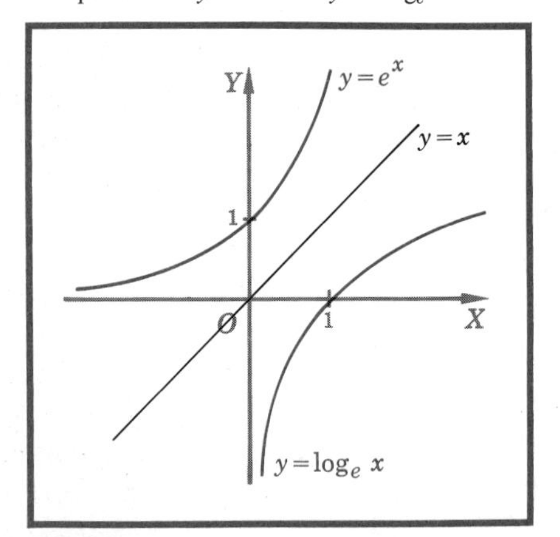

FIGURE 11.7
The graphs of $y = e^x$ and $y = \log_e x$ are symmetric about the line $y = x$.

PROBLEMS 11.4

In Probs. 1 to 14 sketch the graphs.

1 $y = \log_e x$

2 $y = \log_e |x|$

3 $y = \log_e |3x|$

4 $y = |\log_e x|$

5 $y = |\log_e 3x|$

6 $y = 2 \log_e x$

7 $y = 1 + 2 \log_e x$

8 $y = \log_e \sqrt{x}$

9 $y = 10^{\sqrt{x}}$

10 $y = \log_{10} \sqrt{x}$

11 $y = x \log_e x$

12 $y = x \log_e x^2$

13 $y = e^{-x^2}$ (This graph is called the "probability curve".)

14 $y = e^{-2}2^x/x!$, for $x = 0, 1, 2, 3, 4$, etc. (This equation defines an important function in statistics; note that the domain is the positive integers and zero. It is called the "Poisson distribution function".)

In Probs. 15 to 18 solve the equation for x. Use tables in the Appendix. Do not interpolate.

15 $2^x = 10$

16 $2^x = e$

17 $6^{-x} = 36$

18 $3.2^x = e$

19 Sketch $y = e^x$ and $y = 1 + e^{-x}$ on the same axes and to the same scales. From them estimate the number x which satisfies the equation $e^x - e^{-x} = 1$. Check your answer by Table I of the Appendix.

20 Sketch $y = 2e^{-x}$ and $y = 3 - e^x$ on the same axes and to the same scales. From them estimate the number x which satisfies the equation $e^x + 2e^{-x} = 3$. Check your answer by Table I.

11.5 › APPLICATIONS

There are many problems in biology, chemistry, economics, etc., involving growth and decay for which the natural mathematical model is the exponential function. Our basic illustration is taken from the field of economics.

ILLUSTRATION 1

Compound interest

a An amount P dollars (principal) is invested at 100 percent interest (rate), compounded annually. (The accrued interest is to be added to the principal.) Find the total amount A after 1 year.

b Using the same data, find the total amount if interest is compounded monthly.

c Find the total amount with the interest compounded daily (360 days per year).

d Work out the same problem with the interest compounded continuously.

SOLUTION

a $A = P(1 + 1)$

b $A = P(1 + \frac{1}{12})^{12}$

c $A = P(1 + \frac{1}{360})^{360}$

d In order to arrive at something meaningful in this case we should begin with a description of what is meant by compounding "continuously".

At this time we can give only an intuitive explanation since a precise explanation involves the theory of limits. We would have an approximate answer if we compounded each second. A year (360 days) has 31,104,000 sec. The amount, at the end of 1 year, would be

$$A_{31,104,000} = P\left(1 + \frac{1}{31,104,000}\right)^{31,104,000}$$

We should like to know what, if anything, $A = P(1 + 1/n)^n$ would approach with ever-increasing n. The answer (beyond the scope of this text to develop) is Pe. That is, in technical language: "The limit of $(1 + 1/n)^n$, as n grows without bound, is e." Or, in symbols,

$$\lim_{n\to\infty}\left(1 + \frac{1}{n}\right)^n = e$$

If continuous compounding took place over a period of kt years, the amount would be given by

$$A = P\lim_{n\to\infty}\left(1 + \frac{1}{n}\right)^{nkt} = P\lim_{n\to\infty}\left[\left(1 + \frac{1}{n}\right)^n\right]^{kt} = Pe^{kt}$$

The same kind of problem arises in biology, where each of P cells in a given culture splits into two cells in a certain time t.

ILLUSTRATION 2 The number of bacteria in a culture at time t was given by

$$y = N_0e^{5t}$$

What was the number present at time $t = 0$? When was the colony double this initial size?

SOLUTION When $t = 0$, $y = N_0e^0 = N_0$. The colony will be $2N_0$ in size when t satisfies the equation $2N_0 = N_0e^{5t}$, that is, when $5t = \log_e 2$ or when $t = \frac{1}{5}\log_e 2 = 0.6932/5 \approx 0.1386$ unit of time.

In chemistry, certain disintegration problems are explained in a similar manner.

Radioactive decay

ILLUSTRATION 3 Radium decomposes according to the formula $y = k_0 e^{-0.038t}$, where k_0 is the initial amount (corresponding to $t = 0$), and y is the amount undecomposed at time t (in centuries). Find the time when only one-half the original amount will remain. This is known as the "half-life" of radium.

SOLUTION We must solve $\frac{1}{2}k_0 = k_0 e^{-0.038t}$ for t.

$$\begin{aligned} \log_e \tfrac{1}{2} &= -0.038t \\ -0.6932 &= -0.038t \\ t &= \frac{693.2}{38} = 18.24 \text{ centuries} \end{aligned}$$

ILLUSTRATION 4 Given that the half-life of a radioactive substance is 10 min, how much out of a given sample of 5 g will remain undecomposed after 20 min?

SOLUTION The substance decays according to the formula

$$y = 5e^{-kt}$$

First we must find k. From the given data

$$\tfrac{5}{2} = 5e^{-10k} \qquad \text{or} \qquad \tfrac{1}{2} = e^{-10k}$$

Taking natural logarithms of both sides, we have

$$\begin{aligned} -\log_e 2 &= -10k \\ k &= \frac{\log_e 2}{10} \end{aligned}$$

Substituting back, we find

$$y = 5e^{-(\log_e 2)(t/10)} = 5e^{-\frac{\log_e 2}{10}t}$$

When $t = 20$ min,

$$\begin{aligned} y &= 5e^{-\frac{\log_e 2}{10}20} = 5e^{\log_e 0.25} \\ &= 5(0.25) = 1.25 \text{ g} \end{aligned}$$

We could have seen this at once, for half remains after 10 min and so half of a half, or a quarter, remains after 20 min. The above method, however, will give us the answer for any time t.

PROBLEMS 11.5

In Probs. 1 to 6 solve for the unknown.

1 $2 = 3e^{1.2t}$

2 $16 = 5e^{0.2t}$

3 $1.4 = 7e^{-0.5k}$

4 $6.4 = 1.6e^{-2.3x}$

5 $1.7 = 0.7t(3^{0.1})$

6 $5 = 3x(4^{-2.4})$

7 A special case of Newton's Law for the rate at which a hot body cools is $100 = 50e^{-0.25r}$. Find r.

8 An approximation for the pressure p in millimeters of mercury at a height h km above sea level is given by the equation $p = 760e^{-0.144h}$. Find the height for which the pressure is one-quarter the pressure at sea level.

9 One "healing law" for a skin wound is $A = Be^{-n/10}$, where A (in square centimeters) is the unhealed area after n days, and B (in square centimeters) is the original wound area. Find the number of days required to reduce the wound down to one-quarter the original area.

10 Given the half-life of a radioactive substance is 1,000 years, how much out of a given sample of 2 g will remain after 5,000 years?

11 How long will it take a sum to double at 50 percent interest compounded continuously?

12 A radioactive substance decays from 6 to 2 g in 2 days. Find the half-life.

13 Let N be the number of π^0 mesons generated at time $t = 0$, and let

$$y = Ne^{-\left(\frac{10^{16}\log_e 2}{3}\right)t}$$

be the number at any subsequent time t (in seconds). If only $N/2$ are present when $t = t_1$ sec, find t_1.

11.6 › THE LOGARITHMIC SCALE

Slide rule

Ordinary addition can be performed mechanically quite simply by sliding one ruler along another as in Fig. 11.8. We assume that the rulers are graduated in the usual way with linear scales (see Sec. 2.10). A linear scale is one in which the marks 1, 2, 3, . . . are placed 1, 2, 3, . . . units from one end (say the left). Thus with ruler I in its present position we could add $2.5 + 1.5 = 4$, $2.5 + 3 = 5.5$. A logarithmic scale is obtained by changing the coordinates x to new coordinates x' by means of the transformation $x' = \log_{10} x$, $x \geq 1$ (see Sec. 6.13). Examine the scale in Fig. 11.9. The distance from the left end to the mark 3 is not 3 units but is the logarithm (to base 10) of 3 units; that is, the distance is

Slide rule for addition

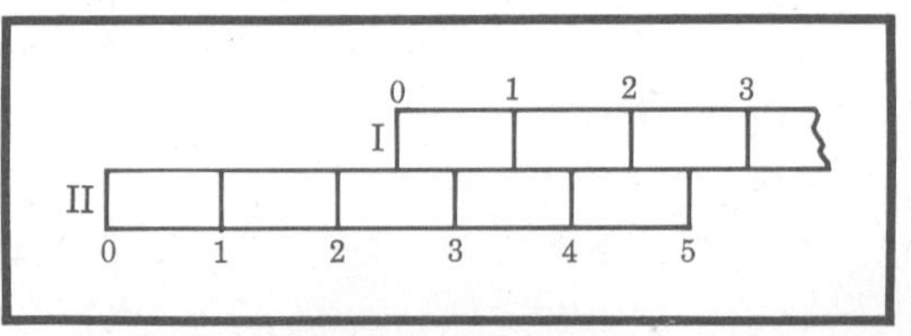

FIGURE 11.8
The two rules have identical linear scales.

Logarithmic scale

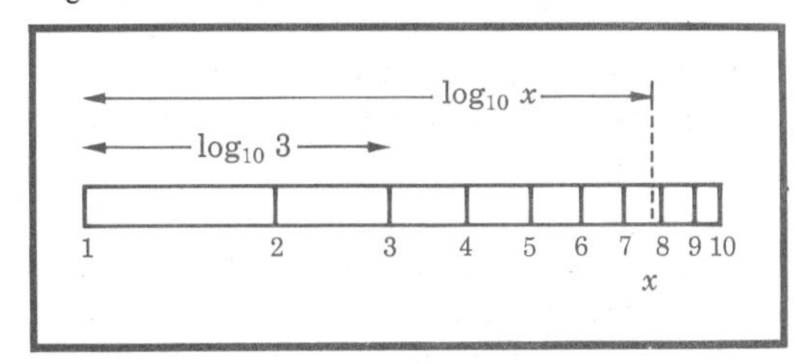

FIGURE 11.9 In a logarithmic scale the distance of the point labeled x from the left end is actually $\log_{10} x$.

$\log_{10} 3 = 0.4771$ of the whole length. Similarly the mark x is placed at a distance of $\log_{10} x$ units from the left end. Note that the left end itself is marked 1 as it should be since $\log_{10} 1 = 0$. If we placed two such scales side by side as in Fig. 11.10, we could add the logarithms of numbers and hence multiply the numbers themselves. Thus $\log_{10} 2 + \log_{10} 3 = \log_{10} 6$. But since the scales are *marked* with units 2, 3, 6, etc., we *read* 2 (on the D scale) $\times$ 3 (on the C scale) $= 6$ (back on the D scale). In the same way we compute $2 \times 3\frac{1}{2} = 7$, $2 \times x = 2x$, etc.

Reading "backward", we perform division; thus 6(on D)/3(on C) = 2 (on D, opposite 1 on C), etc. The usual slide rule also has scales that permit raising to powers and extraction of square and cube roots. A slide rule is a useful aid in calculating where only two- or three-place accuracy is required. Instructions come with a rule.

In all of our graph work up to this point we have described and used but one type of graph paper, called rectangular coordinate paper, in which the rulings are laid out on linear scales. Many other types are available for special purposes. We shall devote Sec. 14.15 to a type called polar coordinate paper.

It is appropriate at this time to mention briefly two other types that are in common use and are available at a bookstore. These are:

Semilog paper
1 Semilogarithmic (semilog) paper in which one axis has a linear scale, the other a logarithmic scale (see Fig. 11.11).

Log-log paper
2 Double-logarithmic (log-log) paper in which each axis is marked with a logarithmic scale (see Fig. 11.12).

EXERCISE A Why is there no zero on a logarithmic scale?

Slide rule for multiplication

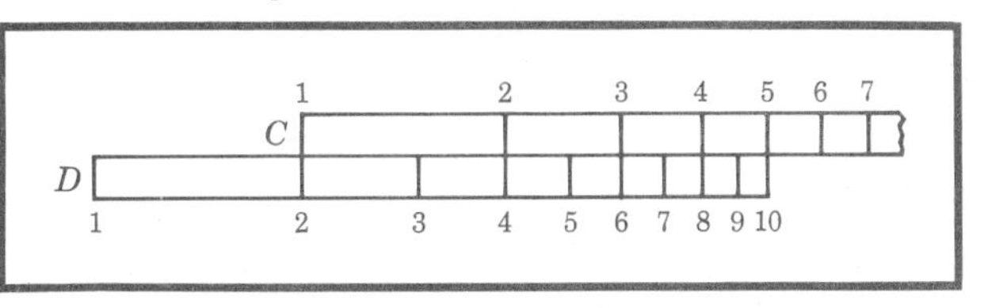

FIGURE 11.10 The two rules have identical logarithmic scales. Since $\log x + \log y = \log xy$, these rules can be used for multiplication.

Semilog paper

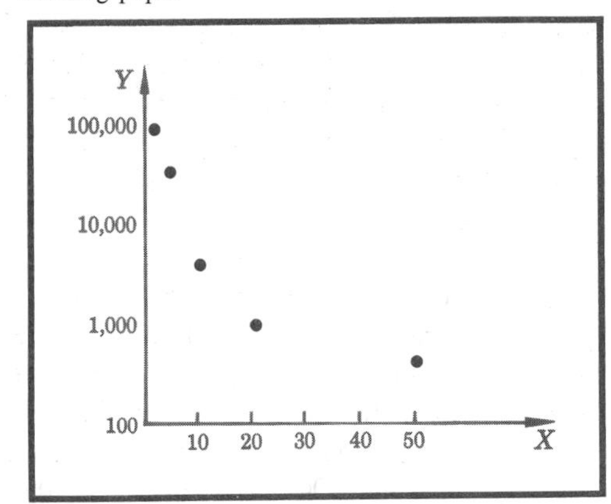

FIGURE 11.11
Semilog paper has a vertical logarithmic scale and a horizontal linear scale. See Illustration 1 for data.

One use for semilog paper is for graphing functions that have both small and relatively large image values, such data as, for example, in the following illustration.

ILLUSTRATION 1 Graph the function whose total set of elements is given by the following table.

x	50	20	8	3	1
y	500	1,000	5,000	25,000	100,000

Log-log paper

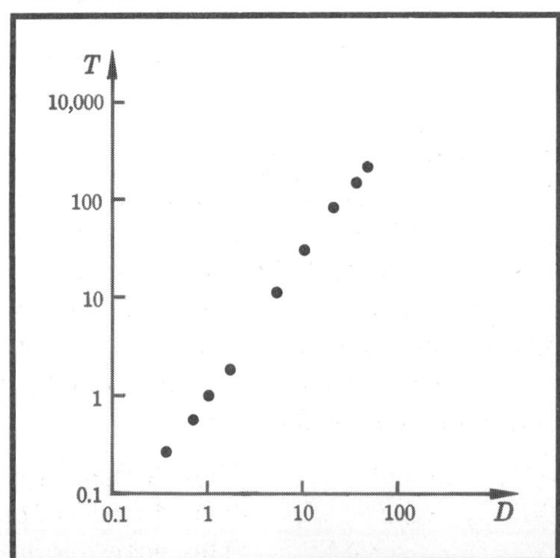

FIGURE 11.12
Both scales of log-log paper are logarithmic. See Illustration 2 for data.

SOLUTION First note that the domain is the set X of the five integers 50, 20, 8, 3, and 1. The image is the set Y of the five corresponding integers 500, 1,000, 5,000, 25,000, and 100,000. You can see that ordinary rectangular paper is inadequate because of the tremendous differences in magnitude of the values of y. These data might refer to the number of bank depositors x each writing checks (for an average month) with a total value y.

In order to get some geometric picture of this function, we resort to semilog paper with the linear scale on the X-axis and the logarithmic scale on the Y-axis. The graph is shown in Fig. 11.11. We do not connect the points, since there are no other elements of this function.

Note that a logarithmic scale goes by repetitive blocks, the set of marks representing 100, 200, 300, . . . , 900 being repeated in the same pattern for 1,000, 2,000, 3,000, . . . , 9,000. This is because of the decimal characteristics of the base 10 and is made clear by the following partial table. By blocks the mantissas repeat; the characteristics increase by 1. Semilog (and log-log) paper comes in several block styles. The one of Fig. 11.11 is three-block paper. In the printed forms the scales in each block run from 1 to 10; you will have to relabel them to fit the given problem.

y	$\log_{10} y$
⋮	⋮
100	2.0000
200	2.3010
300	2.4771
⋮	⋮
900	2.9542
1,000	3.0000
2,000	3.3010
3,000	3.4771
⋮	⋮
9,000	3.9542
⋮	⋮

Curve fitting

Another use for logarithmic paper is in the search for an equation which might be satisfied by a set of data obtained, perhaps, as the result of some kind of an experiment. In running a given experiment, a research worker may feel that there is possibly a simple "law" that the data should or might follow. If he were able to discover this law in the form of an equation, he could then use the equation for purposes of prediction.

Before we illustrate with an example we must make a few remarks about equations of the form

(1) $y = ae^{bx}$ (This defines an *exponential function*)
(2) $y = ax^b$ (This defines a *power function*)
(3) $y = ax + b$ (This defines a *linear function*)

We shall prove in Chap. 14 that the graph of (3) is a straight line when plotted on rectangular coordinate paper. We have considered special cases of (1) and (2) before.

Let us take the logarithms of each side of (1) and (2); we do this in the following double column:

(4) $y = ae^{bx}$	$y = ax^b$
$\log_{10} y = \log_{10} a + bx \log_{10} e$	$\log_{10} y = \log_{10} a + b \log_{10} x$
$= \log_{10} a + 0.43429\, bx$	
Rewrite these in the form	
(5) $Y = A + Bx$	$Y = A + bX$

where $Y = \log_{10} y$, $A = \log_{10} a$, $B = 0.43429b$, $X = \log_{10} x$.

With the introduction of the new variables X and Y, we see that each equation (4) can be written as a linear equation (5). Therefore each equation (5) will plot a straight line. Thus, if we had data such as

x	x_1	x_2	$\cdots$	x_n
y	y_1	y_2	$\cdots$	y_n

and suspected that they followed (approximately) an exponential law (power law), we could look up the logarithms and write them down in a table as follows:

x	x_1	x_2	$\cdots$	x_n	I
y	y_1	y_2	$\cdots$	y_n	
$X = \log_{10} x$	X_1	X_2	$\cdots$	X_n	II
$Y = \log_{10} y$	Y_1	Y_2	$\cdots$	Y_n	III

Plotting I against III on rectangular paper would yield a straight line if the original data followed an exponential law. Plotting II against III on rectangular paper would yield a straight line if the original data followed a power law.

But we can do better. If we plot x against y (original data) on semilog paper, we will get a straight line if the law is exponential since, in effect, the paper looks up the logarithms for us. Similarly x plotted against y on log-log paper will yield a straight line if the data follow a power law.

By plotting data on semilog or log-log paper it is therefore a simple matter to tell whether the law is exponential or power (or approximately so) by determining whether the points lie along a line (or nearly so).

If the points lie along some other curve, we must resort to other methods to find a suitable equation. The general process is called *curve fitting*.

ILLUSTRATION 2 The mean distance D of the planets from the sun and their periods of revolution (T years) are given by the table. (The distance of the earth from the sun is taken as one unit.) Discover the (approximate) law.

	Mercury	*Venus*	*Earth*	*Mars*	*Jupiter*	*Saturn*	*Uranus*	*Neptune*	*Pluto*
T	0.241	0.615	1.00	1.88	11.9	29.5	84.0	165	265
D	0.387	0.723	1.00	1.52	5.20	9.54	19.2	30.1	41.3

SOLUTION We plot the data on rectangular, semilog, and log-log paper. The results are shown in Fig. 11.12 for log-log paper (three-block by four-block). We take D as the independent variable.

The points lie (almost) on a straight line, and thus the law is (approximately) given by $T = aD^b$. It turns out (we will not compute it here) that $a = 1$ and $b = \frac{3}{2}$ (approximately) so that the final answer is $T = D^{3/2}$ or $T^2 = D^3$. Thus the square of the time is the cube of the distance. This is known as *Kepler's Law*.

EXERCISE B For Illustration 2, show the graphs of the data on rectangular and semilog paper.

EXERCISE C Plot the graph of the equation $T = D^{3/2}$ on rectangular paper.

PROBLEMS 11.6

1. Given that the number N of bacteria in a culture at time t is $N = ae^{bt}$, and that $N(0) = 10$, $N(1) = 100$; find $N(2)$.
2. If $y = ae^{bx}$, $y(2) = 4$, $y(3) = 9$, find $y(1)$.
3. Discover an approximate law from the following data:

x	0	2	3	5	7
y	4	12	16	23	33

4. The total adsorption (x cubic units) of a certain gas by another chemical varied with the time (t units) as follows:

x	0	1	2	3
y	0	3	12	26

Discover an appropriate law.

5 The speed s of a certain chemical reaction doubles every time the temperature T° is raised 6°. Make out a table of some of the elements of the function $f:(T,s)$ thus defined, and discover the type of law. Let one element of f be $(0^\circ,1)$.

In Probs. 6 to 15 name the kind of paper on which the graph is a straight line.

6 $y = 2x^3$

7 $y = e^{-6x}$

8 $y = (e^2)^x$

9 $y = x^e$

10 $y = e^2$

11 $2xy = 1$

12 $x^3y = 1$

13 $xy^3 = 1$

14 $y = 2x + 1$

15 $y = a \cdot 2^{bx}$

REFERENCES

In addition to the many standard textbooks on algebra, the reader should consult the following articles in the *American Mathematical Monthly*.

Cairns, W. D.: Napier's Logarithms as He Developed Them, vol. 35, p. 64 (1928).

Cajori, Florian: History of the Exponential and Logarithmic Concepts, vol. 20, pp. 5, 20, 35, 75, 107, 148, 173, 205 (1913).

Huntington, E. V.: An Elementary Theory of the Exponential and Logarithmic Functions, vol. 23, p. 241 (1916).

Lenser, W. T.: Note on Semi-logarithmic Graphs, vol. 49, p. 611 (1942).

Sandham, H. F.: An Approximate Construction for e, vol. 54, p. 215 (1947).

Thomas, J. M.: Pointing Off in Slide Rule Work, vol. 55, p. 567 (1948).

12 TRIGONOMETRIC FUNCTIONS OF ANGLES

12.1 › INTRODUCTION

Trigonometry was originally developed in connection with the study of the relationships between the sides and angles of a triangle. You have probably already met some of the trigonometric functions, such as the sine and cosine, and have applied them to simple problems about triangles. This aspect of trigonometry was investigated extensively by the early Greeks, especially by Hipparchus (circa 180–125 B.C.), who, because of his work in astronomy, actually developed spherical rather than plane trigonometry. The trigonometry of the triangle continues to be of importance in modern technology in such areas as surveying, navigation, and the applications of vectors to mechanics. The present chapter is concerned with those portions of this material which deal with the geometry of the plane. You will need to consult other books for material on spherical trigonometry.

It would be a serious error, however, to limit the study of trigonometry to its applications to triangles. Its modern uses are widespread in many theoretical and applied fields of knowledge. The trigonometric functions force themselves on you in a very surprising fashion when you study the calculus of certain algebraic functions. You will also meet them when you study wave motion, vibrations, alternating current, sound, etc. To treat these subjects adequately we must extend the concept of a trigonometric function so that it is a function of a general real variable, and no longer merely a function of an angle. These more general trigonometric functions become, then, members of our arsenal of functions which have been developed in the previous chapters. Their definitions and properties are given in the following chapter.

The complete set, consisting of the algebraic functions, the exponential function, the logarithmic functions, and the trigonometric functions, is called the set of "elementary functions". Virtually all undergraduate courses in mathematics restrict themselves to these elementary functions. In more advanced work, however, it is necessary to introduce additional functions which carry curious names such as the "gamma function", "Bessel function", "theta function", etc. We shall not need to refer to these hereafter in this book.

12.2 » DISTANCE IN THE PLANE

We begin our study of trigonometry by developing certain properties of plane geometry which we shall need. Naturally we assume that you are already familiar with much of this subject from your study of it in high school. We shall be using all the logical structure of this geometry, including the undefined words, axioms, definitions, and theorems. Of course, we must assume these here, for a review of this material would take us too far afield. As a minimum, you should be familiar with the properties of similar triangles and with the Theorem of Pythagoras.

We shall employ the usual system of rectangular axes which was discussed in Secs. 2.10 to 2.12. In our work so far in this book we have permitted you to use quite different units on the two axes according to your immediate needs. Here, however, we must be more particular. In this chapter, coordinates on the X-axis and the Y-axis will represent *distance* in the *same* units of measurement.

Distance

Let us now consider two points P_1 and P_2 which do not lie on a line parallel to one of the axes (Fig. 12.1). The length of the segment, or the "distance P_1P_2", can be computed from the Theorem of Pythagoras. Construct the right triangle P_1RP_2 with P_1R parallel to the Y-axis and RP_2 parallel to the X-axis. R has coordinates (x_1,y_2). From the Theorem of Pythagoras,

$$(P_1P_2)^2 = (RP_2)^2 + (P_1R)^2$$

Slant distance

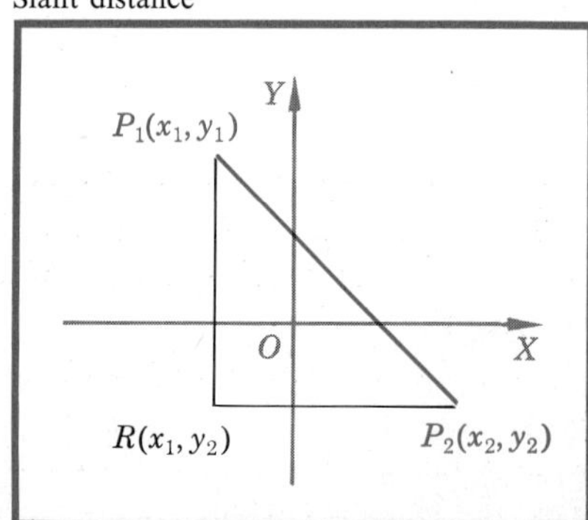

FIGURE 12.1
$(P_1P_2)^2 = (x_2 - x_1)^2 + (y_2 - y_1)^2$

We know (Sec. 2.12) that

$$RP_2 = |x_2 - x_1| \qquad \text{and} \qquad P_1R = |y_2 - y_1|$$

Hence
$$(P_1P_2)^2 = (x_2 - x_1)^2 + (y_2 - y_1)^2$$

We observe that this is also true even if the line P_1P_2 is parallel to one of the axes. We have thus proved the general theorem.

THEOREM 1 The distance d between any two points in the plane $P_1(x_1,y_1)$ and $P_2(x_2,y_2)$ is given by

$$d = \sqrt{(x_2 - x_1)^2 + (y_2 - y_1)^2}$$

ILLUSTRATION 1 Find the distance between $A(4,-3)$ and $B(-2,5)$.

SOLUTION The distance $d = AB$ is given by

$$\begin{aligned} d &= \sqrt{[4 - (-2)]^2 + (-3 - 5)^2} \\ &= \sqrt{36 + 64} \\ &= 10 \end{aligned}$$

ILLUSTRATION 2 Find the lengths of the diagonals of the quadrilateral $A(1,2)$, $B(-2,1)$, $C(-3,-4)$, $D(5,-7)$.

SOLUTION The diagonals are AC and BD, and their lengths are given by

$$\begin{aligned} AC &= \sqrt{4^2 + 6^2} = \sqrt{52} \\ BD &= \sqrt{(-7)^2 + 8^2} = \sqrt{113} \end{aligned}$$

PROBLEMS 12.2

In Probs. 1 to 6 find the distance d between the pairs of points.

1 $(4,0), (0,6)$
2 $(0,-2), (6,0)$
3 $(6,2), (-9,2)$
4 $(2,9), (2,-5)$
5 $(2,-3), (-4,-5)$
6 $(4,16), (-3,6)$

7 Show that the triangle $A(4,1)$, $B(4,11)$, $C(0,3)$ is a right triangle.
8 Show that the triangle $A(4,1)$, $B(3,2)$, $C(6,5)$ is a right triangle.
9 Show that the triangle $A(-\frac{1}{2}, \frac{1}{2}\sqrt{3})$, $B(2,3\sqrt{3})$, $C(\frac{9}{2},\frac{1}{2}\sqrt{3})$ is an equilateral triangle.
10 Show that the triangle $A(0,0)$, $B(\sqrt{3},1)$, $C(0,2)$ is an equilateral triangle.
11 Show that the points $A(1,2)$, $B(2,1)$, $C(0,-3)$, $D(-1,-2)$ are the vertices of a parallelogram.

12 Show that the points $A(0,1)$, $B(-3,-2)$, $C(1,0)$, $D(4,3)$ are the vertices of a parallelogram.

13 Show that $A(2,-2)$, $B(4,2)$, $C(-4,6)$, $D(-6,2)$ are the vertices of a rectangle.

14 Show that $A(5,-4)$, $B(6,-3)$, $C(7,-4)$, $D(6,-5)$ are the vertices of a square.

In Probs. 15 and 16 show that the point P is on the perpendicular bisector of the line segment AB.

15 $P(3,4)$, $A(-2,3)$, $B(4,-1)$ **16** $P(3,\frac{3}{2})$, $A(6,2)$, $B(0,1)$

17 Find the distance between any two points on the X-axis.

18 Show that in the XY-plane the distance between $A(x_1,y_1)$ and $B(x_2,y_2)$ is equal to the distance between $B(x_2,y_2)$ and $A(x_1,y_1)$. HINT: Compute the two distances by the use of the formula.

12.3 » ANGLES

Consider a directed line L and a point O on it. In the spirit of Sec. 2.10 we define a *ray* to be O and the points of L which are beyond O with respect to the given direction. The point O is called the initial point. Further, the *angle AOB* is defined as the union of two rays OA and OB, not lying in the same straight line, which have a common initial point O called the *vertex of the angle* (Fig. 12.2).

Now a line determines two half-planes, one on either side of the line. A line containing ray OA determines a half-plane which contains B (Sec. 8.5). Similarly, a line containing ray OB determines a half-plane which contains A. The intersection of these two half-planes is called the *interior* of angle AOB (Fig. 12.3).

Interior of an angle

In order to assign a measure to angle AOB, we draw a circle of radius 1 with center at the origin O, place OA along the positive X-axis, and let OB fall into whatever position it takes (Fig. 12.4). (A circle of any radius would serve as well; we use radius 1 here for convenience.) Angle AOB is then said to be in *standard position*. OA intersects this circle in the point $P(1,0)$ and OB in a point which we call $Q(x,y)$.

Standard position

Examples of angles

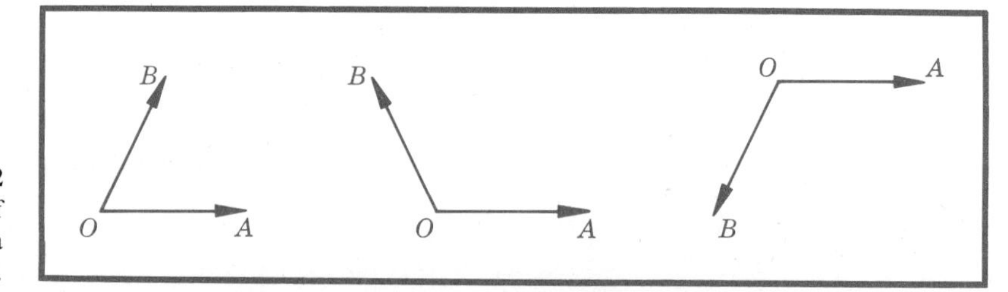

FIGURE 12.2
An angle is the union of two noncollinear rays with a common initial point.

Interior of an angle

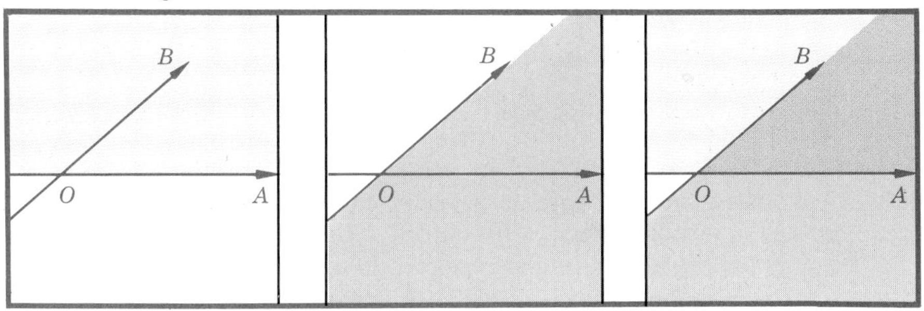

FIGURE 12.3
The interior of angle AOB is in heavy color in the right-hand figure.

Now it is easy to think of many ways of dividing the circumference of a circle into a set of equal arcs. Some of these ways are ruler and compass constructions. Examples are: a set of four equal arcs, and a set of six equal arcs. One very common (but not constructible by ruler and compass) way is to think of the unit circle as being divided into 360 equal arcs, the measure of each of which is defined to be "one degree", written "$1°$". Each degree is further divided into 60 minutes ($60'$) and each minute into 60 seconds ($60''$). In this chapter we use this scale to measure arcs on a unit circle. (See the next chapter for more on this subject.) In terms of this measure for arcs on a unit circle, we can now define a measure for angles.

Measure of an angle

Angles and their intercepted arcs

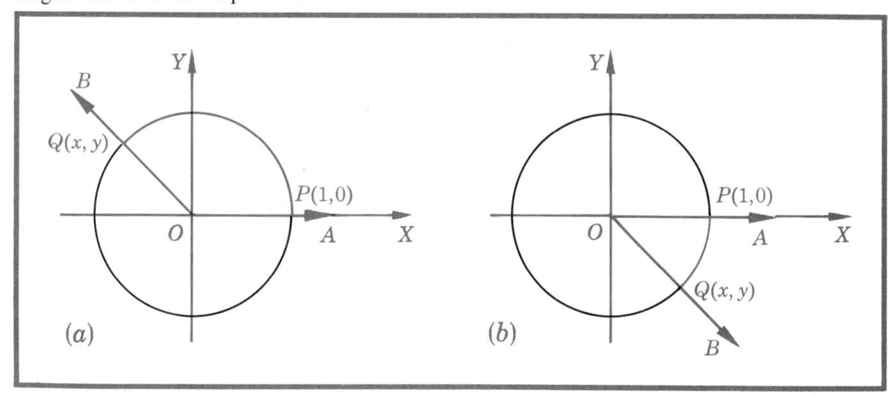

FIGURE 12.4
An angle is measured by the arc which it intercepts on a unit circle.

DEFINITION The measure (in degrees) of angle AOB is the measure (in degrees) of the intercepted arc PQ of the unit circle.

EXERCISE A Show that arc PQ is the intersection of the unit circle with the interior of angle AOB.

Signed angles

This definition of the measure of an angle agrees with that which is used in school geometry. In contrast to the notion of *signed* or *directed* angles about to be developed, we call the above angles *unsigned angles.*

In trigonometry we need the notions of *directed* arcs and angles. Intuitively we may direct an arc by placing an arrow on it, which determines whether it is drawn in the counterclockwise or the clockwise direction around the circle. We then assign a positive measure to a counterclockwise arc and a negative measure to a clockwise arc. These signed measures are then carried over to the corresponding angles.

More specifically, let $\theta°$ be the unsigned measure of arc PQ. Then the measure of angle AOB is (Fig. 12.4):

1 $\theta°$ if B lies in the upper half-plane
2 $-\theta°$ if B lies in the lower half-plane

We now generalize the above definition of angle and permit the rays OA and OB to be collinear. In this case we define the measure of angle AOB to be:

3 $0°$ if B lies on the positive half of the X-axis
4 $180°$ if B lies on the negative half of the X-axis

Thus, there is established a one-to-one correspondence between angles in standard position and their measures in degrees in the interval $-180° < \theta \leq 180°$. Where no confusion arises, we may drop "the measure of" and speak simply of an angle of $\theta°$. At times we may use the same symbol for the angle and for the measure of the angle.

Angle from one line to another

When two undirected lines meet in a point O, four angles are determined, and so it is not proper to refer to *the angle between two undirected lines.* On the other hand, two directed lines meeting at O do determine a unique angle, namely the angle defined by the two rays beginning at O and pointing in the two given directions. Since the two given lines play equivalent roles in this situation, this is an unsigned angle, and we take its measure to be positive. We may, however, be interested in the *directed angle from the first directed line to the second directed line.* In this case we place the directed angle in standard position, with the first directed line coinciding with the positive X-axis, and attach the appropriate sign. Our terms *first directed line* and *second directed line* may be replaced respectively by *initial ray* and *terminal ray* with common initial point O.

Generalized angles

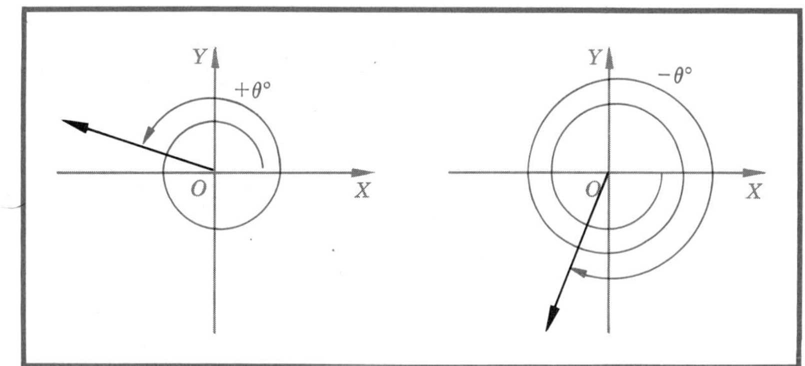

FIGURE 12.5
A generalized angle is determined by two rays and a directed arc.

Generalized angle

In our definition of angle we have restricted ourselves to angles whose measures lie in the interval $-180° < \theta° \leq 180°$. It is frequently desirable, however, to enlarge this concept of an angle to include angles whose measures lie outside of this range. In view of the above discussion, it is natural to use the following definition of a generalized angle.

DEFINITION A generalized angle is the plane figure consisting of (1) two rays, an initial and a terminal, with common initial point O; and (2) a directed arc of a circle (with center at O) whose ends lie on the two given rays. This arc may be of any length and hence may wrap around the circle any number of times in either direction.

The measure in degrees of this arc, together with its sign, is defined to be the measure of the generalized angle. The sign is positive if the wrap is counterclockwise, negative if the wrap is clockwise (Fig. 12.5).

Hereafter, we shall drop the term *generalized angle* and use *angle* to refer to both the angles of plane geometry and to the generalized angles just defined.

EXERCISE B Show that:

1 The measure of a quarter revolution counterclockwise is $90°$.
2 The measure of a half revolution clockwise is $-180°$.
3 The measure of $1\frac{1}{2}$ revolutions counterclockwise is $540°$.
4 The measure of a full revolution clockwise is $-360°$.

PROBLEMS 12.3

In Probs. 1 to 16 sketch roughly the directed angle θ.

1 $\theta = 0°$ **2** $\theta = 45°$ **3** $\theta = 30°$
4 $\theta = 60°$ **5** $\theta = 120°$ **6** $\theta = 135°$

7 $\theta = 150°$ **8** $\theta = 210°$ **9** $\theta = 225°$
10 $\theta = 240°$ **11** $\theta = 300°$ **12** $\theta = 315°$
13 $\theta = 330°$ **14** $\theta = -120°$ **15** $\theta = -240°$
16 $\theta = -300°$

In Probs. 17 to 30 draw, with the aid of a protractor, the directed angle θ.

17 $\theta = 35°$ **18** $\theta = 145°$ **19** $\theta = 205°$
20 $\theta = 340°$ **21** $\theta = -30°$ **22** $\theta = -72°$
23 $\theta = -150°$ **24** $\theta = -200°$ **25** $\theta = 400°$
26 $\theta = 500°$ **27** $\theta = -600°$ **28** $\theta = -500°$
29 $\theta = 1{,}000°$ **30** $\theta = -1{,}500°$

12.4 › POLAR COORDINATES

We have seen earlier how to locate a point in the plane by means of its rectangular coordinates (x,y). For many purposes, however, it is more convenient to locate points by using a different system of coordinates (r,θ), called polar coordinates. The intuitive idea is quite familiar. Suppose we are on a mountain top and wish to locate another peak. We can do so if we are told that it is 50 miles away in a direction 35° east of north. The essentials are that we know the distance r from a fixed point and the angle θ relative to a fixed reference line. The formal definition is as follows (Fig. 12.6).

DEFINITIONS

1 The origin O is called the *pole.*
2 The horizontal (initial) ray, directed to the right from O, is called the *polar axis.*
3 Given a point P in the plane, the directed segment OP is called the *radius vector.* Its length—which we assume to be positive in this chapter—is denoted by r.
4 A directed angle (positive or negative) with the polar axis as initial side and the radius vector on its terminal side is denoted by θ (standard position).
5 In the ordered pair (r,θ), r and θ are called *polar coordinates* of P.

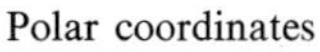

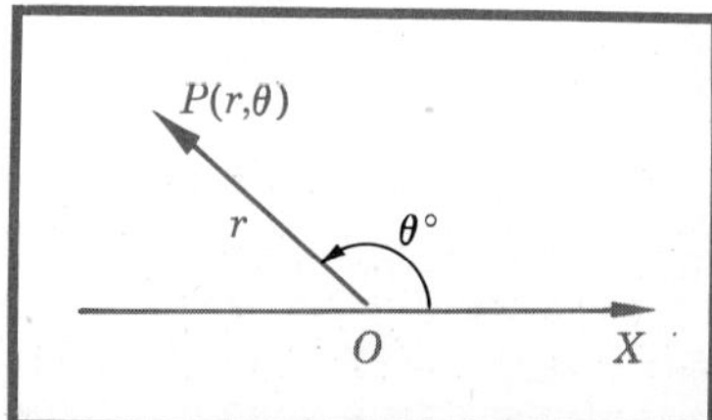

FIGURE 12.6
In polar coordinates a point is determined by a distance and an angle.

Two sets of polar coordinates

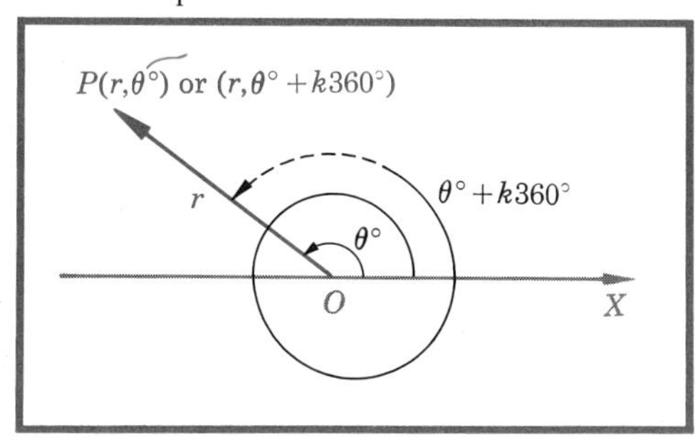

FIGURE 12.7
A point has infinitely many sets of polar coordinates.

Polar and rectangular coordinates

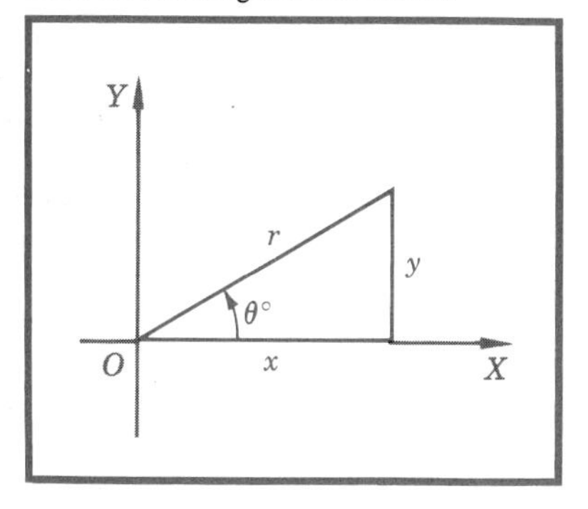

FIGURE 12.8
If either the polar or rectangular coordinates of a point are known, the other can be calculated.

We observe, first of all, that any pair (r,θ) determines a unique point P. On the other hand, a fixed point P has many sets of polar coordinates. Indeed, if P has coordinates (r,θ), then it also has coordinates $(r, \theta° + k360°)$ for every integer k, positive or negative (Fig. 12.7).

We now have two sets of coordinates for locating a point in the plane: (1) rectangular (x,y) and (2) polar (r,θ). So that we can use these interchangeably, we need to be able to go from one to the other at will. We observe at once that $r = \sqrt{x^2 + y^2}$, but, to make further progress, we need to introduce the trigonometric functions sine and cosine. These are treated in the following section.

12.5 » SINE AND COSINE OF A DIRECTED ANGLE

$\sin\theta = y/r$
$\cos\theta = x/r$

Let θ be a directed angle in standard position with initial ray p (polar axis) and terminal ray q. Choose any point Q on q, and let the segment OQ have length r. Then Q has the polar coordinates (r,θ); let its rectangular coordinates be (x,y). Then we define the sine of θ (written $\sin\theta$) and the cosine of θ (written $\cos\theta$) as follows.

DEFINITION $\sin\theta = y/r$; $\cos\theta = x/r$

This definition appears to depend upon our choice of Q, but this is not really the case. Let us choose Q' with coordinates (r',θ) and (x',y'). Then, from the elementary properties of similar triangles, we see that $y/r = y'/r'$ and $x/r = x'/r'$. Thus the definitions of $\sin\theta$ and $\cos\theta$ do not depend upon the choice of the point Q on the terminal side of θ.

Change in the position of Q

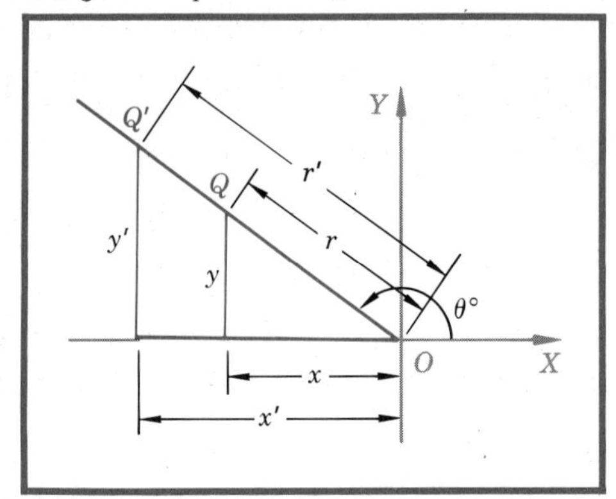

FIGURE 12.9
The values of $\sin\theta$ and $\cos\theta$ are independent of the position of Q.
$\sin\theta = y/r = y'/r'$
$\cos\theta = x/r = x'/r'$

We see in this way that, given a directed angle θ, the real numbers $\sin\theta$ and $\cos\theta$ are uniquely determined. The sets of ordered pairs $\{\theta, \sin\theta\}$ and $\{\theta, \cos\theta\}$ thus *define two functions* which we call "sine" and "cosine", respectively. The domain of each is the set of all directed angles, and their respective images are subsets of the real numbers, which will be discussed in Sec. 13.3.

Let us now return to the problem of determining the rectangular coordinates of a point from its polar coordinates, and vice versa. If we are given the polar coordinates (r,θ), the rectangular coordinates are obtained from the formulas

$$x = r\cos\theta \qquad y = r\sin\theta$$

which are immediate consequences of the definitions of $\sin\theta$ and $\cos\theta$. For a point P we call the number x the *abscissa,* the number y the *ordinate,* and the positive number r the *radius vector,* or, simply, the *distance* (from the origin).

In order to go from (x,y) to (r,θ), we use the formulas

$$r = \sqrt{x^2 + y^2} \qquad \frac{\sin\theta}{\cos\theta} = \frac{y}{x}$$

We shall presently be able to simplify the second of these two formulas.

We can use these formulas to derive an important relation between $\sin\theta$ and $\cos\theta$. First, let us substitute $x = r\cos\theta$ and $y = r\sin\theta$ in $x^2 + y^2 = r^2$. Then we obtain

$$r^2\cos^2\theta + r^2\sin^2\theta = r^2$$

or, dividing by r^2 and rearranging,

$$\sin^2\theta + \cos^2\theta = 1$$

This equation is called an *identity* because it is true for *all* values of θ. It is the most useful identity in the whole subject of trigonometry.

We can also find the distance between two points in terms of their polar coordinates. Suppose that we have two points whose polar coordinates are $P_1(r_1,\theta_1)$ and $P_2(r_2,\theta_2)$ and whose rectangular coordinates are (x_1,y_1) and (x_2,y_2), respectively. We have seen that the distance d between P_1 and P_2 is given by the formula

$$d^2 = (x_2 - x_1)^2 + (y_2 - y_1)^2$$

Substituting in this formula, we find

$$\begin{aligned} d^2 &= (r_2 \cos\theta_2 - r_1 \cos\theta_1)^2 + (r_2 \sin\theta_2 - r_1 \sin\theta_1)^2 \\ &= r_2^2 \cos^2\theta_2 - 2r_1r_2 \cos\theta_2 \cos\theta_1 + r_1^2 \cos^2\theta_1 + r_2^2 \sin^2\theta_2 \\ &\qquad - 2r_1r_2 \sin\theta_2 \sin\theta_1 + r_1^2 \sin^2\theta_1 \\ &= r_1^2(\sin^2\theta_1 + \cos^2\theta_1) + r_2^2(\sin^2\theta_2 + \cos^2\theta_2) \\ &\qquad - 2r_1r_2(\cos\theta_1 \cos\theta_2 + \sin\theta_1 \sin\theta_2) \end{aligned}$$

Hence

$$d^2 = r_1^2 + r_2^2 - 2r_1r_2(\cos\theta_1 \cos\theta_2 + \sin\theta_1 \sin\theta_2)$$

12.6 SINE AND COSINE OF SPECIAL ANGLES

If we are to make any use of the sine and cosine functions, we must know their values for any directed angle. We begin the discussion of this matter by treating some special angles of great importance.

0°

SIN 0° AND COS 0° When $\theta = 0$, it is clear that the terminal side q lies along the initial side p, which we take to be horizontal (the polar axis or the X-axis). Choose Q to be the point (1,0). Then $x = 1, y = 0, r = 1$. Hence

$$\sin 0^\circ = \frac{y}{r} = \frac{0}{1} = 0$$

$$\cos 0^\circ = \frac{x}{r} = \frac{1}{1} = 1$$

90°

SIN 90° AND COS 90° Here we can choose Q to be the point (0,1); so $x = 0$, $y = 1$, $r = 1$, and

$$\sin 90^\circ = \tfrac{1}{1} = 1$$

$$\cos 90^\circ = \tfrac{0}{1} = 0$$

45°

SIN 45° AND COS 45° From Fig. 12.10 we see that we can choose Q to be point (1,1); so $x = 1$, $y = 1$, $r = \sqrt{2}$. Hence

$$\sin 45^\circ = \frac{1}{\sqrt{2}} = \frac{\sqrt{2}}{2} \approx \frac{1.414}{2} \approx 0.707$$

$$\cos 45^\circ = \frac{1}{\sqrt{2}} = \frac{\sqrt{2}}{2} \approx \frac{1.414}{2} \approx 0.707$$

Functions of 45°

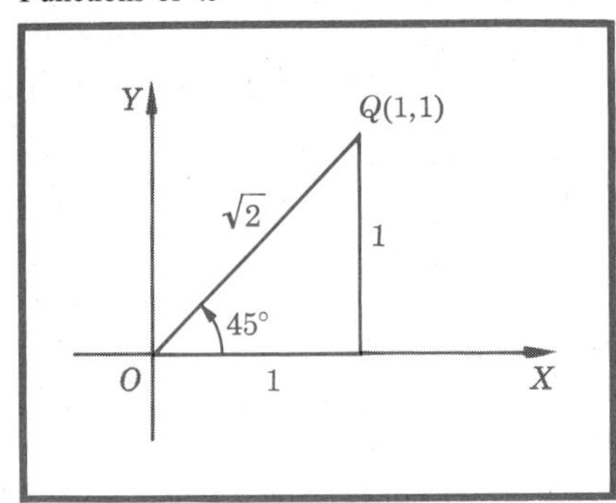

FIGURE 12.10

Functions of 30°

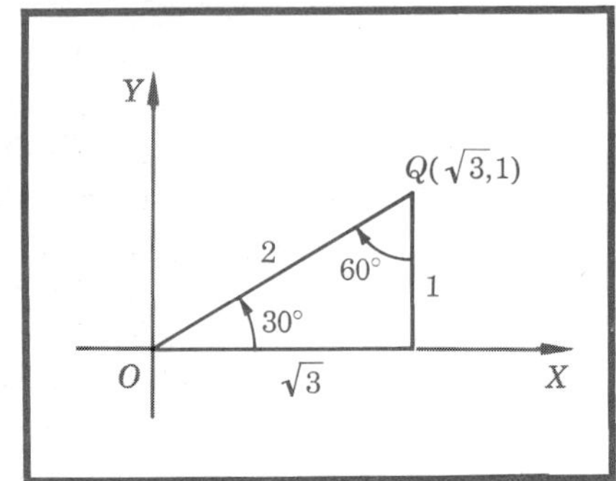

FIGURE 12.11

30° **SIN 30° AND COS 30°** From Fig. 12.11 we see that we can choose Q to be the point $(\sqrt{3},1)$; so $x = \sqrt{3}$, $y = 1$, $r = 2$. Hence

$$\sin 30° = \tfrac{1}{2} = 0.500$$

$$\cos 30° = \frac{\sqrt{3}}{2} \approx \frac{1.732}{2} \approx 0.866$$

60° **SIN 60° AND COS 60°** From Fig. 12.12 we see that we can choose Q to be the point $(1,\sqrt{3})$; so $x = 1$, $y = \sqrt{3}$, $r = 2$. Hence

$$\sin 60° = \frac{\sqrt{3}}{2} \approx 0.866$$

$$\cos 60° = \tfrac{1}{2} = 0.500$$

In the cases just treated every angle lay in the first quadrant. Similar constructions permit us to compute the values of these functions for related angles in other quadrants. The details are contained in the following problems.

Functions of 60°

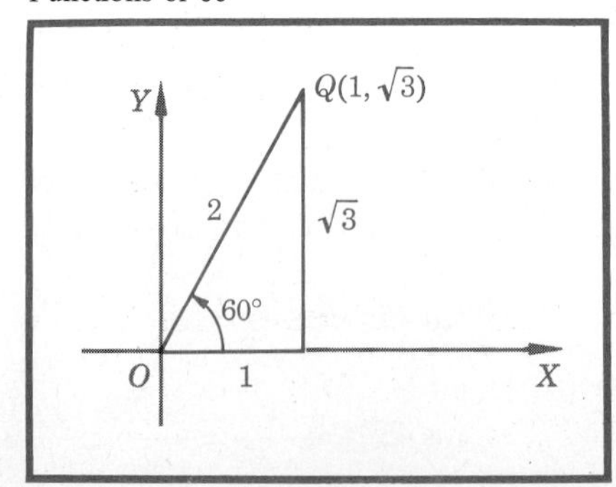

FIGURE 12.12

PROBLEMS 12.6

In Probs. 1 to 30 find $\sin\theta$ and $\cos\theta$.

1 $\theta = 135°$ **2** $\theta = 120°$ **3** $\theta = 180°$
4 $\theta = 150°$ **5** $\theta = 225°$ **6** $\theta = 210°$
7 $\theta = 270°$ **8** $\theta = 240°$ **9** $\theta = 315°$
10 $\theta = 300°$ **11** $\theta = 360°$ **12** $\theta = 330°$
13 $\theta = 600°$ **14** $\theta = 450°$ **15** $\theta = -30°$
16 $\theta = 750°$ **17** $\theta = -60°$ **18** $\theta = -45°$
19 $\theta = -120°$ **20** $\theta = -90°$ **21** $\theta = -150°$
22 $\theta = -135°$ **23** $\theta = -225°$ **24** $\theta = -210°$
25 $\theta = -330°$ **26** $\theta = -300°$ **27** $\theta = -750°$
28 $\theta = -600°$ **29** $\theta = -1575°$ **30** $\theta = -855°$

In Probs. 31 to 46 find x and y for the given values of r and θ.

31 $r = 8, \theta = 30°$ **32** $r = 4, \theta = 45°$
33 $r = 3, \theta = 180°$ **34** $r = 5, \theta = 120°$
35 $r = 10, \theta = 240°$ **36** $r = 6, \theta = 270°$
37 $r = 10, \theta = 300°$ **38** $r = 8, \theta = 315°$
39 $r = \frac{1}{2}\sqrt{2}, \theta = 135°$ **40** $r = \frac{1}{2}\sqrt{2}, \theta = 225°$
41 $r = 3, \theta = -150°$ **42** $r = 1, \theta = -270°$
43 $r = \sqrt{2}, \theta = -45°$ **44** $r = \sqrt{2}, \theta = 45°$
45 $r = 300, \theta = 390°$ **46** $r = 10\sqrt{2}, \theta = 405°$

In Probs. 47 to 62 find r and the least positive value of θ for the given values of x and y.

47 $x = 0, y = -4$ **48** $x = 2, y = 0$
49 $x = 1, y = -\sqrt{3}$ **50** $x = 4\sqrt{3}, y = 4$
51 $x = 50\sqrt{3}, y = -50$ **52** $x = 50\sqrt{2}, y = -50\sqrt{2}$
53 $x = -\frac{3}{2}\sqrt{3}, y = -\frac{3}{2}$ **54** $x = 0, y = 10$
55 $x = -\frac{1}{2}\sqrt{3}, y = -\frac{1}{2}\sqrt{3}$ **56** $x = -\sqrt{2}, y = \sqrt{2}$
57 $x = -8\sqrt{3}, y = -8$ **58** $x = -21, y = 0$
59 $x = -\frac{1}{2}\sqrt{3}, y = \frac{1}{2}$ **60** $x = \sqrt{2}, y = -\sqrt{2}$
61 $x = -6, y = 6\sqrt{3}$ **62** $x = 0, y = -\sqrt{2}$

In Probs. 63 to 70 find r and also find the θ whose absolute value is least.

63 $x = 16, y = 0$ **64** $x = 6\sqrt{3}, y = -6$
65 $x = -1, y = 1$ **66** $x = 4, y = -4\sqrt{3}$
67 $x = 0, y = -7$ **68** $x = -25, y = 25\sqrt{3}$
69 $x = -1, y = -\sqrt{3}$ **70** $x = -2\sqrt{3}, y = -2$

In Probs. 71 to 76 find the distance between the given pairs of points.

71 $(\sqrt{3}, 60°)$ and $(2, 270°)$ **72** $(2, 30°)$ and $(1, 120°)$
73 $(10, 60°)$ and $(1, 135°)$ **74** $(3, 180°)$ and $(2, 390°)$
75 $(1, 45°)$ and $(3, 135°)$ **76** $(3, 90°)$ and $(3, 225°)$

12.7 › OTHER TRIGONOMETRIC FUNCTIONS

Although the sine and cosine are adequate for the study of trigonometry, certain combinations occur so often that they are given special names. We therefore introduce the following four additional trigonometric functions.

DEFINITIONS

$\tan\theta$ 1 The set of ordered pairs $\{\theta, \sin\theta/\cos\theta\}$ defines a function called *tangent,* or simply, *tan.* That is, for a given θ, $\tan\theta = \sin\theta/\cos\theta$, $\cos\theta \neq 0$.

$\cot\theta$ 2 The set of ordered pairs $\{\theta, \cos\theta/\sin\theta\}$ defines a function called *cotangent,* or, simply, *cot.* That is, for a given θ, $\cot\theta = \cos\theta/\sin\theta$, $\sin\theta \neq 0$.

$\sec\theta$ 3 The set of ordered pairs $\{\theta, 1/\cos\theta\}$ defines a function called *secant,* or *sec.* That is, for a given θ, $\sec\theta = 1/\cos\theta$, $\cos\theta \neq 0$.

$\csc\theta$ 4 The set of ordered pairs $\{\theta, 1/\sin\theta\}$ defines a function called *cosecant,* or *csc.* That is, for a given θ, $\csc\theta = 1/\sin\theta$, $\sin\theta \neq 0$.

5 The functions cosine, cotangent, cosecant are called the cofunctions of sine, tangent, secant, respectively.

EXERCISE A What is the domain of definition of each of the functions tangent, cotangent, secant, cosecant?

EXERCISE B Show that

1 $\tan\theta = y/x, x \neq 0$ 2 $\sec\theta = r/x$
3 $\cot\theta = x/y, y \neq 0$ 4 $\csc\theta = r/y$

The signs of the six trigonometric functions of θ depend upon the quadrant in which the terminal side of θ lies. We shall say that "θ is a *second-quadrantal angle* if its terminal side lies in the second quadrant" and shall use similar expressions for the other quadrants. We see that $\sin\theta = y/r$ is positive when θ is a first- or second-quadrantal angle, since y is positive in the first and second quadrants and r is always positive. However, $\sin\theta$ is negative when θ is a third- or fourth-quadrantal angle, for y is negative in these quadrants. The situation is summarized in the following table, which you should verify for yourself.

Quadrant	*Sine*	*Cosine*	*Tangent*	*Cosecant*	*Secant*	*Cotangent*
I	+	+	+	+	+	+
II	+	−	−	+	−	−
III	−	−	+	−	−	+
IV	−	+	−	−	+	−

The following mnemonic scheme will help you to remember the signs of the sine, tangent, and cosine: "All Scholars Take Chemistry", or "ASTC". This says that All functions are positive in the first quadrant, only the Sine in the second quadrant, only the Tangent in the third quadrant, and only the Cosine in the fourth quadrant.

The identity

Basic identity

(1)
$$\sin^2\theta + \cos^2\theta = 1$$

can be written in the form

$$\frac{\sin^2\theta}{\cos^2\theta} + 1 = \frac{1}{\cos^2\theta} \qquad \cos\theta \neq 0$$

From this it follows that

(2)
$$1 + \tan^2\theta = \sec^2\theta$$

for every value of θ for which $\tan\theta$ and $\sec\theta$ have meaning.

EXERCISE C Name one value of θ for which (2) has no meaning.

Similarly we arrive at the identity

(3)
$$1 + \cot^2\theta = \csc^2\theta$$

You should memorize these three basic identities, Eqs. (1), (2), and (3) above.

The values of $\tan\theta$, $\cot\theta$, $\sec\theta$, and $\csc\theta$ for the special angles treated in the previous section can be readily determined. These are of considerable importance both in pure mathematics and in applications in science and engineering. For example, to determine $\tan 30°$, we write the following:

$$\tan 30° = \frac{\sin 30°}{\cos 30°} = \frac{1/2}{\sqrt{3}/2} = \frac{1}{\sqrt{3}} = \frac{\sqrt{3}}{3} = 0.57735$$

Again

$$\sec 60° = \frac{1}{\cos 60°} = \frac{1}{1/2} = 2$$

$$\cot 135° = \frac{\cos 135°}{\sin 135°} = \frac{-1/\sqrt{2}}{1/\sqrt{2}} = -1$$

$$\csc 270° = \frac{1}{\sin 270°} = \frac{1}{-1} = -1$$

Note that $\tan 90° = \sin 90°/\cos 90°$ is not defined since $\cos 90° = 0$ and in defining $\tan\theta$ we excluded the case where $\cos\theta = 0$.

The following problems are designed for a systematic study of these special angles.

PROBLEMS 12.7

In Probs. 1 to 32 compute the values of the six trigonometric functions for the indicated angle. Draw a figure.

1 $0°$	**2** $30°$	**3** $60°$	**4** $45°$
5 $120°$	**6** $90°$	**7** $150°$	**8** $135°$
9 $210°$	**10** $180°$	**11** $240°$	**12** $225°$
13 $300°$	**14** $270°$	**15** $330°$	**16** $315°$
17 $-45°$	**18** $-30°$	**19** $-90°$	**20** $-60°$
21 $-135°$	**22** $-120°$	**23** $-180°$	**24** $-150°$
25 $-225°$	**26** $-210°$	**27** $-270°$	**28** $-240°$
29 $-315°$	**30** $-300°$	**31** $-360°$	**32** $-330°$

12.8 › SOME IMPORTANT IDENTITIES

There are many useful identities in trigonometry, and, as a scientist or mathematician, you will find it necessary at times to make use of some of them. This is particularly true when the angle involved is, in absolute value, greater than $90°$. For example, the tables of values of the natural trigonometric functions—and these we discuss in the next section—are made up for angles ranging between 0 and $90°$ only. We need, therefore, methods of reducing expressions like $\sin(90° + \theta°)$, $\cos(180° + \theta°)$, $\tan(270° - \theta°)$, etc., to some simpler form involving only the angle $\theta°$, where $0° \leq \theta° \leq 90°$.

By way of illustration let us reduce $\sin(90° + \theta°)$. Examine Fig. 12.13. You should be able to show, by methods of plane geometry, that $\triangle OAB$ is congruent to $\triangle OCD$ if, say, $|OA| = CD$.

Functions of $90° + \theta°$

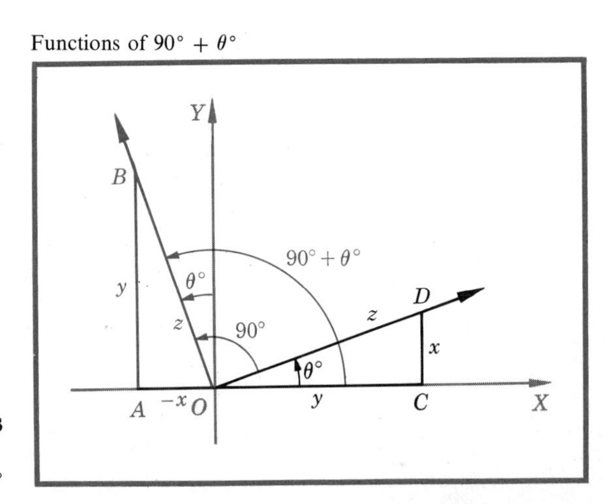

FIGURE 12.13
$\sin(90° + \theta°) = \cos\theta°$
$\cos(90° + \theta°) = -\sin\theta°$

Functions of $180° + \theta°$

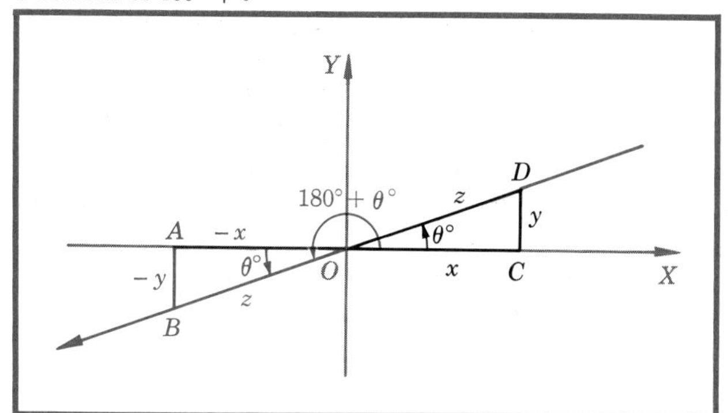

FIGURE 12.14
$\sin(180° + \theta°) = -\sin\theta°$
$\cos(180° + \theta°) = -\cos\theta°$

EXERCISE A Prove that $\triangle OAB$ is congruent to $\triangle OCD$ if $|OA| = CD$.

Now $\sin(90° + \theta°) = y/z = \cos\theta°$, which is the reduction sought. Similarly, keeping in mind that, as pictured, OA is negative, we have $\cos(90° + \theta°) = -x/z = -\sin\theta°$.

Again, from Fig. 12.14 if $AB = -CD$, it follows that

$$\cos(180° + \theta°) = \frac{-x}{z} = -\cos\theta°$$

Similarly, $$\sin(180° + \theta°) = \frac{-y}{z} = -\sin\theta°$$

From Fig. 12.15, with $OA = CD$ (each is negative), you should see that $\tan(270° - \theta°) = -y/-x = y/x = \cot\theta°$.

Functions of $270° - \theta°$

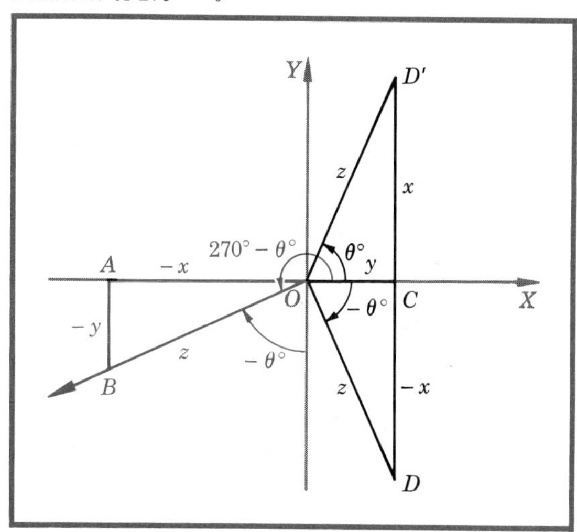

FIGURE 12.15
$\sin(270° - \theta°) = -\cos\theta°$
$\cos(270° - \theta°) = -\sin\theta°$

Functions of $-\theta°$

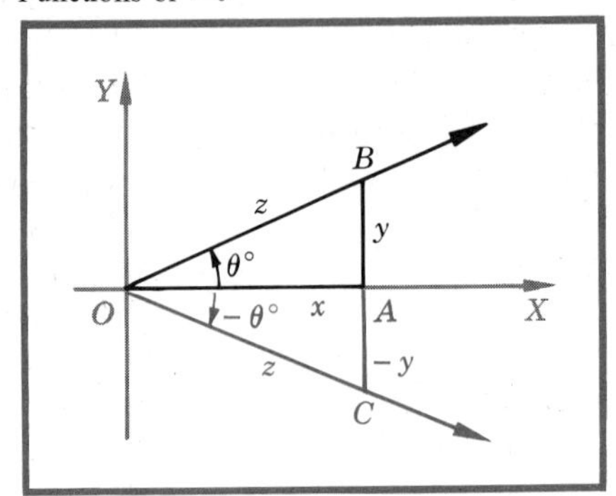

FIGURE 12.16
$\sin(-\theta°) = -\sin\theta°$
$\cos(-\theta°) = \cos\theta°$

These are typical reduction formulas; the general situation is described in the following theorem, which is readily proved.

THEOREM 2 Any trigonometric function of the angle $(k90° \pm \theta°)$ is equal to $(\pm)$ the same function of $\theta°$, if k is an even integer, and is equal to $(\pm)$ the cofunction of $\theta°$, if k is an odd integer. The $(+)$ sign is used if the original function of the original angle $(k90° \pm \theta°)$ is positive; the $(-)$ sign is used if the original function is negative.

The special cases, where $k = 0$ and where $-\theta$ is used, are worthy of further mention. For example, from Fig. 12.16,

$$\sin(-\theta°) = \frac{-y}{z} = -\sin\theta°$$

$$\cos(-\theta°) = \frac{x}{z} = \cos\theta° \qquad \tan(-\theta°) = \frac{-y}{x} = -\tan\theta°$$

and so on.

The following problems are designed for a systematic and detailed study of reduction formulas.

PROBLEMS 12.8 In Probs. 1 to 44 use Theorem 2 to simplify the expression. Assume that $0° < \theta° < 90°$ and that k is an integer. Draw an appropriate figure.

1 $\cos(90° + \theta°)$
2 $\cos(90° - \theta°)$
3 $\sin(90° - \theta°)$
4 $\sin(90° + \theta°)$
5 $\tan(90° - \theta°)$
6 $\tan(90° + \theta°)$
7 $\sin(180° + \theta°)$
8 $\sin(180° - \theta°)$
9 $\tan(180° + \theta°)$
10 $\tan(180° - \theta°)$
11 $\cos(180° - \theta°)$
12 $\cos(180° + \theta°)$

13	$\tan(270° - \theta°)$	**14**	$\tan(270° + \theta°)$
15	$\cos(270° + \theta°)$	**16**	$\sin(270° - \theta°)$
17	$\sin(270° + \theta°)$	**18**	$\cos(270° - \theta°)$
19	$\tan(5 \times 90° + \theta°)$	**20**	$\sin(5 \times 90° - \theta°)$
21	$\cos(5 \times 90° - \theta°)$	**22**	$\tan(5 \times 90° - \theta°)$
23	$\cos(5 \times 90° + \theta°)$	**24**	$\sin(5 \times 90° + \theta°)$
25	$\tan(6 \times 90° - \theta°)$	**26**	$\cos(6 \times 90° + \theta°)$
27	$\sin(6 \times 90° - \theta°)$	**28**	$\tan(6 \times 90° + \theta°)$
29	$\cos(6 \times 90° - \theta°)$	**30**	$\sin(6 \times 90° + \theta°)$
31	$\cos(2k\,90° + \theta°)$	**32**	$\sin(2k\,90° + \theta°)$
33	$\sec(2k\,90° + \theta°)$	**34**	$\csc(2k\,90° + \theta°)$
35	$\cot(2k\,90° + \theta°)$	**36**	$\tan(2k\,90° + \theta°)$
37	$\sin[(2k+1)\,90° - \theta°]$	**38**	$\sin[(2k+1)\,90° + \theta°]$
39	$\cos[(2k+1)\,90° + \theta°]$	**40**	$\cos[(2k+1)\,90° - \theta°]$
41	$\tan[(2k+1)\,90° - \theta°]$	**42**	$\tan[(2k+1)\,90° + \theta°]$
43	$\cot[(2k+1)\,90° + \theta°]$	**44**	$\sec[(2k+1)\,90° + \theta°]$

12.9 › TRIGONOMETRIC TABLES

The values of the six trigonometric functions have been computed for each minute of angle from 0° to 90°, and in the next chapter we show how some of these values were computed. Tables are available in the Appendix, and there are much more extensive tables in handbooks, which also contain other mathematical material. See, for example, R. S. Burlington, "Handbook of Mathematical Tables and Formulas", 4th ed., McGraw-Hill, New York, 1965. Typical portions of these tables are produced on the following page. From it we read $\sin 32° = 0.52992$, $\cos 32°10' = 0.84650$, $\tan 32°16' = 0.63136$. For these we read the *left-hand* column of minutes and the *upper* column headings sin, cos, and tan. To get $\cot 57°5' = 0.64734$, we read the *right-hand* minute column and the *bottom* column heading cot.

Interpolation

Interpolation within any such table is carried out in the following manner: Let us suppose that we want to compute $\cos 32°51.4'$. We first write

$$1.0'\left[\begin{array}{l} 0.4'\left[\begin{array}{l}\cos 32°51' \;\;= 0.84009\\ \cos 32°51.4' = ?\end{array}\right]\Delta \\ \quad\;\; \cos 32°52' \;\;= 0.83994 \end{array}\right]0.00015$$

As the angle increases by 1′, the value of the cosine decreases by 0.00015. In linear interpolation we assume that, if the angle increases by 0.4′, then the value of the cosine decreases by a proportional amount; that is to say, the decrease is $(0.4)(0.00015) = 0.00006$. Formally we write

$$\frac{0.4}{1.0} = \frac{\Delta}{0.00015}$$

32°

′	Sin	Cos	Tan	Cot	′
0	.52992	.84805	.62487	1.6003	**60**
1	.53017	.84789	.62527	1.5993	59
2	.53041	.84774	.62568	1.5983	58
3	.53066	.84759	.62608	1.5972	57
4	.53091	.84743	.62649	1.5962	56
5	.53115	.84728	.62689	1.5952	**55**
6	.53140	.84712	.62730	1.5941	54
7	.53164	.84697	.62770	1.5931	53
8	.53189	.84681	.62811	1.5921	52
9	.53214	.84666	.62852	1.5911	51
10	.53238	.84650	.62892	1.5900	**50**
11	.53263	.84635	.62933	1.5890	49
12	.53288	.84619	.62973	1.5880	48
13	.53312	.84604	.63014	1.5869	47
14	.53337	.84588	.63055	1.5859	46
15	.53361	.84573	.63095	1.5849	**45**
16	.53386	.84557	.63136	1.5839	44
17	.53411	.84542	.63177	1.5829	43
18	.53435	.84526	.63217	1.5818	42
19	.53460	.84511	.63258	1.5808	41
20	.53484	.84495	.63299	1.5798	**40**
21	.53509	.84480	.63340	1.5788	39
22	.53534	.84464	.63380	1.5778	38
23	.53558	.84448	.63421	1.5768	37
24	.53583	.84433	.63462	1.5757	36
25	.53607	.84417	.63503	1.5747	**35**
26	.53632	.84402	.63544	1.5737	34
27	.53656	.84386	.63584	1.5727	33
28	.53681	.84370	.63625	1.5717	32
29	.53705	.84355	.63666	1.5707	31
30	.53730	.84339	.63707	1.5697	**30**
31	.53754	.84324	.63748	1.5687	29
32	.53779	.84308	.63789	1.5677	28
33	.53804	.84292	.63830	1.5667	27
34	.53828	.84277	.63871	1.5657	26
35	.53853	.84261	.63912	1.5647	**25**
36	.53877	.84245	.63953	1.5637	24
37	.53902	.84230	.63994	1.5627	23
38	.53926	.84214	.64035	1.5617	22
39	.53951	.84198	.64076	1.5607	21
40	.53975	.84182	.64117	1.5597	**20**
41	.54000	.84167	.64158	1.5587	19
42	.54024	.84151	.64199	1.5577	18
43	.54049	.84135	.64240	1.5567	17
44	.54073	.84120	.64281	1.5557	16
45	.54097	.84104	.64322	1.5547	**15**
46	.54122	.84088	.64363	1.5537	14
47	.54146	.84072	.64404	1.5527	13
48	.54171	.84057	.64446	1.5517	12
49	.54195	.84041	.64487	1.5507	11
50	.54220	.84025	.64528	1.5497	**10**
51	.54244	.84009	.64569	1.5487	9
52	.54269	.83994	.64610	1.5477	8
53	.54293	.83978	.64652	1.5468	7
54	.54317	.83962	.64693	1.5458	6
55	.54342	.83946	.64734	1.5448	**5**
56	.54366	.83930	.64775	1.5438	4
57	.54391	.83915	.64817	1.5428	3
58	.54415	.83899	.64858	1.5418	2
59	.54440	.83883	.64899	1.5408	1
60	.54464	.83867	.64941	1.5399	**0**
′	Cos	Sin	Cot	Tan	′

57°

from which it follows that $\Delta = 0.00006$. Therefore

$$\cos 32°51.4' = 0.84003$$

Inverse interpolation, where the value of the function is given and the angle is to be determined, is performed in essentially the same manner. For example, if we wish to find θ having been given that $\tan\theta = 0.63530$, we proceed as follows: From the table we find

$$1.0'\left[\Delta\left[\begin{matrix}\tan 32°25' = 0.63503 \\ \tan\theta° = 0.63530\end{matrix}\right]0.00027 \atop \tan 32°26' = 0.63544\right]0.00041$$

Hence

$$\frac{\Delta}{1.0} = \frac{0.00027}{0.00041}$$

$$\Delta = \tfrac{27}{41} \approx 0.658 \approx 0.7$$

Therefore $\theta° = 32°25.7'$.

PROBLEMS 12.9

In Probs. 1 to 40, angles are given. Use tables of the trigonometric functions of angles (Appendix Table V) to find

	sin		*cos*
1	72°32′	**11**	64°19′
2	60°19.6′	**12**	37°31.2′
3	38°14.1′	**13**	116°16.5′
4	110°15.2′	**14**	96°7.7′
5	162°05′	**15**	170°15.2′
6	201°10′	**16**	225°52.2′
7	−16°11′	**17**	−47°18′
8	−36°21′	**18**	−63°33.4′
9	840°	**19**	900°
10	−32°48′	**20**	−1°40.6′

	tan		*cot*
21	26°56′	**31**	109°16′
22	110°42.5′	**32**	50°22.2′
23	322°14.6′	**33**	97°7.7′
24	−13°37′	**34**	200°23.3′
25	−18°16.5′	**35**	−16°34.2′
26	150°6.7′	**36**	−50°16′
27	−86°12.6′	**37**	−12°9′
28	1,500°	**38**	30°5′
29	16°6.1′	**39**	385°33′
30	320°18′	**40**	−37°7.1′

In Probs. 41 to 56, values of the functions are given. Find the indicated angles to the nearest $\frac{1}{10}$ minute. Assume $0° < \theta° < 360°$.

41 $\sin\theta° = 0.64716$ 1st-quadrantal angle
42 $\sin\theta° = 0.94399$ 2d-quadrantal angle
43 $\sin\theta° = -0.34497$ 3d-quadrantal angle
44 $\sin\theta° = -0.73926$ 4th-quadrantal angle
45 $\cos\theta° = 0.69779$ 1st-quadrantal angle
46 $\cos\theta° = -0.12409$ 2d-quadrantal angle
47 $\cos\theta° = -0.32829$ 3d-quadrantal angle
48 $\cos\theta° = 0.63758$ 4th-quadrantal angle
49 $\tan\theta° = 0.37940$ 1st-quadrantal angle
50 $\tan\theta° = -1.5642$ 2d-quadrantal angle
51 $\tan\theta° = 0.38386$ 3d-quadrantal angle
52 $\tan\theta° = -7.3049$ 4th-quadrantal angle
53 $\cot\theta° = 0.93797$ 1st-quadrantal angle
54 $\cot\theta° = -0.85408$ 2d-quadrantal angle
55 $\cot\theta° = 11.888$ 3d-quadrantal angle
56 $\cot\theta° = -1.5809$ 4th-quadrantal angle

12.10 › RIGHT TRIANGLES

The simplest applications of trigonometry involve the solution of right triangles. We are given any two of the following "parts" (including at least one side): the angles A and B, and the sides a, b, and c. Then we are asked to find the remaining three parts.

For a right triangle the definitions of sine, cosine, and tangent given in Secs. 12.5 and 12.7 reduce to the following.

$$\text{The sine of an acute angle} = \frac{\text{opposite side}}{\text{hypotenuse}}$$

$$\text{The cosine of an acute angle} = \frac{\text{adjacent side}}{\text{hypotenuse}}$$

$$\text{The tangent of an acute angle} = \frac{\text{opposite side}}{\text{adjacent side}}$$

As an illustration of the application of these formulas, consider the following problem.

ILLUSTRATION 1 A gable roof has rafters that are 10 ft long. If the eaves are 17 ft apart, how much headroom is there in the center of the attic and how steep is the roof?

SOLUTION Directly from Fig. 12.17 we see that

$$\begin{aligned}\text{Headroom} = x &= \sqrt{100 - 72.25}\\ &= \sqrt{27.75}\\ &= 5.26 \text{ ft} \qquad \text{from a table of square roots}\end{aligned}$$

The steepness of the roof is measured by θ, where

$$\cos\theta = \frac{8.5}{10} = 0.85$$

and, from a table of natural cosines (degree measure),

$$\theta = 31°47.3' \qquad \text{to the nearest } \tfrac{1}{10} \text{ minute}$$

A second example is given in Illustration 2.

ILLUSTRATION 2 A radio tower stands on top of a building 200 ft high. At a point on the ground 500 ft from the base of the building the tower subtends an angle of 10°. How high is the tower?

SOLUTION From Fig. 12.18 we have

$$\begin{aligned}\tan\theta &= \tfrac{200}{500} = 0.40000\\ \theta &= 21°48.1' \qquad \text{to the nearest } \tfrac{1}{10} \text{ minute}\end{aligned}$$

Now $\qquad \tan 31°48.1' = \dfrac{200 + x}{500}$

$$0.62007 = \frac{200 + x}{500}$$

$$\begin{aligned}200 + x &= 310.04\\ x &= 110.04 \text{ ft}\end{aligned}$$

Figure for Illustration 1

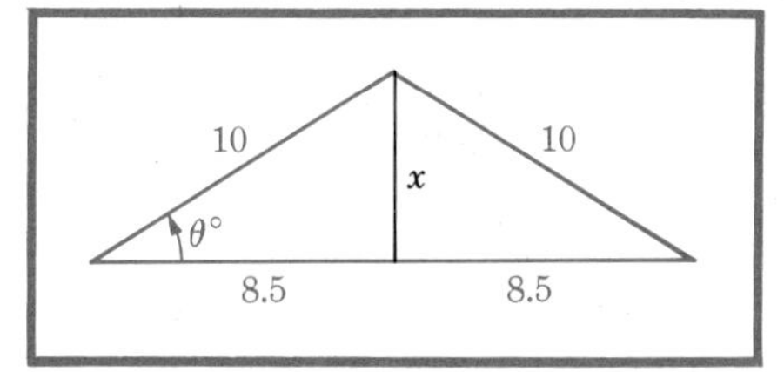

FIGURE 12.17

Figure for Illustration 2

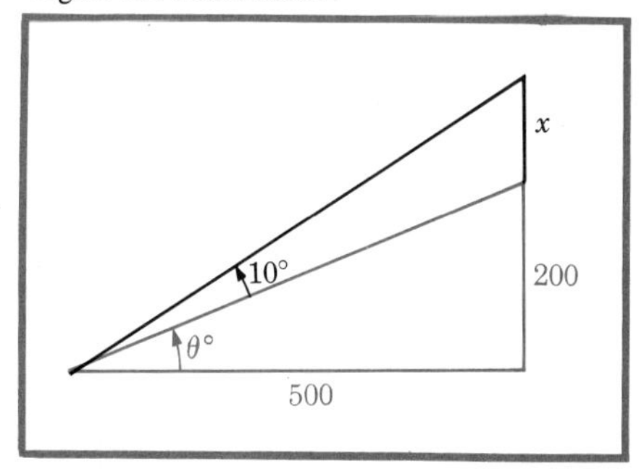

FIGURE 12.18

PROBLEMS 12.10 In Probs. 1 to 10 solve for the unknown parts in the right triangle with angles A, B, $C = 90°$ and opposite sides a, b, c, respectively.

1 $b = 12, c = 13$
2 $a = \sqrt{3}, c = 2$
3 $a = 1, B = 78°41.4'$
4 $a = 1, b = 1$
5 $c = 100, B = 40°$
6 $A = 25°22.6', a = 3$
7 $B = 65°46.9', b = 3$
8 $a = 2, B = 56°18.6'$
9 $b = 14.2, c = 23.7$
10 $a = 29.0, b = 17.4$

In Probs. 11 to 16 find the perimeters of the following regular polygons:

11 A hexagon inscribed in a circle of radius 5 in.
12 A hexagon circumscribed about a circle of radius 5 in.
13 An octagon inscribed in a circle of radius 5 in.
14 An octagon circumscribed about a circle of radius 5 in.
15 A decagon inscribed in a circle of radius 5 in.
16 A decagon circumscribed about a circle of radius 5 in.

17 An object 4 ft tall casts a 3-ft shadow when the angle of elevation of the sun is $\theta°$. Find θ.
18 What is the area of a regular pentagon **(a)** inscribed in a circle of radius 5 in.? **(b)** Circumscribed?
19 What angle does the diagonal of the face of a cube make with a diagonal of the cube (drawn from the same vertex)?
20 A 13-ft ladder, with its foot anchored in an alleyway, will reach 12 ft up a building on one side of the alley and 11 ft up a building on the other. How wide is the alley?
21 Discover a way of measuring the width of a stream by making all the measurements on one bank.
22 A wheel 3 ft in diameter is driven, by means of a noncrossed belt, by a wheel 1 ft in diameter. If the wheel centers are 8 ft apart how long is the belt?
23 Find the area of the traffic island shown in the figure; distances are in feet.

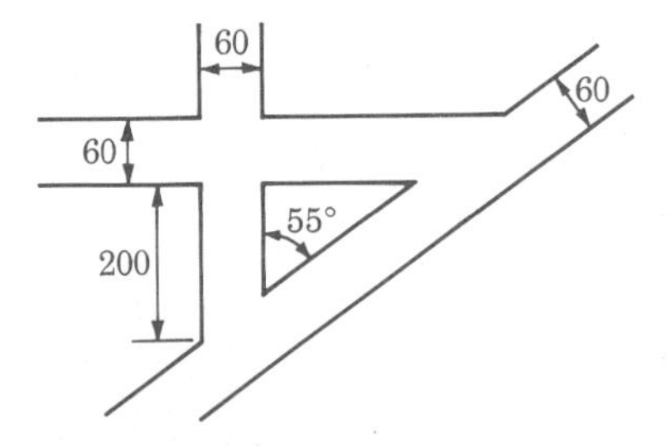

24 At a point P, 8 ft above the ground, the angle of elevation of the top of a lamppost is 60°, and the angle of depression of its base is $30°$. Find the height of the lamppost and the horizontal distance to P.

25 Show the area of a regular polygon of n sides, each of length a, is given by $A = \frac{na^2}{4} \cot \frac{180°}{n}$.

26 Prove that the area S of a right triangle with hypotenuse c is given by $S = \frac{1}{2}c^2 \sin A \cos A$.

12.11 › VECTORS

We have already defined vectors and discussed their algebra (Sec. 7.2). In this section we shall show how trigonometry helps us to use them in practical problems. By way of review, we remind you that a vector whose initial point is at the origin is graphed as a directed line segment $\overrightarrow{OP}$. The vector is determined by the coordinates (a,b) of the endpoint P, and we call a the x-component of $\overrightarrow{OP}$ and b its y-component. Indeed we previously identified $\overrightarrow{OP}$ with the pair (a,b) (Fig. 12.19).

We can, however, describe $\overrightarrow{OP}$ in another important way. We can give its *length* (or *magnitude*) $r = |\overrightarrow{OP}|$ and the angle θ which it makes with the positive X-axis. This pair of numbers (r,θ) are nothing but the polar coordinates of the endpoint P.

Our discussion in Sec. 12.5 now permits us to use these two aspects of a vector interchangeably. If we are given the magnitude r and the direction θ, we find:

The x-component a of the vector (r,θ) is $a = r\cos\theta$.
The y-component b of the vector (r,θ) is $b = r\sin\theta$.

Suppose, on the other hand, that we know the components (a,b) of a vector. Then:

The magnitude r of (a,b) is $r = \sqrt{a^2 + b^2}$.

Components of a vector

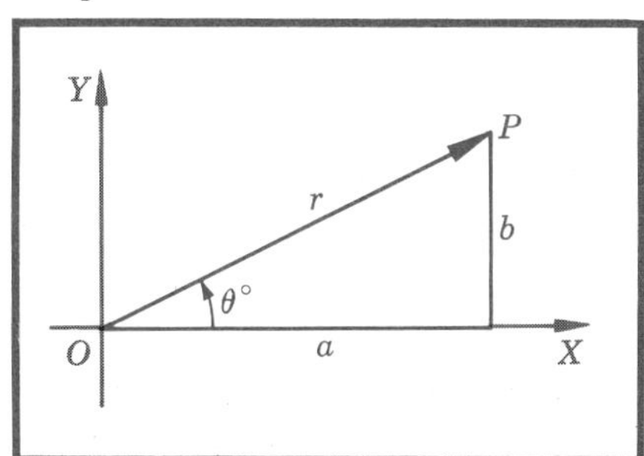

FIGURE 12.19

The vector $\overrightarrow{OP}$ has the horizontal component a and the vertical component b.

The direction θ of (a,b) is given by $\tan\theta = b/a$ (when proper attention is paid to the quadrant).

We can generalize the notion of the x- and y-components of a vector by considering its projections on any pair of perpendicular lines.

DEFINITION

Given a directed line l and a vector $\overrightarrow{AB}$, the *component* of $\overrightarrow{AB}$ on l is $CD = |\overrightarrow{AB}|\cos\theta$ where θ is the angle between the positive directions of $\overrightarrow{AB}$ and l (Fig. 12.20).

Components of a vector

Thus the vector $\overrightarrow{OA}$ (Fig. 12.21) is readily resolved into its components

$$OB = |\overrightarrow{OA}|\cos\alpha \qquad \text{and} \qquad OC = |\overrightarrow{OA}|\cos\beta = |\overrightarrow{OA}|\sin\alpha$$

Let us use these ideas to find the *sum* or *resultant* of two vectors. We recall that the sum of $\overrightarrow{OA_1}$ and $\overrightarrow{OA_2}$ is the vector $\overrightarrow{OB}$, where B completes the parallelogram determined by $\overrightarrow{OA_1}$ and $\overrightarrow{OA_2}$ (Fig. 12.22). We recall that in component language the sum is given by

Addition of vectors

$$(a,b) + (c,d) = (a + c, b + d)$$

Suppose, however, that our two vectors are given in the form (r_1,θ_1) and (r_2,θ_2) and that we wish to find their sum. This is usually the situation in problems in physics and mechanics. First we must compute their components along two perpendicular directions. For convenience we choose the X- and Y-axes. These are

$$\begin{aligned} a_1 &= r_1\cos\theta_1 \qquad & b_1 &= r_1\sin\theta_1 \\ a_2 &= r_2\cos\theta_2 \qquad & b_2 &= r_2\sin\theta_2 \end{aligned}$$

Component of $\overrightarrow{AB}$ in the direction of line l

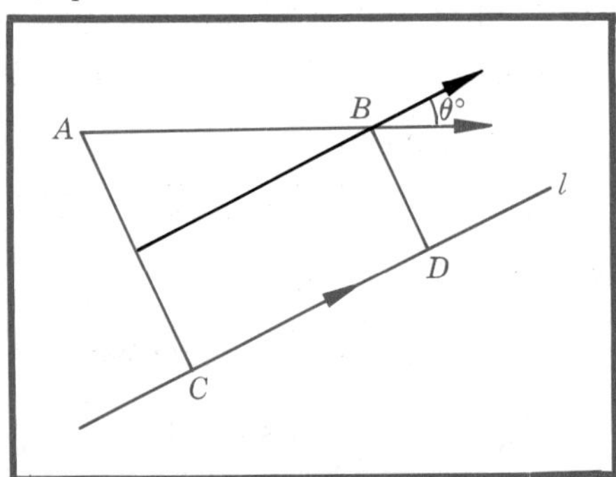

FIGURE 12.20
A vector has a component in any chosen direction.

Components of $\overrightarrow{OA}$

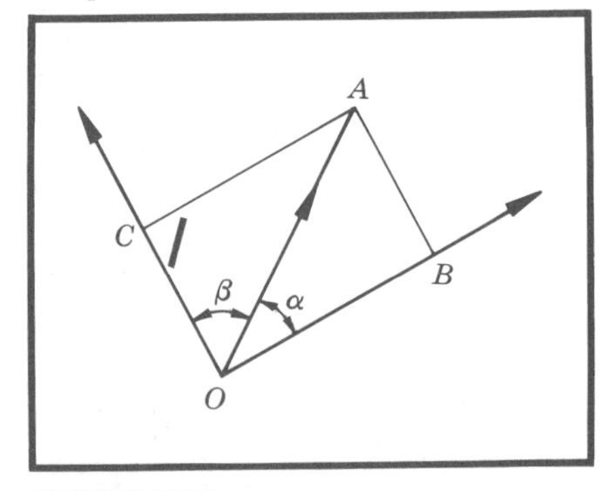

FIGURE 12.21
A vector can be resolved into components corresponding to any two perpendicular directions.

Sum of two vectors

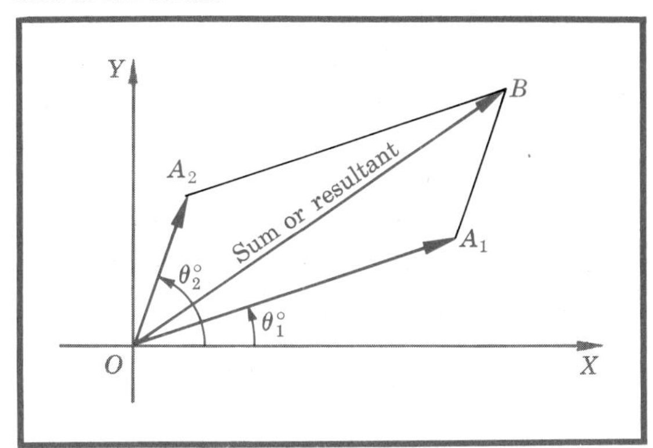

FIGURE 12.22
The sum of two vectors is the diagonal vector of the corresponding parallelogram.

Then the sum is the vector

$$(r_1 \cos \theta_1 + r_2 \cos \theta_2, r_1 \sin \theta_1 + r_2 \sin \theta_2)$$

From this we can find the magnitude and direction of the vector sum by the formulas above.

ILLUSTRATION 1 Two forces F_1 and F_2 of 20 and 10 lb, respectively, act on a body. If F_1 acts at $\theta_1 = 30°$ and F_2 at $\theta_2 = 60°$, find the resultant force F and its direction.

SOLUTION

(F_1,θ_1) has components: $a_1 = 20 \cos 30° = 10\sqrt{3} = 17.32$

$b_1 = 20 \sin 30° = \frac{20}{2} = 10$

(F_2,θ_2) has components: $a_2 = 10 \cos 60° = \frac{10}{2} = 5$

$b_2 = 10 \sin 60° = 5\sqrt{3} = 8.66$

Hence the resultant has components:

$$a_1 + a_2 = 22.32 \qquad b_1 + b_2 = 18.66$$

The magnitude of the resultant is

$$\sqrt{22.32^2 + 18.66^2} = \sqrt{846.4} = 29.1 \text{ lb}$$

The direction of the resultant is given by

$$\tan \theta = \frac{18.66}{22.32} = 0.8360$$

So $\theta = 39°54'$

The resultant is the vector (29.1, 39°54′).

This method applies equally to the sum of three or more vectors. It is particularly useful in analyzing statics problems where the body is at equilibrium. This implies that the vector sum of the forces involved is zero, and hence that the sum of the components in any direction is zero. Good applications of this are given in the next illustrations.

ILLUSTRATION 2 A weight of 10 lb is supported by a rod BP and a rope AP (Fig. 12.23). Find the tension in the rope.

SOLUTION The sum of the horizontal components is

$$-|\overrightarrow{AP}|\sin 60° + |\overrightarrow{BP}|\sin 30° = 0$$

The sum of the vertical components is

$$|\overrightarrow{AP}|\cos 60° + |\overrightarrow{BP}|\cos 30° - 10 = 0$$

We are asked to find $|\overrightarrow{AP}|$ and can do this by solving the above pair of simultaneous equations. This gives

$$|\overrightarrow{AP}| = \frac{10 \sin 30°}{\cos 60° \sin 30° + \sin 60° \cos 30°}$$
$$= 5 \text{ lb}$$

ILLUSTRATION 3 A block of wood weighing 5.0 lb rests on an inclined plane making 20° with the horizontal. Disregarding all forces except that of gravity, determine the force (a vector quantity) required to keep the block from moving (Fig. 12.24).

Figure for Illustration 2

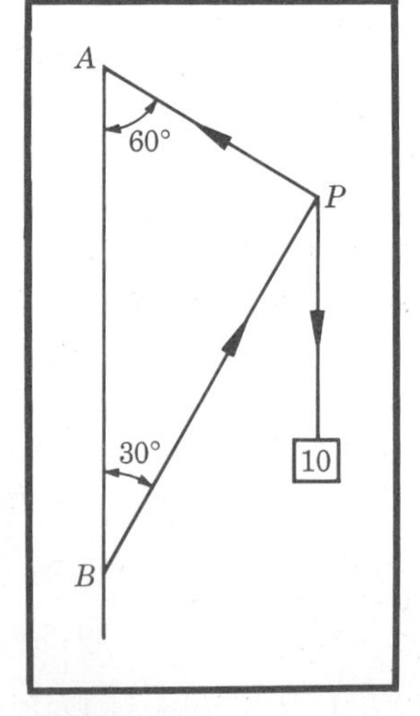

FIGURE 12.23

Figure for Illustration 3

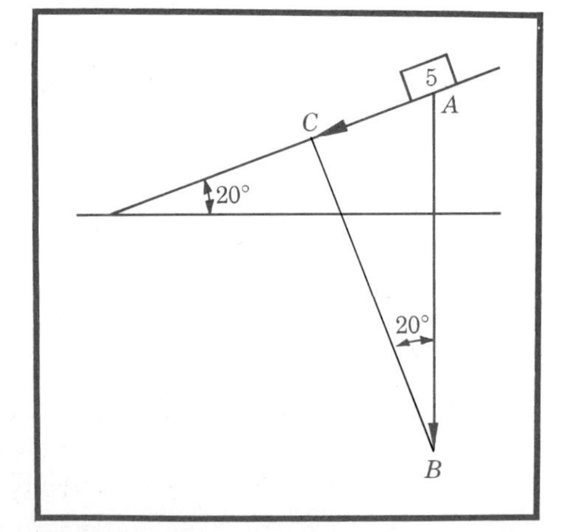

FIGURE 12.24

SOLUTION We draw the vector AB of length 5.0 and downward to represent the force (in pounds). The component of this force in the direction down the inclined plane is $AC = |AB| \sin 20°$. Now

$$\begin{aligned} AC &= 5.0 \sin 20° \\ &= (5.0)(0.34202) \\ &= 1.71010 \approx 1.7 \end{aligned}$$

The answer to the question is given by the vector CA. That is, a force of 1.7 lb acting *up* the plane is required to keep the body from moving.

EXERCISE A Find the component of the force in the upward direction perpendicular to the inclined plane.

PROBLEMS 12.11

1. Two forces of magnitude 6 and 10 act at right angles to each other. Find the resultant, describing the direction of the resultant with respect to the force of magnitude 6.
2. Find the resultant $R(L,\theta)$ of the two forces $F_1(25,40°)$ and $F_2(16,70°)$.
3. Find the resultant $R(L,\theta)$ of the two forces $F_1(14,0°)$ and $F_2(10,70°)$.
4. A stream one mile wide flows between parallel banks at 2 miles/hr. A swimmer, always swimming directly toward the opposite bank, swims $\frac{3}{2}$ miles to reach the opposite bank. What is his speed in still water?
5. An airplane points itself due north, and its northerly speed is 200 miles/hr. An east wind blows the plane west at 20 miles/hr. Find the velocity vector, i.e., |velocity| and direction, of the airplane.
6. A 2-ton truck rests on a 15° incline. What force is required of the brakes to hold the truck at rest?
7. A block weighing x lb is lifted vertically by forces $F_1(10,60°)$ and $F_2(y,120°)$. Find x and y.
8. A 19-lb block is to be lifted vertically by $F_1(x,30°)$ and $F_2(17,\theta)$. Find x and θ.
9. A 10-lb block of wood W (tied to A with a piece of string) rests on scale S. If the whole assembly is rotated 19°26.7′ counterclockwise about O, what will be the reading on the scale?

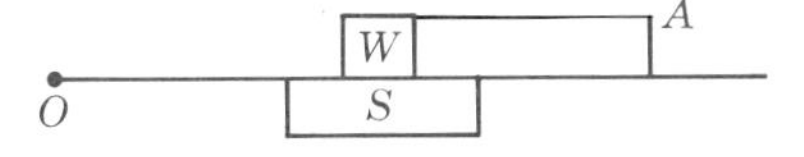

10. A force $F(60,70°)$ is the sum of the two forces $F_1(x,0°)$ and $F_2(y,120°)$. Find x and y.
11. A force of $(1,90°)$ is the sum of two forces, F_1 and F_2, each of magnitude 1 lb. Find the angle between F_1 and F_2.

12 A load of 50 lb is hung from the middle of a rope which is stretched between two walls 40 ft apart. Under the load the rope sags 4 ft in the middle. Find the tension in each half of the rope.

13 A weight W of 10 lb is supported by a rod BP and a rope AP as in the figure. Find the tension in the rope and the compression in the rod.

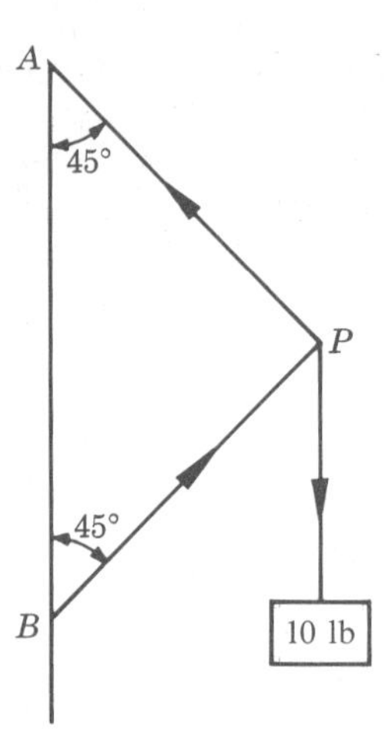

14 A weight of W lb is supported by a rod BP and a rope AP as in the figure. Find general expressions for the tension in the rope and the compression in the rod.

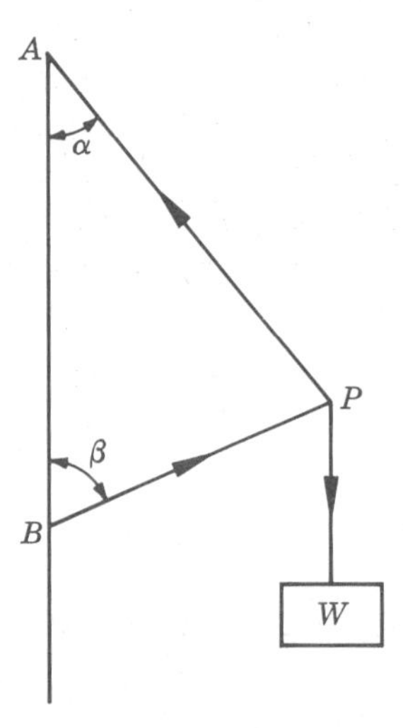

15 Three forces of 4, 16, and 20 lb, respectively, are in equilibrium. What angles do they make with one another?

16 A force $F(a,\theta)$ is the sum of the forces $F_1(30,60°)$, $F_2(60,120°)$, and $F_3(100,150°)$. Find a and θ.

17 Ropes extend from tractors up over a pulley at the top of a haymow and down to a load of hay. If one tractor pulls with a force of 75 lb at 34° with the vertical, and another pulls with a force of 100 lb at 26° with the vertical, what is the weight of the hay they lift?

18 A weight of 60 lb is suspended at the free end of a horizontal bar, which is pivoted at the left end and is held in position by a cord that makes an angle of 40° with the horizontal. Determine the tension in the cord (T lb) and the tension in the bar (P lb).

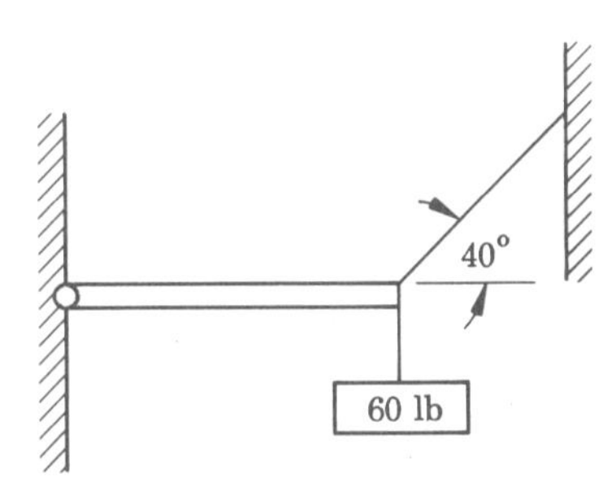

12.12 » LAW OF SINES

So far we have been working essentially with right triangles. In this section and in the next we wish to develop methods for dealing with any triangle. To "solve" a triangle means to find the angles and sides which are not given in the problem.

The solutions of general triangles may be obtained by using the Laws of Sines and Cosines. To derive the Law of Sines, examine Fig. 12.25, in which it is assumed that angle A is acute. We have $h/a = \sin B$ and $h/b = \sin A$ or $h = a \sin B = b \sin A$. Therefore

$$\frac{a}{\sin A} = \frac{b}{\sin B}$$

Similarly

$$a:b:c = \sin A:\sin B:\sin C$$

Law of Sines

FIGURE 12.25
Angle A is assumed to be acute.

This is known as the Law of Sines. The proof of this law when A is obtuse is posed in Prob. 12 of this section. It enables us to solve a triangle when given (1) two sides, one opposite angle, and (2) two angles, one side.

Ambiguous case

When two sides and the angle opposite one of these are given, we have a situation called the *ambiguous case.* Let a, b, and A be given. Then by the Law of Sines

$$\sin B = \frac{b \sin A}{a}$$

Three cases can occur, as follows:

1 $\sin B > 1$. This is impossible, and there is no solution. (See Illustration 1.)
2 $\sin B = 1$. Then $B = 90°$, the triangle is a right triangle, and there is one solution.
3 $\sin B < 1$. Then two values of B must be examined, namely: B_1 in the first quadrant, and B_2 in the second quadrant. B_1 always gives a solution. If $B_2 + A < 180°$, there is a second solution; but if $B_2 + A \geq 180°$, B_2 is impossible and there is only one solution. (See Illustration 2.)

ILLUSTRATION 1 Given $a = 2$, $b = 10$, $A = 30°$; find B.

SOLUTION

$$\sin B = \frac{10 \sin 30°}{2} = \frac{10(0.5000)}{2} = 2.5$$

Since $2.5 > 1$, there is no angle B for which $\sin B = 2.5$. Hence there is no solution.

ILLUSTRATION 2 Given $a = 7.1576$, $b = 8.4632$, $A = 52°14'$; find B.

SOLUTION The Law of Sines is to be used, but the computation is now more difficult. However, with a desk calculator at hand, it is possible for us to determine B very quickly. First, it will be instructive to compute h, the perpendicular distance from C to side c. This is

$$h = 8.4632 \sin 52°14' = (8.4632)(0.79051) = 6.6902$$

Since a is greater than h but less than b, it follows that side a could fall in either position CB or CB' (Fig. 12.26). There are, therefore, two solutions to the problem. Using the Law of Sines, we write

$$\frac{7.1576}{0.79051} = \frac{8.4632}{\sin B}$$

$$\sin B = \frac{(0.79051)(8.4632)}{7.1576}$$
$$= 0.93470$$

Hence $B = 69°10.8'$ and $B' = 110°49.2'$. Either triangle ABC or $AB'C$ will satisfy the given conditions.

If a desk calculator is not available, you can save time by using logarithms for computing sin B. (Review Chap. 11.)

$$\log \sin B = \log 0.79051 + \log 8.4632 - \log 7.1576$$
$$= 9.89791 - 10 + 0.92753 - 0.85477$$
$$= 9.97067 - 10$$

At this point we use a table of the *logarithms* of the trigonometric values and find $B = 69°10.8'$.

SPECIAL NOTE In this book we do not stress the use of logarithms for solving such problems as the above. Desk computers are almost as common as typewriters, and we assume that one of these will be available for your use.

When two angles and a side are given (or computed as in the case just discussed), we can find the third angle immediately by using the fact that the sum of the angles of a triangle is 180°. Then the remaining sides can be found by use of the Law of Sines.

ILLUSTRATION 3 Given $a = 10$, $A = 40°$, $B = 50°$; find b.

SOLUTION From the Law of Sines,

$$b = \frac{a \sin B}{\sin A}$$
$$= \frac{10(0.76604)}{0.64279}$$
$$= 11.917$$

Ambiguous case

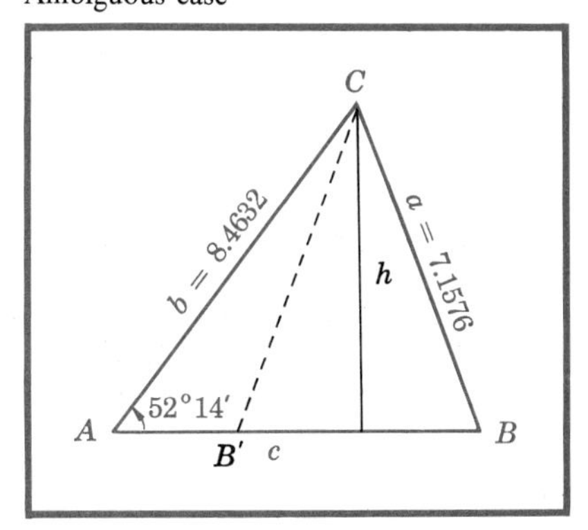

FIGURE 12.26 Given data admit two solutions.

Accuracy of computation

In carrying out such computations we must consider the question of how many significant figures we should keep in the answer. You should always round back to five figures if you are using five-place tables, for the tables place an automatic limit on the accuracy of your result. In addition, you must also consider the accuracy of the given data. When we said in Illustration 3 that $a = 10$, we did not specify the accuracy with which a is measured, and in our solution we assumed that a was exactly 10. If, however, we knew that a was measured only to the nearest integer, we should round the answer for b to the nearest integer, namely 12. In the problems below we assume that all given quantities are exact. In practice, however, you will know how exact your data are and must report your answers to a degree of accuracy compatible with that of your measurements.

PROBLEMS 12.12 Solve the triangle, given the following:

1 $A = 70°, B = 30°, a = 20$
2 $C = 112°, B = 39°, a = 8$
8 $c = 132, a = 190.40, A = 131°26.1'$
4 $B = 45°, C = 105°, c = 3.47$
5 $A = 9°23.7', B = 31°24', b = 15$
6 $B = 13°2', a = 16.013, C = 16°58'$
7 $a = 3, A = 25°, c = 7.5$
8 $C = 55°, c = 8.1, b = 10$
9 $c = 11.750, C = 34°15.3', a = 15.610$
10 $B = 17°16', b = 19.47, c = 28.34$

11 Prove that the area K of triangle ABC is given by the formula $K = \frac{1}{2}bc \sin A$.

12 Develop completely the Law of Sines $a:b:c = \sin A:\sin B:\sin C$, in the case where angle A is obtuse. HINT: $\sin A = \sin (180° - A)$.

13 The magnitudes of two forces are 3 and 4, and their resultant makes an angle of 30° with the first. Describe the system.

14 An observation balloon B and two points A and C on the ground are in the same vertical plane. From A the angle of elevation of the balloon is 70°, and from C 35°. If the balloon is between the two points A and C, what is its elevation if A and C are 1,000 ft apart?

15 If s represents half the perimeter of a triangle and R the radius of the circumscribed circle (see the following figure), show that:

a $R(\sin A + \sin B + \sin C) = s$

b $2R = \dfrac{a}{\sin A} = \dfrac{b}{\sin B} = \dfrac{c}{\sin C}$

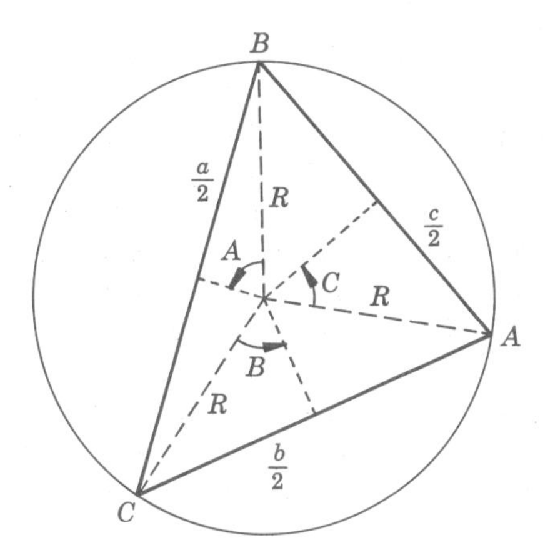

16 Two vectors make an angle of 60° with each other; their resultant is $R(10,40°)$. The vectors issue from the origin, and one makes an angle of 5° with the horizontal. Describe the system.

12.13 › LAW OF COSINES

The Law of Sines is not directly applicable in case three sides of a triangle are given or in case two sides and the included angle are given. To develop a usable formula for these two cases, we proceed as follows (Fig. 12.27): Place the triangle on the axes so that A is at the origin and side c lies along the positive X-axis. Then B has the coordinates $(c,0)$ and C the coordinates $(b \cos A, b \sin A)$. For this construction, A may be acute or obtuse.

The distance BC is equal to a and is given by the distance formula

$$\begin{aligned} a^2 &= (b \cos A - c)^2 + (b \sin A - 0)^2 \\ &= b^2 \cos^2 A - 2bc \cos A + c^2 + b^2 \sin^2 A \\ &= b^2 + c^2 - 2bc \cos A \end{aligned}$$

Law of Cosines

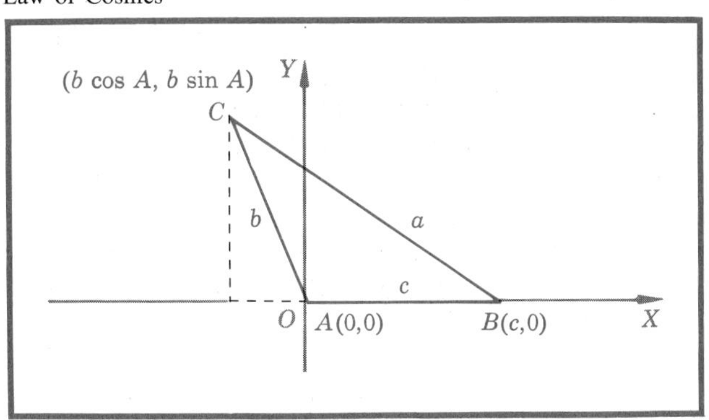

FIGURE 12.27
The angle A may be either acute or obtuse.

Using other letters, this formula may be written

$$b^2 = a^2 + c^2 - 2ac \cos B$$
$$c^2 = a^2 + b^2 - 2ab \cos C$$

Any of these formulas is called the Law of Cosines, and by means of them a triangle can be solved, given (1) three sides and (2) two sides and the included angle.

ILLUSTRATION 1 Given $C = 100°$, $a = 15$, $b = 20$; find c.

SOLUTION Substituting in the third form of the Law of Cosines, we get

$$c^2 = (15)^2 + (20)^2 - 2(15)(20)(-0.17365)$$

recalling that $\cos 100° = -\cos 80° = -0.17365$. Hence

$$\begin{aligned} c^2 &= 729.19 \\ c &= \sqrt{729.19} \\ &= 27.0037 \\ &\approx 27 \end{aligned}$$

EXERCISE A Find A and B in the above illustration.

ILLUSTRATION 2 Given $C = 100°$, $a = 15.277$, $b = 20.593$; find c.

SOLUTION As in Illustration 2 in the preceding section, the problem is now one of more difficult computations. However, with a desk calculator, we compute

$$\begin{aligned} c^2 &= (15.277)^2 + (20.593)^2 - 2(15.277)(20.593)(-0.17365) \\ &= 767.37 \\ c &= 27.701 \approx 27.7 \end{aligned}$$

Note that the Law of Cosines is not well adapted to logarithmic treatment.

Inner product of vectors

We are now in a position to explain the geometric interpretation of the inner product of two vectors (Sec. 7.3). We recall that if $\mathbf{A} = (a_1,a_2)$ and $\mathbf{B} = (b_1,b_2)$, then $\mathbf{A} \cdot \mathbf{B} = a_1b_1 + a_2b_2$.

Construct the triangle OAB as in Fig. 12.28. Then, by the Law of Cosines,

$$(AB)^2 = (OA)^2 + (OB)^2 - 2(OA) \times (OB) \times \cos \theta$$

Using the formula of Sec. 12.2 for the length of a segment, we can write this as

$$(a_1 - b_1)^2 + (a_2 - b_2)^2 = (a_1^2 + a_2^2) + (b_1^2 + b_2^2) - 2\sqrt{a_1^2 + a_2^2}\sqrt{b_1^2 + b_2^2}\cos \theta$$

Interpretation of the inner product

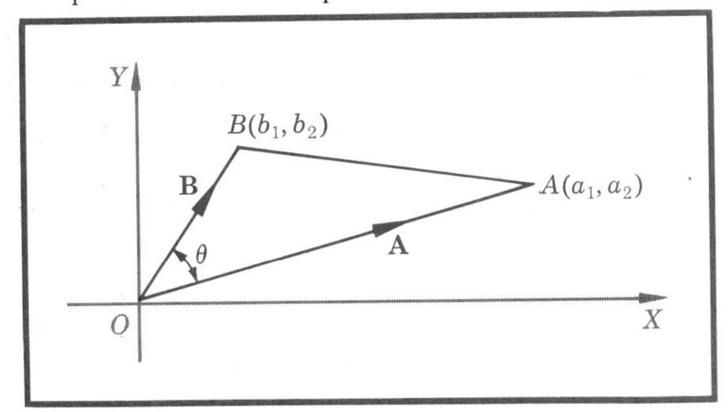

FIGURE 12.28
$\mathbf{A}\cdot\mathbf{B} = |\mathbf{A}|\,|\mathbf{B}| \cos\theta$

or
$$-2(a_1b_1 + a_2b_2) = -2\sqrt{a_1^2 + a_2^2}\sqrt{b_1^2 + b_2^2}\cos\theta$$

Since $|\mathbf{A}| = \sqrt{a_1^2 + a_2^2}$ and $|\mathbf{B}| = \sqrt{b_1^2 + b_2^2}$, this becomes

$$\mathbf{A}\cdot\mathbf{B} = |\mathbf{A}|\,|\mathbf{B}|\cos\theta$$

In words, *the inner product of two vectors is the product of their lengths times the cosine of their included angle.*

EXERCISE B Show that $\mathbf{A}\cdot\mathbf{B} = 0$ if $\mathbf{A}$ and $\mathbf{B}$ are at right angles.

It is now also possible to develop the geometric interpretation of the outer (or vector) product of two vectors in a plane (Sec. 7.8). Let $\mathbf{A} = (a_1,a_2,0)$ and $\mathbf{B} = (b_1,b_2,0)$. Then from its definition

$$\mathbf{A}\wedge\mathbf{B} = \left(0,0,\begin{vmatrix} a_1 & a_2 \\ b_1 & b_2 \end{vmatrix}\right)$$

Then $\mathbf{A}\wedge\mathbf{B}$ is perpendicular to the plane of $\mathbf{A}$ and $\mathbf{B}$. Our problem is to find its length.

First, from the formula above, the length of $\mathbf{A}\wedge\mathbf{B}$ is the absolute value of

$$\begin{vmatrix} a_1 & a_2 \\ b_1 & b_2 \end{vmatrix}$$

Second, consider the triangle in the plane with vertices $O(0,0)$, $A(a_1,a_2)$, and $B(b_1,b_2)$. Its area is given by the formula (Prob. 32, Sec. 7.7)

$$\text{Area} = \frac{1}{2}\begin{vmatrix} 0 & 0 & 1 \\ a_1 & a_2 & 1 \\ b_1 & b_2 & 1 \end{vmatrix} = \frac{1}{2}\begin{vmatrix} a_1 & a_2 \\ b_1 & b_2 \end{vmatrix}$$

Another formula for the area, obtained from Prob. 11, Sec. 12.12, is

$$\text{Area} = \tfrac{1}{2}\sqrt{a_1^2 + a_2^2}\sqrt{b_1^2 + b_2^2}\sin\theta$$

where θ is the angle between OA and OB.

From these two formulas for the area, we find that

$$\begin{aligned}\text{Length of } \mathbf{A} \wedge \mathbf{B} &= \text{absolute value} \begin{vmatrix} a_1 & a_2 \\ b_1 & b_2 \end{vmatrix} \\ &= 2 \times (\text{area of triangle } OAB) \\ &= \sqrt{a_1{}^2 + a_2{}^2}\sqrt{b_1{}^2 + b_2{}^2} \sin\theta\end{aligned}$$

Hence

$$\text{Length of } \mathbf{A} \wedge \mathbf{B} = (\text{length of } \mathbf{A}) \times (\text{length of } \mathbf{B}) \times \sin\theta$$

PROBLEMS 12.13 In Probs. 1 to 12 solve for:

1 Angle C, given $a = 5, b = 7, c = 10$
2 Angle B, given $a = 7, b = 8, c = 11$
3 Angle A, given $a = 5, b = 4, c = 3$
4 Angle B, given $a = 6, b = 5, c = 9$
5 Angle C, given $a = 14.753, b = 19.912, c = 24.384$
6 Angle C, given $a = 9.9171, b = 14.848, c = 16.814$
7 Side c, given $a = 12, b = 14, C = 92°$
8 Side a, given $b = 21.463, c = 6.9897, C = 53°7.4'$
9 Side b, given $a = 9.4671, c = 17.986, B = 86°15'$
10 Side a, given $c = 22.997, b = 6.4374, C = 154°27'$
11 Side a, given $A = 14°, b = 4.9173, c = 6.2057$
12 Side a, given $A = 13°, b = 5.3437, c = 7.9218$

13 The diagonals of a parallelogram are 12 and 13 ft, respectively, and they form an angle of 34°. Find the lengths of the sides a and b.
14 Find a formula for the area of a triangle, given a, b, and C.
15 Show that the perimeter of a regular polygon having n sides is $2nr \sin(180°/n)$, where r is the radius of its circumscribed circle.
16 The magnitudes of two forces and their resultant are 4, 8, and 11, respectively. Describe the system.
17 The magnitudes of two forces and their resultants are 7, 8, and 9, respectively. Describe the system.
18 From an airport at noon one airplane flies northwest at 200 miles/hr and another due south at 350 miles/hr. How far apart are they at 2 P.M.?
19 City X lies 200 miles due west of New York. In a flight from X to New York, a plane, always pointing east, is blown off course by a 30 mile/hr northwest wind. When due south of New York, how far away from New York is the plane if, in still air, it flies at 150 miles/hr?
20 Derive the Law of Cosines from the Law of Sines.
21 Derive the Law of Sines from the Law of Cosines.

12.14 › LAW OF TANGENTS

The Law of Cosines and a desk computer offer the best methods for solving triangles where three sides or two sides and the included angle are given. To solve these two cases with logarithms, it is best to use the Law of Tangents. Even though we do not recommend the latter procedure, it is still a good exercise in trigonometry to develop the Law of Tangents. This is done in the next chapter (Prob. 30, Sec. 13.7).

13 TRIGONOMETRIC FUNCTIONS OF REAL NUMBERS

13.1 » ARC LENGTH

We are now ready to generalize the concepts of the trigonometric functions. In Chap. 12 they were functions of an angle measured in degrees. Here we shall define them as functions of a real number so that sin x, say, has the same domain of definition as e^x, $2x^2 - 4$, etc.

Here, as indeed in the preceding chapter where we defined degree measure, we shall need the concept of the *length of an arc* of a circle, but this concept, as yet undefined, is a very profound one and is not to be passed over lightly. Length measured along a straight line is one thing; but what could be the meaning of length measured along a circle or some other curve? If all of the known history of mathematics is any indication of the truth, Archimedes (287–212 B.C.) was the only one of the ancients who had any clear notion of how to treat arc length, and after his death the subject languished for almost two thousand years.

Circumference of a circle

To find a real number C which he called the length of the circumference of a circle, Archimedes made use of inscribed regular polygons (Fig. 13.1). In his Book I (on the sphere and the cylinder) he assumed (Axiom 1) that a line segment joining two points is shorter than any curve or polygonal line joining these same points. It follows that the perimeter p_n of an inscribed regular polygon of n sides is less than C. Furthermore, if $n < m$, then $p_n < p_m < C$; i.e., as n increases, the length of p_n increases but still stays less than C. When n is very large, the polygon is almost indistinguishable from the circle (at least to the naked eye), and so it is reasonable to define C as follows (see the first few sections of Chap. 15).

DEFINITION The length of the circumference C is the limiting value of the perimeter of an inscribed regular polygon of n sides as the number of sides n is increased indefinitely.

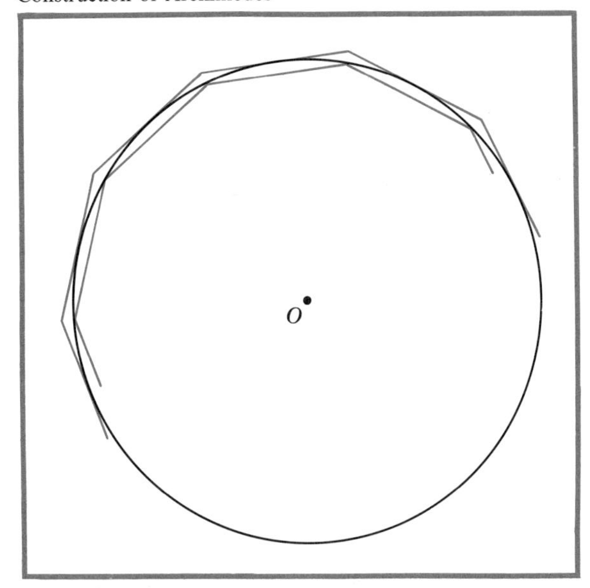

FIGURE 13.1
Approximations to the length of a circle can be obtained by measuring the lengths of inscribed and circumscribed regular polygons.

Archimedes also considered circumscribed regular polygons and assumed (Axiom 2) that the perimeter P_n of a regular polygon of n sides is greater than C. Hence he concluded that $p_n < C < P_n$, that as n increases, P_n decreases but stays greater than C, and finally that $P_n - p_n$ approaches zero as n becomes very large. Thus it follows that C, as defined above, is also the limiting value of the perimeter of a circumscribed regular polygon of n sides as the number of sides increases indefinitely.

As a matter of fact it is not necessary to use regular polygons. It can be proved that the length of the circumference is the limiting value of the perimeter of an inscribed (or circumscribed) polygon of n sides as the number of sides n is increased indefinitely and the length of the longest side is decreased indefinitely.

EXERCISE A Draw a figure which demonstrates that $p_n < p_{2n}$, $n = 4$.

EXERCISE B Draw a figure which demonstrates that $P_{2n} < P_n$, $n = 4$.

EXERCISE C Devise a definition for the length of an arc which is less than a complete circumference.

EXERCISE D Use your definition in Exercise C to show that, in Fig. 13.2,

$$\text{Segment } AB < \text{arc } AB < \text{segment } AC + \text{segment } BC$$

where AC and BC are tangent to the circle at A and B, respectively.

Arc with inscribed and circumscribed polygons

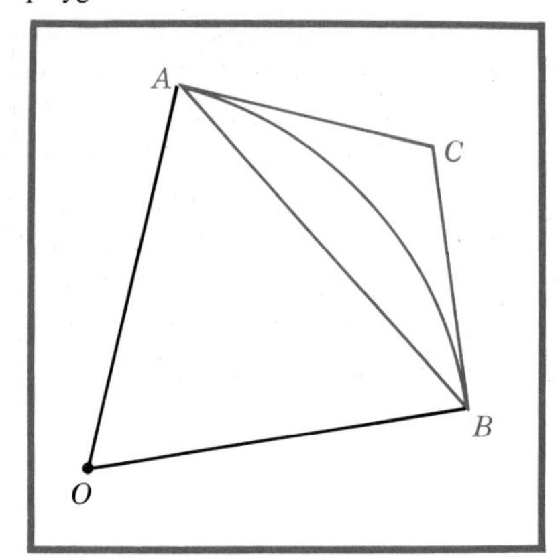

FIGURE 13.2
A polygon inscribed in an arc is shorter than the arc, and a polygon circumscribed about an arc is longer than the arc.

Definition of π

For a circle of radius 1 it is customary to write, exactly, $C = 2\pi$, where, correct to five decimals, $\pi = 3.14159$. This is in fact the definition of π.

DEFINITION π is one-half the length of the circumference of a circle of radius 1.

From this definition one proves by methods of plane geometry that the length of the circumference of a circle of radius r is equal to $2\pi r$.

You may be interested in the calculations in the following table which enable us to find an approximate value for π.

AN APPROXIMATION FOR π

Number of sides of the regular polygon, n	*Perimeter of the inscribed regular polygon in circle of radius 1, p_n*	*Length of circumference of circle of radius 1, C*	*Perimeter of the circumscribed regular polygon about circle of radius 1, P_n*
6	(2)(3.000000)	$< C <$	(2)(3.411017−)
12	(2)(3.105828+)	$< C <$	(2)(3.215391−)
24	(2)(3.132628+)	$< C <$	(2)(3.159660−)
48	(2)(3.139350+)	$< C <$	(2)(3.146087−)
96	(2)(3.141031+)	$< C <$	(2)(3.142715−)
192	(2)(3.141452+)	$< C <$	(2)(3.141874−)
384	(2)(3.141557+)	$< C <$	(2)(3.141664−)
768	(2)(3.141583+)	$< C <$	(2)(3.141611−)
1,536	(2)(3.141590+)	$< C <$	(2)(3.141598−)
3,072	(2)(3.141592+)	$< C <$	(2)(3.141594−)
6,144	(2)(3.141592+)	$< C <$	(2)(3.141593−)

13.2 › NEW DEFINITIONS OF THE TRIGONOMETRIC FUNCTIONS

Consider now a unit circle with center placed at the origin of a rectangular coordinate system (Fig. 13.3). We have already seen that the equation of such a circle is $x^2 + y^2 = 1$, where, of course, the coordinates (x,y) of a point on the circle satisfy the above equation.

Next we lay off on the circle an arc of (positive) length θ beginning at the point (1,0) and running counterclockwise. An arc running clockwise will be called an arc of negative length. The set of all arcs is in one-to-one correspondence with the set of all real numbers. Figure 13.3*e* and *f* shows arcs that lap over more than one circumference. Note that associated with each real number θ there is a unique, ordered pair (x,y) which are the coordinates of the endpoint of the arc θ whose initial point is (1,0).

Various directed arcs

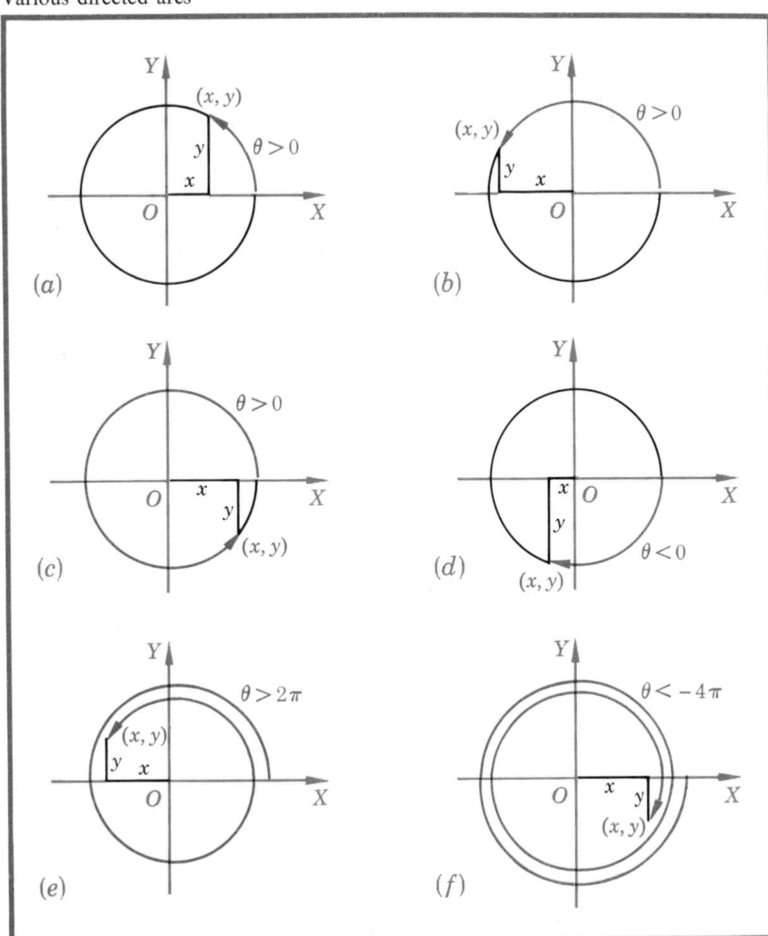

FIGURE 13.3
If θ is the signed length of a directed arc on the unit circle with initial point (1,0) and endpoint (x,y), we define $\sin \theta = y$ and $\cos \theta = x$.

EXERCISE A Is there a unique arc beginning at (1,0) the coordinates of whose endpoint are a given pair (x,y)?

DEFINITIONS

Sine and cosine of a real number

We define x and y, respectively to be $\cos\theta$ and $\sin\theta$ and write

$$x = \cos\theta \qquad y = \sin\theta$$

These are to be read "x is the cosine of the real number θ" and "y is the sine of the real number θ". The sine and cosine are, therefore, functions of the real number θ. The other four trigonometric functions (tan, cot, sec, and csc) of a real number θ are defined in terms of $\sin\theta$ and $\cos\theta$ by the formulas given in Sec. 12.7. Consequently all the identities and formulas developed in Chap. 12 are valid also in the present context.

In order to explain the relationship of these new definitions to those of Chap. 12, where θ was interpreted as an angle measured in degrees, we must first consider how we have measured arcs on the unit circle. There are, indeed, two measures of an arc: (1) its length θ, as defined in Sec. 13.1, (2) its measure in degrees, $\theta°$, as defined in Sec. 12.3. Since an arc whose length is 2π has a degree measure of 360°, these two scales of measurement are related by the linear transformation (Sec. 2.12)

$$\theta° = \frac{360}{2\pi}\theta$$

Some important special cases of this relationship are given in the following table.

SCALES FOR MEASURING ARCS ON THE UNIT CIRCLE

Arc	*Measure in degrees, $\theta°$*	*Length θ*
	0	0
	30	$\pi/6$
	45	$\pi/4$
	60	$\pi/3$
Quarter circle	90	$\pi/2$
Semicircle	180	π
Three-quarter circle	270	$3\pi/2$
Full circle	360	2π

We have also seen (Sec. 12.3) that there is a one-to-one correspondence between angles in standard position and the arcs which they intercept on the unit circle. Indeed, we have used this correspondence to define the *measure of an angle as the measure of the intercepted arc.* In Chap. 12 we confined ourselves to degree measure for both arcs and angles. Here, however, we are concerned with the measure of arcs in terms of length, but so far have not introduced the corresponding measure for angles. It is called *radian measure* and is defined as follows.

Radian measure

DEFINITION The *measure of an angle in radians,* $\theta^{(r)}$, is the length of the arc which this angle intercepts on the unit circle.

Thus an angle which intercepts an arc of length θ has a measure of θ radians. There are, therefore, two scales for the measurement of angles, degrees and radians. The relationship between these is the same as that between the two scales for the measurement of arcs, and is therefore given by the formula

$$\theta^\circ = \frac{360}{2\pi}\theta^{(r)}$$

Now consider an angle whose radian measure is θ. Using the definition of Chap. 12, the sine of this angle is y/r, or y when $r = 1$. From the definition above, the sine of the real number θ is also equal to y. So we have the relation:

The sine of an angle whose radian measure is θ is the sine of the real number θ.

Similar remarks hold true for the other trigonometric functions.

Now let an angle of radian measure θ intercept an arc of length s on a circle of radius r. Then from plane geometry

$$\frac{s}{\theta} = \frac{r}{1}$$

or $$s = r\theta \qquad \text{where } \theta \text{ is in radians}$$

Arc length in a circle of radius r

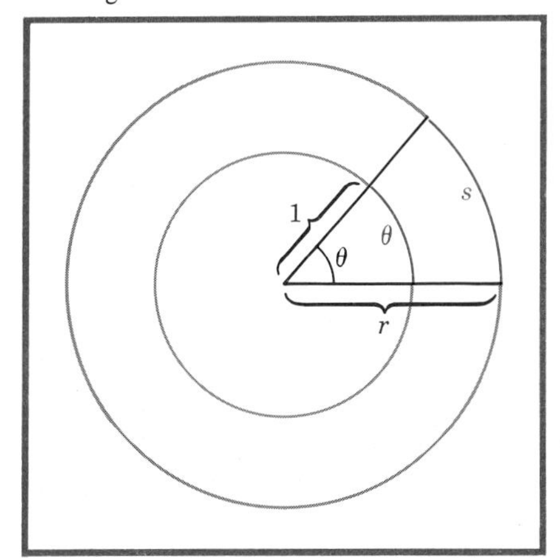

FIGURE 13.4 If an angle of radian measure θ intercepts an arc of length s on a circle of radius r, then $s = r\theta$.

That is, *an angle whose radian measure is θ intercepts an arc of length $r\theta$ on a circle of radius r.*

EXERCISE B Distinguish between sin 2 and sin 2°.

13.3 » COMPUTATIONS

In order to calculate the values of the trigonometric functions of a real number, we can refer to either Table IV or Table V in the Appendix.

ILLUSTRATION 1 Find tan 1.5.

SOLUTION Since $1.5^{(r)} = (180°)(1.5)/\pi = 85°56.6'$, therefore by Table V,

$$\tan 1.5 = \tan 1.5^{(r)} = \tan 85°56.6' = 14.101$$

ILLUSTRATION 2 Find tan 1.5.

SOLUTION From Table IV,

$$\tan 1.5 = 14.101$$

You may be interested in knowing how such a table is prepared. How do we compute $\sin \frac{1}{2}$, $\cos(\pi/7)$, etc., anyway? We cannot give you the details at this time, but at least we can say a few words. By methods of the calculus it can be shown that $\sin x$, where x is any real number, is given by the following "infinite series":

$$\sin x = x - \frac{x^3}{3!} + \frac{x^5}{5!} - \frac{x^7}{7!} + \cdots + (-1)^{n+1}\frac{x^{2n-1}}{(2n-1)!} + \cdots$$

and similarly

$$\cos x = 1 - \frac{x^2}{2!} + \frac{x^4}{4!} - \frac{x^6}{6!} + \cdots + (-1)^{n+1}\frac{x^{2n-2}}{(2n-2)!} + \cdots$$

Approximations for sine and cosine

These hold for every real number x. A table of values of sine and cosine *correct to five decimal places* can be prepared from them by considering the approximations

$$\sin x \approx x - \frac{x^3}{3!} + \frac{x^5}{5!} - \frac{x^7}{7!} \tag{1}$$

$$\cos x \approx 1 - \frac{x^2}{2!} + \frac{x^4}{4!} - \frac{x^6}{6!} \tag{2}$$

This follows from a general theorem which says, for such series, that the absolute value of the error made in taking a finite number of terms as an approximation does not exceed the absolute value of the *first* term omitted. In the case of $\cos x$ this term is $x^8/8!$ and $8! = 40{,}320$. The largest value of x we need to consider is $\pi/4 = 0.785398$. Since

$(0.785398)^8/40{,}320 = 0.0000036$, we see that the error will not affect the fifth decimal.

EXERCISE A Explain why we need only consider $0 \leq x \leq \pi/4$ in order to prepare a complete table for $\sin x$ and $\cos x$.

PROBLEMS 13.3

In Probs. 1 to 8 find the measures in degrees of the angles whose radian measures are:

1 $\frac{\pi}{2}, \frac{\pi}{6}, \frac{\pi}{3}, \frac{\pi}{4}$

2 $\pi, \frac{\pi}{4}, \frac{3\pi}{4}, \frac{5\pi}{4}, \frac{7\pi}{4}$

3 $\frac{\pi}{6}, \frac{4\pi}{6}, \frac{7\pi}{6}, \frac{10\pi}{6}, \frac{11\pi}{6}$

4 $\frac{2\pi}{9}, \frac{5\pi}{9}, \frac{7\pi}{9}, \frac{11\pi}{9}$

5 $-\frac{\pi}{4}, -\frac{\pi}{2}, -\frac{3\pi}{4}, -\pi$

6 $\frac{\pi}{10}, \frac{\pi}{12}, \frac{\pi}{15}, \frac{\pi}{14}, 1$

7 $16\pi, 17\pi, \frac{1}{2}, \frac{1}{3}$

8 $-\frac{\pi}{5}, -\frac{\pi}{7}, -\frac{\pi}{9}, -\frac{\pi}{8}$

In Probs. 9 to 12 find the measures in radians of the angles whose degree measures are:

9 $0°, 30°, 45°, 60°, 90°$

10 $180°, 210°, 225°, 240°$

11 $100°, 217°, 17°, 120°, 12°$

12 $-330°, -21°, -200°, -3°$

For each real number θ, in Probs. 13 to 21, find the value (if it exists) of each of the six trigonometric functions without tables.

13 $\theta = 0$

14 $\theta = \frac{\pi}{2}$

15 $\theta = \frac{\pi}{3}$

16 $\theta = \frac{3\pi}{2}$

17 $\theta = \frac{4\pi}{3}$

18 $\theta = -6\pi$

19 $\theta = 5\pi$

20 $\theta = 37\pi$

21 $\theta = 2n\pi$

In Probs. 22 and 23 draw a figure indicating approximately every arc θ (where $0 \leq \theta \leq 2\pi$) for which:

22 $\tan \theta = 3$

23 $|\tan \theta| = \frac{2}{3}$ and $\cos \theta$ is negative

24 Prove that $-\sin \theta = \sin(-\theta)$, for all θ.

25 Prove that $\cos \theta = \cos(-\theta)$, for all θ.

26 Find a counterexample for the following false statement: For all θ, $\cos 2\theta = 2 \cos \theta$.

27 Find a counterexample for the following false statement: For all ϕ and θ, $\sin(\theta + \phi) = \sin \theta + \sin \phi$.

28 Draw an appropriate figure and prove that, for all θ, $\cos(\theta + 2\pi) = \cos\theta$ and $\cos(\theta + \pi) = -\cos\theta$.

29 Draw an appropriate figure and prove that, for all θ, $\sin(\theta + 2\pi) = \sin\theta$ and $\sin(\theta + \pi) = -\sin\theta$.

30 By making use of the equation of the unit circle, find all θ for which $\tan\theta = \cot\theta$.

31 Find all θ for which $\sin\theta = \cos\theta$.

32 Compute sin 0.3 and cos 0.3 by using the appropriate approximation.

In Probs. 33 to 36 find $\sin\theta$ from a table.

33 $\theta = 1.61$ **34** $\theta = 0.31$
35 $\theta = 0.67$ **36** $\theta = 0.52 + 3\pi$

In Probs. 37 to 40 find $\cos\theta$ from a table.

37 $\theta = 1.38$ **38** $\theta = 0.12$
39 $\theta = 0.76$ **40** $\theta = 1.5 + \pi$

In Probs. 41 to 44 find $\tan\theta$ from a table.

41 $\theta = -1.60$ **42** $\theta = 1.60 + 3\pi$
43 $\theta = 1.60 - \pi$ **44** $\theta = 1.60 - 5\pi$

13.4 › VARIATION AND GRAPHS OF THE FUNCTIONS

As the arc θ increases from 0 to 2π, the trigonometric functions vary. It is relatively simple to see how each varies; we discuss only the cases of sin, cos, and tan. Draw the unit circle, and consider several arcs such that $0 < \theta_1 < \theta_2 < \cdots < \pi/2$ (Fig. 13.5). Remember that for a given

Variation of $\sin\theta$, $\cos\theta$, and $\tan\theta$

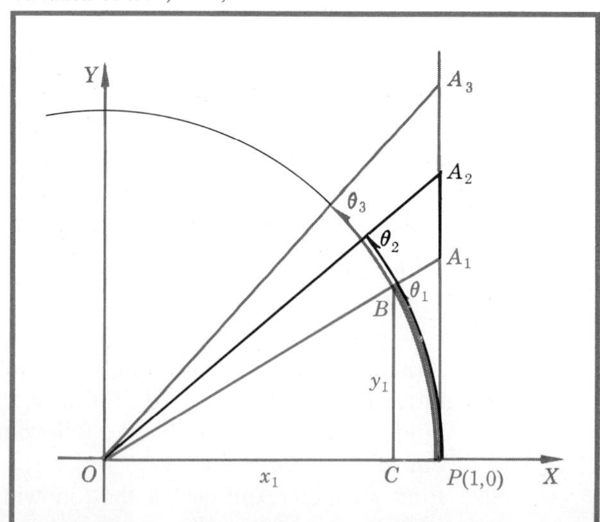

FIGURE 13.5
On a unit circle,
$\sin\theta = BC$,
$\cos\theta = OC$,
$\tan\theta = AP$.

θ, which is a real number, the abscissa x is $\cos\theta$ and the ordinate y is $\sin\theta$. Now erect a line tangent to the circle at $P(1,0)$; also draw the several lines from the origin to the endpoints of $\theta_1, \theta_2, \theta_3, \ldots$, extending these to intersect the tangent line at $A_1, A_2, A_3, \ldots$, respectively. It is seen that the length of the segment of the tangent PA_1 is actually equal to $\tan\theta_1$, by proportional parts of the triangles OPA_1 and OCB. Indeed this is the source of the name "tangent" of θ. By observing the variation of the length of PA as θ varies we can obtain the behavior of $\tan\theta$ as θ varies. From Fig. 13.5 and similar ones for the other quadrants we can read off the variations of $\sin\theta$, $\cos\theta$, and $\tan\theta$. The results are tabulated below.

Quadrant	*As* θ *varies from*	$\sin\theta$ *varies from*	$\cos\theta$ *varies from*	$\tan\theta$ *varies from*
I	0 to $\pi/2$	0 to 1	1 to 0	0 to ∞
II	$\pi/2$ to π	1 to 0	0 to -1	$-\infty$ to 0
III	π to $3\pi/2$	0 to -1	-1 to 0	0 to ∞
IV	$3\pi/2$ to 2π	-1 to 0	0 to 1	$-\infty$ to 0

The entries $\pm\infty$ under $\tan\theta$ need further explanation. It is quite evident that, as θ $(0 < \theta < \pi/2)$ gets nearer and nearer to $\pi/2$, $\tan\theta$ gets larger and larger. When the arc is exactly $\pi/2$, the value of the tan ceases to exist. We indicate this here by writing $\tan(\pi/2) = \infty$. However, in the second quadrant, the tan is negative; hence the entry of $-\infty$ on the second line. Similarly for III and IV quadrant entries.

EXERCISE A Draw a figure similar to Fig. 13.5 but for arcs θ where

$$\frac{\pi}{2} < \theta < \pi$$

Moreover, the variations are "essentially" the same for a given function, quadrant by quadrant: a function may increase or decrease, be positive or turn negative, but the image is the same if sign is disregarded. This will become clearer as we begin to graph the functions. We are already in a position to make up the following detailed table for the first quadrant. Check the following entries.

θ	$\sin\theta$	$\cos\theta$	$\tan\theta$
0	0	1	0
$\pi/6$	$\frac{1}{2} = 0.500$	0.866	$\frac{1}{3}\sqrt{3} = 0.577$
$\pi/4$	$\frac{1}{2}\sqrt{2} = 0.707$	0.707	1
$\pi/3$	$\frac{1}{2}\sqrt{3} = 0.866$	0.5	$\sqrt{3} = 1.732$
$\pi/2$	1	0	∞

For your own understanding you should extend this table through the other three quadrants. Figure 13.6 is helpful. With the aid of this information we sketch the graphs as in Figs. 13.7, 13.8, and 13.9, which also include the graphs of the sec, csc, and cot. To plot a more accurate graph, we could obtain other elements from a table of these functions. Note that sin, cos, sec, csc repeat after 2π but that tan and cot repeat after π, that is, sin, cos, sec, csc are periodic functions of period 2π and tan and cot are periodic functions of period π according to the following definitions.

DEFINITIONS
Periodic functions

A nonconstant function f such that $f(x + a) = f(x)$ for some positive a and all x is said to be a *periodic function*. The least positive a for which this is true is called the *period of the function*.

EXERCISE B Prove that no periodic function can be a rational function. HINT: Use the theorem that if a polynomial of degree n has more than n zeros then the polynomial is identically zero (see Prob. 33, Sec. 10.6).

EXERCISE C Prove that $f(x + 2a) = f(x)$ if f is a periodic function of period a.

Positions of important arcs

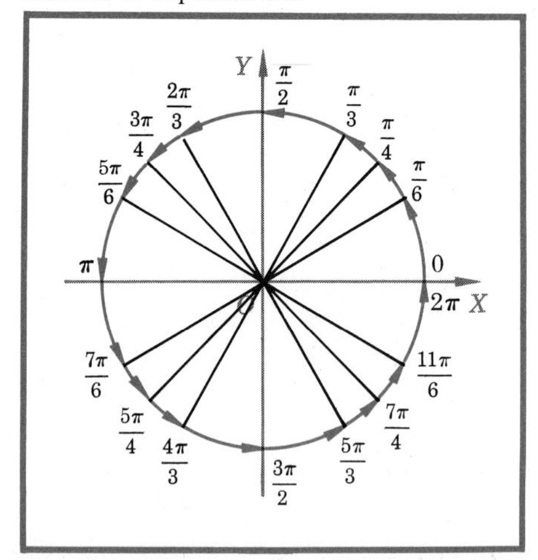

FIGURE 13.6
Observe the behavior of $\sin\theta$, $\cos\theta$, and $\tan\theta$ as θ increases from 0 to 2π.

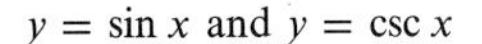

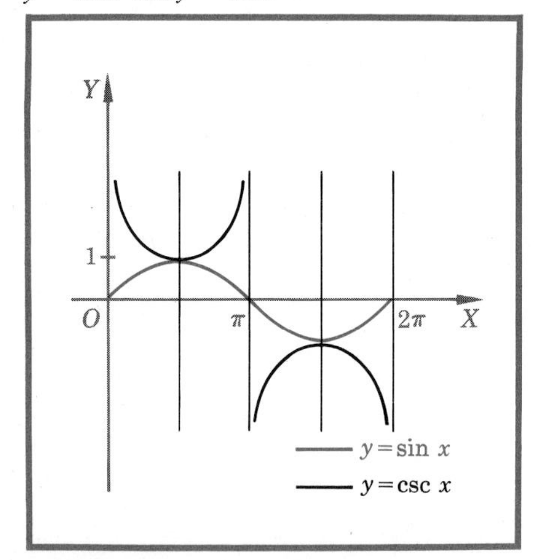

FIGURE 13.7
The graph of $\sin x$ is derived from Figs. 13.5 and 13.6 as the arc increases from 0 to 2π. The length of the arc is written "x" instead of "θ". Remember that $\csc x = 1/\sin x$.

$y = \cos x$ and $y = \sec x$

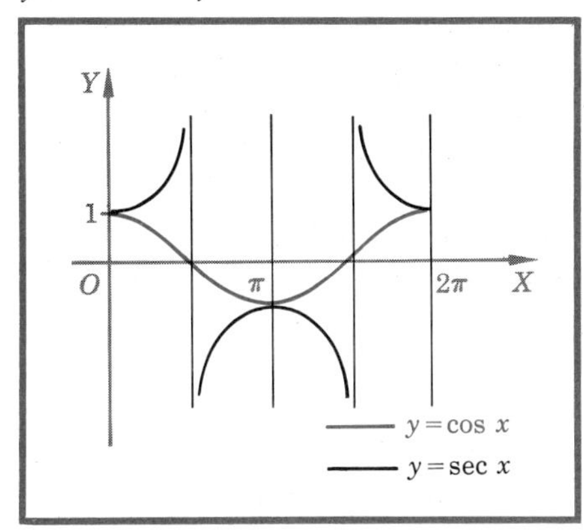

FIGURE 13.8
The graph of $\cos x$ is derived from Figs. 13.5 and 13.6. Remember that $\sec x = 1/\cos x$.

$y = \tan x$ and $y = \cot x$

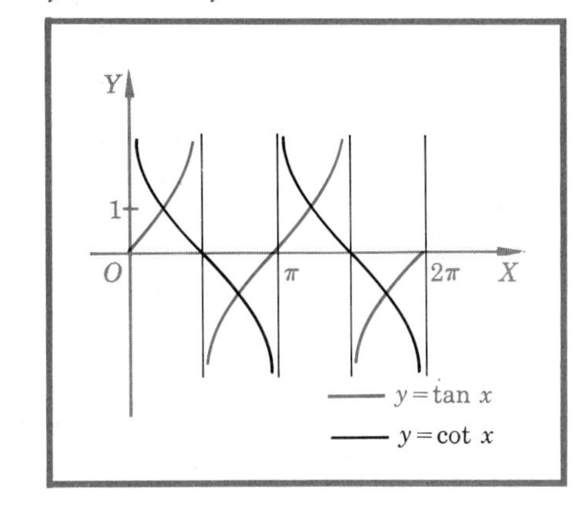

FIGURE 13.9
The graph of $\tan x$ is derived from Figs. 13.5 and 13.6. Remember that $\cot x = 1/\tan x$.

ILLUSTRATION 1 Sketch on the same axes and to the same scale the graphs of $y = \sin x$ and $Y = 3 \sin \frac{1}{2}x$, $0 \leq x \leq 4\pi$.

SOLUTION We compute the following entries, treating $\sin x$ and $3 \sin \frac{1}{2}x$ separately and making use only of the special values we know about. (For a more accurate graph, we should make use of a table.)

x	$\frac{1}{2}x$	$y = \sin x$	$\sin \frac{1}{2}x$	$Y = 3 \sin \frac{1}{2}x$
0	0	0.000	0.000	0.000
$\pi/6$		0.500		
$\pi/4$		0.707		
$\pi/3$	$\pi/6$	0.866	0.500	1.500
$\pi/2$	$\pi/4$	1.000	0.707	2.121
$2\pi/3$	$\pi/3$	0.866	0.866	2.598
$3\pi/4$		0.707		
$5\pi/6$		0.500		
π	$\pi/2$	0.000	1.000	3.000

Now the graph of $y = \sin x$ is "essentially" the same in the second, third, and fourth quadrant as it is in the first quadrant. By this we mean that it is the same shape but placed differently. Also $\sin x$ is periodic of period 2π, while $\sin \frac{1}{2}x$ is periodic of period 4π. With this information we sketch the graphs on page 368. (The whole of a given curve is sketched

$y = 3 \sin \frac{1}{2} x$ and $y = \sin x$

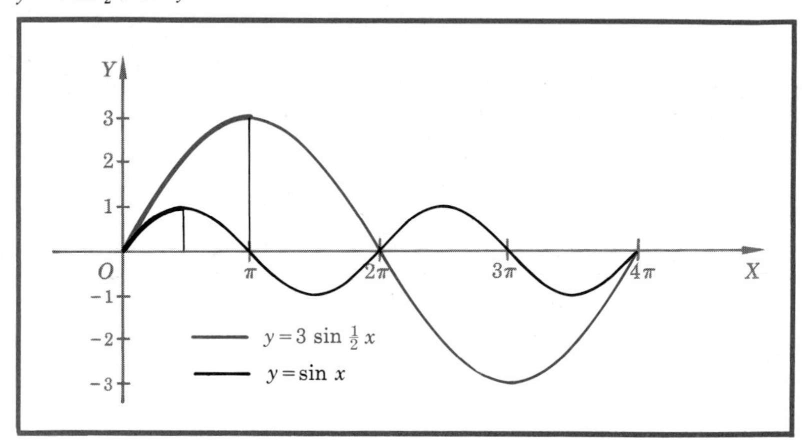

FIGURE 13.10
The graph of $y = A \sin Bx$ is closely related to that of $y = \sin x$.

by means of a template with which the heavy portion was drawn; Fig. 13.10.)

ILLUSTRATION 2 Sketch the graph of $y = \sin x + \frac{1}{3} \sin 3x$.

SOLUTION We should now know enough to sketch

$$Y_1 = \sin x \qquad \text{and} \qquad Y_2 = \tfrac{1}{3} \sin 3x$$

as is shown in Fig. 13.11. Then we sketch $y = Y_1 + Y_2$ by adding the ordinates of the two curves.

ILLUSTRATION 3 Sketch the graph of the function f defined by $y = e^{-x} \sin x$.

$y = \sin x + \frac{1}{3} \sin 3x$

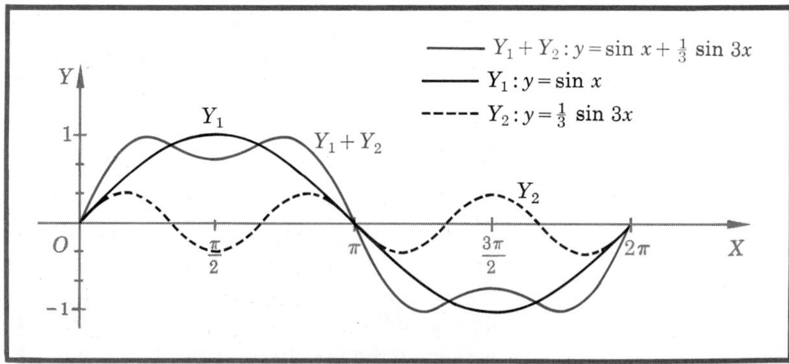

FIGURE 13.11
Graphs of sums of trigonometric functions may be obtained by plotting the graphs of summands and then adding ordinates graphically.

$y = e^{-x} \sin x$

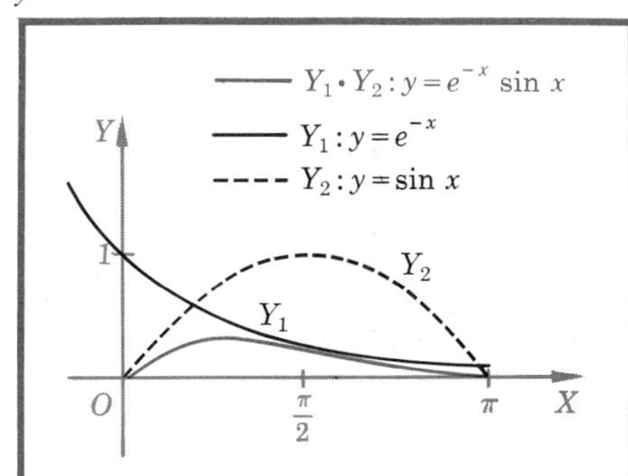

FIGURE 13.12
Graphs of products may be obtained by plotting the graphs of the factors and then multiplying ordinates.

SOLUTION The sketching is best done by considering separately the graphs of $Y_1 = e^{-x}$, $Y_2 = \sin x$, $y = Y_1 \cdot Y_2$. The product $Y_1 \cdot Y_2$ can be estimated from the graphs of $Y_1 = e^{-x}$ and $Y_2 = \sin x$ (Fig. 13.12).

The graph of $y = e^{-x} \sin x$ is called an *exponentially damped sine wave* and is important in the theory of electricity.

PROBLEMS 13.4

1 Figures 13.3 and 13.5 give the so-called line values of the three functions: sin, cos, tan. Prepare a similar figure for sec, csc, and cot.

In Probs. 2 to 13 sketch on the same axes and, to the same scale, the following pairs of graphs (complete period of each).

2 $y = \sin x, y = \sin 2x$
3 $y = \sin x, y = \frac{1}{2} \sin 2x$
4 $y = \sin x, y = 2 \sin x$
5 $y = \sin x, y = 2 \sin \frac{1}{2}x$
6 $y = \cos x, y = \cos 2x$
7 $y = \cos x, y = \frac{1}{2} \cos 2x$
8 $y = \cos x, y = 2 \cos x$
9 $y = \cos x, y = 2 \cos \frac{1}{2}x$
10 $y = \sin x, y = \sin 3x$
11 $y = \cos x, y = \cos 3x$
12 $y = \sin x, y = \sin\left(x + \frac{\pi}{2}\right)$
13 $y = \cos x, y = \frac{1}{2} \cos\left(x - \frac{\pi}{4}\right)$

In Probs. 14 to 22 sketch a complete period.

14 $y = \sin x + \cos x$. (Consider this as a sum $f + g$, plot the graphs of f and g separately, and then add all coordinates.)
15 $y = \sin x - \cos x$
16 $y = \sin x - \frac{1}{3} \sin 3x$
17 $y = \cos x + \frac{1}{3} \cos 3x$
18 $y = 2 \sin x - \cos x$
19 $y = 2 \cos x - \sin x$
20 $y = \cos x + \frac{1}{2} \cos 2x$
21 $y = \sin^2 x + \cos^2 x$
22 $y = \sin x + \frac{1}{2} \sin 2x$

23 What is the image of sec, quadrant by quadrant?
24 What is the image of csc, quadrant by quadrant?
25 What is the image of cot, quadrant by quadrant?

13.5 › AMPLITUDE, PERIOD, PHASE

One of the most important trigonometric concepts is that of a "sinusoidal wave". It occurs in innumerable ways and places in astronomy, mathematics, and all the sciences including the social sciences. It is, simply, the graph of $y = A \sin (Bx + C)$, where A, B, and C are constants.

We begin with a comparison of the graphs of

$$y = \sin x \qquad y = A \sin x$$

These are exhibited superimposed on the same axes and drawn to the same scales, in Fig. 13.13.

Since the maximum value of $\sin x$ is 1 and occurs when

$$x = \frac{\pi}{2} + 2k\pi$$

Amplitude

Period

it is evident that $|A|$ is the maximum value of $A \sin x$. The constant $|A|$ is called the *amplitude* of the sine wave (sinusoidal wave). The *period* p of $y = \sin x$ (and of $y = A \sin x$) is $p = 2\pi$.

Next we compare

$$y = \sin x \qquad y = \sin Bx$$

Now when $Bx = 0$, $x = 0$, and when $Bx = 2\pi$, $x = 2\pi/B$. Therefore the period p of $y = \sin Bx$ is $p = 2\pi/B$ (Fig. 13.14). Combining these two ideas, we have $y = A \sin Bx$ (Fig. 13.15).

Frequency

The frequency of an oscillation is the number of cycles which take place in some standard interval of time, usually one second. Frequency

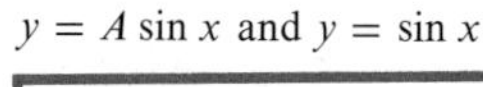

$y = A \sin x$ and $y = \sin x$

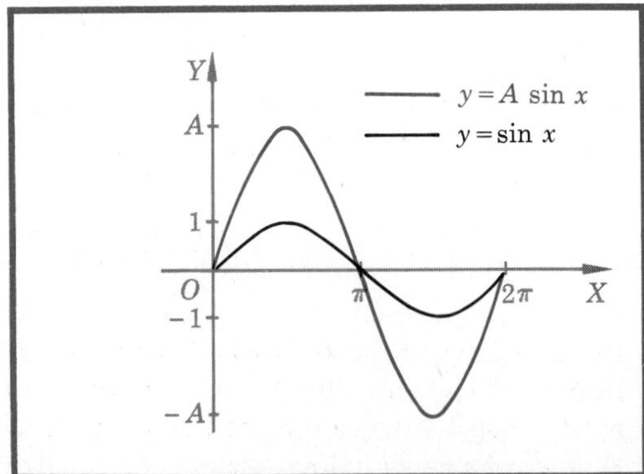

FIGURE 13.13
In $y = A \sin x$, $|A|$ represents the amplitude of the curve.

$y = \sin Bx$ and $y = \sin x$

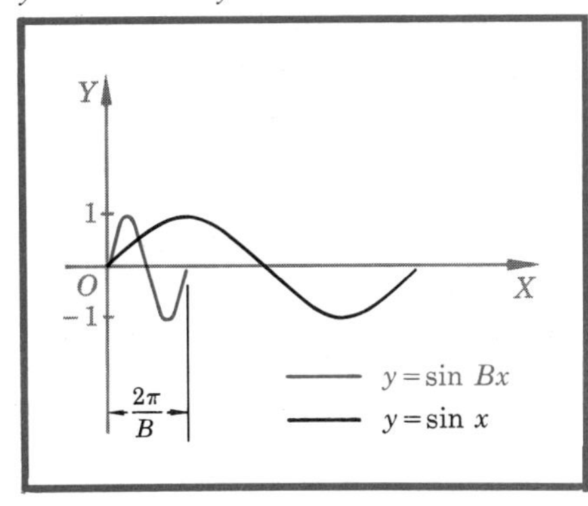

FIGURE 13.14
In $y = \sin Bx$, the period is $p = 2\pi/B$, and the frequency is $\omega = B/2\pi$.

$y = A \sin Bx$ and $y = \sin x$

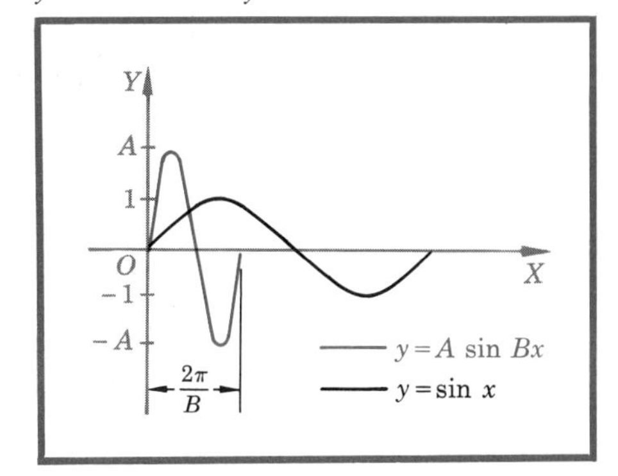

FIGURE 13.15
In $y = A \sin Bx$, both amplitude and period must be discussed.

is measured in cycles per second. A period is the length of a cycle. In radio broadcasting the convenient unit is kilocycles (thousand cycles) per second. These are the numbers on your radio dial. The usual house current is "60-cycle" current, and this means a frequency of 60 cycles/sec.

If an oscillation has a period of p sec, it has a frequency of

$$\omega = \frac{1}{p} \qquad \text{cycles/sec}$$

The frequency of $A \sin Bx$ is $\omega = B/2\pi$, and the frequency of $A \sin (2\pi\omega t)$ is ω cycles/sec. In radio waves the period is usually expressed in terms of the distance which the wave travels in p sec; this distance is called the *wavelength* λ. Since the velocity of a radio wave is 3×10^{10} cm/sec, the wavelength corresponding to one period of p sec is $3p \times 10^{10}$ cm. Thus, in terms of frequency,

$$\lambda = \frac{3 \times 10^{10}}{\omega} \text{cm}$$

Using this formula, one can convert from frequencies to wavelengths and back again. For instance, a broadcast signal with a frequency of 600 kc/sec has a wavelength

$$\lambda = \frac{3 \times 10^{10}}{600 \times 10^{3}} \text{cm} = 500 \text{ meters}$$

As you well know, radio waves are used to transmit the sounds of human voices and musical instruments. To keep things simple, let us restrict ourselves to a pure musical tone. Such pure tones are represented by a sine curve in which the amplitude corresponds to the loudness and

$y = A \sin 2\pi\omega t$

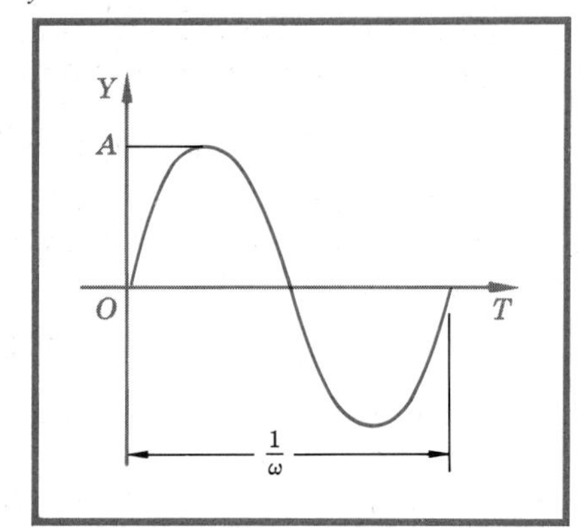

FIGURE 13.16
The graph of a pure tone of frequency ω is a sine wave with period $1/\omega$. Here $y = A \sin 2\pi\omega t$.

the frequency ω to the pitch (Fig. 13.16). For example, the frequency of middle C is 256 cycles/sec. As we have seen, the equation of such a sine curve is

(1) $$y = A \sin 2\pi\omega t$$

The radio station transmits a *carrier wave* which is also a sine curve whose frequency is that assigned to the particular station. These frequencies are much higher than those of sound waves, for example, 800,000 cycles/sec for standard broadcasts and 80,000,000 cycles/sec for the sound portion of a television signal. Let us write the equation of this carrier wave as

(2) $$y = A_0 \sin 2\pi\omega_0 t$$

where $|A_0|$ is its amplitude, and ω_0 its frequency.

Amplitude modulation

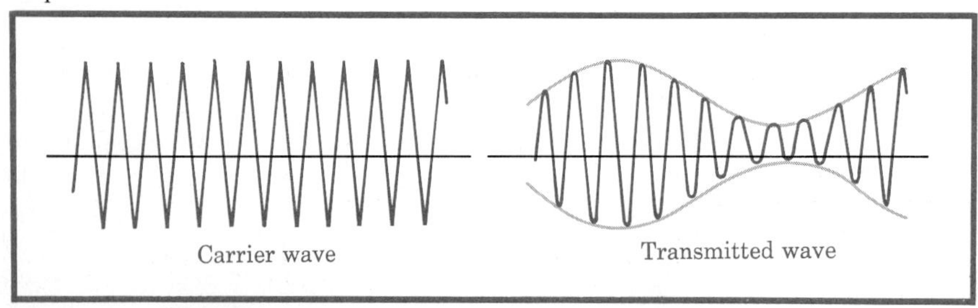

FIGURE 13.17
A carrier wave is a sine wave emitted by the sending station. In amplitude modulation the signal is transmitted by modifying the amplitude of the carrier wave.

The carrier wave can now be used to transmit the pure tone of (1) above by imposing on it a periodic change in its amplitude $|A_0|$. Instead of holding A_0 constant, the station *modulates* A_0 so that it has the value

$$A_0(t) = A_0 + mA_0 \sin 2\pi\omega t$$

where ω is the frequency of the pure tone in Eq. (2), and m is a factor of proportionality called the *degree of modulation*. The transmitted wave then has the equation

$$\textbf{(3)} \qquad y = A_0(1 + m \sin 2\pi\omega t) \sin 2\pi\omega_0 t$$

AM radio

whose graph is sketched in Fig. 13.17. The receiver then converts this signal back into the original pure tone, which is sent out through the loudspeaker. This process is called *amplitude modulation,* or AM.

Alternatively, the frequency of the carrier wave can be modulated in conformity with the tone to be transmitted. In order to do this, a band of frequencies centered on the carrier frequency is selected, say, those in the interval $[\omega_0 - a, \omega_0 + a]$. The frequency ω_0 of the carrier wave is now forced to vary according to the equation

$$\omega_0(t) = \omega_0 + \frac{a}{2\pi\omega t} \sin 2\pi\omega t$$

The transmitted wave then has the equation

$$\begin{aligned} y &= A_0 \sin \left[2\pi t \left(\omega_0 + \frac{a}{2\pi\omega t} \sin 2\pi\omega t \right) \right] \\ &= A_0 \sin \left[2\pi\omega_0 t + \frac{a}{\omega} \sin 2\pi\omega t \right] \end{aligned}$$

FM radio

Its graph is plotted in Fig. 13.18. This process is called *frequency modulation,* or FM.

Consider, now, the graph of $y = \sin(x + C)$. When $x + C = 0$, $x = -C$, and when $x + C = 2\pi$, $x = 2\pi - C$. The graph, indicated in

Frequency modulation

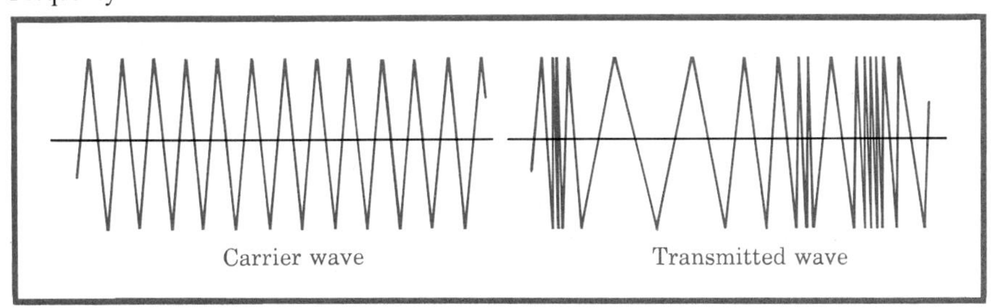

FIGURE 13.18
In frequency modulation the signal is transmitted by modifying the frequency of the carrier wave.

$y = \sin(x + C)$ and $y = \sin x$, $C > 0$

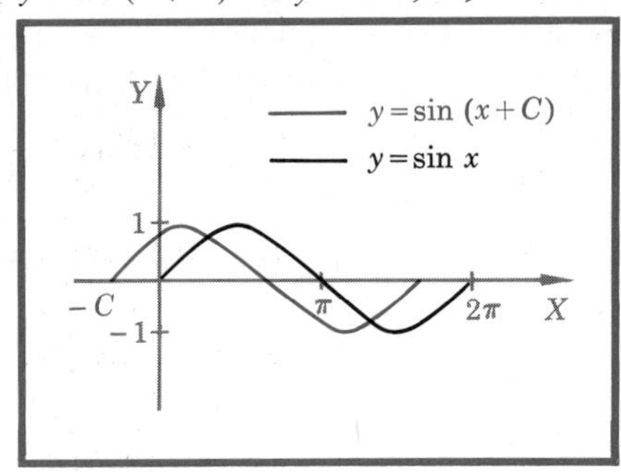

FIGURE 13.19
In $y = \sin(x + C)$, the constant $-C$ is called the phase shift.

$y = \sin(Bx + C)$ and $y = \sin x$, $B > 0$, $C > 0$

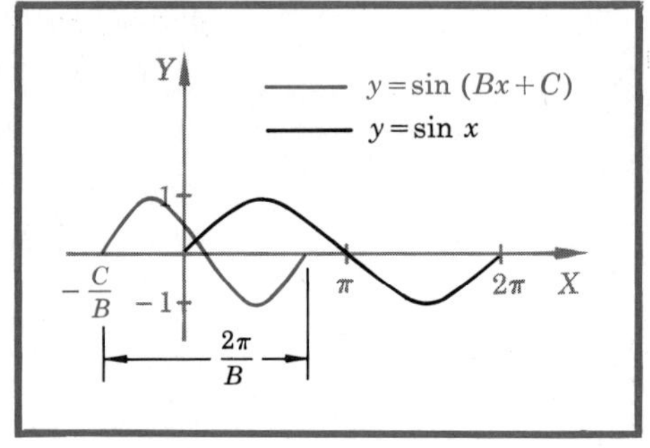

FIGURE 13.20
In $y = \sin(Bx + C)$, the phase shift is $-C/B$.

Fig. 13.19, is therefore a sine wave shifted to the *left* by an amount C. (If C is negative, the shift will be to the *right*.) The constant $-C$ is called the *phase shift*, or *phase angle*. For the wave $y = \sin(Bx + C)$, we note that, when $Bx + C = 0$, $x = -C/B$, and when $Bx + C = 2\pi$, $x = (2\pi - C)/B$ (Fig. 13.20). Here the *phase shift* is represented by the number $-C/B$.

Phase

Finally, we combine all these ideas in the representation of the most general sine wave

$$y = A \sin(Bx + C)$$

The *amplitude* is $|A|$, the *period* is $2\pi/B$, and the *phase shift* is $-C/B$ (Fig. 13.21).

Similar remarks apply to the graphs of the other functions.

$y = A \sin(Bx + C)$ and $y = \sin x$, $A > 0$, $B > 0$, and $C > 0$

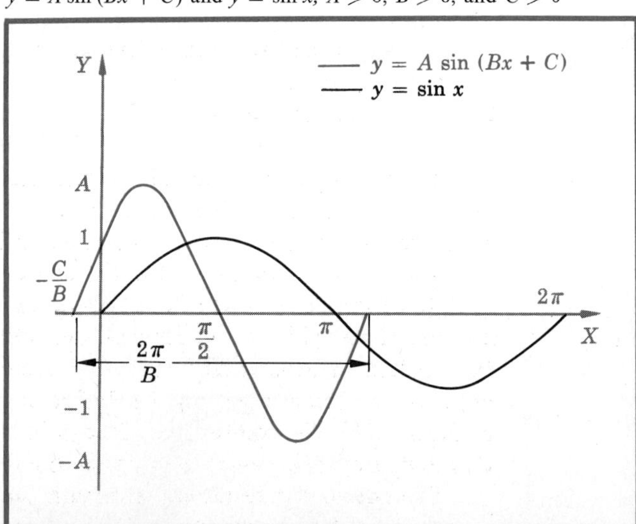

FIGURE 13.21
In the general sine wave, we must discuss amplitude, period, and phase.

PROBLEMS 13.5

In Probs. 1 to 14 sketch the graph. Label amplitude (where meaningful), period, and phase shift.

1 $y = 2\sin\left(x + \frac{\pi}{4}\right)$

2 $y = \frac{1}{2}\sin\left(x - \frac{\pi}{4}\right)$

3 $y = \frac{1}{2}\sin\left(2x + \frac{\pi}{3}\right)$

4 $y = 4\cos\left(x + \frac{\pi}{4}\right)$

5 $y = 3\cos\left(x - \frac{\pi}{4}\right)$

6 $y = 2\cos\left(2x - \frac{\pi}{6}\right)$

7 $y = 2\sin\left(\frac{x}{2} - \frac{\pi}{2}\right)$

8 $y = \frac{1}{2}\cos\left(x - \frac{\pi}{2}\right)$

9 $y = \frac{1}{3}\sin 3x$

10 $y = \frac{1}{3}\cos 3x$

11 $y = \tan\left(x + \frac{\pi}{2}\right)$

12 $y = 2\tan\left(2x + \frac{\pi}{3}\right)$

13 $y = 2\tan\left(x + \frac{\pi}{4}\right)$

14 $y = 3\tan\left(x - \frac{\pi}{4}\right)$

13.6 » ADDITION THEOREMS

Let us consider a function f and its values $f(x_1)$ and $f(x_2)$, where x_1 and x_2 are to be thought of as any two x's in the domain of definition such that $x_1 + x_2$ is also in the domain. The following general question arises: What can we say about $f(x_1 + x_2)$ in terms of $f(x_1)$ and $f(x_2)$ separately? Such a theorem is referred to as an *addition theorem for the function f*. Both classical and modern mathematics place emphasis on the discovery and use of such theorems. They are of special importance in trigonometry where we should like to know the answers to the following questions:

1 Can we express $\sin(\theta \pm \phi)$ and $\cos(\theta \pm \phi)$ in terms of $\sin\theta$, $\sin\phi$, $\cos\theta$, and $\cos\phi$?
2 If so, what are the formulas?

There are many derivations of these formulas, and they all have some degree of artificiality. This can be said about most of the theorems of plane geometry where certain construction lines—once they have been thought of and drawn in—aid in the analysis of the problem. We want to find a formula for one of the four quantities $\sin(\theta \pm \phi)$, $\cos(\theta \pm \phi)$, and therefore we begin by drawing two unit circles as in Fig. 13.22. In Fig. 13.22*a*, we have laid off the arcs θ and ϕ and indicated the resulting arc $\phi - \theta$ as the arc QP. In Fig. 13.22*b* we have laid off the arc $\phi - \theta$ as the arc SR. The coordinates of the points P, Q, R, and S are indicated on the figure; their values follow from the definitions of sine and cosine.

The key to our result is the remark that the segments PQ and RS are equal, for they are corresponding sides of the congruent triangles OPQ

Addition formula

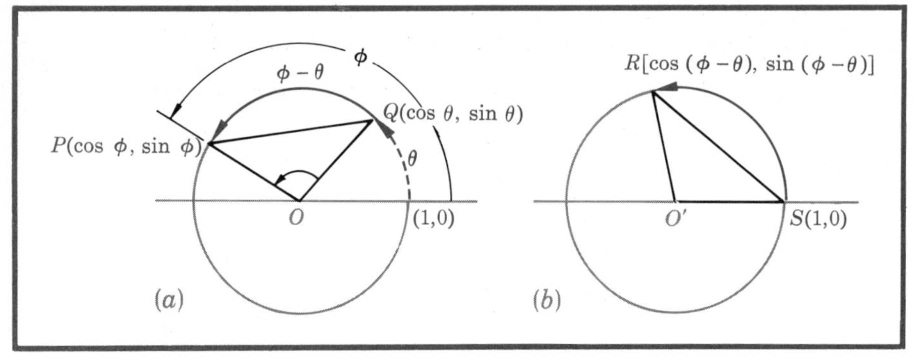

FIGURE 13.22
The "addition formula" for $\cos(\phi - \theta)$ is proved by observing that segment PQ = segment RS.

and $O'RS$. If we express this equality by the use of the distance formula (Sec. 12.2), we find

$$(PQ)^2 = (\cos\phi - \cos\theta)^2 + (\sin\phi - \sin\theta)^2$$

(1)
$$= 2 - 2(\cos\phi\cos\theta + \sin\phi\sin\theta)$$

$$(RS)^2 = [\cos(\phi - \theta) - 1]^2 + [\sin(\phi - \theta) - 0]^2$$

(2)
$$= 2 - 2\cos(\phi - \theta)$$

Equating these, we obtain

(3)
$$2 - 2\cos(\phi - \theta) = 2 - 2(\cos\phi\cos\theta + \sin\phi\sin\theta)$$

Therefore we have the formula

(4)
$$\cos(\phi - \theta) = \cos\phi\cos\theta + \sin\phi\sin\theta$$

In Fig. 13.22 we have implicitly assumed that $\phi - \theta$ is positive. If ϕ and θ are such that $\phi - \theta$ is negative, the figure will be different but the argument remains the same. Moreover, if $\phi - \theta = 2n\pi$, where n is an integer, points P and Q (and hence R and S) will coincide, so that the figure is misleading. Again, however, the algebra is the same and the result is true.

This is exactly the kind of formula we are seeking. It is an identity which expresses the cosine of the difference of two real numbers ϕ and θ in terms of the sines and cosines of ϕ and θ separately. You must memorize formula (4); it is one of the "addition theorems" for the trigonometric functions. We now derive three others, namely, (7), (11), and (12) below.

Directly from the definitions of $\sin\theta$ and $\cos\theta$ it follows that

(5)
$$\sin(-\theta) = -\sin\theta$$

and

(6)
$$\cos(-\theta) = \cos\theta$$

In (6) put $\theta = -\alpha$. [We may do this since (4) is an identity.] We get $\cos(\phi + \alpha) = \cos\phi\cos(-\alpha) + \sin\phi\sin(-\alpha)$. Using (5) and (6), this simplifies to the second important "addition theorem":

(7) $$\cos(\phi + \alpha) = \cos\phi\cos\alpha - \sin\phi\sin\alpha$$

Next, in (4), let us put $\theta = \pi/2$. We get

$$\cos\left(\phi - \frac{\pi}{2}\right) = \cos\phi\cos\frac{\pi}{2} + \sin\phi\sin\frac{\pi}{2}$$

or

(8) $$\cos\left(\phi - \frac{\pi}{2}\right) = \sin\phi$$

If we put $\alpha = \phi - \pi/2$, or $\phi = \alpha + \pi/2$ in (8), we can write (8) in the form

(9) $$\cos\alpha = \sin\left(\alpha + \frac{\pi}{2}\right)$$

Similarly, by putting $\theta = -\pi/2$ in (4), we obtain

(10) $$\cos\left(\phi + \frac{\pi}{2}\right) = -\sin\phi$$

We are now ready to derive the addition theorem for $\sin(\phi - \theta)$. We use (8) to write

$$\sin(\phi - \theta) = \cos\left[(\phi - \theta) - \frac{\pi}{2}\right] = \cos\left[\phi - \left(\theta + \frac{\pi}{2}\right)\right]$$

Now apply (4) to the right-hand expression, and obtain

$$\sin(\phi - \theta) = \cos\phi\cos\left(\theta + \frac{\pi}{2}\right) + \sin\phi\sin\left(\theta + \frac{\pi}{2}\right)$$

Using (9) and (10), we can simplify this to

$$\sin(\phi - \theta) = -\cos\phi\sin\theta + \sin\phi\cos\theta$$

or

(11) $$\sin(\phi - \theta) = \sin\phi\cos\theta - \cos\phi\sin\theta$$

This is the desired result.

Finally, putting $\theta = -\alpha$ in (11), we obtain

(12) $$\sin(\phi + \alpha) = \sin\phi\cos\alpha + \cos\phi\sin\alpha$$

Addition formulas

We now collect (4), (7), (11), and (12) and write them in the form

(I) $$\sin(\phi \pm \theta) = \sin\phi\cos\theta \pm \cos\phi\sin\theta$$

(II) $$\cos(\phi \pm \theta) = \cos\phi\cos\theta \mp \sin\phi\sin\theta$$

To develop a formula for $\tan(\phi \pm \theta)$, write

$$\tan(\phi \pm \theta) = \frac{\sin\phi\cos\theta \pm \cos\phi\sin\theta}{\cos\phi\cos\theta \mp \sin\phi\sin\theta}$$

$$= \frac{\dfrac{\sin\phi\cos\theta}{\cos\phi\cos\theta} \pm \dfrac{\cos\phi\sin\theta}{\cos\phi\cos\theta}}{\dfrac{\cos\phi\cos\theta}{\cos\phi\cos\theta} \mp \dfrac{\sin\phi\sin\theta}{\cos\phi\cos\theta}}$$

(III) $$= \frac{\tan\phi \pm \tan\theta}{1 \mp \tan\phi\tan\theta}$$

You should study this material until you understand it thoroughly. Memorize (I), (II), and (III).

EXERCISE A What restrictions must be placed upon ϕ and θ in (I), (II), and (III)?

PROBLEMS 13.6

In Probs. 1 to 16 make use of (I), (II), (III), to compute without tables. (Leave answers in radical form.)

1 $\sin\left(\frac{\pi}{2} + \frac{\pi}{6}\right)$ **2** $\sin\left(\frac{\pi}{4} - \frac{\pi}{6}\right)$ **3** $\sin\left(\frac{\pi}{2} - \frac{\pi}{6}\right)$

4 $\sin\left(\frac{\pi}{4} - \frac{\pi}{3}\right)$ **5** $\cos\left(\frac{\pi}{4} + \frac{\pi}{3}\right)$ **6** $\cos\left(\frac{\pi}{4} + \frac{\pi}{6}\right)$

7 $\cos\left(\frac{\pi}{4} - \frac{\pi}{6}\right)$ **8** $\cos\left(\frac{\pi}{2} + \frac{\pi}{6}\right)$ **9** $\tan\frac{3\pi}{4}$

10 $\tan\frac{2\pi}{3}$ **11** $\cot\frac{\pi}{12}$ **12** $\cot\frac{5\pi}{12}$

13 $\sec\frac{\pi}{6}$ **14** $\sec\frac{5\pi}{6}$ **15** $\csc\frac{7\pi}{12}$

16 $\csc\frac{5\pi}{12}$

In Probs. 17 to 26 reduce to a function of θ.

17 $\sin(\pi \pm \theta)$ **18** $\sin\left(\frac{\pi}{2} \pm \theta\right)$ **19** $\sin\left(\frac{3\pi}{2} \pm \theta\right)$

20 $\cos\left(\frac{\pi}{2} \pm \theta\right)$ **21** $\cos(\pi \pm \theta)$ **22** $\cos\left(\frac{3\pi}{2} \pm \theta\right)$

23 $\tan\left(\frac{3\pi}{2} \pm \theta\right)$ **24** $\cot(\pi \pm \theta)$ **25** $\sec\left(\frac{3\pi}{2} \pm \theta\right)$

26 $\csc\left(\frac{11\pi}{2} \pm \theta\right)$

27 Show that $\sin \frac{13\pi}{12} = \frac{1}{4}(\sqrt{2} - \sqrt{6})$.

28 Show that $\cos \frac{13\pi}{12} = -\frac{1}{4}(\sqrt{2} + \sqrt{6})$.

29 If $\sin A = -\frac{3}{5}$ and $\cos B = -\frac{12}{13}$, find all possible values of $\cos (A - B)$.

30 Given $\sin \alpha = -\frac{4}{5}$, $\tan \beta = -\frac{5}{12}$, and α is in the fourth quadrant and β is in the second quadrant; find $\sin (\alpha + \beta)$ and $\cos (\alpha + \beta)$.

In Probs. 31 to 34 simplify to a single function of some number.

31 $\sin 6 \cos 2 - \cos 6 \sin 2$

32 $\cos 37 \cos 18 + \sin 37 \sin 18$

33 $\dfrac{\tan 16.2 + \tan 13.4}{1 - \tan 16.2 \tan 13.4}$

34 $\sin 13 \cos 2.1 + \cos 13 \sin (-2.1)$

35 Verify the steps in the following proof of the formula for $\sin (\phi - \theta)$,

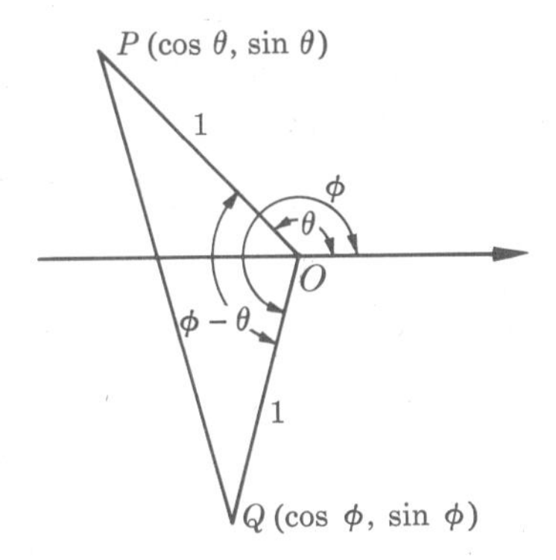

where we assume $0 < \phi - \theta < \pi$. In the figure above, the area K of the triangle POQ is given in the two following ways:

$$K = \tfrac{1}{2} \sin (\phi - \theta)$$

$$K = \frac{1}{2} \begin{vmatrix} \cos \theta & \sin \theta & 1 \\ \cos \phi & \sin \phi & 1 \\ 0 & 0 & 1 \end{vmatrix}$$

$$= \tfrac{1}{2}(\sin \phi \cos \theta - \cos \phi \sin \theta)$$

Therefore $\sin (\phi - \theta) = \sin \phi \cos \theta - \cos \phi \sin \theta$.

36 Derive formula (4) for $\cos (\phi - \theta)$ from the result of Prob. 35.

37 Prove that for any A, B, C, $y = A \sin Cx + B \cos Cx$ represents a sine wave and determine its amplitude, period, and phase. HINT: Put

$$\cos \theta = \frac{A}{\sqrt{A^2 + B^2}} \qquad \sin \theta = \frac{B}{\sqrt{A^2 + B^2}}$$

13.7 » MULTIPLE- AND HALF-ANGLE FORMULAS

The general formulas, or identities, (I), (II), (III) of Sec. 13.6 have some extremely important special cases. These we now derive and we shall mark with Roman numerals the ones you should memorize. They play important roles in all branches of mathematics and in all of the sciences.

From (I), with $\phi = \theta$, it immediately follows that

Double-angle formulas

(IV) $$\sin 2\theta = 2 \sin \theta \cos \theta$$

Similarly from (II)

(V) $$\cos 2\theta = \cos^2 \theta - \sin^2 \theta$$

This can be written in two other ways by making use of the identity $\sin^2 \theta + \cos^2 \theta = 1$. These are

(VI) $$\cos 2\theta = 1 - 2 \sin^2 \theta$$

(VII) $$\cos 2\theta = 2 \cos^2 \theta - 1$$

Formula (IV) expresses the sine of *twice* a number in terms of the sine and cosine of the number itself. Each of the three formulas (V) to (VII) expresses the cosine of *twice* a number in terms of sine and cosine of the number itself.

EXERCISE A Discover a formula for $\tan 2\theta$ in terms of $\tan \theta$.

Now (VI) can be written in the form $\cos x = 1 - 2 \sin^2 (x/2)$ where $2\theta = x$. If we now solve for $\sin^2 (x/2)$ (but now write $\theta/2$ instead of $x/2$), we get

$$\sin^2 \frac{\theta}{2} = \frac{1 - \cos \theta}{2}$$

from which

Half-angle formulas

(VIII) $$\sin \frac{\theta}{2} = \pm \sqrt{\frac{1 - \cos \theta}{2}}$$

The sign before the radical is to be chosen + if $\theta/2$ is a first- or second-quadrantal arc. It is to be chosen − if $\theta/2$ is a third- or fourth-quadrantal arc.

Similarly from (VII) we obtain

(IX) $$\cos \frac{\theta}{2} = \pm \sqrt{\frac{1 + \cos \theta}{2}}$$

EXERCISE B State, for (IX), when the + sign is to be used and when the − sign is to be used.

Formulas (VIII) and (IX) express sine and cosine of *half* a number in terms of the cosine of the number itself. From them we develop three formulas for tangent of *half* a number.

(X) $$\tan\frac{\theta}{2} = \frac{\sin(\theta/2)}{\cos(\theta/2)} = \pm\sqrt{\frac{1-\cos\theta}{1+\cos\theta}}$$

(XI) $$= \frac{1-\cos\theta}{\sin\theta}$$

(XII) $$= \frac{\sin\theta}{1+\cos\theta}$$

EXERCISE C State, for (X), when the $+$ sign is to be used and when the $-$ sign is to be used.

EXERCISE D Derive (XI) and (XII). HINT for (XI):

$$\cos\frac{\theta}{2} = \cos\left(\theta - \frac{\theta}{2}\right) = \cos\theta\cos\frac{\theta}{2} + \sin\theta\sin\frac{\theta}{2}$$

Now solve for $\dfrac{\sin(\theta/2)}{\cos(\theta/2)}$.

We now return to (I) of Sec. 13.6, write

$$\sin(\phi+\theta) = \sin\phi\cos\theta + \cos\phi\sin\theta$$
$$\sin(\phi-\theta) = \sin\phi\cos\theta - \cos\phi\sin\theta$$

and add, getting, after dividing by 2,

(XIII) $$\sin\phi\cos\theta = \tfrac{1}{2}\sin(\phi+\theta) + \tfrac{1}{2}\sin(\phi-\theta)$$

This identity and the two following, which are derived similarly from (II), are most important in a study of the calculus.

Product formulas

(XIV) $$\cos\phi\cos\theta = \tfrac{1}{2}\cos(\phi+\theta) + \tfrac{1}{2}\cos(\phi-\theta)$$

(XV) $$\sin\phi\sin\theta = -\tfrac{1}{2}\cos(\phi+\theta) + \tfrac{1}{2}\cos(\phi-\theta)$$

EXERCISE E Write out the derivation of (XIV) and (XV).

The fifteen formulas (I) to (XV) constitute the basic identities. You should not only memorize them, but you should know how to derive them. They are written in terms of numbers (arcs) ϕ and θ, but, obviously, other symbols could be used, and, of course, degree measure could be used. The following problems, wherein many other letters are used, are based directly on (I) to (XV).

PROBLEMS 13.7

In Probs. 1 to 6 find a counterexample for the false statements which are asserted for all x:

1 $\sin 2x = 2 \sin 2x$ **2** $\cos 2x = 2 \cos x$
3 $2 \sin \frac{1}{2}x = \sin x$ **4** $2 \cos \frac{1}{2}x = \cos x$
5 $\frac{1}{2} \tan 2x = \tan x$ **6** $2 \tan \frac{1}{2}x = \tan x$

7 Derive the formula

$$\cot(\alpha + \beta) = \frac{\cot \alpha \cot \beta - 1}{\cot \alpha + \cot \beta}$$

8 Derive the formula

$$\cot(\alpha - \beta) = \frac{\cot \alpha \cot \beta + 1}{-\cot \alpha + \cot \beta}$$

9 Derive the formula $\sin 3\theta = 3 \sin \theta - 4 \sin^3 \theta$.
10 Derive the formula $\cos 3\theta = 4 \cos^3 \theta - 3 \cos \theta$.
11 Derive a formula for $\sin 4\theta$ in terms of $\sin \theta$ and $\cos \theta$.
12 Derive a formula for $\cos 4\theta$ in terms of $\sin \theta$ and $\cos \theta$.
13 Derive a formula for $\tan 4\theta$ in terms of $\tan \theta$.

In Probs. 14 to 19, let $\sin x = \frac{15}{17}$, $\cos x < 0$, $\tan y = -\frac{12}{5}$, and $\sin y < 0$. Then find:

14 $\sin(x + y)$ **15** $\sin(x - y)$ **16** $\cos(x - y)$
17 $\cos(x + y)$ **18** $\cot(\pi + x)$ **19** $\tan(y - x)$

In Probs. 20 to 23, let $\sin 4\theta = \frac{1}{2}$ and $\tan 4\theta > 0$. Then find:

20 $\sin 2\theta$ **21** $\sin 8\theta$ **22** $\cos 8\theta$ **23** $\cos 2\theta$

24 PROVE: $\cos\left(\frac{\pi}{4} - x\right) = \frac{\cos x + \sin x}{\sqrt{2}}$

25 PROVE: $\sin\left(\frac{\pi}{4} - x\right) = \frac{\cos x - \sin x}{\sqrt{2}}$

Set $x + y = A$ and $x - y = B$, and use the identities (XIII) to (XV) to derive the identities:

26 $\sin A - \sin B = 2 \cos \frac{1}{2}(A + B) \sin \frac{1}{2}(A - B)$
27 $\sin A + \sin B = 2 \sin \frac{1}{2}(A + B) \cos \frac{1}{2}(A - B)$
28 $\cos A - \cos B = -2 \sin \frac{1}{2}(A + B) \sin \frac{1}{2}(A - B)$
29 $\cos A + \cos B = 2 \cos \frac{1}{2}(A + B) \cos \frac{1}{2}(A - B)$

30 Law of Tangents, for triangle ABC, with a = side BC, b = side AC, c = side AB. Verify the steps in the following derivation:

$$\frac{a}{b} = \frac{\sin A}{\sin B}$$

$$\frac{a + b}{b} = \frac{\sin A + \sin B}{\sin B}$$

$$\frac{a-b}{b} = \frac{\sin A - \sin B}{\sin B}$$

$$\begin{aligned}\frac{a+b}{a-b} &= \frac{\sin A + \sin B}{\sin A - \sin B} \\ &= \frac{2 \sin \frac{1}{2}(A+B) \cos \frac{1}{2}(A-B)}{2 \cos \frac{1}{2}(A+B) \sin \frac{1}{2}(A-B)} \\ &= \frac{\tan \frac{1}{2}(A+B)}{\tan \frac{1}{2}(A-B)} \qquad \text{Law of Tangents}\end{aligned}$$

Show how this formula can be used to solve a triangle, given two sides and the included angle. Compare the advantages and disadvantages of this solution with those of the solution in terms of the Law of Cosines.

31 Show that Eq. (3) of Sec. 13.5 (amplitude modulation) can be written as

$$y = A_0 \sin 2\pi\omega_0 t + \frac{A_0 m}{2} \cos 2\pi(\omega - \omega_0)t - \frac{A_0 m}{2} \cos 2\pi(\omega + \omega_0)t$$

The frequencies in the last two terms, $\omega - \omega_0$ and $\omega + \omega_0$, are called *sideband frequencies*. Because of their presence, a band of frequencies is needed to transmit the given signal.

13.8 › IDENTITIES

From the basic identities given above it is possible to establish the truth of a whole host of other identities. These are often useful in applications of mathematics, for they permit us to reduce a formidable-looking trigonometric expression to something simpler and more manageable. Before entering into the details, we must say a few words about the general idea.

Let us consider the example:

ILLUSTRATION 1 Prove that

$$\frac{\sin^2 x}{1 + \cos x} = 1 - \cos x$$

is an identity.

SOLUTION In the first place this is an open sentence whose variable is x. This variable must have a universal set U, which in this case consists of all real numbers except those for which $1 + \cos x = 0$. That is, we must exclude $x = (2n+1)\pi$ from U.

Since $1 + \cos x \neq 0$ for all x in U, we may multiply both sides of the equation by $1 + \cos x$ and obtain the sequence of equivalent equations:

$$\sin^2 x = (1 - \cos x)(1 + \cos x)$$
$$\sin^2 x = 1 - \cos^2 x$$
$$\sin^2 x = \sin^2 x$$

Since $\sin^2 x = \sin^2 x$ for all x in U, we have shown that

$$\frac{\sin^2 x}{1 + \cos x} = 1 - \cos x$$

for all x in U. This proves that it is an identity.

There are a great variety of ways in which such a proof can be arranged. Here is another one.

The equation

$$\frac{\sin^2 x}{1 + \cos x} = 1 - \cos x$$

is equivalent to

$$\frac{1 - \cos^2 x}{1 + \cos x} = 1 - \cos x$$

This is equivalent to

$$\frac{(1 + \cos x)(1 - \cos x)}{1 + \cos x} = 1 - \cos x$$

and to

$$1 - \cos x = 1 - \cos x$$

since we may divide by the nonzero expression $1 + \cos x$. But this last equation is true for all x in U, and so the original equation, which is equivalent to it, is an identity.

For the general situation we have the following definitions and procedures:

DEFINITION An identity is an open sentence (with a given variable and corresponding universal set U) which is true for all elements of U.

METHOD OF PROVING AN IDENTITY

1 Determine U as the set of reals except for those values for which any term in the equation is undefined.
2 Prove that the given equation is equivalent to an obvious identity, by a sequence of steps involving the operations:
   a Adding to both sides any expression which is defined for all elements of U. (The proof that this leads to an equivalent equation is similar to that of Theorem 1, Chap. 6.)
   b Multiplying both sides by any expression which is defined for all elements of U and which is nonzero for each element of U. (The

proof that this leads to an equivalent equation is similar to that of Theorem 2, Chap. 6.)

c Replacing any term in the equation by another expression which is equal to it for all values of the variable in U.

REMARK You must be careful not to use operations which do not lead to an equivalent equation. In particular, avoid squaring both sides. In Illustration 2 we show the dangers of such a procedure.

ILLUSTRATION 2 Prove the (false) identity

$$\sqrt{1 - \cos^2 x} = \sin x$$

SOLUTION U is the set of reals with no exceptions. Squaring both sides, we obtain

$$1 - \cos^2 x = \sin^2 x$$

which is equivalent to the identity

$$\sin^2 x = \sin^2 x$$

Surely, however, this cannot be a valid proof, for

$$\sqrt{1 - \cos^2 x} = \sin x$$

is true only when $\sin x \geq 0$, that is, when x lies in the first or second quadrant. The error occurs because the equations

$$\sqrt{1 - \cos^2 x} = \sin x \qquad \text{and} \qquad 1 - \cos^2 x = \sin^2 x$$

are not equivalent.

EXERCISE A Prove the (correct) identity

$$\sqrt{1 - \cos^2 x} = |\sin x|$$

ILLUSTRATION 3 Prove the identity

$$\frac{\csc x + 1}{\cot x} = \frac{\cot x}{\csc x - 1}$$

SOLUTION U is the set of reals except for values of x for which $\cot x$ or $\csc x$ are undefined and for which $\cot x = 0$ or $\csc x = 1$. Thus we must exclude $x = n\pi$, $(\pi/2) + n\pi$.

Multiplying both sides by $\cot x(\csc x - 1)$, we obtain

$$\begin{aligned}(\csc x + 1)(\csc x - 1) &= \cot^2 x\\ \csc^2 x - 1 &= \cot^2 x\\ \cot^2 x &= \cot^2 x\end{aligned}$$

The last equation is true for all x in U, and so the given equation is an identity.

ILLUSTRATION 4 Prove the identity

$$\frac{\tan^3 x - \cot^3 x}{\tan x - \cot x} = \tan^2 x + \csc^2 x$$

Before attempting a proof, we must choose U such that $\tan x - \cot x \neq 0$. This gives us $U = R$ except $x = (\pi/4) + n(\pi/2)$ where n is an integer.

SOLUTION 1 We recall that $\csc^2 x = 1 + \cot^2 x$; so we have

$$\frac{\tan^3 x - \cot^3 x}{\tan x - \cot x} = \tan^2 x + 1 + \cot^2 x$$

Multiplying both sides by $\tan x - \cot x$, we obtain

$$\begin{aligned}\tan^3 x - \cot^3 x &= (\tan^2 x + 1 + \cot^2 x)(\tan x - \cot x)\\ &= \tan^3 x + \tan x + \tan x \cot^2 x - \cot x \tan^2 x\\ &\qquad - \cot x - \cot^3 x\end{aligned}$$

But

$$\tan x = \frac{1}{\cot x}$$

and so this becomes

$$\begin{aligned}\tan^3 x - \cot^3 x &= \tan^3 x + \tan x + \cot x - \tan x - \cot x - \cot^3 x\\ &= \tan^3 x - \cot^3 x\end{aligned}$$

SOLUTION 2 The numerator on the left factors; so we can write

$$\frac{(\tan x - \cot x)(\tan^2 x + \tan x \cot x + \cot^2 x)}{\tan x - \cot x} = \tan^2 x + 1 + \cot^2 x$$

Dividing out $\tan x - \cot x$ on the left, we have

$$\tan^2 x + \tan x \cot x + \cot^2 x = \tan^2 x + 1 + \cot^2 x$$

or

$$\tan^2 x + 1 + \cot^2 x = \tan^2 x + 1 + \cot^2 x$$

PROBLEMS 13.8

Prove the following identities:

1 $\dfrac{\sin x \sec x}{\tan x + \cot x} = 1 - \cos^2 x$

2 $\tan x + \cot x = \sec x \csc x$

3 $\sin x(\cot x + \csc x) = 1 + \cos x$

4 $\dfrac{1 + \tan x}{1 - \tan x} + \dfrac{1 + \cot x}{1 - \cot x} = 0$

5 $\dfrac{\csc x}{\sec x} = \dfrac{1 + \cot x}{1 + \tan x}$

6 $\dfrac{1 - \sin x}{\cos x} = \dfrac{\cos x}{1 + \sin x}$

7 $\dfrac{\sin x + \cos x}{\sec x + \csc x} = \dfrac{\sin x}{\sec x}$

8 $\dfrac{1 + \cos x}{\sin x} + \dfrac{\sin x}{1 + \cos x} = 2 \csc x$

9 $\dfrac{1 + \sin 2x + \cos 2x}{1 + \sin 2x - \cos 2x} = \cot x$

10 $(\sin \frac{1}{2}\theta - \cos \frac{1}{2}\theta)^2 = 1 - \sin \theta$

11 $\csc x - \cot x = \tan \frac{1}{2}x$

12 $\dfrac{1 - \cos 2x}{\sin 2x} = \tan x$

13 $\sin x = \cos x \tan x = \dfrac{\tan x}{\sec x}$

14 $\cos (x + y) \cos y + \sin (x + y) \sin y = \cos x$

15 $\tan \left(\dfrac{x}{2} + \dfrac{\pi}{4}\right) = \sec x + \tan x$

16 $\csc x = \cot x + \tan \dfrac{x}{2}$

17 $\sin 5x \cos 3x + \cos 5x \sin 3x = 2 \sin 4x \cos 4x$

18 $\cos \frac{3}{5}x \cos \frac{2}{5}x - \sin \frac{2}{5}x \sin \frac{3}{5}x = \cos x$

19 $\cos^6 \theta - \sin^6 \theta = \cos 2\theta - \frac{1}{8} \sin 4\theta \sin 2\theta$

20 $\cos \theta + \cos 2\theta + \cos 3\theta = \cos 2\theta(1 + 2 \cos \theta)$

13.9 › EQUATIONS

In finding the universal set U for the illustrations in Sec. 13.8 we encountered equations such as

$$1 + \cos x = 0$$
$$\csc x - 1 = 0$$
$$\tan x - \cot x = 0$$

and found it necessary to solve these for x. Equations like these are called *trigonometric equations.* In the examples above, each equation had infinitely many solutions. But some equations such as $\sin x = 3$ have no solutions. If the equation is an identity, all x in U are solutions.

There are practically no general rules which, if followed, will lead to the roots of a trigonometric equation. You might try to factor or to solve by quadratic formula where appropriate. Or, again, you might reduce each and every trigonometric function present to one and the same function of one and the same independent variable. In this section we exhibit some of the obvious ways of solving such an equation.

ILLUSTRATION 1 Solve the equation

$$2 \sin^2 x + \sin x - 1 = 0$$

for all roots.

SOLUTION The left-hand member is quadratic in the quantity $\sin x$; that is, it is a polynomial of the second degree in $\sin x$. It is factorable:

$$(2 \sin x - 1)(\sin x + 1) = 0$$

Thus from the first factor we get

$$2 \sin x - 1 = 0$$
$$\sin x = \tfrac{1}{2}$$
$$x = \frac{\pi}{6} + 2n\pi \qquad \text{1st-quadrantal arc}$$
$$x = \frac{5\pi}{6} + 2n\pi \qquad \text{2d-quadrantal arc}$$

EXERCISE A Draw figures for these arcs.

The second factor yields

$$\sin x + 1 = 0$$
$$\sin x = -1$$
$$x = \frac{3\pi}{2} + 2n\pi$$

There are no other roots.

ILLUSTRATION 2 Find all values of x in the interval 0 to 2π satisfying the equation

$$\cos^2 2x + 3 \sin 2x - 3 = 0$$

SOLUTION This appears to offer some difficulty at first thought because of the presence of both sine and cosine. We use the identity

$$\sin^2 2x + \cos^2 2x = 1$$

and rewrite the equation in the form

$$1 - \sin^2 2x + 3 \sin 2x - 3 = 0$$

which factors into

$$(1 - \sin 2x)(2 - \sin 2x) = 0$$

The first factor yields

$$1 - \sin 2x = 0$$
$$\sin 2x = 1$$
$$2x = \frac{\pi}{2} + 2n\pi$$

whence

$$x = \frac{\pi}{4} + n\pi$$

The second factor leads to the equation

$$\sin 2x = 2$$

which has no solution.

ILLUSTRATION 3 Solve the equation $\tan x + 2 \sec x = 1$.

SOLUTION You should be able to follow each of the steps:

$$\frac{\sin x}{\cos x} + \frac{2}{\cos x} = 1$$

$$\frac{\sin x + 2}{\cos x} = 1$$

$$\sin x + 2 = \cos x \qquad \text{provided } \cos x \neq 0$$

$$= \pm\sqrt{1 - \sin^2 x}$$

$$(\sin x + 2)^2 = 1 - \sin^2 x$$

$$\sin^2 x + 4 \sin x + 4 = 1 - \sin^2 x$$

$$2 \sin^2 x + 4 \sin x + 3 = 0$$

$$\sin x = \frac{-4 \pm \sqrt{16 - 24}}{4}$$

Since we are dealing with the real numbers, we conclude that the original equation is satisfied by no real number.

ILLUSTRATION 4 Solve the equation $\sin 2x + \sin x = 0$.

SOLUTION We first write $\sin 2x = 2 \sin x \cos x$.

$$2 \sin x \cos x + \sin x = 0$$

$$\sin x(2 \cos x + 1) = 0$$

$$\sin x = 0 \qquad \cos x = -\tfrac{1}{2}$$

$$x = n\pi \qquad x = \frac{2\pi}{3} + 2n\pi \qquad \text{2d-quadrantal arc}$$

$$x = \frac{4\pi}{3} + 2n\pi \qquad \text{3d-quadrantal arc}$$

EXERCISE B Draw figures for these arcs.

ILLUSTRATION 5 Solve the equation $\tan^2 x - 5 \tan x - 4 = 0$.

SOLUTION This is a quadratic equation in the quantity $\tan x$. Solving this by formula, we get

$$\tan x = \frac{5 \pm \sqrt{25 + 16}}{2}$$

$$= \tfrac{5}{2} \pm \tfrac{1}{2}\sqrt{41}$$

$$= 2.50000 \pm 3.20156$$

$$= 5.70656 \text{ and } -0.70156$$

Since these values do not correspond to any of the special arcs, we use Table IV in the Appendix.

From $\tan x = 5.70156$ we find that

$$x = 1.3971 + 2n\pi \qquad \text{1st-quadrantal arc}$$
$$x = (1.3971 + \pi) + 2n\pi \qquad \text{3d-quadrantal arc}$$
$$= 4.5387 + 2n\pi$$

To solve $\tan x = -0.70156$, we first solve

$$\tan x' = +0.70156$$

This gives

$$x' = 0.6118$$

But now we must use the minus sign since at present what we have is

$$\tan 0.6118 = 0.70156$$

whereas we seek x such that $\tan x = -0.70156$. This means that either

$$x = \pi - x' \qquad \text{or} \qquad x = 2\pi - x'$$

Finally, therefore, we have

$$x = 2.5298 + 2n\pi \qquad \text{2d-quadrantal arc}$$
$$x = 5.6714 + 2n\pi \qquad \text{4th-quadrantal arc}$$

PROBLEMS 13.9 Solve the following equations for all roots:

1 $\sin x = \sin 2x$

2 $\cos 2x = \frac{1}{2}$

3 $\sin 2x = \sqrt{3}/2$

4 $\csc^2 x = \frac{4}{3}$

5 $\cos 3x = \frac{1}{2}$

6 $2 \sin^2 x - \sin x - 1 = 0$

7 $2 \sin^2 x - \sin x = 0$

8 $\sec x + \tan x = 0$

9 $\cos 2x + \cos x = -1$

10 $4 \sin^2 x \tan x - 4 \sin^2 x - 3 \tan x + 3 = 0$

11 $\sin^2 x - \sin x = \frac{1}{4}$

12 $2 \sin x = \csc x + 1$

13 $\sin 4x + \sin 2x = 0$

14 $\tan 2x \tan x = 1$

15 $\csc x = \tan x \cot x$

16 $\tan x + 3 \cot x = 4$

17 $\csc x + \cot x = \sqrt{3}$

18 $2 \cos x = 1 - \sin x$

19 $2 + \sqrt{3} \sec x - 4 \cos x = 2\sqrt{3}$

20 $\sin 3x = \cos 2x$

21 $4 \cos 2x + 3 \cos x = 1$

22 $\cos x + \cos 2x + \cos 3x = 0$

23 $\tan\left(x + \dfrac{\pi}{4}\right) - 1 + \sin 2x = 0$

24 $2 \cos^2 \frac{1}{2}x = \cos^2 x$

25 $\sin 5x - \sin 3x - \sin x = 0$

26 $\tan x + 2 \cot x = \frac{5}{2} \csc x$

13.10 › INVERSE TRIGONOMETRIC FUNCTIONS

If we are given the function $y = \sin x$, we may well ask, For what real numbers x is $\sin x$ equal to a given real number y? We recognize this as the problem of finding the function which is the inverse of $\sin x$ (Sec. 9.8). In order to approach this question, let us reexamine the graph of $y = \sin x$ (Fig. 13.23).

We see at once that $y = \sin x$ is not one-to-one and hence that it does not have an inverse. We have seen, however, in Sec. 9.8, that when a function does not have an inverse, we can sometimes restrict its domain so that the restricted function does have an inverse. To do this we should restrict the function to an interval in which it is one-to-one. There are many choices for such an interval for $y = \sin x$, but it is customary to choose the one in which $|x|$ is as small as possible and in which $\sin x$ varies over its maximum interval, $-1 \leq \sin x \leq 1$. This is the interval $-\pi/2 \leq x \leq \pi/2$ (Fig. 13.24). In this interval $\sin x$ is one-to-one, and so the restricted function has an inverse. So that we may be specific, let us call $y = \text{Sin}\, x$ (read "Cap Sine x") the restriction of $y = \sin x$ to

Does $y = \sin x$ have an inverse?

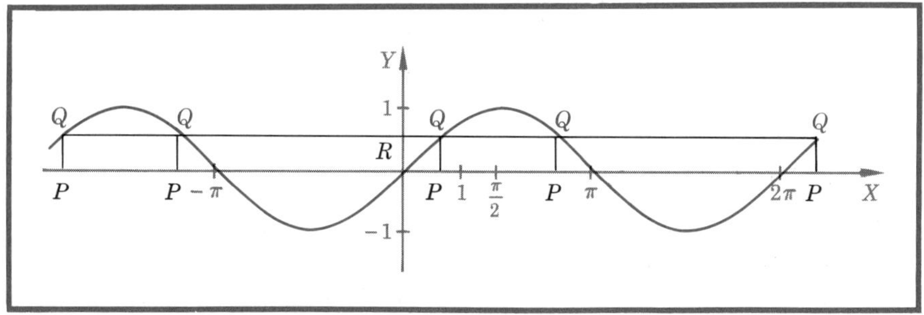

FIGURE 13.23
Since the horizontal line containing points labeled Q cuts the graph in more than one point, $y = \sin x$ does not have an inverse.

$y = \text{Sin}\, x$

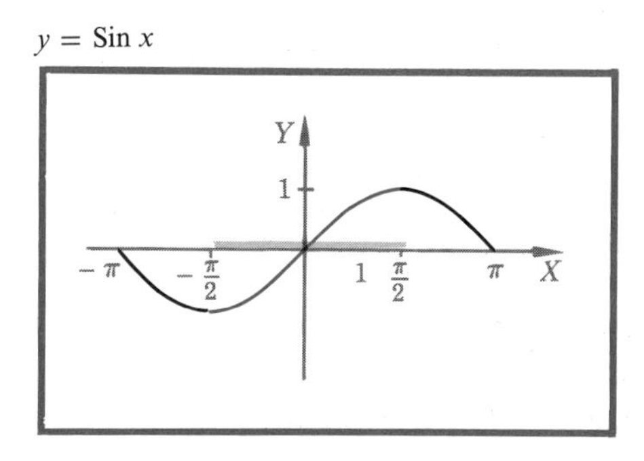

FIGURE 13.24
If the domain of $y = \sin x$ is restricted to the interval $-\pi/2 \leq x \leq \pi/2$, the resulting "restricted sine" function, $y = \text{Sin}\, x$, does have an inverse.

Restricted sine function, Sin x

the interval $-\pi/2 \leq x \leq \pi/2$. The image of Sin x is clearly $-1 \leq y \leq 1$.

According to our procedure, we should now solve $y = \text{Sin}\, x$ for x, but we have no direct way of doing so. Hence the inverse of $y = \text{Sin}\, x$ is a new function to which we must assign a name. Two names are used for this function, namely:

$$x = \text{Sin}^{-1} y \qquad \text{and} \qquad x = \text{arc Sin}\, y$$

Each may be read: "x is the inverse Cap-Sine of y", or "x is the real number whose Cap-Sine is y". The domain of this function is $-1 \leq y \leq 1$, and the image is $-\pi/2 \leq x \leq \pi/2$. As we discussed in Sec. 9.8, we wish to write x as the independent variable and y as the dependent variable, so finally we have

$$y = \text{Sin}^{-1} x \qquad \text{or} \qquad y = \text{arc Sin}\, x$$

whose domain is $-1 \leq x \leq 1$ and whose image is $-\pi/2 \leq y \leq \pi/2$. Its graph is given in Fig. 13.25. Similarly, we define other restricted trigonometric functions:

Other restricted functions

$y = \text{Cos}\, x$, $0 \leq x \leq \pi$, $-1 \leq y \leq 1$, as $\cos x$ restricted to this domain
$y = \text{arc Cos}\, x$, $-1 \leq x \leq 1$, $0 \leq y \leq \pi$, as the inverse of $y = \text{Cos}\, x$
$y = \text{Tan}\, x$, $-\pi/2 < x < \pi/2$, $-\infty < y < \infty$, as $\tan x$ restricted to this domain
$y = \text{arc Tan}\, x$, $-\infty < x < \infty$, $-\pi/2 < y < \pi/2$, as the inverse of $y = \text{Tan}\, x$

EXERCISE A Define arc Cot x.

ILLUSTRATION 1 Find $y = \text{arc Sin}\, \sqrt{3}/2$.

$y = \text{Sin}\, x$ and $y = \text{Sin}^{-1} x$

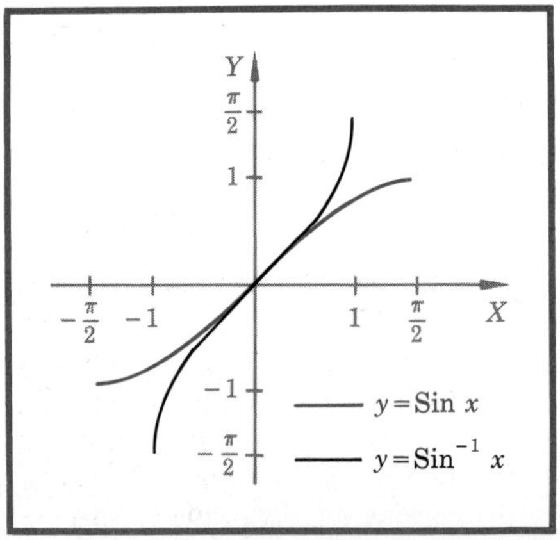

FIGURE 13.25
The curves are symmetrical about $y = x$.

SOLUTION We seek the number y, where $-\pi/2 \le y \le \pi/2$, such that the sine of it is $\sqrt{3}/2$. That is, $\text{Sin}\, y = \sqrt{3}/2$, and, from previous knowledge, we know that $y = \pi/3$.

ILLUSTRATION 2 Find $y = \text{arc Cos}\,(-0.87531)$.

SOLUTION This is the same thing as saying $\text{Cos}\, y = -0.87531$ and, of course, $0 \le y \le \pi$. We may use either Table IV or Table V of the Appendix. To use these tables let $z = \pi - y$. Then $\text{Cos}\, z = +0.87531$. First we find z from the tables.

Radian measure (Table IV)

$$0.01\left[\Delta\left[\begin{array}{l}\cos 0.50 = 0.87758 \\ \cos z \quad\;\; = 0.87531\end{array}\right]227 \atop \cos 0.51 = 0.87274\right]484$$

(We write 227 instead of 0.00227, etc.)

$$\frac{\Delta}{0.01} = \frac{227}{484}$$

$$\Delta = 0.0047$$

Therefore $$z = 0.5047$$

Since $y = \pi - z$, we get

$$y = \pi - 0.5047$$

or $$y = 2.6369$$

Degree measure (Table V)

$$1'\left[\Delta\left[\begin{array}{l}\cos 28°55' = 0.87532 \\ \cos z \qquad\; = 0.87531\end{array}\right]1 \atop \cos 28°56' = 0.87518\right]14$$

(We write 14 instead of 0.00014, etc.)

$$\frac{\Delta}{1} = \frac{1}{14} \qquad \Delta = 0.07 \approx 0.1$$

Therefore $$z = 28°55.1'$$

Since $y = 180° - z$, we have

$$y = 180° - 28°55.1' = 151°4.9'$$

We can reduce $151°4.9'$ to radian measure by using Table VI of the Appendix:

$$\begin{aligned} 4.9' &= 0.00145 \text{ radian} \\ 1° &= 0.01745 \text{ radian} \\ 50° &= 0.87266 \text{ radian} \\ 100° &= 1.74533 \text{ radians} \\ y = 151°4.9' &= 2.63689 \text{ radians} \end{aligned}$$

ILLUSTRATION 3 Evaluate sin arc Sin 0.25837.

SOLUTION Keep in mind that arc Sin 0.25837 means "the number z whose sine is 0.25837, $0 \le z \le \pi/2$". The problem then can be stated as: "What is the sine of the number whose sine is 0.25837?" This is obviously 0.25837.

ILLUSTRATION 4 Evaluate arc Tan $\left(\text{Tan}\frac{\pi}{7}\right)$.

SOLUTION Now the tangent of $\pi/7$ is a number z. The problem is to find arc Tan z. The whole problem can be stated as: "What is the number whose tangent is tangent of $\pi/7$?" The answer is $\pi/7$.

Illustrations 3 and 4 are special cases of $f(f^{-1}) = f^{-1}(f) = E$ (see Sec. 9.8).

ILLUSTRATION 5 Find $y = \cos(\text{arc Tan}\, x + \pi/3)$.

SOLUTION We use $\cos(\phi + \theta) = \cos\phi\cos\theta - \sin\phi\sin\theta$.

$$\begin{aligned}\cos\left(\text{arc Tan}\, x + \frac{\pi}{3}\right) &= \cos(\text{arc Tan}\, x)\cdot\cos\frac{\pi}{3} - \sin(\text{arc Tan}\, x)\cdot\sin\frac{\pi}{3}\\ &= \frac{1}{\sqrt{1+x^2}}\cdot\frac{1}{2} - \frac{x}{\sqrt{1+x^2}}\cdot\frac{\sqrt{3}}{2}\\ &= \frac{1-\sqrt{3}\,x}{2\sqrt{1+x^2}}\end{aligned}$$

PROBLEMS 13.10

In Probs. 1 to 20 compute:

1 $\text{Sin}^{-1}\frac{1}{2}$

2 arc Sin $\frac{1}{2}\sqrt{3}$

3 $\text{Sin}^{-1}(-1)$

4 sin arc Sin 0.21583

5 arc Cos $\frac{1}{2}$

6 $\text{Cos}^{-1}\frac{1}{2}\sqrt{2}$

7 $\text{Cos}^{-1}(-\frac{1}{2}\sqrt{3})$

8 cos Cos^{-1} 0.72134

9 $\text{Tan}^{-1}\left(-\frac{1}{\sqrt{3}}\right)$

10 arc Tan $\sqrt{3}$

11 arc Cos $\left[\tan\left(-\frac{5\pi}{4}\right)\right]$

12 arc Tan $\left(\cos\frac{3\pi}{4}\right)$

13 $\text{Sin}^{-1}(\tan 2)$

14 $\cos(2\,\text{Tan}^{-1} 1)$

15 $\tan[\text{Tan}^{-1}\frac{1}{2} + \text{Tan}^{-1}(-\frac{2}{3})]$

16 sin (arc Sin $\frac{12}{13}$ + arc Sin $\frac{4}{5}$)

17 $\text{Tan}^{-1}[\tan(\text{Cos}^{-1}\frac{1}{2})]$

18 $\tan(2\,\text{Tan}^{-1}\frac{3}{4} + \text{Tan}^{-1}\frac{5}{12})$

19 arc Cos (arc Sin 0)

20 $\cos\left[\left(\text{arc Cos}\frac{2}{3}\right) + \frac{\pi}{2}\right]$

In Probs. 21 to 24 verify:

21 $2\ \mathrm{Tan}^{-1}\frac{1}{3} + \mathrm{Tan}^{-1}\frac{1}{7} = \pi/4$

22 $\mathrm{arc\ Sin}\ \frac{77}{85} - \mathrm{arc\ Sin}\ \frac{3}{5} = \mathrm{arc\ Cos}\ \frac{15}{17}$

23 $\mathrm{Cos}^{-1} x = \mathrm{Tan}^{-1}\dfrac{\sqrt{1-x^2}}{x}$ where $0 < x \leq 1$

24 $\mathrm{Tan}^{-1}\frac{1}{11} + \mathrm{Tan}^{-1}\frac{5}{6} = \mathrm{Tan}^{-1}\frac{1}{3} + \mathrm{Tan}^{-1}\frac{1}{2}$

In Probs. 25 to 28 sketch the graph.

25 $y = \frac{1}{2}\ \mathrm{arc\ Sin}\ 2x$

26 $y = 2\ \mathrm{arc\ Sin}\ \dfrac{x}{2}$

27 $y = \mathrm{Cos}\ x, y = \mathrm{arc\ Cos}\ x$

28 $y = \mathrm{Tan}\ x, y = \mathrm{arc\ Tan}\ x$

29 PROVE: The area K of the segment cut from a circle of radius r by a chord at a distance d from the center is given by

$$K = \left(r^2\ \mathrm{arc\ Cos}\ \frac{d}{r}\right) - \left(d\sqrt{r^2 - d^2}\right)$$

13.11 › COMPLEX NUMBERS

We have already met (Secs. 2.13 and 2.14) a complex number represented in rectangular form: $a + ib$ (or $a + bi$). There is a one-to-one correspondence between such numbers and points in the plane. Now since $a = r\cos\theta$ and $b = r\sin\theta$ (Fig. 13.26),

(1) $$a + ib = r(\cos\theta + i\sin\theta)$$

where $$r = |a + ib| = \sqrt{a^2 + b^2}$$

and $$\tan\theta = \frac{b}{a}$$

The real, nonnegative number $r\ (= \sqrt{a^2 + b^2})$ is called the *absolute value* (or *modulus*) of the complex number and is written $|a + ib|$. The angle

Polar form of $a + ib$

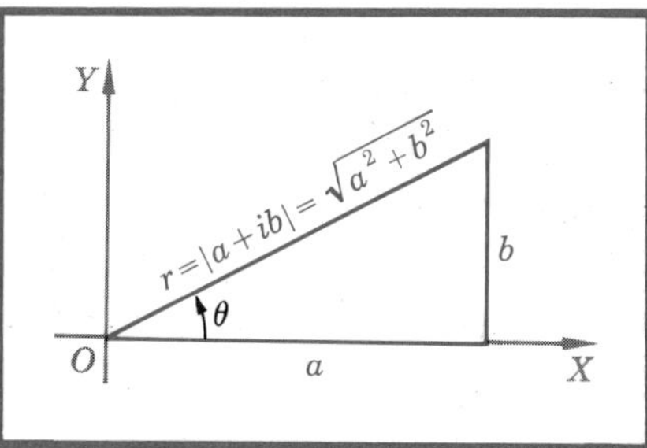

FIGURE 13.26 The graphical representation of complex numbers was discussed in Sec. 2.14. This figure illustrates how complex numbers can be represented in polar coordinates.

θ associated with the number $a + ib$ is called the *argument* (or *amplitude*) of $a + ib$. The left-hand side of (1) is the *rectangular form* and the right-hand side is the *polar form* of a complex number. A complex number is therefore a vector having both magnitude (absolute value) and direction (argument).

Polar form of a complex number

Addition (and subtraction) of complex numbers is best accomplished in rectangular ($a + ib$) form. Thus

$$(a + ib) \pm (c + id) = (a \pm c) + i(b \pm d)$$

But multiplication and division are conveniently treated in polar $[r(\cos\theta + i\sin\theta)]$ form. Often, to simplify the notation, we write $r(\cos\theta + i\sin\theta)$ in the form $r \operatorname{cis}\theta$.

MULTIPLICATION Consider $r_1 \operatorname{cis}\theta_1$ and $r_2 \operatorname{cis}\theta_2$, two complex numbers. Their product (Fig. 13.27) is given by

$$\begin{aligned} r_1 \operatorname{cis}\theta_1 \cdot r_2 \operatorname{cis}\theta_2 &= r_1r_2 \operatorname{cis}\theta_1 \operatorname{cis}\theta_2 \\ &= r_1r_2(\cos\theta_1 + i\sin\theta_1)(\cos\theta_2 + i\sin\theta_2) \\ &= r_1r_2[(\cos\theta_1\cos\theta_2 - \sin\theta_1\sin\theta_2) \\ &\qquad + i(\sin\theta_1\cos\theta_2 + \cos\theta_1\sin\theta_2)] \\ &= r_1r_2[\cos(\theta_1 + \theta_2) + i\sin(\theta_1 + \theta_2)] \\ &= r_1r_2 \operatorname{cis}(\theta_1 + \theta_2) \end{aligned}$$

Therefore the absolute value of the product is the product of the absolute values and the argument of the product is the sum of the arguments (plus or minus a multiple of 2π).

By similar reasoning

$$r_1 \operatorname{cis}\theta_1 \cdot r_2 \operatorname{cis}\theta_2 \cdot r_3 \operatorname{cis}\theta_3 = r_1r_2r_3 \operatorname{cis}(\theta_1 + \theta_2 + \theta_3)$$

Product of complex numbers

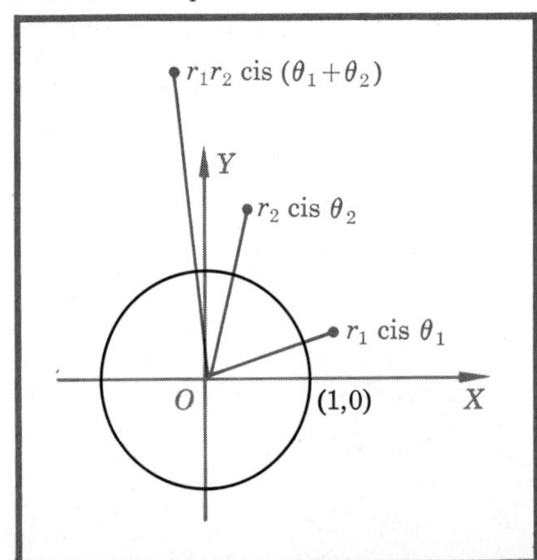

FIGURE 13.27
To multiply two complex numbers in polar form, add their arguments and multiply their absolute values.

If the three numbers θ_1, θ_2, θ_3 are all equal to θ, and if r_1, r_2, r_3 are all equal to r, we have

$$(r \operatorname{cis} \theta)^3 = r^3 \operatorname{cis} 3\theta$$

And similarly

(2) $$(r \operatorname{cis} \theta)^n = r^n \operatorname{cis} n\theta \qquad n \text{ a positive integer}$$

With proper interpretations, (2) is true for any real number n, but we shall not give the proof. This is known as De Moivre's Theorem.

THEOREM 1 **DE MOIVRE'S THEOREM** $(r \operatorname{cis} \theta)^n = r^n \operatorname{cis} n\theta$, n real

EXERCISE A Prove De Moivre's Theorem for the case where n is a positive integer. HINT: Use induction.

DIVISION To find the quotient of two complex numbers, write

$$\frac{r_1 \operatorname{cis} \theta_1}{r_2 \operatorname{cis} \theta_2} = \frac{r_1 \operatorname{cis} \theta_1}{r_2 \operatorname{cis} \theta_2} \times \frac{r_2 \operatorname{cis}(-\theta_2)}{r_2 \operatorname{cis}(-\theta_2)}$$
$$= \frac{r_1 r_2 \operatorname{cis}(\theta_1 - \theta_2)}{r_2^2 \operatorname{cis} 0}$$
$$= \frac{r_1}{r_2} \operatorname{cis}(\theta_1 - \theta_2)$$

Thus we see that the absolute value of the quotient of two complex numbers is the quotient of their absolute values and the argument of the quotient is the argument of the numerator minus the argument of the denominator (Fig. 13.28).

Roots of complex numbers

ROOTS OF COMPLEX NUMBERS First, we note that the argument of a complex number is not uniquely defined. If

$$a + ib = r \operatorname{cis} \theta$$

Quotient of complex numbers

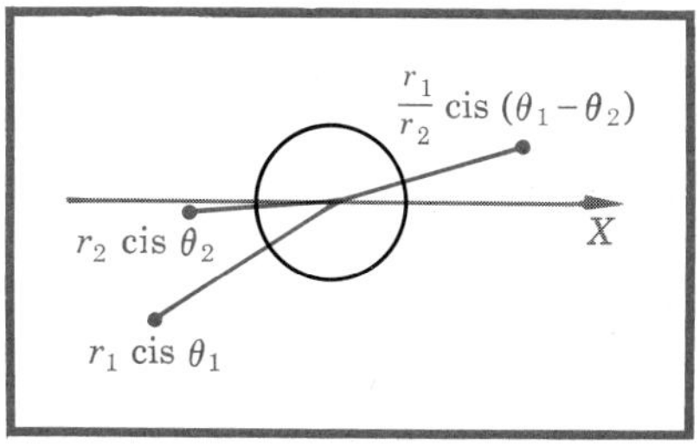

$\frac{r_1}{r_2} \operatorname{cis}(\theta_1 - \theta_2)$

$r_2 \operatorname{cis} \theta_2$

X

$r_1 \operatorname{cis} \theta_1$

FIGURE 13.28
To divide two complex numbers in polar form, take the difference of their arguments and the quotient of their absolute values.

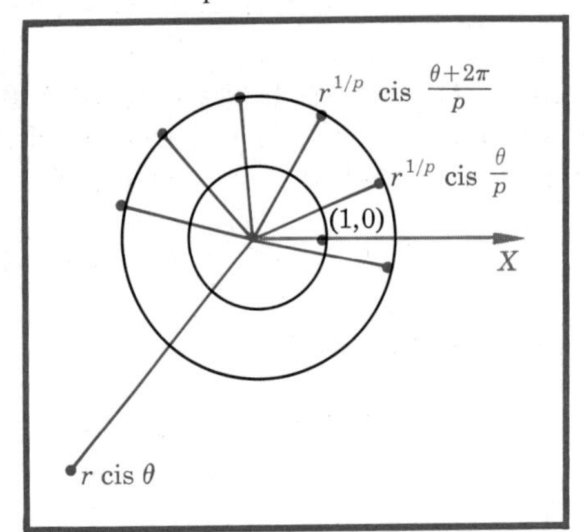

FIGURE 13.29
The p "pth roots" of a complex number are evenly spaced on a circle of radius $r^{1/p}$.

it is also equal to $r[\text{cis}\,(\theta + 2\pi n)]$ for integer n. Up to now this was not important, but we must use it here.

Given the complex number $r \text{ cis}\,\theta$, we seek to find all complex numbers whose pth powers are equal to $r \text{ cis}\,\theta$. These are called its pth roots. From De Moivre's Theorem we see at once that for every n

$$\left(r^{1/p} \text{ cis}\,\frac{\theta + 2\pi n}{p}\right)^p = r \text{ cis}\,(\theta + 2\pi n) = r \text{ cis}\,\theta$$

Therefore the numbers (Fig. 13.29)

$$\textbf{(3)} \qquad r^{1/p} \cos \frac{\theta + 2\pi n}{p}$$

are pth roots of $r \text{ cis}\,\theta$. It can be shown that these comprise all the pth roots.

EXERCISE B If $(R \text{ cis}\,\phi)^p = r \text{ cis}\,\theta$, show that $R \text{ cis}\,\phi$ must have the form (3) for some value of n.

Let us examine (3) for various values of n. Letting $n = 0$, we have

$$r^{1/p} \text{ cis}\,\frac{\theta}{p}$$

In more advanced books this is called the *principal* pth root of $r \text{ cis}\,\theta$ and is denoted by the symbol $(r \text{ cis}\,\theta)^{1/p}$. Letting $n = 1$, we have

$$r^{1/p} \text{ cis}\,\frac{\theta + 2\pi}{p}$$

Each of these two (distinct) numbers is a pth root of $r \text{ cis}\,\theta$. By letting $n = 2, 3, \ldots, p - 1$, we obtain $p - 2$ other distinct pth roots. Letting $n = p$, we have

$$r^{1/p} \operatorname{cis} \frac{\theta + 2\pi p}{p} = r^{1/p} \operatorname{cis} \frac{\theta}{p}$$

which yields the same result as did $n = 0$. And $n = p + 1$ yields the same result as $n = 1$, etc. Therefore there are p (distinct), and only p, pth roots of a complex number, $a + ib = r \operatorname{cis} \theta$. These are given by

$$\textbf{(4)} \qquad r^{1/p} \operatorname{cis} \frac{\theta + 2\pi n}{p} \qquad n = 0, 1, 2, \ldots, p - 1$$

You should memorize (4).

ILLUSTRATION 1 Find the three cube roots of 1 (Fig. 13.30). Since

$$1 = 1 + i0$$

$r = 1$, and $\theta = 0$. Thus $1 = \operatorname{cis}(0 + 2\pi n)$. The cube roots are

$$1^{1/3} \operatorname{cis} \frac{0 + 2\pi n}{3} \qquad n = 0, 1, 2$$

or

$$1 \operatorname{cis} 0 = 1$$

$$1 \operatorname{cis} \frac{2\pi}{3} = \cos \frac{2\pi}{3} + i \sin \frac{2\pi}{3} = -\frac{1}{2} + i\frac{\sqrt{3}}{2}$$

$$1 \operatorname{cis} \frac{4\pi}{3} = \cos \frac{4\pi}{3} + i \sin \frac{4\pi}{3} = -\frac{1}{2} - i\frac{\sqrt{3}}{2}$$

To check the result, multiply out $\left(-\frac{1}{2} + i\frac{\sqrt{3}}{2}\right)^3$ and $\left(-\frac{1}{2} - i\frac{\sqrt{3}}{2}\right)^3$. The results should be 1.

This example is equivalent to solving the equation

$$x^3 - 1 = 0$$

or

$$(x - 1)(x^2 + x + 1) = 0$$

Cube roots of 1

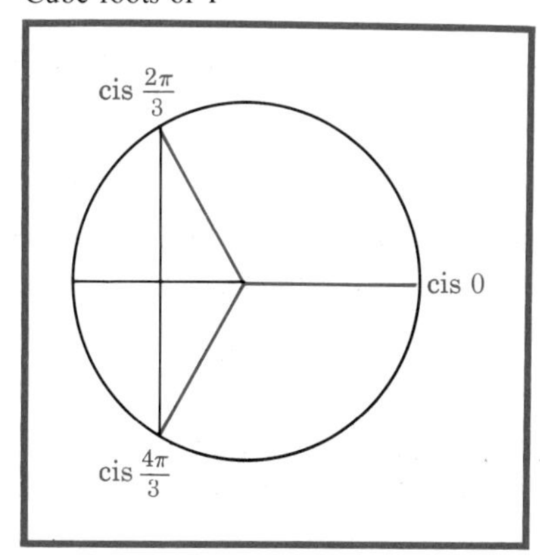

FIGURE 13.30

The roots are

$$x = 1$$

$$x = \frac{-1 \pm \sqrt{1-4}}{2} = -\frac{1}{2} \pm i\frac{\sqrt{3}}{2}$$

PROBLEMS 13.11 In Probs. 1 to 10 change the following to polar form:

1 $\sqrt{2} + \sqrt{2}i$ **2** $3 + 3\sqrt{3}i$

3 $-30i$ **4** $-2\sqrt{2} + 2\sqrt{2}i$

5 3 **6** -20

7 $-\frac{1}{2}\sqrt{3} + \frac{1}{2}i$ **8** $\sqrt{3} + i$

9 $-3 + 4i$ **10** $\dfrac{1-i}{\sqrt{2}}$

In Probs. 11 to 20 change the following to rectangular form:

11 $4 \text{ cis } 30°$ **12** $2 \text{ cis } 210°$

13 $10 \text{ cis } 240°$ **14** $5 \text{ cis } 315°$

15 $\sqrt{3} \text{ cis } 45°$ **16** $2\sqrt{2} \text{ cis } 45°$

17 $2\sqrt{2} \text{ cis } 225°$ **18** $2 \text{ cis } (-270°)$

19 $6 \text{ cis } (-\pi)$ **20** $4 \text{ cis } 90° + 12 \text{ cis } 120°$

21 Change to rectangular form and polar form:

$$14 \text{ cis } 120° + 6 \text{ cis } (-60°)$$

In Probs. 22 to 25 find the product of:

22 $3 \text{ cis } 60°$ and $4 \text{ cis } 80°$ **23** $\frac{1}{2} \text{ cis } 20°$ and $2 \text{ cis } 60°$

24 $4 \text{ cis } \frac{\pi}{3}$ and $3 \text{ cis } \left(-\frac{\pi}{3}\right)$ **25** $3 \text{ cis } \frac{\pi}{4}$ and $4 \text{ cis } \frac{\pi}{2}$

In Probs. 26 to 29 find the quotient of:

26 $2 \text{ cis } 120°$ by $3 \text{ cis } 60°$ **27** $16 \text{ cis } \frac{\pi}{8}$ by $4 \text{ cis } \frac{\pi}{4}$

28 $10 \text{ cis } 3\pi$ by $2 \text{ cis } \pi$ **29** $11 \text{ cis } 60°$ by $6 \text{ cis } 15°$

In Probs. 30 to 33 find the fourth power of:

30 $2 \text{ cis } \frac{\pi}{4}$ **31** $\frac{1}{2} \text{ cis } 15°$

32 $\sqrt{3} \text{ cis } 225°$ **33** $2 \text{ cis } \left(-\frac{\pi}{16}\right)$

34 Find the three cube roots of 27 cis $\frac{3\pi}{2}$. Plot them and the original number.
35 Find the four fourth roots of 1. Plot them and the original number.
36 Find the four fourth roots of -1. Plot them and the original number.
37 Find the four fourth roots of i. Plot them and the original number.
38 Find the three cube roots of 64 cis π. Plot them and the original number.

REFERENCES

Dubisch, Roy: "Trigonometry", Ronald, New York (1955).
Terman, F. E.: "Electronic and Radio Engineering", McGraw-Hill, New York (1955).

In addition to the many standard textbooks on trigonometry, the reader should consult the following articles in the *American Mathematical Monthly:*

Burton, L. J., and E. A. Hedberg: Proofs of the Addition Formulae for Sines and Cosines, vol. 56, p. 471 (1949).
Burton, L. J.: The Laws of Sines and Cosines, vol. 56, p. 550 (1949).
Carver, W. B.: Trigonometric Functions—of What?, vol. 26, p. 243 (1919).
Householder, A. S.: The Addition Formulas in Trigonometry, vol. 49, p. 326 (1942).
McShane, E. J.: The Addition Formulas for the Sine and Cosine, vol. 48, p. 688 (1941).
Vance, E. P.: Teaching Trigonometry, vol. 54, p. 36 (1947).

Also see:

Allendoerfer, C. B.: Angles, Arcs, and Archimedes, *The Mathematics Teacher,* vol. 58 (February, 1965).
———: The Method of Equivalence, *The Mathematics Teacher,* vol. 59, pp. 531–535 (1966).

14 ANALYTIC GEOMETRY

14.1 » INTRODUCTION

René Descartes (1596–1650) introduced the subject of analytic geometry with the publishing of his "La Géométrie" in 1637. Accordingly it is often referred to as *cartesian* geometry; it is, essentially, merely a method of studying geometry by means of a coordinate system and an associated algebra. The application of this basic idea enabled the mathematicians of the seventeenth century to make the first noteworthy advances in the field of geometry since the days of Euclid. The next great advance came with the invention of the calculus (see Chaps. 15 and 16).

There are two central problems in plane analytic geometry:

1. Given an equation in x and y, to plot its graph, or to represent it geometrically as a set of points in the plane.
2. Given a set of points in the plane, defined by certain geometric conditions, to find an equation whose graph will consist wholly of this set of points.

Locus

The second problem is frequently called a *locus problem*. A locus is the geometric counterpart of a relation, and we define it as follows:

DEFINITION A *locus* is a subset of the set of points in the plane.

A locus is defined by some geometric conditions, usually expressed in words. Let P represent an arbitrary point in the plane; then the following are examples of loci:

1. $\{P \mid P$ is at a fixed distance r from a point $C\}$; this locus is then a circle with radius r and center C.
2. $\{P \mid PA = PB$, where A and B are fixed points$\}$; this locus is the perpendicular bisector of the segment AB.

3 $\{P \mid P$ is a fixed point on the rim of a wheel which rolls along a line$\}$; this locus is called a cycloid.

Many loci are defined in terms of a physical notion like example 3. For this reason you may run across statements like: "The locus of a point which moves so that . . .". Since there is no motion in geometry, we prefer to avoid this language except in applications to mechanics.

When we are given such a locus, the problem before us is to find the corresponding relation. That is, we seek an equation whose graph is the given locus. We call this an *equation of the locus*. Having found such an equation, we study its properties by algebraic means and thus derive properties of the locus.

14.2 THE STRAIGHT LINE

We begin our discussion of analytic geometry by treating straight lines in the plane. There are many geometric methods for determining the position of such a line, and from among these we shall choose the following as a starting point for our discussion: A line in the plane is determined by (1) a point P through which it passes and (2) the angle θ which is formed by the line and the X-axis. Since *the* angle between two undirected lines is ambiguous (Sec. 12.3), it is necessary to define this angle more carefully as follows:

DEFINITION

Inclination

The inclination θ of a line in the plane is the angle whose measure $\theta°$ satisfies the inequality $0° \leq \theta° < 180°$, whose initial ray is along the X-axis in the positive direction and whose terminal ray lies on the given line and is directed upward (Fig. 14.1). The inclination of a horizontal line is $0°$.

If we are given a point $P(x_0,y_0)$ and an angle θ such that $0° \leq \theta° < 180°$, we can therefore draw the line through P with inclination θ, and now our task is to find an equation whose graph is this line. To do so we construct a new set of axes X' and Y' whose new origin is the point $P'(x_0,y_0)$ (Fig. 14.2). Each point P in the plane now has two sets of coordinates (x,y) and (x',y'). From the figure it follows that

$$x' = x - x_0 \qquad \text{and} \qquad y' = y - y_0$$

In Fig. 14.3 we have drawn a line through P_0, which is not parallel to the Y-axis. From elementary geometry we know that the inclination θ of this line can be taken to be either the appropriate angle between the line and the X-axis or the corresponding angle between the line and the X'-axis. Now choose an arbitrary point P on the line with coordinates

Inclination of a line

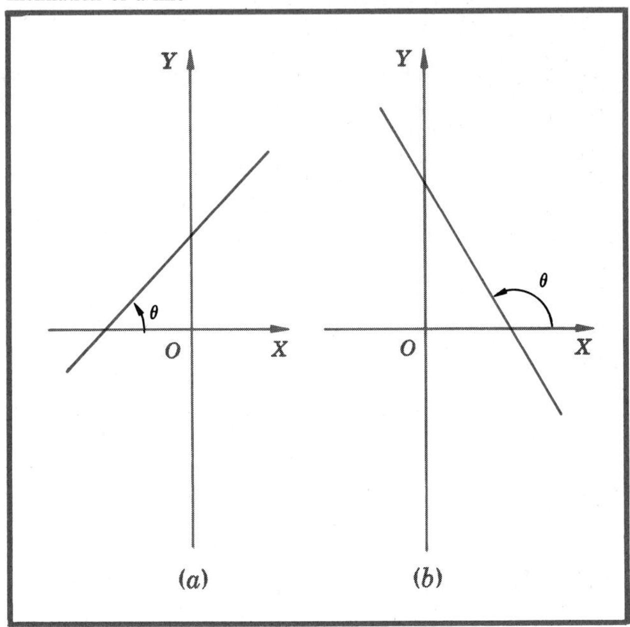

FIGURE 14.1
The inclination θ satisfies the inequality $0° \leq \theta° < 180°$.

The axes X' and Y'

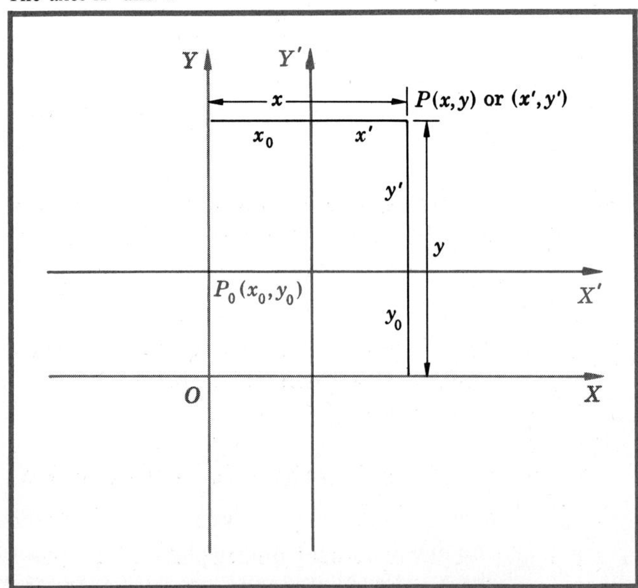

FIGURE 14.2
The new axes are drawn parallel to the old axes and pass through $P_0(x_0,y_0)$. The old and new coordinates of P are related by the equations

$$x' = x - x_0,$$
$$y' = y - y_0.$$

Line referred to both sets of axes

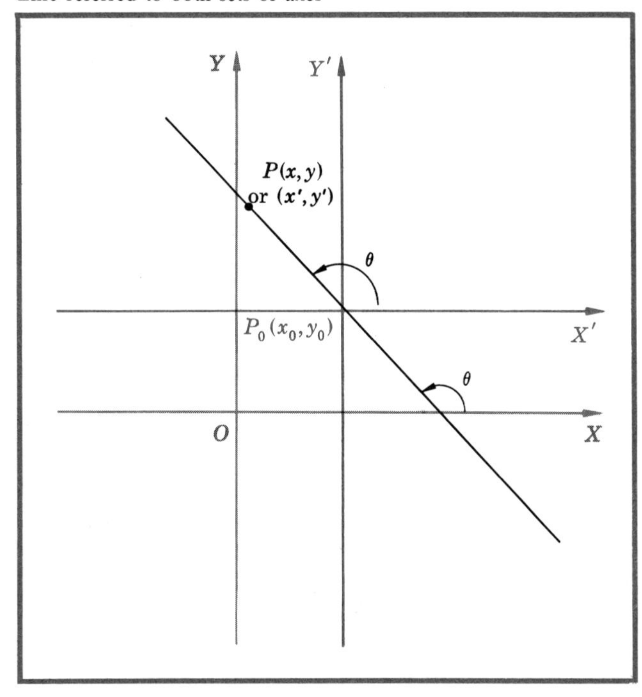

FIGURE 14.3
The inclination is angle θ. For any point P on the line $y' = x' \tan\theta$, and $y - y_0 = (x - x_0)\tan\theta$.

(x,y) or (x',y'). Using the X'- and Y'-axes and elementary trigonometry, we see that regardless of the position of P on the line,

$$\frac{y'}{x'} = \tan\theta \qquad \text{or} \qquad y' = x' \tan\theta$$

In terms of x and y this can be rewritten

$$y - y_0 = (x - x_0)\tan\theta$$

Slope

The expression $\tan\theta$ occurs so often in this chapter that we give it a special name, the *slope* of the line, and simplify our notation by writing $m = \tan\theta$.

DEFINITION The slope m of any nonvertical line in the plane is defined to be the tangent of the angle of inclination of this line. Slope is not defined for vertical lines.

In terms of these notations we have proved the following theorem:

THEOREM 1 Let a nonvertical line l in the plane pass through point $P_0(x_0,y_0)$ and have slope m. Then an arbitrary point $P(x,y)$ in the plane lies on this

In Probs. 11 to 18 find the slope and the intercepts of the given line. Write the equation of the line in the two-intercept form and in the slope-intercept form.

11 $x + 3y - 6 = 0$ **12** $2x - y + 4 = 0$
13 $-3x + 4y + 12 = 0$ **14** $5x - 3y - 15 = 0$
15 $x + 7y - 5 = 0$ **16** $2x + 5y + 8 = 0$
17 $3x - 6 = 0$ **18** $4y + 2 = 0$

In Probs. 19 to 24 find an equation of the line satisfying the given conditions.

19 Passing through $(4,-3)$; inclination $30°$
20 Passing through $(2,-3)$; inclination $60°$
21 Passing through $(-2,3)$; x-intercept is twice y-intercept.
22 Passing through (x_1,y_1); equal intercepts
23 Passing through the origin and the intersection of $x = a$ and $y + b = 0$
24 Passing through $(1,2)$ and the point of intersection of $2x - 3y = 0$ and $x + 5y - 4 = 0$

25 Show that every horizontal line has an equation of the form $y - y_0 = 0$.
26 If $A = 0$ and $B = 0$, what is the graph of $Ax + By + C = 0$? Consider the two cases $C = 0$ and $C \neq 0$.

14.5 › OTHER PROPERTIES OF LINES IN THE PLANE

Parallel lines

PARALLEL LINES Two nonvertical lines in the plane are parallel if they are distinct and have the same inclination. Thus they have the same slope. Since the slope of a nonvertical line whose equation is $Ax + By + C = 0$ is $m = -A/B$, it follows that the equations

$$\left.\begin{aligned} Ax + By + C_1 = 0 \\ Ax + By + C_2 = 0 \end{aligned}\right\} \qquad \begin{aligned} &C_1 \neq C_2 \text{ and not both} \\ &A \text{ and } B \text{ are zero} \end{aligned}$$

represent parallel lines.

ILLUSTRATION 1 Find an equation of a line parallel to $3x - 4y + 5 = 0$ and passing through $(1,-2)$.

SOLUTION Any line whose equation is of the form $3x - 4y + c = 0$ is parallel to $3x - 4y + 5 = 0$. To find c, substitute $(x,y) = (1,-2)$. Then $3 + 8 + c = 0$, or $c = -11$. The answer is $3x - 4y - 11 = 0$.

ILLUSTRATION 2 Find an equation of a line parallel to $x = 5$ and passing through $(2,-7)$.

SOLUTION Since the given line is vertical, the required line must also be vertical and pass through (2,−7). An equation for it is then $x = 2$.

ILLUSTRATION 3 Show that the line joining (1,3) and (2,−4) is parallel to the line joining (2,−3) and (3,−10).

SOLUTION They are parallel since

$$m_1 = \frac{-4-3}{2-1} = -7 \qquad \text{and} \qquad m_2 = \frac{-10+3}{3-2} = -7$$

PERPENDICULAR LINES If a line l_1 has inclination θ_1, then any line l_2 perpendicular to l_1 has an inclination θ_2 of $\theta_1 + 90°$ or else $\theta_1 - 90°$. From Theorem 2, Chap. 12, it follows that

$$m_2 = \tan(\theta_1 \pm 90°) = -\cot\theta_1 = -\frac{1}{m_1}$$

We, therefore, have proved Theorem 4.

THEOREM 4 Perpendicular lines

Two lines with slopes m_1 and m_2 are perpendicular if and only if $m_1 m_2 + 1 = 0$.

A useful consequence of this theorem is the corollary:

COROLLARY

If $Ax + By + C = 0$ is an equation of the nonvertical line l_1, then any line l_2 perpendicular to l_1 has an equation of the form

$$Bx - Ay + D = 0$$

PROOF The slope of l_1 is $m_1 = -A/B$. The slope of l_2 is $m_2 = B/A$. Hence $m_1 m_2 + 1 = 0$.

ILLUSTRATION 4 Find an equation of the line passing through (1,3) and perpendicular to the line $2x + y - 7 = 0$.

SOLUTION The equation of a line perpendicular to $2x + y - 7 = 0$ has the form $x - 2y + c = 0$. To find c put $(x,y) = (1,3)$. Then $1 - 6 + c = 0$ or $c = 5$. The answer is $x - 2y + 5 = 0$.

ILLUSTRATION 5 Find an equation of the line l_2 passing through (2,−5) and perpendicular to the line l_1 whose equation is $x = 7$.

SOLUTION Since l_1 is vertical, l_2 must be horizontal, and so its equation is $y = -5$.

ILLUSTRATION 6 Show that the line joining (3,1) and (−2,2) is perpendicular to the line joining (3,4) and (4,9).

SOLUTION We find that $m_1 = -\frac{1}{5}$ and $m_2 = 5$. Hence $m_1 m_2 + 1 = 0$, and the lines are perpendicular.

MIDPOINT OF A LINE SEGMENT Consider a line segment $P_1(x_1,y_1)$, $P_2(x_2,y_2)$. We seek the coordinates $(\bar{x},\bar{y})$ of the midpoint P in terms of x_1, y_1, x_2, and y_2. From Fig. 14.5 it is evident that

(1) $$\frac{x_2 - x_1}{P_1P_2} = \frac{\bar{x} - x_1}{P_1P}$$

But $P_1P_2 = 2P_1P$. Therefore (1) becomes

$$\frac{x_2 - x_1}{2} = \frac{\bar{x} - x_1}{1}$$

from which we get

$$2\bar{x} - 2x_1 = x_2 - x_1$$
$$2\bar{x} = x_1 + x_2$$

Midpoint formula

or $$\bar{x} = \frac{x_1 + x_2}{2}$$

Similarly, $$\bar{y} = \frac{y_1 + y_2}{2}$$

EXERCISE A From Fig. 14.5 derive the expression for $\bar{y}$.

Thus the x-coordinate of the midpoint is the average of the x-coordinates of the endpoints; the y-coordinate is the average of the y-coordinates of the endpoints. For example, the midpoint of the segment whose endpoints are (−1,5), (4,−7) has coordinates $(\frac{3}{2},-1)$.

Midpoint of a segment

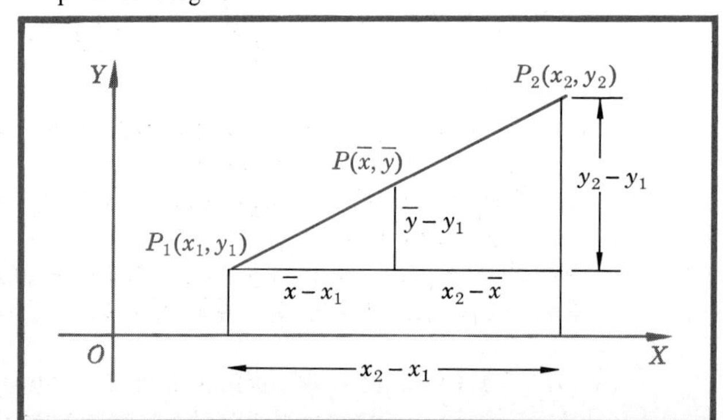

FIGURE 14.5 The midpoint of P_1P_2 has coordinates $\bar{x} = (x_1 + x_2)/2$, $\bar{y} = (y_1 + y_2)/2$.

EXERCISE B Find the coordinates of the midpoints of the sides of the triangle whose vertices are $A(4,7)$, $B(-3,-3)$, $C(2,-5)$.

PROBLEMS 14.5

In Probs. 1 to 6 find an equation of the line through the given point and parallel to the given line.

1 $(1,-3)$, $2x - 4y + 7 = 0$
2 $(4,2)$, $x + 3y - 5 = 0$
3 $(1,4)$, $-3x + 2y + 9 = 0$
4 $(0,2)$, $5x - 3 = 0$
5 $(2,-1)$, line with slope 3
6 $(-3,5)$, line with slope -4

In Probs. 7 to 12 find an equation of the line through the given point and perpendicular to the given line.

7 $(4,2)$, $x - 3y + 7 = 0$
8 $(1,-3)$, $2x + 5y + 3 = 0$
9 $(5,4)$, $-3x + 2y - 8 = 0$
10 $(3,3)$, $7x - y + 1 = 0$
11 $(1,4)$, line with slope $\frac{1}{2}$
12 $(-2,5)$, line with slope -4

In Probs. 13 to 16 find the midpoint of the segment whose endpoints are:

13 $(1,-3)$, $(2,5)$
14 $(2,7)$, $(1,-4)$
15 $(4,0)$, $(0,9)$
16 (x_1,y_1), (x_2,y_2)

In Probs. 17 to 20 the point P is the midpoint of P_1P_2. Find the coordinates of:

17 P_2, given $P_1(3,4)$, $P(5,6)$
18 P_2, given $P_1(-3,10)$, $P(2,2)$
19 P_1, given $P(2,3)$, $P_2(1,-5)$
20 P_1, given $P(2,4)$, $P_2(10,16)$

21 Given a triangle with vertices $P_1(x_1,y_1)$, $P_2(x_2,y_2)$, $P_3(x_3,y_3)$:
   a Find the midpoints of the sides.
   b Find the equations of the medians.
   c Prove that the medians intersect in the point

$$\left(\frac{x_1 + x_2 + x_3}{3}, \frac{y_1 + y_2 + y_3}{3}\right)$$

22 Find an equation of the perpendicular bisector of the segment with endpoints $A(2,-4)$ and $B(-4,8)$.
23 Prove that the triangle $A(2,1)$, $B(1,3)$, $C(-2,5)$ is not isosceles and not equilateral.
24 Prove that the triangle $A(3,2)$, $B(2,4)$, $C(1,1)$ is a right triangle.
25 Show that $A(-3,5)$, $B(4,9)$, and $C(-17,-3)$ are on the same straight line.
26 Find the slope of each side of the triangle $A(5,1)$, $B(-8,2)$, $C(0,3)$.
27 Find the slopes of the medians of the triangle $A(0,0)$, $B(-2,6)$, $C(8,4)$.

28 Prove that $A(0,-1)$, $B(-3,2)$, $C(1,9)$, and $D(4,6)$ are the vertices of a parallelogram.

29 Write an equation which states that $P(x,y)$ is 2 units from the point $(-3,2)$.

30 Write an equation which states that $P(x,y)$ is two times as far from $(4,0)$ as it is from $(-4,0)$.

31 The points $A(1,3)$, $B(3,1)$, $C(1,-1)$ are vertices of a square. Find the coordinates of the fourth vertex.

32 Plot the four points $A(2,6)$, $B(4,2)$, $C(-6,-8)$, $D(-8,2)$. Find the coordinates of the midpoints of the sides of the quadrilateral $ABCD$. Prove that the midpoints are the vertices of a parallelogram.

33 For the line segment P_1P_2, where $P_1(x_1,y_1)$ and $P_2(x_2,y_2)$, find the coordinates of $P(x,y)$ such that $P_1P/PP_2 = r_1/r_2$ where r_1 and r_2 are two given real numbers.

34 Find the coordinates of the point P such that $P_1P = 3PP_2$ where P_1 is $(2,3)$ and P_2 is $(8,12)$.

14.6 » DIRECTED LINES AND VECTORS

Directed lines

So far in this chapter all lines have been undirected, but now we must introduce the concept of a directed line. We *direct* a line by putting an arrow on it to indicate which direction is to be considered positive. This has already been done for the coordinate axes, and now we are extending this concept to other lines.

Vector along a line

Consider a directed line l through the point $P_1(x_1,y_1)$ and another point $P_2(x_2,y_2)$ on l such that the directed segment $\overrightarrow{P_1P_2}$ is in the positive direction of l (Fig. 14.6). Then $\overrightarrow{P_1P_2}$ can be considered to be a *vector* with components $x_2 - x_1$ and $y_2 - y_1$. The length of the vector $(x_2 - x_1, y_2 - y_1)$ is given by

$$d = \sqrt{(x_2 - x_1)^2 + (y_2 - y_1)^2}$$

and so is the same as the distance between P_1 and P_2.

A unit vector $\mathbf{L}$ (that is, a vector of length one) in the same direction as the line is, therefore, given by the expression

Unit vector

$$\mathbf{L} = \left(\frac{x_2 - x_1}{d}, \frac{y_2 - y_1}{d}\right)$$

EXERCISE A Verify that the length of $\mathbf{L}$ is one.

From Fig. 14.6 it is apparent that

$$\frac{x_2 - x_1}{d} = \cos\alpha \qquad \text{and} \qquad \frac{y_2 - y_1}{d} = \cos\beta$$

Direction cosines of a directed line

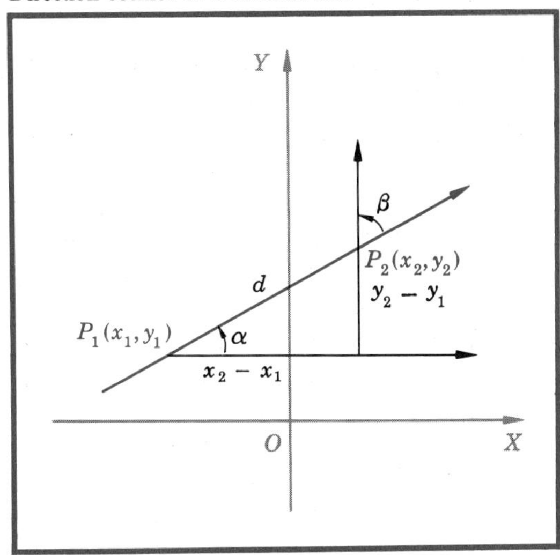

FIGURE 14.6

$$\cos \alpha = \frac{x_2 - x_1}{d}$$

$$\cos \beta = \frac{y_2 - y_1}{d}$$

where α is the unsigned angle between the positive direction of l and the positive X-axis, and β is the unsigned angle between the positive direction of l and the positive Y-axis. Thus **L** has the expression

$$\mathbf{L} = (\cos \alpha, \cos \beta)$$

Direction cosines

The components of **L**, namely $\cos \alpha$ and $\cos \beta$, are called the *direction cosines* of the directed line l.

EXERCISE B Prove that $\cos^2 \alpha + \cos^2 \beta = 1$. See Exercise A.

EXERCISE C Prove that $|\sin \alpha| = |\cos \beta|$ and $|\sin \beta| = |\cos \alpha|$.

ILLUSTRATION 1 Find the direction cosines of the line through $P_1(3,5)$ and $P_2(2,-6)$ directed from P_1 to P_2.

SOLUTION $x_2 - x_1 = -1$, $y_2 - y_1 = -11$. So $d = \sqrt{(-11)^2 + (-1)^2} = \sqrt{122}$.

Therefore, $\cos \alpha = -1/\sqrt{122}$ and $\cos \beta = -11/\sqrt{122}$.

ILLUSTRATION 2 Given the direction cosines $\cos \alpha$ and $\cos \beta$ of a directed line, find its slope.

SOLUTION $$m = \frac{y_2 - y_1}{x_2 - x_1} = \frac{(y_2 - y_1)/d}{(x_2 - x_1)/d} = \frac{\cos \beta}{\cos \alpha}$$

ILLUSTRATION 3 Given the slope m of a directed line, find its direction cosines.

SOLUTION From Fig. 14.6, $\theta = \alpha$. So $\tan\alpha = \tan\theta = m$. But $\sec^2\alpha = 1 + \tan^2\alpha = 1 + m^2$. So $\cos\alpha = \pm 1/\sqrt{1+m^2}$. From Illustration 2, $\cos\beta = m\cos\alpha = \pm m/\sqrt{1+m^2}$. The $\pm$ sign is ambiguous since the slope is a property of an undirected line. Depending on the direction chosen, we have either

$$\cos\alpha = \frac{1}{\sqrt{1+m^2}} \qquad \cos\beta = \frac{m}{\sqrt{1+m^2}}$$

or

$$\cos\alpha = \frac{-1}{\sqrt{1+m^2}} \qquad \cos\beta = \frac{-m}{\sqrt{1+m^2}}$$

ILLUSTRATION 4 Given the equation of a line

$$Ax + By + C = 0 \qquad \text{not both } A \text{ and } B \text{ zero}$$

find its direction cosines. Since $m = -A/B$, it follows from Illustration 3 that

$$\cos\alpha = \frac{B}{\sqrt{A^2+B^2}} \qquad \cos\beta = \frac{-A}{\sqrt{A^2+B^2}}$$

or

$$\cos\alpha = \frac{-B}{\sqrt{A^2+B^2}} \qquad \cos\beta = \frac{A}{\sqrt{A^2+B^2}}$$

depending on the chosen direction of the line.

ILLUSTRATION 5 If $Ax + By + C = 0$ is an equation for l, prove that the vector (A,B) is perpendicular to l.

SOLUTION Two vectors are perpendicular if their inner product is zero (Secs. 7.3 and 12.13). Hence in this case perpendicularity requires that $(\cos\alpha, \cos\beta)\cdot(A,B) = 0$. From Illustration 4, it follows that this is the same condition as:

$$\left(\frac{\pm B}{\sqrt{A^2+B^2}}, \frac{\mp A}{\sqrt{A^2+B^2}}\right)\cdot(A,B) = 0$$

which can be immediately verified to be true.

ILLUSTRATION 6 Let l pass through $P_0(x_0,y_0)$ and have direction cosines $\cos\alpha$ and $\cos\beta$. Find an equation for l.

SOLUTION From the point-slope form, an equation for l is

$$y - y_0 = m(x - x_0)$$

From Illustration 3, $m = \cos\beta/\cos\alpha$. So an equation for l is

$$y - y_0 = \frac{\cos\beta}{\cos\alpha}(x - x_0)$$

or $$(x - x_0)\cos\beta - (y - y_0)\cos\alpha = 0$$

We now employ direction cosines to obtain three important geometrical results.

ANGLE BETWEEN TWO DIRECTED LINES From our discussion of the inner product of two vectors in Secs. 7.3 and 12.13 we know that for any two vectors **A** and **B**,

$$\mathbf{A} \cdot \mathbf{B} = |\mathbf{A}| \times |\mathbf{B}| \times \cos\theta$$

where $|\mathbf{A}|$ and $|\mathbf{B}|$ are the lengths of **A** and **B**, respectively, and θ is the angle between them. Suppose, then, that we have two directed lines, l_1 and l_2 with their associated unit vectors:

$$\mathbf{L}_1 = (\cos\alpha_1, \cos\beta_1) \qquad \text{and} \qquad \mathbf{L}_2 = (\cos\alpha_2, \cos\beta_2)$$

It is then an immediate conclusion that the angle θ between these directed lines can be computed from the equation

$$\mathbf{L}_1 \cdot \mathbf{L}_2 = 1 \times 1 \times \cos\theta$$

This can be rewritten as

Angle between directed lines

$$\cos\theta = \cos\alpha_1 \cos\alpha_2 + \cos\beta_1 \cos\beta_2$$

or as

(1) $$\theta = \text{Cos}^{-1}(\cos\alpha_1 \cos\alpha_2 + \cos\beta_1 \cos\beta_2)$$

ILLUSTRATION 7 Find the angle between the directed lines l_1 and l_2 where l_1 passes through $P_1(3,-1)$ and $Q_1(1,4)$ and is directed from P_1 to Q_1, and l_2 passes through $P_2(2,5)$ and $Q_2(-2,1)$ and is directed from P_2 to Q_2.

SOLUTION $d_1 = \sqrt{5^2 + 2^2} = \sqrt{29}$, $\cos\alpha_1 = -2/\sqrt{29}$, $\cos\beta_1 = 5/\sqrt{29}$. Also $d_2 = \sqrt{4^2 + 4^2} = 4\sqrt{2}$, $\cos\alpha_2 = -4/4\sqrt{2} = -1/\sqrt{2}$, $\cos\beta_2 = -4/4\sqrt{2} = -1/\sqrt{2}$. Hence

$$\cos\theta = \left(\frac{-2}{\sqrt{29}}\right)\left(\frac{-1}{\sqrt{2}}\right) + \left(\frac{5}{\sqrt{29}}\right)\left(\frac{-1}{\sqrt{2}}\right) = \frac{-3}{\sqrt{58}}$$

Hence $$\theta = \text{Cos}^{-1}\left(\frac{-3}{\sqrt{58}}\right)$$

ILLUSTRATION 8 Find the cosine of the angle B of the triangle $A(0,0)$, $B(2,-1)$, $C(9,2)$.

SOLUTION In order to obtain an angle (interior) of a triangle, we *think* of the sides as being directed *away* from that particular vertex. To obtain angle B, therefore, we impose the directions $\overrightarrow{BA}$ and $\overrightarrow{BC}$. We compute

$$d_{BA} = \sqrt{5} \qquad\qquad d_{BC} = \sqrt{58}$$

$$\cos\alpha_{BA} = -\frac{2}{\sqrt{5}} \qquad\qquad \cos\alpha_{BC} = \frac{7}{\sqrt{58}}$$

$$\cos\beta_{BA} = \frac{1}{\sqrt{5}} \qquad\qquad \cos\beta_{BC} = \frac{3}{\sqrt{58}}$$

Therefore

$$\begin{aligned}\cos\theta = \cos B &= \cos\alpha_{BA}\cos\alpha_{BC} + \cos\beta_{BA}\cos\beta_{BC} \\ &= \frac{-14}{\sqrt{5}\sqrt{58}} + \frac{3}{\sqrt{5}\sqrt{58}} \\ &= \frac{-11}{\sqrt{5}\sqrt{58}}\end{aligned}$$

The angle B is obtuse.

ILLUSTRATION 9 Find the cosine of the acute angle between the undirected lines

$$A_1x + B_1y + C_1 = 0$$
$$A_2x + B_2y + C_2 = 0$$

SOLUTION From Illustration 4,

$$\cos\alpha_1 = \frac{\pm B_1}{\sqrt{A_1^2 + B_1^2}} \qquad \cos\beta_1 = \frac{\mp A_1}{\sqrt{A_1^2 + B_1^2}}$$

and

$$\cos\alpha_2 = \frac{\pm B_2}{\sqrt{A_2^2 + B_2^2}} \qquad \cos\beta_2 = \frac{\mp A_2}{\sqrt{A_2^2 + B_2^2}}$$

Therefore

$$\cos\theta = \frac{\pm(A_1A_2 + B_1B_2)}{\sqrt{A_1^2 + B_1^2}\sqrt{A_2^2 + B_2^2}}$$

where the sign in front must still be determined.

Since θ is required to be acute, $\cos\theta$ must be positive. Hence

$$\cos\theta = \frac{|A_1A_2 + B_1B_2|}{\sqrt{A_1^2 + B_1^2}\sqrt{A_2^2 + B_2^2}}$$

DISTANCE BETWEEN TWO PARALLEL LINES Let

$$Ax + By + C_1 = 0$$
$$Ax + By + C_2 = 0$$

be two parallel lines which are not parallel to the Y-axis (Fig. 14.7). Their respective y-intercepts are $b_1 = -C_1/B$ and $b_2 = -C_2/B$. From the figure, the distance d between the two lines is

$$d = |(b_1 - b_2)\sin\beta|$$

Distance between parallel lines

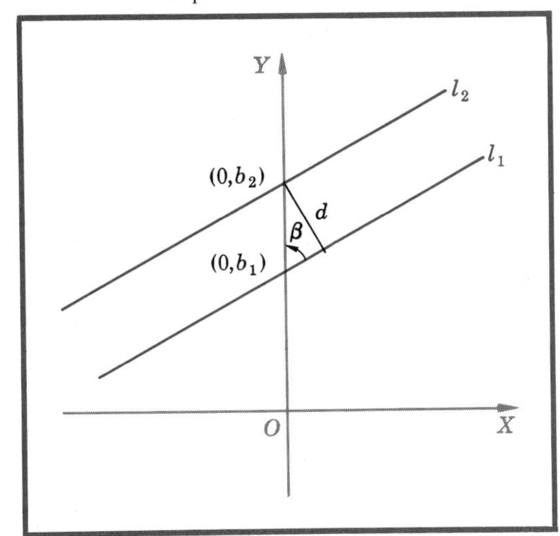

FIGURE 14.7 The distance $d = |(b_2 - b_1)\sin\beta| = |(b_2 - b_1)\cos\alpha|$. If the lines have the equations $A_1x + B_1y + C_1 = 0$ and $A_2x + B_2y + C_2 = 0$, then $d = \dfrac{|C_2 - C_1|}{\sqrt{A^2 + B^2}}$.

Since

$$|\sin\beta| = |\cos\alpha| = \frac{|B|}{\sqrt{A^2 + B^2}} \qquad \text{(Illustration 4)}$$

it follows that

Distance between parallel lines

$$d = \left|\frac{-C_1}{B} + \frac{C_2}{B}\right| \times \frac{|B|}{\sqrt{A^2 + B^2}} = \frac{|C_2 - C_1|}{\sqrt{A^2 + B^2}}$$

EXERCISE D Show that this formula is still true for the vertical lines $Ax + C_1 = 0$ and $Ax + C_2 = 0$.

DISTANCE BETWEEN A POINT AND A LINE We are seeking the (perpendicular) distance between $P_0(x_0,y_0)$ and the line l whose equation is $Ax + By + C = 0$. This is clearly the same as the distance between l and the line l_0 through P_0 parallel to l, whose equation is

$$Ax + By - (Ax_0 + By_0) = 0$$

Hence

Distance between point and line

$$d = \frac{|-(Ax_0 + By_0) - C|}{\sqrt{A^2 + B^2}} = \frac{|Ax_0 + By_0 + C|}{\sqrt{A^2 + B^2}}$$

ILLUSTRATION 10 The distance d between the parallel lines $3x - 4y + 7 = 0$ and $3x - 4y - 3 = 0$ is

$$d = \frac{|7 - (-3)|}{5} = 2$$

The distance d between the point $(2,-3)$ and the line $3x - 4y + 2 = 0$ is

$$d = \frac{|(3)(2) - 4(-3) + 2|}{5} = 4$$

PROBLEMS 14.6 In Probs. 1 to 8 find the direction cosines of the line from the first point to the second point.

1 $(1,3)$, $(2,4)$

2 $(-1,4)$, $(2,3)$

3 $(-3,-5)$, $(2,-1)$

4 $(4,-3)$, $(6,-3)$

5 $(4,-9)$, $(2,3)$

6 $(5,-25)$, $(-20,-10)$

7 $(5,6)$, $(-2,6)$

8 $(4,1)$, $(4,-10)$

In Probs. 9 to 12 find the cosine of the angle between the two lines, where the first is directed from A to B and the second from C to D.

9 $A(0,0)$, $B(1,3)$; $C(3,-1)$, $D(2,-2)$

10 $A(1,2)$, $B(-3,5)$; $C(5,3)$, $D(-4,-1)$

11 $A(3,3)$, $B(-2,1)$; $C(-3,0)$, $D(3,1)$

12 $A(1,6)$, $B(-2,8)$; $C(1,-2)$, $D(-1,-3)$

In Probs. 13 to 16 find (to within a common sign) the direction cosines of the lines whose equations are:

13 $2x + 5y - 9 = 0$

14 $3x - 7y + 12 = 0$

15 $-4x + 5y - 7 = 0$

16 $3x - 5 = 0$

In Probs. 17 to 20 find the slope of the line whose direction cosines are:

17 $\frac{2}{3}, -\frac{\sqrt{5}}{3}$

18 $\frac{3}{7}, \frac{2\sqrt{10}}{7}$

19 $\frac{\sqrt{2}}{2}, \frac{\sqrt{2}}{2}$

20 $-\frac{3}{5}, -\frac{4}{5}$

In Probs. 21 to 24 find an equation of the line with the given direction cosines and passing through the given point.

21 $\cos\alpha = \frac{\sqrt{5}}{5}$, $\cos\beta = \frac{2\sqrt{5}}{5}$, $P(2,-1)$

22 $\cos\alpha = \frac{3\sqrt{11}}{11}$, $\cos\beta = \frac{-2\sqrt{11}}{11}$, $P(1,3)$

23 $\cos\alpha = \frac{\sqrt{10}}{10}$, $\cos\beta = \frac{3\sqrt{10}}{10}$, $P(-1,4)$

24 $\cos\alpha = 0$, $\cos\beta = -1$, $P(4,5)$

In Probs. 25 to 30 find the cosine of the angle A of the triangle.

25 $A(0,0)$, $B(-2,4)$, $C(10,5)$
26 $A(0,0)$, $B(5,2)$, $C(-4,10)$
27 $A(3,7)$, $B(-3,-1)$, $C(0,-2)$
28 $A(-6,1)$, $B(0,-2)$, $C(1,15)$
29 $A(3,-1)$, $B(4,3)$, $C(2,1)$
30 $A(-1,1)$, $B(-3,-3)$, $C(3,0)$

In Probs. 31 to 34 find the cosine of the acute angle between the given pair of lines.

31 $2x - 7y + 3 = 0;\ 5x + y + 1 = 0$
32 $2x - 3y + 2 = 0;\ 5x - y + 2 = 0$
33 $3x + y + 9 = 0;\ 3x + y - 10 = 0$
34 $7x - 2y + 1 = 0;\ 2x + 7y - 13 = 0$

In Probs. 35 to 38 find the distance between the given parallel lines.

35 $3x - 4y + 5 = 0$
$3x - 4y + 7 = 0$

36 $5x + 12y - 5 = 0$
$5x + 12y + 2 = 0$

37 $3x - 6y + 3 = 0$
$3x - 6y - 11 = 0$

38 $x - 5y + 10 = 0$
$x - 5y - 3 = 0$

In Probs. 39 to 42 find the distance between the given point and the given line.

39 $(4,-1)$, $4x + 3y - 6 = 0$
40 $(1,5)$, $5x - 12y + 7 = 0$
41 $(-1,2)$, $3x + y + 8 = 0$
42 $(0,3)$, $-2x + 5y - 6 = 0$

43 Find equations for the bisectors of the angles made by the pair of lines:

$$3x - y = 0$$
$$x - 2y = 0$$

44 Find equations for the bisectors of the angles made by the pair of lines:

$$6x + 8y - 41 = 0$$
$$12x - 5y - 30 = 0$$

45 Prove that (2,3) is the center of a circle which touches the three lines:

$$4x + 3y - 7 = 0$$
$$5x + 12y - 20 = 0$$
$$3x + 4y - 8 = 0$$

***46** Consider the point $P_1(x_1,y_1)$ on a circle of radius 1 whose center is at the origin. **(a)** Find the direction cosines of the radius vector $\overrightarrow{OP_1}$.

(b) Prove that the line $xx_1 + yy_1 = 1$ passes through P_1 and is tangent to the circle at P_1.

***47** In the triangle $P_1(x_1,y_1)$, $P_2(x_2,y_2)$, $P_3(x_3,y_3)$ find the length of the altitude from P_1 to side P_2P_3. Then prove that the area of triangle $P_1P_2P_3$ is the absolute value of

$$\frac{1}{2}\begin{vmatrix} x_1 & y_1 & 1 \\ x_2 & y_2 & 1 \\ x_3 & y_3 & 1 \end{vmatrix}$$

***48** Using the formula from trigonometry for the tangent of the difference of two angles, show that the acute angle between a line with slope m_1 and a line with slope m_2 is given by

$$\tan\theta = \left|\frac{m_2 - m_1}{1 + m_1m_2}\right|$$

Normal form ***49** The *normal form* of the equation of a line given by $Ax + By + C = 0$ is defined to be

$$\frac{A}{\sqrt{A^2 + B^2}}x + \frac{B}{\sqrt{A^2 + B^2}}y + \frac{C}{\sqrt{A^2 + B^2}} = 0$$

Find the geometrical interpretation of the vector:

$$\left(\frac{A}{\sqrt{A^2 + B^2}}, \frac{B}{\sqrt{A^2 + B^2}}\right) \qquad \text{(See Illustration 5.)}$$

***50** In the normal form of the line of Prob. 49 find a geometrical interpretation of the scalar $|C|/\sqrt{A^2 + B^2}$.

14.7 › APPLICATIONS TO PLANE GEOMETRY

The properties of a given geometric configuration usually found in Euclidean plane geometry do not in any way depend upon a related coordinate system. It often happens, however, that the introduction of a coordinate system will help to simplify the work of proving a theorem, especially if axes are chosen properly. But the axes must be chosen so that there will be no loss in generality. For example, if the problem is to prove some proposition relating to a triangle, then a coordinate axis can be chosen coincident with a side and one vertex can then be taken as the origin. Consider the following illustration.

ILLUSTRATION 1 PROVE: The line segment joining the midpoints of two sides of a triangle is parallel to the third side and equal to one-half its length.

SOLUTION We choose axes as in Fig. 14.8. The midpoints D and E have the coordinates $D\left(\frac{b}{2}, \frac{c}{2}\right)$ and $E\left(\frac{a + b}{2}, \frac{c}{2}\right)$. The slope of DE is

Figure for Illustration 1

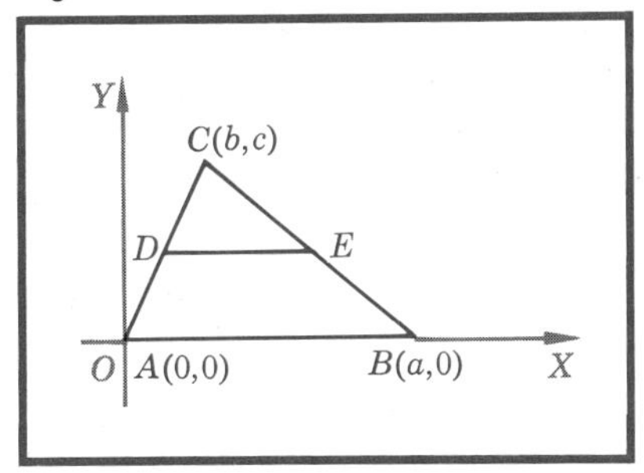

FIGURE 14.8

$$m_{DE} = \frac{(c/2) - (c/2)}{(a+b)/2 - (b/2)} = \frac{0}{a/2} = 0$$

Since AB also has slope zero, it follows that DE is parallel to AB. The length AB is a. The length of DE is

$$\begin{aligned} DE &= \sqrt{\left(\frac{b}{2} - \frac{a+b}{2}\right)^2 + \left(\frac{c}{2} - \frac{c}{2}\right)^2} \\ &= \sqrt{\frac{a^2}{4}} \\ &= \frac{a}{2} \end{aligned}$$

Thus the theorem is proved.

ILLUSTRATION 2 PROVE: The diagonals of a parallelogram bisect each other.

SOLUTION Choose axes as in Fig. 14.9, letting the coordinates of three vertices be $A(0,0)$, $B(a,0)$, and $C(b,c)$. Then, since the figure $ABCD$ is a parallelogram, the coordinates of D are determined. It is easy to see

Figure for Illustration 2

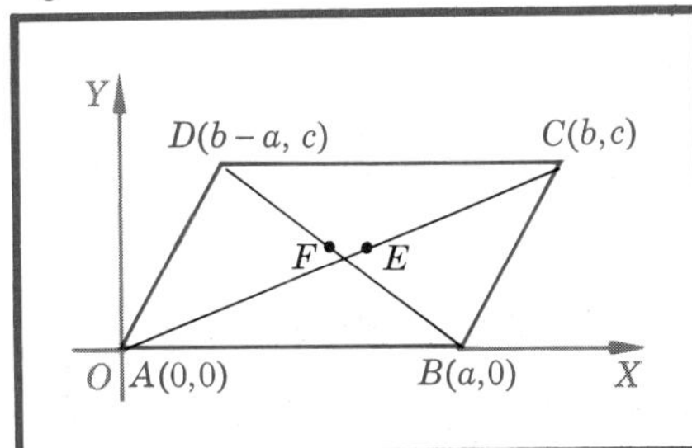

FIGURE 14.9

that they must be $D(b - a, c)$. The midpoint of AC has coordinates $E\left(\frac{b}{2}, \frac{c}{2}\right)$. Let F be the midpoint of BD. The coordinates of F are

$$F\left(\frac{b - a + a}{2}, \frac{c}{2}\right)$$

that is, $F\left(\frac{b}{2}, \frac{c}{2}\right)$. Since E and F have the same coordinates, they must coincide. Hence the proposition is proved.

PROBLEMS 14.7 Draw a figure and prove by analytic geometry:

1. The midpoint of the hypotenuse of a right triangle is equidistant from the vertices.
2. The diagonals of a rectangle have equal lengths.
3. The diagonals of a rhombus are perpendicular.
4. If the diagonals of a parallelogram are equal, then the parallelogram is a rectangle.
5. In a triangle the perpendicular bisectors of the sides meet in a point.
6. The midpoints of the sides of any quadrilateral are the vertices of a parallelogram.
7. The base angles of an isosceles triangle are equal.
8. If two parallel lines are cut by a transversal, the alternate interior angles are equal.
9. The distance between the midpoints of the nonparallel sides of a trapezoid is one-half the sum of the parallel sides.
10. The lines joining the midpoints of the sides of a triangle divide it into four congruent triangles.
11. The sum of the squares of the lengths of sides of a parallelogram equals the sum of the squares of the lengths of the diagonals.
12. The locus of points P equidistant from two points A and B is the perpendicular bisector of segment AB.
13. If two medians of a triangle are equal, the triangle is isosceles.
14. In any triangle four times the sum of the squares of the lengths of the medians is equal to three times the sum of the squares of the lengths of the sides.

14.8 › CONIC SECTIONS

One way of generalizing $Ax + By + C = 0$, which represents a straight line, is to add all possible quadratic terms (terms of the second degree in x and y). Where an obvious shift has been made in renaming the coefficients, such an equation can be written in the form

$$Ax^2 + Bxy + Cy^2 + Dx + Ey + F = 0 \tag{1}$$

It is the general equation of the second degree in each variable (provided it is not true that $A = B = C = 0$).

We shall consider some special cases of (1). The treatment of the general case is complicated; but the total set of points corresponding to the ordered pairs (x,y) satisfying the relation defined by (1) is called a *conic section*. This is because, geometrically, the curve can be obtained by cutting a cone with a plane. This fact was known to the Greek mathematicians of 300 B.C.; we shall give the appropriate geometric illustration as we treat each case.

14.9 › CASE I. THE CIRCLE

DEFINITION A *circle* is the locus of points P which are at a fixed distance from a fixed point.

Thus consider a fixed point $C(h,k)$. Now the point $P(x,y)$ will be r units from C if and only if the distance PC equals r, that is, if and only if (Fig. 14.10)

$$\sqrt{(x-h)^2 + (y-k)^2} = r$$

This becomes, upon squaring,

$$(x-h)^2 + (y-k)^2 = r^2 \tag{1}$$

which is the equation whose graph is the circle with center at $C(h,k)$ and with radius r since (1) expresses the condition that the point P, with coordinates x and y, shall always be exactly r units from C.

Equation (1) reduces, after a little rearranging, to

$$x^2 + y^2 - 2hx - 2ky + h^2 + k^2 - r^2 = 0 \tag{2}$$

This is a special case of Eq. (1), Sec. 14.8, where $A = C$ and $B = 0$ [which indeed constitutes a necessary condition that Eq. (1), Sec. 14.8, represent a circle].

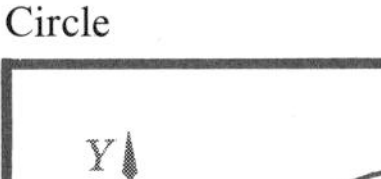

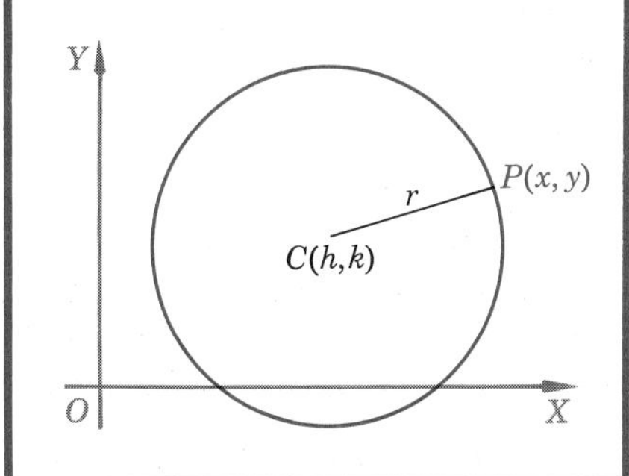

FIGURE 14.10 The center is (h,k), and the radius is r.

A circle is a conic section

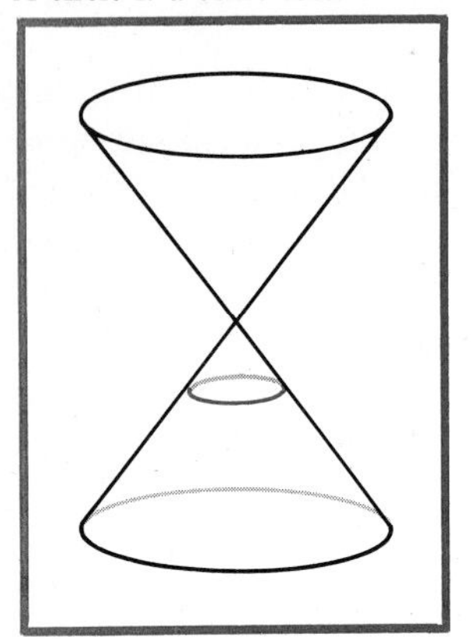

FIGURE 14.11
A plane parallel to the base of a right circular cone cuts the cone in a circle.

EXERCISE A In (2) above, what coefficients correspond to A, B, C, D, E, and F in Eq. (1), Sec. 14.8?

EXERCISE B Is the necessary condition that Eq. (1), Sec. 14.8, represent a circle, namely, $A = C$ and $B = 0$, also sufficient?

The circle is a conic section. Geometrically it is obtained by cutting a right circular cone with a plane parallel to the base (Fig. 14.11).

ILLUSTRATION 1 Write the equation of the circle with center at $(-2,1)$ and with radius 3.

SOLUTION It is

$$(x + 2)^2 + (y - 1)^2 = 9$$

ILLUSTRATION 2 Plot the curve given by

$$x^2 + y^2 - 3x + 6y - 5 = 0$$

SOLUTION We complete the square separately on the x terms and on the y terms as follows:

$$x^2 - 3x + [\tfrac{9}{4}] + y^2 + 6y + [9] = 5 + [\tfrac{9}{4}] + [9]$$

The "5" was transposed, and the brackets merely indicate the terms added to complete the square. (To complete the square on the x terms, we must add the square of one-half the coefficient of x; and similarly for the y terms.) This can be rewritten as

$$(x^2 - 3x + \tfrac{9}{4}) + (y^2 + 6y + 9) = \tfrac{65}{4}$$

or, again, as

$$(x - \tfrac{3}{2})^2 + (y + 3)^2 = \tfrac{65}{4}$$

This is precisely in the form (1) so that the graph is a circle with center at $(\frac{3}{2},-3)$ and $r = \sqrt{\frac{65}{4}} = \frac{1}{2}\sqrt{65}$.

The equation

$$x^2 + y^2 = r^2 \tag{3}$$

is that of a circle of radius r with center at the origin.

EXERCISE C In (3) above, what coefficients correspond to A, B, C, D, E, and F in Eq. (1), Sec. 14.8?

PROBLEMS 14.9 In Probs. 1 to 12 sketch and find the equation of the circle:

1 Center at $(-2,-1)$, radius 2
2 Center at $(3,-3)$, radius 6
3 Center at $(-1,4)$, radius 4
4 Center at $(2,3)$, radius 3
5 Ends of diameter at $(-1,-2)$, $(2,2)$
6 Ends of diameter at $(3,-2)$, $(-5,4)$
7 Tangent to the X-axis, center at $(3,4)$
8 Tangent to the Y-axis, center at $(-2,1)$
9 Has for diameter the portion of $4x + 3y - 12 = 0$ lying in first quadrant.
10 Has radius 2 and is concentric with $x^2 + y^2 - 2x - 4y = 0$.
11 Has radius 3 and is tangent to both axes.
12 Tangent to both axes, and center lies on $y = -x$.

In Probs. 13 to 20 find center and radius. Sketch.

13 $x^2 + y^2 - 6x + 8y = 0$
14 $-14 + 10x - 10y - x^2 - y^2 = 0$
15 $x^2 + y^2 - 6x + 4y = 0$
16 $2x^2 + 2y^2 - 7x + 9y + 7 = 0$
17 $-x^2 - y^2 + 2x - 4y - 2 = 0$
18 $-16x + x^2 + y^2 = 0$
19 $x^2 + y^2 - 6x + 8y + 30 = 0$
20 $5x^2 + 5y^2 + ay = 0$

21 Find the locus of points P such that the sum of squares of the distances from P to $(2,5)$ and to $(-3,1)$ is 33.
22 Find the locus of points P such that the sum of squares of the distances from P to $(-1,5)$ and to $(-5,7)$ is 82.
23 Write the equation of every circle of radius 1.
24 Write the equation of every circle of radius 1 with center on $y = -x$.
25 Write the equation of every circle of radius 1 tangent to the circle $x^2 + y^2 = 1$.
*26 The locus of points P, such that the sum of the squares of the distances from P to n fixed points (x_1,y_1), (x_2,y_2), . . . , (x_n,y_n) is (a sufficiently large) constant k, is a circle. Prove it.

14.10 › CASE II. THE PARABOLA

We are already somewhat familiar with the parabola (Sec. 6.5).

DEFINITIONS A *parabola* is the locus of points P such that the distance of P from a fixed point is always equal to its distance from a fixed line. The fixed

Parabola with focus and directrix

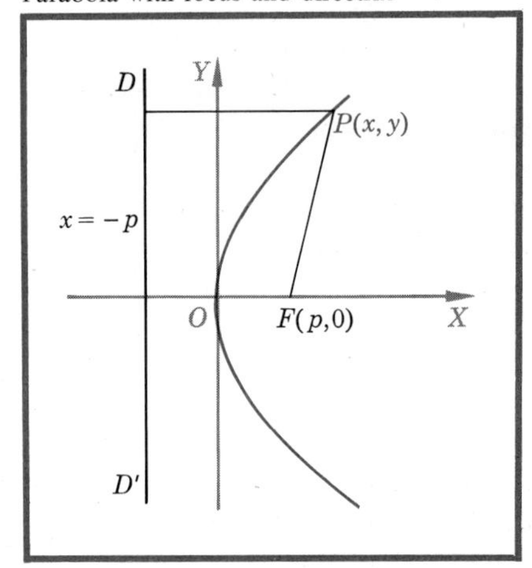

FIGURE 14.12
For a parabola, F is the focus, DD' is the directrix, and $PD = PF$.

Focus and directrix

point is called the *focus;* the fixed line is called the *directrix*. The line perpendicular to the directrix and passing through the focus is called the *axis of the parabola.*

In order to arrive at an equation for this locus, we choose the coordinate axes so that the focus F has the coordinates $F(p,0)$ and the directrix line DD' has the equation $x = -p$ (Fig. 14.12). (This choice of axes leads to the simplest equation, although this is not immediately apparent.) By definition, the distance PF must equal the (perpendicular) distance from P to DD'. The distance from P to DD' is $|x + p|$. We have

A parabola is a conic section

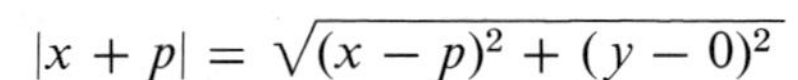

$$|x + p| = \sqrt{(x - p)^2 + (y - 0)^2}$$

which yields, upon squaring,

$$x^2 + 2px + p^2 = x^2 - 2px + p^2 + y^2$$

This reduces to

$$y^2 = 4px \qquad \textbf{(1)}$$

This is the equation sought; it defines a relation. It is a special case of Eq. (1), Sec. 14.8.

EXERCISE A In (1) above, what coefficients correspond to A, B, C, D, E, and F in Eq. (1), Sec. 14.8?

FIGURE 14.13
A plane parallel to a generator of a right circular cone cuts the cone in a parabola.

The parabola is a conic section (Fig. 14.13). Geometrically, the parabola can be obtained by cutting a right circular cone with a plane parallel to a generator.

Focus and directrix of $y^2 = -7x$

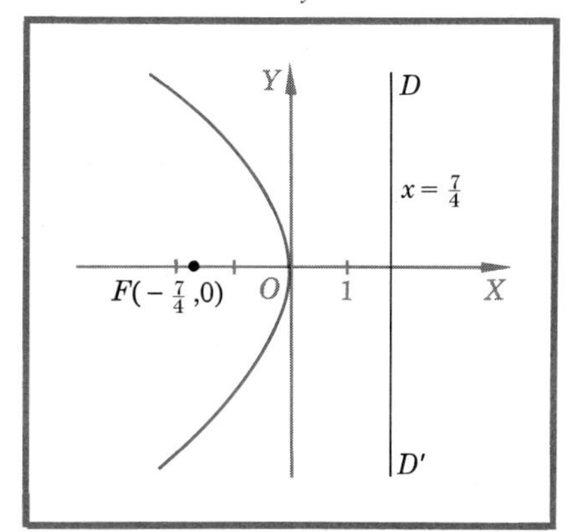

FIGURE 14.14
The focus and directrix of a parabola can be in positions quite different from those in Fig. 14.12.

ILLUSTRATION 1 Write the equation of the parabola with $F(3,0)$ and directrix $x = -3$.

SOLUTION In this case $p = 3$, and the equation is consequently $y^2 = 12x$.

ILLUSTRATION 2 Sketch the parabola whose equation is $y^2 = -7x$. Find the coordinates of the focus and the equation of the directrix.

SOLUTION Here $4p = -7$. Hence the focus has coordinates $F(-\frac{7}{4},0)$, and the equation of the directrix DD' is $x = \frac{7}{4}$ (Fig. 14.14).

PROBLEMS 14.10

In Probs. 1 to 10 find the coordinates of the focus and the equation of the directrix, and sketch.

1 $y^2 = 16x$
2 $x^2 = 4y$
3 $x^2 = 14y$
4 $x^2 = -2y$
5 $y^2 = x$
6 $y^2 = -7x$
7 $y^2 = -4x$
8 $x^2 = -y$
9 $(y + 2)^2 = -4(x - 3)$. HINT: Plot the lines $y + 2 = 0$ and $x - 3 = 0$, and think of these as new axes. That is, make the transformations $y' = y + 2$, $x' = x - 3$.
10 $(x + 1)^2 = 12(y - 5)$. (See hint, Prob. 9.)

In Probs. 11 to 22 find the equation of the parabola, and sketch.

11 Focus at $(0,-4)$, directrix $y = -2$
12 Focus at $(2,0)$, directrix $x = -4$
13 Focus at $(0,-1)$, directrix $y = 1$
14 Focus at $(0,-4)$, directrix $y = 4$

15 Focus at $(0,\frac{3}{4})$, directrix $y = -\frac{3}{4}$
16 Focus at $(0,\frac{3}{2})$, directrix $y = -\frac{3}{2}$
17 Focus at $(4,0)$, directrix $x = -4$
18 Focus at $(2,0)$, directrix $x = -2$
19 Focus at $(-\frac{1}{2},0)$, directrix $x = \frac{1}{2}$
20 Focus at $(-1,0)$, directrix $x = 1$
21 Focus at $(-1,-1)$, directrix $y = -3$
22 Focus at $(2a,b)$, vertex at (a,b), where $a \neq 0$

23 Find the points of intersection of $y^2 = -4x$, $x^2 = -4y$.
24 Find the points of intersection of $x^2 = -2y$, $x^2 = y + 6$.
25 A point has the property that the sum of its distances from $F(6,-1)$, $F'(-6,-1)$ is 20. Find the equation of the locus of such points.
26 A point has the property that the sum of its distances from $F(1,-3)$, $F'(-1,-3)$ is $2\sqrt{3}$. Find the equation of the locus of such points.
27 Each circle of a set of circles passes through $(1,0)$ and is tangent to the line $x = -5$. Find the equation of the locus of the centers of the circles.
28 Sketch and discuss: $(y - k)^2 = 4p(x - h)$, where $p \neq 0$.
29 Sketch on the same axes: $y = x^2$, $y = 2mx - 1$. Find the x-coordinates of the points of intersection of this parabola and straight line. Find the condition that there is only one point of intersection. Discuss the geometry for the lines $y = 2x - 1$ and $y = -2x - 1$.

14.11 › CASE III. THE ELLIPSE

DEFINITIONS An *ellipse* is the locus of points P such that the sum of the distances from P to two fixed points is a constant. The two fixed points are called *foci*.

A very simple equation results from choosing the axes and scales so that the foci F and F' have the coordinates $F(c,0)$, $F'(-c,0)$. We let the sum of the distances be the constant $2a$. Note that $2a > 2c$; hence $a > c$ (Fig. 14.15). The definition requires that

$$\text{(1)} \qquad \sqrt{(x + c)^2 + y^2} + \sqrt{(x - c)^2 + y^2} = 2a$$

We transpose the second radical and square, getting

$$x^2+2cx+c^2+y^2=4a^2-4a\sqrt{(x-c)^2+y^2}+x^2-2cx+c^2+y^2$$

which simplifies to

$$4cx - 4a^2 = -4a\sqrt{(x - c)^2 + y^2}$$

We can now cast out the 4, and the reason for choosing $2a$ as the sum of the distances instead of a is now apparent. Square again. Thus

$$c^2x^2 - 2a^2cx + a^4 = a^2(x^2 - 2cx + c^2 + y^2)$$

Ellipse with foci F and F'

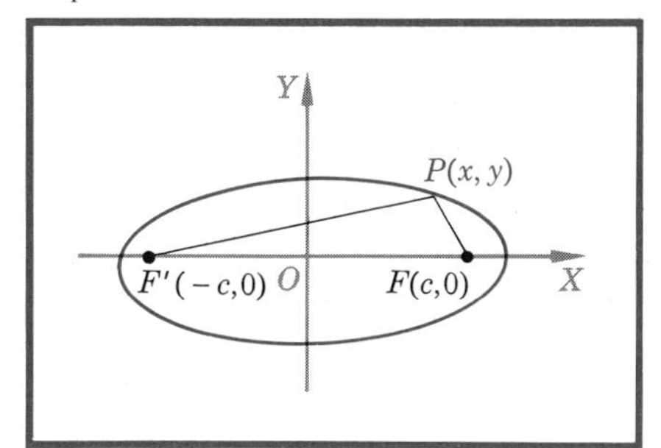

FIGURE 14.15 For an ellipse: $PF + PF' = 2a$.

which reduces to

$$(2) \qquad (a^2 - c^2)x^2 + a^2y^2 = a^4 - a^2c^2 = a^2(a^2 - c^2)$$

Since $a > c$, it follows that $a^2 > c^2$ and $a^2 - c^2 > 0$. Let us call $a^2 - c^2 = b^2$ (a positive number). We can then write (2) in the form

$$b^2x^2 + a^2y^2 = a^2b^2$$

or, finally,

$$(3) \qquad \frac{x^2}{a^2} + \frac{y^2}{b^2} = 1$$

This is the equation of the ellipse.

EXERCISE A In (3) above, what coefficients correspond to A, B, C, D, E, and F in Eq. (1), Sec. 14.8?

EXERCISE B Show that the points $V(a,0)$ and $V'(-a,0)$ are on the ellipse.

EXERCISE C Show that the points $(0,b)$ and $(0,-b)$ are on the ellipse.

DEFINITIONS

Major axis and minor axis

The points V and V' are called the *vertices* of the ellipse. The segment joining V and V' is called the *major axis;* its length is $2a$. The segment joining $(0,b)$ and $(0,-b)$ is called the *minor axis;* its length is $2b$. The *center* of the ellipse is the midpoint of the major axis. The foci are $F(c,0)$ and $F'(-c,0)$, where $a^2 - b^2 = c^2$.

The ellipse is a conic section (Fig. 14.16). Geometrically, the ellipse can be obtained by cutting a right circular cone with a plane inclined (but not parallel to a generator) so that it cuts only one nappe of the cone. This permits an ellipse to reduce to a circle if the cutting plane

An ellipse is a conic section

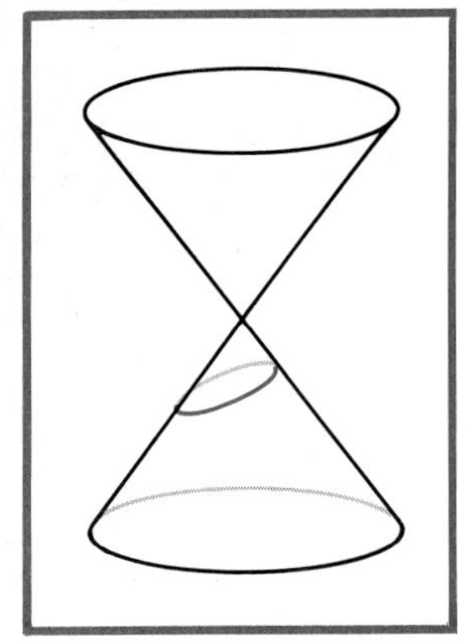

FIGURE 14.16 A plane which intersects only one nappe of a right circular cone cuts this cone in an ellipse.

is perpendicular to the axis of the cone. Algebraically, this is the case where $a = b$ and where, therefore, Eq. (3) reduces to

$$\frac{x^2}{a^2} + \frac{y^2}{a^2} = 1$$

which represents a circle of radius a.

ILLUSTRATION 1 Plot the graph of

$$\textbf{(4)} \qquad \frac{x^2}{9} + \frac{y^2}{4} = 1$$

SOLUTION The total graph (Fig. 14.17) will be made up of the graphs of the two algebraic functions f and g derived from (4) and defined by the equations

$$y = \tfrac{2}{3}\sqrt{9 - x^2} \qquad \begin{cases} \text{Domain: } -3 \le x \le 3 \\ \text{Image: } \quad 0 \le y \le 2 \end{cases}$$

$$y = -\tfrac{2}{3}\sqrt{9 - x^2} \qquad \begin{cases} \text{Domain: } -3 \le x \le 3 \\ \text{Image: } \quad -2 \le y \le 0 \end{cases}$$

The zeros of both f and g are $x = \pm 3$; that is, the vertices of the ellipse are $V(3,0)$ and $V'(-3,0)$. The point (0,2) is on the graph of f, (0,−2) is on the graph of g; each graph is a semiellipse. The coordinates of the foci are $F(\sqrt{5},0)$ and $F'(-\sqrt{5},0)$ since $c^2 = a^2 - b^2 = 5$. You should compute a few elements of f and g.

ILLUSTRATION 2 Write the equation of the ellipse with $F(1,0)$, $F'(-1,0)$, and major axis 5.

Figure for Illustration 1

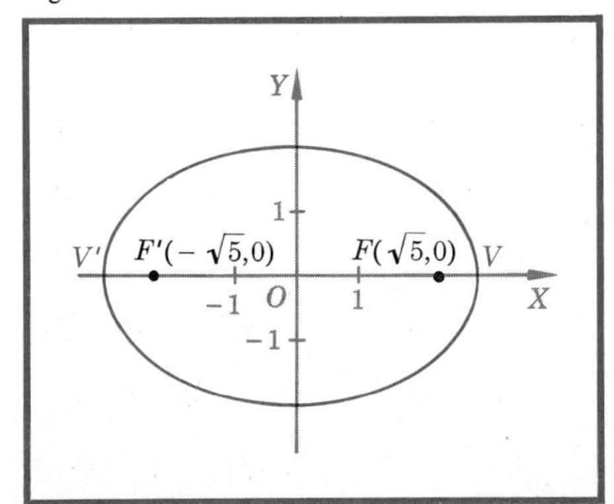

FIGURE 14.17

Figure for Illustration 2

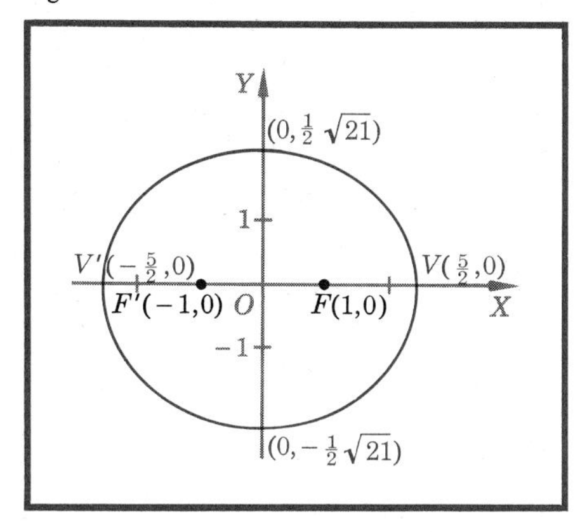

FIGURE 14.18

SOLUTION Now $c = 1$ and $2a = 5$. Therefore $a = \frac{5}{2}$; and since $b^2 = a^2 - c^2$, we have $b^2 = \frac{25}{4} - 1 = \frac{21}{4}$. Therefore the equation is

$$\frac{x^2}{\frac{25}{4}} + \frac{y^2}{\frac{21}{4}} = 1$$

The graph is drawn in Fig. 14.18.

PROBLEMS 14.11 In Probs. 1 to 10 sketch and find the coordinates of the foci and of the vertices.

1 $\dfrac{x^2}{12} + \dfrac{y^2}{9} = 1$

2 $\dfrac{x^2}{4} + \dfrac{y^2}{1} = 1$

3 $\dfrac{x^2}{25} + \dfrac{y^2}{16} = 1$

4 $\dfrac{x^2}{4} + \dfrac{y^2}{3} = 1$

5 $\dfrac{x^2}{144} + \dfrac{y^2}{36} = 1$

6 $3x^2 + 5y^2 = 15$

7 $\dfrac{x^2}{36} + \dfrac{y^2}{100} = 1$

8 $\dfrac{x^2}{16} + \dfrac{y^2}{25} = 1$

9 $\dfrac{(x-3)^2}{4} + \dfrac{(y+3)^2}{1} = 1$. HINT: Plot the lines $x - 3 = 0$ and $y + 3 = 0$, and think of these as new axes.

10 $\dfrac{(x-1)^2}{25} + \dfrac{(y+2)^2}{9} = 1$. (See hint for Prob. 9.)

In Probs. 11 to 22 sketch and find the equation of the ellipse.

11 $F(2,0)$, $F'(-2,0)$, and major axis 8
12 $F(2,0)$, $F'(-2,0)$, and minor axis 8
13 $F(3,0)$, $F'(-3,0)$, and minor axis 20
14 $F(3,0)$, $F'(-3,0)$, and major axis 20
15 Major axis (along X-axis) 6, minor axis 4, and center at (0,0)
16 Major axis (along X-axis) 10, minor axis 8, and center at (1,0)
17 $V(-3,0)$, $V'(3,0)$, $F(-1,0)$, $F'(1,0)$
18 $V(-5,0)$, $V'(5,0)$, $F(-4,0)$, $F'(4,0)$
19 $V(-4,2)$, $V'(4,2)$, $F(-3,2)$, $F'(3,2)$
20 $V(-6,-2)$, $V'(6,-2)$, $F(-2,-2)$, $F'(2,-2)$
21 $V(0,-3)$, $V'(0,3)$, $F(0,-1)$, $F'(0,1)$
22 $V(0,7)$, $V'(0,-7)$, $F(0,3)$, $F'(0,-3)$

23 A point has the property that the absolute value of the difference of its distances from $F(3,-1)$ and $F'(-3,-1)$ is 4. Find the equation of the locus of such points.

24 The hypotenuse of each of a set of right triangles is the segment joining (2,2) and (2,−2). Find the equation of the locus of the third vertices.

25 For the ellipse $\frac{x^2}{a^2} + \frac{y^2}{b^2} = 1$, show that the line segment drawn from $(0,b)$ to a focus is of length a.

26 Determine the coordinates of the foci of the ellipse

$$\frac{x^2}{4} + \frac{y^2}{4} = 1$$

27 Sketch and discuss:

$$\frac{(x-h)^2}{a^2} + \frac{(y-k)^2}{b^2} = 1$$

Assume $a > b$.

14.12 › CASE IV. THE HYPERBOLA

DEFINITIONS A *hyperbola* is the locus of points P such that the absolute value of the difference of the distances from P to two fixed points is a constant. The two fixed points are called *foci*.

We choose axes as we did for the ellipse, writing $F(c,0)$ and $F'(-c,0)$ (Fig. 14.19). We let the absolute value of the difference of the distances be the constant $2a$. The definition requires

$$\left|\sqrt{(x+c)^2 + y^2} - \sqrt{(x-c)^2 + y^2}\right| = 2a$$

Hyperbola with foci and asymptotes

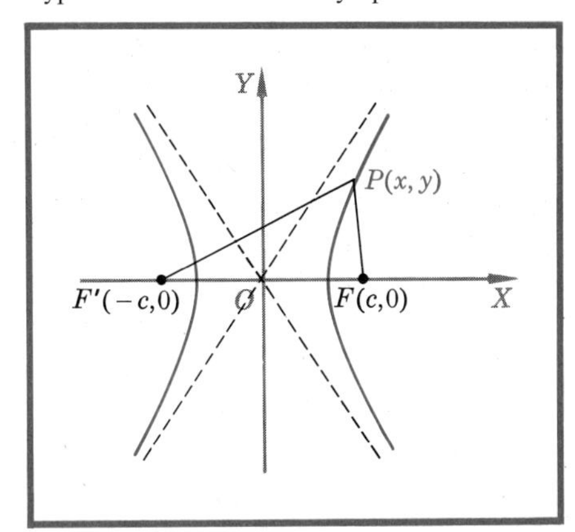

FIGURE 14.19
For a hyperbola:
$|PF - PF'| = 2a$.

This is equivalent to

(1) $$\sqrt{(x+c)^2+y^2} - \sqrt{(x-c)^2+y^2} = +2a$$

if $$\sqrt{(x+c)^2+y^2} > \sqrt{(x-c)^2+y^2}$$

and to

(2) $$\sqrt{(x+c)^2+y^2} - \sqrt{(x-c)^2+y^2} = -2a$$

if $$\sqrt{(x+c)^2+y^2} < \sqrt{(x-c)^2+y^2}$$

In either case, (1) or (2), if we square (twice) and simplify as we did in the case of the ellipse, we shall arrive again at Eq. (2), Sec. 14.11. For the hyperbola, however, $2a < 2c$, as can be seen directly from the figure. This means that $a^2 - c^2 < 0$; we set $a^2 - c^2 = -b^2$ (a negative number). Continuing the simplification of Eq. (2), Sec. 14.12, we get, on this basis,

$$b^2x^2 - a^2y^2 = a^2b^2$$

or, finally,

(3) $$\frac{x^2}{a^2} - \frac{y^2}{b^2} = 1$$

This is the equation of the hyperbola.

EXERCISE A In (3) above, what coefficients correspond to A, B, C, D, E, and F in Eq. (1), Sec. 14.8?

EXERCISE B Show that the points $V(a,0)$, $V'(-a,0)$ are on the hyperbola.

DEFINITIONS The points $V(a,0)$ and $V'(-a,0)$ are called the *vertices* of the hyperbola. The segment VV' is called the *transverse axis;* its length is $2a$. The

Transverse axis and conjugate axis

segment joining $(0,b)$ and $(0,-b)$ is called the *conjugate axis;* its length is $2b$. The *center* of the hyperbola is the midpoint of the transverse axis. The foci are $F(c,0)$ and $F'(-c,0)$, where $a^2 + b^2 = c^2$.

If we divide (3) by x^2, we get, after a little simplification,

$$\frac{y^2}{x^2b^2} = \frac{1}{a^2} - \frac{1}{x^2}$$

$$\frac{y^2}{x^2} = \frac{b^2}{a^2} - \frac{b^2}{x^2}$$

$$\left|\frac{y}{x}\right| = \sqrt{\frac{b^2}{a^2} - \frac{b^2}{x^2}}$$

Asymptotes

This says that, for large x (since b^2/x^2 is then small), the positive ratio $|y/x|$ is just a very little less than b/a. (See Chap. 15, where the notion of *limit* is treated.) With this information at hand we sketch in the two lines (called the *asymptotes* of the *hyperbola*) whose equations are $y = \pm(b/a)x$ to act as guides in sketching the graph of the hyperbola itself. The asymptotes are *not* part of the locus; they merely serve as aids in plotting.

The hyperbola is a conic section (Fig. 14.20). Geometrically, the hyperbola can be obtained by cutting a right circular cone with a plane that is inclined so as to cut both nappes of the cone but not placed so as to pass through the vertex of the cone.

ILLUSTRATION 1 Sketch the graph of the equation

A hyperbola is a conic section

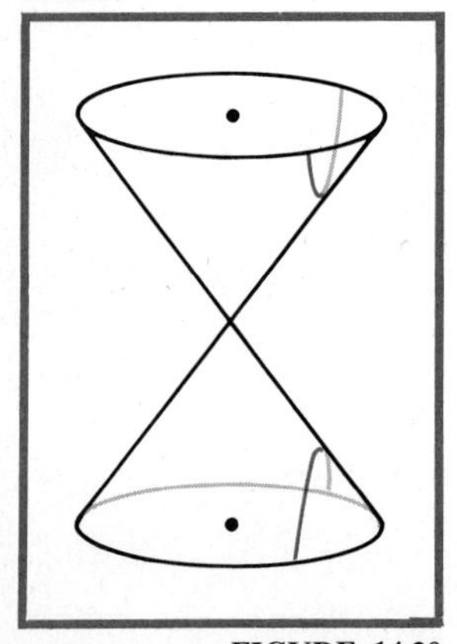

FIGURE 14.20
A plane which intersects both nappes of a right circular cone but which does not pass through the vertex cuts this cone in a hyperbola.

$$\frac{x^2}{9} - \frac{y^2}{4} = 1 \tag{4}$$

SOLUTION The total graph will be made up of the graphs of the two algebraic functions f and g derived from (4) and defined by the equations

$$y = \tfrac{2}{3}\sqrt{x^2 - 9} \qquad \begin{cases} \text{Domain:} & -\infty < x \le -3,\ 3 \le x < \infty \\ \text{Image:} & 0 \le y < \infty \end{cases}$$

$$y = -\tfrac{2}{3}\sqrt{x^2 - 9} \qquad \begin{cases} \text{Domain:} & -\infty < x \le -3,\ 3 \le x < \infty \\ \text{Image:} & -\infty < y \le 0 \end{cases}$$

The zeros are $x = \pm 3$ for each function; that is, the vertices of the hyperbola are $V(3,0)$ and $V'(-3,0)$. The coordinates of the foci are $F(\sqrt{13},0)$ and $F'(-\sqrt{13},0)$, since $c^2 = b^2 + a^2 = 13$. You should compute a few elements of f and g.

The length of the transverse axis is 6; the length of the conjugate axis is 4. The equations of the asymptotes are $y = \pm\frac{2}{3}x$ (Fig. 14.21).

ILLUSTRATION 2 Write the equation of the hyperbola with vertices $V(2,0)$, $V'(-2,0)$ and with foci $F(3,0)$, $F'(-3,0)$.

Figure for Illustration 1

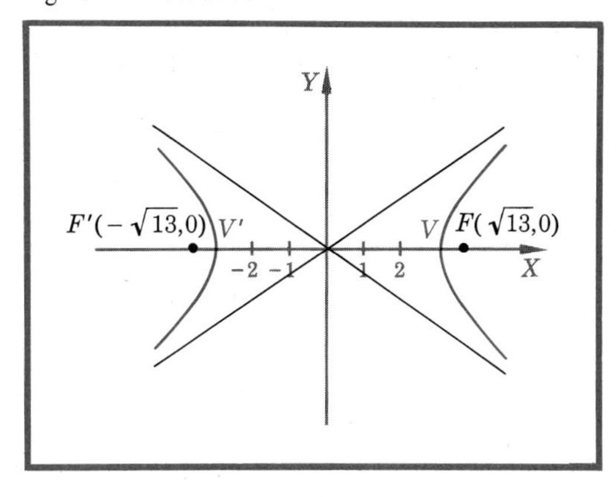

FIGURE 14.21

SOLUTION We are given $a = 2$ and $c = 3$. Therefore, since $c^2 = b^2 + a^2$, we find $b^2 = c^2 - a^2 = 9 - 4 = 5$. The equation is

$$\frac{x^2}{4} - \frac{y^2}{5} = 1$$

The equations of the asymptotes are

$$y = \pm\frac{\sqrt{5}}{2}x \qquad \text{(Fig. 14.22)}$$

PROBLEMS 14.12 In Probs. 1 to 10 sketch and find the coordinates of the foci and the equations of the asymptotes.

1 $\dfrac{x^2}{25} - \dfrac{y^2}{9} = 1$

2 $\dfrac{x^2}{9} - \dfrac{y^2}{4} = 1$

Figure for Illustration 2

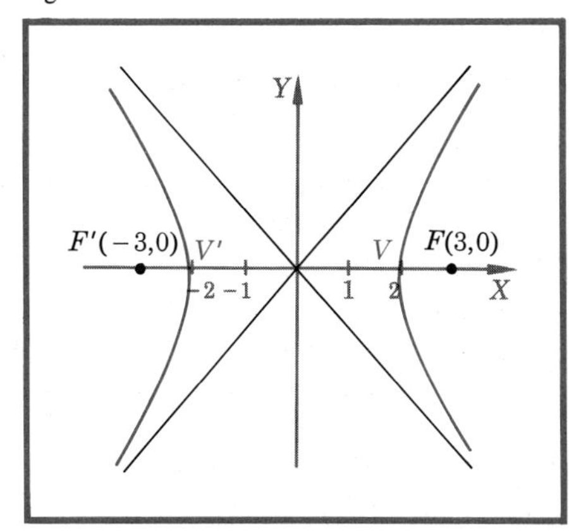

FIGURE 14.22

3 $-\dfrac{x^2}{144} + \dfrac{y^2}{16} = 1$ **4** $-\dfrac{x^2}{100} + \dfrac{y^2}{25} = 1$

5 $-x^2 + 4y^2 = 4$ **6** $\dfrac{x^2}{5} - \dfrac{y^2}{2} = 1$

7 $\dfrac{x^2}{9} - \dfrac{y^2}{25} = -1$ **8** $\dfrac{x^2}{16} - \dfrac{y^2}{36} = -1$

9 $\dfrac{(x+3)^2}{49} - \dfrac{(y-2)^2}{16} = 1$ **10** $\dfrac{(x-2)^2}{4} - \dfrac{(y+1)^2}{2} = 1$

In Probs. 11 to 28 find the equation of the hyperbola, and the equations of the asymptotes.

11 $F(4,0)$, $F'(-4,0)$ and transverse axis 4
12 $F(5,0)$, $F'(-5,0)$ and transverse axis 6
13 $F(4,0)$, $F'(-4,0)$ and conjugate axis 2
14 $F(5,0)$, $F'(-5,0)$ and conjugate axis 6
15 $F(0,5)$, $F'(0,-5)$ and transverse axis 8
16 $F(0,\sqrt{41})$, $F'(0,-\sqrt{41})$ and conjugate axis 10
17 $F(8,2)$, $F'(-2,2)$ and transverse axis 8
18 $F(6,3)$, $F'(-2,3)$ and transverse axis 4
19 $F(-3,15)$, $F'(-3,-5)$ and transverse axis 12
20 $F(4,9)$, $F'(4,-7)$ and transverse axis 14
21 Transverse axis (along X-axis) 10, conjugate axis 6, center at $(0,0)$
22 Transverse axis (along X-axis) 10, conjugate axis 6, center at $(3,0)$
23 $V(6,0)$, $V'(-6,0)$, $F(9,0)$, $F'(-9,0)$
24 $V(5,0)$, $V'(-5,0)$, $F(8,0)$, $F'(-8,0)$
25 $V(6,-1)$, $V'(-6,-1)$, $F(9,-1)$, $F'(-9,-1)$
26 $V(5,2)$, $V'(-5,2)$, $F(7,2)$, $F'(-7,2)$
27 $V(3,7)$, $V'(3,-13)$, $F(3,11)$, $F'(3,-17)$
28 $V(-2,10)$, $V'(-2,-10)$, $F(-2,14)$, $F'(-2,-14)$

29 Sketch and discuss: $\dfrac{(x-h)^2}{a^2} - \dfrac{(y-k)^2}{b^2} = 1.$
30 Sketch $x^2 - y^2 - x + y = 0$. HINT: Factor.
31 Sketch $x^2 + 2xy + y^2 - 3x - 3y + 2 = 0$. HINT: Factor.
32 What geometric configurations, other than ellipse, parabola, and hyperbola, can be obtained by cutting a cone with a plane? Illustrate with a figure.

14.13 › APPLICATIONS

In order to treat in detail many of the scientific applications of the theory of conic sections, we need especially the methods of the calculus (Chaps. 15 and 16). Therefore at this time we shall just mention some of them briefly.

Parabola

1 Path of a projectile (neglecting air resistance)
2 Cable of a suspension bridge (uniformly loaded along the bridge)
3 Parabolic reflector [surface generated by revolving a parabola about its axis has the property that each light ray coming in parallel to the axis will be reflected to (through) the focus]
4 Graphs of many equations in physics are parabolas
5 The antenna of a radio telescope

Ellipse

1 Orbit of a planet (sun at one focus)
2 Orbits of planetary moons, satellites, some comets
3 Elliptic gears for certain machine tools
4 Focal property: a ray emanating at one focus is reflected to the other
5 Many scientific formulas are equations which plot into ellipses

Hyperbola

1 Used in the construction of certain telescopic lenses
2 Some comets trace hyperbolas
3 Formulas taken from the field of the physical sciences are often of hyperbolic type

14.14 › LOCUS PROBLEMS

The conic sections—parabola, ellipse, and hyperbola—have been defined above as loci satisfying certain geometric conditions. In this section we present you with a variety of similar locus problems. To solve such problems let an arbitrary point on the locus be $P(x,y)$. Then write an equation of the condition on P given in the problem and simplify. This is then the equation of the required locus.

ILLUSTRATION 1 Find an equation satisfied by a point P whose distance from the point $A(1,3)$ is equal to its distance from the point $B(-2,4)$.

SOLUTION Let P be the point $P(x,y)$. Then the distance $PA = \sqrt{(x-1)^2 + (y-3)^2}$, and the distance $PB = \sqrt{(x+2)^2 + (y-4)^2}$. The condition on P is that distance $PA = PB$. So the equation is

$$\sqrt{(x-1)^2 + (y-3)^2} = \sqrt{(x+2)^2 + (y-4)^2}$$

This can be simplified as follows:

$$(x-1)^2 + (y-3)^2 = (x+2)^2 + (y-4)^2$$
$$x^2 - 2x + 1 + y^2 - 6y + 9 = x^2 + 4x + 4 + y^2 - 8y + 16$$
$$-6x + 2y - 10 = 0$$

or finally,

$$3x - y + 5 = 0$$

PROBLEMS 14.14 In Probs. 1 to 4 find the equation of the given locus.

1 The locus of points P such that the distance from P to $(-3,2)$ is always 4 units.
2 The locus of points P such that the distance from P to the line $x = -2$ is always equal to its distance from the point $(2,0)$.
3 The locus of points P such that the sum of the distances of P from the two points $(-4,0)$ and $(4,0)$ is always 10 units.
4 The locus of points P such that the absolute value of the difference of the distances of P from $(0,5)$ and $(0,-5)$ is 4 units.

5 Show that an ellipse is the locus of points P such that the ratio of the distances of P from a fixed point and from a fixed line is a constant e less than unity. HINT: Take the fixed point $F(ae,0)$ and the fixed line $x = a/e$. Show that the equation of the locus is then

$$\frac{x^2}{a^2} + \frac{y^2}{a^2(1 - e^2)} = 1$$

Eccentricity

The constant e is called the *eccentricity* of the ellipse. What is the eccentricity of a circle?

6 Show that a hyperbola is the locus of points P such that the ratio of the distances of P from a fixed point and from a fixed line is a constant e greater than unity. HINT: Take the fixed point $F(ae,0)$ and the fixed line $x = a/e$. Show that the equation of the locus is then

$$\frac{x^2}{a^2} - \frac{y^2}{a^2(e^2 - 1)} = 1$$

The constant e is called the *eccentricity* of the hyperbola. (See the definition of the parabola where e, defined similarly, would be equal to unity.)

In Probs. 7 to 16 find the equation of the stated locus and identify the curve. For convenience choose the given points and lines as indicated in each problem. This choice does not affect the generality of your conclusions.

7 Find the locus of a point whose distances from two fixed points are in a constant ratio $r \neq 1$. Let the points be $(-a,0)$ and $(a,0)$.
8 Find the locus of a point the square of whose distance from a fixed point is a constant r times its distance from a fixed line not passing through the fixed point. Let the point be $(-a,0)$ and the line $x = a$.
9 Find the locus of a point which is the center of a circle passing through a fixed point and tangent to a fixed line. Let the fixed point be $(0,0)$ and the line be $x = a$.

10 Find the locus of a point such that the square of its distance from the base of an isosceles triangle is equal to the product of its distances from the other two sides. Let the triangle have vertices $(-a,0)$, $(a,0)$, $(0,b)$.

11 Find the locus of a point the sum of whose distances from two fixed perpendicular lines is equal to the square of its distance from the point of intersection of the lines. Let the lines be $x = 0$ and $y = 0$.

12 Find the locus of a point the product of whose distances from two fixed intersecting lines is a constant r. Let the lines be $y = \pm ax$.

13 Find the locus of a point the sum of the squares of whose distances from two intersecting lines is a constant r. Let the lines be $y = \pm ax$.

14 A variable line makes with two fixed perpendicular lines a triangle of constant area A. Find the locus of the midpoint of the segment of the variable line whose endpoints are on the two fixed lines. Let the two fixed lines be $x = 0$ and $y = 0$.

15 Given two parallel lines L_1 and L_2 and a third line L_3 perpendicular to the first two, find the locus of a point the product of whose distances from L_1 and L_2 is a constant r times the square of its distance from L_3. Let L_1 and L_2 be $y = 0$, $y = a$; and L_3 be $x = 0$.

16 A variable line is drawn parallel to the base BC of a triangle ABC meeting AB and AC in the points D and E, respectively. Find the locus of the intersection of BE and CD. Let $A = (0,a)$, $B = (b,0)$, $C = (c,0)$.

14.15 › POLAR COORDINATES

In discussing the trigonometric functions of angles we introduced in Sec. 12.4 the ideas of polar coordinates. Our interest at that time was trigonometry, and certain discussions were simplified by assuming that the radius vector r was positive. For purposes of analytic geometry it is highly desirable to remove this restriction. This we now do according to the following definitions (which are just slightly modified versions of Definitions 1 to 5 in Sec. 12.4).

DEFINITIONS

Consider a (horizontal) line called the *polar axis* and a point O on it called the *pole,* or *origin*. From the pole to an arbitrary point P, draw the line segment r, called the *radius vector;* the radius vector makes a directed angle θ with the positive direction of the polar axis, which is taken to the right. We assign to P the ordered pair (r,θ); call them the *polar coordinates of* P, and write $P(r,\theta)$.

We permit θ to be positive (counterclockwise) or negative (clockwise). If no restrictions are imposed upon us, we may use any convenient angular

Polar coordinates

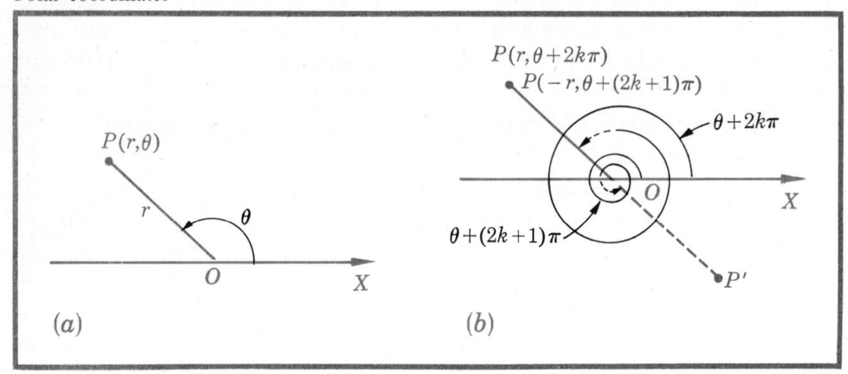

FIGURE 14.23
In contrast to our agreement in Chap. 12, we now permit r to be negative.

unit, radian measure and degree measure being the most common. Likewise we permit r to be positive or negative, as we shall explain.

First, note that a fixed point P has several sets of polar coordinates. Indeed, if P has coordinates (r,θ), then it also has coordinates $(r, \theta + 2k\pi)$ for every integer k. Regardless of what integer k is used in the second element of the number pair $(r, \theta + 2k\pi)$, r itself is positive. It is desirable to permit r to be negative. We agree to call the number pair $(-r, \theta + \pi)$ coordinates of P; likewise for the pair $(-r, \theta + (2k + 1)\pi)$, k an integer. The geometric interpretation is evident from Fig. 14.23*b*: we extend PO in the direction OP' and measure θ to the extension. For this case r is negative.

In summary, the polar coordinates of the point P are

(1) $$(r, \theta + 2k\pi) \qquad k \text{ an integer}$$

or

(2) $$(-r, \theta + (2k + 1)\pi) \qquad k \text{ an integer}$$

The pole itself is a very special point since, when $r = 0$, there is no unique angle θ.

There is no one-to-one correspondence between points in the plane and polar number pairs. To a given pair (r,θ) there corresponds a unique point, but to a given point there corresponds no unique pair (r,θ). This is in contrast to the situation in rectangular coordinates.

There is evidently a connection between the rectangular coordinates and the polar coordinates of a point P. By superposition of the two systems (Fig. 14.24) we find that

(I) $$x = r\cos\theta \qquad y = r\sin\theta$$

EXERCISE A Figure 14.24 assumes that r is positive. Draw a figure with r negative, and use it to establish formulas (I).

Polar and rectangular coordinates

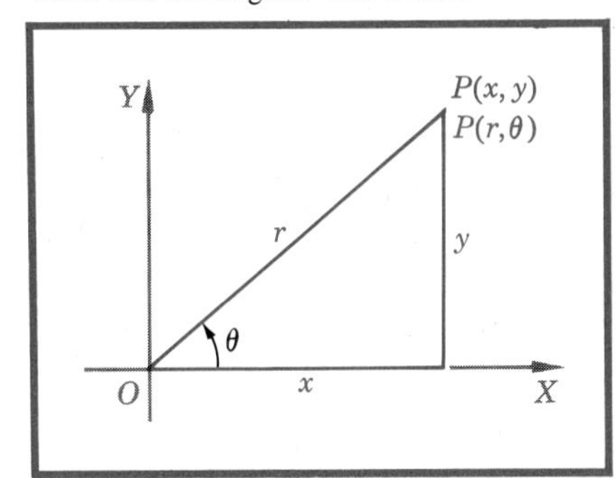

FIGURE 14.24
This figure shows that
$x = r \cos \theta$,
$y = r \sin \theta$.

From these equations we can readily find the rectangular coordinates when we know the polar coordinates. To arrive at polar coordinates when we are first given rectangular coordinates is harder because of the ambiguity in the former. We generally settle upon the least angle θ, where $0 \leq \theta < 360°$, to be associated with P. In this case r is positive. But there are times when we need to consider expressions (1) and (2) from each of which we will then pick out appropriate coordinates (perhaps more than one set) to suit our purpose. Thus, usually,

$$\textbf{(II)} \qquad r = \sqrt{x^2 + y^2} \qquad \tan \theta = \frac{y}{x}$$

where θ is determined as the least angle $\theta \geq 0$, which satisfies the above equation and the equations

$$\textbf{(III)} \qquad \sin \theta = \frac{y}{\sqrt{x^2 + y^2}} \qquad \cos \theta = \frac{x}{\sqrt{x^2 + y^2}}$$

By means of (I), (II), and (III) we can transform equations from one system to another. Sometimes one system is more suitable to a given problem than another.

ILLUSTRATION 1 Transform the polar equation

$$Ar \cos \theta + Br \sin \theta + C = 0$$

to rectangular coordinates.

SOLUTION We make use of (I) and write the transformed equation

$$Ax + By + C = 0$$

In either system the graph is a straight line.

EXERCISE A Show that $r \cos (\theta - a) = b$ is the equation of a straight line and compare with $r(A \cos \theta + B \sin \theta) + C = 0$.

EXERCISE B Sketch the graph of $r = \sec\theta$.

EXERCISE C Write the polar equation of an arbitrary line passing through the pole.

ILLUSTRATION 2 Transform the rectangular equation $x^2 + y^2 = a^2$ (of a circle of radius a with center at the origin) to polar coordinates.

SOLUTION Using (I), we write $x^2 + y^2 = a^2$ in the form

$$\begin{aligned}(r\cos\theta)^2 + (r\sin\theta)^2 &= a^2\\ r^2(\cos^2\theta + \sin^2\theta) &= a^2\\ r^2 &= a^2\end{aligned}$$

The graph of $r = a$ is a circle of radius a with center at the pole; $r = -a$ plots the same circle. Hence there are two different equations in polar coordinates for this circle. This is not an isolated example; certain curves may have several distinct polar equations. This is due to the fact that the polar coordinates of a point are not unique. It is important to note that the *coordinates* (a,θ), satisfying $r = a$, do *not* satisfy $r = -a$.

EXERCISE D Write down the polar equation of the circle with unit radius and center at the point $(r = \frac{1}{2}, \theta = \pi/2)$.

PROBLEMS 14.15

In Probs. 1 to 12 transform to polar coordinates.

1 $3x + y = 4$
2 $3x - 4y = 0$
3 $x^2 + y^2 = 16$
4 $x^2 + y^2 + 4x - 6y = 0$
5 $x^2 = y$
6 $y^2 = 4x$
7 $\dfrac{x^2}{16} + \dfrac{y^2}{25} = 1$
8 $\dfrac{x^2}{4} + \dfrac{y^2}{9} = 1$
9 $\dfrac{x^2}{16} - \dfrac{y^2}{9} = 1$
10 $\dfrac{y^2}{25} - \dfrac{x^2}{36} = 1$
11 $(x - x^2 - y^2)^2 = x^2 + y^2$
12 $(x^2 + y^2 + y)^2 = x^2 + y^2$

In Probs. 13 to 28 transform to rectangular coordinates.

13 $r = 4\sin\theta$
14 $r = 3\cos\theta$
15 $r = -4\cos\theta$
16 $r = -5\sin\theta$
17 $r = 3 + \cos\theta$
18 $r = 4 - \cos\theta$
19 $r = 1 - \sin\theta$
20 $r = 1 - 2\sin\theta$
21 $r = \dfrac{1}{1 + \sin\theta}$
22 $r = \dfrac{1}{1 + \cos\theta}$

23 $r = -2$	**24** $\theta = 45°$
25 $\theta = 30°$	**26** $r\theta = 5$
27 $r^2\theta = 2$	**28** $r^2 = 4r$

14.16 » POLAR COORDINATES (Continued)

To derive the equations of conic sections in polar coordinates we shall proceed from the following definition which should be compared with those given in Sec. 14.10 and Sec. 14.14, Probs. 5 and 6.

DEFINITIONS

Conic section

The locus of points P such that the ratio of the distances from P to a fixed point F and to a fixed line DD' is a constant e is called a *conic section.* The point F is called the *focus,* DD' is called the *directrix,* and e is called the *eccentricity.* If

$e = 1$, the locus is a *parabola*
$e < 1$, the locus is an *ellipse*
$e > 1$, the locus is a *hyperbola*

To derive the equation of a conic in polar coordinates is quite simple if we choose the focus F for the pole and the line through F and perpendicular to DD' for the polar axis. Consult Fig. 14.25; we let p be the distance from F to DD'. By definition, for every point $P(r,\theta)$ it must be true that

$$\frac{\text{Dist. } PF}{\text{Dist. } P \text{ to } DD'} = e$$

Focus and directrix in polar coordinates

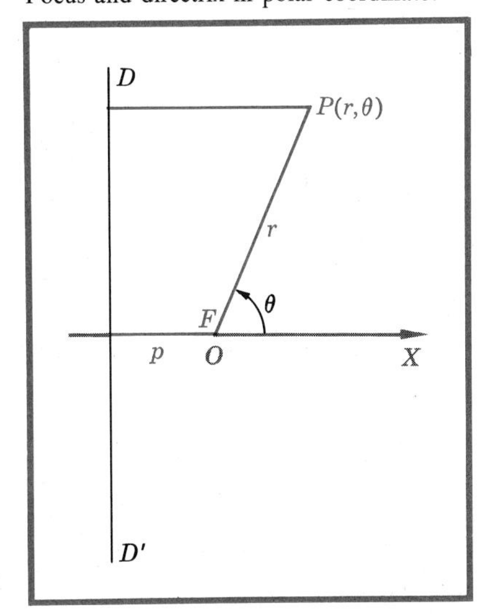

FIGURE 14.25
A general conic section is given by a focus F, a directrix DD', and the relation $PF/PD = e$.

that is,

$$\frac{r}{p + r\cos\theta} = e$$

This reduces to

(1)
$$r = \frac{ep}{1 - e\cos\theta}$$

which is the equation of the conic.

ILLUSTRATION 1 Sketch the graph of the parabola $r = 1/(1 - \cos\theta)$.

SOLUTION We make out a table of values.

θ	0	$\pi/6$	$\pi/4$	$\pi/3$	$\pi/2$	π
$1 - \cos\theta$	0	0.134	0.293	0.500	1	2
r	...	7.47	3.41	2	1	$\frac{1}{2}$

You can easily make out another table for third and fourth quadrants (for the lower half of the curve). It is important to be careful in plotting points for values of θ near 0 since r is not defined for $\theta = 0$ (Fig. 14.26).

EXERCISE A When the directrix is to the right of the pole, show that the equation of the conic is

$$r = \frac{ep}{1 + e\cos\theta}$$

Parabola in polar coordinates

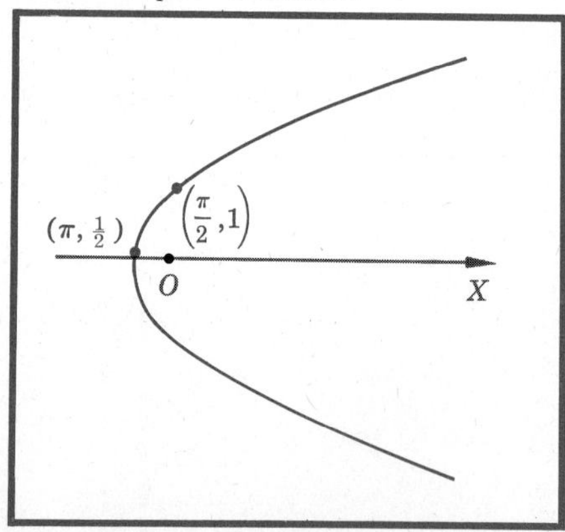

FIGURE 14.26
$r = \dfrac{1}{1 - \cos\theta}$

EXERCISE B When the directrix is horizontal and at a distance p from the pole, show that the equation of the conic is

$$r = \frac{ep}{1 \pm e \sin \theta}$$

We may obtain the graphs of other equations in polar coordinates by the methods of the following illustrations.

ILLUSTRATION 2 Sketch the graph of $r = \sin 2\theta$.

SOLUTION Again we prepare a table of values. This time, for practice, we shall use degree measure.

θ	0°	15°	22.5°	30°	45°	60°	67.5°	75°	90°
2θ	0°	30°	45°	60°	90°	120°	135°	150°	180°
r	0	0.500	0.707	0.866	1	0.866	0.707	0.500	0

The graph of this much is given in Fig. 14.27*a*.

We continue the table.

θ	105°	112.5°	120°	135°	150°	157.5°	165°	180°
2θ	210°	225°	240°	270°	300°	315°	330°	360°
r	−0.500	−0.707	−0.866	−1	−0.866	−0.707	−0.500	0

The graph of the preceding two tables is given in Fig. 14.27*b*. So far, no portion of the graph has repeated. Indeed θ must run the full

Partial graph of $r = \sin 2\theta$

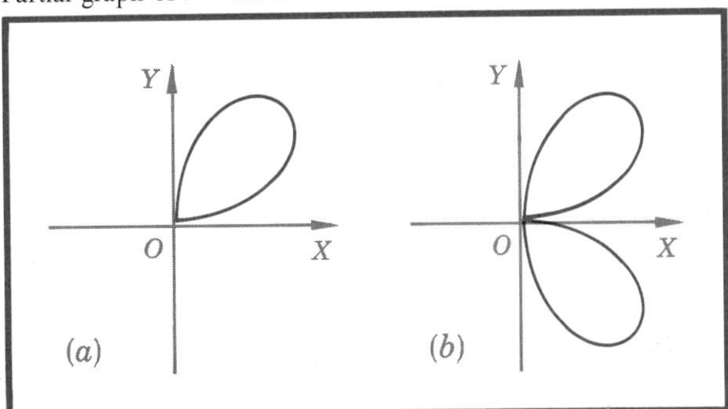

FIGURE 14.27 This graph has the same form in each quadrant, so we need plot only one quadrant in detail.

Complete graph of $r = \sin 2\theta$

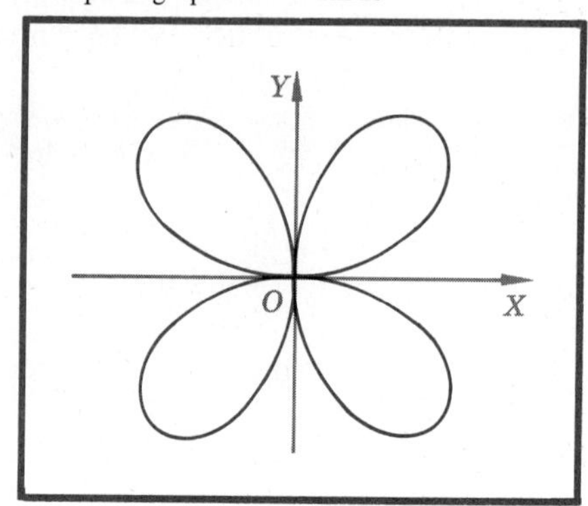

FIGURE 14.28
Four-leafed rose.

course of 360° before repetition. The total graph is the "four-leaved rose" exhibited in Fig. 14.28.

EXERCISE C Make out the remaining portion of the above table for the complete graph (drawn in Fig. 14.28).

You should study this illustration to see that really all we need is the set of the first five entries in the first portion of the table. With a knowledge of the behavior of the trigonometric function sine, we can draw the total graph. The essential variations of $\sin 2\theta$ take place in the first half of the first quadrant. This is typical of much of the graph work in polar coordinates involving the trigonometric functions where periodicity plays an important role.

ILLUSTRATION 3 Plot the graph of $r = \theta$.

SOLUTION Here we must use radian measure; but there is no need of making out a table. Take note of the portion corresponding to negative values of θ in Fig. 14.29. The curve is known as the Spiral of Archimedes.

Graph of $r = \theta$

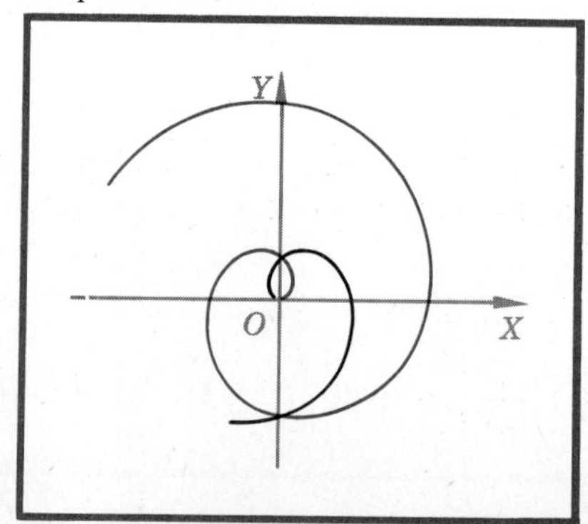

FIGURE 14.29
Spiral of Archimedes. The colored portion corresponds to $\theta > 0$; the black to $\theta < 0$.

Graph of $r = \cos\theta$

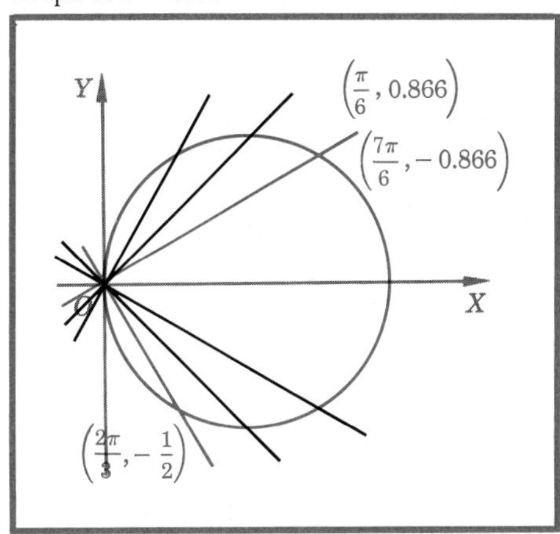

FIGURE 14.30
Circle. The curve is traversed twice as θ varies from 0 to 2π.

ILLUSTRATION 4 Sketch the graph of $r = \cos\theta$.

SOLUTION We prepare the following table.

θ	0	$\pi/6$	$\pi/4$	$\pi/3$	$\pi/2$	$2\pi/3$	$3\pi/4$	$5\pi/6$	π	$7\pi/6$
r	1	0.866	0.707	0.500	0	−0.500	−0.707	−0.866	−1	−0.866

The geometric point whose coordinates are $(\pi/6, 0.866)$ is the same as that with coordinates $(7\pi/6, -0.866)$. These are the second and last entries in the preceding table. Extension of the table through third and fourth quadrantal angles shows that the curve is being traced a second time. Therefore the description of the curve is complete after θ runs through the first two quadrants (Fig. 14.30). The curve is actually a circle, as you can immediately verify by transforming $r = \cos\theta$ to the rectangular form

$$(x - \tfrac{1}{2})^2 + y^2 = \tfrac{1}{4}$$

PROBLEMS 14.16

In Probs. 1 to 28 transform to rectangular coordinates. Sketch, using either polar coordinates or rectangular coordinates.

1 $r = \dfrac{8}{3 - \cos\theta}$

2 $r = \dfrac{3}{2 - \cos\theta}$

3 $r = -\dfrac{8}{1 - 3\cos\theta}$

4 $r = \dfrac{3}{1 + 2\cos\theta}$

5 $r = \dfrac{1}{1 - \sin\theta}$

6 $r = \dfrac{1}{1 + \sin\theta}$

7 $r^2 + r\sin(\theta - 30°) = 0$
8 $r^2 + r\cos(\theta + 30°) = 0$
9 $r = -\sin 2\theta$
10 $r = \sin\theta$
11 $r = \sin 4\theta$
12 $r = \sin 3\theta$ Three-leaved rose
13 $r = \cos 2\theta$ Four-leaved rose
14 $r = \cos\theta$
15 $r = \cos 4\theta$
16 $r = \cos 3\theta$ Three-leaved rose
17 $r = 1 + 2\sin\theta$
18 $r = 1 + 2\cos\theta$ Limaçon
19 $r = 1 + \sin\theta$
20 $r = 1 + \cos\theta$ Cardioid
21 $r = 2 + \sin\theta$
22 $r = 2 + \cos\theta$
23 $r = \pi/\theta$ Hyperbolic spiral
24 $r = -\pi/\theta$
25 $r^2 = -\pi/\theta$
26 $r^2 = \pi/\theta$ The lituus
27 $r^2 = \sin 2\theta$
28 $r^2 = \cos 2\theta$ Lemniscate

29 Sketch and find the points of intersection:

$$r = \cos\theta \qquad \text{and} \qquad r = 1 - \cos\theta$$

30 Sketch and find the points of intersection:

$$r = \sin\theta \qquad \text{and} \qquad r = \sqrt{3}\cos\theta$$

31 Find the equation of the locus of the midpoints of chords of a circle of radius a drawn from a fixed point Q on the circle.
32 Find the distance between $P_1(r_1,\theta_1)$ and $P_2(r_2,\theta_2)$.
33 Find the equation of the locus of points P such that the length of the radius vector of P is proportional to the square of its vectorial angle.

14.17 › PARAMETRIC EQUATIONS

In order to explain the meaning of the parametric equations of a curve, we shall begin by discussing a function somewhat different from those treated in Chap. 9. Here we are interested in a function F whose domain is the real line R (or some subset of it) and whose range is the plane, that is, $R \times R$. Hence we have a function F which may be written

$$F{:}R \to R \times R$$

A particular function of this type can be defined in terms of two functions f and g: $f{:}R \to R$ and $g{:}R \to R$ as follows:

(1) $$x = f(t) \quad \text{and} \quad y = g(t)$$

In this notation t is a point of the real line which is mapped onto the point (x,y) in the plane. As t varies on the line, (x,y) moves in the plane and traces out a curve which is the image of the function F.

REMARKS

Parameter

1 We refer to Eq. (1) as the *parametric equations* of the curve and call t the *parameter*.
2 Since the parameter can be chosen in many ways, we expect to find a great variety of parametric equations representing the same curve. For example,

$$x = t, y = t \quad \text{and} \quad x = t^3, y = t^3$$

both represent the line $x = y$.
3 Cartesian equations of the curve can sometimes be obtained by eliminating t from $x = f(t)$ and $y = g(t)$. The resulting equation (often difficult or impossible to obtain) has the form $F(x,y) = 0$, but its graph may not be identical with the graph of the given parametric equations. For example, eliminating t from $x = t^2$ and $y = t^2$, we obtain $y = x$. The parametric equations, however, give only the portion of this line above the origin. See Illustration 5 for another example.

ILLUSTRATION 1 Write the equation of a straight line in parametric form.

SOLUTION In Sec. 14.6, we saw that

$$x_2 - x_1 = d \cos \alpha \qquad y_2 - y_1 = d \cos \beta$$

where (x_1,y_1) and (x_2,y_2) are points on the line and d is the distance between them. If we write (x,y) for (x_2,y_2) and t for the distance between (x,y) and (x_1,y_1), these can be written

$$x = x_1 + t \cos \alpha \qquad y = y_1 + t \cos \beta$$

These are parametric equations of the line. They may also be written in the form

$$x = x_1 + u(x_2 - x_1) \qquad y = y_1 + u(y_2 - y_1)$$

where $u = t/d$. The graph of them is the whole line.

ILLUSTRATION 2 Plot the line whose parametric equations are $x = 1 + 3u, y = 2 - 2u$. Then eliminate u and find the cartesian equation of the line.

SOLUTION Construct a table of values:

u	-2	-1	0	1	2
x	-5	-2	1	4	7
y	6	4	2	0	-2

Then plot the points (x, y) as in Fig. 14.31. To find the cartesian equation, first solve $x = 1 + 3u$ for u. The result is $u = (x - 1)/3$. Substitute this expression in $y = 2 - 2u$. Hence we obtain

$$y = 2 - \frac{2(x - 1)}{3}$$

which simplifies to $2x + 3y - 8 = 0$.

ILLUSTRATION 3 Find parametric equations for the circle $x^2 + y^2 = a^2$. These are:

$$x = a \cos t \qquad y = a \sin t$$

where t represents an angle as in polar coordinates. When the point (x,y) is traversing the circle with angular velocity ω, its motion is described by the parametric equations,

$$x = a \cos \omega t \qquad y = a \sin \omega t$$

where t now represents time.

Graph of line from its parametric equations

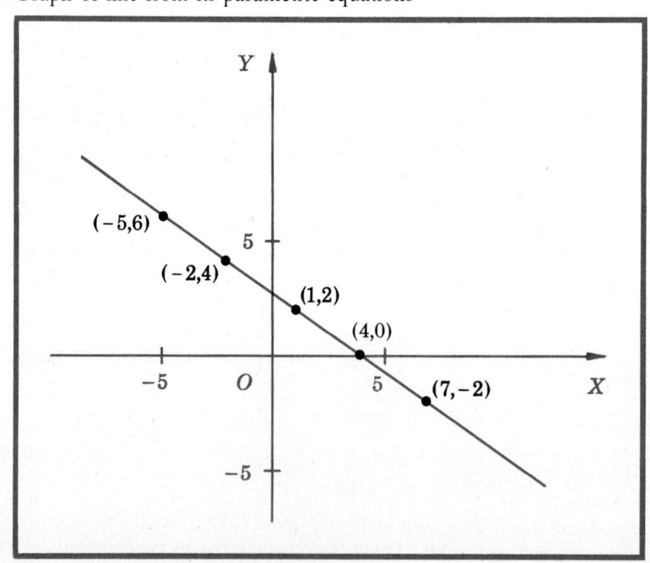

FIGURE 14.31
$x = 1 + 3u$,
$y = 2 - 2u$,
or $2x + 3y - 8 = 0$.

Parametric equations of an ellipse

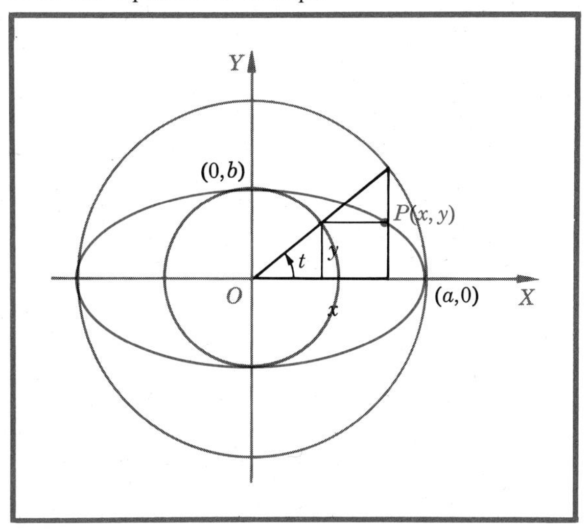

FIGURE 14.32 The parametric equations of the ellipse $x^2/a^2 + y^2/b^2 = 1$ are $x = a \cos t$ and $y = b \sin t$.

ILLUSTRATION 4 Find parametric equations for the ellipse

$$b^2x^2 + a^2y^2 = a^2b^2$$

SOLUTION We choose the parameter t as the angle shown in Fig. 14.32. In terms of the angle t we can write down the equations immediately since $x/a = \cos t$ and $y/b = \sin t$. They are therefore

(2) $$x = a \cos t \qquad y = b \sin t$$

and the graph is the complete ellipse.

EXERCISE A Eliminate t from Eq. (2).

Graph of $x = \sin^2 t$; $y = 2 \cos t$

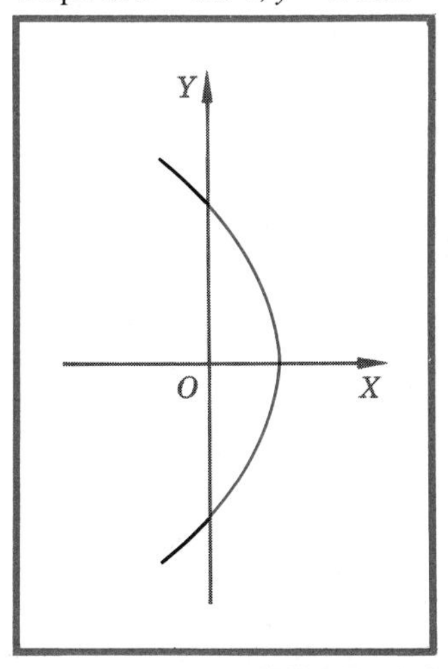

FIGURE 14.33 The parametric equations represent only the colored portion of the curve.

ILLUSTRATION 5 Show that $x = \sin^2 t$, $y = 2 \cos t$ represents only a portion of the parabola whose cartesian equation is $y^2 = 4(1 - x)$ (Fig. 14.33).

SOLUTION The given parametric equations permit x to vary from 0 to 1 only and y to vary from -2 to $+2$. We eliminate t as follows:

$$x = \sin^2 t$$
$$\tfrac{1}{4}y^2 = \cos^2 t$$

Adding, we get

$$x + \tfrac{1}{4}y^2 = 1$$

or

$$y^2 = 4(1 - x)$$

ILLUSTRATION 6 A circle of radius a rolls along a line. Find the locus described by a point on the circumference.

Graph of $x = a(t - \sin t)$; $y = a(1 - \cos t)$

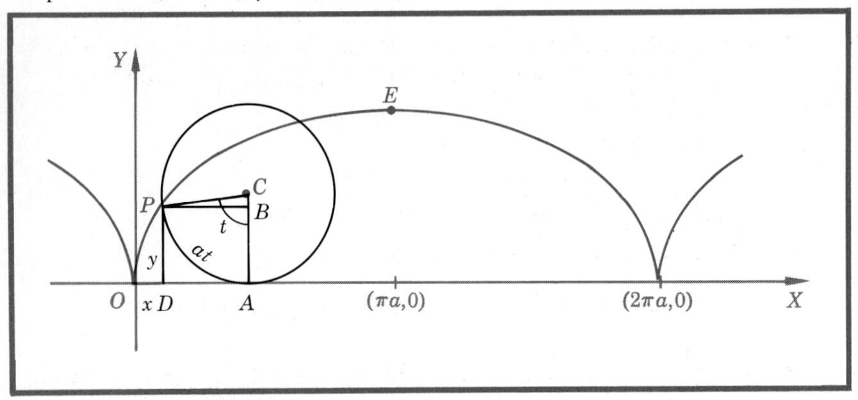

FIGURE 14.34
A cycloid is the path of a fixed point on a rolling circle.

SOLUTION Take the X-axis as coincident with the line and the initial position of the tracing point as the origin (Fig. 14.34). The positive angle

$$PCA = t \text{ radians}$$

will be chosen as the parameter. The arc $PA = at$ (Sec. 13.2). Clearly

$$OA = PA = at$$

Hence

$$\begin{aligned} x &= OA - DA \\ &= OA - PB \\ &= at - a \sin t \end{aligned}$$

Further

$$\begin{aligned} y &= PD = CA - CB \\ &= a - a \cos t \end{aligned}$$

Cycloid

Parametric equations of the locus, called a *cycloid*, are therefore

$$\begin{aligned} x &= a(t - \sin t) \\ y &= a(1 - \cos t) \end{aligned}$$

This curve is very important in physics, where it is called the *brachistochrone*, or the *curve of quickest descent*. This means that, if we think of the curve as turned upside down, then, out of all possible curves connecting O and E, the brachistochrone is the one down which a frictionless particle will slide in least time. As a matter of fact this time is independent of the point on the curve from which the particle is released; the particle will slide from O to E in the same time that it will slide from any other point (such as P) to E. The cartesian equation obtained by eliminating t is troublesome; therefore we do not consider it.

PROBLEMS 14.17

In Probs. 1 to 26 eliminate the parameter t, and identify the curve if possible.

1 $x = 2t + 5, y = 3t - 7$ **2** $x = t + 1, y = t - 2$

3 $x = t^2, y = t - 2$

4 $x = t, y = 2$

5 $x = t^2 + 3t + 1, y = t$

6 $x = 2t, y = t^2 + t + 1$

7 $x = \cos t, y = \sin^2 t$

8 $x = 1 - \cos t, y = 1 + \sin t$

9 $x = \sin^2 t, y = 3 \cos t$

10 $x = 2 \sin t, y = \cos^2 t$

11 $x = a \sin t + b \cos t, y = a \cos t - b \sin t$

12 $x = \cos t + \sin t, y = \cos t - \sin t$

13 $x = \dfrac{1}{1 - t}, y = \dfrac{1}{1 - t}$

14 $x = a \cos t + b \sin t, y = a \sin t - b \cos t$

15 $x = 1 - t, y = t^2 + 2t$

16 $x = \dfrac{1}{2t}, y = 1 - t$

17 $x = 2 \sin t, y = \cos t$

18 $x = \dfrac{1}{\sqrt{1 + t^2}}, y = \dfrac{2t}{\sqrt{1 + t^2}}$

19 $x = 4 - t, y = \sqrt{-t}$

20 $x = t^2 - t, y = 1 + t$

21 $x = \dfrac{t}{1 + t^3}, y = \dfrac{t^2}{1 + t^3}$ Folium of Descartes

22 $x = 1 - 3 \cos t, y = 2 + 2 \sin t$

23 $x = \sin 2t, y = \cos t$

24 $x = \tan t, y = \sec t$

25 $x = -\dfrac{2at}{1 + t^2}, y = a\dfrac{1 - t^2}{1 + t^2}$

26 $x = \cos 2t, y = \sin t$

27 Find the parametric equations of the parabola whose equation is $y^2 = 4px$, in terms of the parameter t, which is the slope of the line $y = tx$.

28 A circle of radius $a/4$ rolls inside a circle of radius a. Show that parametric equations of the locus described by a point on the circumference of the rolling circle are $x = a \cos^3 t, y = a \sin^3 t$. [The parameter t is the angle through which the line of centers turns, the center of the stationary circle being placed at the origin. The initial position of the line of centers coincides with the X-axis, and $(a,0)$ is the initial position of the tracing point.] The curve is called the hypocycloid.

REFERENCES

In addition to the many standard textbooks on analytic geometry, the reader should consult the following articles in the *American Mathematical Monthly:*

Boyer, C. B.: The Equation of an Ellipse, vol. 54, p. 410 (1947).
Boyer, C. B.: Newton as an Originator of Polar Coordinates, vol. 56, p. 73 (1949).

Hammer, D. C.: Plotting Curves in Polar Coordinates, vol. 48, p. 397 (1941).

Hawthorne, Frank: Derivation of the Equations of Conics, vol. 54, p. 219 (1947).

Johns, A. E.: The Reduced Equation of the General Conic, vol. 54, p. 100 (1947).

Wagner, R. W.: Equations and Loci in Polar Coordinates, vol. 55, p. 360 (1948).

15 INTUITIVE INTEGRATION

15.1 » INTRODUCTION

In this chapter and in the next we shall be concerned with a branch of mathematics known as the "calculus". The treatment is in two parts: integral calculus (integration) and differential calculus (differentiation). This body of material was developed by Newton, Leibniz, and others in the seventeenth century; before their time only Archimedes seems to have had any clear notion of what was involved. That which is involved is the *theory of limits*. Our approach will be an intuitive one inasmuch as a detailed study of limits is beyond the scope of this book.

To begin with, we recall that the early Greeks gave a definition of the (measure of the) length of the circumference of a circle (see Sec. 13.1). They also defined the (measure of the) area of a rectangle as the product of length times width. That is, $A = L \times W$ (Fig. 15.1). From this it follows that the area of a right triangle of legs L and W is $\frac{1}{2}LW$, which can be read "one-half the base times the altitude". The notion is readily extended by trigonometry to cover the case of any triangle ($A =$ one-half the product of two sides times the sine of the angle included by those two sides). Since any polygon can be broken up into triangles, the above definition has led to a method of determining the area of a polygon (Fig. 15.2).

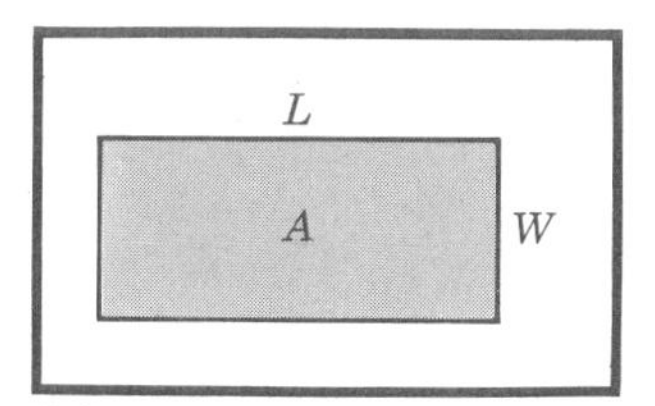

FIGURE 15.1

EXERCISE A Find the area of Fig. 15.3.

This procedure is obviously applicable only to polygons—plane figures bounded by straight-line segments—and not to figures bounded by curves. Yet, intuitively, we feel that a rope stretched out on the floor in a simple closed curve without kinks, crossings, or knots (Fig. 15.4) encloses something we should like to call "area". This means two things: first, we must give a definition of "area bounded by a closed curve", and, second, if area is to be a fruitful idea, we must develop a way of com-

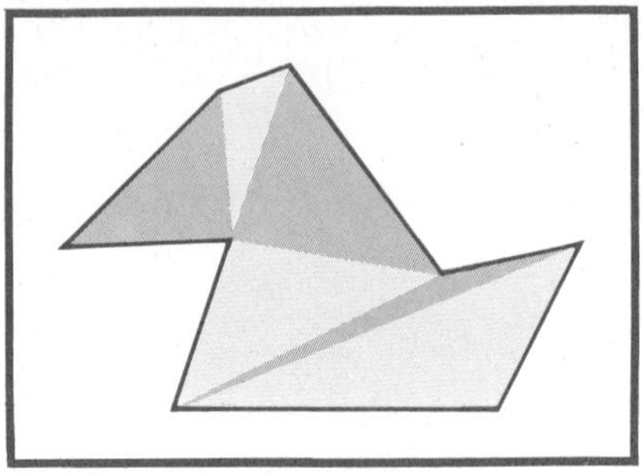

Triangular subdivision

FIGURE 15.2
The area of a polygonal region can be calculated by subdividing it into triangles.

Polygonal region

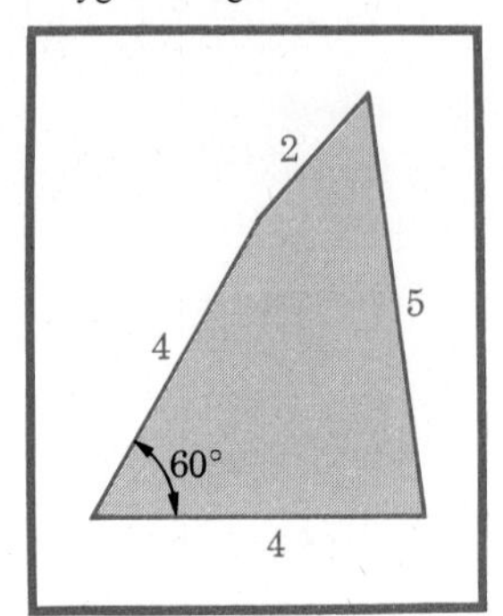

FIGURE 15.3
To find the area, subdivide into two triangles.

Nonpolygonal region

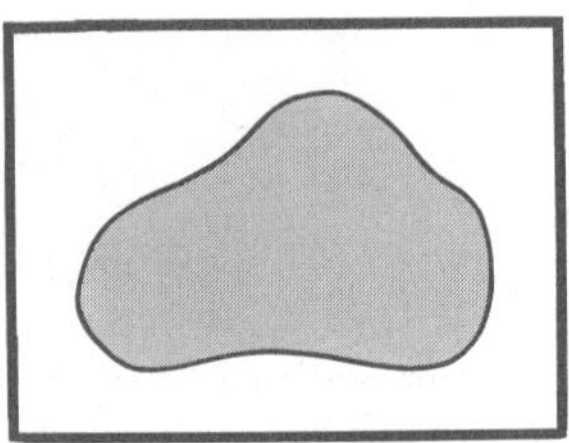

FIGURE 15.4
New methods must be devised to find areas of nonpolygonal regions.

puting it. That is, we should like to be able to find, at least theoretically, a real number which will be called, indifferently, the "measure of the area", or, simply, the "area".

As a matter of fact, Euclid gave such a definition for the area of a circle, and, using it and a circle of radius one unit, Archimedes (287–212 B.C.) was able to approximate π to within the inequality $3\frac{10}{71} < \pi < 3\frac{10}{70}$.

15.2 ➤ AREA OF A CIRCLE

Euclid defined the area of a circle as follows.

DEFINITION The *area of a circle* is the limiting value of the area of an inscribed (or circumscribed) regular polygon of n sides as the number of sides n is increased indefinitely.

Let us try to find the area of a circle of radius r by using this definition. We begin by determining the area of an inscribed regular polygon of n sides. Radii are drawn from the center of the circle to the vertices of the polygon, dividing the polygon into n isosceles triangles. A typical triangle OPQ, with central angle $\theta = 2\pi/n$, is indicated in Fig. 15.5. The inscribed polygon has n times the area of this triangle since there are n such triangles. Now the area of OPQ is $\frac{1}{2}r^2 \sin(2\pi/n)$, and the area of the polygon is $(n/2)r^2 \sin(2\pi/n)$, which, in turn, is less than the area of the circle. By Euclid's definition

(1) Area of circle = limiting value of $\dfrac{n}{2}r^2 \sin\dfrac{2\pi}{n}$

as n increases indefinitely

Circle and an inscribed polygon

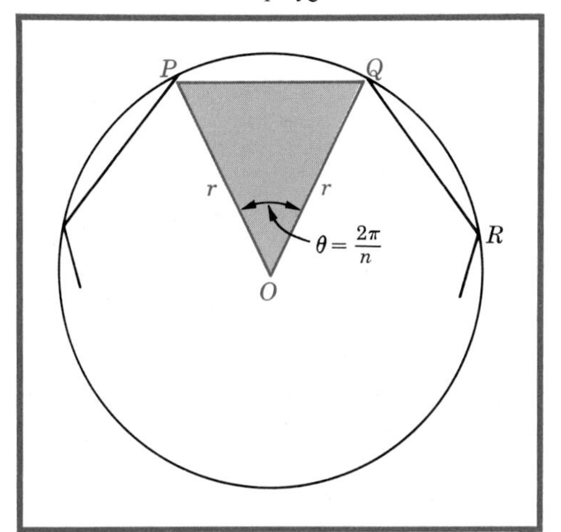

FIGURE 15.5
An approximation to the area of a circle is given by the area of an inscribed polygon.

Circle and a circumscribed polygon

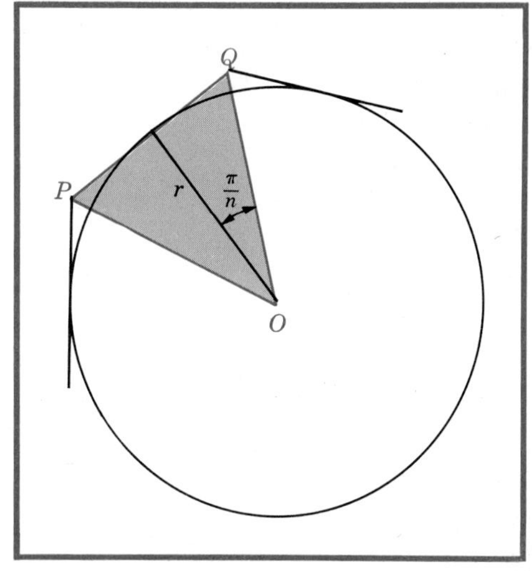

FIGURE 15.6
Another approximation to the area of a circle is given by the area of a circumscribed polygon.

If we had used a circumscribed regular polygon of n sides (Fig. 15.6), we would find that the area of a typical triangle (with central angle $\theta = 2\pi/n$) is $r^2 \tan(\pi/n)$, and so the area of the polygon is $nr^2 \tan(\pi/n)$. This is greater than the area of the circle, but by Euclid's definition

$$\text{(2)} \quad \text{Area of circle} = \text{limiting value of } nr^2 \tan\frac{\pi}{n} \quad \text{as } n \text{ increases indefinitely}$$

Archimedes used (1) and (2) and actually carried out the computations for the cases $n = 6, 12, 24, 48, 96$. It was inscribed and circumscribed regular polygons of 96 sides which produced the famous inequality referred to in Sec. 15.1.

EXERCISE A Set $r = 1$, and compute from (1) $\frac{96}{2}\sin(360°/96) = 3.1392$ and from (2) $96\tan(180°/96) = 3.1420$.

EXERCISE B From Exercise A above show that the area πr^2 of a circle of radius r satisfies the inequality

$$3.1392r^2 < \pi r^2 < 3.1420r^2$$

EXERCISE C From Figs. 15.5 and 15.6, show that the perimeters of the inscribed and circumscribed polygons are respectively $2nr\sin(\pi/n)$ and $2nr\tan(\pi/n)$. (See Sec. 13.1.)

Inscribed polygon with large number of sides

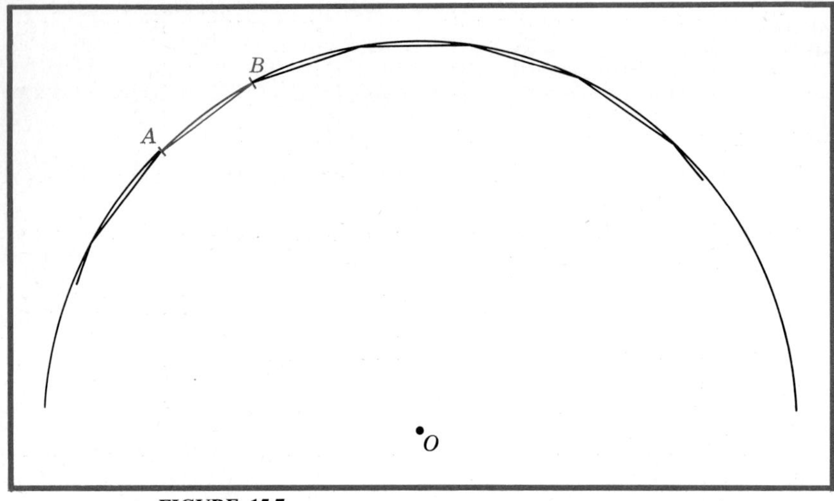

FIGURE 15.7
The approximation improves as the number of sides increases.

Now the idea of the "limit" expressed in the definition is intuitively clear: we feel that we can "see" what happens geometrically as the number n increases. For very large n an inscribed side AB (which is the chord AB) becomes indistinguishable from the arc AB (Fig. 15.7), and the area of the inscribed polygon differs from the area of the circle by a very, very small amount. On the other hand, when we try to compute the limits in (1) and (2), which we write in the notation

(3) $$\lim_{n\to\infty} \frac{n}{2} r^2 \sin \frac{2\pi}{n}$$

(4) $$\lim_{n\to\infty} nr^2 \tan \frac{\pi}{n}$$

we meet some technical difficulties.

It is not the purpose of this book to study in great detail the theory of limits, but in the next section we make a few remarks which should help you see that

$$\lim_{n\to\infty} \frac{n}{2} r^2 \sin \frac{2\pi}{n} = \pi r^2$$

PROBLEMS 15.2 In Probs. 1 and 2 compute the values indicated. (Change to degrees and use Table V of the Appendix.)

	n	6	12	24	48
1	$n \tan \frac{\pi}{n}$				
2	$\frac{n}{2} \sin \frac{2\pi}{n}$				

In Probs. 3 to 9 compute the values indicated using Table IV of the Appendix.

	x	1.00	0.50	0.05	0.01
3	$\sin x$				
4	$\cos x$				
5	$\tan x$				
6	$\frac{1}{\cos x}$				
7	$1 - \cos x$				
8	$\frac{\tan x}{x}$				
9	$\frac{\sin x}{x}$				

10 Paraphrase the language of the definition of the *area* of a circle, and give a precise definition of the *circumference* of a circle.
11 Define "length" of the parabolic arc $y = +\sqrt{x}$, from $x = 0$ to $x = 1$.
12 Define the area of the ellipse whose equation is $x^2 + 4y^2 = 4$.

15.3 › SOME LIMITS

From Table IV of the Appendix, we find:

x	1.00	0.50	0.05	0.01	0.00
$\sin x$	0.84147	0.47943	0.04998	0.01000	0.00000
$\frac{\sin x}{x}$	0.84147	0.95886	0.9996	1.00000	?

We are able to write $\sin 0 = 0.00000$, but we are unable to write a number in place of the question mark (we never divide by zero). However, the entries for $(\sin x)/x$ seem to indicate that $(\sin x)/x$ approaches 1 as x approaches 0. This is a correct guess, and we shall prove it later in Theorem 6.

First, we should know what a limit is. The statement

$$\text{“}\lim_{x\to 0} \frac{\sin x}{x} = 1\text{”}$$

means geometrically that, as x gets close to 0, $(\sin x)/x$ gets close to 1. Or, more generally, the statement "$\lim_{x\to a} f(x) = L$" means that as x gets close to a, $f(x)$ gets close to L. This is said more precisely and elegantly in the following definition.

DEFINITION
Definition of limit

The value $f(x)$ of the function f is said to *approach* the constant L as a limit as x approaches a if to each positive number ϵ, there corresponds a number $\delta(\epsilon)$ such that, if $0 < |x - a| < \delta(\epsilon)$ then

$$|f(x) - L| < \epsilon\dagger$$

† Similarly we have the definition: The value of the function f is said to approach the limit $+\infty$ as x approaches a if to each positive number A there corresponds a number $\delta(\epsilon)$ such that, if $0 < |x - a| < \delta(\epsilon)$, then $f(x) > A$.

In other words the statement that $f(x)$ has L as a limit when x approaches a means that: Given any vertical interval $[L - \epsilon, L + \epsilon]$, we can find a corresponding horizontal interval $[a - \delta(\epsilon), a + \delta(\epsilon)]$ such that whenever x $(\neq a)$ is in this horizontal interval, then $f(x)$ is in the vertical interval.

You should examine Fig. 15.8 until you think you understand the wording of the definition even though the process of obtaining limits from

Limit of a function

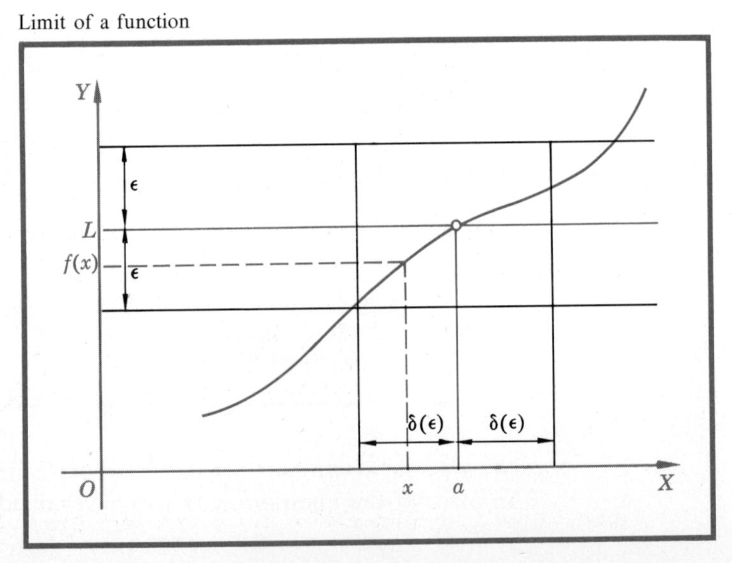

FIGURE 15.8
To show that $\lim_{x\to a} f(x) = L$, first choose a horizontal strip of width 2ϵ about $y = L$. Then we must find an interval of width 2δ (where δ depends on ϵ) about $x = a$ such that, for each x (different from a) in this interval, the point $(x, f(x))$ lies in the chosen horizontal strip. If we can do this for each choice of the positive number ϵ, then $\lim_{x\to a} f(x) = L$.

Theorems about limits

it is very complicated. The following theorems, which we state without proof, will seem quite reasonable to you, and we shall make use of them. Let $\lim_{x\to a} f(x) = F$ and $\lim_{x\to a} g(x) = G$.

THEOREM 1 For the constant function f, where $f(x) = C$, it is true that $\lim_{x\to a} f(x) = C$, for all a.

THEOREM 2 $\lim_{x\to a} (f(x) \pm g(x)) = \lim_{x\to a} f(x) \pm \lim_{x\to a} g(x) = F \pm G$

THEOREM 3 $\lim_{x\to a} f(x) \cdot g(x) = \lim_{x\to a} f(x) \cdot \lim_{x\to a} g(x) = F \cdot G$

COROLLARY $\lim_{x\to a} k \cdot f(x) = k \cdot \lim_{x\to a} f(x) = kF$, k constant

THEOREM 4 $\lim_{x\to a} \dfrac{f(x)}{g(x)} = \dfrac{\lim_{x\to a} f(x)}{\lim_{x\to a} g(x)} = \dfrac{F}{G}$, if $G \neq 0$

ILLUSTRATION 1 Let $f(x) = x$, $-\infty < x < \infty$. Find $\lim_{x\to 2} f(x)$.

SOLUTION From an inspection of $f(x)$ for x near 2, we guess that $f(x) \to 2$ as $x \to 2$. Note that we must not set $x = 2$, for this is not the idea of a limit. Now we prove that our guess is correct.

Using the definition, we choose an ϵ (>0) and try to find a $\delta(\epsilon)$ such that

$$|f(x) - 2| < \epsilon$$

when $0 < |x - 2| < \delta(\epsilon)$. Since $f(x) = x$, this is an easy matter: just choose $\delta(\epsilon) = \epsilon$.

Similarly

$$\lim_{x\to a} f(x) = a$$

ILLUSTRATION 2 Find $\lim_{x\to a} x^2$. From Illustration 1 we know that $\lim_{x\to a} x = a$. Now $x^2 = x \cdot x$. Therefore Theorem 3 tells us that

$$\lim_{x\to a} x^2 = a^2$$

EXERCISE A Find $\lim_{x\to a} x^n$.

EXERCISE B Find $\lim_{x\to a} Cx^n$, where C is a constant.

ILLUSTRATION 3 Find $\lim_{x\to 2} (3x^3 - 2x^2 + 4)$. From Theorem 2,

$$\lim_{x\to 2} (3x^3 - 2x^2 + 4) = \lim_{x\to 2} 3x^3 - \lim_{x\to 2} (2x^2) + \lim_{x\to 2} 4$$

Finally, from Exercise B,

$$\lim_{x\to 2} (3x^3 - 2x^2 + 4) = 24 - 8 + 4 = 20$$

ILLUSTRATION 4 Find $\lim_{x\to 1} \frac{x^2+3}{x+2}$. From Theorem 4,

$$\lim_{x\to 1} \frac{x^2+3}{x+2} = \frac{\lim_{x\to 1}(x^2+3)}{\lim_{x\to 1}(x+2)} = \frac{4}{3}$$

Domination principle

An important aid in some proofs is the "domination principle", which, in this case, has the following statement.

THEOREM 5 If $F(x) \le f(x) \le G(x)$ for all x in an interval containing $x = a$, except possibly at $x = a$, and if $\lim_{x\to a} F(x) = L$ and $\lim_{x\to a} G(x) = L$, then $\lim_{x\to a} f(x) = L$.

We use this principle in the next two illustrations.

ILLUSTRATION 5 Show that $\lim_{x\to 0} \sin x = 0$.

SOLUTION We recall the definition of $\sin x$ and note that, for small x, $|\sin x| \le |x|$. This follows from the fact in Fig. 15.9

$$2|\sin x| = PP'$$
$$2|x| = \text{arc } PP'$$

Construction to show that $|\sin x| \le |x|$

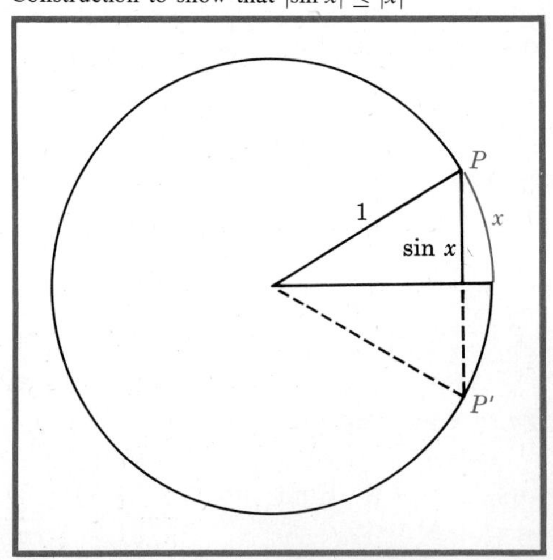

FIGURE 15.9
The fundamental fact is that a chord of a circle is shorter than the corresponding arc.

Since a chord has length less than the corresponding arc, $PP' <$ arc PP' or $|\sin x| \leq |x|$. The equality occurs at $x = 0$. Therefore

$$-|x| \leq \sin x \leq |x|$$

in a small interval about $x = 0$. We apply the domination principle with $F(x) = -|x|$ and $G(x) = |x|$. Since $\lim_{x \to 0} |x| = 0$, we conclude that

$$\lim_{x \to 0} \sin x = 0$$

ILLUSTRATION 6 Show that $\lim_{x \to 0} \cos x = 1$.

SOLUTION Consider $1 - \cos x$. Then

$$1 - \cos x = \frac{1 - \cos^2 x}{1 + \cos x} < 1 - \cos^2 x$$

for x near zero. Hence, for x near zero,

$$0 \leq 1 - \cos x \leq 1 - \cos^2 x = \sin^2 x$$

Therefore

$$\lim_{x \to 0} (1 - \cos x) = 0 \qquad \text{Theorem 5 and Illustration 5}$$

and

$$\lim_{x \to 0} \cos x = 1$$

$\lim_{x \to 0} \frac{\sin x}{x} = 1$

We shall now prove Theorem 6 and, by means of it, find the area of a circle.

THEOREM 6 $\lim_{x \to 0} \frac{\sin x}{x} = 1$

Consider Fig. 15.10 which is drawn with $|x| < \pi/2$. We observe that

$$\text{Arc } PSP' = 2|x|$$
$$\text{Chord } PQP' = 2 \sin |x| = 2|\sin x|$$
$$\text{Segment } PR = \text{segment } P'R = \tan |x| = |\tan x|$$

From Sec. 13.1, Exercise D, we know that

$$PQP' < PSP' < PR + P'R$$

or

$$2|\sin x| < 2|x| < 2|\tan x|$$

Dividing by $2|\sin x|$, we obtain

$$1 < \left| \frac{x}{\sin x} \right| < \frac{1}{|\cos x|}$$

Construction to show that $|\sin x| \le |x| \le |\tan x|$

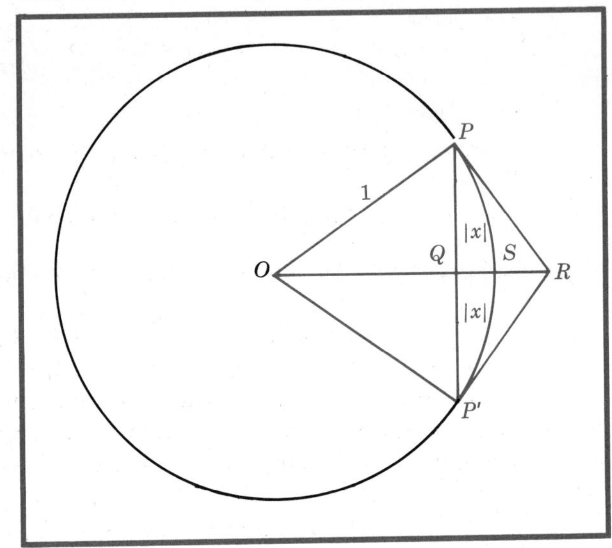

FIGURE 15.10
A circular arc is longer than any inscribed polygon and shorter than any circumscribed polygon.

Since $\lim_{x\to 0} \cos x = 1$ (Illustration 6), it follows from Theorem 5 that

$$\lim_{x\to 0} \left| \frac{x}{\sin x} \right| = 1$$

or, from Theorem 4,

$$\lim_{x\to 0} \left| \frac{\sin x}{x} \right| = 1$$

Now x and $\sin x$ have the same signs for small x, and hence $(\sin x)/x$ is positive. Therefore for small x ($x \neq 0$),

$$\frac{\sin x}{x} = \left| \frac{\sin x}{x} \right|$$

and thus

$$\lim_{x\to 0} \frac{\sin x}{x} = 1$$

For an alternate proof of this important limit see Prob. 27 at the end of this section.

We are finally in a position to find the area of a circle of radius r by Euclid's definition. This area is given by

$$\text{Area} = \lim_{n\to\infty} \frac{n}{2} r^2 \sin \frac{2\pi}{n}$$

$$= r^2 \lim_{n\to\infty} \pi \frac{\sin (2\pi/n)}{2\pi/n}$$

Now set $n = 1/t$, and we get (note that, as $n \to \infty$, $t \to 0$)

$$\text{Area} = \pi r^2 \lim_{t \to 0} \frac{\sin 2\pi t}{2\pi t}$$
$$= \pi r^2$$

since $2\pi t$ plays the same role as x in Theorem 6.

EXERCISE C Prove $\lim_{x \to 0} \frac{\tan x}{x} = 1$, and hence find the area of a circle by applying Euclid's definition with circumscribed regular polygons.

EXERCISE D From the results of Exercise C, Sec. 15.2, and from the definition of circumference (Sec. 13.1), and from Theorem 6, find the circumference of a circle of radius r. Note that this is not circular reasoning. We have defined π to be half the circumference of a unit circle. We must now find the circumference of a circle of radius r.

EXERCISE E Argue that the "π" in the area formula above is the same as the "π" previously used in the circumference formula.

PROBLEMS 15.3

In Probs. 1 to 23 find the limit indicated.

1 $\lim_{x \to 2} (4x - 9)$

2 $\lim_{x \to 4} (x - x^2 + 4)$

3 $\lim_{x \to 0} \frac{x^2 - 1}{x - 1}$

4 $\lim_{x \to -1} \frac{x^2 + 4x - 2}{x - 2}$

5 $\lim_{x \to 1/2} \frac{x^2}{x^4}$

6 $\lim_{x \to 0} \frac{x}{x^4}$

7 $\lim_{x \to -4} |x|$

8 $\lim_{x \to -1/2} \frac{|x - 1|}{2}$

9 $\lim_{x \to \pi} \cos x$

10 $\lim_{x \to \pi/4} (2x^2 + \sin x)$

11 $\lim_{x \to 0} \frac{\sin x}{x}$

12 $\lim_{x \to \pi/2} \frac{x}{\sin x}$

13 $\lim_{x \to 0} (3 \sin x + 2 \cos x)$

14 $\lim_{x \to 2} \frac{\sin (x - 2)}{(x - 2)}$

15 $\lim_{x \to 3\pi/2} |\sin x|$

16 $\lim_{x \to 0} \frac{\sin x}{\cos x}$

17 $\lim_{x \to 1} \frac{(x + \frac{1}{2}) \sin x}{1 + \sin x}$

18 $\lim_{x \to 2\pi} \frac{x}{2 \cos x}$

19 $\lim_{x \to \pi/4} \frac{1}{\sin^2 x + \cos^2 x}$

20 $\lim_{x \to 0} \frac{x \sin^2 x}{x^2}$

21 $\lim_{x \to \pi/3} \sec x$

22 $\lim_{x \to \pi/4} \tan x$

23 $\lim_{x \to -3} \dfrac{x}{|x|}$

24 Sketch $y = \sin(1/x), x \neq 0$.

25 Sketch $y = x \sin(1/x), x \neq 0$.

26 Find (when they exist):

a $\lim_{x \to \infty} \sin x$ **b** $\lim_{x \to \infty} \dfrac{\sin x}{x}$ **c** $\lim_{x \to 0} \dfrac{\sin x}{x}$

d $\lim_{x \to 0} x \sin \dfrac{1}{x}$ **e** $\lim_{x \to \infty} x \sin \dfrac{1}{x}$ **f** $\lim_{x \to \infty} \dfrac{1}{x} \sin \dfrac{1}{x}$

27 Prove Theorem 6. In the figure below, chord BC (associated with arc x) continually bisected produces equal chords (and arcs) P_0P_1, $P_1P_2, \ldots, P_{n-1}P_n$. For small $|x|$ argue that

a
$$\begin{aligned} a_1 + a_2 + \cdots + a_n \le P_0P_1 + P_1P_2 + \cdots + P_{n-1}P_n \\ \le (a_1 + a_2 + \cdots + a_n) \\ + (b_1 + b_2 + \cdots + b_n) \end{aligned}$$

In the limit, as $n \to \infty$, (a) yields (with Illustration 6):

b $|\sin x| \le |x| \le |\sin x| + (1 - \cos x) \le |\sin x| + \sin^2 x$

Hence $$1 \le \left| \frac{x}{\sin x} \right| < 1 + |\sin x|$$

Now use the domination principle and Illustration 5.

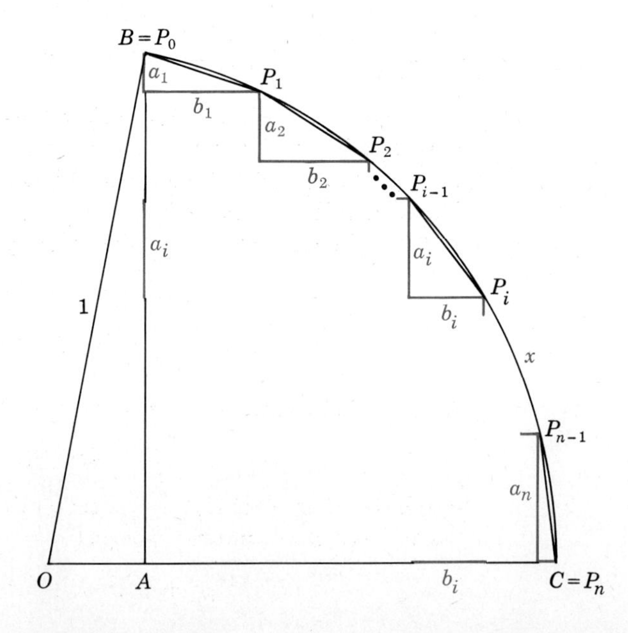

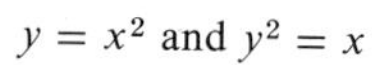

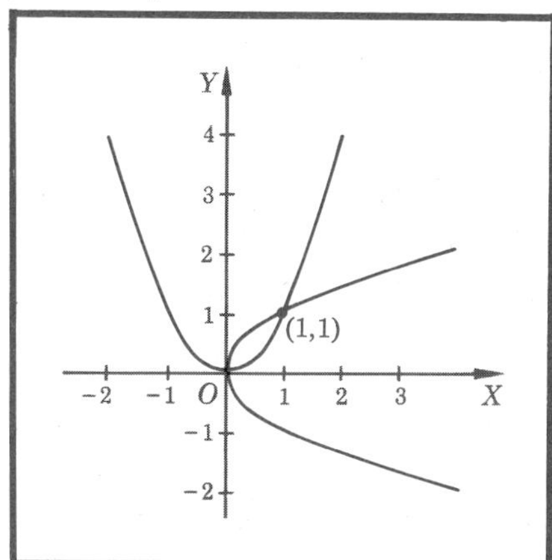

FIGURE 15.11
New methods are needed to find the area between two

Area under $y = x^2$ from $x = 0$ to $x = 1$

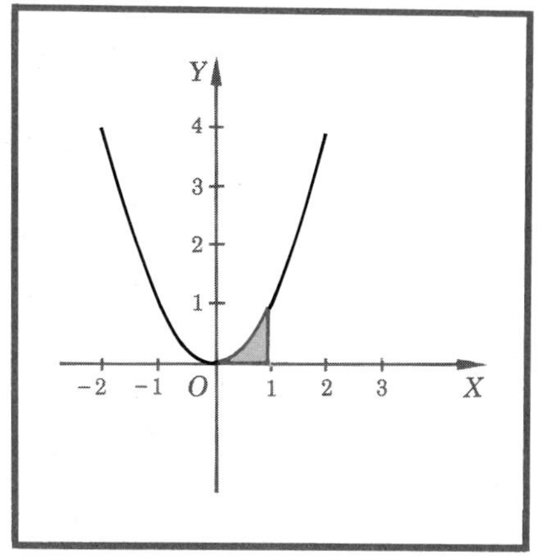

FIGURE 15.12
The problem is to define (and then calculate) the area of the colored region bounded by $y = x^2$, $y = 0$, and $x = 1$.

15.4 › AREA UNDER $y = x^2$

Euclid's definition of area works quite well for the circle, but we need a different definition for other types of areas. For example, if the area were that enclosed by the two curves $y = x^2$ and $y^2 = x$, it would not be possible to make sensible use of inscribed (or circumscribed) *regular* polygons (Fig. 15.11).

EXERCISE A Give a definition of this area, using inscribed polygons.

Area as the limit of a sum

Again, consider the area enclosed by the *curves* $y = x^2, y = 0, x = 1$ (Fig. 15.12). To find this area, we use the method which is due, essentially, to Archimedes, who reasoned as follows. The area sought, call it A, is larger than the combined areas of the rectangles formed as in Fig. 15.13, where ordinates have been erected at the quarter marks. That is,

$$\tfrac{1}{4}(\tfrac{1}{4})^2 + \tfrac{1}{4}(\tfrac{2}{4})^2 + \tfrac{1}{4}(\tfrac{3}{4})^2 < A$$

or

$$\frac{1}{4^3}[1^2 + 2^2 + 3^2] < A$$

This reduces, numerically, to

$$\tfrac{14}{64} < A$$

Similarly, from Fig. 15.14, A is smaller than the sum of the rectangles which have been drawn. That is,

$$\tfrac{1}{4}(\tfrac{1}{4})^2 + \tfrac{1}{4}(\tfrac{2}{4})^2 + \tfrac{1}{4}(\tfrac{3}{4})^2 + \tfrac{1}{4}(\tfrac{4}{4})^2 > A$$

Internal approximation of desired area

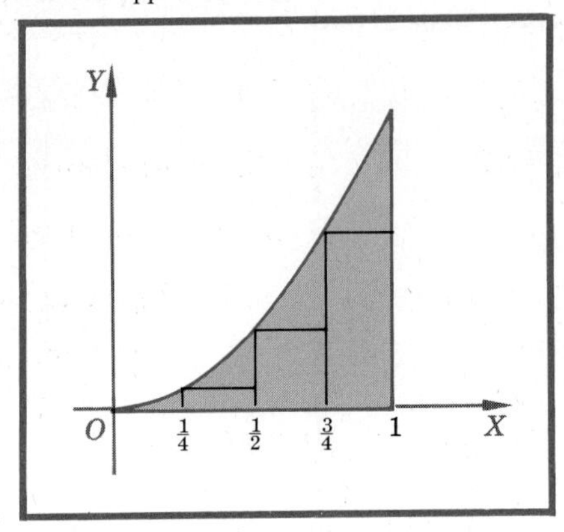

FIGURE 15.13

External approximation of desired area

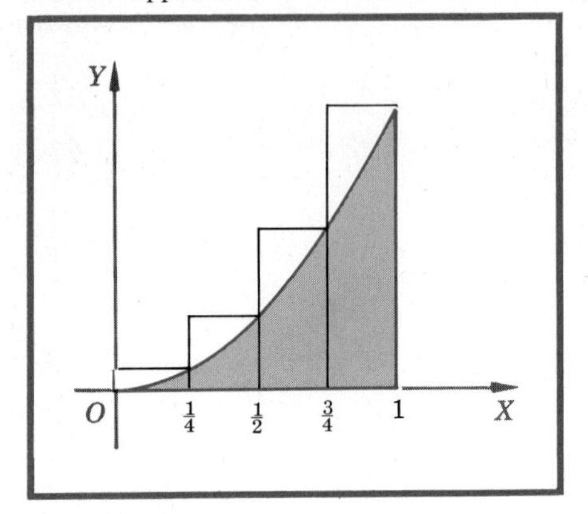

FIGURE 15.14

or $$\frac{1}{4^3}[1^2 + 2^2 + 3^2 + 4^2] > A$$

that is, $$\tfrac{30}{64} > A$$

Hence $$\tfrac{14}{64} < A < \tfrac{30}{64}$$

If ordinates had been erected at the eighth marks, the corresponding inequalities would have been (you should draw a figure for each case)

$$\tfrac{1}{8}(\tfrac{1}{8})^2 + \tfrac{1}{8}(\tfrac{2}{8})^2 + \cdots + \tfrac{1}{8}(\tfrac{7}{8})^2 < A$$

$$\frac{1}{8^3}[1^2 + 2^2 + \cdots + 7^2] < A$$

$$\tfrac{140}{512} < A$$

and $$\tfrac{1}{8}(\tfrac{1}{8})^2 + \tfrac{1}{8}(\tfrac{2}{8})^2 + \cdots + \tfrac{1}{8}(\tfrac{8}{8})^2 > A$$

$$\frac{1}{8^3}[1^2 + 2^2 + \cdots + 8^2] > A$$

$$\tfrac{204}{512} > A$$

or $$\tfrac{140}{512} < A < \tfrac{204}{512}$$

Observe that the bounding interval in the second case lies wholly within that of the first case; i.e.,

$$\tfrac{14}{64} < \tfrac{140}{512} < A < \tfrac{204}{512} < \tfrac{30}{64}$$

Now let us *imagine* that ordinates are erected at abscissa marks which are multiples of $1/n$ (Figs. 15.15 and 15.16). Then we would have A "boxed in" by

(1) $$\frac{1}{n^3}[1^2 + 2^2 + \cdots + (n-1)^2] < A < \frac{1}{n^3}[1^2 + 2^2 + \cdots + (n-1)^2 + n^2]$$

Let us write L_n for the left term of (1) and R_n for its right term. Hence (1) is rewritten as

$$L_n < A < R_n$$

From Prob. 12, Sec. 2.7, we learn by mathematical induction that

$$1^2 + 2^2 + \cdots + n^2 = \tfrac{1}{6}n(n+1)(2n+1)$$

Hence
$$L_n = \frac{(n-1)n(2n-1)}{6n^3} = \frac{(n-1)(2n-1)}{6n^2}$$

$$R_n = \frac{(n+1)(2n+1)}{6n^2}$$

Also
$$L_{n+1} = \frac{n(2n+1)}{6(n+1)^2}$$

$$R_{n+1} = \frac{(n+2)(2n+3)}{6(n+1)^2}$$

From these it follows that

$$L_{n+1} - L_n = \frac{n(2n+1)}{6(n+1)^2} - \frac{(n-1)(2n-1)}{6n^2}$$

Finer internal approximation

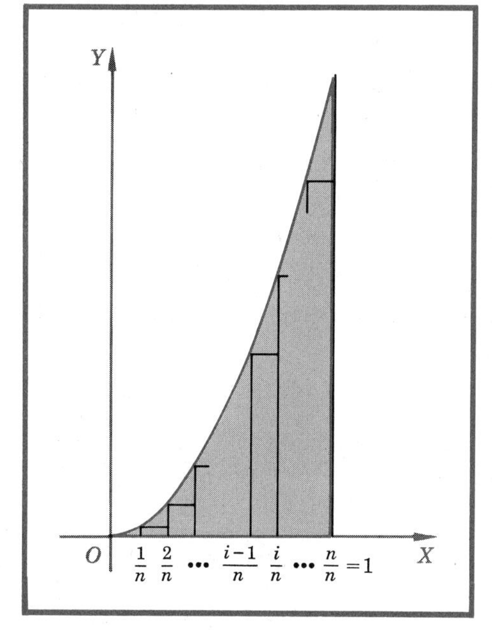

FIGURE 15.15

Finer external approximation

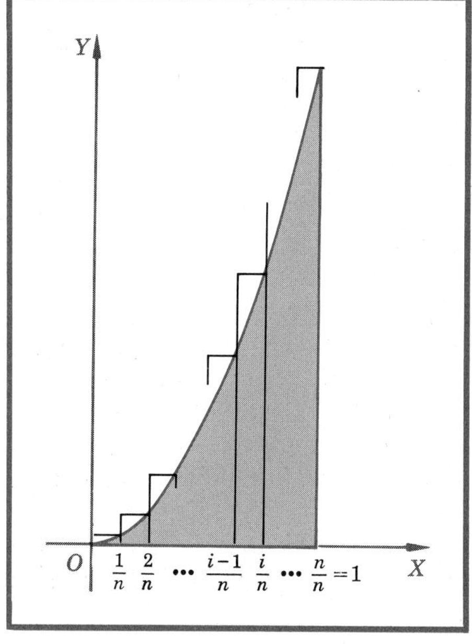

FIGURE 15.16

$$= \frac{n^3(2n+1) - (n+1)^2(n-1)(2n-1)}{6(n+1)^2n^2}$$

$$= \frac{3n^2 + n - 1}{6(n+1)^2n^2} > 0$$

or

$$L_n < L_{n+1}$$

and similarly $R_{n+1} < R_n$. Hence we have

$$L_n < L_{n+1} < A < R_{n+1} < R_n$$

Further, $R_n - L_n = 1/n$, which tends to zero as n tends to ∞. Hence it appears that we can approach the true value of A by taking larger and larger values of n, and A should equal the limiting value of L_n or R_n as n gets larger and larger. [Archimedes did not quite say it this way, but he must have had some such notion in his mind; it was Cavalieri (1598–1647) who first carried out the details in the year 1630.] Pictorially also, this seems reasonable, although you are cautioned against relying too heavily on your geometric intuition. The question now is to find this limiting value.

We have seen that

$$R_n = \frac{(n+1)(2n+1)}{6n^2}$$

$$= \frac{1}{6}\left(1 + \frac{1}{n}\right)\left(2 + \frac{1}{n}\right)$$

It is reasonable to say that $1/n$ tends to zero as n tends to ∞.† From this and from Theorem 2 we conclude that $(1 + 1/n)$ tends to 1 as n tends to ∞ and that $(2 + 1/n)$ tends to 2 as n tends to ∞. Consequently, R_n tends to $\frac{1}{6}(1)(2) = \frac{1}{3}$.

† More specifically $\lim_{x \to \infty} f(x) = L$ means: The value $f(x)$ of the function f is said to approach the constant L as a limit as x tends to $+\infty$ if, for every positive ϵ, there exists a positive number A such that, if $x > A$, then $|f(x) - L| < \epsilon$.

Similarly, we have seen that

$$L_n = \frac{(n-1)(2n-1)}{6n^2}$$

$$= \frac{1}{6}\left(1 - \frac{1}{n}\right)\left(2 - \frac{1}{n}\right)$$

By the same argument it follows that L_n tends to $\frac{1}{3}$ as n tends to ∞. Since A is between L_n and R_n, we note that $A = \frac{1}{3}$.

We conclude, therefore, that we did not need to work with both R_n and L_n. From either one the answer follows.

PROBLEMS 15.4

In Probs. 1 to 8 make use of the above results to sketch and find the area enclosed by:

1 $y = \sqrt{x}, y = 0, x = 1$
2 $y^2 = x, x = 1$

3 $y = x^2, y^2 = x$
4 $y = 3x^2,\ y = 0,\ x = 1$
5 $y = kx^2,\ y = 0,\ x = 1$
6 $y = x^2,\ y = x$
7 $(y + 1)^2 = x - 2,\ x = 3$
8 $(y - k)^2 = x - h,\ x = h + 1$

9 Consider a unit square, its diagonal AB, and a zigzag path from A to B made up of segments parallel to the sides, as in the figure. Now the sum of the lengths of the zigzag is surely just two units.

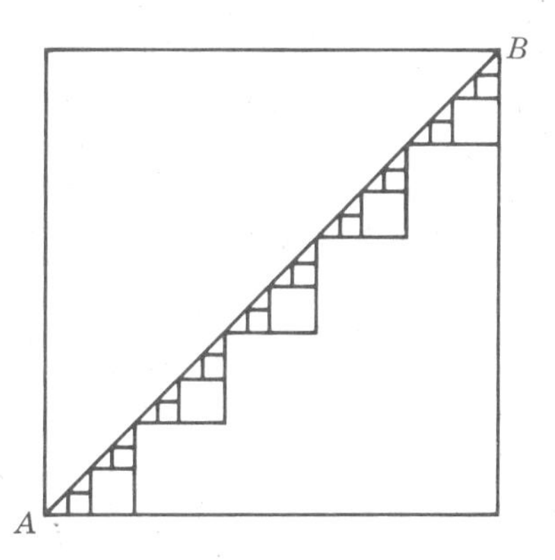

Making the zigs and the zags smaller and smaller indefinitely, we approach the diagonal closer and closer; thus we see that the length of the diagonal of the square is two units. Point out the flaw in the reasoning.

10 The following "argument" is ascribed to Zeno† (495–435 B.C.). See if you can detect a flaw in the reasoning, and write it down in a few words. Achilles cannot catch the tortoise in a race if the tortoise has a head start; for, before he catches the tortoise, he must get up to the place from which the tortoise started. But in the meantime, the tortoise has gone ahead and so has another headstart. Repeating this argument indefinitely, we see that the tortoise will never be caught.

† Look up a fine account of the history of Zeno's Paradoxes by Florian Cajori in the *American Mathematical Monthly,* vol. 22, pp. 1, 39, 77, 109, 143, 179, 215, 253, 292 (1915).

15.5 › AREA UNDER $y = x^n$

The area bounded by $y = x^3$, $y = 0$, $x = 0$, $x = 1$ is set up in the same way in which it was set up for $y = x^2$. Divide the interval [0, 1] up into n equal intervals by the points $x_0 = 0$, $x_1 = 1/n$, $x_2 = 2/n$, . . . , $x_n = n/n = 1$ (Fig. 15.17). On each interval as base form a rectangle using the ordinate at the right-hand point of that interval as altitude. The areas

Area under $y = x^3$ from $x = 0$ to $x = 1$

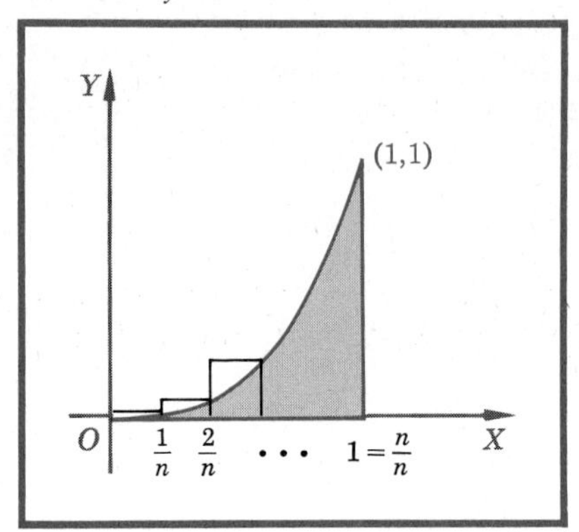

FIGURE 15.17

of the first few rectangles are $\left(\frac{1}{n}\right)^3 \cdot \frac{1}{n}, \left(\frac{2}{n}\right)^3 \cdot \frac{1}{n}, \left(\frac{3}{n}\right)^3 \cdot \frac{1}{n}$, and the area A under the curve is approximately

$$A \approx \left[\left(\frac{1}{n}\right)^3 + \left(\frac{2}{n}\right)^3 + \cdots + \left(\frac{n}{n}\right)^3\right] \cdot \frac{1}{n}$$

$$\approx \frac{1}{n^4}[1^3 + 2^3 + \cdots + n^3]$$

The area is exactly

$$A = \lim_{n\to\infty} \frac{1}{n^4}[1^3 + 2^3 + \cdots + n^3]$$

We are unable to proceed until we have found how to reduce $1^3 + 2^3 + \cdots + n^3$ to some other form because we cannot take the limit as it stands. To discover this "on our own" would be quite difficult, and here we shall assume the answer,† which is

† See mathematical induction in Sec 2.6 and Prob. 13, Sec. 2.6.

$$1^3 + 2^3 + \cdots + n^3 = \frac{n^2(n+1)^2}{4}$$

Therefore

$$A = \lim_{n\to\infty} \frac{n^2(n+1)^2}{4n^4} = \lim_{n\to\infty} \frac{1}{4} \cdot \frac{n^2 + 2n + 1}{n^2}$$

$$= \lim_{n\to\infty} \frac{1}{4}\left(1 + \frac{2}{n} + \frac{1}{n^2}\right)$$

$$= \tfrac{1}{4}$$

EXERCISE A By these methods find the area of the triangle defined by $y = x$, $y = 0$, $x = 1$.

In a similar way we could set up the area bounded by $y = x^4$, $y = 0$, $x = 1$. It would be

$$A = \lim_{n\to\infty} \left[\left(\frac{1}{n}\right)^4 + \left(\frac{2}{n}\right)^4 + \cdots + \left(\frac{n}{n}\right)^4\right] \cdot \frac{1}{n}$$

$$= \lim_{n\to\infty} \frac{1}{n^5}[1^4 + 2^4 + \cdots + n^4]$$

and again we are faced with the difficult problem of reducing the sum of the fourth powers of the integers 1, 2, . . . , n to some simpler form. It is

$$1^4 + 2^4 + \cdots + n^4 = \frac{6n^5 + 15n^4 + 10n^3 - n}{30}$$

Now A becomes

$$A = \lim_{n\to\infty} \frac{1}{30}\left[6 + \frac{15}{n} + \frac{10}{n^2} - \frac{1}{n^4}\right]$$

$$= \tfrac{1}{5}$$

You should see that there is a pattern developing.

TABLE 1

Enclosing curves	*Area enclosed*
$y = x$, $x = 0$, $x = 1$	$\frac{1}{2}$
$y = x^2$, $x = 0$, $x = 1$	$\frac{1}{3}$
$y = x^3$, $x = 0$, $x = 1$	$\frac{1}{4}$
$y = x^4$, $x = 0$, $x = 1$	$\frac{1}{5}$

It is true that the area enclosed by $y = x^n$, $x = 0$, $x = 1$ is $1/(n + 1)$, for $n = 0, 1, 2, \ldots$, but we defer further study of this until we have met with some of the ideas of the next chapter.

In order to find the area enclosed by $y = x^2$, $y = 0$, $x = b$, we divide the interval $[0,b]$ into n equal intervals by the points $x_0 = 0$, $x_1 = \frac{1}{n}(b)$, $x_2 = \frac{2}{n}(b), \ldots, x_n = \frac{n}{n}(b)$ (Fig. 15.18). For $y = x^2$ the area A is therefore

$$A = \lim_{n\to\infty} \left\{\left[\frac{1}{n}(b)\right]^2 \cdot \frac{1}{n}(b) + \left[\frac{2}{n}(b)\right]^2 \cdot \frac{1}{n}(b) + \cdots + \left[\frac{n}{n}(b)\right]^2 \cdot \frac{1}{n}(b)\right\}$$

$$= \lim_{n\to\infty} \frac{b^3}{n^3}[1^2 + 2^2 + \cdots + n^2]$$

$$= \lim_{n\to\infty} \frac{b^3}{n^3}\left[\frac{n(n + 1)(2n + 1)}{6}\right]$$

$$= \frac{b^3}{3}$$

Area under $y = x^2$ from $x = 0$ to $x = b$

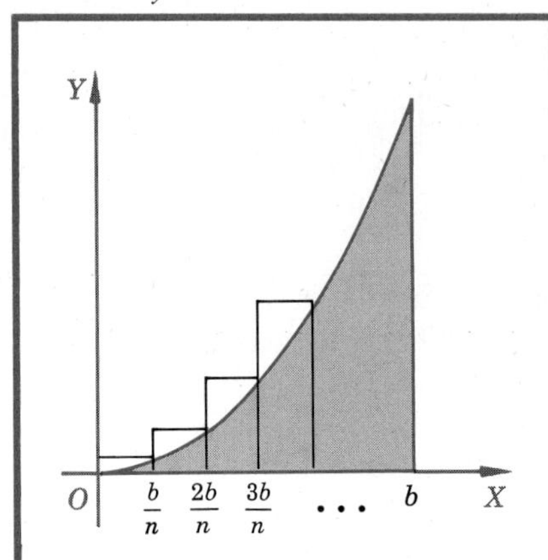

FIGURE 15.18

Area under $y = x^2$ from $x = a$ to $x = b$

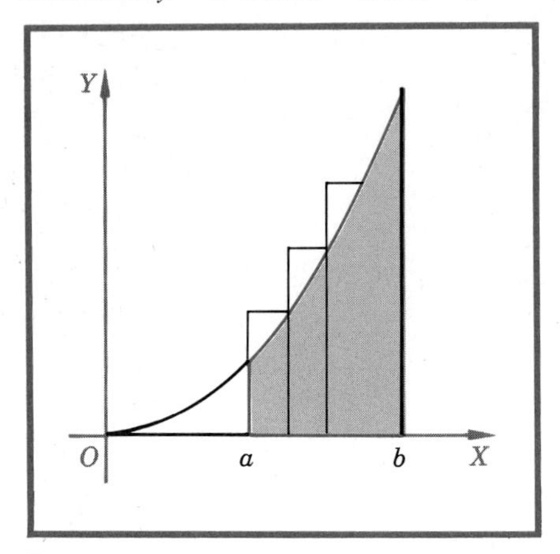

FIGURE 15.19

EXERCISE B Show that the area bounded by $y = x^3, x = 0, x = b$ is $b^4/4$.

By the addition of areas we can readily find the area enclosed by $y = x^2, x = a, x = b$ where $a > 0$ and $b > 0$ (Fig. 15.19). Since the area (under $y = x^2$) from 0 to b is $b^3/3$, the area from 0 to a is $a^3/3$ and the area from a to b is $b^3/3 - a^3/3$.

Again there is a pattern ($b > a$):

TABLE 2

Enclosing curves	*Area enclosed*
$y = 1, y = 0, x = a, x = b$	$(b - a)$
$y = x, y = 0, x = a, x = b$	$\frac{1}{2}(b^2 - a^2)$
$y = x^2, y = 0, x = a, x = b$	$\frac{1}{3}(b^3 - a^3)$
$y = x^3, y = 0, x = a, x = b$	$\frac{1}{4}(b^4 - a^4)$
$y = x^4, y = 0, x = a, x = b$	$\frac{1}{5}(b^5 - a^5)$
. . . .	. . . .
$y = x^n, y = 0, x = a, x = b$ $n \geq 0$	$\frac{1}{n+1}(b^{n+1} - a^{n+1})$

EXERCISE C Argue that if $b < a$, then the answers above should be written $\frac{1}{2}(a^2 - b^2)$, etc.

PROBLEMS 15.5

In Probs. 1 to 8 sketch and find the enclosed area by summing and taking the limit.

1 $y = -x^2 + 1, y = \frac{1}{2}, y = 0$
2 $y + 2 = (x - 1)^2, y = -1$

3 $y - \frac{1}{2} = (x - \frac{1}{2})^2, x = 1, y = 1, x = 0$
4 $y = x^2 - 2, y = \frac{1}{2}, x = -1, x = 1$
5 $y = x^3, y = 0, x = 0, x = 1$
6 $y = 1 - x^3, y = 0, x = 0, x = 1$
7 $y = |x^3|, y = 0, x = 0, x = 1$
8 $y = x^4, x = -1, x = -2, y = 0$

In Probs. 9 to 18 sketch and find the enclosed area by using the general results of this section as formulas. All exponents are assumed to be ≥ 0.

9 $y = -x^4, y = -1$
10 $y = x^4, y = 0, x = -1, x = 1$
11 $y = 3x^5, y = 0, x = 1$
12 $y = 2x^3, y = 0, x = 3$
13 $y = x^6, y = 0, x = -2, x = 3$
14 $y = x^6, y = 2$
15 $y = nx^{n-1}, y = 0, x = 0, x = 1$
16 $y = x^{n-1}, y = 0, x = 0, x = a$
17 $y = 3nx^{3n-1}, y = 0, x = a, x = b$
18 $y = x^{n-3}, y = 0, x = a, x = b$

15.6 › AREA UNDER GRAPH OF A POLYNOMIAL FUNCTION

With a few exceptions the problems of Sec. 15.5 are such that the enclosed area is above the X-axis. Such areas are positive. The physical area enclosed by $y = -x^2$, $y = 0$, $x = 1$ lies *below* the X-axis. Such an area is negative when computed as above. It is obvious that it should be since the ordinates (heights) of the small rectangles are themselves negative. Regardless of whether the area lies above or below the X-axis, we still speak of the area "under" a curve.

It is our usual convention to read from left to right on the horizontal axis and thus to choose $b > a$ when we develop formulas like those in Table 2, Sec. 15.5. There are occasions, however, when we wish to read in the opposite direction and thus choose $b < a$. When we do this we will, by convention, maintain the formulas of Table 2 without change. In certain cases this introduces negative area. With these two situations in mind we point out that the area under $y = x^n$, $y = 0$, from $x = a$ to $x = b$ may be positive, negative, or zero. A similar situation arises in the case of polynomials.

Algebraic area

The graph of a typical polynomial equation $y = P(x)$ is much like that in Fig. 15.20. In computing the area A under this from $x = a$ to $x = b$, we would arrive at $A = B + C$, where B is the *lightly* colored areas (positive) and where C is the *heavily* colored areas (negative). Therefore what we get is the *algebraic* sum, and this number A may be positive, negative, or zero.

Area "under" the graph of a general polynomial

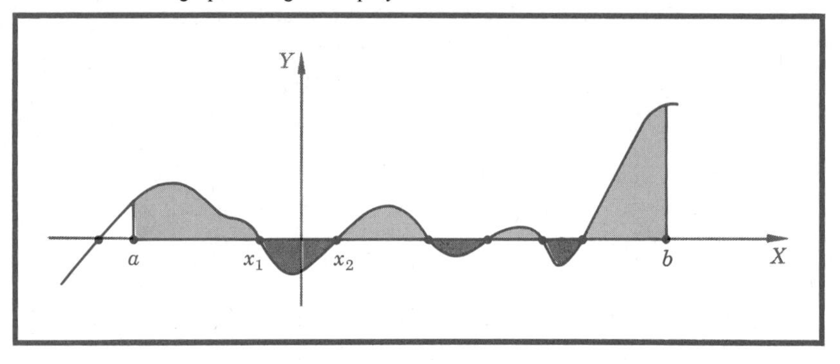

FIGURE 15.20
Our method of computing area gives a positive answer if the region is above the X-axis and a negative answer if the region is below the X-axis. When applied to a general polynomial it gives the algebraic sum of the positive and negative parts.

Physical area

REMARK If we wish to compute the total colored area—as if it were land along the banks of a river—then it is necessary first of all to find $\{x \mid P(x) = 0\}$, that is, the roots $x_1, x_2, \ldots$ of the polynomial equation. We would then write for the physical area A

$$A = |A_a^{x_1}| + |A_{x_1}^{x_2}| + \cdots$$

We have used "$A_a^{x_1}$" to represent the area from a to x_1, etc.

To compute the area under a polynomial we need the facts that

$$A_a^b \text{ under } kf(x) = k[A_a^b \text{ under } f(x)]$$
$$A_a^b \text{ under } f(x) + g(x) = [A_a^b \text{ under } f(x)] + [A_a^b \text{ under } g(x)]$$

Since A_a^b is given as a limit, these facts are consequences of the corresponding theorems on limits, namely Theorem 2 and the Corollary to Theorem 3, Sec. 15.3. Proceed as in the illustration below.

ILLUSTRATION 1 Find the algebraic area under $y = \frac{1}{6}x(x^2 - 9)$ from $x = -1$ to $x = 4$.

SOLUTIONS We first sketch the graph (Fig. 15.21). We write $y = \frac{1}{6}x^3 - \frac{3}{2}x$, consider the terms separately, and make use of Table 2, Sec. 15.5.

$$A_{-1}^0 = \tfrac{1}{6}[\tfrac{1}{4}(0^4 - (-1)^4)] - \tfrac{3}{2}[\tfrac{1}{2}(0^2 - (-1)^2)]$$
$$A_0^3 = \tfrac{1}{6}[\tfrac{1}{4}(3^4 - 0^4)] - \tfrac{3}{2}[\tfrac{1}{2}(3^2 - 0^2)]$$
$$A_3^4 = \tfrac{1}{6}[\tfrac{1}{4}(4^4 - 3^4)] - \tfrac{3}{2}[\tfrac{1}{2}(4^2 - 3^2)]$$

Reducing, we have $A_{-1}^0 = \frac{17}{24}, A_0^3 = -\frac{81}{24}, A_3^4 = \frac{49}{24}$ so that we can find the algebraic area by adding the separate areas as they stand. Thus the algebraic area $A_{-1}^4 = -\frac{15}{24}$, and this is exactly what we get by evaluating A_{-1}^4 directly.

Area "under" $y = \frac{1}{6}x(x^2 - 9)$ from $x = -1$ to $x = 4$

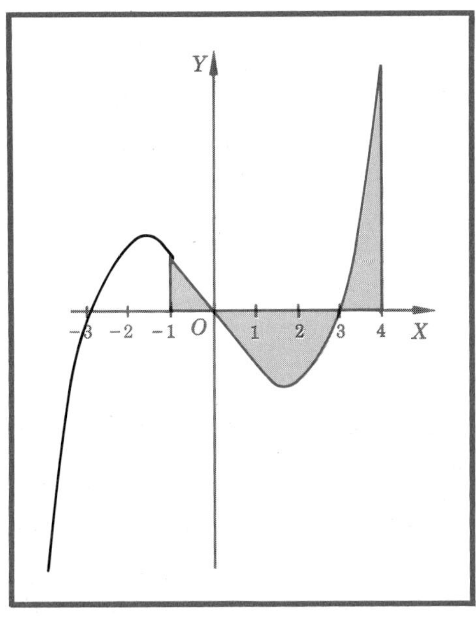

FIGURE 15.21

$$A_{-1}^{4} = \tfrac{1}{6}[\tfrac{1}{4}(4^4 - (-1)^4)] - \tfrac{3}{2}[\tfrac{1}{2}(4^2 - (-1)^2)]$$
$$= \frac{255}{24} - \frac{3(15)}{4}$$
$$= -\tfrac{15}{24}$$

If, on the other hand, we want the physical area, whether it is above or below the X-axis, then we need to write

$$|A_{-1}^{0}| + |A_{0}^{3}| + |A_{3}^{4}| = \tfrac{147}{24}$$

Unless explicitly stated to the contrary, algebraic area is implied when "area" is used by itself. In general the context will be clear.

PROBLEMS 15.6

In Probs. 1 to 16 sketch and find the enclosed physical area by using the formulas of Table 2, Sec. 15.5, and the ideas of Sec. 15.6.

1 $y = x^2 - 4, y = 0$ **2** $y = -(x - 1)(x + 5), y = 0$
3 $y = x(3 - x), y = 0$ **4** $y = -x^2, y = -1$
5 $y = -x + 2, y = 0, x = 0, x = 2$
6 $y = x^2 - x, y = 0, x = 0, x = 1$
7 $y = x^3 - x^2, y = 0$
8 $y = (x + 1)(x - 1)(x), y = 0, x = 1, x = -1$

9 $y = x^2(x-2), y = 0, x = 0, x = 1$
10 $y = x(x-2)(x-4), y = 0, x = 0, x = 2$
11 $y = x^2(x^2-9), y = 0, x = -3, x = 3$
12 $y = (x+1)(x-1)(x+2)(x-2), y = 0, x = -2, x = 2$
13 $y = (x-2)^2(x-1)^2, y = 0, x = 0, x = 2$
14 $y = -x^5, y = 0, x = -1, x = 1$
15 $y = x^5 - x^3, y = 0, x = -1, x = 2$
16 $y = x^3(x-1)(x-1), y = 0, x = -2, x = 2$

In Probs. 17 to 22 sketch and find the algebraic area enclosed; i.e., find A_a^b as indicated.

17 $y = 2x$; find A_{-2}^2.
18 $y = 3x^2$; find A_{-1}^0.
19 $y = 16x^3$; find A_{-1}^0.
20 $y = (x-1)(x-2)(x-4)$; find A_0^4.
21 $y = (x)(x-\frac{1}{2})(x+\frac{1}{2})$; find A_{-2}^2.
22 $y = 5x^4 + 3x^2 + 2x + 1$; find A_{-1}^2.

23 For $y = a + bx^2 + cx^4 + dx^6$; argue that $A_{-100}^0 = A_0^{100}$.
24 For $y = ax + bx^3 + cx^5 + dx^7$; argue that $A_{-100}^{100} = 0$.

15.7 ➤ AREA UNDER $y = f(x)$

We are now in a position to define "area enclosed by $y = f(x), y = 0, x = a, x = b$", at least for some simple functions f (Fig. 15.22). First, divide the interval $[a,b]$ up into n *equal* intervals by the points $x_0 = a, x_1, x_2, \ldots, x_{i-1}, x_i, \ldots, x_n = b$. Let us call the common interval width Δx, read "delta x". As we have been doing, use x_1 to determine the height $f(x)$ of a rectangle, use x_2 to determine the height $f(x_2)$ of the next rectangle, etc. Form the rectangles whose areas are $f(x_1)\,\Delta x, f(x_2)\,\Delta x, \ldots, f(x_n)\,\Delta x$. The enclosed area A_a^b—often referred to simply as the

Area under $y = f(x)$ from $x = a$ to $x = b$

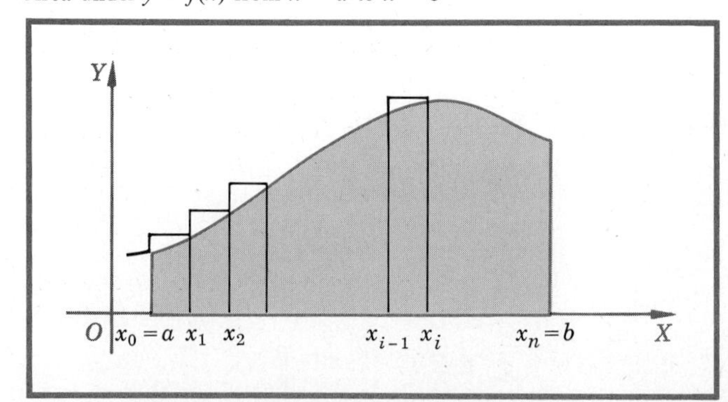

FIGURE 15.22
We now apply this method to areas bounded by more general curves.

"area under the curve from a to b"— is approximately

$$A_a^b \approx f(x_1)\,\Delta x + f(x_2)\,\Delta x + \cdots + f(x_n)\,\Delta x$$

or

$$A_a^b \approx [f(x_1) + f(x_2) + \cdots + f(x_n)]\,\Delta x \tag{1}$$

At this point we should like to introduce a useful shorthand notation, called the Σ notation, for the sum $f(x_1) + f(x_2) + \cdots + f(x_n)$, but we shall first explain the notation with simpler examples. The Greek Σ (sigma, corresponding to S) is used to indicate the "sum of". A dummy letter, such as i or j, is used with Σ. Thus $\sum_{i=1}^{3} x^i$ is read "the sum of x to the ith power, i running from 1 to 3". When written out, this becomes

$$\sum_{i=1}^{3} x^i = x + x^2 + x^3$$

Summation notation

Similarly,

$$\sum_{j=0}^{2} 7x_j = 7x_0 + 7x_1 + 7x_2$$

$$\sum_{i=1}^{n} i = 1 + 2 + 3 + \cdots + n$$

$$\sum_{j=1}^{n} i^2 = 1^2 + 2^2 + \cdots + n^2$$

$$\sum_{j=1}^{k} (2 + 3j) = (2 + 3) + (2 + 3 \cdot 2) + \cdots + (2 + 3k)$$

In this notation we can therefore write (1) in the form

$$A_a^b \approx \sum_{i=1}^{n} f(x_i)\,\Delta x$$

EXERCISE A Express in Σ (sigma) notation:

1 $2x^2 + 4x^4 + 8x^8 + \cdots + 32x^{32}$
2 $y + y^2 + y^3 + \cdots + y^n$
3 $1 + \frac{1}{2} + \frac{1}{3} + \cdots + \frac{1}{n}$

EXERCISE B Write out (expand) the given Σ notation:

1 $\sum_{i=1}^{5} (2i + x_i)$

2 $\sum_{i=1}^{n} (f_i - f_{i-1})$

3 $\sum_{j=0}^{n} (3 - j)$

DEFINITION The *area* A_a^b *under* $y = f(x)$, from $x = a$ to $x = b$, is given by

$$A_a^b = \lim_{n\to\infty} [f(x_1) + f(x_2) + \cdots + f(x_n)]\,\Delta x$$

$$= \lim_{n\to\infty} \sum_{i=1}^{n} f(x_i)\,\Delta x$$

if this limit exists.

Although we shall find better ways of computing A_a^b than by working directly with the limit, yet it is important that you understand the definition.

15.8 › INTEGRATION

The limit in the definition above appears so often in mathematical literature (where it has many, many other interpretations) that a very special notation and name have been assigned to it. It is

$$\lim_{n\to\infty} \sum_{i=1}^{n} f(x_i)\,\Delta x = \int_a^b f(x)\,dx$$

Definite integral

which is read "the integral of f, with respect to x, from a to b". (Some authors refer to this as the *definite integral.*) The function f is called the *integrand;* a and b are called the lower and upper limits† of integration, respectively. The dx says that, regardless of the other variables or parameters that might appear in f, the operations of summing and limit taking are to be performed with respect to the variable x, the *variable of integration.* The elongated S, that is, the symbol "$\int$", is to remind us of the sum and limit operations involved. Remember: the integral is *not* a sum; it is the *limit.* The integral is a real number, which depends upon f, a, and b but not upon what we call the independent variable x. The letter x could be called t or z, etc.; for this reason the variable of integration is often referred to as a *dummy* variable. Hence we have the following theorem.

† The word "limit" here is used in the sense of "bound" and has no connection with the limit of the sequence or of a function. This regrettable confusion is so well established in mathematics that the student will just have to keep alert to be sure of the sense in which "limit" is used.

THEOREM 7 $\int_a^b f(x)\,dx = \int_a^b f(t)\,dt = \cdots = \int_a^b f(z)\,dz = \cdots$

To include the case where $a = b$, we give the following intuitively reasonable definition.

DEFINITION $\int_a^a f(x)\,dx = 0$

The following theorems (which we actually used, for polynomials, in Sec. 15.6) are readily proved directly from the definition of integral.

THEOREM 8 $\int_a^b k \cdot f(x)\,dx = k \cdot \int_a^b f(x)\,dx$, k constant

THEOREM 9 $\int_a^b [f(x) \pm g(x)]\,dx = \int_a^b f(x)\,dx \pm \int_a^b g(x)\,dx$

THEOREM 10 $\int_a^b f(x)\,dx = -\int_b^a f(x)\,dx$

THEOREM 11 $\int_a^c f(x)\,dx = \int_a^b f(x)\,dx + \int_b^c f(x)\,dx$

EXERCISE A Prove Theorem 8.

EXERCISE B Prove Theorem 9.

EXERCISE C Prove Theorem 10.

EXERCISE D Prove Theorem 11.

EXERCISE E Show that

$$\int_a^b (px^2 + qx + r)\,dx = p\int_a^b x^2\,dx + q\int_a^b x\,dx + r\int_a^b 1\,dx$$

The integral $\int_a^b f(x)\,dx$ depends on f, a, and b. In order to study this dependence more fully, let us think of f as a given function, a as being fixed, and b as a variable. We should perhaps change the notation; therefore consider $y = f(u)$, a fixed and x, a variable, as the upper limit.

DEFINITION The function F whose values are given by

$$F(x,a) = \int_a^x f(u)\,du$$

is called the *integral of f,* with respect to u, from a to x.

By changing a to some other number b we get $F(x,b) = \int_b^x f(u)\,du$, another integral. How do $F(x,a)$ and $F(x,b)$ differ? To answer this, we consider the difference $F(x,a) - F(x,b)$. We have

$$\begin{aligned} F(x,a) - F(x,b) &= \int_a^x f(u)\,du - \int_b^x f(u)\,du \\ &= \int_a^x f(u)\,du + \int_x^b f(u)\,du \\ &= \int_a^b f(u)\,du = k \qquad \text{a constant} \end{aligned}$$

Hence any two integrals of one and the same function differ only by an additive constant; that is, $F(x,a) = F(x,b) + k$.

EXERCISE F Give a geometric interpretation of $\int_a^x f(u)\,du$.

Look now at Table 2, Sec. 15.5, above. In terms of an integral, the last line would read $\int_a^b x^n\,dx = \frac{1}{n+1}(b^{n+1} - a^{n+1})$, but there we were thinking of n as an integer ≥ 0. As a matter of fact, the equality holds for every real number $n \neq -1$ according to the following theorem, which we merely state.

THEOREM 12 $\int_a^x u^n\,du = \frac{x^{n+1}}{n+1} - \frac{a^{n+1}}{n+1}$, for any real number $n \neq -1$, provided u^n is continuous in the interval $[a,x]$.

Note that when n is negative, u^n is discontinuous at $u = 0$, so that, in this case, 0 must not lie in $[a,x]$.

ILLUSTRATION 1 Compute the value of the integral $\int_2^4 2x^2 + 3x\,dx$.

SOLUTION

$$\begin{aligned} \int_2^4 2x^2 + 3x\,dx &= \left[\frac{2x^3}{3} + \frac{3x^2}{2}\right]_2^4 \\ &= \left(\frac{128}{3} + \frac{48}{2}\right) - \left(\frac{16}{3} + \frac{12}{2}\right) \end{aligned}$$

$$= \frac{112}{3} + \frac{36}{2}$$
$$= 45\tfrac{1}{3}$$

ILLUSTRATION 2 Compute the value of the integral $\int_0^1 x^3y^2x^4\,dy$.

SOLUTION The notation indicates that x and z are constant during the process of integration. So

$$\int_0^1 x^3y^2z^4\,dy = \left[\frac{x^3y^3z^4}{3}\right]_{y=0}^{y=1} = \frac{x^3z^4}{3}$$

PROBLEMS 15.8

Compute the value of the integral where possible.

1 $\int_0^1 (6x + 9)\,dx$

2 $\int_1^2 (2z + 6)\,dz$

3 $\int_1^2 \frac{h^2}{2}\,dh$

4 $\int_0^1 (m^3 + m)\,dm$

5 $\int_{-1}^0 (y^4 - y^2)\,dy$

6 $\int_0^1 v^{1/2}\,dv$

7 $\int_2^3 3(t^{1/3})\,dt$

8 $\int_0^1 (x^{1/2} + x^{1/3})\,dx$

9 $\int_{-1}^2 \left(x - \frac{1}{x^2}\right)dx$

10 $\int_1^2 4x\left(x + \frac{1}{x^3}\right)dx$

11 $\int_0^1 (ax^2 + bx + c)\,dx$

12 $\int_1^0 (at + b)\,dt$

13 $\int_1^2 (ax + b)^2\,dx$

14 $\int_0^c dx$

15 $\int_1^5 2py\,dy$

16 $\int_0^2 \frac{2p}{y}\,dp$

17 $\int_0^1 xy^2z\,dz$

18 $\int_0^1 x^2yz^2\,dz$

19 $\int_0^1 x^4y^4z\,dy$

20 $\int_1^3 (t^{1/2})z\,dt$

21 $\int_0^1 \frac{t^4x^5y^3}{xy}\,dt$

22 $\int_0^t (y - z)\,dy$

23 $\int_0^a q(x^{1/4})\,dx$

24 $\int_a^b (x^2 + x^2y)\,dx$

25 $\int_{b/2}^{b} (bx - 2b)\,dx$

26 $\int_{-2}^{1} x^{-1/2}\,dx$

27 $\int_{0}^{b+6} (x - 2b)\,dx$

28 $\int_{1}^{1} (t - 3)x\,dx$

29 $\int_{0}^{1} \left(\int_{0}^{y} x\,dx\right) dy$

30 $\int_{-1}^{1} \left(\int_{0}^{y} 3x^2\,dx\right) dy$

15.9 › SETTING UP PROBLEMS; APPLICATIONS

Let us emphasize that in this chapter we have tried to give you some ideas of integration without going into complicated proofs. The subject of integral calculus is large and difficult and will occupy the serious student for many years. We have seen how to find areas under polynomial curves by integration, i.e., by setting up an *approximating sum* and then taking the *limit of the sum*. Actually we made use of the simple formula $\int_a^b x^n\,dx = \frac{1}{n+1}(b^{n+1} - a^{n+1})$ in working with polynomials. In this text we shall not discuss the integration of other functions, but there are obviously a host of others to study.

In discussing a few other applications of the integral calculus it will be our object also to show how the basic definition furnishes us with a powerful method of formulating a given problem. To this end, we consider once more the problem of the area under a curve.

ILLUSTRATION 1 Find the area bounded by the curves $y = f(x)$, $y = 0$, $x = a$, $x = b$.

SOLUTION We *think* (Fig. 15.23): Divide the interval $[a,b]$ into n equal subintervals by the points whose x-coordinates are $a = x_0, x_1, x_2, \ldots, x_i, \ldots, x_n = b$, and set $\Delta x = x_i - x_{i-1}$. Form the sum $\sum_{i=1}^{n} f(x_i)\,\Delta x$.

Area under $y = f(x)$ from $x = a$ to $x = b$

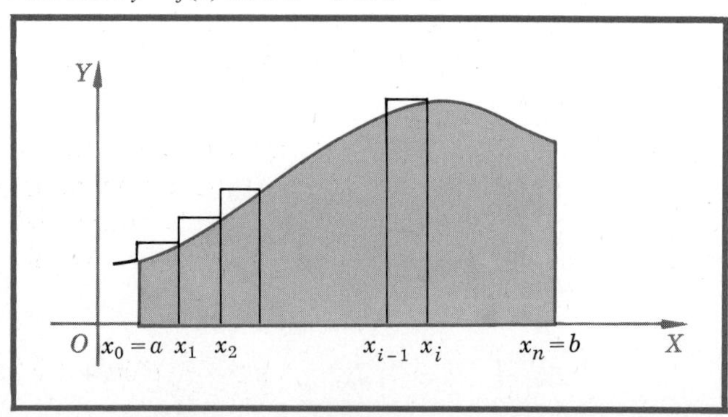

FIGURE 15.23

Element of area

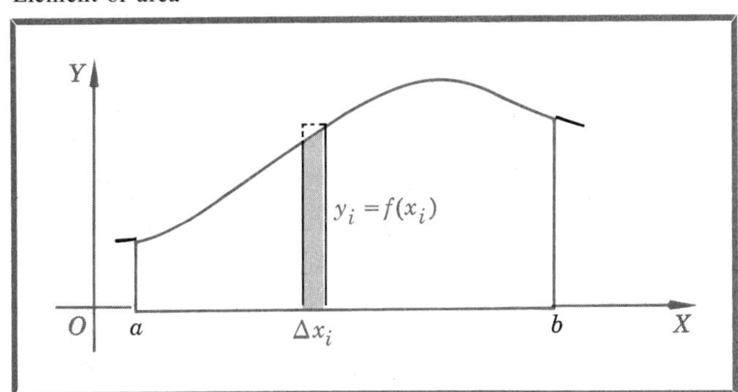

FIGURE 15.24
Instead of drawing all the approximating rectangles, we pick a typical one and call it the "element of area".

The area will then be given by

$$\lim_{n\to\infty} \sum_{i=1}^{n} f(x_i)\,\Delta x = \int_a^b f(x)\,dx$$

At this point we would pay no further attention to the left-hand side of the equation *if* we could calculate $\int_a^b f(x)\,dx$.

To shorten the work, we *write* as follows (Fig. 15.24): One small rectangle used in the definition has the area $f(x_i)\,\Delta x$. Because of the end result we are seeking, we actually write this as $f(x)\,dx$. Then the total area will be the limiting value of the sum of all such small rectangles, or simply, $\int_a^b f(x)\,dx$. Indeed this is where the symbol $\int$ came from: it is something of an elongated S which is to remind us of both *sum* and *limit*.

Work

ILLUSTRATION 2 A bag of sand is raised 10 ft but steadily loses sand at the rate of 3 lb/ft of elevation. Find the work done if at the beginning the full bag weighed 400 lb.

SOLUTION We *think:* Divide the interval (10 ft) into equal subintervals of width Δx (Fig. 15.25). From the definition of *work* = *force* × *distance* we compute, approximately, the work done in lifting the bag through the ith interval $[x_{i-1},x_i]$. It is $(400 - 3x_i)\,\Delta x$. The total work would be, exactly,

$$\lim_{n\to\infty} \sum_{i=1}^{n} (400 - 3x_i)\,\Delta x$$

But we *write* (Fig. 15.26): At height x the bag weighs $400 - 3x$ lb. In lifting this through a distance dx the work done is $(400 - 3x)\,dx$, and the total work is $\int_0^{10} (400 - 3x)\,dx$.

At this point we compute

$$\text{Work} = \int_0^{10} (400 - 3x)\,dx = 400(10) - \tfrac{3}{2}(10)^2$$
$$= 4{,}000 - 150 = 3{,}850 \text{ ft-lb}$$

Total force

ILLUSTRATION 3 A rectangular fish tank is 3 ft long, 2 ft wide, and 1 ft deep. It is filled with water weighing 62.4 lb/ft³. Find the total force on one end of the tank.

Actual subdivision

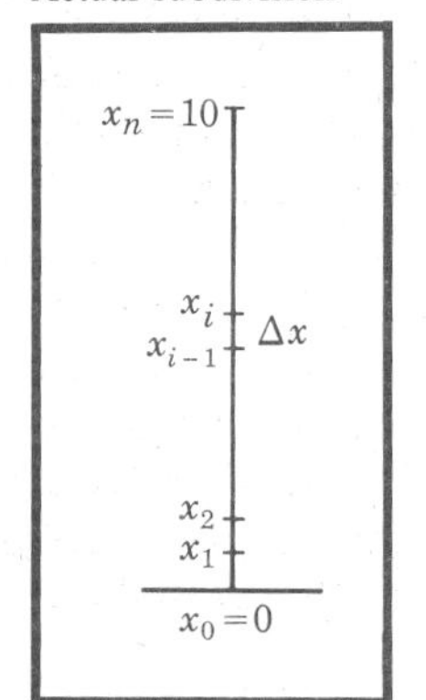

FIGURE 15.25
See Illustration 2.

SOLUTION We *think* (Fig. 15.27): Divide the interval corresponding to the depth of the tank into equal subintervals of width Δx. From the definition of *total force* $= 62.4 \times A \times h$, where $A =$ area at depth h, we compute, approximately, the total force on the strip between the depth x_{i-1} and x_i. It is $62.4(2)x_i\,\Delta x$. Hence the total force on the end is exactly

$$\lim_{n\to\infty} \sum_{i=1}^{n} 62.4(2)x_i\,\Delta x$$

But we *write* (Fig. 15.28): The force on the small strip at depth x is $62.4(2)x\,dx$, and the total force is $2(62.4)\int_0^1 x\,dx$. This yields total force $= 2(62.4)\frac{1}{2} = 62.4$ lb.

Symbolic element of distance

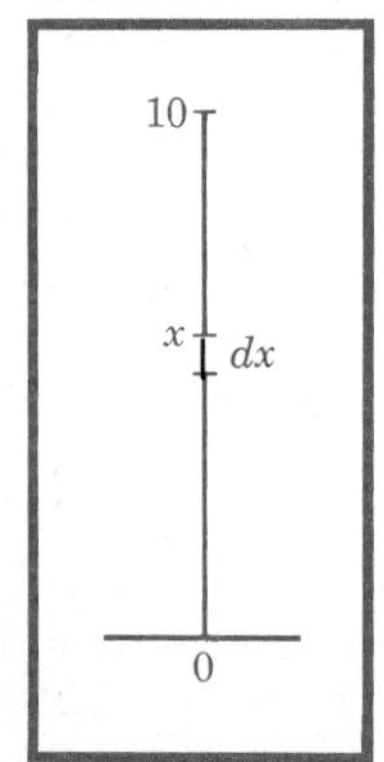

FIGURE 15.26
See Illustration 2.

ILLUSTRATION 4 A vertical cylindrical tank of radius 5 ft and height 20 ft is filled with a liquid weighing w lb/ft³. Find the work done in pumping the liquid out over the rim of the top (Fig. 15.29).

SOLUTION We *think:* Divide the interval corresponding to height into equal subintervals of width Δx. The weight contained between x_{i-1} and x_i is $25\pi w\,\Delta x$. This must be lifted x_i ft; so the work done is $25\pi w x_i\,\Delta x$. The total work would be $\lim_{n\to\infty} \sum_{i=1}^{n} 25\pi w x_i\,\Delta x$.

But we *write:* The typical small weight is $25\pi w\,dx$, and this must be lifted x ft. The work, therefore, is

$$W = 25\pi w \int_0^{20} x\,dx = 25\pi w \frac{20^2}{2} = 5{,}000\pi w \text{ ft-lb}$$

Volume of revolution

ILLUSTRATION 5 Find the volume enclosed by a surface of revolution.

Actual subdivision

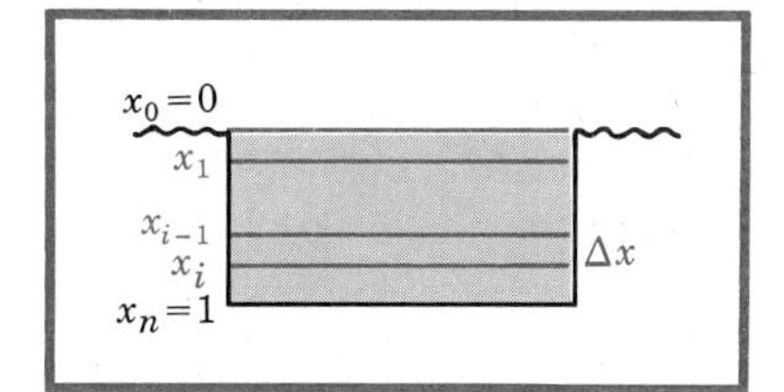

FIGURE 15.27
See Illustration 3.

Symbolic element of area

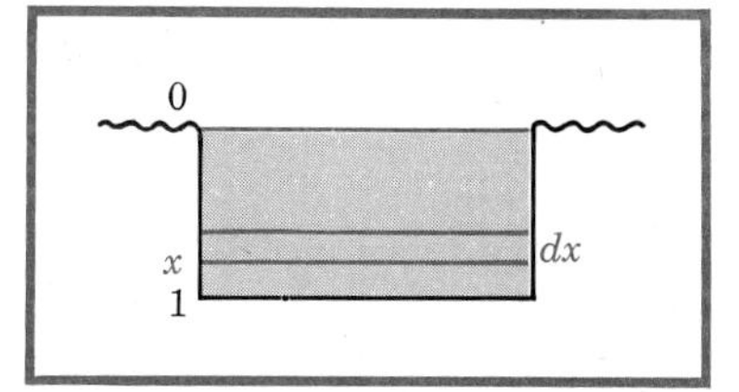

FIGURE 15.28
See Illustration 3.

SOLUTION Consider the curve $y = f(x)$, between a and b; revolve it around the X-axis, thus generating a surface of revolution (Fig. 15.30). This surface, together with the planes $x = a$, $x = b$, enclose something we would like to call "volume". This, once again, is an intuitive notion. We shall give a definition of this intuitive notion in a moment. First form n intervals on the X-axis from a to b by the points $x_0 = a, x_1, x_2, \ldots,$ $x_n = b$ such that $x_i - x_{i-1} = \Delta x$. Cut the surface by the planes $x = x_i$, $i = 0, 1, \ldots, n$. Construct the cylinders as indicated in the figure; the general one will have a base of radius $f(x_i)$ and a height of $\Delta x_i = x_i - x_{i-1}$. We assume that the volume of a cylinder is known from elementary geometry to be area of base times height. Hence the general elementary cylinder will have a volume of $\pi[f(x_i)]^2 \cdot \Delta x$ and the sum of all such

$$\sum_{i=1}^{n} \pi[f(x_i)]^2 \cdot \Delta x$$

Symbolic element of volume

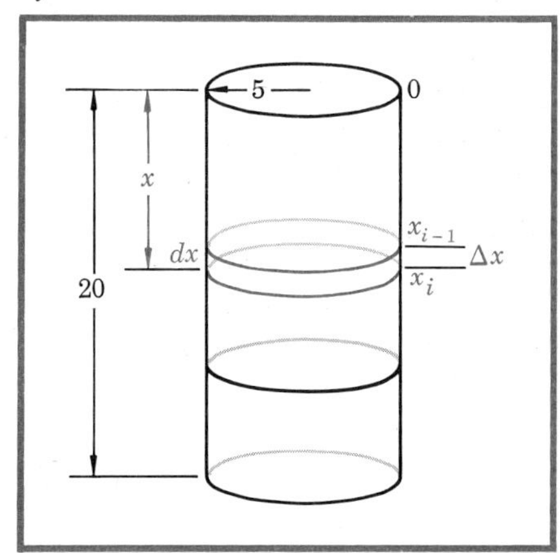

FIGURE 15.29
See Illustration 4.

Symbolic element of volume

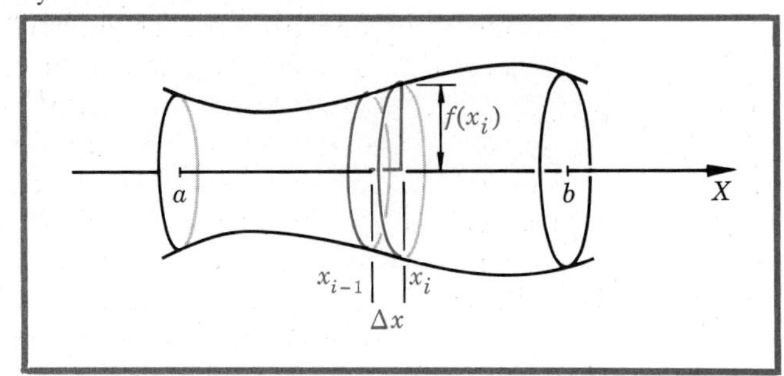

FIGURE 15.30
See Illustration 5.

will be an approximation to the "volume" being considered. If we take the limit of this sum as $n \to \infty$, we shall have the volume by definition.

DEFINITION The *volume* contained between the planes $x = a$ and $x = b$ and the surface generated by revolving $y = f(x)$ about the X-axis is

(1)
$$V_a^b = \lim_{n\to\infty} \sum_{i=1}^{n} \pi[f(x_i)]^2 \cdot \Delta x$$

if this limit exists.

By now you should be able to translate this definition directly into the integral

(2)
$$V_a^b = \pi \int_a^b [f(x)]^2 \, dx$$

Recall the mental process we go through in writing down such an expression. We *write* (2) but we *think* and *talk* (1) because (1) is the basic one. We say to ourselves something like this: "A typical little cylinder in a given subinterval will have the volume approximately equal to $\pi[f(x)]^2 \cdot \Delta x$." Then we continue: "If we sum up all such little cylinders, we will get, in the limit, the total volume." And we write $\int$ for limit of sum; so we have

$$\pi \int_a^b [f(x)]^2 \, dx$$

ILLUSTRATION 6 As a special example we now find the volume generated by revolving $y = x^2$ about the X-axis between $x = 0$ and $x = 2$.

SOLUTION
$$V_0^2 = \pi \int_0^2 [x^2]^2 \, dx$$
$$= \pi \frac{2^5}{5}$$
$$= \tfrac{32}{5}\pi \text{ cubic units}$$

PROBLEMS 15.9

1 A small boat anchor chain weighs 2 lb/ft. What is the work done in pulling up anchor if the anchor itself weighs 110 lb and 22 ft of chain is out. Assume the lift is vertical.

2 The force required to stretch a certain spring x in. is $3x$ lb. Find the work done in stretching the spring 4 in. beyond its natural length.

3 The natural length of a spring is 4 in., and the force required to compress it x in. is $8x$ lb. Find the work done in compressing it to half its natural length.

4 A 16- by 40-ft rectangular floodgate is placed vertically in water with the 40-ft edge in the surface of the water. Find the force on the side in contact with the water.

5 A plate in the form of an equilateral triangle, of side 10, is submerged vertically until one edge is just in the surface of the water. Find the total force on one side of such a plate.

6 A plate in the form of the parabola $y = 4x^2$ is lowered vertically into water to a depth of 1 ft, vertex down. Find the force on one side.

7 A tank, in the form of a square right pyramid 15 ft on a side, is 6 ft deep, vertex down. Find the work required to empty the tank by pumping the water to a point 2 ft above the top of the tank. Assume that water weighs w lb/ft^3.

In Probs. 8 to 13 find the volume generated as indicated. Sketch.

8 $y = \sqrt{4 - x^2}$ about the X-axis from $x = -2$ to $x = +2$

9 $x^2 = py$ about the Y-axis from $x = 0$ to $x = p/2$

10 $y = ax$ about the Y-axis from $x = 0$ to $x = t$

11 $y = b\sqrt{a^2 - x^2}$ about the X-axis from $x = -a$ to $x = +a$

12 $y = x^{3/2}$ about the X-axis from $x = 0$ to $x = 1$

13 $y = Ax^2 + Bx$ about the X-axis from $x = 0$ to $x = x_0$, where $x_0 > 0$, $A > 0$, $B \geq 0$, and $C \geq 0$

In Probs. 14 to 17 set up each problem as the limit of a sum and as an integral, but do not attempt to evaluate.

14 A circular water main 10 ft in diameter is full of water. Find the pressure on the gate valve when closed.

15 Calculate the work done in pumping out the water from a filled hemispherical reservoir 50 ft deep.

16 The natural length of a spring is 5 in., and the force required to stretch it x in. is $25x$ lb. Find the work done in extending the length of the spring from 6 to 7 in.

17 Find the volume generated by revolving $y = \cos x$ about the X-axis from $x = 0$ to $x = \pi/4$. Can you find this volume approximately if it is known that

$$\cos^2 x \approx 1 - x^2 + \frac{x^4}{3}$$

16 INTUITIVE DIFFERENTIATION

16.1 » INTRODUCTION

In Chap. 15 we were concerned with the problem of area. There we indicated that Euclid gave a definition of the area of a circle and that Archimedes, using this definition, arrived at the approximation $\pi \approx 3\frac{1}{7}$—a value that is in current use. Archimedes also found, in effect, by methods of sums and limits, the area enclosed by the parabola $y^2 = x$ and $x = 1$. It is reasonable to say that the early Greeks had some insight in that branch of mathematics now called *integral calculus*. They seem to have had no notion of *differential calculus*, the subject of this chapter.

16.2 » NOTION OF A TANGENT

While the concept of *integral* grew out of the problem of the area under a curve, the concept of *derivative* arose in connection with the *geometric tangent* to a plane curve and also in connection with the physical quantity *velocity*. We shall treat these in turn but begin with Euclid's definition of a tangent to a circle, an idea you met in plane geometry. Euclid said:

The tangent to a circle at point D, one endpoint of a diameter DD', is the line passing through D and perpendicular to DD'.

You can readily see that such a definition will be of no use to us if we try to apply it to curves other than circles.

Intuitively we feel that a curve, though bending and turning, should have some sort of "nearly constant direction" in a very small interval, and our intuition furnishes us with a clue as to how to define a tangent line.

Consider a curve C, such as is pictured in Fig. 16.1, and draw the line PQ, called a secant line. If Q were made to trace the curve until it approached the point P, the secant PQ would take different positions therewith.

Secants approximating a tangent

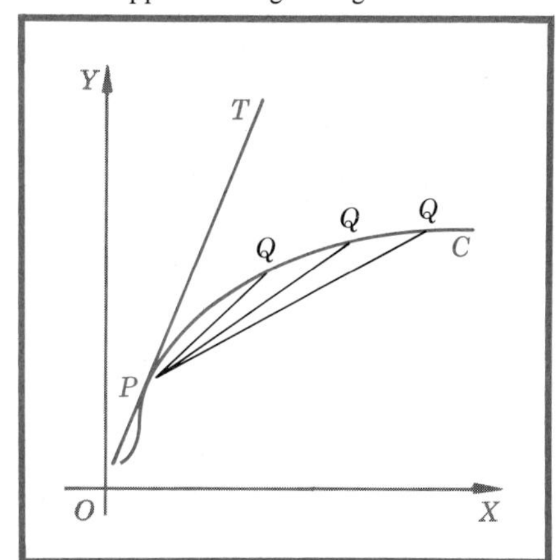

FIGURE 16.1
We can approximate a tangent at P by drawing a secant PQ where Q is close to P.

DEFINITION
Tangent to a curve

The line, if one exists, whose position is the unique limiting position PT of the secant line PQ, as $Q \to P$ along the curve is called the *tangent line* to the curve at the point P.

For some curves there is no unique limiting position (for some points P), in which case there is no tangent at P. But for graphs of polynomials, the tangent line is well defined for each and every point P on the graph. In this chapter we deal mainly with polynomials, but we shall develop our basic concepts for a more general function f.

The definition and the discussion of a tangent so far have been geometric in nature. Now let us translate the geometric wording into an equation for this line. This is indeed quite essential. For our definition of the tangent to a curve we used the phrase "limiting position of a secant". Although this makes intuitive sense, we have not defined the meaning of such a limit and hence cannot proceed deductively here. Instead, we must use our knowledge of analytic geometry to translate this intuitive idea into a sharp, clear one.

We know that a line is completely determined by a point and a slope, and in this case the point is given. Therefore we must seek the slope of the tangent. From our early discussion we might well infer that the slope of the tangent should equal the limit of the slopes of the secants. The slope of a secant which cuts the curve $y = f(x)$ in the points $(x_1, f(x_1))$, $(x_1 + h, f(x_1 + h))$ is

$$\text{Slope of secant} = \frac{f(x_1 + h) - f(x_1)}{h}$$

Hence we state the following definition (Fig. 16.2).

Construction for finding slope of a tangent

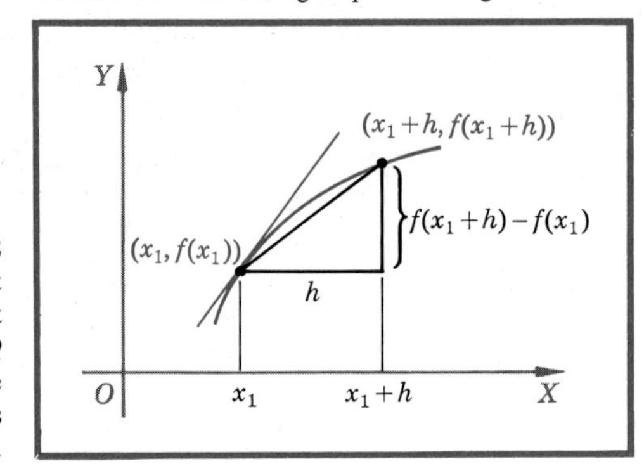

FIGURE 16.2
To find the slope of a tangent at P, find the slope of a secant PQ and take the limit as Q approaches P. In the figure P is $(x_1,f(x_1))$, and Q is $(x_1 + h, f(x_1 + h))$.

DEFINITION The tangent to the curve $y = f(x)$ at the point $(x_1,f(x_1))$ is the line passing through this point whose *slope* $m(x_1)$ is given by

(1)
$$m(x_1) = \lim_{h \to 0} \frac{f(x_1 + h) - f(x_1)}{h}$$

provided this limit exists.

We recall that the equation of such a line is

(2)
$$y - y_1 = m(x_1)(x - x_1)$$

Equation of the tangent

Finally, therefore, we write down the equation of this tangent line by computing $m(x_1)$ and substituting in (2) above.

ILLUSTRATION 1 Find the equation of the line tangent to the curve

$$y = x^2 - x + 1$$

at the point (0,1) (Fig. 16.3).

Tangent to $y = x^2 - x + 1$ at (0,1)

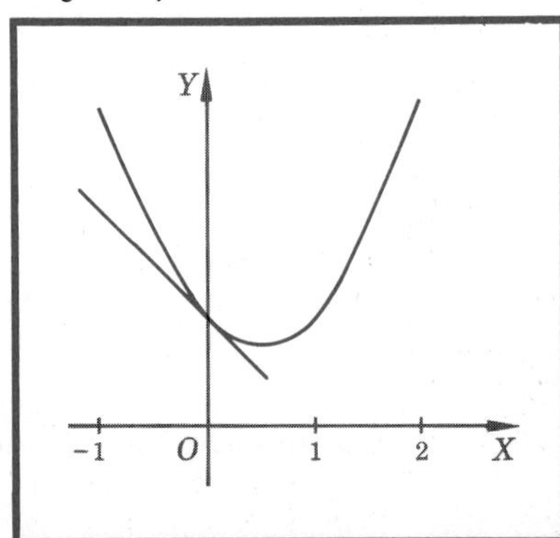

FIGURE 16.3
See Illustration 1.

SOLUTION

$$\begin{aligned}
f(x_1) &= f(0) = 1 \\
f(x_1 + h) &= (x_1 + h)^2 - (x_1 + h) + 1 \\
&= (0 + h)^2 - (0 + h) + 1 \\
&= h^2 - h + 1 \\
m(x_1) &= \lim_{h \to 0} \frac{f(x_1 + h) - f(x_1)}{h} \\
&= \lim_{h \to 0} \frac{h^2 - h}{h} \\
&= \lim_{h \to 0} (h - 1) \qquad h \neq 0 \\
&= -1
\end{aligned}$$

The equation of the tangent is, therefore,

$$y - 1 = -1(x)$$

ILLUSTRATION 2 Find the equation of the tangent to $y = x^2$ at the point (x_1, y_1) (Fig. 16.4).

SOLUTION

$$\begin{aligned}
f(x_1) &= x_1^2 \\
f(x_1 + h) &= (x_1 + h)^2 = x_1^2 + 2x_1h + h^2 \\
m(x_1) &= \lim_{h \to 0} \frac{f(x_1 + h) - f(x_1)}{h} \\
&= \lim_{h \to 0} \frac{x_1^2 + 2x_1h + h^2 - x_1^2}{h} \\
&= 2x_1
\end{aligned}$$

The equation of the tangent is, therefore,

$$y - y_1 = 2x_1(x - x_1)$$

Tangent to $y = x^2$ at (x_1, y_1)

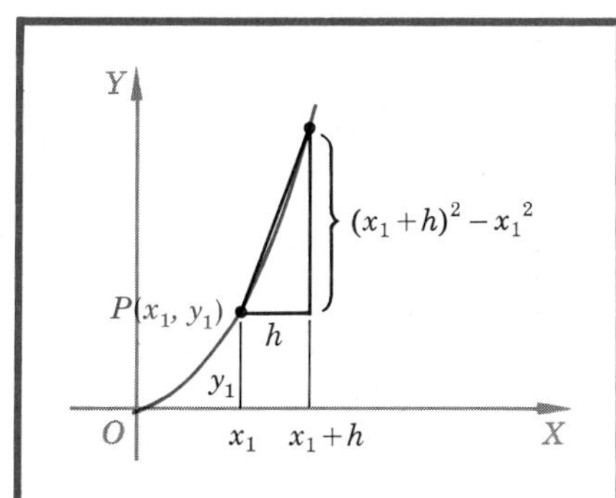

FIGURE 16.4
See Illustration 2.

PROBLEMS 16.2 Find the equation of the tangent line as indicated, and sketch.

1 $y = 3x - 5$ at $(1,-2)$
2 $y = 6x - 2$ at $(2,10)$
3 $y = 4x^2 - 3$ at $(1,1)$
4 $y = 2 + 9x^2$ at $(0,2)$
5 $y = 4x^2 - 2x$ at $(1,2)$
6 $y = x^2 - 4x + 4$ $(2,0)$
7 $y = 3x^2 + 4x + 5$ at $(0,5)$
8 $y = x^2 - 9$ at $(1,-8)$
9 $y = x^3 - 1$ at $(2,7)$
10 $y = 4x^2 - 4x^3$ at $(2,16)$
11 $y = 2x^4 + 3$ at $(1,5)$
12 $y = x^6 - 64$ at $(2,0)$
13 $y = mx + b$ at $(x_0, mx_0 + b)$
14 $y = 3x^4$ at the point on the curve where $x = x_1$
15 $y = x^2$, with positive slope and passing through the point $(2,10)$
16 $y = Ax^2 + Bx + C$, at the point on the curve where $x = x_0$

16.3 › VELOCITY AND ACCELERATION

When a particle moves, there are associated with the motion certain quantities such as time, distance, velocity, and acceleration. We shall restrict ourselves to the case where the motion takes place on a straight line, since we are unprepared at this time to consider general curvilinear motion.

Let $y = f(t)$ give the position of the particle on the Y-axis at any time t. The time variable is measured continuously by a clock and is usually thought of as positive or zero, although on occasion we may want to assign a negative value in order to describe a past event. The y-coordinate is a linear distance positive, negative, or zero from some fixed point on the line called the origin (Fig. 16.5). Suppose the particle to be at $y = f(t_1)$ and $y = f(t_1 + h)$, when t is t_1 and $t_1 + h$, respectively. Then the particle has moved $f(t_1 + h) - f(t_1)$ units of distance in $h > 0$ units of time.

DEFINITION *Average velocity*

If a particle moves a distance of

$$f(t_1 + h) - f(t_1)$$

in time h, then the ratio (Fig. 16.6*a*)

(1)
$$\bar{v} = \frac{f(t_1 + h) - f(t_1)}{h}$$

is called the *average velocity during the time interval h.*

Average velocity is thus the change in distance per unit change in time. Units often encountered are miles per hour, centimeters per second, etc. These are abbreviated miles/hr, cm/sec, etc. Since distance may be negative, so also velocity may be negative. The absolute value of the average velocity is called average speed.

Distance in terms of time

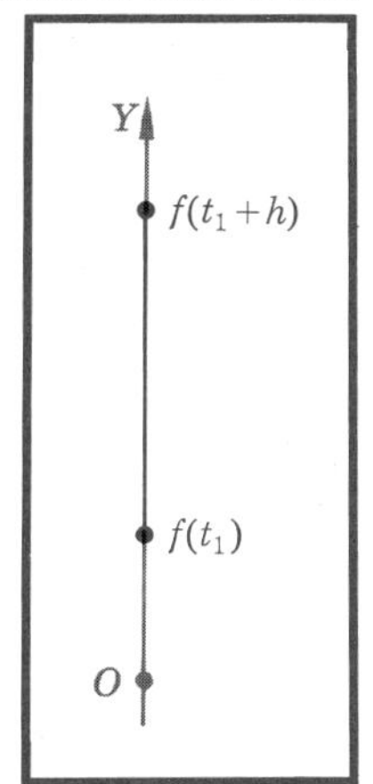

FIGURE 16.5
The average velocity is the ratio of the elapsed distance to the elapsed time.

Now average velocity (and also average speed) is an interval property since it describes what happens in an interval of time. Hence it cannot directly explain such a statement as "exactly at that instant the plane was traveling at 500 miles per hour", because there is no interval of time involved in this observation. And yet, intuitively the statement does have some sense. It seems to say that, if the plane had continued at the same (constant) speed as it was traveling at that instant, then it would have covered 500 miles every hour thereafter. But this does not supply an answer to the inherent difficulty in the notion of traveling at 500 miles/hr *at* a certain (clock) value, say t_1, of the time variable. But let us think of a small interval of time $[t_1, t_1 + h]$, where $h > 0$, and the average velocity $\bar{v}$ during this interval. Then consider the average velocity $\bar{v}$ for smaller and smaller intervals of time h. Intuitively we feel that there should be something, $v(t_1)$, called *instantaneous velocity at* $t = t_1$ which the average velocity $\bar{v}$ would approach as h approaches zero. We lay down the following definition.

DEFINITION
Velocity

Given distance y as a function f of t, then the *instantaneous velocity,* or, simply, *velocity for a particular value of t, say* t_1, is defined to be

(2)
$$v(t_1) = \lim_{h \to 0} \frac{f(t_1 + h) - f(t_1)}{h}$$

provided this limit exists.

REMARK If we substitute $h = 0$ in the expression $\dfrac{f(t_1 + h) - f(t_1)}{h}$, it takes the meaningless form 0/0. However, the limit of this expression may still have meaning and be of great value. We shall discuss this further in Sec. 16.4.

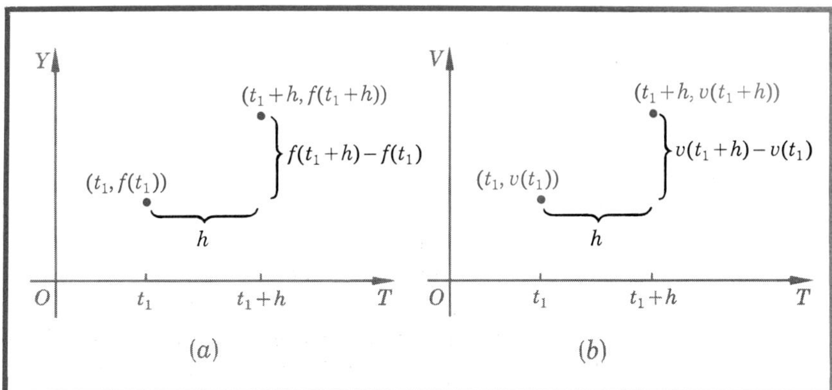

FIGURE 16.6
(*a*) Average velocity $= \dfrac{f(t_1 + h) - f(t_1)}{h}$
(*b*) Average acceleration $= \dfrac{v(t_1 + h) - v(t_1)}{h}$

The concept of acceleration is no more difficult to grasp mathematically than that of velocity. It is known to be the rate at which velocity is changing. To describe this precisely, let us compute the two values of instantaneous velocity $v(t_1)$ and $v(t_1 + h)$ corresponding to the two values of t, namely, t_1 and $t_1 + h$.

DEFINITION The ratio

Average acceleration **(3)**

$$\bar{a} = \frac{v(t_1 + h) - v(t_1)}{h}$$

is called the *average acceleration during the interval h.* It may be positive, negative, or zero (Fig. 16.6*b*).

DEFINITION The *instantaneous acceleration,* or, simply, *acceleration at* t_1, is defined by

Acceleration **(4)**

$$a(t_1) = \lim_{h \to 0} \frac{v(t_1 + h) - v(t_1)}{h}$$

provided this limit exists.

Average acceleration is an interval property. Instantaneous acceleration is a point property; it is a limit. The unit of acceleration is "units of velocity per unit of time", such as feet per second, per second; miles per hour, per minute; etc. These are abbreviated ft/sec/sec, or ft/sec^2, miles/hr/min, etc.

To summarize, velocity is the rate of change of distance with respect to time. Acceleration is the rate of change of velocity with respect to time.

ILLUSTRATION 1 A particle moves vertically (up and down) in a straight line under the following law of motion: $y = 8t - t^2$, where t is in seconds and y is in feet. Find (1) the velocity at any time t_1; (2) the acceleration at any time t_1; (3) the domain of values of $t > 0$ for which velocity is positive; (4) maximum value of y.

SOLUTION

1 $$v = \lim_{h \to 0} \frac{f(t_1 + h) - f(t_1)}{h}$$

$$= \lim_{h \to 0} \frac{[8(t_1 + h) - (t_1 + h)^2] - [8t_1 - t_1^2]}{h}$$

$$= \lim_{h \to 0} \frac{8t_1 + 8h - t_1^2 - 2t_1h - h^2 - 8t_1 + t_1^2}{h}$$

$$= \lim_{h \to 0} \frac{8h - 2t_1h - h^2}{h}$$

$$= \lim_{h \to 0} (8 - 2t_1 - h) \qquad h \neq 0$$

$$= 8 - 2t_1 \text{ ft/sec}$$

2 $$a = \lim_{h \to 0} \frac{v(t_1 + h) - v(t_1)}{h} = \lim_{h \to 0} \frac{[8 - 2(t_1 + h)] - [8 - 2t_1]}{h}$$

$$a = \lim_{h \to 0} \frac{8 - 2t_1 - 2h - 8 + 2t_1}{h}$$

$$= \lim_{h \to 0} \frac{-2h}{h} = \lim_{h \to 0} (-2)$$

$$= -2 \text{ ft/sec/sec}$$

3 $v = 8 - 2t > 0$, or $t < 4$ sec. Since also $t > 0$, the answer is: $0 < t < 4$.

4 The particle is at the origin when $t = 0$ and again when $t = 8$. Since it helps to sketch the graph of $y = 8t - t^2$, we do so in Fig. 16.7; however, this graph does not represent the curve traversed by the particle—it simply shows more clearly how high the particle is at time t. The particle evidently rises to some maximum height, then falls back down again, reaching the "ground" in 8 sec. The velocity is positive ($0 < t < 4$) going up, negative ($4 < t < 8$) coming down. It therefore reached its maximum height at $t = 4$, when the velocity was zero. Its maximum height was $8 \cdot 4 - 4^2 = 16$ ft.

Graph of height versus time

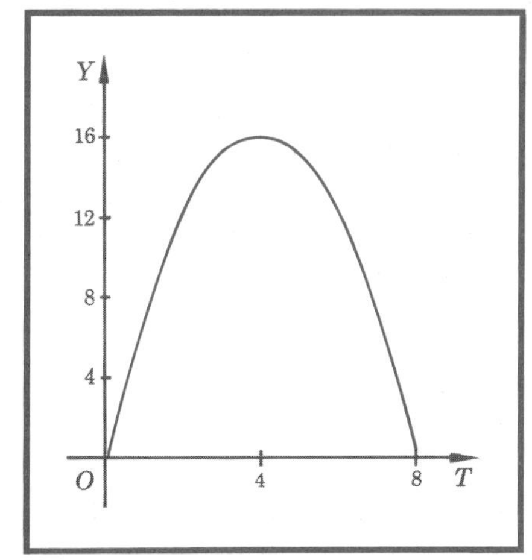

FIGURE 16.7
$y = 8t - t^2$.
See Illustration 1.

PROBLEMS 16.3

	GIVEN:	FIND:		
	The height y, ft, of a particle at time t, sec	**(a)** *Average velocity $\bar{v}$ during interval from*	**(b)** *Average velocity $\bar{v}$ during interval from*	**(c)** *Velocity $\bar{v}$ at*
1	$y = 4t^2 + 1$	$t = 0$ to $t = 2$	$t = 0$ to $t = 1$	$t = 7$
2	$y = t^2 - 4t$	$t = 0$ to $t = 3$	$t = 1$ to $t = 3$	$t = 4$
3	$y = \frac{t}{2} - 16t^2$	$t = 0$ to $t = 1$	$t = 2$ to $t = 4$	$t = 6$
4	$y = \frac{t^2}{2}$	$t = 0$ to $t = \frac{1}{2}$	$t = 1$ to $t = 2$	$t = 0$
5	$y = 4t^2 + 12t + 16$	$t = 0$ to $t = 2$	$t = 0$ to $t = 1$	$t = 2$
6	$y = 4t^2 - 16t - 2$	$t = 0$ to $t = 3$	$t = \frac{1}{2}$ to $t = 1$	$t = 3$
7	$y = 2t^3 + 1$	$t = 0$ to $t = 1$	$t = 1$ to $t = 2$	$t = 1$
8	$y = \frac{t^2}{2} + t - 1$	$t = 0$ to $t = 2$	$t = 2$ to $t = 3$	$t = 1$

	GIVEN:	FIND:	
	The velocity v, ft/sec, at time t, sec	**(a)** *Average acceleration $\bar{a}$ during interval from*	**(b)** *Acceleration a at*
9	$v = t + 1$	$t = 0$ to $t = 3$	$t = 1$
10	$v = 2t^2 + t$	$t = 1$ to $t = 2$	$t = 2$
11	$v = 32t^2 - 8t$	$t = 0$ to $t = 2$	$t = \frac{1}{2}$
12	$v = t^3 + 1$	$t = 0$ to $t = 1$	$t = 3$

	GIVEN:	FIND:	
	The height y, ft, at time t, sec	**(a)** *Velocity v at*	**(b)** *Acceleration a at*
13	$y = t^2 - 4$	$t = 0$	$t = t_1$
14	$y = t^2 - 2t + 3$	$t = 1$	$t = 0$
15	$y = 10t^2 - 5$	$t = 1$	$t = 2$
16	$y = t^3 - t^2 - t$	$t = \frac{1}{2}$	$t = 1$
17	$y = t^3 - t$	$t = 2$	$t = 7$
18	$y = 4t^3$	$t = t_1$	$t = 3$
19	$y = t^2$	$t = t_1$	$t = t_1$
20	$y = at^n$, n a positive integer	$t = t_1$	$t = t_1$

16.4 › DERIVATIVE

In determining the slope $m(x_1)$ of the tangent to the curve $y = f(x)$ at a point $x = x_1$, we were led to formula (1), Sec. 16.2,

$$m(x_1) = \lim_{h \to 0} \frac{f(x_1 + h) - f(x_1)}{h} \tag{1}$$

Formula (2), Sec. 16.3, for velocity at $t = t_1$ [where $y = f(t)$ relates distance and time] was

$$v(t_1) = \lim_{h \to 0} \frac{f(t_1 + h) - f(t_1)}{h} \tag{2}$$

and formula (4), Sec. 16.3, for acceleration at $t = t_1$ [where $y = v(t)$ relates velocity and time] was

$$a(t_1) = \lim_{h \to 0} \frac{v(t_1 + h) - v(t_1)}{h} \tag{3}$$

It is a phenomenon worth recording that these three processes are abstractly identical: each is the same limit operation on a function. A name has been given to this operation.

DEFINITION The limit

$$\lim_{h \to 0} \frac{f(x_1 + h) - f(x_1)}{h} \tag{4}$$

if it exists, is called the *derivative of f with respect to x at the point $x = x_1$*.

The derivative is therefore a new function which is the result of an operation on a function at a point. The process is called *differentiation*.

EXERCISE A Are the domains of definition of a function and its derivative necessarily the same? Explain.

Other notations are often used. Instead of h let us write Δx (read "delta x"), which stands for a change in x. Also let us write Δf for the quantity

$$f(x_1 + \Delta x) - f(x_1)$$

(Fig. 16.8). Then the derivative is

$$\lim_{\Delta x \to 0} \frac{\Delta f}{\Delta x} \tag{5}$$

if we make the appropriate substitutions in (4).

The notion of a derivative is more subtle than might appear at first sight. Be sure that you understand the following remarks.

Derivative of f with respect to x

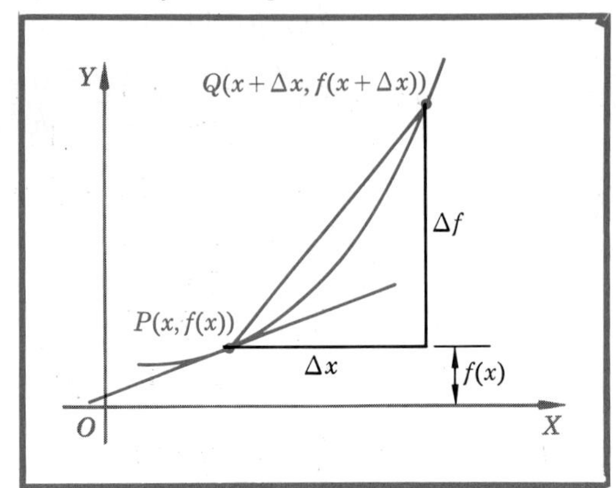

FIGURE 16.8 The derivative of f with respect to x equals $\lim_{\Delta x \to 0} \dfrac{\Delta f}{\Delta x}$.

REMARKS

1 The derivative is *not* the value of $\Delta f/\Delta x$ when $\Delta x = 0$; for when $\Delta x = 0$, $\Delta f/\Delta x = 0/0$, which is indeterminate.

2 The derivative is the *limit* of $\Delta f/\Delta x$ as $\Delta x \to 0$ if this limit exists. However, the limit may fail to exist, and in this case the derivative is not defined. For polynomials, however, the derivative exists for every value of x. An example of a curve which has a point at which there is no derivative is given in the following illustration.

ILLUSTRATION 1 Find the derivative at the point (0,0), if it exists, of the function f whose values are $f(x) = \sqrt{x}$.

$$\begin{aligned} \Delta f &= \sqrt{x + \Delta x} - \sqrt{x} \\ &= \sqrt{\Delta x} \qquad \text{when } x = 0 \end{aligned}$$

$$\begin{aligned} \frac{\Delta f}{\Delta x} &= \frac{\sqrt{\Delta x}}{\Delta x} \\ &= \frac{1}{\sqrt{\Delta x}} \qquad \text{provided } \Delta x > 0 \end{aligned}$$

But $\lim_{\Delta x \to 0} \Delta f/\Delta x$ does not exist, for $1/\sqrt{\Delta x}$ can be made as large as we please by choosing Δx sufficiently small; i.e., it "tends to infinity".

3 Where f is a polynomial, we shall evaluate $\lim_{\Delta x \to 0} \Delta f/\Delta x$ by a process which may seem to contradict our first two remarks. Let us examine it closely.

ILLUSTRATION 2 Find the derivative at the point where $x = 0$ of the function f whose values are $f(x) = x^2 - x + 1$. (See Illustration 1, Sec. 16.2.)

SOLUTION We have

$$f(0) = 1$$
$$f(0 + \Delta x) = (0 + \Delta x)^2 - (0 + \Delta x) + 1$$
$$= \overline{\Delta x}^2 - \Delta x + 1$$

So
$$\Delta f = f(0 + \Delta x) - f(0)$$
$$= \overline{\Delta x}^2 - \Delta x$$

Then
$$\frac{\Delta f}{\Delta x} = \frac{\overline{\Delta x}^2 - \Delta x}{\Delta x}$$
$$= \Delta x - 1 \qquad \text{provided } \Delta x \neq 0$$

Finally
$$\lim_{\Delta x \to 0} \frac{\Delta f}{\Delta x} = \lim_{\Delta x \to 0} (\Delta x - 1)$$
$$= -1$$

In the above illustration we first computed Δf at the required point and then found $\Delta f/\Delta x$. In its original form

$$\frac{\Delta f}{\Delta x} = \frac{\overline{\Delta x}^2 - \Delta x}{\Delta x}$$

and we cannot find its limit without another step. We simplify this fraction by dividing numerator and denominator by Δx, a permissible process as long as $\Delta x \neq 0$. This gives

$$\frac{\Delta f}{\Delta x} = \Delta x - 1 \qquad \text{provided } \Delta x \neq 0$$

Now we take the limit of $\Delta x - 1$ as $\Delta x \to 0$ and assert that the answer is -1.

In evaluating $\lim_{\Delta x \to 0} (\Delta x - 1)$ we *do not put* $\Delta x = 0$; we let Δx *approach* zero. As Δx approaches zero, $\Delta x - 1$ approaches -1, the answer. The point of confusion is that we also get the right answer if we put $\Delta x = 0$ in $\Delta x - 1$. Thus we get the right answer by a process which is apparently illegitimate. Actually this process can be justified whenever $\Delta f/\Delta x$ can be reduced (by division) to a polynomial in Δx. The justification follows from the following theorem on polynomials, which we give without proof.

THEOREM 1 If $f(x)$ is a polynomial in x, then for any a,

$$\lim_{x \to a} f(x) = f(a)$$

In other words, we may find the limit of a polynomial by the simple device of substitution. This is *not* true of functions in general. Since, in Illustration 2, $\Delta x - 1$ is a polynomial in Δx, we can find $\lim_{\Delta x \to 0} (\Delta x - 1)$ by putting $\Delta x = 0$, and hence the apparent confusion is resolved.

The symbols in (4) and (5), while standard, are still long to write down, and thus we devise other symbols which are also quite standard. Let f be the function whose values are given by $y = f(x)$. Then the derivative of f with respect to x is written as

(6)	$\lim_{\Delta x \to 0} \dfrac{f(x + \Delta x) - f(x)}{\Delta x}$	Definition
(7)	$D_x y = D_x f(x) = D_x f$	After Cauchy, 1789–1857
(8)	$f'(x)$	After Lagrange, 1736–1813
(9)	$\dfrac{dy}{dx} = \dfrac{df(x)}{dx}$	After Leibniz, 1646–1716

$D_x y$

In this book we shall use the notations (7) and (8) but shall avoid (9). You will often run across (9) in books on science and engineering, but it is basically misleading. It gives every appearance of being a fraction, but it is *not* a fraction—it is the limit of a fraction. The various terms dy, dx, $df(x)$ have no separate meanings, but students (and others who should know better) sometimes are misled and ascribe meaning to them.

The following statements are easy to prove.

THEOREM 2 $D_x cf = c \cdot D_x f$, c constant

THEOREM 3 $D_x(f \pm g) = D_x f \pm D_x g$

Derivative of a sum

EXERCISE B Prove Theorems 2 and 3.

EXERCISE C Show that

$$D(px^2 + qx + r) = p\ Dx^2 + q\ Dx + rD1$$

We now turn to the problem of differentiating systematically the various nonnegative integral powers of x, namely, $1 = x^0, x, x^2, \ldots, x^n, \ldots$. Draw the associated figure, and remember that we are calling

$$\Delta f = f(x + \Delta x) - f(x)$$

THEOREM 4 $D_x 1 = 0$

PROOF

$$f(x) = 1$$
$$f(x + \Delta x) = 1$$

$$\Delta f = 0$$

$$\frac{\Delta f}{\Delta x} = 0 \qquad \Delta x \neq 0$$

$$D_x f = \lim_{\Delta x \to 0} \frac{\Delta f}{\Delta x} = 0$$

EXERCISE D Prove that for a constant C,

$$D_x C = 0$$

THEOREM 5 $D_x x = 1$

PROOF

$$f(x) = x$$
$$f(x + \Delta x) = x + \Delta x$$
$$\Delta f = \Delta x$$
$$\frac{\Delta f}{\Delta x} = \frac{\Delta x}{\Delta x} = 1 \qquad \Delta x \neq 0$$
$$D_x f = \lim_{\Delta x \to 0} \frac{\Delta f}{\Delta x} = 1$$

THEOREM 6 $D_x x^2 = 2x$

PROOF

$$f(x) = x^2$$
$$f(x + \Delta x) = (x + \Delta x)^2 = x^2 + 2x\,\overline{\Delta x} + \overline{\Delta x}^2$$
$$\Delta f = 2x\,\overline{\Delta x} + \overline{\Delta x}^2$$
$$\frac{\Delta f}{\Delta x} = 2x + \Delta x \qquad \Delta x \neq 0$$
$$D_x f = \lim_{\Delta x \to 0} \frac{\Delta f}{\Delta x} = 2x$$

THEOREM 7 $D_x x^3 = 3x^2$

PROOF

$$f(x) = x^3$$
$$f(x + \Delta x) = (x + \Delta x)^3 = x^3 + 3x^2\,\overline{\Delta x} + 3x\,\overline{\Delta x}^2 + \overline{\Delta x}^3$$
$$\Delta f = 3x^2\,\overline{\Delta x} + 3x\,\overline{\Delta x}^2 + \overline{\Delta x}^3$$
$$\frac{\Delta f}{\Delta x} = 3x^2 + 3x\,\overline{\Delta x} + \overline{\Delta x}^2 \qquad \Delta x \neq 0$$
$$D_x f = \lim_{\Delta x \to 0} \frac{\Delta f}{\Delta x} = 3x^2$$

THEOREM 8 $D_x x^4 = 4x^3$

PROOF $f(x) = x^4$

$$f(x + \Delta x) = (x + \Delta x)^4 = x^4 + 4x^3\overline{\Delta x} + 6x^2\overline{\Delta x}^2 + 4x\overline{\Delta x}^3 + \overline{\Delta x}^4$$

$$\Delta f = 4x^3\overline{\Delta x} + 6x^2\overline{\Delta x}^2 + 4x\overline{\Delta x}^3 + \overline{\Delta x}^4$$

$$\frac{\Delta f}{\Delta x} = 4x^3 + 6x^2\overline{\Delta x} + 4x\overline{\Delta x}^2 + \overline{\Delta x}^3 \qquad \Delta x \neq 0$$

$$D_x f = \lim_{\Delta x \to 0} \frac{\Delta f}{\Delta x} = 4x^3$$

We have passed over such questions as the $\lim_{\Delta x \to 0} 6x^2\overline{\Delta x}$, $4x\overline{\Delta x}^2$, etc. By now it must be clear that these go to zero with Δx for any given fixed x.

You might want to guess what the derivatives of the higher powers would be. A reasonable guess is that $D_x x^n = nx^{n-1}$ for n a positive integer. We shall prove this below.

THEOREM 9 $D_x x^n = nx^{n-1}$, when n is a positive integer.

$D_x x^n$

PROOF We proceed by induction. We know that, for $n = 1$,

$$D_x x^1 = D_x x = 1$$

Thus the formula is verified. We must now prove: for all $k \geq 1$, if

$$D_x x^k = kx^{k-1}$$

then

$$D_x x^{k+1} = (k + 1)x^k$$

Now $x^{k+1} = x^k \cdot x$. To find its derivative, we consider

$$\begin{aligned}
\lim_{\Delta x \to 0} \frac{(x + \Delta x)^{k+1} - x^{k+1}}{\Delta x} &= \lim_{\Delta x \to 0} \frac{(x + \Delta x)^k(x + \Delta x) - x^k x}{\Delta x} \\
&= \lim_{\Delta x \to 0} \frac{[(x + \Delta x)^k - x^k]x}{\Delta x} \\
&\qquad + \lim_{\Delta x \to 0} \frac{(x + \Delta x)^k \cdot \Delta x}{\Delta x} \\
&= (D_x x^k)x + x^k \\
&= kx^{k-1} \cdot x + x^k \\
&= (k + 1)x^k
\end{aligned}$$

From the induction axiom it then follows that the theorem is true for all positive integers n. As a matter of fact, we state without proof that Theorem 9 holds for any real value of the exponent n. Memorize:

(10) $$D_x x^n = nx^{n-1} \qquad n \text{ a real number}$$

EXERCISE E Prove Theorem 9 by the Δ-process, using the Binomial Theorem (Sec. 3.4). Model your proof on that of Theorem 8.

PROBLEMS 16.4 In Probs. 1 to 12, find $D_x y$.

1 $y = 4x^6 + 2x + x^{1/2},\ x \geq 0$

2 $y = \frac{3}{4}x^7 + 2x^2 + \frac{3}{x^2},\ x \neq 0$

3 $y = \frac{-3}{x^3} - \frac{x^3}{3},\ x \neq 0$

4 $y = (Ax^{\sqrt{2}} + B)$

5 $y = (x + 1)(x + 2)$

6 $y = \frac{(2x^2 - 10x)}{3x^4},\ x \neq 0$

7 $y = \frac{-x^3 - x^2 + x + 1}{1 - x},\ x \neq 1$

8 $y = x + \frac{x^2}{2} + \frac{x^3}{3} + \frac{x^4}{4}$

9 $y = (\frac{1}{2} - x)^3$

10 $y = \sum_{k=0}^{n} a_k x^k,\ a_n \neq 0$

11 $y = \frac{x^4 + x^3 + 2x^2 + 3x + 1}{1 + x},\ x \neq -1$

12 $y = \frac{x^2 - 3x - 10}{x - 5},\ x \neq 5$

In Probs. 13 to 20, find the slope of the tangent to each curve at the point indicated and the equation of the tangent at that point.

13 $y = cx^{\pi} - (x^{\pi})^2$, at (0,0)

14 $y = x^4 - 3x^2$, at (1,−2)

15 $y = 4(\sqrt[4]{x})$, at (1,4)

16 $y = \frac{x^2}{2} - 2x^{1/2}$, at (4,4)

17 $y = ax^2 + bx + c$, at (0,c)

18 $y = 4x^2 - \frac{1}{x^2}$, at (1,3)

19 $y = x^2 - 5x + 6$, at (3,0)

20 $y = x^{2/3} + 5$, at (1,6)

21 Find the equation of the tangent to $y = \frac{x^3}{3} + \frac{x^2}{2} + x + 1$ where $x = 2$.

22 Find the equation of the tangent to $y = -x + \frac{x^3}{6} - \frac{x^5}{120}$ where $x = 6$.

23 Find **(a)** the equation of the tangent and **(b)** the equation of the line perpendicular to the tangent at (0,0) on $y = x^2$.

24 Find **(a)** the equation of the tangent and **(b)** the equation of the line perpendicular to the tangent at (2,25) on $y = 3x^3 + x^2 - x - 1$.

In Probs. 25 to 30, given distance y ft as a function of time t sec, find the velocity at the time indicated.

25 $y = \frac{At^3}{3} + \frac{Bt^2}{2} + Ct,\ t = x_0$ **26** $y = 2t^3 - t^5 + 6,\ t = 4$

27 $y = 18t^6 - 64,\ t = t_0$ **28** $y = t^m + x_0,\ t = x_0$

29 $y = 3t^5 - t + 3,\ t = 1$ **30** $y = t^{1/2} - t^2,\ t = 9$

In Probs. 31 to 36 given velocity v ft/sec as a function of time t, find the acceleration at the time indicated.

31 $v = t^2 + t + 1,\ t = x_0$ **32** $v = 2t^{10} - t^3,\ t = 1$

33 $v = k,\ t = t_0,\ k$ a constant **34** $v = 6t^5 - t,\ t = 2$

35 $v = m - t^n,\ t = m^{1/(n-1)}$ **36** $v = t - 3t^{16},\ t = 0$

37 From the definition of derivative, find $D_x y$ for $y = x^{-3}$. Hence show that the general rule applies.

38 From the definition of derivative, find $D_x y$ for $y = x^{-n}$, where n is a positive integer. Hence show that the same rule applies for either positive or negative powers of x. HINT: Use mathematical induction.

39 Illustrate with an example to show that, in general, $D(f \cdot g) \neq Df \cdot dG$. (The derivative of a product is not the product of the derivatives.)

***40** From first principles derive the formula for the derivative of a product: $D(f \cdot g) = f\,Dg + g\,Df$.

41 Illustrate with an example to show that, in general, $D(f/g) \neq Df/Dg$. (The derivative of a quotient is not the quotient of the derivatives.)

***42** From first principles derive the formula for the derivative of a quotient: $D(f/g) = (g\,Df - f\,Dg)/g^2$.

16.5 › SECOND DERIVATIVE

Since $D_x f(x) = f'(x)$ is itself a function f' of x it has a derivative, namely,

$$D_x(D_x f(x)) = \lim_{\Delta x \to 0} \frac{f'(x + \Delta x) - f'(x)}{\Delta x}$$

provided this limit exists. We write

$$D_x^2 f(x) = \lim_{\Delta x \to 0} \frac{f'(x + \Delta x) - f'(x)}{\Delta x} = f''(x)$$

and call this the second derivative of f with respect to x at the point x. The superscript 2 on D is not a square; it stands for the *second* derivative.

Where $y = f(x)$, we may write $\dfrac{d^2y}{dx^2}$ for $f''(x)$. Still higher derivatives could be written:

$$D_x^3 f, \ldots, D_x^n f \quad \text{or} \quad \frac{d^3 f}{dx^3}, \ldots, \frac{d^n f}{dx^n} \quad \text{or} \quad f'''(x), \ldots, f^{(n)}(x)$$

We have already seen that for motion in a straight line, velocity is the derivative of distance with respect to time and that acceleration is the derivative of velocity with respect to time. Therefore acceleration is the second derivative of distance with respect to time. Thus if $y = f(t)$ is the distance from the origin at any time t:

Distance: $y = f(t)$
Velocity: $v(t) = D_t y = f'(t)$
Acceleration: $a(t) = D_t^2 y = f''(t)$

ILLUSTRATION 1 If the distance y from the origin at time t is given by $y = -16t^2 + 3{,}000t + 50{,}000$ find:

a The initial distance, i.e., the value of y when $t = 0$
b The velocity at any time t and the initial velocity
c The acceleration at any time t and the initial acceleration

SOLUTION
a $y(0) = 50{,}000$
b $v(t) = -32t + 3{,}000$
$v(0) = 3{,}000$
c $a(t) = -32$
$a(0) = -32$

16.6 » THE CHAIN RULE

We now know how to differentiate a monomial term of the form x^n, where n is any real number, and a polynomial function provided it is given, in form, by $y = a_0x^n + a_1x^{n-1} + \cdots + a_n$. If, however, it were given in some other form, we might not know, at this stage, how to differentiate it—at least without a lot of work. For example, $y = (x^2 + 3)^{37}$ defines a polynomial function (expand by the Binomial Theorem), but we should like to find the derivative directly without carrying out this expansion. We can do so by using the following theorem which is a special case of a more general theorem known as the *Chain Rule.*

THEOREM 10 Let u be a function of x, and let n be any real number. Let $y = [u(x)]^n$. Then

(1) $$D_x y = D_x[u(x)]^n = n[u(x)]^{n-1} \cdot D_x u(x)$$

PROOF We give the proof only for the case where n is a positive integer, and proceed by induction. The theorem is trivially true for $n = 1$. We must show that for all $k \geq 1$, if

$$D_x[u(x)]^k = k[u(x)]^{k-1} \cdot D_x u(x)$$

then

$$D_x[u(x)]^{k+1} = (k+1)[u(x)]^k \cdot D_x u(x)$$

Now

$$\begin{aligned} D_x[u(x)]^{k+1} &= D_x[[u(x)]^k \cdot u(x)] \\ &= D_x[u(x)]^k \cdot u(x) + [u(x)]^k \cdot D_x u(x) \quad \text{[Prob. 40, Sec. 16.4]} \\ &= [k[u(x)]^{k-1} \cdot D_x u(x)] \cdot u(x) + [u(x)]^k \cdot D_x u(x) \\ &= k[u(x)]^k \cdot D_x u(x) + [u(x)]^k \cdot D_x u(x) \\ &= (k+1)[u(x)]^k \cdot D_x u(x) \end{aligned}$$

This completes the proof. Memorize (1).

ILLUSTRATION 1 Find the derivative of $(x^3 + 6x - 1)^{17}$ with respect to x.

SOLUTION Think of $x^3 + 6x - 1$ as u; then

$$D_x(x^3 + 6x - 1)^{17} = 17(x^3 + 6x - 1)^{16}(3x^2 + 6)$$

ILLUSTRATION 2 Find the derivative $D_x y$ of $y = u^3 + u - 5$, where $u = x^2 + 6x$.

SOLUTION We are asked to find

$$\begin{aligned} D_x[(x^2 + 6x)^3 + (x^2 + 6x) - 5] &= D_x(x^2 + 6x)^3 + D_x(x^2 + 6x) - D_x(5) \\ &= 3(x^2 + 6x)^2(2x + 6) + (2x + 6) - 0 \\ &= [3(x^2 + 6x)^2 + 1](2x + 6) \end{aligned}$$

ILLUSTRATION 3 Find the derivative of $\sqrt{4 - x^2}$ with respect to x.

SOLUTION Since $\sqrt{4 - x^2} = (4 - x^2)^{1/2}$, we have from Theorem 10, where $u(x) = 4 - x^2$, that

$$\begin{aligned} D_x\sqrt{4 - x^2} &= \tfrac{1}{2}(4 - x^2)^{-1/2} \cdot (-2x) \\ &= \frac{-x}{\sqrt{4 - x^2}} \end{aligned}$$

ILLUSTRATION 4 Find the equation of the tangent to $(x^2/a^2) - (y^2/b^2) = 1$ (a hyperbola) at the point (x_1, y_1). Note that this implies that $|x_1| \geq a$.

SOLUTION This equation defines a relation, and not a function. In order to differentiate, we must consider one of the functions which can be derived from this relation. We may choose either

$$y = +\frac{b}{a}\sqrt{x^2 - a^2} \qquad \text{or} \qquad y = -\frac{b}{a}\sqrt{x^2 - a^2}$$

The domain of definition of each function is $|x| \geq a$. The given point on the hyperbola (x_1,y_1) will satisfy exactly one of these equations, and we then select this one as the definition of a function:

$$y = f(x)$$

With this definition of $f(x)$, the function given by

$$F(x) = \frac{x^2}{a^2} - \frac{[f(x)]^2}{b^2} - 1$$

has the value zero for all x such that $|x| \geq a$. Its derivative, $F'(x)$, must also be zero. Hence

$$F'(x) = \frac{2x}{a^2} - \frac{2f(x)f'(x)}{b^2} = 0$$

Solving, we find

$$f'(x) = \frac{b^2}{a^2}\frac{x}{f(x)} = \frac{b^2}{a^2}\frac{x}{y}$$

Hence the slope of the tangent to the hyperbola at (x_1,y_1) is

$$m = \frac{b^2}{a^2}\frac{x_1}{y_1}$$

Hence the equation of the tangent is

$$y - y_1 = \frac{b^2}{a^2}\frac{x_1}{y_1}(x - x_1)$$

or, simplifying,

$$a^2yy_1 - a^2{y_1}^2 = b^2xx_1 - b^2{x_1}^2$$

or

$$b^2xx_1 - a^2yy_1 = b^2{x_1}^2 - a^2{y_1}^2$$

Since the right-hand member is a^2b^2 [the point (x_1,y_1) is on the hyperbola, and therefore the coordinates satisfy its equation], we have

$$b^2xx_1 - a^2yy_1 = a^2b^2$$

or, finally,

$$\frac{xx_1}{a^2} - \frac{yy_1}{b^2} = 1$$

PROBLEMS 16.6 In Probs. 1 to 8 given distance y ft as a function of time t sec, find the velocity and acceleration at any time t_0:

1 $y = t^3 - t + t^{1/2}$

2 $y = (8t + 1)^2$

3 $y = (t^2 + 2)(t + 1)$

4 $y = (t^2 + 1)(t^2 - 1)$

5 $y = (t + 2)^2 + (t - 2)^2$

6 $y = (t^2 + 3)^2 + 1$

7 $y = (t - 3)^{1/2} + 2t^2$

8 $y = (t^2 - 3)^{3/2}$

In Probs. 9 to 18 find the first, second, and third derivatives with respect to t.

9 $y = t^{1/2} + t^2$

10 $y = (t + 1)^3$

11 $y = (t^2 + 1)^3$

12 $y = [(3t)^2 + 1]^2$

13 $y = t - 1$

14 $y = 1 - 2t$

15 $y = (t + t^2)^2$

16 $y = t + t^2 + t^3$

17 $y = (2t + 3)^{1/4}$

18 $y = t - (t + 1)^{1/2}$

In Probs. 19 to 26 find $D_x y$ by the Chain Rule.

19 $y = (x^2 + 5x + 25)^3$

20 $y = (x^{-2} + x^2)^2$

21 $y = 2t^3$ where $t = 6x^2 - 3x$

22 $y = v^2$ where $v = \dfrac{1}{x^2}$

23 $y = (x^2 + x^3)^{2/5}$

24 $y = \left(x^2 - \dfrac{1}{x^3}\right)^{7/3}$

25 $y = (b - x^2)^{1/4}$

26 $y = (2x + x^2)^{2/3}$

27 Find $D_t^2 y$ and $D_t^3 y$ where $y = 13t^2 - 7t + 8$.

28 Find $D_t^2 y$ and $D_t^3 y$ at $x = 1$ where $y = 4t^3 + 6t^2 - 8t + 5$.

29 Find $D_x^2 y$ and $D_x^3 y$ at $x = 1$ where $y = -3x^3 - 8x^2 + 2x - 3$.

30 Find $D_x y$ and $D_x^2 y$ at $x = a$ where $y = x^2 + 4x - 7$.

16.7 › MAXIMA AND MINIMA

In this section we apply the ideas of the calculus to help us draw the graphs of certain functions.

DEFINITION A function f is said to be *increasing* at the point x_0 if, for all $|\Delta x|$ sufficiently small,

Increasing function

(1)
$$f(x_0 + \Delta x) < f(x_0) \qquad \text{when } \Delta x < 0$$
$$f(x_0 + \Delta x) > f(x_0) \qquad \text{when } \Delta x > 0$$

A function is increasing in an interval if it is increasing at each point of the interval. As x traces such an interval in the positive direction, the graph of $y = f(x)$ rises.

THEOREM 11 If $f'(x_0) > 0$, then f is increasing at x_0.

PROOF Given

$$f'(x_0) = \lim_{\Delta x \to 0} \frac{f(x_0 + \Delta x) - f(x_0)}{\Delta x} > 0$$

If in the limit the ratio $\frac{f(x_0 + \Delta x) - f(x_0)}{\Delta x}$ is positive, then, for sufficiently small $\Delta x < 0$, it must be true that $f(x_0 + \Delta x) - f(x_0) < 0$. That is, $f(x_0 + \Delta x) < f(x_0)$, which is the first part of condition (1). Again if $\Delta x > 0$ and is small, it must be true that

$$f(x_0 + \Delta x) - f(x_0) > 0$$

and the second condition of (1), $f(x_0 + \Delta x) > f(x_0)$, is satisfied. Hence the theorem is proved.

EXERCISE A State and prove a partial converse of Theorem 11 for a differentiable function f. [If f is increasing at x_0, then $f'(x_0) \geq 0$.]

EXERCISE B Write out a definition of decreasing function and a theorem (and a partial converse) corresponding to Theorem 11.

Consider the curve $y = f(x)$, where f is a differentiable function (Fig. 16.9). The value $f(x_1)$ is the largest that the function f assumes in a small interval containing x_1. Such a value of the function f is called a relative maximum of f. Similarly $f(x_2)$ is called a relative minimum of f. We often omit the adjective "relative", but it will still be understood. At each point of a suitably small interval to the left of x_1, the derivative $f'(x) > 0$ (Exercise A). At x_1, the derivative $f'(x_1) = 0$. At each point of a small interval to the right of x_1, the derivative $f'(x) < 0$ (Exercise B).

EXERCISE C What are the corresponding facts for small intervals to the left and to the right of x_2?

Relative maxima and minima and stationary points

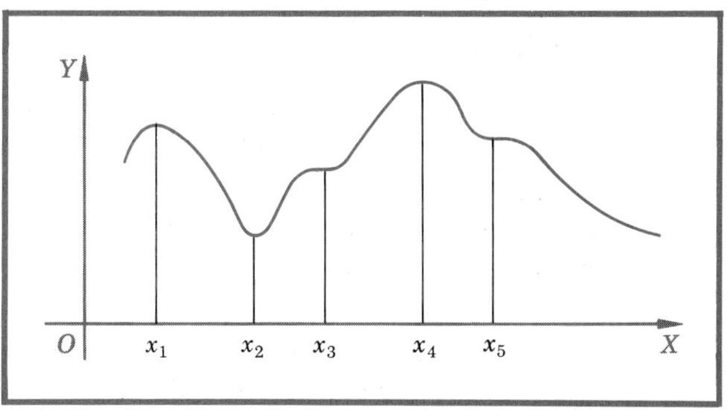

FIGURE 16.9 A point like x_1 and x_4 which is higher than its immediate neighbors is a relative maximum. A point like x_2 is a relative minimum if it is lower than its immediate neighbors. A point like x_3 and x_5 at which the tangent is horizontal but which is neither a relative maximum nor minimum is a stationary point.

DEFINITIONS
Relative maximum, minimum

The point where $x = x_0$ is called a (relative) *maximum* of the function f if and only if $f(x_0 \pm \Delta x) < f(x_0)$ for all sufficiently small values of $\Delta x \neq 0$. The point where $x = x_0$ is called a (relative) *minimum* of the function f if and only if $f(x_0 \pm \Delta x) > f(x_0)$ for all sufficiently small values of $\Delta x \neq 0$.

As an aid to finding (relative) maxima and minima we prove the next theorem.

THEOREM 12 If f is differentiable for all values of x, then $f'(x) = 0$ at the (relative) maxima and minima of f.

PROOF At any point, $f'(x) > 0$, $= 0$, or < 0. If $f'(x) > 0$ at $x = x_0$, Theorem 11 tells us that $f(x)$ is increasing at this point. Hence x_0 is neither a maximum nor a minimum. Similarly, by Exercise B, $f'(x)$ cannot be < 0 at a maximum or minimum. Therefore it must be zero.

The converse of Theorem 12 is false, for $f'(x)$ can be zero at points which are neither maxima nor minima. Such points are called stationary points. In Fig. 16.9, x_3 and x_5 are such points.

DEFINITIONS
Stationary, critical points

The point $x = x_0$ is called a *stationary point* of f if $f'(x_0) = 0$ and either $f'(x_0 \pm \Delta x) > 0$ or $f'(x_0 \pm \Delta x) < 0$ for all small Δx. Any point $x = x_0$ for which $f'(x_0) = 0$ is called a *critical point* of f.

In Fig. 16.9, x_3 and x_5 are stationary points, while x_1, x_2, x_3, x_4, and x_5 are critical points.

EXERCISE D For the differentiable function f to have a maximum at $x = x_0$, it is necessary and sufficient that . . . . Complete so as to form a meaningful theorem.

EXERCISE E In order for the differentiable function f to have a minimum at $x = x_0$, it is necessary and sufficient that . . . . Complete so as to form a meaningful theorem.

EXERCISE F On what basis have we ruled out relative maxima and minima as exhibited in the graph in Fig. 16.10?

We summarize with a rule as follows.

RULE FOR FINDING THE MAXIMUM (MINIMUM) VALUE OF A FUNCTION First, find the function to be maximized! This function may be given. Again

Points at which derivative does not exist

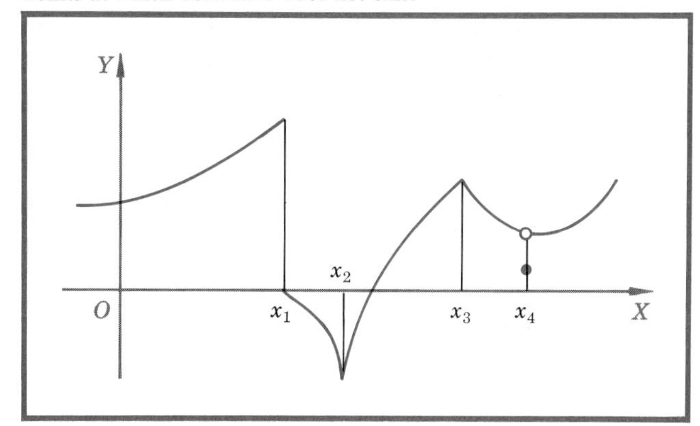

FIGURE 16.10
If f is not differentiable for some values of x, there may be relative maxima and minima at points where $f'(x)$ does not exist.

the statement of the original problem may be in words, and you will then have to translate these into the appropriate mathematical expressions. You may have to differentiate some given function several times, or you may have to perform other operations on given quantities, but, regardless of what the operations are, you must first find the function whose maximum (minimum) is sought. Call this function f.

Second, find $f'(x)$. The solutions of $f'(x) = 0$ are the critical values, and they must be tested in order to determine whether a certain one yields a maximum value of f, a minimum value of f, or a stationary value of f.

In Table 1 the test may be made by using $f(x)$ or by using $f'(x)$ as indicated. In this table α and β are used to designate certain positive constants; each plays the role of a Δx to be chosen so as to simplify the test.

EXERCISE G Make out a similar table for a stationary point.

TABLE 1
TESTING FOR MAXIMA AND MINIMA

x	$f(x)$	$f'(x)$	*Comments*
x_0	$f(x_0)$	Given $f'(x_0) = 0$	Testing $f(x_0)$ for a maximum;
$x_0 - \alpha$ $x_0 + \beta$	$f(x_0 - \alpha) < f(x_0)$ $f(x_0 + \beta) < f(x_0)$	$f'(x_0 - \alpha) > 0$ $f'(x_0 + \beta) < 0$	$\therefore f(x_0)$ is a relative maximum [because of the inequalities in either the $f(x)$ or the $f'(x)$ column]
x_0	$f(x_0)$	Given $f'(x_0) = 0$	Testing $f(x_0)$ for a minimum;
$x_0 - \alpha$ $x_0 + \beta$	$f(x_0 - \alpha) > f(x_0)$ $f(x_0 + \beta) > f(x_0)$	$f'(x_0 - \alpha) < 0$ $f'(x_0 + \beta) > 0$	$\therefore f(x_0)$ is a relative minimum [because of the inequalities in either the $f(x)$ or the $f'(x)$ column]

CAUTION: The interval $[x_0 - \alpha, x_0 + \beta]$ must not be so large as to include other critical values of f or of f'

Let us now apply the above reasoning to the function f', whose values give the slopes of the various tangents to $y = f(x)$. When $f'(x)$ is increasing, $f''(x) > 0$; and geometrically the tangents are rotating counterclockwise as x increases. When $f'(x)$ is decreasing, $f''(x) < 0$; and geometrically the tangents are rotating clockwise as x increases. When $f'(x)$ has a (relative) maximum or minimum, there is a change in the direction of rotation of the tangents. Such a point is called a point of inflection.

Inflection

DEFINITION A point at which f' has a (relative) maximum or minimum is called a *point of inflection.*

THEOREM 13 If f is twice differentiable for all values of x, then $f''(x) = 0$ at its points of inflection.

The proof is an immediate consequence of the definition and Theorem 12. The converse of Theorem 13 is false, for $f''(x)$ may be zero at points of other types as well.

In Fig. 16.11, x_1, x_2, x_3 are points of inflection. At x_1 the slope $f'(x_1)$ is a minimum, while at x_2 the slope $f'(x_2)$ is a maximum. At x_3 the slope $f'(x_3)$ has a minimum value of 0, and hence x_3 is also a stationary point.

When we have gained information about the points where f is stationary, about the maximum and minimum values of f, and about the points of inflection, we are in a better position to plot the curve. Hence the calculus is a powerful tool indeed in curve tracing.

ILLUSTRATION 1 Sketch the graph of $y = 2x^3 + 3x^2 - 12x$.

Inflection points

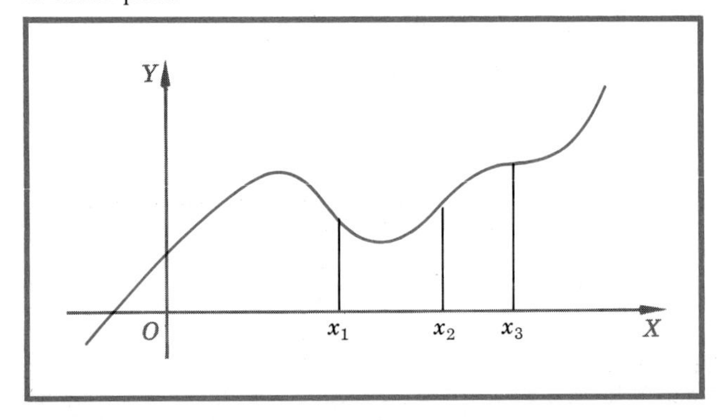

FIGURE 16.11
At a point of inflection, such as x_1, x_2, and x_3, f' has a relative maximum or minimum.

SOLUTION The zeros of the polynomial $2x^3 + 3x^2 - 12x$ are

$$x = 0 \qquad x = -\frac{3}{4} \pm \frac{\sqrt{105}}{4}$$

The domain of definition is $-\infty < x < \infty$. There is no symmetry, and the function is everywhere continuous. We find $f'(x)$ and set $f'(x) = 0$.

$$f'(x) = 6x^2 + 6x - 12 = 0$$

that is,

$$x^2 + x - 2 = (x - 1)(x + 2) = 0$$
$$x = 1, \ -2$$

These are the critical points which must be tested.

Further, $f''(x) = 12x + 6 = 0$ when solved for $x = -\frac{1}{2}$ gives the point of inflection which is $(-\frac{1}{2}, 6\frac{1}{2})$. We compute a few more values of the function (Table 2) and sketch in Fig. 16.12.

TABLE 2

x	$f(x)$	$f'(x)$	*Comments*
1	$f(1) = -7$	$f'(1) = 0$	Testing $f(1)$;
Set $x_0 - \alpha = 1 - 1 = 0$	$f(0) = 0 > -7$	$f'(0) = -12 < 0$	$\therefore f(1) = -7$
Set $x_0 + \beta = 1 + 1 = 2$	$f(2) = 4 > -7$	$f'(2) = 24 > 0$	is a relative minimum
-2	$f(-2) = 20$	$f'(-2) = 0$	Testing $f(-2)$;
Set $x_0 - \alpha = -2 - 1 = -3$	$f(-3) = 9 < 20$	$f'(-3) = 24 > 0$	$\therefore f(-2) = 20$
Set $x_0 + \beta = -2 + 2 = 0$	$f(0) = 0 < 20$	$f'(0) = -12 < 0$	is a relative maximum

$y = 2x^3 + 3x^2 - 12x$

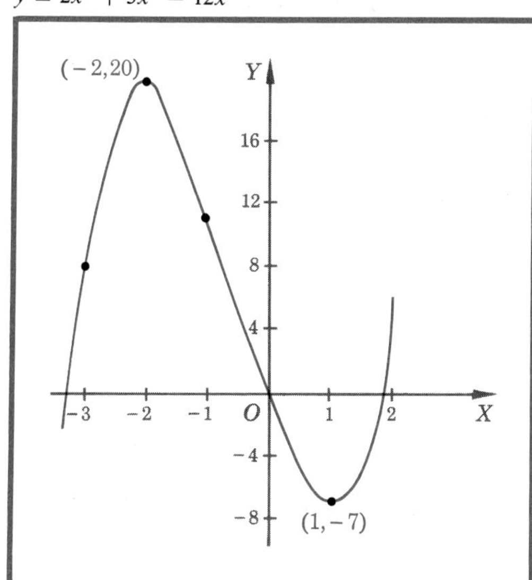

FIGURE 16.12
See Illustration 1.

ILLUSTRATION 2 Prove that among all rectangles with fixed perimeter P the square is the one with maximum area.

SOLUTION Call the sides of the general rectangle x and y. Then

$$P = 2x + 2y \tag{2}$$

The quantity to be maximized is the area A, where

$$A = xy$$

We cannot yet proceed to differentiate A, however, because there are two variables momentarily present, namely, x and y. But, using relation (2), we may eliminate either x or y. From (2),

$$y = \frac{P}{2} - x$$

so that $A = xy$ becomes

$$A = x\left(\frac{P}{2} - x\right)$$

Now we may proceed.

$$A'(x) = \frac{P}{2} - 2x$$

The critical value is obtained by setting

$$A'(x) = \frac{P}{2} - 2x = 0$$

and solving; hence $x = P/4$. Using this, we find, from (2), that $y = P/4$; also that the rectangle has equal sides, i.e., is a square.

We still do not know (except intuitively) that these values correspond to a maximum; we must test (Table 3).

TABLE 3

x	$A(x)$	$A'(x)$	*Comments*
$\frac{P}{4}$	This test will not be used in this problem	$A'\left(\frac{P}{4}\right) = 0$	Testing $A\left(\frac{P}{4}\right)$;
$\frac{P}{4} - \alpha = 0$		$A'(0) = \frac{P}{2} > 0$	$A\left(\frac{P}{4}\right) = \frac{P^2}{16}$ is a relative maximum given by $x = y = \frac{P}{4}$;
$\frac{P}{4} + \beta = \frac{P}{2}$		$A'\left(\frac{P}{2}\right) = -\frac{P}{2} < 0$	$\therefore$ rectangle of maximum area is a square

PROBLEMS 16.7 In Probs. 1 to 24 find all maxima, minima, and points of inflection; and plot the curve.

1 $y = x^2 + 4x + 4$
2 $y = 5x^2 - 8x + 2$
3 $y = -5x^2 + 8x - 2$
4 $y = x^2 + 4$
5 $y = 2x^2 - 4x + 1$
6 $y = x^2 + x + 1$
7 $y = Ax^3, A > 0$
8 $y = 16x^4$
9 $y = 2x^2 - 3x^3$
10 $y = 4x + 2x^2 + x^3$
11 $y = x^4 - 2x^2$
12 $y = -x^3 + x^4$
13 $y = x^5$
14 $y = x^4 + x^3 + x^2$
15 $y = \sqrt{2 + x^2}$
16 $y = \sqrt{x^2 - 4}$
17 $y = (x + 3)^{1/3}$
18 $y = (2x + 1)^{7/3}$
19 $y = \frac{1}{4}\sqrt{16 - x^2}$
20 $y = \frac{1}{16}\sqrt{x^2 - x}$
21 $y = (x - 1)(x - 2)(x - 3)$
22 $y = 2x^3 - 7x^2 + 3x - 2$
23 $y = 3x^3 - 9x^2 + 9x - 3$
24 $y = 3x^3 + 18x^2 + 6$

In Probs. 25 to 30 find all points of inflection.

25 $y = x^3 - 6x^2 - 8x + 2$
26 $y = 2x^3 - 6x^2 + 6x - 1$
27 $y = 4x^3 - 4x + 1$
28 $y = 3x^4 + 18x^2 - 2$
29 $y = -x^4 + 2x^3 - 3x$
30 $y = 3x^3 - 2x^2 - 5x$

In Probs. 31 to 34, find all minimum points.

31 $y = \dfrac{x^3}{3} + \dfrac{x^2}{2} + x + 1$
32 $y = x^4 + x^2 + 4$
33 $y = 3x^5 + 4x^4 + 5x^3$
34 $y = x^6 + 4x^3 + 4$

35 If the velocity of a particle is given by $v = t^2 + 4$, find the minimum velocity.

36 If the acceleration of a particle is given by $a = t^2 - t - 30$, find the minimum acceleration.

37 If the height y of a particle is given by $3t^4 + 2t^3 + t^2 - t + 1$, find the minimum acceleration.

38 If the height y of a particle is given by $y = \dfrac{2t^3}{3} - t^2 - 12t$, find:

- **a** The time t_1, when it stops for the first time
- **b** The time t_2, when it stops for the second time
- **c** The velocity at t_1 and t_2
- **d** The acceleration at t_1 and t_2
- **e** The time when the velocity is minimum
- **f** The time when the acceleration is zero

39 Prove that among all rectangles with fixed area A, the square is the one with minimum perimeter.

40 A man has P running feet of chicken wire, and with it wishes to form a rectangular pen, making use of an existing stone wall as one side. Find the dimensions so that the pen will have maximum area.

41 A watermelon grower can ship 15 tons now at a profit of \$5 per ton. By waiting, he can add 3 tons/week, but his profit will be reduced 50 cents/ton/week. When should he ship for maximum profit?

42 **(a)** Find the relative dimensions of a closed tin can (cylindrical) to be made from a given amount of metal (without losses in cutting, etc.) that will have maximum volume. **(b)** Find the same for an open tin can.

43 A man in a boat 3 miles from the nearest point P on the shore (a straight line) wishes to reach a point Q also on the shore 5 miles from P. Where should he land (between P and Q) to minimize his travel time, if he can row 2 miles/hr and walk 4 miles/hr? Now do the problem again, considering that Q is 50 miles along the shore from P. Again, for one mile from P.

44 Find the absolute maximum and absolute minimum of f where $f(x) = x^3 + 10x^2 + 25x$ in the interval $0 \leq x \leq 3$. For what values of x do these occur?

45 Find the absolute maximum and absolute minimum of f where $f(x) = x^3 + 10x^2 + 25x$ in the interval $-5 \leq x \leq 0$. For what values of x do these occur?

46 Let f be a function such that $f'(x)$ exists for all x in the interval $a < x < b$. Will every function which satisfies this property have an absolute maximum and an absolute minimum? Give reasons for your answer.

16.8 › RELATED RATES

Theorem 10 gave the formula for $D_x y$, where $y = [u(x)]^n$. Since this has many applications in rate problems involving time as independent variable, we shall replace x by the letter t and write

$$D_t y = n[u(t)]^{n-1} D_t u \tag{1}$$

Usually the variable t does not enter explicitly: we are given either $D_t y$ or $D_t u$ and are asked to solve for the other by using formula (1).

ILLUSTRATION 1 The radius of a circle is increasing at the rate of 2 ft/min. How fast is the area increasing when $r = r$ ft? when $r = 3$ ft?

SOLUTION Evidently we have

$$A = \pi r^2$$

where r is such a function of t that $D_t r$ (given) $= 2$ ft/min. We are asked to find $D_t A$; we therefore differentiate A with respect to t, getting

$$\begin{aligned} D_t A &= 2\pi r\, D_t r \\ &= 4\pi r \text{ ft}^2/\text{min} \end{aligned} \tag{2}$$

which is the first part of the answer. For the second part we substitute $r = 3$ in (2), getting

$$D_t A = 12\pi \text{ ft}^2/\text{min}$$

when $r = 3$ ft.

ILLUSTRATION 2 A man, 100 ft away from the base of a flagpole, starts walking toward the base at 10 ft/sec just as a flag at the top of the pole is lowered at the rate of 5 ft/sec. If the pole is 70 ft tall, find how the distance between the man and the flag is changing per unit of time at the end of 2 sec.

Relationship among x, y, and z

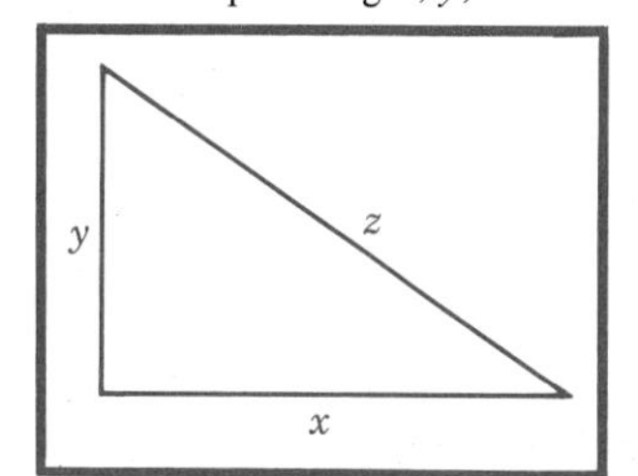

FIGURE 16.13
See Illustration 2.

SOLUTION Call x the distance the man is from the base, y the height of flag, and z the distance between man and flag at any time t (Fig. 16.13). Then we are given

$$D_t x = -10 \text{ ft/sec}$$
$$D_t y = -5 \text{ ft/sec}$$

(The minus sign is present because x and y are decreasing.) Moreover,

(3) $$x = -10t + 100 \qquad y = -5t + 70$$

Now, always we have

(4) $$z^2 = x^2 + y^2$$

and differentiating with respect to time gives

(5) $$2z\, D_t z = 2x\, D_t x + 2y\, D_t y$$

[z^2, x^2, and y^2, each is a u^n problem, Sec. 16.6, formula (1)]. From (3) above we compute $x = 80$ ft, $y = 60$ ft, when $t = 2$; and then, from (4),

$$z = \sqrt{x^2 + y^2} = \sqrt{(80)^2 + (60)^2} = 100 \text{ ft}$$

Hence (5) becomes, canceling the multiplicative factor 2,

$$100\, D_t z = 80(-10) + 60(-5) = -1{,}100$$

or $D_t z = -11$ ft/sec. The minus sign says that the distance z is decreasing at $t = 2$.

PROBLEMS 16.8

1 The edge of a cube is decreasing at 3 in./sec. When the edge is 2 in., find **(a)** how fast the volume is decreasing and **(b)** how fast the surface area is decreasing.

2 The volume of a cube increases at 4 in.^{3}/sec. When the edge of the cube is 3 in., find **(a)** how fast the edge is increasing and **(b)** how fast the surface area is increasing.

3 A baseball diamond is a square 90 ft on each side. If a batter runs down the first-base line at 25 ft/sec, how fast is his distance from third base increasing as he passes first base?

4 A floating man is pulled toward a pier 50 ft above water. The rope is pulled in at a rate of 3 ft/sec. How fast is the man approaching the base of the pier when 130 ft of rope remains to be pulled in?

5 A private jet flying due east at 500 miles/hr passes over a ship traveling due south at 5 miles/hr. How fast are they separating 1 hr later if the jet is 1 mile high?

6 Gas is being pumped into a spherical balloon at 200 ft^3/min. When $r = 6$ ft, find **(a)** how fast r increases and **(b)** how fast the surface area increases.

7 A conical filter has a 15-in. depth and a 6-in. radius. Liquid passes through it at 3 in.^{3}/sec. How fast is the level of the liquid falling when it is 9 in.?

8 A man 6 ft tall walks at 4 miles/hr away from a light which hangs 30 ft above the ground. Find:

- **a** How fast his shadow lengthens.
- **b** How fast the head of his shadow moves away from the light.

9 A rectangle is 6 in. long and 3 in. wide. Its length is increasing at 1 in./sec, and its width is decreasing at 2 in./sec. Find how fast the area is changing.

10 If the variable x is increasing, for what value of x is the rate of increase of x^4 thirty-two times that of x?

16.9 FUNDAMENTAL THEOREM OF CALCULUS

In this chapter we have been concerned so far with differentiation and its applications but there is obviously an inverse problem—that of anti-differentiation. For example, we might be given the derivative $D_x f$ and asked to find f.

ILLUSTRATION 1 Given $D_x f(x) = x^4$; find $f(x)$.

SOLUTION From our knowledge of differentiation we know that, if $f(x) = x^5$, then $D_x f(x) = 5x^4$. Hence $f(x) = \frac{1}{5}x^5$ is an $f(x)$ such that $D_x f(x) = x^4$; but so is $f(x) = \frac{1}{5}x^5 + 7$. As a matter of fact $f(x) = \frac{1}{5}x^5 + C$, where C is a constant, is such that $D_x f(x) = x^4$. The function $f:(x, \frac{1}{5}x^5 + C)$, for any constant C, is called a *primitive* of x^4 according to the following general definition.

Primitive

DEFINITION Any function G defined in an interval I and such that $D_x G(x) = f(x)$ in I is called a *primitive* of f.

Some authors use the phrase "indefinite integral" instead of "primitive" in this connection. For the polynomials the interval I is usually chosen to be the entire real line.

ILLUSTRATION 2 Find a primitive of x^n.

SOLUTION By differentiating x^{n+1} we get $(n+1)x^n$, and therefore a primitive of x^n is $\dfrac{x^{n+1}}{n+1} + C$, for any constant C, provided $n + 1 \neq 0$.

From the illustrations above it should be clear that, if G is a primitive of f, then $G + C$ is another primitive of f. You may well wonder if the set of functions $G + C$ includes all primitives of F, and this is answered in the affirmative by the next theorem, which we give without proof.

THEOREM 14 If F and G are two primitives of a function f for the same interval I, then $F(x) = G(x) + C$ in I.

EXERCISE A Let

$$f(x) = 0 \text{ in } (1,2) \cup (3,4) \qquad \text{The union of two open intervals}$$
$$G_1(x) = 3 \text{ in } (1,2) \cup (3,4)$$
$$G_2(x) = \begin{cases} 4 \text{ in } (1,2) \\ 7 \text{ in } (3,4) \end{cases}$$

Show that $DG_1(x) = f(x)$ and $DG_2(x) = f(x)$, but that

$$G_1(x) \neq G_2(x) + C$$

It is because of examples of this type that primitives are defined only for an interval I.

If the thought has not come to you already it should fairly pop into your mind now: *Antidifferentiation resembles integration!* There seems to be no offhand reason under the sun why there should be a connection between these seemingly independent ideas:

$$\text{(1)} \qquad \text{Integral} = \int_a^x f(u)\,du = \lim_{n\to\infty} \sum_{i=1}^{n} f(u_i)(u_i - u_{i-1})$$

$$\text{(2)} \qquad \text{Derivative} = D_x f = \lim_{\Delta x \to 0} \frac{\Delta f}{\Delta x}$$

But look at the following table for a good comparison.

The function f	*The integral* $F(x) = \int_a^x f(u)\,du$	*The derivative* D_xF
x	$\frac{x^2}{2} - \frac{a^2}{2}$	x
x^2	$\frac{x^3}{3} - \frac{a^3}{3}$	x^2
x^3	$\frac{x^4}{4} - \frac{a^4}{4}$	x^3
x^4	$\frac{x^5}{5} - \frac{a^5}{5}$	x^4
$x^n,\ n \geq 1$	$\frac{x^{n+1}}{n+1} - \frac{a^{n+1}}{n+1}$	x^n

EXERCISE B Integrate in the first column from a to x, and show that the second column is obtained.

EXERCISE C Differentiate in the second column, and show that the third column is obtained.

Thus it turns out that integration and differentiation are essentially inverse to one another, at least for the special functions used. That this situation obtains for an extensive set of functions was an important discovery made by Isaac Newton (1642–1727) in the year 1665. [The set is the set of all *continuous functions.* A function f is said to be *continuous at the point* $x = a$ if (1) $\lim_{x \to a} f(x)$ exists, (2) $f(a)$ exists, and (3) $\lim_{x \to a} f(x) = f(a)$, but we do not pursue the matter.] We state without proof Newton's theorem, now called the Fundamental Theorem of Integral Calculus.

Fundamental Theorem

THEOREM 15 Every integral of a continuous function f is a primitive of f. In other words, if $F(x) = \int_a^x f(u)\,du$, then $D_xF(x) = f(x)$.

The practical importance of this theorem should now become clear. It is a difficult task to find integrals directly from the basic definition since limits of complicated sums must be evaluated. Often it is easier to find a primitive by looking for a function G whose derivative is the given function $f(x)$. We now show that, *if a primitive can be found, then an integral can be obtained from it.*

We are seeking to find $F(x) = \int_a^x f(u)\,du$. Let G be any primitive of f whose domain of definition is an interval I containing a and x. Since

F is also a primitive of f by the above theorem, it follows from Theorem 14 that

$$F(x) = G(x) + C \qquad \text{in } I \tag{3}$$

for a suitable constant C. To find the value of C, put $x = a$ in Eq. (3). Hence $F(a) = G(a) + C$. We note, however, that

$$F(a) = \int_a^a f(u)\,du = 0$$

Hence $\qquad C = -G(a)$

Therefore $\qquad F(x) = G(x) - G(a)$

This gives us a rule for evaluating any integral

$$F(b) = \int_a^b f(u)\,du$$

RULE First, find a primitive $G(x)$ such that $D_x G(x) = f(x)$ for all points in the interval $[a,b]$. Then

$$F(b) = \int_a^b f(u)\,du = G(b) - G(a)$$

In computation we usually write this in the form

$$\int_a^b f(u)\,du = G(x)\Big]_a^b = G(b) - G(a)$$

One final remark should be made about a primitive of f. Our definition says that G is a primitive of f if $D_x G(x) = f(x)$. A common notation for G is

$$G(x) = \int f(x)\,dx + C$$

where C is to be thought of as an arbitrary constant.

The fundamental problem of finding integrals and primitives of given functions can occupy the serious student for several years. The process is known simply as *integration*.

ILLUSTRATION 3 Find the value of $\int_1^2 (x^3 - \frac{1}{2}x^4 + x - 3)\,dx$.

SOLUTION

$$\begin{aligned}\int_1^2 (x^3 - \tfrac{1}{2}x^4 + x - 3)\,dx &= \Big[\tfrac{1}{4}x^4 - \tfrac{1}{10}x^5 + \tfrac{1}{2}x^2 - 3x\Big]_1^2 \\ &= (\tfrac{16}{4} - \tfrac{32}{10} + \tfrac{4}{2} - 6) - (\tfrac{1}{4} - \tfrac{1}{10} + \tfrac{1}{2} - 3) \\ &= -\tfrac{17}{20}\end{aligned}$$

ILLUSTRATION 4 Find a primitive of $6x^2 - 2$.

SOLUTION
$$G(x) = \int (6x^2 - 2)\,dx$$
$$= 2x^3 - 2x + C$$

We can apply this method of integration to solve problems involving distance and velocity.

ILLUSTRATION 5 The velocity of a particle moving on the X-axis is given by $v = 2t - 3t^2 + 1$. At $t = 0$, the particle is at the origin. Where is it when $t = 1$?

SOLUTION We know that $D_t x = v = 2t - 3t^2 + 1$. Thus x is some primitive of $2t - 3t^2 + 1$, or

$$x = t^2 - t^3 + t + C$$

The value of C is obtained by putting $x = 0$, $t = 0$ in this equation and solving for C. We find that $C = 0$; thus

$$x = t^2 - t^3 + t$$

At $t = 1$, $x = 1 - 1 + 1 = 1$ unit from the origin.

ILLUSTRATION 6 Find the area under the curve $y = 3x - x^2$ from $x = 1$ to $x = 2$.

SOLUTION
$$A_1^2 = \int_1^2 (3x - x^2)\,dx$$
$$= \frac{3x^2}{2} - \frac{x^3}{3}\Big]_1^2$$
$$= \frac{3 \cdot 4}{2} - \frac{8}{3} - \left(\frac{3}{2} - \frac{1}{3}\right)$$
$$= \tfrac{13}{6} \text{ square units}$$

PROBLEMS 16.9

In Probs. 1 to 12 a primitive of a certain function is given. Find the function.

1 $2x^3 - 4x + 3x^2$
2 $x^4 + 2x^3 + 4$
3 $9x^2 + 10x^4 - 2x^3$
4 $3x^2 + 2x^3$
5 $4x^{-3} + 2x^{-2}$
6 $x^2 - x^{-2}$
7 $6x^3 - 2x^{-4} + 8$
8 $ax^2 + bx + c$
9 $ax^{-3} + bx^{-2}$
10 $ax^{-4} + bx^3 + cx^{-2} + dx$
11 $Ax^{n+4} + Bx^{n+3} + Cx$
12 $Ax^p + Bx^q$

In Probs. 13 to 26 a function is given. Find a primitive of the function.

13 $3x^2 + 2x - 4$

14 $7x^2 - 2x$

15 $x^4 + 3x^2 - 2$

16 $5x^4 + 3x^2 + 2x$

17 $-x^{-2} + (-x)^2$

18 $3x^{-2} + 4x^{-6} + 2x^{-4}$

19 $x^{-5} + x^5 + 4x^3 - 2x$

20 $4x^3 - 2x^{-3/2} + x^{-3}$

21 $ax^4 + bx^2 - cx^{-3}$

22 $4x^{-3} + bx^2 - cx^{-2}$

23 $ax^{-6} + bx^{-7}$

24 $-3x^{-2} + x^{1/2} + \dfrac{x}{2}$

25 $(p + 1)x^p + B(q + 1)x^q,\ (p, q \neq -1)$

26 $(Ax^3)(Bx^2)$

In Probs. 27 to 42, evaluate:

27 $\int_0^1 (2x + 4x^3)\,dx$

28 $\int_0^1 (x^2 + x^4)\,dx$

29 $\int_{-1}^1 2x^3\,dx$

30 $\int_1^3 (3x^2 + 2x + 1)\,dx$

31 $\int_0^3 (ax^2 + bx + c)\,dx$

32 $\int_0^3 (6x^2 + 3x - 4)\,dx$

33 $\int_a^{a+1} (x - 1)\,dx$

34 $\int_{-b}^b (u^{-3} + u)\,du$

35 $\int_a^{2a} (t^2 - 2t)\,dt$

36 $\int_0^2 (4y^3 + 3y^2)\,dy$

37 $\int_2^6 x^3\,dx$

38 $\int_{-1}^1 (3x + 4)\,dx$

39 $\int_1^6 5t^{-4}\,dt$

40 $\int_1^2 At^{-3}\,dt$

41 $\int_2^4 (-3z^{-2} + z^{-3})\,dz$

42 $\int_1^{10} t + 1\,dt$

In Probs. 43 to 50 find the area bounded by:

43 $y = 3x + 2$
$x = 0,\ y = 3$

44 $y = 3x + 2$
$x = 0,\ x = 1,\ y = 0$

45 $y = 3x^2$
$x = 0,\ x = 1,\ y = 0$

46 $y = 3x^2$
$y = 3x$

47 $y = 4x^3$
$x = -1,\ x = 1,\ y = 0$

48 $y = 4x^3$
$x = \frac{1}{2},\ x = 0,\ y = -\frac{3}{2}x + 3$

49 $y = \dfrac{1}{x^2},\ y = 0$
$x = -2,\ x = -1$

50 $y = \dfrac{-3}{x^3},\ y = 0$
$x = -3,\ x = -2$

51 The distance y from the origin at time t of a particle is given by $y = 3t^2 + t + 1$. Find **(a)** the velocity at time t_0 and **(b)** the velocity at time $t = 0$.

52 The distance y from the origin of a particle at time t is given by $y = 4t^3 + 5t^4 + 3t^2$. Find **(a)** the velocity at time t_0 and **(b)** the acceleration at time t.

53 The velocity v of a particle at time t is given by $v = (3t^2/2) + 2$. Find **(a)** the acceleration at time t_0, **(b)** the acceleration at $t = 0$, and **(c)** the distance traversed between $t = 0$ and $t = 1$.

54 The velocity v of a particle at time t is given by $v = 3t + 2t^2 + t^3$. If the particle is at the origin when $t = 0$, find the distance traversed from $t = 0$ to $t = 1$.

55 PROVE: The rate of change of area [enclosed by $y = f(x)$, $x = a$, $x = x_0$, $y = 0$] per unit change in x, at $x = x_0$, is $f(x_0)$.

56 If the graph of $y = f(x)$ is a curve passing through the point $(-1,2)$ and if $f'(x) = (x - 1) + (x - 2)^2$, what is the exact expression for $f(x)$?

16.10 › FALLING BODIES

As a first approximation to the theory of falling bodies it is customary to disregard the retarding forces due to air friction, etc., and to assume that the only force acting is that of gravity. Under these circumstances, the acceleration will be -32 ft/sec/sec. This figure has been determined empirically by physicists, and the minus sign is supplied so that to a falling body is assigned a negative velocity. Now if we call y the height (in feet) of a particle at time t (in seconds), then we may write acceleration $= D_t^2 y = -32$. One integration yields (the primitive)

(1) $$\text{Velocity} = D_t y = -32t + C$$

If now we know that the particle was fired from a height of y_0 ft with *initial velocity* v_0, then we may write

$$v_0 = D_t y|_{t=0} = -32(0) + C_1$$

Hence
$$C_1 = v_0$$

and (1) becomes

(2) $$D_t y = -32t + v_0$$

which gives the velocity at any time t. The integration of (2) yields

$$y = -16t^2 + v_0 t + C_2$$

Since we were told that the particle was fired from a height of y_0 ft (from the ground), we can write

$$y_0 = -16(0) + v_0(0) + C_2$$

Therefore $C_2 = y_0$, and, for this problem, the height at any time t is given by

(3) $$y = -16t^2 + v_0 t + y_0$$

This is the general equation which applies to falling bodies under our assumptions.

ILLUSTRATION 1 A bomb was dropped from an airplane 16,000 ft high. When did it strike the ground?

SOLUTION We are given $v_0 = 0$, $y_0 = 16{,}000$. From (3), the height at any time t is given by

$$y = -16t^2 + 16{,}000$$

The bomb struck the ground at the time when $y = 0$. Thus

$$0 = -16t^2 + 16{,}000$$

and

$$t = \pm 10\sqrt{10}$$

The minus sign is of no interest here, and so the answer is: The bomb struck the ground $10\sqrt{10}$ sec later.

ILLUSTRATION 2 At 12 noon the motors of a certain rocket burned out at 6,400 ft elevation when the rocket was still traveling straight up and at 8.0×10^4 ft/sec. When was the rocket highest? What was its maximum height? When did it strike the ground, and with what velocity did it strike?

SOLUTION

$$\begin{aligned} D_t^2 y &= -32 \\ D_t y &= -32t + 8.0 \times 10^4 \\ y &= -16t^2 + 8.0 \times 10^4 t + 6.4 \times 10^3 \end{aligned}$$

It was highest when $D_t y = -32t + 8.0 \times 10^4 = 0$, that is to say, when $t = 2{,}500$ sec (past noon, or at 12:42). The maximum height was

$$\begin{aligned} y_{\max} &= -16(625 \times 10^4) + 20{,}000 \times 10^4 + 0.64 \times 10^4 \\ &= 10^4(10^4 + 0.64) \\ &= 100{,}006{,}400 \text{ ft} \end{aligned}$$

Since $Dy\big|_{t<2{,}500} > 0$ and $Dy\big|_{t>2{,}500} < 0$, the test for maximum is satisfied.

It struck the ground when $y = 0$, that is, when

$$-16t^2 + 8.0 \times 10^4 t + 6.4 \times 10^3 = 0$$

$$\begin{aligned} t &= \frac{-8.0 \times 10^4 \pm \sqrt{64 \times 10^8 + (64)^2 10^2}}{-32} \\ &= 0.25 \times 10^4 \pm 2.5\sqrt{10^6 + 64} \\ &\approx 2{,}500 \pm 2{,}502 \\ &= 5{,}002 \text{ sec} \quad \text{(past noon, or 1:23)} \end{aligned}$$

It struck the ground with velocity

$$Dy\Big|_{t=5,002} = -32(5002) + 8.0 \times 10^4$$
$$= -80{,}064 \text{ ft/sec}$$

PROBLEMS 16.10 In Probs. 1 to 8 you are given the height y ft after t sec of a body moving straight up and down. Find **(a)** the velocity and acceleration at any time t, **(b)** the initial velocity, **(c)** the maximum height, **(d)** the time when it struck the ground.

1 $y = 160t - 8t^2$
2 $y = 25t - 5t^2$
3 $y = 64t - t^2$
4 $y = 16 - t$
5 $y = 400 + 160t - 8t^2$
6 $y = 100 + 100t - 50t^2$
7 $y = 32 + 16t - 4t^2$
8 $y = -24 + 10t - t^2$

In Probs. 9 to 12 vertical motion is assumed. Find the height y ft at time t sec from the conditions given.

9 Particle thrown up from an altitude of 200 ft with an initial velocity of 20 ft/sec
10 Particle thrown upward from ground with initial velocity of 20 ft/sec
11 Particle fired upward from an altitude of 5 miles and with an initial velocity of 400 miles/hr
12 Particle dropped from 1,000 ft

17 HYPERBOLIC FUNCTIONS

17.1 » HYPERBOLIC FUNCTIONS

In many areas of pure and applied mathematics and engineering there are functions, closely related to sine and cosine, that are of very great importance. These are $f\colon\left(\theta,\dfrac{e^{\theta}-e^{-\theta}}{2}\right)$ and $g\colon\left(\theta,\dfrac{e^{\theta}+e^{-\theta}}{2}\right)$; the domain of each is the set of real numbers.

Although these are just simple combinations of the exponential functions given by $y = e^{\theta}$ and $y = e^{-\theta}$, they are used so extensively that tables have been prepared for them and names given to them. For reasons that will be made clear in the next section, they are called the "hyperbolic sine of the number θ" and the "hyperbolic cosine of the number θ", respectively. These are written "$\sinh\theta$" and "$\cosh\theta$". Thus we write

(1)
$$\sinh\theta = \frac{e^{\theta}-e^{-\theta}}{2}$$

(2)
$$\cosh\theta = \frac{e^{\theta}+e^{-\theta}}{2}$$

17.2 » HYPERBOLIC AND CIRCULAR TRIGONOMETRIC FUNCTIONS

In order to make clear the connection between the hyperbolic and the circular functions, we first reconsider the latter. These ($\sin\phi$, $\cos\phi$, etc.) were defined with respect to the circle $x^2 + y^2 = 1$. For this reason we refer to them as the "circular trigonometric functions" (Fig. 17.1).

Let $P(x,y)$ be a point on the circle in the first quadrant, and set $\phi = \text{arc}\,AP$. The area of the sector OAP is equal to $\phi/2$ square units.† Since the area of triangle OAB is $\frac{1}{2}$ square unit, the number ϕ may be thought of as the ratio of the area of the sector to the area of the triangle.

† In your study of plane geometry you should have met with the more general result that, in a circle of radius a, the area of a sector of central angle ϕ radians is $a^2\phi/2$.

Area definition of the ordinary (circular) measure of an angle

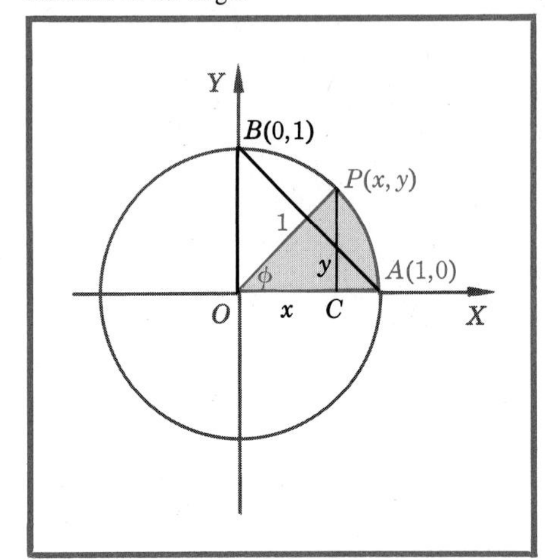

FIGURE 17.1
The measure of an angle ϕ can be defined as twice the area of the circular sector OAP which it determines in the unit circle.

That is,

Angle in terms of sector

(3) $$\phi = \frac{\text{area of sector } OAP}{\text{area of } \triangle\, OAB} = \frac{\phi/2}{\frac{1}{2}}$$

As in Chap. 13,

(4) $$\sin\phi = PC = y \qquad \cos\phi = OC = x$$

EXERCISE A Extend these ideas to the case of the circle $x^2 + y^2 = a^2$.

It is this new way of looking at ϕ that shows us how to develop a trigonometry based upon the hyperbola $x^2 - y^2 = 1$. Let $P(x,y)$ be a point on this hyperbola in the first quadrant, and set (Fig. 17.2)

$$\theta = \frac{\text{area of sector } OAP}{\text{area of } \triangle\, OAB} = \frac{\text{area of sector } OAP}{\frac{1}{2}}$$

Now

Area of sector OAP = (area OCP) − (area under hyperbola from A to C)

The area *under* the hyperbola can be found by calculus as follows: The equation of the curve is $y = +\sqrt{x^2 - 1}$, and the area is given by

$$\text{Area } ACP = \int_1^x \sqrt{z^2 - 1}\, dz$$

(We changed to the dummy variable of integration z because we have already used x as the abscissa of P.) The evaluation of this integral

Area definition of the hyperbolic measure of an angle

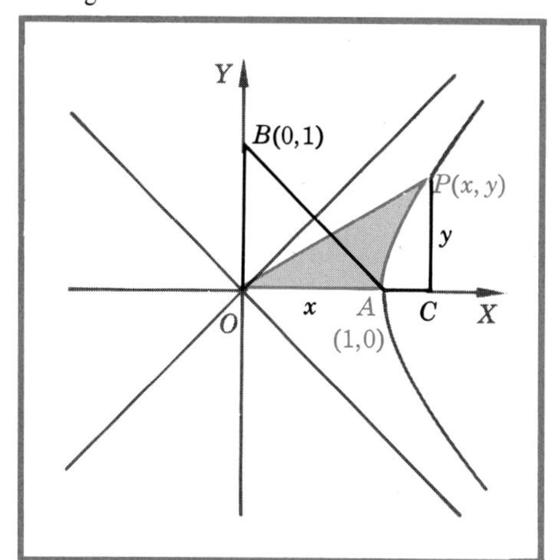

FIGURE 17.2 By analogy with Fig. 17.1 define the real number θ as twice the area of the hyperbolic sector OAP, where $x^2 - y^2 = 1$. This θ is definitely not the ordinary measure of the angle AOP; call it the "hyperbolic measure" of this angle.

involves more calculus than we have studied in this text, and so we can only write down the results without derivation. We write $\log_e \alpha$ simply as $\ln \alpha$.

$$\int_1^x \sqrt{z^2 - 1}\, dz = \left[\frac{z}{2}\sqrt{z^2 - 1} - \tfrac{1}{2}\ln\left(z + \sqrt{z^2 - 1}\right)\right]_1^x$$

$$= \frac{x}{2}\sqrt{x^2 - 1} - \tfrac{1}{2}\ln\left(x + \sqrt{x^2 - 1}\right)$$

From this we get

$$\text{Area of sector } OAP = \tfrac{1}{2}xy - \tfrac{1}{2}x\sqrt{x^2 - 1} + \tfrac{1}{2}\ln\left(x + \sqrt{x^2 - 1}\right)$$
$$= \tfrac{1}{2}xy - \tfrac{1}{2}xy + \tfrac{1}{2}\ln(x + y) = \tfrac{1}{2}\ln(x + y)$$

since $y = \sqrt{x^2 - 1}$. Therefore

(5) $$\theta = \ln(x + y)$$

We now define

(6) $$\sinh\theta = PC = y \qquad \cosh\theta = OC = x$$

to correspond to $\sin\theta$ and $\cos\theta$ in (4). In order to obtain (1) and (2) from (6), we solve (5) for x and for y by the following device: From (5) it follows at once that

(7) $$e^{\theta} = x + y \qquad e^{-\theta} = \frac{1}{x + y}$$

From (6) and (7) we have

$$\frac{e^{\theta} - e^{-\theta}}{2} = \frac{x + y - 1/(x + y)}{2}$$

$$= \frac{x^2 + 2xy + y^2 - 1}{2(x + y)}$$

$$= \frac{2xy + 2y^2}{2(x + y)} \qquad \text{since } x^2 = 1 + y^2$$

$$= y$$

Hence
$$\sinh \theta = y = \frac{e^{\theta} + e^{-\theta}}{2}$$

Similarly from (6) and (7) we have

$$\frac{e^{\theta} + e^{-\theta}}{2} = \frac{x + y + 1/(x + y)}{2}$$

$$= \frac{x^2 + 2xy + y^2 + 1}{2(x + y)}$$

$$= x$$

Hence
$$\cosh \theta = x = \frac{e^{\theta} + e^{-\theta}}{2}$$

EXERCISE B Extend these ideas to the case of the hyperbola $x^2 - y^2 = a^2$.

Although these geometric derivations assume that $P(x,y)$ lies in the first quadrant, the analytic definitions of $\sinh \theta$ and $\cosh \theta$ in (1) and (2) are subject to no such restrictions. These hyperbolic functions are hence defined for all real values of θ. Note that θ is *not* the usual measure, ϕ, of the angle AOP; it can be considered to be a new measure of this angle, its *hyperbolic* measure.

17.3 › HYPERBOLIC TRIGONOMETRY

Other hyperbolic functions are defined by the following rules:

$$\tanh \theta = \frac{\sinh \theta}{\cosh \theta} = \frac{e^{\theta} - e^{-\theta}}{e^{\theta} + e^{-\theta}}$$

$$\coth \theta = \frac{e^{\theta} + e^{-\theta}}{e^{\theta} - e^{-\theta}} \qquad \theta \neq 0$$

$$\operatorname{sech} \theta = \frac{1}{\cosh \theta}$$

$$\operatorname{csch} \theta = \frac{1}{\sinh \theta} \qquad \theta \neq 0$$

A trigonometry of hyperbolic functions can be developed comparable with that of the circular functions.

Identities **ILLUSTRATION 1** Show that $\cosh^2\theta - \sinh^2\theta = 1$.

SOLUTION

$$\cosh^2\theta = \frac{(e^\theta + e^{-\theta})^2}{4}$$

$$= \frac{e^{2\theta} + 2 + e^{-2\theta}}{4}$$

$$\sinh^2\theta = \frac{(e^\theta - e^{-\theta})^2}{4}$$

$$= \frac{e^{2\theta} - 2 + e^{-2\theta}}{4}$$

By subtraction the result follows.

ILLUSTRATION 2 Verify that

$$\sinh(x + y) = \sinh x \cosh y + \cosh x \sinh y$$

SOLUTION For the verification we work with the right-hand side.

$$\sinh(x + y) = \frac{e^x - e^{-x}}{2} \cdot \frac{e^y + e^{-y}}{2} + \frac{e^x + e^{-x}}{2} \cdot \frac{e^y - e^{-y}}{2}$$

$$= \frac{e^{x+y}}{4} - \frac{e^{-x+y}}{4} + \frac{e^{x-y}}{4} - \frac{e^{-(x+y)}}{4}$$

$$+ \frac{e^{x+y}}{4} + \frac{e^{-x+y}}{4} - \frac{e^{x-y}}{4} - \frac{e^{-(x+y)}}{4}$$

$$= \frac{e^{x+y}}{2} - \frac{e^{-(x+y)}}{2} = \sinh(x + y)$$

ILLUSTRATION 3 Given $y = \sinh x = \dfrac{e^x - e^{-x}}{2}$ and hence the function $f{:}\left(x, \dfrac{e^x - e^{-x}}{2}\right)$; find the inverse function f^{-1}.

SOLUTION First we wish to solve for x in terms of y.

$$y = \frac{e^x - e^{-x}}{2}$$

$$e^x - e^{-x} = 2y$$

Adding $-2y$ to both sides and then multiplying by e^x, we obtain

$$e^{2x} - 2ye^x - 1 = 0$$

This quadratic equation (in e^x) has the two solutions

$$e^x = \frac{2y \pm \sqrt{4y^2 + 4}}{2}$$
$$= y \pm \sqrt{y^2 + 1}$$

but the one with the minus sign is impossible since e^x is always positive. Therefore

$$e^x = y + \sqrt{y^2 + 1}$$

Taking logarithms on both sides to base e, we have

$$x = \ln (y + \sqrt{y^2 + 1})$$

The inverse function is therefore

$$f^{-1}:(y, \ln (y + \sqrt{y^2 + 1}))$$

or, in our usual x notation

$$f^{-1}:(x, \ln (x + \sqrt{x^2 + 1}))$$

ILLUSTRATION 4 Sketch $y = \cosh x$. (This is the *catenary:* that curve assumed by a power cable between poles.)

SOLUTION To compute ordered pairs of this function f:(x,cosh x), we use Table I in the Appendix. There is symmetry with respect to the Y-axis.

x	0	1	−1	1.5	2	−2	2.5	3
e^x	1	2.72	0.37	4.48	7.39	0.14	12.18	20.09
e^{-x}	1	0.37	2.72	0.22	0.14	7.39	0.08	0.05
$e^x + e^{-x}$	2	3.09	3.09	4.70	7.53	7.53	12.26	20.14
cosh x	1	1.54	1.54	2.35	3.76	3.76	6.13	10.07

The graph is plotted in Fig. 17.3.

PROBLEMS 17.3

In Probs. 1 to 11 prove the identity:

1 $\sinh (x - y) = \sinh x \cosh y - \cosh x \sinh y$
2 $\cosh (x + y) = \cosh x \cosh y + \sinh x \sinh y$
3 $\cosh (x - y) = \cosh x \cosh y - \sinh x \sinh y$
4 $\tanh (x + y) = \dfrac{\tanh x + \tanh y}{1 + \tanh x \tanh y}$

Catenary

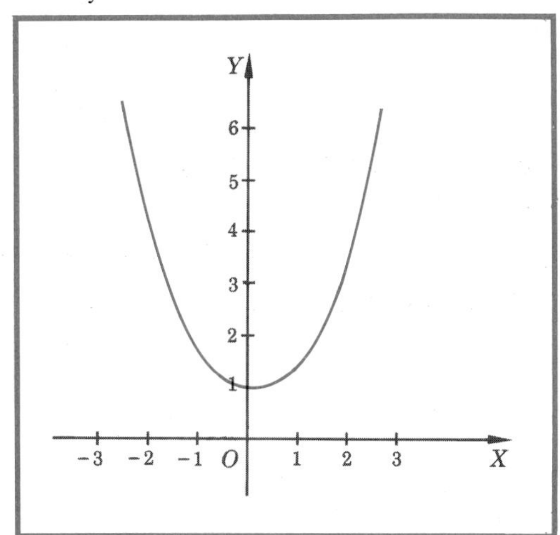

FIGURE 17.3
$y = \cosh x$.
See Illustration 4.

5 $\operatorname{sech}^2 x = 1 - \tanh^2 x$

6 $\operatorname{csch}^2 x = \coth^2 x - 1$

7 $\sinh 2x = 2 \sinh x \cosh x$

8 $\cosh 2x = \cosh^2 x + \sinh^2 x$

9 $\tanh 2x = \dfrac{2 \tanh x}{1 + \tanh^2 x}$

10 $\sinh \dfrac{x}{2} = \begin{cases} +\sqrt{\dfrac{\cosh x - 1}{2}} & x \geq 0 \\ -\sqrt{\dfrac{\cosh x - 1}{2}} & x \leq 0 \end{cases}$

11 $\cosh \dfrac{x}{2} = \sqrt{\dfrac{\cosh x + 1}{2}}$

In Probs. 12 to 17 discover the relation between:

12 $\sinh(-x)$ and $\sinh x$

13 $\cosh(-x)$ and $\cosh x$

14 $\tanh(-x)$ and $\tanh x$

15 $\operatorname{sech}(-x)$ and $\operatorname{sech} x$

16 $\operatorname{csch}(-x)$ and $\operatorname{csch} x$

17 $\coth(-x)$ and $\coth x$

In Probs. 18 to 22 determine $f^{-1}(x)$. State domain.

18 $f(x) = \tanh x$

19 $f(x) = \coth x$

20 $f(x) = \operatorname{csch} x$

21 $f(x) = \operatorname{Cosh} x$ ($\cosh x$ restricted by $x \geq 0$)

22 $f(x) = \operatorname{Sech} x$ ($\operatorname{sech} x$ restricted by $x \geq 0$)

In Probs. 23 to 27 sketch the curve.

23 $y = \sinh x$ **24** $y = \tanh x$
25 $y = \coth x$ **26** $y = \text{sech}\, x$
27 $y = \text{csch}\, x$

28 In the transmission of electric power the formula $I = \cosh x + \sinh x$ occurs. Sketch the graph.
29 Sketch $y = \cosh x - \sinh x$.
30 To the same scales and on the same axes sketch $y = \sinh x$ and $y = \sinh 2x$.
31 A parachute with unit mass is falling. Under certain conditions its velocity v and height s above ground are given by
a $v = -\sqrt{32} \tanh \sqrt{32}\, t$
b $s = -\ln \cosh \sqrt{32}\, t + 4{,}000$
Sketch these curves.

17.4 » EULER'S FORMULA

There is a remarkable formula due to Euler which permits us to write $\sin \theta$ and $\cos \theta$ in terms of the values of the exponential function in forms very similar to those of the definitions of $\sinh \theta$ and $\cosh \theta$. This formula involves the expression $e^{i\theta}$ and hence introduces imaginary exponents. Since these have not been defined so far, we must define them before proceeding.

DEFINITION (EULER) $e^{i\theta} = \cos \theta + i \sin \theta$

This may seem to be a very artificial definition; so let us motivate it as follows. It can be shown that

$$\cos \theta = 1 - \frac{\theta^2}{2!} + \frac{\theta^4}{4!} - \frac{\theta^6}{6!} + \cdots \tag{1}$$

$$\sin \theta = \theta - \frac{\theta^3}{3!} + \frac{\theta^5}{5!} - \frac{\theta^7}{7!} + \cdots \tag{2}$$

$$e^{\theta} = 1 + \theta + \frac{\theta^2}{2!} + \frac{\theta^3}{3!} + \frac{\theta^4}{4!} + \cdots \tag{3}$$

A reasonable definition of $e^{i\theta}$ is therefore

$$\begin{aligned} e^{i\theta} &= 1 + (i\theta) + \frac{(i\theta)^2}{2!} + \frac{(i\theta)^3}{3!} + \frac{(i\theta)^4}{4!} + \cdots \\ &= \left(1 - \frac{\theta^2}{2!} + \frac{\theta^4}{4!} \cdots\right) + i\left(\theta - \frac{\theta^3}{3!} + \frac{\theta^5}{5!} + \cdots\right) \\ &= \cos \theta + i \sin \theta \end{aligned} \tag{4}$$

Formulas (1) to (4) involve infinite series, which we have not discussed in this book, and so we cannot base a rigorous definition of $e^{i\theta}$ on these. They serve us here, however, as motivation for the rigorous definition given above.

$e^{i\theta} = \text{cis}\,\theta$

EXERCISE A Note that $e^{i\theta}$ is now the same as the $\text{cis}\,\theta$ of Sec. 13.11. Hence show that $e^{i\theta}e^{i\phi} = e^{i(\theta+\phi)}$.

EXERCISE B Show that $1/e^{i\theta} = e^{-i\theta} = \cos\theta - i\sin\theta$.

EXERCISE C Show that $e^{i\pi} = -1$.

From Euler's Formula we find that

$$e^{i\theta} = \cos\theta + i\sin\theta \tag{5}$$

$$e^{-i\theta} = \cos\theta - i\sin\theta \tag{6}$$

Adding (5) and (6) and simplifying, we see that

$$\cos\theta = \frac{e^{i\theta} + e^{-i\theta}}{2} \tag{7}$$

Subtracting (6) from (5) and simplifying, we obtain

$$\sin\theta = \frac{e^{i\theta} - e^{-i\theta}}{2i} \tag{8}$$

Formulas (7) and (8) are the ones to which we referred at the beginning of this section.

APPENDIX

TABLE I VALUES OF e^x AND e^{-x}

x	e^x	e^{-x}	x	e^x	e^{-x}
0.00	1.0000	1.00000	2.10	8.1662	0.12246
0.01	1.0101	0.99005	2.20	9.0250	0.11080
0.02	1.0202	0.98020	2.30	9.9742	0.10026
0.03	1.0305	0.97045	2.40	11.023	0.09072
0.04	1.0408	0.96079	2.50	12.182	0.08208
0.05	1.0513	0.95123	2.60	13.464	0.07427
0.06	1.0618	0.94176	2.70	14.880	0.06721
0.07	1.0725	0.93239	2.80	16.445	0.06081
0.08	1.0833	0.92312	2.90	18.174	0.05502
0.09	1.0942	0.91393	3.00	20.086	0.04979
0.10	1.1052	0.90484	3.10	22.198	0.04505
0.20	1.2214	0.81873	3.20	24.533	0.04076
0.30	1.3499	0.74082	3.30	27.113	0.03688
0.40	1.4918	0.67032	3.40	29.964	0.03337
0.50	1.6487	0.60653	3.50	33.115	0.03020
0.60	1.8221	0.54881	3.60	36.598	0.02732
0.70	2.0138	0.49659	3.70	40.447	0.02472
0.80	2.2255	0.44933	3.80	44.701	0.02237
0.90	2.4596	0.40657	3.90	49.402	0.02024
1.00	2.7183	0.36788	4.00	54.598	0.01832
1.10	3.0042	0.33287	4.10	60.340	0.01657
1.20	3.3201	0.30119	4.20	66.686	0.01500
1.30	3.6693	0.27253	4.30	73.700	0.01357
1.40	4.0552	0.24660	4.40	81.451	0.01228
1.50	4.4817	0.22313	4.50	90.017	0.01111
1.60	4.9530	0.20190	4.60	99.484	0.01005
1.70	5.4739	0.18268	4.70	109.95	0.00910
1.80	6.0496	0.16530	4.80	121.51	0.00823
1.90	6.6859	0.14957	4.90	134.29	0.00745
2.00	7.3891	0.13534	5.00	148.41	0.00674

TABLE II **COMMON LOGARITHMS (BASE 10)**

N.	0	1	2	3	4	5	6	7	8	9
10	0000	0043	0086	0128	0170	0212	0253	0294	0334	0374
11	0414	0453	0492	0531	0569	0607	0645	0682	0719	0755
12	0792	0828	0864	0899	0934	0969	1004	1038	1072	1106
13	1139	1173	1206	1239	1271	1303	1335	1367	1399	1430
14	1461	1492	1523	1553	1584	1614	1644	1673	1703	1732
15	1761	1790	1818	1847	1875	1903	1931	1959	1987	2014
16	2041	2068	2095	2122	2148	2175	2201	2227	2253	2279
17	2304	2330	2355	2380	2405	2430	2455	2480	2504	2529
18	2553	2577	2601	2625	2648	2672	2695	2718	2742	2765
19	2788	2810	2833	2856	2878	2900	2923	2945	2967	2989
20	3010	3032	3054	3075	3096	3118	3139	3160	3181	3201
21	3222	3243	3263	3284	3304	3324	3345	3365	3385	3404
22	3424	3444	3464	3483	3502	3522	3541	3560	3579	3598
23	3617	3636	3655	3674	3692	3711	3729	3747	3766	3784
24	3802	3820	3838	3856	3874	3892	3909	3927	3945	3962
25	3979	3997	4014	4031	4048	4065	4082	4099	4116	4133
26	4150	4166	4183	4200	4216	4232	4249	4265	4281	4298
27	4314	4330	4346	4362	4378	4393	4409	4425	4440	4456
28	4472	4487	4502	4518	4533	4548	4564	4579	4594	4609
29	4624	4639	4654	4669	4683	4698	4713	4728	4742	4757
30	4771	4786	4800	4814	4829	4843	4857	4871	4886	4900
31	4914	4928	4942	4955	4969	4983	4997	5011	5024	5038
32	5051	5065	5079	5092	5105	5119	5132	5145	5159	5172
33	5185	5198	5211	5224	5237	5250	5263	5276	5289	5302
34	5315	5328	5340	5353	5366	5378	5391	5403	5416	5428
35	5441	5453	5465	5478	5490	5502	5514	5527	5539	5551
36	5563	5575	5587	5599	5611	5623	5635	5647	5658	5670
37	5682	5694	5705	5717	5729	5740	5752	5763	5775	5786
38	5798	5809	5821	5832	5843	5855	5866	5877	5888	5899
39	5911	5922	5933	5944	5955	5966	5977	5988	5999	6010
40	6021	6031	6042	6053	6064	6075	6085	6096	6107	6117
41	6128	6138	6149	6160	6170	6180	6191	6201	6212	6222
42	6232	6243	6253	6263	6274	6284	6294	6304	6314	6325
43	6335	6345	6355	6365	6375	6385	6395	6405	6415	6425
44	6435	6444	6454	6464	6474	6484	6493	6503	6513	6522
45	6532	6542	6551	6561	6571	6580	6590	6599	6609	6618
46	6628	6637	6646	6656	6665	6675	6684	6693	6702	6712
47	6721	6730	6739	6749	6758	6767	6776	6785	6794	6803
48	6812	6821	6830	6839	6848	6857	6866	6875	6884	6893
49	6902	6911	6920	6928	6937	6946	6955	6964	6972	6981
50	6990	6998	7007	7016	7024	7033	7042	7050	7059	7067
51	7076	7084	7093	7101	7110	7118	7126	7135	7143	7152
52	7160	7168	7177	7185	7193	7202	7210	7218	7226	7235
53	7243	7251	7259	7267	7275	7284	7292	7300	7308	7316
54	7324	7332	7340	7348	7356	7364	7372	7380	7388	7396
N.	0	1	2	3	4	5	6	7	8	9

TABLE II **COMMON LOGARITHMS (Continued)**

N.	0	1	2	3	4	5	6	7	8	9
55	7404	7412	7419	7427	7435	7443	7451	7459	7466	7474
56	7482	7490	7497	7505	7513	7520	7528	7536	7543	7551
57	7559	7566	7574	7582	7589	7597	7604	7612	7619	7627
58	7634	7642	7649	7657	7664	7672	7679	7686	7694	7701
59	7709	7716	7723	7731	7738	7745	7752	7760	7767	7774
60	7782	7789	7796	7803	7810	7818	7825	7832	7839	7846
61	7853	7860	7868	7875	7882	7889	7896	7903	7910	7917
62	7924	7931	7938	7945	7952	7959	7966	7973	7980	7987
63	7993	8000	8007	8014	8021	8028	8035	8041	8048	8055
64	8062	8069	8075	8082	8089	8096	8102	8109	8116	8122
65	8129	8136	8142	8149	8156	8162	8169	8176	8182	8189
66	8195	8202	8209	8215	8222	8228	8235	8241	8248	8254
67	8261	8267	8274	8280	8287	8293	8299	8306	8312	8319
68	8325	8331	8338	8344	8351	8357	8363	8370	8376	8382
69	8388	8395	8401	8407	8414	8420	8426	8432	8439	8445
70	8451	8457	8463	8470	8476	8482	8488	8494	8500	8506
71	8513	8519	8525	8531	8537	8543	8549	8555	8561	8567
72	8573	8579	8585	8591	8597	8603	8609	8615	8621	8627
73	8633	8639	8645	8651	8657	8663	8669	8675	8681	8686
74	8692	8698	8704	8710	8716	8722	8727	8733	8739	8745
75	8751	8756	8762	8768	8774	8779	8785	8791	8797	8802
76	8808	8814	8820	8825	8831	8837	8842	8848	8854	8859
77	8865	8871	8876	8882	8887	8893	8899	8904	8910	8915
78	8921	8927	8932	8938	8943	8949	8954	8960	8965	8971
79	8976	8982	8987	8993	8998	9004	9009	9015	9020	9025
80	9031	9036	9042	9047	9053	9058	9063	9069	9074	9079
81	9085	9090	9096	9101	9106	9112	9117	9122	9128	9133
82	9138	9143	9149	9154	9159	9165	9170	9175	9180	9186
83	9191	9196	9201	9206	9212	9217	9222	9227	9232	9238
84	9243	9248	9253	9258	9263	9269	9274	9279	9284	9289
85	9294	9299	9304	9309	9315	9320	9325	9330	9335	9340
86	9345	9350	9355	9360	9365	9370	9375	9380	9385	9390
87	9395	9400	9405	9410	9415	9420	9425	9430	9435	9440
88	9445	9450	9455	9460	9465	9469	9474	9479	9484	9489
89	9494	9499	9504	9509	9513	9518	9523	9528	9533	9538
90	9542	9547	9552	9557	9562	9566	9571	9576	9581	9586
91	9590	9595	9600	9605	9609	9614	9619	9624	9628	9633
92	9638	9643	9647	9652	9657	9661	9666	9671	9675	9680
93	9685	9689	9694	9699	9703	9708	9713	9717	9722	9727
94	9731	9736	9741	9745	9750	9754	9759	9763	9768	9773
95	9777	9782	9786	9791	9795	9800	9805	9809	9814	9818
96	9823	9827	9832	9836	9841	9845	9850	9854	9859	9863
97	9868	9872	9877	9881	9886	9890	9894	9899	9903	9908
98	9912	9917	9921	9926	9930	9934	9939	9943	9948	9952
99	9956	9961	9965	9969	9974	9978	9983	9987	9991	9996
N.	0	1	2	3	4	5	6	7	8	9

TABLE III NATURAL LOGARITHMS (BASE e)

	.00	.01	.02	.03	.04	.05	.06	.07	.08	.09
1.0	0.0000	0.0100	0.0198	0.0296	0.0392	0.0488	0.0583	0.0677	0.0770	0.0862
1.1	0.0953	0.1044	0.1133	0.1222	0.1310	0.1398	0.1484	0.1570	0.1655	0.1740
1.2	0.1823	0.1906	0.1989	0.2070	0.2151	0.2231	0.2311	0.2390	0.2469	0.2546
1.3	0.2624	0.2700	0.2776	0.2852	0.2927	0.3001	0.3075	0.3148	0.3221	0.3293
1.4	0.3365	0.3436	0.3507	0.3577	0.3646	0.3716	0.3784	0.3853	0.3920	0.3988
1.5	0.4055	0.4121	0.4187	0.4253	0.4318	0.4383	0.4447	0.4511	0.4574	0.4637
1.6	0.4700	0.4762	0.4824	0.4886	0.4947	0.5008	0.5068	0.5128	0.5188	0.5247
1.7	0.5306	0.5365	0.5423	0.5481	0.5539	0.5596	0.5653	0.5710	0.5766	0.5822
1.8	0.5878	0.5933	0.5988	0.6043	0.6098	0.6152	0.6206	0.6259	0.6313	0.6366
1.9	0.6419	0.6471	0.6523	0.6575	0.6627	0.6678	0.6729	0.6780	0.6831	0.6881
2.0	0.6932	0.6981	0.7031	0.7080	0.7129	0.7178	0.7227	0.7275	0.7324	0.7372
2.1	0.7419	0.7467	0.7514	0.7561	0.7608	0.7655	0.7701	0.7747	0.7793	0.7839
2.2	0.7885	0.7930	0.7975	0.8020	0.8065	0.8109	0.8154	0.8198	0.8242	0.8286
2.3	0.8329	0.8373	0.8416	0.8459	0.8502	0.8544	0.8587	0.8629	0.8671	0.8713
2.4	0.8755	0.8796	0.8838	0.8879	0.8920	0.8961	0.9002	0.9042	0.9083	0.9123
2.5	0.9163	0.9203	0.9243	0.9282	0.9322	0.9361	0.9400	0.9439	0.9478	0.9517
2.6	0.9555	0.9594	0.9632	0.9670	0.9708	0.9746	0.9783	0.9821	0.9858	0.9895
2.7	0.9933	0.9969	1.0006	1.0043	1.0080	1.0116	1.0152	1.0188	1.0225	1.0260
2.8	1.0296	1.0332	1.0367	1.0403	1.0438	1.0473	1.0508	1.0543	1.0578	1.0613
2.9	1.0647	1.0682	1.0716	1.0750	1.0784	1.0818	1.0852	1.0886	1.0919	1.0953
3.0	1.0986	1.1019	1.1053	1.1086	1.1119	1.1151	1.1184	1.1217	1.1249	1.1282
3.1	1.1314	1.1346	1.1378	1.1410	1.1442	1.1474	1.1506	1.1537	1.1569	1.1600
3.2	1.1632	1.1663	1.1694	1.1725	1.1756	1.1787	1.1817	1.1848	1.1878	1.1909
3.3	1.1939	1.1969	1.2000	1.2030	1.2060	1.2090	1.2119	1.2149	1.2179	1.2208
3.4	1.2238	1.2267	1.2296	1.2326	1.2355	1.2384	1.2413	1.2442	1.2470	1.2499
3.5	1.2528	1.2556	1.2585	1.2613	1.2641	1.2669	1.2698	1.2726	1.2754	1.2782
3.6	1.2809	1.2837	1.2865	1.2892	1.2920	1.2947	1.2975	1.3002	1.3029	1.3056
3.7	1.3083	1.3110	1.3137	1.3164	1.3191	1.3218	1.3244	1.3271	1.3297	1.3324
3.8	1.3350	1.3376	1.3403	1.3429	1.3455	1.3481	1.3507	1.3533	1.3558	1.3584
3.9	1.3610	1.3635	1.3661	1.3686	1.3712	1.3737	1.3762	1.3788	1.3813	1.3838
4.0	1.3863	1.3888	1.3913	1.3938	1.3962	1.3987	1.4012	1.4036	1.4061	1.4085
4.1	1.4110	1.4134	1.4159	1.4183	1.4207	1.4231	1.4255	1.4279	1.4303	1.4327
4.2	1.4351	1.4375	1.4398	1.4422	1.4446	1.4469	1.4493	1.4516	1.4540	1.4563
4.3	1.4586	1.4609	1.4633	1.4656	1.4679	1.4702	1.4725	1.4748	1.4771	1.4793
4.4	1.4816	1.4839	1.4861	1.4884	1.4907	1.4929	1.4951	1.4974	1.4996	1.5019
4.5	1.5041	1.5063	1.5085	1.5107	1.5129	1.5151	1.5173	1.5195	1.5217	1.5239
4.6	1.5261	1.5282	1.5304	1.5326	1.5347	1.5369	1.5390	1.5412	1.5433	1.5454
4.7	1.5476	1.5497	1.5518	1.5539	1.5560	1.5581	1.5602	1.5623	1.5644	1.5665
4.8	1.5686	1.5707	1.5728	1.5748	1.5769	1.5790	1.5810	1.5831	1.5851	1.5872
4.9	1.5892	1.5913	1.5933	1.5953	1.5974	1.5994	1.6014	1.6034	1.6054	1.6074
5.0	1.6094	1.6114	1.6134	1.6154	1.6174	1.6194	1.6214	1.6233	1.6253	1.6273
5.1	1.6292	1.6312	1.6332	1.6351	1.6371	1.6390	1.6409	1.6429	1.6448	1.6467
5.2	1.6487	1.6506	1.6525	1.6544	1.6563	1.6582	1.6601	1.6620	1.6639	1.6658
5.3	1.6677	1.6696	1.6715	1.6734	1.6752	1.6771	1.6790	1.6808	1.6827	1.6845
5.4	1.6864	1.6882	1.6901	1.6919	1.6938	1.6956	1.6974	1.6993	1.7011	1.7029

TABLE III NATURAL LOGARITHMS (Continued)

	.00	.01	.02	.03	.04	.05	.06	.07	.08	.09
5.5	1.7047	1.7066	1.7084	1.7102	1.7120	1.7138	1.7156	1.7174	1.7192	1.7210
5.6	1.7228	1.7246	1.7263	1.7281	1.7299	1.7317	1.7334	1.7352	1.7370	1.7387
5.7	1.7405	1.7422	1.7440	1.7457	1.7475	1.7492	1.7509	1.7527	1.7544	1.7561
5.8	1.7579	1.7596	1.7613	1.7630	1.7647	1.7664	1.7681	1.7699	1.7716	1.7733
5.9	1.7750	1.7766	1.7783	1.7800	1.7817	1.7834	1.7851	1.7868	1.7884	1.7901
6.0	1.7918	1.7934	1.7951	1.7967	1.7984	1.8001	1.8017	1.8034	1.8050	1.8066
6.1	1.8083	1.8099	1.8116	1.8132	1.8148	1.8165	1.8181	1.8197	1.8213	1.8229
6.2	1.8245	1.8262	1.8278	1.8294	1.8310	1.8326	1.8342	1.8358	1.8374	1.8390
6.3	1.8405	1.8421	1.8437	1.8453	1.8469	1.8485	1.8500	1.8516	1.8532	1.8547
6.4	1.8563	1.8579	1.8594	1.8610	1.8625	1.8641	1.8656	1.8672	1.8687	1.8703
6.5	1.8718	1.8733	1.8749	1.8764	1.8779	1.8795	1.8810	1.8825	1.8840	1.8856
6.6	1.8871	1.8886	1.8901	1.8916	1.8931	1.8946	1.8961	1.8976	1.8991	1.9006
6.7	1.9021	1.9036	1.9051	1.9066	1.9081	1.9095	1.9110	1.9125	1.9140	1.9155
6.8	1.9169	1.9184	1.9199	1.9213	1.9228	1.9242	1.9257	1.9272	1.9286	1.9301
6.9	1.9315	1.9330	1.9344	1.9359	1.9373	1.9387	1.9402	1.9416	1.9430	1.9445
7.0	1.9459	1.9473	1.9488	1.9502	1.9516	1.9530	1.9544	1.9559	1.9573	1.9587
7.1	1.9601	1.9615	1.9629	1.9643	1.9657	1.9671	1.9685	1.9699	1.9713	1.9727
7.2	1.9741	1.9755	1.9769	1.9782	1.9796	1.9810	1.9824	1.9838	1.9851	1.9865
7.3	1.9879	1.9892	1.9906	1.9920	1.9933	1.9947	1.9961	1.9974	1.9988	2.0001
7.4	2.0015	2.0028	2.0042	2.0055	2.0069	2.0082	2.0096	2.0109	2.0122	2.0136
7.5	2.0149	2.0162	2.0176	2.0189	2.0202	2.0215	2.0229	2.0242	2.0255	2.0268
7.6	2.0281	2.0295	2.0308	2.0321	2.0334	2.0347	2.0360	2.0373	2.0386	2.0399
7.7	2.0412	2.0425	2.0438	2.0451	2.0464	2.0477	2.0490	2.0503	2.0516	2.0528
7.8	2.0541	2.0554	2.0567	2.0580	2.0592	2.0605	2.0618	2.0631	2.0643	2.0656
7.9	2.0669	2.0681	2.0694	2.0707	2.0719	2.0732	2.0744	2.0757	2.0769	2.0782
8.0	2.0794	2.0807	2.0819	2.0832	2.0844	2.0857	2.0869	2.0882	2.0894	2.0906
8.1	2.0919	2.0931	2.0943	2.0956	2.0968	2.0980	2.0992	2.1005	2.1017	2.1029
8.2	2.1041	2.1054	2.1066	2.1078	2.1090	2.1102	2.1114	2.1126	2.1138	2.1150
8.3	2.1163	2.1175	2.1187	2.1199	2.1211	2.1223	2.1235	2.1247	2.1259	2.1270
8.4	2.1282	2.1294	2.1306	2.1318	2.1330	2.1342	2.1353	2.1365	2.1377	2.1389
8.5	2.1401	2.1412	2.1424	2.1436	2.1448	2.1459	2.1471	2.1483	2.1494	2.1506
8.6	2.1518	2.1529	2.1541	2.1552	2.1564	2.1576	2.1587	2.1599	2.1610	2.1622
8.7	2.1633	2.1645	2.1656	2.1668	2.1679	2.1691	2.1702	2.1713	2.1725	2.1736
8.8	2.1748	2.1759	2.1770	2.1782	2.1793	2.1804	2.1815	2.1827	2.1838	2.1849
8.9	2.1861	2.1872	2.1883	2.1894	2.1905	2.1917	2.1928	2.1939	2.1950	2.1961
9.0	2.1972	2.1983	2.1994	2.2006	2.2017	2.2028	2.2039	2.2050	2.2061	2.2072
9.1	2.2083	2.2094	2.2105	2.2116	2.2127	2.2138	2.2148	2.2159	2.2170	2.2181
9.2	2.2192	2.2203	2.2214	2.2225	2.2235	2.2246	2.2257	2.2268	2.2279	2.2289
9.3	2.2300	2.2311	2.2322	2.2332	2.2343	2.2354	2.2364	2.2375	2.2386	2.2396
9.4	2.2407	2.2418	2.2428	2.2439	2.2450	2.2460	2.2471	2.2481	2.2492	2.2502
9.5	2.2513	2.2523	2.2534	2.2544	2.2555	2.2565	2.2576	2.2586	2.2597	2.2607
9.6	2.2618	2.2628	2.2638	2.2649	2.2659	2.2670	2.2680	2.2690	2.2701	2.2711
9.7	2.2721	2.2732	2.2742	2.2752	2.2762	2.2773	2.2783	2.2793	2.2803	2.2814
9.8	2.2824	2.2834	2.2844	2.2854	2.2865	2.2875	2.2885	2.2895	2.2905	2.2915
9.9	2.2925	2.2935	2.2946	2.2956	2.2966	2.2976	2.2986	2.2996	2.3006	2.3016

TABLE III NATURAL LOGARITHMS (Continued)

N	Nat Log	N	Nat Log	N	Nat Log	N	Nat Log	N	Nat Log
0	$-\infty$	40	3.68 888	80	4.38 203	120	4.78 749	160	5.07 517
1	0.00 000	41	3.71 357	81	4.39 445	121	4.79 579	161	5.08 140
2	0.69 315	42	3.73 767	82	4.40 672	122	4.80 402	162	5.08 760
3	1.09 861	43	3.76 120	83	4.41 884	123	4.81 218	163	5.09 375
4	1.38 629	44	3.78 419	84	4.43 082	124	4.82 028	164	5.09 987
5	1.60 944	45	3.80 666	85	4.44 265	125	4.82 831	165	5.10 595
6	1.79 176	46	3.82 864	86	4.45 435	126	4.83 628	166	5.11 199
7	1.94 591	47	3.85 015	87	4.46 591	127	4.84 419	167	5.11 799
8	2.07 944	48	3.87 120	88	4.47 734	128	4.85 203	168	5.12 396
9	2.19 722	49	3.89 182	89	4.48 864	129	4.85 981	169	5.12 990
10	2.30 259	50	3.91 202	90	4.49 981	130	4.86 753	170	5.13 580
11	2.39 790	51	3.93 183	91	4.51 086	131	4.87 520	171	5.14 166
12	2.48 491	52	3.95 124	92	4.52 179	132	4.88 280	172	5.14 749
13	2.56 495	53	3.97 029	93	4.53 260	133	4.89 035	173	5.15 329
14	2.63 906	54	3.98 898	94	4.54 329	134	4.89 784	174	5.15 906
15	2.70 805	55	4.00 733	95	4.55 388	135	4.90 527	175	5.16 479
16	2.77 259	56	4.02 535	96	4.56 435	136	4.91 265	176	5.17 048
17	2.83 321	57	4.04 305	97	4.57 471	137	4.91 998	177	5.17 615
18	2.89 037	58	4.06 044	98	4.58 497	138	4.92 725	178	5.18 178
19	2.94 444	59	4.07 754	99	4.59 512	139	4.93 447	179	5.18 739
20	2.99 573	60	4.09 434	100	4.60 517	140	4.94 164	180	5.19 296
21	3.04 452	61	4.11 087	101	4.61 512	141	4.94 876	181	5.19 850
22	3.09 104	62	4.12 713	102	4.62 497	142	4.95 583	182	5.20 401
23	3.13 549	63	4.14 313	103	4.63 473	143	4.96 284	183	5.20 949
24	3.17 805	64	4.15 888	104	4.64 439	144	4.96 981	184	5.21 494
25	3.21 888	65	4.17 439	105	4.65 396	145	4.97 673	185	5.22 036
26	3.25 810	66	4.18 965	106	4.66 344	146	4.98 361	186	5.22 575
27	3.29 584	67	4.20 469	107	4.67 283	147	4.99 043	187	5.23 111
28	3.33 220	68	4.21 951	108	4.68 213	148	4.99 721	188	5.23 644
29	3.36 730	69	4.23 411	109	4.69 135	149	5.00 395	189	5.24 175
30	3.40 120	70	4.24 850	110	4.70 048	150	5.01 064	190	5.24 702
31	3.43 399	71	4.26 268	111	4.70 953	151	5.01 728	191	5.25 227
32	3.46 574	72	4.27 667	112	4.71 850	152	5.02 388	192	5.25 750
33	3.49 651	73	4.29 046	113	4.72 739	153	5.03 044	193	5.26 269
34	3.52 636	74	4.30 407	114	4.73 620	154	5.03 695	194	5.26 786
35	3.55 535	75	4.31 749	115	4.74 493	155	5.04 343	195	5.27 300
36	3.58 352	76	4.33 073	116	4.75 359	156	5.04 986	196	5.27 811
37	3.61 092	77	4.34 381	117	4.76 217	157	5.05 625	197	5.28 320
38	3.63 759	78	4.35 671	118	4.77 068	158	5.06 260	198	5.28 827
39	3.66 356	79	4.36 945	119	4.77 912	159	5.06 890	199	5.29 330
40	3.68 888	80	4.38 203	120	4.78 749	160	5.07 517	200	5.29 832

TABLE IV **TRIGONOMETRIC FUNCTIONS OF REAL NUMBERS**

x	Sin x	Tan x	Cot x	Cos x	x	Sin x	Tan x	Cot x	Cos x
.00	.00000	.00000	∞	1.00000	**.50**	.47943	.54630	1.8305	.87758
.01	.01000	.01000	99.997	0.99995	.51	.48818	.55936	1.7878	.87274
.02	.02000	.02000	49.993	.99980	.52	.49688	.57256	1.7465	.86782
.03	.03000	.03001	33.323	.99955	.53	.50553	.58592	1.7067	.86281
.04	.03999	.04002	24.987	.99920	.54	.51414	.59943	1.6683	.85771
.05	.04998	.05004	19.983	.99875	.55	.52269	.61311	1.6310	.85252
.06	.05996	.06007	16.647	.99820	.56	.53119	.62695	1.5950	.84726
.07	.06994	.07011	14.262	.99755	.57	.53963	.64097	1.5601	.84190
.08	.07991	.08017	12.473	.99680	.58	.54802	.65517	1.5263	.83646
.09	.08988	.09024	11.081	.99595	.59	.55636	.66956	1.4935	.83094
.10	.09983	.10033	9.9666	.99500	**.60**	.56464	.68414	1.4617	.82534
.11	.10978	.11045	9.0542	.99396	.61	.57287	.69892	1.4308	.81965
.12	.11971	.12058	8.2933	.99281	.62	.58104	.71391	1.4007	.81388
.13	.12963	.13074	7.6489	.99156	.63	.58914	.72911	1.3715	.80803
.14	.13954	.14092	7.0961	.99022	.64	.59720	.74454	1.3431	.80210
.15	.14944	.15144	6.6166	.98877	.65	.60519	.76020	1.3154	.79608
.16	.15932	.16138	6.1966	.98723	.66	.61312	.77610	1.2885	.78999
.17	.16918	.17166	5.8256	.98558	.67	.62099	.79225	1.2622	.78382
.18	.17903	.18197	5.4954	.98384	.68	.62879	.80866	1.2366	.77757
.19	.18886	.19232	5.1997	.98200	.69	.63654	.82534	1.2116	.77125
.20	.19867	.20271	4.9332	.98007	**.70**	.64422	.84229	1.1872	.76484
.21	.20846	.21314	4.6917	.97803	.71	.65183	.85953	1.1634	.75836
.22	.21823	.22362	4.4719	.97590	.72	.65938	.87707	1.1402	.75181
.23	.22798	.23414	4.2709	.97367	.73	.66687	.89492	1.1174	.74517
.24	.23770	.24472	4.0864	.97134	.74	.67429	.91309	1.0952	.73847
.25	.24740	.25534	3.9163	.96891	.75	.68164	.93160	1.0734	.73169
.26	.25708	.26602	3.7591	.96639	.76	.68892	.95045	1.0521	.72484
.27	.26673	.27676	3.6133	.96377	.77	.69614	.96967	1.0313	.71791
.28	.27636	.28755	3.4776	.96106	.78	.70328	.98926	1.0109	.71091
.29	.28595	.29841	3.3511	.95824	.79	.71035	1.0092	.99084	.70385
.30	.29552	.30934	3.2327	.95534	**.80**	.71736	1.0296	.97121	.69671
.31	.30506	.32033	3.1218	.95233	.81	.72429	1.0505	.95197	.68950
.32	.31457	.33139	3.0176	.94924	.82	.73115	1.0717	.93309	.68222
.33	.32404	.34252	2.9195	.94604	.83	.73793	1.0934	.91455	.67488
.34	.33349	.35374	2.8270	.94275	.84	.74464	1.1156	.89635	.66746
.35	.34290	.36503	2.7395	.93937	.85	.75128	1.1383	.87848	.65998
.36	.35227	.37640	2.6567	.93590	.86	.75784	1.1616	.86091	.65244
.37	.36162	.38786	2.5782	.93233	.87	.76433	1.1853	.84365	.64483
.38	.37092	.39941	2.5037	.92866	.88	.77074	1.2097	.82668	.63715
.39	.38019	.41105	2.4328	.92491	.89	.77707	1.2346	.80998	.62941
.40	.38942	.42279	2.3652	.92106	**.90**	.78333	1.2602	.79355	.62161
.41	.39861	.43463	2.3008	.91712	.91	.78950	1.2864	.77738	.61375
.42	.40776	.44657	2.2393	.91309	.92	.79560	1.3133	.76146	.60582
.43	.41687	.45862	2.1804	.90897	.93	.80162	1.3409	.74578	.59783
.44	.42594	.47078	2.1241	.90475	.94	.80756	1.3692	.73034	.58979
.45	.43497	.48306	2.0702	.90045	.95	.81342	1.3984	.71511	.58168
.46	.44395	.49545	2.1084	.89605	.96	.81919	1.4284	.70010	.57352
.47	.45289	.50797	1.9686	.89157	.97	.82489	1.4592	.68531	.56530
.48	.46178	.52061	1.9208	.88699	.98	.83050	1.4910	.67071	.55702
.49	.47063	.53339	1.8748	.88233	.99	.83603	1.5237	.65631	.54869
.50	.47943	.54630	1.8305	.87758	**1.00**	.84147	1.5574	.64209	.54030
x	Sin x	Tan x	Cot x	Cos x	x	Sin x	Tan x	Cot x	Cos x

TABLE IV **TRIGONOMETRIC FUNCTIONS OF REAL NUMBERS (Continued)**

x	Sin x	Tan x	Cot x	Cos x	x	Sin x	Tan x	Cot x	Cos x
1.00	.84147	1.5574	.64209	.54030	**1.50**	.99749	14.101	.07091	.07074
1.01	.84683	1.5922	.62806	.53186	1.51	.99815	16.428	.06087	.06076
1.02	.85211	1.6281	.61420	.52337	1.52	.99871	19.670	.05084	.05077
1.03	.85730	1.6652	.60051	.51482	1.53	.99917	24.498	.04082	.04079
1.04	.86240	1.7036	.58699	.50622	1.54	.99953	32.461	.03081	.03079
1.05	.86742	1.7433	.57362	.49757	1.55	.99978	48.078	.02080	.02079
1.06	.87236	1.7844	.56040	.48887	1.56	.99994	92.621	.01080	.01080
1.07	.87720	1.8270	.54734	.48012	1.57	1.00000	1255.8	.00080	.00080
1.08	.88196	1.8712	.53441	.47133	1.58	.99996	−108.65	−.00920	−.00920
1.09	.88663	1.9171	.52162	.46249	1.59	.99982	−52.067	−.01921	−.01920
1.10	.89121	1.9648	.50897	.45360	**1.60**	.99957	−34.233	−.02921	−.02920
1.11	.89570	2.0143	.49644	.44466	1.61	.99923	−25.495	−.03922	−.03919
1.12	.90010	2.0660	.48404	.43568	1.62	.99879	−20.307	−.04924	−.04918
1.13	.90441	2.1198	.47175	.42666	1.63	.99825	−16.871	−.05927	−.05917
1.14	.90863	2.1759	.45959	.41759	1.64	.99761	−14.427	−.06931	−.06915
1.15	.91276	2.2345	.44753	.40849	1.65	.99687	−12.599	−.07937	−.07912
1.16	.91680	2.2958	.43558	.39934	1.66	.99602	−11.181	−.08944	−.08909
1.17	.92075	2.3600	.42373	.39015	1.67	.99508	−10.047	−.09953	−.09904
1.18	.92461	2.4273	.41199	.38092	1.68	.99404	− 9.1208	−.10964	−.10899
1.19	.92837	2.4979	.40034	.37166	1.69	.99290	− 8.3492	−.11977	−.11892
1.20	.93204	2.5722	.38878	.36236	**1.70**	.99166	− 7.6966	−.12993	−.12884
1.21	.93562	2.6503	.37731	.35302	1.71	.99033	− 7.1373	−.14011	−.13875
1.22	.93910	2.7328	.36593	.34365	1.72	.98889	− 6.6524	−.15032	−.14865
1.23	.94249	2.8198	.35463	.33424	1.73	.98735	− 6.2281	−.16056	−.15853
1.24	.94578	2.9119	.34341	.32480	1.74	.98572	− 5.8535	−.17084	−.16840
1.25	.94898	3.0096	.33227	.31532	1.75	.98399	− 5.5204	−.18115	−.17825
1.26	.95209	3.1133	.32121	.30582	1.76	.98215	− 5.2221	−.19149	−.18808
1.27	.95510	3.2236	.31021	.29628	1.77	.98022	− 4.9534	−.20188	−.19789
1.28	.95802	3.3413	.29928	.28672	1.78	.97820	− 4.7101	−.21231	−.20768
1.29	.96084	3.4672	.28842	.27712	1.79	.97607	− 4.4887	−.22278	−.21745
1.30	.96356	3.6021	.27762	.26750	**1.80**	.97385	− 4.2863	−.23330	−.22720
1.31	.96618	3.7471	.26687	.25785	1.81	.97153	− 4.1005	−.24387	−.23693
1.32	.96872	3.9033	.25619	.24818	1.82	.96911	− 3.9294	−.25449	−.24663
1.33	.97115	4.0723	.24556	.23848	1.83	.96659	− 3.7712	−.26517	−.25631
1.34	.97348	4.2556	.23498	.22875	1.84	.96398	− 3.6245	−.27590	−.26596
1.35	.97572	4.4552	.22446	.21901	1.85	.96128	− 3.4881	−.28669	−.27559
1.36	.97786	4.6734	.21398	.20924	1.86	.95847	− 3.3608	−.29755	−.28519
1.37	.97991	4.9131	.20354	.19945	1.87	.95557	− 3.2419	−.30846	−.29476
1.38	.98185	5.1774	.19315	.18964	1.88	.95258	− 3.1304	−.31945	−.30430
1.39	.98370	5.4707	.18279	.17981	1.89	.94949	− 3.0257	−.33051	−.31381
1.40	.98545	5.7979	.17248	.16997	**1.90**	.94630	− 2.9271	−.34164	−.32329
1.41	.98710	6.1654	.16220	.16010	1.91	.94302	− 2.8341	−.35284	−.33274
1.42	.98865	6.5811	.15195	.15023	1.92	.93965	− 2.7463	−.36413	−.34215
1.43	.99010	7.0555	.14173	.14033	1.93	.93618	− 2.6632	−.37549	−.35153
1.44	.99146	7.6018	.13155	.13042	1.94	.93262	− 2.5843	−.38695	−.36087
1.45	.99271	8.2381	.12139	.12050	1.95	.92896	− 2.5095	−.39849	−.37018
1.46	.99387	8.9886	.11125	.11057	1.96	.92521	− 2.4383	−.41012	−.37945
1.47	.99492	9.8874	.10114	.10063	1.97	.92137	− 2.3705	−.42185	−.38868
1.48	.99588	10.983	.09105	.09067	1.98	.91744	− 2.3058	−.43368	−.39788
1.49	.99674	12.350	.08097	.08071	1.99	.91341	− 2.2441	−.44562	−.40703
1.50	.99749	14.101	.07081	.07074	**2.00**	.90930	− 2.1850	−.45766	−.41615
x	Sin x	Tan x	Cot x	Cos x	x	Sin x	Tan x	Cot x	Cos x

TABLE V TRIGONOMETRIC FUNCTIONS OF ANGLES

0° (180°) (359°) 179°

′	Sin	Tan	Cot	Cos	′
0	.00000	.00000	——	1.0000	60
1	.00029	.00029	3437.7	1.0000	59
2	.00058	.00058	1718.9	1.0000	58
3	.00087	.00087	1145.9	1.0000	57
4	.00116	.00116	859.44	1.0000	56
5	.00145	.00145	687.55	1.0000	55
6	.00175	.00175	572.96	1.0000	54
7	.00204	.00204	491.11	1.0000	53
8	.00233	.00233	429.72	1.0000	52
9	.00262	.00262	381.97	1.0000	51
10	.00291	.00291	343.77	1.0000	50
11	.00320	.00320	312.52	.99999	49
12	.00349	.00349	286.48	.99999	48
13	.00378	.00378	264.44	.99999	47
14	.00407	.00407	245.55	.99999	46
15	.00436	.00436	229.18	.99999	45
16	.00465	.00465	214.86	.99999	44
17	.00495	.00495	202.22	.99999	43
18	.00524	.00524	190.98	.99999	42
19	.00553	.00553	180.93	.99998	41
20	.00582	.00582	171.89	.99998	40
21	.00611	.00611	163.70	.99998	39
22	.00640	.00640	156.26	.99998	38
23	.00669	.00669	149.47	.99998	37
24	.00698	.00698	143.24	.99998	36
25	.00727	.00727	137.51	.99997	35
26	.00756	.00756	132.22	.99997	34
27	.00785	.00785	127.32	.99997	33
28	.00814	.00815	122.77	.99997	32
29	.00844	.00844	118.54	.99996	31
30	.00873	.00873	114.59	.99996	30
31	.00902	.00902	110.89	.99996	29
32	.00931	.00931	107.43	.99996	28
33	.00960	.00960	104.17	.99995	27
34	.00989	.00989	101.11	.99995	26
35	.01018	.01018	98.218	.99995	25
36	.01047	.01047	95.489	.99995	24
37	.01076	.01076	92.908	.99994	23
38	.01105	.01105	90.463	.99994	22
39	.01134	.01135	88.144	.99994	21
40	.01164	.01164	85.940	.99993	20
41	.01193	.01193	83.844	.99993	19
42	.01222	.01222	81.847	.99993	18
43	.01251	.01251	79.943	.99992	17
44	.01280	.01280	78.126	.99992	16
45	.01309	.01309	76.390	.99991	15
46	.01338	.01338	74.729	.99991	14
47	.01367	.01367	73.139	.99991	13
48	.01396	.01396	71.615	.99990	12
49	.01425	.01425	70.153	.99990	11
50	.01454	.01455	68.750	.99989	10
51	.01483	.01484	67.402	.99989	9
52	.01513	.01513	66.105	.99989	8
53	.01542	.01542	64.858	.99988	7
54	.01571	.01571	63.657	.99988	6
55	.01600	.01600	62.499	.99987	5
56	.01629	.01629	61.383	.99987	4
57	.01658	.01658	60.306	.99986	3
58	.01687	.01687	59.266	.99986	2
59	.01716	.01716	58.261	.99985	1
60	.01745	.01746	57.290	.99985	0
′	Cos	Cot	Tan	Sin	′

90° (270°) (269°) 89°

1° (181°) (358°) **178°**

′	Sin	Tan	Cot	Cos	′
0	.01745	.01746	57.290	.99985	60
1	.01774	.01775	56.351	.99984	59
2	.01803	.01804	55.442	.99984	58
3	.01832	.01833	54.561	.99983	57
4	.01862	.01862	53.709	.99983	56
5	.01891	.01891	52.882	.99982	55
6	.01920	.01920	52.081	.99982	54
7	.01949	.01949	51.303	.99981	53
8	.01978	.01978	50.549	.99980	52
9	.02007	.02007	49.816	.99980	51
10	.02036	.02036	49.104	.99979	50
11	.02065	.02066	48.412	.99979	49
12	.02094	.02095	47.740	.99978	48
13	.02123	.02124	47.085	.99977	47
14	.02152	.02153	46.449	.99977	46
15	.02181	.02182	45.829	.99976	45
16	.02211	.02211	45.226	.99976	44
17	.02240	.02240	44.639	.99975	43
18	.02269	.02269	44.066	.99974	42
19	.02298	.02298	43.508	.99974	41
20	.02327	.02328	42.964	.99973	40
21	.02356	.02357	42.433	.99972	39
22	.02385	.02386	41.916	.99972	38
23	.02414	.02415	41.411	.99971	37
24	.02443	.02444	40.917	.99970	36
25	.02472	.02473	40.436	.99969	35
26	.02501	.02502	39.965	.99969	34
27	.02530	.02531	39.506	.99968	33
28	.02560	.02560	39.057	.99967	32
29	.02589	.02589	38.618	.99966	31
30	.02618	.02619	38.188	.99966	30
31	.02647	.02648	37.769	.99965	29
32	.02676	.02677	37.358	.99964	28
33	.02705	.02706	36.956	.99963	27
34	.02734	.02735	36.563	.99963	26
35	.02763	.02764	36.178	.99962	25
36	.02792	.02793	35.801	.99961	24
37	.02821	.02822	35.431	.99960	23
38	.02850	.02851	35.070	.99959	22
39	.02879	.02881	34.715	.99959	21
40	.02908	.02910	34.368	.99958	20
41	.02938	.02939	34.027	.99957	19
42	.02967	.02968	33.694	.99956	18
43	.02996	.02997	33.366	.99955	17
44	.03025	.03026	33.045	.99954	16
45	.03054	.03055	32.730	.99953	15
46	.03083	.03084	32.421	.99952	14
47	.03112	.03114	32.118	.99952	13
48	.03141	.03143	31.821	.99951	12
49	.03170	.03172	31.528	.99950	11
50	.03199	.03201	31.242	.99949	10
51	.03228	.03230	30.960	.99948	9
52	.03257	.03259	30.683	.99947	8
53	.03286	.03288	30.412	.99946	7
54	.03316	.03317	30.145	.99945	6
55	.03345	.03346	29.882	.99944	5
56	.03374	.03376	29.624	.99943	4
57	.03403	.03405	29.371	.99942	3
58	.03432	.03434	29.122	.99941	2
59	.03461	.03463	28.877	.99940	1
60	.03490	.03492	28.636	.99939	0
′	Cos	Cot	Tan	Sin	′

91° (271°) (268°) **88°**

For degrees indicated at the top (bottom) of the page use the column headings at the top (bottom). With degrees at the left (right) of each block (top or bottom), use the minute column at the left (right). The correct sign (plus or minus) must be prefixed in accordance with Sec. 12.7.

TABLE V TRIGONOMETRIC FUNCTIONS OF ANGLES (Continued)

2° (182°) (357°) **177°**

′	Sin	Tan	Cot	Cos	′
0	.03490	.03492	28.636	.99939	60
1	.03519	.03521	28.399	.99938	59
2	.03548	.03550	28.166	.99937	58
3	.03577	.03579	27.937	.99936	57
4	.03606	.03609	27.712	.99935	56
5	.03635	.03638	27.490	.99934	55
6	.03664	.03667	27.271	.99933	54
7	.03693	.03696	27.057	.99932	53
8	.03723	.03725	26.845	.99931	52
9	.03752	.03754	26.637	.99930	51
10	.03781	.03783	26.432	.99929	50
11	.03810	.03812	26.230	.99927	49
12	.03839	.03842	26.031	.99926	48
13	.03868	.03871	25.835	.99925	47
14	.03897	.03900	25.642	.99924	46
15	.03926	.03929	25.452	.99923	45
16	.03955	.03958	25.264	.99922	44
17	.03984	.03987	25.080	.99921	43
18	.04013	.04016	24.898	.99919	42
19	.04042	.04046	24.719	.99918	41
20	.04071	.04075	24.542	.99917	40
21	.04100	.04104	24.368	.99916	39
22	.04129	.04133	24.196	.99915	38
23	.04159	.04162	24.026	.99913	37
24	.04188	.04191	23.859	.99912	36
25	.04217	.04220	23.695	.99911	35
26	.04246	.04250	23.532	.99910	34
27	.04275	.04279	23.372	.99909	33
28	.04304	.04308	23.214	.99907	32
29	.04333	.04337	23.058	.99906	31
30	.04362	.04366	22.904	.99905	30
31	.04391	.04395	22.752	.99904	29
32	.04420	.04424	22.602	.99902	28
33	.04449	.04454	22.454	.99901	27
34	.04478	.04483	22.308	.99900	26
35	.04507	.04512	22.164	.99898	25
36	.04536	.04541	22.022	.99897	24
37	.04565	.04570	21.881	.99896	23
38	.04594	.04599	21.743	.99894	22
39	.04623	.04628	21.606	.99893	21
40	.04653	.04658	21.470	.99892	20
41	.04682	.04687	21.337	.99890	19
42	.04711	.04716	21.205	.99889	18
43	.04740	.04745	21.075	.99888	17
44	.04769	.04774	20.946	.99886	16
45	.04798	.04803	20.819	.99885	15
46	.04827	.04833	20.693	.99883	14
47	.04856	.04862	20.569	.99882	13
48	.04885	.04891	20.446	.99881	12
49	.04914	.04920	20.325	.99879	11
50	.04943	.04949	20.206	.99878	10
51	.04972	.04978	20.087	.99876	9
52	.05001	.05007	19.970	.99875	8
53	.05030	.05037	19.855	.99873	7
54	.05059	.05066	19.740	.99872	6
55	.05088	.05095	19.627	.99870	5
56	.05117	.05124	19.516	.99869	4
57	.05146	.05153	19.405	.99867	3
58	.05175	.05182	19.296	.99866	2
59	.05205	.05212	19.188	.99864	1
60	.05234	.05241	19.081	.99863	0
′	Cos	Cot	Tan	Sin	′

92° (272°) (267°) **87°**

3° (183°) (356°) **176°**

′	Sin	Tan	Cot	Cos	′
0	.05234	.05241	19.081	.99863	60
1	.05263	.05270	18.976	.99861	59
2	.05292	.05299	18.871	.99860	58
3	.05321	.05328	18.768	.99858	57
4	.05350	.05357	18.666	.99857	56
5	.05379	.05387	18.564	.99855	55
6	.05408	.05416	18.464	.99854	54
7	.05437	.05445	18.366	.99852	53
8	.05466	.05474	18.268	.99851	52
9	.05495	.05503	18.171	.99849	51
10	.05524	.05533	18.075	.99847	50
11	.05553	.05562	17.980	.99846	49
12	.05582	.05591	17.886	.99844	48
13	.05611	.05620	17.793	.99842	47
14	.05640	.05649	17.702	.99841	46
15	.05669	.05678	17.611	.99839	45
16	.05698	.05708	17.521	.99838	44
17	.05727	.05737	17.431	.99836	43
18	.05756	.05766	17.343	.99834	42
19	.05785	.05795	17.256	.99833	41
20	.05814	.05824	17.169	.99831	40
21	.05844	.05854	17.084	.99829	39
22	.05873	.05883	16.999	.99827	38
23	.05902	.05912	16.915	.99826	37
24	.05931	.05941	16.832	.99824	36
25	.05960	.05970	16.750	.99822	35
26	.05989	.05999	16.668	.99821	34
27	.06018	.06029	16.587	.99819	33
28	.06047	.06058	16.507	.99817	32
29	.06076	.06087	16.428	.99815	31
30	.06105	.06116	16.350	.99813	30
31	.06134	.06145	16.272	.99812	29
32	.06163	.06175	16.195	.99810	28
33	.06192	.06204	16.119	.99808	27
34	.06221	.06233	16.043	.99806	26
35	.06250	.06262	15.969	.99804	25
36	.06279	.06291	15.895	.99803	24
37	.06308	.06321	15.821	.99801	23
38	.06337	.06350	15.748	.99799	22
39	.06366	.06379	15.676	.99797	21
40	.06395	.06408	15.605	.99795	20
41	.06424	.06438	15.534	.99793	19
42	.06453	.06467	15.464	.99792	18
43	.06482	.06496	15.394	.99790	17
44	.06511	.06525	15.325	.99788	16
45	.06540	.06554	15.257	.99786	15
46	.06569	.06584	15.189	.99784	14
47	.06598	.06613	15.122	.99782	13
48	.06627	.06642	15.056	.99780	12
49	.06656	.06671	14.990	.99778	11
50	.06685	.06700	14.924	.99776	10
51	.06714	.06730	14.860	.99774	9
52	.06743	.06759	14.795	.99772	8
53	.06773	.06788	14.732	.99770	7
54	.06802	.06817	14.669	.99768	6
55	.06831	.06847	14.606	.99766	5
56	.06860	.06876	14.544	.99764	4
57	.06889	.06905	14.482	.99762	3
58	.06918	.06934	14.421	.99760	2
59	.06947	.06963	14.361	.99758	1
60	.06976	.06993	14.301	.99756	0
′	Cos	Cot	Tan	Sin	′

93° (273°) (266°) **86°**

TABLE V TRIGONOMETRIC FUNCTIONS OF ANGLES (Continued)

4° (184°) (355°) **175°**

′	Sin	Tan	Cot	Cos	′
0	.06976	.06993	14.301	.99756	60
1	.07005	.07022	14.241	.99754	59
2	.07034	.07051	14.182	.99752	58
3	.07063	.07080	14.124	.99750	57
4	.07092	.07110	14.065	.99748	56
5	.07121	.07139	14.008	.99746	55
6	.07150	.07168	13.951	.99744	54
7	.07179	.07197	13.894	.99742	53
8	.07208	.07227	13.838	.99740	52
9	.07237	.07256	13.782	.99738	51
10	.07266	.07285	13.727	.99736	50
11	.07295	.07314	13.672	.99734	49
12	.07324	.07344	13.617	.99731	48
13	.07353	.07373	13.563	.99729	47
14	.07382	.07402	13.510	.99727	46
15	.07411	.07431	13.457	.99725	45
16	.07440	.07461	13.404	.99723	44
17	.07469	.07490	13.352	.99721	43
18	.07498	.07519	13.300	.99719	42
19	.07527	.07548	13.248	.99716	41
20	.07556	.07578	13.197	.99714	40
21	.07585	.07607	13.146	.99712	39
22	.07614	.07636	13.096	.99710	38
23	.07643	.07665	13.046	.99708	37
24	.07672	.07695	12.996	.99705	36
25	.07701	.07724	12.947	.99703	35
26	.07730	.07753	12.898	.99701	34
27	.07759	.07782	12.850	.99699	33
28	.07788	.07812	12.801	.99696	32
29	.07817	.07841	12.754	.99694	31
30	.07846	.07870	12.706	.99692	30
31	.07875	.07899	12.659	.99689	29
32	.07904	.07929	12.612	.99687	28
33	.07933	.07958	12.566	.99685	27
34	.07962	.07987	12.520	.99683	26
35	.07991	.08017	12.474	.99680	25
36	.08020	.08046	12.429	.99678	24
37	.08049	.08075	12.384	.99676	23
38	.08078	.08104	12.339	.99673	22
39	.08107	.08134	12.295	.99671	21
40	.08136	.08163	12.251	.99668	20
41	.08165	.08192	12.207	.99666	19
42	.08194	.08221	12.163	.99664	18
43	.08223	.08251	12.120	.99661	17
44	.08252	.08280	12.077	.99659	16
45	.08281	.08309	12.035	.99657	15
46	.08310	.08339	11.992	.99654	14
47	.08339	.08368	11.950	.99652	13
48	.08368	.08397	11.909	.99649	12
49	.08397	.08427	11.867	.99647	11
50	.08426	.08456	11.826	.99644	10
51	.08455	.08485	11.785	.99642	9
52	.08484	.08514	11.745	.99639	8
53	.08513	.08544	11.705	.99637	7
54	.08542	.08573	11.664	.99635	6
55	.08571	.08602	11.625	.99632	5
56	.08600	.08632	11.585	.99630	4
57	.08629	.08661	11.546	.99627	3
58	.08658	.08690	11.507	.99625	2
59	.08687	.08720	11.468	.99622	1
60	.08716	.08749	11.430	.99619	0
′	Cos	Cot	Tan	Sin	′

94° (274°) (265°) **85°**

5° (185°) (354°) **174°**

′	Sin	Tan	Cot	Cos	′
0	.08716	.08749	11.430	.99619	60
1	.08745	.08778	11.392	.99617	59
2	.08774	.08807	11.354	.99614	58
3	.08803	.08837	11.316	.99612	57
4	.08831	.08866	11.279	.99609	56
5	.08860	.08895	11.242	.99607	55
6	.08889	.08925	11.205	.99604	54
7	.08918	.08954	11.168	.99602	53
8	.08947	.08983	11.132	.99599	52
9	.08976	.09013	11.095	.99596	51
10	.09005	.09042	11.059	.99594	50
11	.09034	.09071	11.024	.99591	49
12	.09063	.09101	10.988	.99588	48
13	.09092	.09130	10.953	.99586	47
14	.09121	.09159	10.918	.99583	46
15	.09150	.09189	10.883	.99580	45
16	.09179	.09218	10.848	.99578	44
17	.09208	.09247	10.814	.99575	43
18	.09237	.09277	10.780	.99572	42
19	.09266	.09306	10.746	.99570	41
20	.09295	.09335	10.712	.99567	40
21	.09324	.09365	10.678	.99564	39
22	.09353	.09394	10.645	.99562	38
23	.09382	.09423	10.612	.99559	37
24	.09411	.09453	10.579	.99556	36
25	.09440	.09482	10.546	.99553	35
26	.09469	.09511	10.514	.99551	34
27	.09498	.09541	10.481	.99548	33
28	.09527	.09570	10.449	.99545	32
29	.09556	.09600	10.417	.99542	31
30	.09585	.09629	10.385	.99540	30
31	.09614	.09658	10.354	.99537	29
32	.09642	.09688	10.322	.99534	28
33	.09671	.09717	10.291	.99531	27
34	.09700	.09746	10.260	.99528	26
35	.09729	.09776	10.229	.99526	25
36	.09758	.09805	10.199	.99523	24
37	.09787	.09834	10.168	.99520	23
38	.09816	.09864	10.138	.99517	22
39	.09845	.09893	10.108	.99514	21
40	.09874	.09923	10.078	.99511	20
41	.09903	.09952	10.048	.99508	19
42	.09932	.09981	10.019	.99506	18
43	.09961	.10011	9.9893	.99503	17
44	.09990	.10040	9.9601	.99500	16
45	.10019	.10069	9.9310	.99497	15
46	.10048	.10099	9.9021	.99494	14
47	.10077	.10128	9.8734	.99491	13
48	.10106	.10158	9.8448	.99488	12
49	.10135	.10187	9.8164	.99485	11
50	.10164	.10216	9.7882	.99482	10
51	.10192	.10246	9.7601	.99479	9
52	.10221	.10275	9.7322	.99476	8
53	.10250	.10305	9.7044	.99473	7
54	.10279	.10334	9.6768	.99470	6
55	.10308	.10363	9.6493	.99467	5
56	.10337	.10393	9.6220	.99464	4
57	.10366	.10422	9.5949	.99461	3
58	.10395	.10452	9.5679	.99458	2
59	.10424	.10481	9.5411	.99455	1
60	.10453	.10510	9.5144	.99452	0
′	Cos	Cot	Tan	Sin	′

95° (275°) (264°) **84°**

TABLE V **TRIGONOMETRIC FUNCTIONS OF ANGLES (Continued)**

6° (186°) (353°) **173°**

′	Sin	Tan	Cot	Cos	′
0	.10453	.10510	9.5144	.99452	60
1	.10482	.10540	9.4878	.99449	59
2	.10511	.10569	9.4614	.99446	58
3	.10540	.10599	9.4352	.99443	57
4	.10569	.10628	9.4090	.99440	56
5	.10597	.10657	9.3831	.99437	55
6	.10626	.10687	9.3572	.99434	54
7	.10655	.10716	9.3315	.99431	53
8	.10684	.10746	9.3060	.99428	52
9	.10713	.10775	9.2806	.99424	51
10	.10742	.10805	9.2553	.99421	50
11	.10771	.10834	9.2302	.99418	49
12	.10800	.10863	9.2052	.99415	48
13	.10829	.10893	9.1803	.99412	47
14	.10858	.10922	9.1555	.99409	46
15	.10887	.10952	9.1309	.99406	45
16	.10916	.10981	9.1065	.99402	44
17	.10945	.11011	9.0821	.99399	43
18	.10973	.11040	9.0579	.99396	42
19	.11002	.11070	9.0338	.99393	41
20	.11031	.11099	9.0098	.99390	40
21	.11060	.11128	8.9860	.99386	39
22	.11089	.11158	8.9623	.99383	38
23	.11118	.11187	8.9387	.99380	37
24	.11147	.11217	8.9152	.99377	36
25	.11176	.11246	8.8919	.99374	35
26	.11205	.11276	8.8686	.99370	34
27	.11234	.11305	8.8455	.99367	33
28	.11263	.11335	8.8225	.99364	32
29	.11291	.11364	8.7996	.99360	31
30	.11320	.11394	8.7769	.99357	30
31	.11349	.11423	8.7542	.99354	29
32	.11378	.11452	8.7317	.99351	28
33	.11407	.11482	8.7093	.99347	27
34	.11436	.11511	8.6870	.99344	26
35	.11465	.11541	8.6648	.99341	25
36	.11494	.11570	8.6427	.99337	24
37	.11523	.11600	8.6208	.99334	23
38	.11552	.11629	8.5989	.99331	22
39	.11580	.11659	8.5772	.99327	21
40	.11609	.11688	8.5555	.99324	20
41	.11638	.11718	8.5340	.99320	19
42	.11667	.11747	8.5126	.99317	18
43	.11696	.11777	8.4913	.99314	17
44	.11725	.11806	8.4701	.99310	16
45	.11754	.11836	8.4490	.99307	15
46	.11783	.11865	8.4280	.99303	14
47	.11812	.11895	8.4071	.99300	13
48	.11840	.11924	8.3863	.99297	12
49	.11869	.11954	8.3656	.99293	11
50	.11898	.11983	8.3450	.99290	10
51	.11927	.12013	8.3245	.99286	9
52	.11956	.12042	8.3041	.99283	8
53	.11985	.12072	8.2838	.99279	7
54	.12014	.12101	8.2636	.99276	6
55	.12043	.12131	8.2434	.99272	5
56	.12071	.12160	8.2234	.99269	4
57	.12100	.12190	8.2035	.99265	3
58	.12129	.12219	8.1837	.99262	2
59	.12158	.12249	8.1640	.99258	1
60	.12187	.12278	8.1443	.99255	0
′	Cos	Cot	Tan	Sin	′

96° (276°) (263°) **83°**

7° (187°) (352°) **172°**

′	Sin	Tan	Cot	Cos	′
0	.12187	.12278	8.1443	.99255	60
1	.12216	.12308	8.1248	.99251	59
2	.12245	.12338	8.1054	.99248	58
3	.12274	.12367	8.0860	.99244	57
4	.12302	.12397	8.0667	.99240	56
5	.12331	.12426	8.0476	.99237	55
6	.12360	.12456	8.0285	.99233	54
7	.12389	.12485	8.0095	.99230	53
8	.12418	.12515	7.9906	.99226	52
9	.12447	.12544	7.9718	.99222	51
10	.12476	.12574	7.9530	.99219	50
11	.12504	.12603	7.9344	.99215	49
12	.12533	.12633	7.9158	.99211	48
13	.12562	.12662	7.8973	.99208	47
14	.12591	.12692	7.8789	.99204	46
15	.12620	.12722	7.8606	.99200	45
16	.12649	.12751	7.8424	.99197	44
17	.12678	.12781	7.8243	.99193	43
18	.12706	.12810	7.8062	.99189	42
19	.12735	.12840	7.7882	.99186	41
20	.12764	.12869	7.7704	.99182	40
21	.12793	.12899	7.7525	.99178	39
22	.12822	.12929	7.7348	.99175	38
23	.12851	.12958	7.7171	.99171	37
24	.12880	.12988	7.6996	.99167	36
25	.12908	.13017	7.6821	.99163	35
26	.12937	.13047	7.6647	.99160	34
27	.12966	.13076	7.6473	.99156	33
28	.12995	.13106	7.6301	.99152	32
29	.13024	.13136	7.6129	.99148	31
30	.13053	.13165	7.5958	.99144	30
31	.13081	.13195	7.5787	.99141	29
32	.13110	.13224	7.5618	.99137	28
33	.13139	.13254	7.5449	.99133	27
34	.13168	.13284	7.5281	.99129	26
35	.13197	.13313	7.5113	.99125	25
36	.13226	.13343	7.4947	.99122	24
37	.13254	.13372	7.4781	.99118	23
38	.13283	.13402	7.4615	.99114	22
39	.13312	.13432	7.4451	.99110	21
40	.13341	.13461	7.4287	.99106	20
41	.13370	.13491	7.4124	.99102	19
42	.13399	.13521	7.3962	.99098	18
43	.13427	.13550	7.3800	.99094	17
44	.13456	.13580	7.3639	.99091	16
45	.13485	.13609	7.3479	.99087	15
46	.13514	.13639	7.3319	.99083	14
47	.13543	.13669	7.3160	.99079	13
48	.13572	.13698	7.3002	.99075	12
49	.13600	.13728	7.2844	.99071	11
50	.13629	.13758	7.2687	.99067	10
51	.13658	.13787	7.2531	.99063	9
52	.13687	.13817	7.2375	.99059	8
53	.13716	.13846	7.2220	.99055	7
54	.13744	.13876	7.2066	.99051	6
55	.13773	.13906	7.1912	.99047	5
56	.13802	.13935	7.1759	.99043	4
57	.13831	.13965	7.1607	.99039	3
58	.13860	.13995	7.1455	.99035	2
59	.13889	.14024	7.1304	.99031	1
60	.13917	.14054	7.1154	.99027	0
′	Cos	Cot	Tan	Sin	′

97° (277°) (262°) **82°**

TABLE V TRIGONOMETRIC FUNCTIONS OF ANGLES (Continued)

8° (188°) — (351°) **171°**

′	Sin	Tan	Cot	Cos	′
0	.13917	.14054	7.1154	.99027	60
1	.13946	.14084	7.1004	.99023	59
2	.13975	.14113	7.0855	.99019	58
3	.14004	.14143	7.0706	.99015	57
4	.14033	.14173	7.0558	.99011	56
5	.14061	.14202	7.0410	.99006	55
6	.14090	.14232	7.0264	.99002	54
7	.14119	.14262	7.0117	.98998	53
8	.14148	.14291	6.9972	.98994	52
9	.14177	.14321	6.9827	.98990	51
10	.14205	.14351	6.9682	.98986	50
11	.14234	.14381	6.9538	.98982	49
12	.14263	.14410	6.9395	.98978	48
13	.14292	.14440	6.9252	.98973	47
14	.14320	.14470	6.9110	.98969	46
15	.14349	.14499	6.8969	.98965	45
16	.14378	.14529	6.8828	.98961	44
17	.14407	.14559	6.8687	.98957	43
18	.14436	.14588	6.8548	.98953	42
19	.14464	.14618	6.8408	.98948	41
20	.14493	.14648	6.8269	.98944	40
21	.14522	.14678	6.8131	.98940	39
22	.14551	.14707	6.7994	.98936	38
23	.14580	.14737	6.7856	.98931	37
24	.14608	.14767	6.7720	.98927	36
25	.14637	.14796	6.7584	.98923	35
26	.14666	.14826	6.7448	.98919	34
27	.14695	.14856	6.7313	.98914	33
28	.14723	.14886	6.7179	.98910	32
29	.14752	.14915	6.7045	.98906	31
30	.14781	.14945	6.6912	98902	30
31	.14810	.14975	6.6779	.98897	29
32	.14838	.15005	6.6646	.98893	28
33	.14867	.15034	6.6514	.98889	27
34	.14896	.15064	6.6383	.98884	26
35	.14925	.15094	6.6252	.98880	25
36	.14954	.15124	6.6122	.98876	24
37	.14982	.15153	6.5992	.98871	23
38	.15011	.15183	6.5863	.98867	22
39	.15040	.15213	6.5734	.98863	21
40	.15069	.15243	6.5606	.98858	20
41	.15097	.15272	6.5478	.98854	19
42	.15126	.15302	6.5350	.98849	18
43	.15155	.15332	6.5223	.98845	17
44	.15184	.15362	6.5097	.98841	16
45	.15212	.15391	6.4971	.98836	15
46	.15241	.15421	6.4846	.98832	14
47	.15270	.15451	6.4721	.98827	13
48	.15299	.15481	6.4596	.98823	12
49	.15327	.15511	6.4472	.98818	11
50	.15356	.15540	6.4348	.98814	10
51	.15385	.15570	6.4225	.98809	9
52	.15414	.15600	6.4103	.98805	8
53	.15442	.15630	6.3980	.98800	7
54	.15471	.15660	6.3859	.98796	6
55	.15500	.15689	6.3737	.98791	5
56	.15529	.15719	6.3617	.98787	4
57	.15557	.15749	6.3496	.98782	3
58	.15586	.15779	6.3376	.98778	2
59	.15615	.15809	6.3257	.98773	1
60	.15643	.15838	6.3138	.98769	0
′	Cos	Cot	Tan	Sin	′

98° (278°) — (261°) **81°**

9° (189°) — (350°) **170°**

′	Sin	Tan	Cot	Cos	′
0	.15643	.15838	6.3138	.98769	60
1	.15672	.15868	6.3019	.98764	59
2	.15701	.15898	6.2901	.98760	58
3	.15730	.15928	6.2783	.98755	57
4	.15758	.15958	6.2666	.98751	56
5	.15787	.15988	6.2549	.98746	55
6	.15816	.16017	6.2432	.98741	54
7	.15845	.16047	6.2316	.98737	53
8	.15873	.16077	6.2200	.98732	52
9	.15902	.16107	6.2085	.98728	51
10	.15931	.16137	6.1970	.98723	50
11	.15959	.16167	6.1856	.98718	49
12	.15988	.16196	6.1742	.98714	48
13	.16017	.16226	6.1628	.98709	47
14	.16046	.16256	6.1515	.98704	46
15	.16074	.16286	6.1402	.98700	45
16	.16103	.16316	6.1290	.98695	44
17	.16132	.16346	6.1178	.98690	43
18	.16160	.16376	6.1066	.98686	42
19	.16189	.16405	6.0955	.98681	41
20	.16218	.16435	6.0844	.98676	40
21	.16246	.16465	6.0734	.98671	39
22	.16275	.16495	6.0624	.98667	38
23	.16304	.16525	6.0514	.98662	37
24	.16333	.16555	6.0405	.98657	36
25	.16361	.16585	6.0296	.98652	35
26	.16390	.16615	6.0188	.98648	34
27	.16419	.16645	6.0080	.98643	33
28	.16447	.16674	5.9972	.98638	32
29	.16476	.16704	5.9865	.98633	31
30	.16505	.16734	5.9758	.98629	30
31	.16533	.16764	5.9651	.98624	29
32	.16562	.16794	5.9545	.98619	28
33	.16591	.16824	5.9439	.98614	27
34	.16620	.16854	5.9333	.98609	26
35	.16648	.16884	5.9228	.98604	25
36	.16677	.16914	5.9124	.98600	24
37	.16706	.16944	5.9019	.98595	23
38	.16734	.16974	5.8915	.98590	22
39	.16763	.17004	5.8811	.98585	21
40	.16792	.17033	5.8708	.98580	20
41	.16820	.17063	5.8605	.98575	19
42	.16849	.17093	5.8502	.98570	18
43	.16878	.17123	5.8400	.98565	17
44	.16906	.17153	5.8298	.98561	16
45	.16935	.17183	5.8197	.98556	15
46	.16964	.17213	5.8095	.98551	14
47	.16992	.17243	5.7994	.98546	13
48	.17021	.17273	5.7894	.98541	12
49	.17050	.17303	5.7794	.98536	11
50	.17078	.17333	5.7694	.98531	10
51	.17107	.17363	5.7594	.98526	9
52	.17136	.17393	5.7495	.98521	8
53	.17164	.17423	5.7396	.98516	7
54	.17193	.17453	5.7297	.98511	6
55	.17222	.17483	5.7199	.98506	5
56	.17250	.17513	5.7101	.98501	4
57	.17279	.17543	5.7004	.98496	3
58	.17308	.17573	5.6906	.98491	2
59	.17336	.17603	5.6809	.98486	1
60	.17365	.17633	5.6713	.98481	0
′	Cos	Cot	Tan	Sin	′

99° (279°) — (260°) **80°**

TABLE V TRIGONOMETRIC FUNCTIONS OF ANGLES (Continued)

10° (190°) (349°) **169°**

′	Sin	Tan	Cot	Cos	′
0	.17365	.17633	5.6713	.98481	60
1	.17393	.17663	5.6617	.98476	59
2	.17422	.17693	5.6521	.98471	58
3	.17451	.17723	5.6425	.98466	57
4	.17479	.17753	5.6329	.98461	56
5	.17508	.17783	5.6234	.98455	55
6	.17537	.17813	5.6140	.98450	54
7	.17565	.17843	5.6045	.98445	53
8	.17594	.17873	5.5951	.98440	52
9	.17623	.17903	5.5857	.98435	51
10	.17651	.17933	5.5764	.98430	50
11	.17680	.17963	5.5671	.98425	49
12	.17708	.17993	5.5578	.98420	48
13	.17737	.18023	5.5485	.98414	47
14	.17766	.18053	5.5393	.98409	46
15	.17794	.18083	5.5301	.98404	45
16	.17823	.18113	5.5209	.98399	44
17	.17852	.18143	5.5118	.98394	43
18	.17880	.18173	5.5026	.98389	42
19	.17909	.18203	5.4936	.98383	41
20	.17937	.18233	5.4845	.98378	40
21	.17966	.18263	5.4755	.98373	39
22	.17995	.18293	5.4665	.98368	38
23	.18023	.18323	5.4575	.98362	37
24	.18052	.18353	5.4486	.98357	36
25	.18081	.18384	5.4397	.98352	35
26	.18109	.18414	5.4308	.98347	34
27	.18138	.18444	5.4219	.98341	33
28	.18166	.18474	5.4131	.98336	32
29	.18195	.18504	5.4043	.98331	31
30	.18224	.18534	5.3955	.98325	30
31	.18252	.18564	5.3868	.98320	29
32	.18281	.18594	5.3781	.98315	28
33	.18309	.18624	5.3694	.98310	27
34	.18338	.18654	5.3607	.98304	26
35	.18367	.18684	5.3521	.98299	25
36	.18395	.18714	5.3435	.98294	24
37	.18424	.18745	5.3349	.98288	23
38	.18452	.18775	5.3263	.98283	22
39	.18481	.18805	5.3178	.98277	21
40	.18509	.18835	5.3093	.98272	20
41	.18538	.18865	5.3008	.98267	19
42	.18567	.18895	5.2924	.98261	18
43	.18595	.18925	5.2839	.98256	17
44	.18624	.18955	5.2755	.98250	16
45	.18652	.18986	5.2672	.98245	15
46	.18681	.19016	5.2588	.98240	14
47	.18710	.19046	5.2505	.98234	13
48	.18738	.19076	5.2422	.98229	12
49	.18767	.19106	5.2339	.98223	11
50	.18795	.19136	5.2257	.98218	10
51	.18824	.19166	5.2174	.98212	9
52	.18852	.19197	5.2092	.98207	8
53	.18881	.19227	5.2011	.98201	7
54	.18910	.19257	5.1929	.98196	6
55	.18938	.19287	5.1848	.98190	5
56	.18967	.19317	5.1767	.98185	4
57	.18995	.19347	5.1686	.98179	3
58	.19024	.19378	5.1606	.98174	2
59	.19052	.19408	5.1526	.98168	1
60	.19081	.19438	5.1446	.98163	0
′	Cos	Cot	Tan	Sin	′

100° (280°) (259°) **79°**

11° (191°) (348°) **168°**

′	Sin	Tan	Cot	Cos	′
0	.19081	.19438	5.1446	.98163	60
1	.19109	.19468	5.1366	.98157	59
2	.19138	.19498	5.1286	.98152	58
3	.19167	.19529	5.1207	.98146	57
4	.19195	.19559	5.1128	.98140	56
5	.19224	.19589	5.1049	.98135	55
6	.19252	.19619	5.0970	.98129	54
7	.19281	.19649	5.0892	.98124	53
8	.19309	.19680	5.0814	.98118	52
9	.19338	.19710	5.0736	.98112	51
10	.19366	.19740	5.0658	.98107	50
11	.19395	.19770	5.0581	.98101	49
12	.19423	.19801	5.0504	.98096	48
13	.19452	.19831	5.0427	.98090	47
14	.19481	.19861	5.0350	.98084	46
15	.19509	.19891	5.0273	.98079	45
16	.19538	.19921	5.0197	.98073	44
17	.19566	.19952	5.0121	.98067	43
18	.19595	.19982	5.0045	.98061	42
19	.19623	.20012	4.9969	.98056	41
20	.19652	.20042	4.9894	.98050	40
21	.19680	.20073	4.9819	.98044	39
22	.19709	.20103	4.9744	.98039	38
23	.19737	.20133	4.9669	.98033	37
24	.19766	.20164	4.9594	.98027	36
25	.19794	.20194	4.9520	.98021	35
26	.19823	.20224	4.9446	.98016	34
27	.19851	.20254	4.9372	.98010	33
28	.19880	.20285	4.9298	.98004	32
29	.19908	.20315	4.9225	.97998	31
30	.19937	.20345	4.9152	.97992	30
31	.19965	.20376	4.9078	.97987	29
32	.19994	.20406	4.9006	.97981	28
33	.20022	.20436	4.8933	.97975	27
34	.20051	.20466	4.8860	.97969	26
35	.20079	.20497	4.8788	.97963	25
36	.20108	.20527	4.8716	.97958	24
37	.20136	.20557	4.8644	.97952	23
38	.20165	.20588	4.8573	.97946	22
39	.20193	.20618	4.8501	.97940	21
40	.20222	.20648	4.8430	.97934	20
41	.20250	.20679	4.8359	.97928	19
42	.20279	.20709	4.8288	.97922	18
43	.20307	.20739	4.8218	.97916	17
44	.20336	.20770	4.8147	.97910	16
45	.20364	.20800	4.8077	.97905	15
46	.20393	.20830	4.8007	.97899	14
47	.20421	.20861	4.7937	.97893	13
48	.20450	.20891	4.7867	.97887	12
49	.20478	.20921	4.7798	.97881	11
50	.20507	.20952	4.7729	.97875	10
51	.20535	.20982	4.7659	.97869	9
52	.20563	.21013	4.7591	.97863	8
53	.20592	.21043	4.7522	.97857	7
54	.20620	.21073	4.7453	.97851	6
55	.20649	.21104	4.7385	.97845	5
56	.20677	.21134	4.7317	.97839	4
57	.20706	.21164	4.7249	.97833	3
58	.20734	.21195	4.7181	.97827	2
59	.20763	.21225	4.7114	.97821	1
60	.20791	.21256	4.7046	.97815	0
′	Cos	Cot	Tan	Sin	′

101° (281°) (258°) **78°**

TABLE V TRIGONOMETRIC FUNCTIONS OF ANGLES (Continued)

12° (192°) — (347°) **167°**

′	Sin	Tan	Cot	Cos	′
0	.20791	.21256	4.7046	.97815	60
1	.20820	.21286	4.6979	.97809	59
2	.20848	.21316	4.6912	.97803	58
3	.20877	.21347	4.6845	.97797	57
4	.20905	.21377	4.6779	.97791	56
5	.20933	.21408	4.6712	.97784	55
6	.20962	.21438	4.6646	.97778	54
7	.20990	.21469	4.6580	.97772	53
8	.21019	.21499	4.6514	.97766	52
9	.21047	.21529	4.6448	.97760	51
10	.21076	.21560	4.6382	.97754	50
11	.21104	.21590	4.6317	.97748	49
12	.21132	.21621	4.6252	.97742	48
13	.21161	.21651	4.6187	.97735	47
14	.21189	.21682	4.6122	.97729	46
15	.21218	.21712	4.6057	.97723	45
16	.21246	.21743	4.5993	.97717	44
17	.21275	.21773	4.5928	.97711	43
18	.21303	.21804	4.5864	.97705	42
19	.21331	.21834	4.5800	.97698	41
20	.21360	.21864	4.5736	.97692	40
21	.21388	.21895	4.5673	.97686	39
22	.21417	.21925	4.5609	.97680	38
23	.21445	.21956	4.5546	.97673	37
24	.21474	.21986	4.5483	.97667	36
25	.21502	.22017	4.5420	.97661	35
26	.21530	.22047	4.5357	.97655	34
27	.21559	.22078	4.5294	.97648	33
28	.21587	.22108	4.5232	.97642	32
29	.21616	.22139	4.5169	.97636	31
30	.21644	.22169	4.5107	.97630	30
31	.21672	.22200	4.5045	.97623	29
32	.21701	.22231	4.4983	.97617	28
33	.21729	.22261	4.4922	.97611	27
34	.21758	.22292	4.4860	.97604	26
35	.21786	.22322	4.4799	.97598	25
36	.21814	.22353	4.4737	.97592	24
37	.21843	.22383	4.4676	.97585	23
38	.21871	.22414	4.4615	.97579	22
39	.21899	.22444	4.4555	.97573	21
40	.21928	.22475	4.4494	.97566	20
41	.21956	.22505	4.4434	.97560	19
42	.21985	.22536	4.4373	.97553	18
43	.22013	.22567	4.4313	.97547	17
44	.22041	.22597	4.4253	.97541	16
45	.22070	.22628	4.4194	.97534	15
46	.22098	.22658	4.4134	.97528	14
47	.22126	.22689	4.4075	.97521	13
48	.22155	.22719	4.4015	.97515	12
49	.22183	.22750	4.3956	.97508	11
50	.22212	.22781	4.3897	.97502	10
51	.22240	.22811	4.3838	.97496	9
52	.22268	.22842	4.3779	.97489	8
53	.22297	.22872	4.3721	.97483	7
54	.22325	.22903	4.3662	.97476	6
55	.22353	.22934	4.3604	.97470	5
56	.22382	.22964	4.3546	.97463	4
57	.22410	.22995	4.3488	.97457	3
58	.22438	.23026	4.3430	.97450	2
59	.22467	.23056	4.3372	.97444	1
60	.22495	.23087	4.3315	.97437	0
′	Cos	Cot	Tan	Sin	′

102° (282°) — (257°) **77°**

13° (193°) — (346°) **166°**

′	Sin	Tan	Cot	Cos	′
0	.22495	.23087	4.3315	.97437	60
1	.22523	.23117	4.3257	.97430	59
2	.22552	.23148	4.3200	.97424	58
3	.22580	.23179	4.3143	.97417	57
4	.22608	.23209	4.3086	.97411	56
5	.22637	.23240	4.3029	.97404	55
6	.22665	.23271	4.2972	.97398	54
7	.22693	.23301	4.2916	.97391	53
8	.22722	.23332	4.2859	.97384	52
9	.22750	.23363	4.2803	.97378	51
10	.22778	.23393	4.2747	.97371	50
11	.22807	.23424	4.2691	.97365	49
12	.22835	.23455	4.2635	.97358	48
13	.22863	.23485	4.2580	.97351	47
14	.22892	.23516	4.2524	.97345	46
15	.22920	.23547	4.2468	.97338	45
16	.22948	.23578	4.2413	.97331	44
17	.22977	.23608	4.2358	.97325	43
18	.23005	.23639	4.2303	.97318	42
19	.23033	.23670	4.2248	.97311	41
20	.23062	.23700	4.2193	.97304	40
21	.23090	.23731	4.2139	.97298	39
22	.23118	.23762	4.2084	.97291	38
23	.23146	.23793	4.2030	.97284	37
24	.23175	.23823	4.1976	.97278	36
25	.23203	.23854	4.1922	.97271	35
26	.23231	.23885	4.1868	.97264	34
27	.23260	.23916	4.1814	.97257	33
28	.23288	.23946	4.1760	.97251	32
29	.23316	.23977	4.1706	.97244	31
30	.23345	.24008	4.1653	.97237	30
31	.23373	.24039	4.1600	.97230	29
32	.23401	.24069	4.1547	.97223	28
33	.23429	.24100	4.1493	.97217	27
34	.23458	.24131	4.1441	.97210	26
35	.23486	.24162	4.1388	.97203	25
36	.23514	.24193	4.1335	.97196	24
37	.23542	.24223	4.1282	.97189	23
38	.23571	.24254	4.1230	.97182	22
39	.23599	.24285	4.1178	.97176	21
40	.23627	.24316	4.1126	.97169	20
41	.23656	.24347	4.1074	.97162	19
42	.23684	.24377	4.1022	.97155	18
43	.23712	.24408	4.0970	.97148	17
44	.23740	.24439	4.0918	.97141	16
45	.23769	.24470	4.0867	.97134	15
46	.23797	.24501	4.0815	.97127	14
47	.23825	.24532	4.0764	.97120	13
48	.23853	.24562	4.0713	.97113	12
49	.23882	.24593	4.0662	.97106	11
50	.23910	.24624	4.0611	.97100	10
51	.23938	.24655	4.0560	.97093	9
52	.23966	.24686	4.0509	.97086	8
53	.23995	.24717	4.0459	.97079	7
54	.24023	.24747	4.0408	.97072	6
55	.24051	.24778	4.0358	.97065	5
56	.24079	.24809	4.0308	.97058	4
57	.24108	.24840	4.0257	.97051	3
58	.24136	.24871	4.0207	.97044	2
59	.24164	.24902	4.0158	.97037	1
60	.24192	.24933	4.0108	.97030	0
′	Cos	Cot	Tan	Sin	′

103° (283°) — (256°) **76°**

TABLE V TRIGONOMETRIC FUNCTIONS OF ANGLES (Continued)

14° (194°) (345°) **165°**

′	Sin	Tan	Cot	Cos	′
0	.24192	.24933	4.0108	.97030	60
1	.24220	.24964	4.0058	.97023	59
2	.24249	.24995	4.0009	.97015	58
3	.24277	.25026	3.9959	.97008	57
4	.24305	.25056	3.9910	.97001	56
5	.24333	.25087	3.9861	.96994	55
6	.24362	.25118	3.9812	.96987	54
7	.24390	.25149	3.9763	.96980	53
8	.24418	.25180	3.9714	.96973	52
9	.24446	.25211	3.9665	.96966	51
10	.24474	.25242	3.9617	.96959	50
11	.24503	.25273	3.9568	.96952	49
12	.24531	.25304	3.9520	.96945	48
13	.24559	.25335	3.9471	.96937	47
14	.24587	.25366	3.9423	.96930	46
15	.24615	.25397	3.9375	.96923	45
16	.24644	.25428	3.9327	.96916	44
17	.24672	.25459	3.9279	.96909	43
18	.24700	.25490	3.9232	.96902	42
19	.24728	.25521	3.9184	.96894	41
20	.24756	.25552	3.9136	.96887	40
21	.24784	.25583	3.9089	.96880	39
22	.24813	.25614	3.9042	.96873	38
23	.24841	.25645	3.8995	.96866	37
24	.24869	.25676	3.8947	.96858	36
25	.24897	.25707	3.8900	.96851	35
26	.24925	.25738	3.8854	.96844	34
27	.24954	.25769	3.8807	.96837	33
28	.24982	.25800	3.8760	.96829	32
29	.25010	.25831	3.8714	.96822	31
30	.25038	.25862	3.8667	.96815	30
31	.25066	.25893	3.8621	.96807	29
32	.25094	.25924	3.8575	.96800	28
33	.25122	.25955	3.8528	.96793	27
34	.25151	.25986	3.8482	.96786	26
35	.25179	.26017	3.8436	.96778	25
36	.25207	.26048	3.8391	.96771	24
37	.25235	.26079	3.8345	.96764	23
38	.25263	.26110	3.8299	.96756	22
39	.25291	.26141	3.8254	.96749	21
40	.25320	.26172	3.8208	.96742	20
41	.25348	.26203	3.8163	.96734	19
42	.25376	.26235	3.8118	.96727	18
43	.25404	.26266	3.8073	.96719	17
44	.25432	.26297	3.8028	.96712	16
45	.25460	.26328	3.7983	.96705	15
46	.25488	.26359	3.7938	.96697	14
47	.25516	.26390	3.7893	.96690	13
48	.25545	.26421	3.7848	.96682	12
49	.25573	.26452	3.7804	.96675	11
50	.25601	.26483	3.7760	.96667	10
51	.25629	.26515	3.7715	.96660	9
52	.25657	.26546	3.7671	.96653	8
53	.25685	.26577	3.7627	.96645	7
54	.25713	.26608	3.7583	.96638	6
55	.25741	.26639	3.7539	.96630	5
56	.25769	.26670	3.7495	.96623	4
57	.25798	.26701	3.7451	.96615	3
58	.25826	.26733	3.7408	.96608	2
59	.25854	.26764	3.7364	.96600	1
60	.25882	.26795	3.7321	.96593	0
′	Cos	Cot	Tan	Sin	′

104° (284°) (255°) **75°**

15° (195°) (344°) **164°**

′	Sin	Tan	Cot	Cos	′
0	.25882	.26795	3.7321	.96593	60
1	.25910	.26826	3.7277	.96585	59
2	.25938	.26857	3.7234	.96578	58
3	.25966	.26888	3.7191	.96570	57
4	.25994	.26920	3.7148	.96562	56
5	.26022	.26951	3.7105	.96555	55
6	.26050	.26982	3.7062	.96547	54
7	.26079	.27013	3.7019	.96540	53
8	.26107	.27044	3.6976	.96532	52
9	.26135	.27076	3.6933	.96524	51
10	.26163	.27107	3.6891	.96517	50
11	.26191	.27138	3.6848	.96509	49
12	.26219	.27169	3.6806	.96502	48
13	.26247	.27201	3.6764	.96494	47
14	.26275	.27232	3.6722	.96486	46
15	.26303	.27263	3.6680	.96479	45
16	.26331	.27294	3.6638	.96471	44
17	.26359	.27326	3.6596	.96463	43
18	.26387	.27357	3.6554	.96456	42
19	.26415	.27388	3.6512	.96448	41
20	.26443	.27419	3.6470	.96440	40
21	.26471	.27451	3.6429	.96433	39
22	.26500	.27482	3.6387	.96425	38
23	.26528	.27513	3.6346	.96417	37
24	.26556	.27545	3.6305	.96410	36
25	.26584	.27576	3.6264	.96402	35
26	.26612	.27607	3.6222	.96394	34
27	.26640	.27638	3.6181	.96386	33
28	.26668	.27670	3.6140	.96379	32
29	.26696	.27701	3.6100	.96371	31
30	.26724	.27732	3.6059	.96363	30
31	.26752	.27764	3.6018	.96355	29
32	.26780	.27795	3.5978	.96347	28
33	.26808	.27826	3.5937	.96340	27
34	.26836	.27858	3.5897	.96332	26
35	.26864	.27889	3.5856	.96324	25
36	.26892	.27921	3.5816	.96316	24
37	.26920	.27952	3.5776	.96308	23
38	.26948	.27983	3.5736	.96301	22
39	.26976	.28015	3.5696	.96293	21
40	.27004	.28046	3.5656	.96285	20
41	.27032	.28077	3.5616	.96277	19
42	.27060	.28109	3.5576	.96269	18
43	.27088	.28140	3.5536	.96261	17
44	.27116	.28172	3.5497	.96253	16
45	.27144	.28203	3.5457	.96246	15
46	.27172	.28234	3.5418	.96238	14
47	.27200	.28266	3.5379	.96230	13
48	.27228	.28297	3.5339	.96222	12
49	.27256	.28329	3.5300	.96214	11
50	.27284	.28360	3.5261	.96206	10
51	.27312	.28391	3.5222	.96198	9
52	.27340	.28423	3.5183	.96190	8
53	.27368	.28454	3.5144	.96182	7
54	.27396	.28486	3.5105	.96174	6
55	.27424	.28517	3.5067	.96166	5
56	.27452	.28549	3.5028	.96158	4
57	.27480	.28580	3.4989	.96150	3
58	.27508	.28612	3.4951	.96142	2
59	.27536	.28643	3.4912	.96134	1
60	.27564	.28675	3.4874	.96126	0
′	Cos	Cot	Tan	Sin	′

105° (285°) (254°) **74°**

TABLE V **TRIGONOMETRIC FUNCTIONS OF ANGLES (Continued)**

16° (196°) (343°) **163°**

′	Sin	Tan	Cot	Cos	′
0	.27564	.28675	3.4874	.96126	60
1	.27592	.28706	3.4836	.96118	59
2	.27620	.28738	3.4798	.96110	58
3	.27648	.28769	3.4760	.96102	57
4	.27676	.28801	3.4722	.96094	56
5	.27704	.28832	3.4684	.96086	55
6	.27731	.28864	3.4646	.96078	54
7	.27759	.28895	3.4608	.96070	53
8	.27787	.28927	3.4570	.96062	52
9	.27815	.28958	3.4533	.96054	51
10	.27843	.28990	3.4495	.96046	50
11	.27871	.29021	3.4458	.96037	49
12	.27899	.29053	3.4420	.96029	48
13	.27927	.29084	3.4383	.96021	47
14	.27955	.29116	3.4346	.96013	46
15	.27983	.29147	3.4308	.96005	45
16	.28011	.29179	3.4271	.95997	44
17	.28039	.29210	3.4234	.95989	43
18	.28067	.29242	3.4197	.95981	42
19	.28095	.29274	3.4160	.95972	41
20	.28123	.29305	3.4124	.95964	40
21	.28150	.29337	3.4087	.95956	39
22	.28178	.29368	3.4050	.95948	38
23	.28206	.29400	3.4014	.95940	37
24	.28234	.29432	3.3977	.95931	36
25	.28262	.29463	3.3941	.95923	35
26	.28290	.29495	3.3904	.95915	34
27	.28318	.29526	3.3868	.95907	33
28	.28346	.29558	3.3832	.95898	32
29	.28374	.29590	3.3796	.95890	31
30	.28402	.29621	3.3759	.95882	30
31	.28429	.29653	3.3723	.95874	29
32	.28457	.29685	3.3687	.95865	28
33	.28485	.29716	3.3652	.95857	27
34	.28513	.29748	3.3616	.95849	26
35	.28541	.29780	3.3580	.95841	25
36	.28569	.29811	3.3544	.95832	24
37	.28597	.29843	3.3509	.95824	23
38	.28625	.29875	3.3473	.95816	22
39	.28652	.29906	3.3438	.95807	21
40	.28680	.29938	3.3402	.95799	20
41	.28708	.29970	3.3367	.95791	19
42	.28736	.30001	3.3332	.95782	18
43	.28764	.30033	3.3297	.95774	17
44	.28792	.30065	3.3261	.95766	16
45	.28820	.30097	3.3226	.95757	15
46	.28847	.30128	3.3191	.95749	14
47	.28875	.30160	3.3156	.95740	13
48	.28903	.30192	3.3122	.95732	12
49	.28931	.30224	3.3087	.95724	11
50	.28959	.30255	3.3052	.95715	10
51	.28987	.30287	3.3017	.95707	9
52	.29015	.30319	3.2983	.95698	8
53	.29042	.30351	3.2948	.95690	7
54	.29070	.30382	3.2914	.95681	6
55	.29098	.30414	3.2879	.95673	5
56	.29126	.30446	3.2845	.95664	4
57	.29154	.30478	3.2811	.95656	3
58	.29182	.30509	3.2777	.95647	2
59	.29209	.30541	3.2743	.95639	1
60	.29237	.30573	3.2709	.95630	0
′	Cos	Cot	Tan	Sin	′

106° (286°) (253°) **73°**

17° (197°) (342°) **162°**

′	Sin	Tan	Cot	Cos	′
0	.29237	.30573	3.2709	.95630	60
1	.29265	.30605	3.2675	.95622	59
2	.29293	.30637	3.2641	.95613	58
3	.29321	.30669	3.2607	.95605	57
4	.29348	.30700	3.2573	.95596	56
5	.29376	.30732	3.2539	.95588	55
6	.29404	.30764	3.2506	.95579	54
7	.29432	.30796	3.2472	.95571	53
8	.29460	.30828	3.2438	.95562	52
9	.29487	.30860	3.2405	.95554	51
10	.29515	.30891	3.2371	.95545	50
11	.29543	.30923	3.2338	.95536	49
12	.29571	.30955	3.2305	.95528	48
13	.29599	.30987	3.2272	.95519	47
14	.29626	.31019	3.2238	.95511	46
15	.29654	.31051	3.2205	.95502	45
16	.29682	.31083	3.2172	.95493	44
17	.29710	.31115	3.2139	.95485	43
18	.29737	.31147	3.2106	.95476	42
19	.29765	.31178	3.2073	.95467	41
20	.29793	.31210	3.2041	.95459	40
21	.29821	.31242	3.2008	.95450	39
22	.29849	.31274	3.1975	.95441	38
23	.29876	.31306	3.1943	.95433	37
24	.29904	.31338	3.1910	.95424	36
25	.29932	.31370	3.1878	.95415	35
26	.29960	.31402	3.1845	.95407	34
27	.29987	.31434	3.1813	.95398	33
28	.30015	.31466	3.1780	.95389	32
29	.30043	.31498	3.1748	.95380	31
30	.30071	.31530	3.1716	.95372	30
31	.30098	.31562	3.1684	.95363	29
32	.30126	.31594	3.1652	.95354	28
33	.30154	.31626	3.1620	.95345	27
34	.30182	.31658	3.1588	.95337	26
35	.30209	.31690	3.1556	.95328	25
36	.30237	.31722	3.1524	.95319	24
37	.30265	.31754	3.1492	.95310	23
38	.30292	.31786	3.1460	.95301	22
39	.30320	.31818	3.1429	.95293	21
40	.30348	.31850	3.1397	.95284	20
41	.30376	.31882	3.1366	.95275	19
42	.30403	.31914	3.1334	.95266	18
43	.30431	.31946	3.1303	.95257	17
44	.30459	.31978	3.1271	.95248	16
45	.30486	.32010	3.1240	.95240	15
46	.30514	.32042	3.1209	.95231	14
47	.30542	.32074	3.1178	.95222	13
48	.30570	.32106	3.1146	.95213	12
49	.30597	.32139	3.1115	.95204	11
50	.30625	.32171	3.1084	.95195	10
51	.30653	.32203	3.1053	.95186	9
52	.30680	.32235	3.1022	.95177	8
53	.30708	.32267	3.0991	.95168	7
54	.30736	.32299	3.0961	.95159	6
55	.30763	.32331	3.0930	.95150	5
56	.30791	.32363	3.0899	.95142	4
57	.30819	.32396	3.0868	.95133	3
58	.30846	.32428	3.0838	.95124	2
59	.30874	.32460	3.0807	.95115	1
60	.30902	.32492	3.0777	.95106	0
′	Cos	Cot	Tan	Sin	′

107° (287°) (252°) **72°**

TABLE V TRIGONOMETRIC FUNCTIONS OF ANGLES (Continued)

18° (198°) (341°) **161°**

′	Sin	Tan	Cot	Cos	′
0	.30902	.32492	3.0777	.95106	60
1	.30929	.32524	3.0746	.95097	59
2	.30957	.32556	3.0716	.95088	58
3	.30985	.32588	3.0686	.95079	57
4	.31012	.32621	3.0655	.95070	56
5	.31040	.32653	3.0625	.95061	55
6	.31068	.32685	3.0595	.95052	54
7	.31095	.32717	3.0565	.95043	53
8	.31123	.32749	3.0535	.95033	52
9	.31151	.32782	3.0505	.95024	51
10	.31178	.32814	3.0475	.95015	50
11	.31206	.32846	3.0445	.95006	49
12	.31233	.32878	3.0415	.94997	48
13	.31261	.32911	3.0385	.94988	47
14	.31289	.32943	3.0356	.94979	46
15	.31316	.32975	3.0326	.94970	45
16	.31344	.33007	3.0296	.94961	44
17	.31372	.33040	3.0267	.94952	43
18	.31399	.33072	3.0237	.94943	42
19	.31427	.33104	3.0208	.94933	41
20	.31454	.33136	3.0178	.94924	40
21	.31482	.33169	3.0149	.94915	39
22	.31510	.33201	3.0120	.94906	38
23	.31537	.33233	3.0090	.94897	37
24	.31565	.33266	3.0061	.94888	36
25	.31593	.33298	3.0032	.94878	35
26	.31620	.33330	3.0003	.94869	34
27	.31648	.33363	2.9974	.94860	33
28	.31675	.33395	2.9945	.94851	32
29	.31703	.33427	2.9916	.94842	31
30	.31730	.33460	2.9887	.94832	30
31	.31758	.33492	2.9858	.94823	29
32	.31786	.33524	2.9829	.94814	28
33	.31813	.33557	2.9800	.94805	27
34	.31841	.33589	2.9772	.94795	26
35	.31868	.33621	2.9743	.94786	25
36	.31896	.33654	2.9714	.94777	24
37	.31923	.33686	2.9686	.94768	23
38	.31951	.33718	2.9657	.94758	22
39	.31979	.33751	2.9629	.94749	21
40	.32006	.33783	2.9600	.94740	20
41	.32034	.33816	2.9572	.94730	19
42	.32061	.33848	2.9544	.94721	18
43	.32089	.33881	2.9515	.94712	17
44	.32116	.33913	2.9487	.94702	16
45	.32144	.33945	2.9459	.94693	15
46	.32171	.33978	2.9431	.94684	14
47	.32199	.34010	2.9403	.94674	13
48	.32227	.34043	2.9375	.94665	12
49	.32254	.34075	2.9347	.94656	11
50	.32282	.34108	2.9319	.94646	10
51	.32309	.34140	2.9291	.94637	9
52	.32337	.34173	2.9263	.94627	8
53	.32364	.34205	2.9235	.94618	7
54	.32392	.34238	2.9208	.94609	6
55	.32419	.34270	2.9180	.94599	5
56	.32447	.34303	2.9152	.94590	4
57	.32474	.34335	2.9125	.94580	3
58	.32502	.34368	2.9097	.94571	2
59	.32529	.34400	2.9070	.94561	1
60	.32557	.34433	2.9042	.94552	0
′	Cos	Cot	Tan	Sin	′

108° (288°) (251°) **71°**

19° (199°) (340°) **160°**

′	Sin	Tan	Cot	Cos	′
0	.32557	.34433	2.9042	.94552	60
1	.32584	.34465	2.9015	.94542	59
2	.32612	.34498	2.8987	.94533	58
3	.32639	.34530	2.8960	.94523	57
4	.32667	.34563	2.8933	.94514	56
5	.32694	.34596	2.8905	.94504	55
6	.32722	.34628	2.8878	.94495	54
7	.32749	.34661	2.8851	.94485	53
8	.32777	.34693	2.8824	.94476	52
9	.32804	.34726	2.8797	.94466	51
10	.32832	.34758	2.8770	.94457	50
11	.32859	.34791	2.8743	.94447	49
12	.32887	.34824	2.8716	.94438	48
13	.32914	.34856	2.8689	.94428	47
14	.32942	.34889	2.8662	.94418	46
15	.32969	.34922	2.8636	.94409	45
16	.32997	.34954	2.8609	.94399	44
17	.33024	.34987	2.8582	.94390	43
18	.33051	.35020	2.8556	.94380	42
19	.33079	.35052	2.8529	.94370	41
20	.33106	.35085	2.8502	.94361	40
21	.33134	.35118	2.8476	.94351	39
22	.33161	.35150	2.8449	.94342	38
23	.33189	.35183	2.8423	.94332	37
24	.33216	.35216	2.8397	.94322	36
25	.33244	.35248	2.8370	.94313	35
26	.33271	.35281	2.8344	.94303	34
27	.33298	.35314	2.8318	.94293	33
28	.33326	.35346	2.8291	.94284	32
29	.33353	.35379	2.8265	.94274	31
30	.33381	.35412	2.8239	.94264	30
31	.33408	.35445	2.8213	.94254	29
32	.33436	.35477	2.8187	.94245	28
33	.33463	.35510	2.8161	.94235	27
34	.33490	.35543	2.8135	.94225	26
35	.33518	.35576	2.8109	.94215	25
36	.33545	.35608	2.8083	.94206	24
37	.33573	.35641	2.8057	.94196	23
38	.33600	.35674	2.8032	.94186	22
39	.33627	.35707	2.8006	.94176	21
40	.33655	.35740	2.7980	.94167	20
41	.33682	.35772	2.7955	.94157	19
42	.33710	.35805	2.7929	.94147	18
43	.33737	.35838	2.7903	.94137	17
44	.33764	.35871	2.7878	.94127	16
45	.33792	.35904	2.7852	.94118	15
46	.33819	.35937	2.7827	.94108	14
47	.33846	.35969	2.7801	.94098	13
48	.33874	.36002	2.7776	.94088	12
49	.33901	.36035	2.7751	.94078	11
50	.33929	.36068	2.7725	.94068	10
51	.33956	.36101	2.7700	.94058	9
52	.33983	.36134	2.7675	.94049	8
53	.34011	.36167	2.7650	.94039	7
54	.34038	.36199	2.7625	.94029	6
55	.34065	.36232	2.7600	.94019	5
56	.34093	.36265	2.7575	.94009	4
57	.34120	.36298	2.7550	.93999	3
58	.34147	.36331	2.7525	.93989	2
59	.34175	.36364	2.7500	.93979	1
60	.34202	.36397	2.7475	.93969	0
′	Cos	Cot	Tan	Sin	′

109° (289°) (250°) **70°**

TABLE V TRIGONOMETRIC FUNCTIONS OF ANGLES (Continued)

20° (200°) (339°) **159°**

′	Sin	Tan	Cot	Cos	′
0	.34202	.36397	2.7475	.93969	60
1	.34229	.36430	2.7450	.93959	59
2	.34257	.36463	2.7425	.93949	58
3	.34284	.36496	2.7400	.93939	57
4	.34311	.36529	2.7376	.93929	56
5	.34339	.36562	2.7351	.93919	55
6	.34366	.36595	2.7326	.93909	54
7	.34393	.36628	2.7302	.93899	53
8	.34421	.36661	2.7277	.93889	52
9	.34448	.36694	2.7253	.93879	51
10	.34475	.36727	2.7228	.93869	50
11	.34503	.36760	2.7204	.93859	49
12	.34530	.36793	2.7179	.93849	48
13	.34557	.36826	2.7155	.93839	47
14	.34584	.36859	2.7130	.93829	46
15	.34612	.36892	2.7106	.93819	45
16	.34639	.36925	2.7082	.93809	44
17	.34666	.36958	2.7058	.93799	43
18	.34694	.36991	2.7034	.93789	42
19	.34721	.37024	2.7009	.93779	41
20	.34748	.37057	2.6985	.93769	40
21	.34775	.37090	2.6961	.93759	39
22	.84803	.37123	2.6937	.93748	38
23	.34830	.37157	2.6913	.93738	37
24	.34857	.37190	2.6889	.93728	36
25	.34884	.37223	2.6865	.93718	35
26	.34912	.37256	2.6841	.93708	34
27	.34939	.37289	2.6818	.93698	33
28	.34966	.37322	2.6794	.93688	32
29	.34993	.37355	2.6770	.93677	31
30	.35021	.37388	2.6746	.93667	30
31	.35048	.37422	2.6723	.93657	29
32	.35075	.37455	2.6699	.93647	28
33	.35102	.37488	2.6675	.93637	27
34	.35130	.37521	2.6652	.93626	26
35	.35157	.37554	2.6628	.93616	25
36	.35184	.37588	2.6605	.93606	24
37	.35211	.37621	2.6581	.93596	23
38	.35239	.37654	2.6558	.93585	22
39	.35266	.37687	2.6534	.93575	21
40	.35293	.37720	2.6511	.93565	20
41	.35320	.37754	2.6488	.93555	19
42	.35347	.37787	2.6464	.93544	18
43	.35375	.37820	2.6441	.93534	17
44	.35402	.37853	2.6418	.93524	16
45	.35429	.37887	2.6395	.93514	15
46	.35456	.37920	2.6371	.93503	14
47	.35484	.37953	2.6348	.93493	13
48	.35511	.37986	2.6325	.93483	12
49	.35538	.38020	2.6302	.93472	11
50	.35565	.38053	2.6279	.93462	10
51	.35592	.38086	2.6256	.93452	9
52	.35619	.38120	2.6233	.93441	8
53	.35647	.38153	2.6210	.93431	7
54	.35674	.38186	2.6187	.93420	6
55	.35701	.38220	2.6165	.93410	5
56	.35728	.38253	2.6142	.93400	4
57	.35755	.38286	2.6119	.93389	3
58	.35782	.38320	2.6096	.93379	2
59	.35810	.38353	2.6074	.93368	1
60	.35837	.38386	2.6051	.93358	0
′	Cos	Cot	Tan	Sin	′

110° (290°) (249°) **69°**

21° (201°) (338°) **158°**

′	Sin	Tan	Cot	Cos	′
0	.35837	.38386	2.6051	.93358	60
1	.35864	.38420	2.6028	.93348	59
2	.35891	.38453	2.6006	.93337	58
3	.35918	.38487	2.5983	.93327	57
4	.35945	.38520	2.5961	.93316	56
5	.35973	.38553	2.5938	.93306	55
6	.36000	.38587	2.5916	.93295	54
7	.36027	.38620	2.5893	.93285	53
8	.36054	.38654	2.5871	.93274	52
9	.36081	.38687	2.5848	.93264	51
10	.36108	.38721	2.5826	.93253	50
11	.36135	.38754	2.5804	.93243	49
12	.36162	.38787	2.5782	.93232	48
13	.36190	.38821	2.5759	.93222	47
14	.36217	.38854	2.5737	.93211	46
15	.36244	.38888	2.5715	.93201	45
16	.36271	.38921	2.5693	.93190	44
17	.36298	.38955	2.5671	.93180	43
18	.36325	.38988	2.5649	.93169	42
19	.36352	.39022	2.5627	.93159	41
20	.36379	.39055	2.5605	.93148	40
21	.36406	.39089	2.5583	.93137	39
22	.36434	.39122	2.5561	.93127	38
23	.36461	.39156	2.5539	.93116	37
24	.36488	.39190	2.5517	.93106	36
25	.36515	.39223	2.5495	.93095	35
26	.36542	.39257	2.5473	.93084	34
27	.36569	.39290	2.5452	.93074	33
28	.36596	.39324	2.5430	.93063	32
29	.36623	.39357	2.5408	.93052	31
30	.36650	.39391	2.5386	.93042	30
31	.36677	.39425	2.5365	.93031	29
32	.36704	.39458	2.5343	.93020	28
33	.36731	.39492	2.5322	.93010	27
34	.36758	.39526	2.5300	.92999	26
35	.36785	.39559	2.5279	.92988	25
36	.36812	.39593	2.5257	.92978	24
37	.36839	.39626	2.5236	.92967	23
38	.36867	.39660	2.5214	.92956	22
39	.36894	.39694	2.5193	.92945	21
40	.36921	.39727	2.5172	.92935	20
41	.36948	.39761	2.5150	.92924	19
42	.36975	.39795	2.5129	.92913	18
43	.37002	.39829	2.5108	.92902	17
44	.37029	.39862	2.5086	.92892	16
45	.37056	.39896	2.5065	.92881	15
46	.37083	.39930	2.5044	.92870	14
47	.37110	.39963	2.5023	.92859	13
48	.37137	.39997	2.5002	.92849	12
49	.37164	.40031	2.4981	.92838	11
50	.37191	.40065	2.4960	.92827	10
51	.37218	.40098	2.4939	.92816	9
52	.37245	.40132	2.4918	.92805	8
53	.37272	.40166	2.4897	.92794	7
54	.37299	.40200	2.4876	.92784	6
55	.37326	.40234	2.4855	.92773	5
56	.37353	.40267	2.4834	.92762	4
57	.37380	.40301	2.4813	.92751	3
58	.37407	.40335	2.4792	.92740	2
59	.37434	.40369	2.4772	.92729	1
60	.37461	.40403	2.4751	.92718	0
′	Cos	Cot	Tan	Sin	′

111° (291°) (248°) **68°**

TABLE V TRIGONOMETRIC FUNCTIONS OF ANGLES (Continued)

22° (202°) (337°) **157°**

′	Sin	Tan	Cot	Cos	′
0	.37461	.40403	2.4751	.92718	60
1	.37488	.40436	2.4730	.92707	59
2	.37515	.40470	2.4709	.92697	58
3	.37542	.40504	2.4689	.92686	57
4	.37569	.40538	2.4668	.92675	56
5	.37595	.40572	2.4648	.92664	55
6	.37622	.40606	2.4627	.92653	54
7	.37649	.40640	2.4606	.92642	53
8	.37676	.40674	2.4586	.92631	52
9	.37703	.40707	2.4566	.92620	51
10	.37730	.40741	2.4545	.92609	50
11	.37757	.40775	2.4525	.92598	49
12	.37784	.40809	2.4504	.92587	48
13	.37811	.40843	2.4484	.92576	47
14	.37838	.40877	2.4464	.92565	46
15	.37865	.40911	2.4443	.92554	45
16	.37892	.40945	2.4423	.92543	44
17	.37919	.40979	2.4403	.92532	43
18	.37946	.41013	2.4383	.92521	42
19	.37973	.41047	2.4362	.92510	41
20	.37999	.41081	2.4342	.92499	40
21	.38026	.41115	2.4322	.92488	39
22	.38053	.41149	2.4302	.92477	38
23	.38080	.41183	2.4282	.92466	37
24	.38107	.41217	2.4262	.92455	36
25	.38134	.41251	2.4242	.92444	35
26	.38161	.41285	2.4222	.92432	34
27	.38188	.41319	2.4202	.92421	33
28	.38215	.41353	2.4182	.92410	32
29	.38241	.41387	2.4162	.92399	31
30	.38268	.41421	2.4142	.92388	30
31	.38295	.41455	2.4122	.92377	29
32	.38322	.41490	2.4102	.92366	28
33	.38349	.41524	2.4083	.92355	27
34	.38376	.41558	2.4063	.92343	26
35	.38403	.41592	2.4043	.92332	25
36	.38430	.41626	2.4023	.92321	24
37	.38456	.41660	2.4004	.92310	23
38	.38483	.41694	2.3984	.92299	22
39	.38510	.41728	2.3964	.92287	21
40	.38537	.41763	2.3945	.92276	20
41	.38564	.41797	2.3925	.92265	19
42	.38591	.41831	2.3906	.92254	18
43	.38617	.41865	2.3886	.92243	17
44	.38644	.41899	2.3867	.92231	16
45	.38671	.41933	2.3847	.92220	15
46	.38698	.41968	2.3828	.92209	14
47	.38725	.42002	2.3808	.92198	13
48	.38752	.42036	2.3789	.92186	12
49	.38778	.42070	2.3770	.92175	11
50	.38805	.42105	2.3750	.92164	10
51	.38832	.42139	2.3731	.92152	9
52	.38859	.42173	2.3712	.92141	8
53	.38886	.42207	2.3693	.92130	7
54	.38912	.42242	2.3673	.92119	6
55	.38939	.42276	2.3654	.92107	5
56	.38966	.42310	2.3635	.92096	4
57	.38993	.42345	2.3616	.92085	3
58	.39020	.42379	2.3597	.92073	2
59	.39046	.42413	2.3578	.92062	1
60	.39073	.42447	2.3559	.92050	0
′	Cos	Cot	Tan	Sin	′

112° (292°) (247°) **67°**

23° (203°) (336°) **156°**

′	Sin	Tan	Cot	Cos	′
0	.39073	.42447	2.3559	.92050	60
1	.39100	.42482	2.3539	.92039	59
2	.39127	.42516	2.3520	.92028	58
3	.39153	.42551	2.3501	.92016	57
4	.39180	.42585	2.3483	.92005	56
5	.39207	.42619	2.3464	.91994	55
6	.39234	.42654	2.3445	.91982	54
7	.39260	.42688	2.3426	.91971	53
8	.39287	.42722	2.3407	.91959	52
9	.39314	.42757	2.3388	.91948	51
10	.39341	.42791	2.3369	.91936	50
11	.39367	.42826	2.3351	.91925	49
12	.39394	.42860	2.3332	.91914	48
13	.39421	.42894	2.3313	.91902	47
14	.39448	.42929	2.3294	.91891	46
15	.39474	.42963	2.3276	.91879	45
16	.39501	.42998	2.3257	.91868	44
17	.39528	.43032	2.3238	.91856	43
18	.39555	.43067	2.3220	.91845	42
19	.39581	.43101	2.3201	.91833	41
20	.39608	.43136	2.3183	.91822	40
21	.39635	.43170	2.3164	.91810	39
22	.39661	.43205	2.3146	.91799	38
23	.39688	.43239	2.3127	.91787	37
24	.39715	.43274	2.3109	.91775	36
25	.39741	.43308	2.3090	.91764	35
26	.39768	.43343	2.3072	.91752	34
27	.39795	.43378	2.3053	.91741	33
28	.39822	.43412	2.3035	.91729	32
29	.39848	.43447	2.3017	.91718	31
30	.39875	.43481	2.2998	.91706	30
31	.39902	.43516	2.2980	.91694	29
32	.39928	.43550	2.2962	.91683	28
33	.39955	.43585	2.2944	.91671	27
34	.39982	.43620	2.2925	.91660	26
35	.40008	.43654	2.2907	.91648	25
36	.40035	.43689	2.2889	.91636	24
37	.40062	.43724	2.2871	.91625	23
38	.40088	.43758	2.2853	.91613	22
39	.40115	.43793	2.2835	.91601	21
40	.40141	.43828	2.2817	.91590	20
41	.40168	.43862	2.2799	.91578	19
42	.40195	.43897	2.2781	.91566	18
43	.40221	.43932	2.2763	.91555	17
44	.40248	.43966	2.2745	.91543	16
45	.40275	.44001	2.2727	.91531	15
46	.40301	.44036	2.2709	.91519	14
47	.40328	.44071	2.2691	.91508	13
48	.40355	.44105	2.2673	.91496	12
49	.40381	.44140	2.2655	.91484	11
50	.40408	.44175	2.2637	.91472	10
51	.40434	.44210	2.2620	.91461	9
52	.40461	.44244	2.2602	.91449	8
53	.40488	.44279	2.2584	.91437	7
54	.40514	.44314	2.2566	.91425	6
55	.40541	.44349	2.2549	.91414	5
56	.40567	.44384	2.2531	.91402	4
57	.40594	.44418	2.2513	.91390	3
58	.40621	.44453	2.2496	.91378	2
59	.40647	.44488	2.2478	.91366	1
60	.40674	.44523	2.2460	.91355	0
′	Cos	Cot	Tan	Sin	′

113° (293°) (246°) **66°**

TABLE V **TRIGONOMETRIC FUNCTIONS OF ANGLES (Continued)**

24° (204°) (335°) **155°**

′	Sin	Tan	Cot	Cos	′
0	.40674	.44523	2.2460	.91355	60
1	.40700	.44558	2.2443	.91343	59
2	.40727	.44593	2.2425	.91331	58
3	.40753	.44627	2.2408	.91319	57
4	.40780	.44662	2.2390	.91307	56
5	.40806	.44697	2.2373	.91295	55
6	.40833	.44732	2.2355	.91283	54
7	.40860	.44767	2.2338	.91272	53
8	.40886	.44802	2.2320	.91260	52
9	.40913	.44837	2.2303	.91248	51
10	.40939	.44872	2.2286	.91236	50
11	.40966	.44907	2.2268	.91224	49
12	.40992	.44942	2.2251	.91212	48
13	.41019	.44977	2.2234	.91200	47
14	.41045	.45012	2.2216	.91188	46
15	.41072	.45047	2.2199	.91176	45
16	.41098	.45082	2.2182	.91164	44
17	.41125	.45117	2.2165	.91152	43
18	.41151	.45152	2.2148	.91140	42
19	.41178	.45187	2.2130	.91128	41
20	.41204	.45222	2.2113	.91116	40
21	.41231	.45257	2.2096	.91104	39
22	.41257	.45292	2.2079	.91092	38
23	.41284	.45327	2.2062	.91080	37
24	.41310	.45362	2.2045	.91068	36
25	.41337	.45397	2.2028	.91056	35
26	.41363	.45432	2.2011	.91044	34
27	.41390	.45467	2.1994	.91032	33
28	.41416	.45502	2.1977	.91020	32
29	.41443	.45538	2.1960	.91008	31
30	.41469	.45573	2.1943	.90996	30
31	.41496	.45608	2.1926	.90984	29
32	.41522	.45643	2.1909	.90972	28
33	.41549	.45678	2.1892	.90960	27
34	.41575	.45713	2.1876	.90948	26
35	.41602	.45748	2.1859	.90936	25
36	.41628	.45784	2.1842	.90924	24
37	.41655	.45819	2.1825	.90911	23
38	.41681	.45854	2.1808	.90899	22
39	.41707	.45889	2.1792	.90887	21
40	.41734	.45924	2.1775	.90875	20
41	.41760	.45960	2.1758	.90863	19
42	.41787	.45995	2.1742	.90851	18
43	.41813	.46030	2.1725	.90839	17
44	.41840	.46065	2.1708	.90826	16
45	.41866	.46101	2.1692	.90814	15
46	.41892	.46136	2.1675	.90802	14
47	.41919	.46171	2.1659	.90790	13
48	.41945	.46206	2.1642	.90778	12
49	.41972	.46242	2.1625	.90766	11
50	.41998	.46277	2.1609	.90753	10
51	.42024	.46312	2.1592	.90741	9
52	.42051	.46348	2.1576	.90729	8
53	.42077	.46383	2.1560	.90717	7
54	.42104	.46418	2.1543	.90704	6
55	.42130	.46454	2.1527	.90692	5
56	.42156	.46489	2.1510	.90680	4
57	.42183	.46525	2.1494	.90668	3
58	.42209	.46560	2.1478	.90655	2
59	.42235	.46595	2.1461	.90643	1
60	.42262	.46631	2.1445	.90631	0
′	Cos	Cot	Tan	Sin	′

114° (294°) (245°) **65°**

25° (205°) (334°)**154°**

′	Sin	Tan	Cot	Cos	′
0	.42262	.46631	2.1445	.90631	60
1	.42288	.46666	2.1429	.90618	59
2	.42315	.46702	2.1413	.90606	58
3	.42341	.46737	2.1396	.90594	57
4	.42367	.46772	2.1380	.90582	56
5	.42394	.46808	2.1364	.90569	55
6	.42420	.46843	2.1348	.90557	54
7	.42446	.46879	2.1332	.90545	53
8	.42473	.46914	2.1315	.90532	52
9	.42499	.46950	2.1299	.90520	51
10	.42525	.46985	2.1283	.90507	50
11	.42552	.47021	2.1267	.90495	49
12	.42578	.47056	2.1251	.90483	48
13	.42604	.47092	2.1235	.90470	47
14	.42631	.47128	2.1219	.90458	46
15	.42657	.47163	2.1203	.90446	45
16	.42683	.47199	2.1187	.90433	44
17	.42709	.47234	2.1171	.90421	43
18	.42736	.47270	2.1155	.90408	42
19	.42762	.47305	2.1139	.90396	41
20	.42788	.47341	2.1123	.90383	40
21	.42815	.47377	2.1107	.90371	39
22	.42841	.47412	2.1092	.90358	38
23	.42867	.47448	2.1076	.90346	37
24	.42894	.47483	2.1060	.90334	36
25	.42920	.47519	2.1044	.90321	35
26	.42946	.47555	2.1028	.90309	34
27	.42972	.47590	2.1013	.90296	33
28	.42999	.47626	2.0997	.90284	32
29	.43025	.47662	2.0981	.90271	31
30	.43051	.47698	2.0965	.90259	30
31	.43077	.47733	2.0950	.90246	29
32	.43104	.47769	2.0934	.90233	28
33	.43130	.47805	2.0918	.90221	27
34	.43156	.47840	2.0903	.90208	26
35	.43182	.47876	2.0887	.90196	25
36	.43209	.47912	2.0872	.90183	24
37	.43235	.47948	2.0856	.90171	23
38	.43261	.47984	2.0840	.90158	22
39	.43287	.48019	2.0825	.90146	21
40	.43313	.48055	2.0809	.90133	20
41	.43340	.48091	2.0794	.90120	19
42	.43366	.48127	2.0778	.90108	18
43	.43392	.48163	2.0763	.90095	17
44	.43418	.48198	2.0748	.90082	16
45	.43445	.48234	2.0732	.90070	15
46	.43471	.48270	2.0717	.90057	14
47	.43497	.48306	2.0701	.90045	13
48	.43523	.48342	2.0686	.90032	12
49	.43549	.48378	2.0671	.90019	11
50	.43575	.48414	2.0655	.90007	10
51	.43602	.48450	2.0540	.89994	9
52	.43628	.48486	2.0625	.89981	8
53	.43654	.48521	2.0609	.89968	7
54	.43680	.48557	2.0594	.89956	6
55	.43706	.48593	2.0579	.89943	5
56	.43733	.48629	2.0564	.89930	4
57	.43759	.48665	2.0549	.89918	3
58	.43785	.48701	2.0533	.89905	2
59	.43811	.48737	2.0518	.89892	1
60	.43837	.48773	2.0503	.89879	0
′	Cos	Cot	Tan	Sin	′

115° (295°) (244°) **64°**

TABLE V TRIGONOMETRIC FUNCTIONS OF ANGLES (Continued)

26° (206°) (333°) **153°**

′	Sin	Tan	Cot	Cos	′
0	.43837	.48773	2.0503	.89879	60
1	.43863	.48809	2.0488	.89867	59
2	.43889	.48845	2.0473	.89854	58
3	.43916	.48881	2.0458	.89841	57
4	.43942	.48917	2.0443	.89828	56
5	.43968	.48953	2.0428	.89816	55
6	.43994	.48989	2.0413	.89803	54
7	.44020	.49026	2.0398	.89790	53
8	.44046	.49062	2.0383	.89777	52
9	.44072	.49098	2.0368	.89764	51
10	.44098	.49134	2.0353	.89752	50
11	.44124	.49170	2.0338	.89739	49
12	.44151	.49206	2.0323	.89726	48
13	.44177	.49242	2.0308	.89713	47
14	.44203	.49278	2.0293	.89700	46
15	.44229	.49315	2.0278	.89687	45
16	.44255	.49351	2.0263	.89674	44
17	.44281	.49387	2.0248	.89662	43
18	.44307	.49423	2.0233	.89649	42
19	.44333	.49459	2.0219	.89636	41
20	.44359	.49495	2.0204	.89623	40
21	.44385	.49532	2.0189	.89610	39
22	.44411	.49568	2.0174	.89597	38
23	.44437	.49604	2.0160	.89584	37
24	.44464	.49640	2.0145	.89571	36
25	.44490	.49677	2.0130	.89558	35
26	.44516	.49713	2.0115	.89545	34
27	.44542	.49749	2.0101	.89532	33
28	.44568	.49786	2.0086	.89519	32
29	.44594	.49822	2.0072	.89506	31
30	.44620	.49858	2.0057	.89493	30
31	.44646	.49894	2.0042	.89480	29
32	.44672	.49931	2.0028	.89467	28
33	.44698	.49967	2.0013	.89454	27
34	.44724	.50004	1.9999	.89441	26
35	.44750	.50040	1.9984	.89428	25
36	.44776	.50076	1.9970	.89415	24
37	.44802	.50113	1.9955	.89402	23
38	.44828	.50149	1.9941	.89389	22
39	.44854	.50185	1.9926	.89376	21
40	.44880	.50222	1.9912	.89363	20
41	.44906	.50258	1.9897	.89350	19
42	.44932	.50295	1.9883	.89337	18
43	.44958	.50331	1.9868	.89324	17
44	.44984	.50368	1.9854	.89311	16
45	.45010	.50404	1.9840	.89298	15
46	.45036	.50441	1.9825	.89285	14
47	.45062	.50477	1.9811	.89272	13
48	.45088	.50514	1.9797	.89259	12
49	.45114	.50550	1.9782	.89245	11
50	.45140	.50587	1.9768	.89232	10
51	.45166	.50623	1.9754	.89219	9
52	.45192	.50660	1.9740	.89206	8
53	.45218	.50696	1.9725	.89193	7
54	.45243	.50733	1.9711	.89180	6
55	.45269	.50769	1.9697	.89167	5
56	.45295	.50806	1.9683	.89153	4
57	.45321	.50843	1.9669	.89140	3
58	.45347	.50879	1.9654	.89127	2
59	.45373	.50916	1.9640	.89114	1
60	.45399	.50953	1.9626	.89101	0
′	Cos	Cot	Tan	Sin	′

116° (296°) (243°) **63°**

27° (207°) (332°) **152°**

′	Sin	Tan	Cot	Cos	′
0	.45399	.50953	1.9626	.89101	60
1	.45425	.50989	1.9612	.89087	59
2	.45451	.51026	1.9598	.89074	58
3	.45477	.51063	1.9584	.89061	57
4	.45503	.51099	1.9570	.89048	56
5	.45529	.51136	1.9556	.89035	55
6	.45554	.51173	1.9542	.89021	54
7	.45580	.51209	1.9528	.89008	53
8	.45606	.51246	1.9514	.88995	52
9	.45632	.51283	1.9500	.88981	51
10	.45658	.51319	1.9486	.88968	50
11	.45684	.51356	1.9472	.88955	49
12	.45710	.51393	1.9458	.88942	48
13	.45736	.51430	1.9444	.88928	47
14	.45762	.51467	1.9430	.88915	46
15	.45787	.51503	1.9416	.88902	45
16	.45813	.51540	1.9402	.88888	44
17	.45839	.51577	1.9388	.88875	43
18	.45865	.51614	1.9375	.88862	42
19	.45891	.51651	1.9361	.88848	41
20	.45917	.51688	1.9347	.88835	40
21	.45942	.51724	1.9333	.88822	39
22	.45968	.51761	1.9319	.88808	38
23	.45994	.51798	1.9306	.88795	37
24	.46020	.51835	1.9292	.88782	36
25	.46046	.51872	1.9278	.88768	35
26	.46072	.51909	1.9265	.88755	34
27	.46097	.51946	1.9251	.88741	33
28	.46123	.51983	1.9237	.88728	32
29	.46149	.52020	1.9223	.88715	31
30	.46175	.52057	1.9210	.88701	30
31	.46201	.52094	1.9196	.88688	29
32	.46226	.52131	1.9183	.88674	28
33	.46252	.52168	1.9169	.88661	27
34	.46278	.52205	1.9155	.88647	26
35	.46304	.52242	1.9142	.88634	25
36	.46330	.52279	1.9128	.88620	24
37	.46355	.52316	1.9115	.88607	23
38	.46381	.52353	1.9101	.88593	22
39	.46407	.52390	1.9088	.88580	21
40	.46433	.52427	1.9074	.88566	20
41	.46458	.52464	1.9061	.88553	19
42	.46484	.52501	1.9047	.88539	18
43	.46510	.52538	1.9034	.88526	17
44	.46536	.52575	1.9020	.88512	16
45	.46561	.52613	1.9007	.88499	15
46	.46587	.52650	1.8993	.88485	14
47	.46613	.52687	1.8980	.88472	13
48	.46639	.52724	1.8967	.88458	12
49	.46664	.52761	1.8953	.88445	11
50	.46690	.52798	1.8940	.88431	10
51	.46716	.52836	1.8927	.88417	9
52	.46742	.52873	1.8913	.88404	8
53	.46767	.52910	1.8900	.88390	7
54	.46793	.52947	1.8887	.88377	6
55	.46819	.52985	1.8873	.88363	5
56	.46844	.53022	1.8860	.88349	4
57	.46870	.53059	1.8847	.88336	3
58	.46896	.53096	1.8834	.88322	2
59	.46921	.53134	1.8820	.88308	1
60	.46947	.53171	1.8807	.88295	0
′	Cos	Cot	Tan	Sin	′

117° (297°) (242°) **62°**

TABLE V TRIGONOMETRIC FUNCTIONS OF ANGLES (Continued)

28° (208°) (331°) **151°**

′	Sin	Tan	Cot	Cos	′
0	.46947	.53171	1.8807	.88295	60
1	.46973	.53208	1.8794	.88281	59
2	.46999	.53246	1.8781	.88267	58
3	.47024	.53283	1.8768	.88254	57
4	.47050	.53320	1.8755	.88240	56
5	.47076	.53358	1.8741	.88226	55
6	.47101	.53395	1.8728	.88213	54
7	.47127	.53432	1.8715	.88199	53
8	.47153	.53470	1.8702	.88185	52
9	.47178	.53507	1.8689	.88172	51
10	.47204	.53545	1.8676	.88158	50
11	.47229	.53582	1.8663	.88144	49
12	.47255	.53620	1.8650	.88130	48
13	.47281	.53657	1.8637	.88117	47
14	.47306	.53694	1.8624	.88103	46
15	.47332	.53732	1.8611	.88089	45
16	.47358	.53769	1.8598	.88075	44
17	.47383	.53807	1.8585	.88062	43
18	.47409	.53844	1.8572	.88048	42
19	.47434	.53882	1.8559	.88034	41
20	.47460	.53920	1.8546	.88020	40
21	.47486	.53957	1.8533	.88006	39
22	.47511	.53995	1.8520	.87993	38
23	.47537	.54032	1.8507	.87979	37
24	.47562	.54070	1.8495	.87965	36
25	.47588	.54107	1.8482	.87951	35
26	.47614	.54145	1.8469	.87937	34
27	.47639	.54183	1.8456	.87923	33
28	.47665	.54220	1.8443	.87909	32
29	.47690	.54258	1.8430	.87896	31
30	.47716	.54296	1.8418	.87882	30
31	.47741	.54333	1.8405	.87868	29
32	.47767	.54371	1.8392	.87854	28
33	.47793	.54409	1.8379	.87840	27
34	.47818	.54446	1.8367	.87826	26
35	.47844	.54484	1.8354	.87812	25
36	.47869	.54522	1.8341	.87798	24
37	.47895	.54560	1.8329	.87784	23
38	.47920	.54597	1.8316	.87770	22
39	.47946	.54635	1.8303	.87756	21
40	.47971	.54673	1.8291	.87743	20
41	.47997	.54711	1.8278	.87729	19
42	.48022	.54748	1.8265	.87715	18
43	.48048	.54786	1.8253	.87701	17
44	.48073	.54824	1.8240	.87687	16
45	.48099	.54862	1.8228	.87673	15
46	.48124	.54900	1.8215	.87659	14
47	.48150	.54938	1.8202	.87645	13
48	.48175	.54975	1.8190	.87631	12
49	.48201	.55013	1.8177	.87617	11
50	.48226	.55051	1.8165	.87603	10
51	.48252	.55089	1.8152	.87589	9
52	.48277	.55127	1.8140	.87575	8
53	.48303	.55165	1.8127	.87561	7
54	.48328	.55203	1.8115	.87546	6
55	.48354	.55241	1.8103	.87532	5
56	.48379	.55279	1.8090	.87518	4
57	.48405	.55317	1.8078	.87504	3
58	.48430	.55355	1.8065	.87490	2
59	.48456	.55393	1.8053	.87476	1
60	.48481	.55431	1.8040	.87462	0
′	Cos	Cot	Tan	Sin	′

118° (298°) (241°) **61°**

29° (209°) (330°) **150°**

′	Sin	Tan	Cot	Cos	′
0	.48481	.55431	1.8040	.87462	60
1	.48506	.55469	1.8028	.87448	59
2	.48532	.55507	1.8016	.87434	58
3	.48557	.55545	1.8003	.87420	57
4	.48583	.55583	1.7991	.87406	56
5	.48608	.55621	1.7979	.87391	55
6	.48634	.55659	1.7966	.87377	54
7	.48659	.55697	1.7954	.87363	53
8	.48684	.55736	1.7942	.87349	52
9	.48710	.55774	1.7930	.87335	51
10	.48735	.55812	1.7917	.87321	50
11	.48761	.55850	1.7905	.87306	49
12	.48786	.55888	1.7893	.87292	48
13	.48811	.55926	1.7881	.87278	47
14	.48837	.55964	1.7868	.87264	46
15	.48862	.56003	1.7856	.87250	45
16	.48888	.56041	1.7844	.87235	44
17	.48913	.56079	1.7832	.87221	43
18	.48938	.56117	1.7820	.87207	42
19	.48964	.56156	1.7808	.87193	41
20	.48989	.56194	1.7796	.87178	40
21	.49014	.56232	1.7783	.87164	39
22	.49040	.56270	1.7771	.87150	38
23	.49065	.56309	1.7759	.87136	37
24	.49090	.56347	1.7747	.87121	36
25	.49116	.56385	1.7735	.87107	35
26	.49141	.56424	1.7723	.87093	34
27	.49166	.56462	1.7711	.87079	33
28	.49192	.56501	1.7699	.87064	32
29	.49217	.56539	1.7687	.87050	31
30	.49242	.56577	1.7675	.87036	30
31	.49268	.56616	1.7663	.87021	29
32	.49293	.56654	1.7651	.87007	28
33	.49318	.56693	1.7639	.86993	27
34	.49344	.56731	1.7627	.86978	26
35	.49369	.56769	1.7615	.86964	25
36	.49394	.56808	1.7603	.86949	24
37	.49419	.56846	1.7591	.86935	23
38	.49445	.56885	1.7579	.86921	22
39	.49470	.56923	1.7567	.86906	21
40	.49495	.56962	1.7556	.86892	20
41	.49521	.57000	1.7544	.86878	19
42	.49546	.57039	1.7532	.86863	18
43	.49571	.57078	1.7520	.86849	17
44	.49596	.57116	1.7508	.86834	16
45	.49622	.57155	1.7496	.86820	15
46	.49647	.57193	1.7485	.86805	14
47	.49672	.57232	1.7473	.86791	13
48	.49697	.57271	1.7461	.86777	12
49	.49723	.57309	1.7449	.86762	11
50	.49748	.57348	1.7437	.86748	10
51	.49773	.57386	1.7426	.86733	9
52	.49798	.57425	1.7414	.86719	8
53	.49824	.57464	1.7402	.86704	7
54	.49849	.57503	1.7391	.86690	6
55	.49874	.57541	1.7379	.86675	5
56	.49899	.57580	1.7367	.86661	4
57	.49924	.57619	1.7355	.86646	3
58	.49950	.57657	1.7344	.86632	2
59	.49975	.57696	1.7332	.86617	1
60	.50000	.57735	1.7321	.86603	0
′	Cos	Cot	Tan	Sin	′

119° (299°) (240°) **60°**

TABLE V TRIGONOMETRIC FUNCTIONS OF ANGLES (Continued)

30° (210°) (329°) **149°**

′	Sin	Tan	Cot	Cos	′
0	.50000	.57735	1.7321	.86603	60
1	.50025	.57774	1.7309	.86588	59
2	.50050	.57813	1.7297	.86573	58
3	.50076	.57851	1.7286	.86559	57
4	.50101	.57890	1.7274	.86544	56
5	.50126	.57929	1.7262	.86530	55
6	.50151	.57968	1.7251	.86515	54
7	.50176	.58007	1.7239	.86501	53
8	.50201	.58046	1.7228	.86486	52
9	.50227	.58085	1.7216	.86471	51
10	.50252	.58124	1.7205	.86457	50
11	.50277	.58162	1.7193	.86442	49
12	.50302	.58201	1.7182	.86427	48
13	.50327	.58240	1.7170	.86413	47
14	.50352	.58279	1.7159	.86398	46
15	.50377	.58318	1.7147	.86384	45
16	.50403	.58357	1.7136	.86369	44
17	.50428	.58396	1.7124	.86354	43
18	.50453	.58435	1.7113	.86340	42
19	.50478	.58474	1.7102	.86325	41
20	.50503	.58513	1.7090	.86310	40
21	.50528	.58552	1.7079	.86295	39
22	.50553	.58591	1.7067	.86281	38
23	.50578	.58631	1.7056	.86266	37
24	.50603	.58670	1.7045	.86251	36
25	.50628	.58709	1.7033	.86237	35
26	.50654	.58748	1.7022	.86222	34
27	.50679	.58787	1.7011	.86207	33
28	.50704	.58826	1.6999	.86192	32
29	.50729	.58865	1.6988	.86178	31
30	.50754	.58905	1.6977	.86163	30
31	.50779	.58944	1.6965	.86148	29
32	.50804	.58983	1.6954	.86133	28
33	.50829	.59022	1.6943	.86119	27
34	.50854	.59061	1.6932	.86104	26
35	.50879	.59101	1.6920	.86089	25
36	.50904	.59140	1.6909	.86074	24
37	.50929	.59179	1.6898	.86059	23
38	.50954	.59218	1.6887	.86045	22
39	.50979	.59258	1.6875	.86030	21
40	.51004	.59297	1.6864	.86015	20
41	.51029	.59336	1.6853	.86000	19
42	.51054	.59376	1.6842	.85985	18
43	.51079	.59415	1.6831	.85970	17
44	.51104	.59454	1.6820	.85956	16
45	.51129	.59494	1.6808	.85941	15
46	.51154	.59533	1.6797	.85926	14
47	.51179	.59573	1.6786	.85911	13
48	.51204	.59612	1.6775	.85896	12
49	.51229	.59651	1.6764	.85881	11
50	.51254	.59691	1.6753	.85866	10
51	.51279	.59730	1.6742	.85851	9
52	.51304	.59770	1.6731	.85836	8
53	.51329	.59809	1.6720	.85821	7
54	.51354	.59849	1.6709	.85806	6
55	.51379	.59888	1.6698	.85792	5
56	.51404	.59928	1.6687	.85777	4
57	.51429	.59967	1.6676	.85762	3
58	.51454	.60007	1.6665	.85747	2
59	.51479	.60046	1.6654	.85732	1
60	.51504	.60086	1.6643	.85717	0
′	Cos	Cot	Tan	Sin	′

120° (300°) (239°) **59°**

31° (211°) (328°) **148°**

′	Sin	Tan	Cot	Cos	′
0	.51504	.60086	1.6643	.85717	60
1	.51529	.60126	1.6632	.85702	59
2	.51554	.60165	1.6621	.85687	58
3	.51579	.60205	1.6610	.85672	57
4	.51604	.60245	1.6599	.85657	56
5	.51628	.60284	1.6588	.85642	55
6	.51653	.60324	1.6577	.85627	54
7	.51678	.60364	1.6566	.85612	53
8	.51703	.60403	1.6555	.85597	52
9	.51728	.60443	1.6545	.85582	51
10	.51753	.60483	1.6534	.85567	50
11	.51778	.60522	1.6523	.85551	49
12	.51803	.60562	1.6512	.85536	48
13	.51828	.60602	1.6501	.85521	47
14	.51852	.60642	1.6490	.85506	46
15	.51877	.60681	1.6479	.85491	45
16	.51902	.60721	1.6469	.85476	44
17	.51927	.60761	1.6458	.85461	43
18	.51952	.60801	1.6447	.85446	42
19	.51977	.60841	1.6436	.85431	41
20	.52002	.60881	1.6426	.85416	40
21	.52026	.60921	1.6415	.85401	39
22	.52051	.60960	1.6404	.85385	38
23	.52076	.61000	1.6393	.85370	37
24	.52101	.61040	1.6383	.85355	36
25	.52126	.61080	1.6372	.85340	35
26	.52151	.61120	1.6361	.85325	34
27	.52175	.61160	1.6351	.85310	33
28	.52200	.61200	1.6340	.85294	32
29	.52225	.61240	1.6329	.85279	31
30	.52250	.61280	1.6319	.85264	30
31	.52275	.61320	1.6308	.85249	29
32	.52299	.61360	1.6297	.85234	28
33	.52324	.61400	1.6287	.85218	27
34	.52349	.61440	1.6276	.85203	26
35	.52374	.61480	1.6265	.85188	25
36	.52399	.61520	1.6255	.85173	24
37	.52423	.61561	1.6244	.85157	23
38	.52448	.61601	1.6234	.85142	22
39	.52473	.61641	1.6223	.85127	21
40	.52498	.61681	1.6212	.85112	20
41	.52522	.61721	1.6202	.85096	19
42	.52547	.61761	1.6191	.85081	18
43	.52572	.61801	1.6181	.85066	17
44	.52597	.61842	1.6170	.85051	16
45	.52621	.61882	1.6160	.85035	15
46	.52646	.61922	1.6149	.85020	14
47	.52671	.61962	1.6139	.85005	13
48	.52696	.62003	1.6128	.84989	12
49	.52720	.62043	1.6118	.84974	11
50	.52745	.62083	1.6107	.84959	10
51	.52770	.62124	1.6097	.84943	9
52	.52794	.62164	1.6087	.84928	8
53	.52819	.62204	1.6076	.84913	7
54	.52844	.62245	1.6066	.84897	6
55	.52869	.62285	1.6055	.84882	5
56	.52893	.62325	1.6045	.84866	4
57	.52918	.62366	1.6034	.84851	3
58	.52943	.62406	1.6024	.84836	2
59	.52967	.62446	1.6014	.84820	1
60	.52992	.62487	1.6003	.84805	0
′	Cos	Cot	Tan	Sin	′

121° (301°) (238°) **58°**

TABLE V **TRIGONOMETRIC FUNCTIONS OF ANGLES (Continued)**

32° (212°) (327°) **147°**

′	Sin	Tan	Cot	Cos	′
0	.52992	.62487	1.6003	.84805	60
1	.53017	.62527	1.5993	.84789	59
2	.53041	.62568	1.5983	.84774	58
3	.53066	.62608	1.5972	.84759	57
4	.53091	.62649	1.5962	.84743	56
5	.53115	.62689	1.5952	.84728	55
6	.53140	.62730	1.5941	.84712	54
7	.53164	.62770	1.5931	.84697	53
8	.53189	.62811	1.5921	.84681	52
9	.53214	.62852	1.5911	.84666	51
10	.53238	.62892	1.5900	.84650	50
11	.53263	.62933	1.5890	.84635	49
12	.53288	.62973	1.5880	.84619	48
13	.53312	.63014	1.5869	.84604	47
14	.53337	.63055	1.5859	.84588	46
15	.53361	.63095	1.5849	.84573	45
16	.53386	.63136	1.5839	.84557	44
17	.53411	.63177	1.5829	.84542	43
18	.53435	.63217	1.5818	.84526	42
19	.53460	.63258	1.5808	.84511	41
20	.53484	.63299	1.5798	.84495	40
21	.53509	.63340	1.5788	.84480	39
22	.53534	.63380	1.5778	.84464	38
23	.53558	.63421	1.5768	.84448	37
24	.53583	.63462	1.5757	.84433	36
25	.53607	.63503	1.5747	.84417	35
26	.53632	.63544	1.5737	.84402	34
27	.53656	.63584	1.5727	.84386	33
28	.53681	.63625	1.5717	.84370	32
29	.53705	.63666	1.5707	.84355	31
30	.53730	.63707	1.5697	.84339	30
31	.53754	.63748	1.5687	.84324	29
32	.53779	.63789	1.5677	.84308	28
33	.53804	.63830	1.5667	.84292	27
34	.53828	.63871	1.5657	.84277	26
35	.53853	.63912	1.5647	.84261	25
36	.53877	.63953	1.5637	.84245	24
37	.53902	.63994	1.5627	.84230	23
38	.53926	.64035	1.5617	.84214	22
39	.53951	.64076	1.5607	.84198	21
40	.53975	.64117	1.5597	.84182	20
41	.54000	.64158	1.5587	.84167	19
42	.54024	.64199	1.5577	.84151	18
43	.54049	.64240	1.5567	.84135	17
44	.54073	.64281	1.5557	.84120	16
45	.54097	.64322	1.5547	.84104	15
46	.54122	.64363	1.5537	.84088	14
47	.54146	.64404	1.5527	.84072	13
48	.54171	.64446	1.5517	.84057	12
49	.54195	.64487	1.5507	.84041	11
50	.54220	.64528	1.5497	.84025	10
51	.54244	.64569	1.5487	.84009	9
52	.54269	.64610	1.5477	.83994	8
53	.54293	.64652	1.5468	.83978	7
54	.54317	.64693	1.5458	.83962	6
55	.54342	.64734	1.5448	.83946	5
56	.54366	.64775	1.5438	.83930	4
57	.54391	.64817	1.5428	.83915	3
58	.54415	.64858	1.5418	.83899	2
59	.54440	.64899	1.5408	.83883	1
60	.54464	.64941	1.5399	.83867	0
′	Cos	Cot	Tan	Sin	′

122° (302°) (237°) **57°**

33° (213°) (326°) **146°**

′	Sin	Tan	Cot	Cos	′
0	.54464	.64941	1.5399	.83867	60
1	.54488	.64982	1.5389	.83851	59
2	.54513	.65024	1.5379	.83835	58
3	.54537	.65065	1.5369	.83819	57
4	.54561	.65106	1.5359	.83804	56
5	.54586	.65148	1.5350	.83788	55
6	.54610	.65189	1.5340	.83772	54
7	.54635	.65231	1.5330	.83756	53
8	.54659	.65272	1.5320	.83740	52
9	.54683	.65314	1.5311	.83724	51
10	.54708	.65355	1.5301	.83708	50
11	.54732	.65397	1.5291	.83692	49
12	.54756	.65438	1.5282	.83676	48
13	.54781	.65480	1.5272	.83660	47
14	.54805	.65521	1.5262	.83645	46
15	.54829	.65563	1.5253	.83629	45
16	.54854	.65604	1.5243	.83613	44
17	.54878	.65646	1.5233	.83597	43
18	.54902	.65688	1.5224	.83581	42
19	.54927	.65729	1.5214	.83565	41
20	.54951	.65771	1.5204	.83549	40
21	.54975	.65813	1.5195	.83533	39
22	.54999	.65854	1.5185	.83517	38
23	.55024	.65896	1.5175	.83501	37
24	.55048	.65938	1.5166	.83485	36
25	.55072	.65980	1.5156	.83469	35
26	.55097	.66021	1.5147	.83453	34
27	.55121	.66063	1.5137	.83437	33
28	.55145	.66105	1.5127	.83421	32
29	.55169	.66147	1.5118	.83405	31
30	.55194	.66189	1.5108	.83389	30
31	.55218	.66230	1.5099	.83373	29
32	.55242	.66272	1.5089	.83356	28
33	.55266	.66314	1.5080	.83340	27
34	.55291	.66356	1.5070	.83324	26
35	.55315	.66398	1.5061	.83308	25
36	.55339	.66440	1.5051	.83292	24
37	.55363	.66482	1.5042	.83276	23
38	.55388	.66524	1.5032	.83260	22
39	.55412	.66566	1.5023	.83244	21
40	.55436	.66608	1.5013	.83228	20
41	.55460	.66650	1.5004	.83212	19
42	.55484	.66692	1.4994	.83195	18
43	.55509	.66734	1.4985	.83179	17
44	.55533	.66776	1.4975	.83163	16
45	.55557	.66818	1.4966	.83147	15
46	.55581	.66860	1.4957	.83131	14
47	.55605	.66902	1.4947	.83115	13
48	.55630	.66944	1.4938	.83098	12
49	.55654	.66986	1.4928	.83082	11
50	.55678	.67028	1.4919	.83066	10
51	.55702	.67071	1.4910	.83050	9
52	.55726	.67113	1.4900	.83034	8
53	.55750	.67155	1.4891	.83017	7
54	.55775	.67197	1.4882	.83001	6
55	.55799	.67239	1.4872	.82985	5
56	.55823	.67282	1.4863	.82969	4
57	.55847	.67324	1.4854	.82953	3
58	.55871	.67366	1.4844	.82936	2
59	.55895	.67409	1.4835	.82920	1
60	.55919	.67451	1.4826	.82904	0
′	Cos	Cot	Tan	Sin	′

123° (303°) (236°) **56°**

TABLE V **TRIGONOMETRIC FUNCTIONS OF ANGLES (Continued)**

34° (214°) — (325°) **145°**

′	Sin	Tan	Cot	Cos	′
0	.55919	.67451	1.4826	.82904	60
1	.55943	.67493	1.4816	.82887	59
2	.55968	.67536	1.4807	.82871	58
3	.55992	.67578	1.4798	.82855	57
4	.56016	.67620	1.4788	.82839	56
5	.56040	.67663	1.4779	.82822	55
6	.56064	.67705	1.4770	.82806	54
7	.56088	.67748	1.4761	.82790	53
8	.56112	.67790	1.4751	.82773	52
9	.56136	.67832	1.4742	.82757	51
10	.56160	.67875	1.4733	.82741	50
11	.56184	.67917	1.4724	.82724	49
12	.56208	.67960	1.4715	.82708	48
13	.56232	.68002	1.4705	.82692	47
14	.56256	.68045	1.4696	.82675	46
15	.56280	.68088	1.4687	.82659	45
16	.56305	.68130	1.4678	.82643	44
17	.56329	.68173	1.4669	.82626	43
18	.56353	.68215	1.4659	.82610	42
19	.56377	.68258	1.4650	.82593	41
20	.56401	.68301	1.4641	.82577	40
21	.56425	.68343	1.4632	.82561	39
22	.56449	.68386	1.4623	.82544	38
23	.56473	.68429	1.4614	.82528	37
24	.56497	.68471	1.4605	.82511	36
25	.56521	.68514	1.4596	.82495	35
26	.56545	.68557	1.4586	.82478	34
27	.56569	.68600	1.4577	.82462	33
28	.56593	.68642	1.4568	.82446	32
29	.56617	.68685	1.4559	.82429	31
30	.56641	.68728	1.4550	.82413	30
31	.56665	.68771	1.4541	.82396	29
32	.56689	.68814	1.4532	.82380	28
33	.56713	.68857	1.4523	.82363	27
34	.56736	.68900	1.4514	.82347	26
35	.56760	.68942	1.4505	.82330	25
36	.56784	.68985	1.4496	.82314	24
37	.56808	.69028	1.4487	.82297	23
38	.56832	.69071	1.4478	.82281	22
39	.56856	.69114	1.4469	.82264	21
40	.56880	.69157	1.4460	.82248	20
41	.56904	.69200	1.4451	.82231	19
42	.56928	.69243	1.4442	.82214	18
43	.56952	.69286	1.4433	.82198	17
44	.56976	.69329	1.4424	.82181	16
45	.57000	.69372	1.4415	.82165	15
46	.57024	.69416	1.4406	.82148	14
47	.57047	.69459	1.4397	.82132	13
48	.57071	.69502	1.4388	.82115	12
49	.57095	.69545	1.4379	.82098	11
50	.57119	.69588	1.4370	.82082	10
51	.57143	.69631	1.4361	.82065	9
52	.57167	.69675	1.4352	.82048	8
53	.57191	.69718	1.4344	.82032	7
54	.57215	.69761	1.4335	.82015	6
55	.57238	.69804	1.4326	.81999	5
56	.57262	.69847	1.4317	.81982	4
57	.57286	.69891	1.4308	.81965	3
58	.57310	.69934	1.4299	.81949	2
59	.57334	.69977	1.4290	.81932	1
60	.57358	.70021	1.4281	.81915	0
′	Cos	Cot	Tan	Sin	′

124° (304°) — (235°) **55°**

35° (215°) — (324°) **144°**

′	Sin	Tan	Cot	Cos	′
0	.57358	.70021	1.4281	.81915	60
1	.57381	.70064	1.4273	.81899	59
2	.57405	.70107	1.4264	.81882	58
3	.57429	.70151	1.4255	.81865	57
4	.57453	.70194	1.4246	.81848	56
5	.57477	.70238	1.4237	.81832	55
6	.57501	.70281	1.4229	.81815	54
7	.57524	.70325	1.4220	.81798	53
8	.57548	.70368	1.4211	.81782	52
9	.57572	.70412	1.4202	.81765	51
10	.57596	.70455	1.4193	.81748	50
11	.57619	.70499	1.4185	.81731	49
12	.57643	.70542	1.4176	.81714	48
13	.57667	.70586	1.4167	.81698	47
14	.57691	.70629	1.4158	.81681	46
15	.57715	.70673	1.4150	.81664	45
16	.57738	.70717	1.4141	.81647	44
17	.57762	.70760	1.4132	.81631	43
18	.57786	.70804	1.4124	.81614	42
19	.57810	.70848	1.4115	.81597	41
20	.57833	.70891	1.4106	.81580	40
21	.57857	.70935	1.4097	.81563	39
22	.57881	.70979	1.4089	.81546	38
23	.57904	.71023	1.4080	.81530	37
24	.57928	.71066	1.4071	.81513	36
25	.57952	.71110	1.4063	.81496	35
26	.57976	.71154	1.4054	.81479	34
27	.57999	.71198	1.4045	.81462	33
28	.58023	.71242	1.4037	.81445	32
29	.58047	.71285	1.4028	.81428	31
30	.58070	.71329	1.4019	.81412	30
31	.58094	.71373	1.4011	.81395	29
32	.58118	.71417	1.4002	.81378	28
33	.58141	.71461	1.3994	.81361	27
34	.58165	.71505	1.3985	.81344	26
35	.58189	.71549	1.3976	.81327	25
36	.58212	.71593	1.3968	.81310	24
37	.58236	.71637	1.3959	.81293	23
38	.58260	.71681	1.3951	.81276	22
39	.58283	.71725	1.3942	.81259	21
40	.58307	.71769	1.3934	.81242	20
41	.58330	.71813	1.3925	.81225	19
42	.58354	.71857	1.3916	.81208	18
43	.58378	.71901	1.3908	.81191	17
44	.58401	.71946	1.3899	.81174	16
45	.58425	.71990	1.3891	.81157	15
46	.58449	.72034	1.3882	.81140	14
47	.58472	.72078	1.3874	.81123	13
48	.58496	.72122	1.3865	.81106	12
49	.58519	.72167	1.3857	.81089	11
50	.58543	.72211	1.3848	.81072	10
51	.58567	.72255	1.3840	.81055	9
52	.58590	.72299	1.3831	.81038	8
53	.58614	.72344	1.3823	.81021	7
54	.58637	.72388	1.3814	.81004	6
55	.58661	.72432	1.3806	.80987	5
56	.58684	.72477	1.3798	.80970	4
57	.58708	.72521	1.3789	.80953	3
58	.58731	.72565	1.3781	.80936	2
59	.58755	.72610	1.3772	.80919	1
60	.58779	.72654	1.3764	.80902	0
′	Cos	Cot	Tan	Sin	′

125° (305°) — (234°) **54°**

TABLE V **TRIGONOMETRIC FUNCTIONS OF ANGLES (Continued)**

36° (216°)				(323°) **143°**	
′	Sin	Tan	Cot	Cos	′
0	.58779	.72654	1.3764	.80902	60
1	.58802	.72699	1.3755	.80885	59
2	.58826	.72743	1.3747	.80867	58
3	.58849	.72788	1.3739	.80850	57
4	.58873	.72832	1.3730	.80833	56
5	.58896	.72877	1.3722	.80816	55
6	.58920	.72921	1.3713	.80799	54
7	.58943	.72966	1.3705	.80782	53
8	.58967	.73010	1.3697	.80765	52
9	.58990	.73055	1.3688	.80748	51
10	.59014	.73100	1.3680	.80730	50
11	.59037	.73144	1.3672	.80713	49
12	.59061	.73189	1.3663	.80696	48
13	.59084	.73234	1.3655	.80679	47
14	.59108	.73278	1.3647	.80662	46
15	.59131	.73323	1.3638	.80644	45
16	.59154	.73368	1.3630	.80627	44
17	.59178	.73413	1.3622	.80610	43
18	.59201	.73457	1.3613	.80593	42
19	.59225	.73502	1.3605	.80576	41
20	.59248	.73547	1.3597	.80558	40
21	.59272	.73592	1.3588	.80541	39
22	.59295	.73637	1.3580	.80524	38
23	.59318	.73681	1.3572	.80507	37
24	.59342	.73726	1.3564	.80489	36
25	.59365	.73771	1.3555	.80472	35
26	.59389	.73816	1.3547	.80455	34
27	.59412	.73861	1.3539	.80438	33
28	.59436	.73906	1.3531	.80420	32
29	.59459	.73951	1.3522	.80403	31
30	.59482	.73996	1.3514	.80386	30
31	.59506	.74041	1.3506	.80368	29
32	.59529	.74086	1.3498	.80351	28
33	.59552	.74131	1.3490	.80334	27
34	.59576	.74176	1.3481	.80316	26
35	.59599	.74221	1.3473	.80299	25
36	.59622	.74267	1.3465	.80282	24
37	.59646	.74312	1.3457	.80264	23
38	.59669	.74357	1.3449	.80247	22
39	.59693	.74402	1.3440	.80230	21
40	.59716	.74447	1.3432	.80212	20
41	.59739	.74492	1.3424	.80195	19
42	.59763	.74538	1.3416	.80178	18
43	.59786	.74583	1.3408	.80160	17
44	.59809	.74628	1.3400	.80143	16
45	.59832	.74674	1.3392	.80125	15
46	.59856	.74719	1.3384	.80108	14
47	.59879	.74764	1.3375	.80091	13
48	.59902	.74810	1.3367	.80073	12
49	.59926	.74855	1.3359	.80056	11
50	.59949	.74900	1.3351	.80038	10
51	.59972	.74946	1.3343	.80021	9
52	.59995	.74991	1.3335	.80003	8
53	.60019	.75037	1.3327	.79986	7
54	.60042	.75082	1.3319	.79968	6
55	.60065	.75128	1.3311	.79951	5
56	.60089	.75173	1.3303	.79934	4
57	.60112	.75219	1.3295	.79916	3
58	.60135	.75264	1.3287	.79899	2
59	.60158	.75310	1.3278	.79881	1
60	.60182	.75355	1.3270	.79864	0
′	Cos	Cot	Tan	Sin	′
126° (306°)				(233°) **53°**	

37° (217°)				(322°) **142°**	
′	Sin	Tan	Cot	Cos	′
0	.60182	.75355	1.3270	.79864	60
1	.60205	.75401	1.3262	.79846	59
2	.60228	.75447	1.3254	.79829	58
3	.60251	.75492	1.3246	.79811	57
4	.60274	.75538	1.3238	.79793	56
5	.60298	.75584	1.3230	.79776	55
6	.60321	.75629	1.3222	.79758	54
7	.60344	.75675	1.3214	.79741	53
8	.60367	.75721	1.3206	.79723	52
9	.60390	.75767	1.3198	.79706	51
10	.60414	.75812	1.3190	.79688	50
11	.60437	.75858	1.3182	.79671	49
12	.60460	.75904	1.3175	.79653	48
13	.60483	.75950	1.3167	.79635	47
14	.60506	.75996	1.3159	.79618	46
15	.60529	.76042	1.3151	.79600	45
16	.60553	.76088	1.3143	.79583	44
17	.60576	.76134	1.3135	.79565	43
18	.60599	.76180	1.3127	.79547	42
19	.60622	.76226	1.3119	.79530	41
20	.60645	.76272	1.3111	.79512	40
21	.60668	.76318	1.3103	.79494	39
22	.60691	.76364	1.3095	.79477	38
23	.60714	.76410	1.3087	.79459	37
24	.60738	.76456	1.3079	.79441	36
25	.60761	.76502	1.3072	.79424	35
26	.60784	.76548	1.3064	.79406	34
27	.60807	.76594	1.3056	.79388	33
28	.60830	.76640	1.3048	.79371	32
29	.60853	.76686	1.3040	.79353	31
30	.60876	.76733	1.3032	.79335	30
31	.60899	.76779	1.3024	.79318	29
32	.60922	.76825	1.3017	.79300	28
33	.60945	.76871	1.3009	.79282	27
34	.60968	.76918	1.3001	.79264	26
35	.60991	.76964	1.2993	.79247	25
36	.61015	.77010	1.2985	.79229	24
37	.61038	.77057	1.2977	.79211	23
38	.61061	.77103	1.2970	.79193	22
39	.61084	.77149	1.2962	.79176	21
40	.61107	.77196	1.2954	.79158	20
41	.61130	.77242	1.2946	.79140	19
42	.61153	.77289	1.2938	.79122	18
43	.61176	.77335	1.2931	.79105	17
44	.61199	.77382	1.2923	.79087	16
45	.61222	.77428	1.2915	.79069	15
46	.61245	.77475	1.2907	.79051	14
47	.61268	.77521	1.2900	.79033	13
48	.61291	.77568	1.2892	.79016	12
49	.61314	.77615	1.2884	.78998	11
50	.61337	.77661	1.2876	.78980	10
51	.61360	.77708	1.2869	.78962	9
52	.61383	.77754	1.2861	.78944	8
53	.61406	.77801	1.2853	.78926	7
54	.61429	.77848	1.2846	.78908	6
55	.61451	.77895	1.2838	.78891	5
56	.61474	.77941	1.2830	.78873	4
57	.61497	.77988	1.2822	.78855	3
58	.61520	.78035	1.2815	.78837	2
59	.61543	.78082	1.2807	.78819	1
60	.61566	.78129	1.2799	.78801	0
′	Cos	Cot	Tan	Sin	′
127° (307°)				(232°) **52°**	

TABLE V TRIGONOMETRIC FUNCTIONS OF ANGLES (Continued)

38° (218°) (321°) **141°**

′	Sin	Tan	Cot	Cos	′
0	.61566	.78129	1.2799	.78801	60
1	.61589	.78175	1.2792	.78783	59
2	.61612	.78222	1.2784	.78765	58
3	.61635	.78269	1.2776	.78747	57
4	.61658	.78316	1.2769	.78729	56
5	.61681	.78363	1.2761	.78711	55
6	.61704	.78410	1.2753	.78694	54
7	.61726	.78457	1.2746	.78676	53
8	.61749	.78504	1.2738	.78658	52
9	.61772	.78551	1.2731	.78640	51
10	.61795	.78598	1.2723	.78622	50
11	.61818	.78645	1.2715	.78604	49
12	.61841	.78692	1.2708	.78586	48
13	.61864	.78739	1.2700	.78568	47
14	.61887	.78786	1.2693	.78550	46
15	.61909	.78834	1.2685	.78532	45
16	.61932	.78881	1.2677	.78514	44
17	.61955	.78928	1.2670	.78496	43
18	.61978	.78975	1.2662	.78478	42
19	.62001	.79022	1.2655	.78460	41
20	.62024	.79070	1.2647	.78442	40
21	.62046	.79117	1.2640	.78424	39
22	.62069	.79164	1.2632	.78405	38
23	.62092	.79212	1.2624	.78387	37
24	.62115	.79259	1.2617	.78369	36
25	.62138	.79306	1.2609	.78351	35
26	.62160	.79354	1.2602	.78333	34
27	.62183	.79401	1.2594	.78315	33
28	.62206	.79449	1.2587	.78297	32
29	.62229	.79496	1.2579	.78279	31
30	.62251	.79544	1.2572	.78261	30
31	.62274	.79591	1.2564	.78243	29
32	.62297	.79639	1.2557	.78225	28
33	.62320	.79686	1.2549	.78206	27
34	.62342	.79734	1.2542	.78188	26
35	.62365	.79781	1.2534	.78170	25
36	.62388	.79829	1.2527	.78152	24
37	.62411	.79877	1.2519	.78134	23
38	.62433	.79924	1.2512	.78116	22
39	.62456	.79972	1.2504	.78098	21
40	.62479	.80020	1.2497	.78079	20
41	.62502	.80067	1.2489	.78061	19
42	.62524	.80115	1.2482	.78043	18
43	.62547	.80163	1.2475	.78025	17
44	.62570	.80211	1.2467	.78007	16
45	.62592	.80258	1.2460	.77988	15
46	.62615	.80306	1.2452	.77970	14
47	.62638	.80354	1.2445	.77952	13
48	.62660	.80402	1.2437	.77934	12
49	.62683	.80450	1.2430	.77916	11
50	.62706	.80498	1.2423	.77897	10
51	.62728	.80546	1.2415	.77879	9
52	.62751	.80594	1.2408	.77861	8
53	.62774	.80642	1.2401	.77843	7
54	.62796	.80690	1.2393	.77824	6
55	.62819	.80738	1.2386	.77806	5
56	.62842	.80786	1.2378	.77788	4
57	.62864	.80834	1.2371	.77769	3
58	.62887	.80882	1.2364	.77751	2
59	.62909	.80930	1.2356	.77733	1
60	.62932	.80978	1.2349	.77715	0
′	Cos	Cot	Tan	Sin	′

128° (308°) (231°) **51°**

39° (219°) (320°) **140°**

′	Sin	Tan	Cot	Cos	′
0	.62932	.80978	1.2349	.77715	60
1	.62955	.81027	1.2342	.77696	59
2	.62977	.81075	1.2334	.77678	58
3	.63000	.81123	1.2327	.77660	57
4	.63022	.81171	1.2320	.77641	56
5	.63045	.81220	1.2312	.77623	55
6	.63068	.81268	1.2305	.77605	54
7	.63090	.81316	1.2298	.77586	53
8	.63113	.81364	1.2290	.77568	52
9	.63135	.81413	1.2283	.77550	51
10	.63158	.81461	1.2276	.77531	50
11	.63180	.81510	1.2268	.77513	49
12	.63203	.81558	1.2261	.77494	48
13	.63225	.81606	1.2254	.77476	47
14	.63248	.81655	1.2247	.77458	46
15	.63271	.81703	1.2239	.77439	45
16	.63293	.81752	1.2232	.77421	44
17	.63316	.81800	1.2225	.77402	43
18	.63338	.81849	1.2218	.77384	42
19	.63361	.81898	1.2210	.77366	41
20	.63383	.81946	1.2203	.77347	40
21	.63406	.81995	1.2196	.77329	39
22	.63428	.82044	1.2189	.77310	38
23	.63451	.82092	1.2181	.77292	37
24	.63473	.82141	1.2174	.77273	36
25	.63496	.82190	1.2167	.77255	35
26	.63518	.82238	1.2160	.77236	34
27	.63540	.82287	1.2153	.77218	33
28	.63563	.82336	1.2145	.77199	32
29	.63585	.82385	1.2138	.77181	31
30	.63608	.82434	1.2131	.77162	30
31	.63630	.82483	1.2124	.77144	29
32	.63653	.82531	1.2117	.77125	28
33	.63675	.82580	1.2109	.77107	27
34	.63698	.82629	1.2102	.77088	26
35	.63720	.82678	1.2095	.77070	25
36	.63742	.82727	1.2088	.77051	24
37	.63765	.82776	1.2081	.77033	23
38	.63787	.82825	1.2074	.77014	22
39	.63810	.82874	1.2066	.76996	21
40	.63832	.82923	1.2059	.76977	20
41	.63854	.82972	1.2052	.76959	19
42	.63877	.83022	1.2045	.76940	18
43	.63899	.83071	1.2038	.76921	17
44	.63922	.83120	1.2031	.76903	16
45	.63944	.83169	1.2024	.76884	15
46	.63966	.83218	1.2017	.76866	14
47	.63989	.83268	1.2009	.76847	13
48	.64011	.83317	1.2002	.76828	12
49	.64033	.83366	1.1995	.76810	11
50	.64056	.83415	1.1988	.76791	10
51	.64078	.83465	1.1981	.76772	9
52	.64100	.83514	1.1974	.76754	8
53	.64123	.83564	1.1967	.76735	7
54	.64145	.83613	1.1960	.76717	6
55	.64167	.83662	1.1953	.76698	5
56	.64190	.83712	1.1946	.76679	4
57	.64212	.83761	1.1939	.76661	3
58	.64234	.83811	1.1932	.76642	2
59	.64256	.83860	1.1925	.76623	1
60	.64279	.83910	1.1918	.76604	0
′	Cos	Cot	Tan	Sin	′

129° (309°) (230°) **50°**

TABLE V TRIGONOMETRIC FUNCTIONS OF ANGLES (Continued)

40° (220°) (319°) **139°**

′	Sin	Tan	Cot	Cos	′
0	.64279	.83910	1.1918	.76604	60
1	.64301	.83960	1.1910	.76586	59
2	.64323	.84009	1.1903	.76567	58
3	.64346	.84059	1.1896	.76548	57
4	.64368	.84108	1.1889	.76530	56
5	.64390	.84158	1.1882	.76511	55
6	.64412	.84208	1.1875	.76492	54
7	.64435	.84258	1.1868	.76473	53
8	.64457	.84307	1.1861	.76455	52
9	.64479	.84357	1.1854	.76436	51
10	.64501	.84407	1.1847	.76417	50
11	.64524	.84457	1.1840	.76398	49
12	.64546	.84507	1.1833	.76380	48
13	.64568	.84556	1.1826	.76361	47
14	.64590	.84606	1.1819	.76342	46
15	.64612	.84656	1.1812	.76323	45
16	.64635	.84706	1.1806	.76304	44
17	.64657	.84756	1.1799	.76286	43
18	.64679	.84806	1.1792	.76267	42
19	.64701	.84856	1.1785	.76248	41
20	.64723	.84906	1.1778	.76229	40
21	.64746	.84956	1.1771	.76210	39
22	.64768	.85006	1.1764	.76192	38
23	.64790	.85057	1.1757	.76173	37
24	.64812	.85107	1.1750	.76154	36
25	.64834	.85157	1.1743	.76135	35
26	.64856	.85207	1.1736	.76116	34
27	.64878	.85257	1.1729	.76097	33
28	.64901	.85308	1.1722	.76078	32
29	.64923	.85358	1.1715	.76059	31
30	.64945	.85408	1.1708	.76041	30
31	.64967	.85458	1.1702	.76022	29
32	.64989	.85509	1.1695	.76003	28
33	.65011	.85559	1.1688	.75984	27
34	.65033	.85609	1.1681	.75965	26
35	.65055	.85660	1.1674	.75946	25
36	.65077	.85710	1.1667	.75927	24
37	.65100	.85761	1.1660	.75908	23
38	.65122	.85811	1.1653	.75889	22
39	.65144	.85862	1.1647	.75870	21
40	.65166	.85912	1.1640	.75851	20
41	.65188	.85963	1.1633	.75832	19
42	.65210	.86014	1.1626	.75813	18
43	.65232	.86064	1.1619	.75794	17
44	.65254	.86115	1.1612	.75775	16
45	.65276	.86166	1.1606	.75756	15
46	.65298	.86216	1.1599	.75738	14
47	.65320	.86267	1.1592	.75719	13
48	.65342	.86318	1.1585	.75700	12
49	.65364	.86368	1.1578	.75680	11
50	.65386	.86419	1.1571	.75661	10
51	.65408	.86470	1.1565	.75642	9
52	.65430	.86521	1.1558	.75623	8
53	.65452	.86572	1.1551	.75604	7
54	.65474	.86623	1.1544	.75585	6
55	.65496	.86674	1.1538	.75566	5
56	.65518	.86725	1.1531	.75547	4
57	.65540	.86776	1.1524	.75528	3
58	.65562	.86827	1.1517	.75509	2
59	.65584	.86878	1.1510	.75490	1
60	.65606	.86929	1.1504	.75471	0
′	Cos	Cot	Tan	Sin	′

130° (310°) (229°) **49°**

41° (221°) (318°) **138°**

′	Sin	Tan	Cot	Cos	′
0	.65606	.86929	1.1504	.75471	60
1	.65628	.86980	1.1497	.75452	59
2	.65650	.87031	1.1490	.75433	58
3	.65672	.87082	1.1483	.75414	57
4	.65694	.87133	1.1477	.75395	56
5	.65716	.87184	1.1470	.75375	55
6	.65738	.87236	1.1463	.75356	54
7	.65759	.87287	1.1456	.75337	53
8	.65781	.87338	1.1450	.75318	52
9	.65803	.87389	1.1443	.75299	51
10	.65825	.87441	1.1436	.75280	50
11	.65847	.87492	1.1430	.75261	49
12	.65869	.87543	1.1423	.75241	48
13	.65891	.87595	1.1416	.75222	47
14	.65913	.87646	1.1410	.75203	46
15	.65935	.87698	1.1403	.75184	45
16	.65956	.87749	1.1396	.75165	44
17	.65978	.87801	1.1389	.75146	43
18	.66000	.87852	1.1383	.75126	42
19	.66022	.87904	1.1376	.75107	41
20	.66044	.87955	1.1369	.75088	40
21	.66066	.88007	1.1363	.75069	39
22	.66088	.88059	1.1356	.75050	38
23	.66109	.88110	1.1349	.75030	37
24	.66131	.88162	1.1343	.75011	36
25	.66153	.88214	1.1336	.74992	35
26	.66175	.88265	1.1329	.74973	34
27	.66197	.88317	1.1323	.74953	33
28	.66218	.88369	1.1316	.74934	32
29	.66240	.88421	1.1310	.74915	31
30	.66262	.88473	1.1303	.74896	30
31	.66284	.88524	1.1296	.74876	29
32	.66306	.88576	1.1290	.74857	28
33	.66327	.88628	1.1283	.74838	27
34	.66349	.88680	1.1276	.74818	26
35	.66371	.88732	1.1270	.74799	25
36	.66393	.88784	1.1263	.74780	24
37	.66414	.88836	1.1257	.74760	23
38	.66436	.88888	1.1250	.74741	22
39	.66458	.88940	1.1243	.74722	21
40	.66480	.88992	1.1237	.74703	20
41	.66501	.89045	1.1230	.74683	19
42	.66523	.89097	1.1224	.74664	18
43	.66545	.89149	1.1217	.74644	17
44	.66566	.89201	1.1211	.74625	16
45	.66588	.89253	1.1204	.74606	15
46	.66610	.89306	1.1197	.74586	14
47	.66632	.89358	1.1191	.74567	13
48	.66653	.89410	1.1184	.74548	12
49	.66675	.89463	1.1178	.74528	11
50	.66697	.89515	1.1171	.74509	10
51	.66718	.89567	1.1165	.74489	9
52	.66740	.89620	1.1158	.74470	8
53	.66762	.89672	1.1152	.74451	7
54	.66783	.89725	1.1145	.74431	6
55	.66805	.89777	1.1139	.74412	5
56	.66827	.89830	1.1132	.74392	4
57	.66848	.89883	1.1126	.74373	3
58	.66870	.89935	1.1119	.74353	2
59	.66891	.89988	1.1113	.74334	1
60	.66913	.90040	1.1106	.74314	0
′	Cos	Cot	Tan	Sin	′

131° (311°) (228°) **48°**

TABLE V **TRIGONOMETRIC FUNCTIONS OF ANGLES (Continued)**

42° (222°) (317°) **137°**

′	Sin	Tan	Cot	Cos	′
0	.66913	.90040	1.1106	.74314	60
1	.66935	.90093	1.1100	.74295	59
2	.66956	.90146	1.1093	.74276	58
3	.66978	.90199	1.1087	.74256	57
4	.66999	.90251	1.1080	.74237	56
5	.67021	.90304	1.1074	.74217	55
6	.67043	.90357	1.1067	.74198	54
7	.67064	.90410	1.1061	.74178	53
8	.67086	.90463	1.1054	.74159	52
9	.67107	.90516	1.1048	.74139	51
10	.67129	.90569	1.1041	.74120	50
11	.67151	.90621	1.1035	.74100	49
12	.67172	.90674	1.1028	.74080	48
13	.67194	.90727	1.1022	.74061	47
14	.67215	.90781	1.1016	.74041	46
15	.67237	.90834	1.1009	.74022	45
16	.67258	.90887	1.1003	.74002	44
17	.67280	.90940	1.0996	.73983	43
18	.67301	.90993	1.0990	.73963	42
19	.67323	.91046	1.0983	.73944	41
20	.67344	.91099	1.0977	.73924	40
21	.67366	.91153	1.0971	.73904	39
22	.67387	.91206	1.0964	.73885	38
23	.67409	.91259	1.0958	.73865	37
24	.67430	.91313	1.0951	.73846	36
25	.67452	.91366	1.0945	.73826	35
26	.67473	.91419	1.0939	.73806	34
27	.67495	.91473	1.0932	.73787	33
28	.67516	.91526	1.0926	.73767	32
29	.67538	.91580	1.0919	.73747	31
30	.67559	.91633	1.0913	.73728	30
31	.67580	.91687	1.0907	.73708	29
32	.67602	.91740	1.0900	.73688	28
33	.67623	.91794	1.0894	.73669	27
34	.67645	.91847	1.0888	.73649	26
35	.67666	.91901	1.0881	.73629	25
36	.67688	.91955	1.0875	.73610	24
37	.67709	.92008	1.0869	.73590	23
38	.67730	.92062	1.0862	.73570	22
39	.67752	.92116	1.0856	.73551	21
40	.67773	.92170	1.0850	.73531	20
41	.67795	.92224	1.0843	.73511	19
42	.67816	.92277	1.0837	.73491	18
43	.67837	.92331	1.0831	.73472	17
44	.67859	.92385	1.0824	.73452	16
45	.67880	.92439	1.0818	.73432	15
46	.67901	.92493	1.0812	.73413	14
47	.67923	.92547	1.0805	.73393	13
48	.67944	.92601	1.0799	.73373	12
49	.67965	.92655	1.0793	.73353	11
50	.67987	.92709	1.0786	.73333	10
51	.68008	.92763	1.0780	.73314	9
52	.68029	.92817	1.0774	.73294	8
53	.68051	.92872	1.0768	.73274	7
54	.68072	.92926	1.0761	.73254	6
55	.68093	.92980	1.0755	.73234	5
56	.68115	.93034	1.0749	.73215	4
57	.68136	.93088	1.0742	.73195	3
58	.68157	.93143	1.0736	.73175	2
59	.68179	.93197	1.0730	.73155	1
60	.68200	.93252	1.0724	.73135	0
	Cos	Cot	Tan	Sin	

132° (312°) (227°) **47°**

43° (223°) (316°) **136°**

′	Sin	Tan	Cot	Cos	′
0	.68200	.93252	1.0724	.73135	60
1	.68221	.93306	1.0717	.73116	59
2	.68242	.93360	1.0711	.73096	58
3	.68264	.93415	1.0705	.73076	57
4	.68285	.93469	1.0699	.73056	56
5	.68306	.93524	1.0692	.73036	55
6	.68327	.93578	1.0686	.73016	54
7	.68349	.93633	1.0680	.72996	53
8	.68370	.93688	1.0674	.72976	52
9	.68391	.93742	1.0668	.72957	51
10	.68412	.93797	1.0661	.72937	50
11	.68434	.93852	1.0655	.72917	49
12	.68455	.93906	1.0649	.72897	48
13	.68476	.93961	1.0643	.72877	47
14	.68497	.94016	1.0637	.72857	46
15	.68518	.94071	1.0630	.72837	45
16	.68539	.94125	1.0624	.72817	44
17	.68561	.94180	1.0618	.72797	43
18	.68582	.94235	1.0612	.72777	42
19	.68603	.94290	1.0606	.72757	41
20	.68624	.94345	1.0599	.72737	40
21	.68645	.94400	1.0593	.72717	39
22	.68666	.94455	1.0587	.72697	38
23	.68688	.94510	1.0581	.72677	37
24	.68709	.94565	1.0575	.72657	36
25	.68730	.94620	1.0569	.72637	35
26	.68751	.94676	1.0562	.72617	34
27	.68772	.94731	1.0556	.72597	33
28	.68793	.94786	1.0550	.72577	32
29	.68814	.94841	1.0544	.72557	31
30	.68835	.94896	1.0538	.72537	30
31	.68857	.94952	1.0532	.72517	29
32	.68878	.95007	1.0526	.72497	28
33	.68899	.95062	1.0519	.72477	27
34	.68920	.95118	1.0513	.72457	26
35	.68941	.95173	1.0507	.72437	25
36	.68962	.95229	1.0501	.72417	24
37	.68983	.95284	1.0495	.72397	23
38	.69004	.95340	1.0489	.72377	22
39	.69025	.95395	1.0483	.72357	21
40	.69046	.95451	1.0477	.72337	20
41	.69067	.95506	1.0470	.72317	19
42	.69088	.95562	1.0464	.72297	18
43	.69109	.95618	1.0458	.72277	17
44	.69130	.95673	1.0452	.72257	16
45	.69151	.95729	1.0446	.72236	15
46	.69172	.95785	1.0440	.72216	14
47	.69193	.95841	1.0434	.72196	13
48	.69214	.95897	1.0428	.72176	12
49	.69235	.95952	1.0422	.72156	11
50	.69256	.96008	1.0416	.72136	10
51	.69277	.96064	1.0410	.72116	9
52	.69298	.96120	1.0404	.72095	8
53	.69319	.96176	1.0398	.72075	7
54	.69340	.96232	1.0392	.72055	6
55	.69361	.96288	1.0385	.72035	5
56	.69382	.96344	1.0379	.72015	4
57	.69403	.96400	1.0373	.71995	3
58	.69424	.96457	1.0367	.71974	2
59	.69445	.96513	1.0361	.71954	1
60	.69466	.96569	1.0355	.71934	0
	Cos	Cot	Tan	Sin	′

133° (313°) (226°) **46°**

TABLE V TRIGONOMETRIC FUNCTIONS OF ANGLES (Continued)

44° (224°) (315°) **135°**

′	Sin	Tan	Cot	Cos	′
0	.69466	.96569	1.0355	.71934	60
1	.69487	.96625	1.0349	.71914	59
2	.69508	.96681	1.0343	.71894	58
3	.69529	.96738	1.0337	.71873	57
4	.69549	.96794	1.0331	.71853	56
5	.69570	.96850	1.0325	.71833	55
6	.69591	.96907	1.0319	.71813	54
7	.69612	.96963	1.0313	.71792	53
8	.69633	.97020	1.0307	.71772	52
9	.69654	.97076	1.0301	.71752	51
10	.69675	.97133	1.0295	.71732	50
11	.69696	.97189	1.0289	.71711	49
12	.69717	.97246	1.0283	.71691	48
13	.69737	.97302	1.0277	.71671	47
14	.69758	.97359	1.0271	.71650	46
15	.69779	.97416	1.0265	.71630	45
16	.69800	.97472	1.0259	.71610	44
17	.69821	.97529	1.0253	.71590	43
18	.69842	.97586	1.0247	.71569	42
19	.69862	.97643	1.0241	.71549	41
20	.69883	.97700	1.0235	.71529	40
21	.69904	.97756	1.0230	.71508	39
22	.69925	.97813	1.0224	.71488	38
23	.69946	.97870	1.0218	.71468	37
24	.69966	.97927	1.0212	.71447	36
25	.69987	.97984	1.0206	.71427	35
26	.70008	.98041	1.0200	.71407	34
27	.70029	.98098	1.0194	.71386	33
28	.70049	.98155	1.0188	.71366	32
29	.70070	.98213	1.0182	.71345	31
30	.70091	.98270	1.0176	.71325	30
31	.70112	.98327	1.0170	.71305	29
32	.70132	.98384	1.0164	.71284	28
33	.70153	.98441	1.0158	.71264	27
34	.70174	.98499	1.0152	.71243	26
35	.70195	.98556	1.0147	.71223	25
36	.70215	.98613	1.0141	.71203	24
37	.70236	.98671	1.0135	.71182	23
38	.70257	.98728	1.0129	.71162	22
39	.70277	.98786	1.0123	.71141	21
40	.70298	.98843	1.0117	.71121	20
41	.70319	.98901	1.0111	.71100	19
42	.70339	.98958	1.0105	.71080	18
43	.70360	.99016	1.0099	.71059	17
44	.70381	.99073	1.0094	.71039	16
45	.70401	.99131	1.0088	.71019	15
46	.70422	.99189	1.0082	.70998	14
47	.70443	.99247	1.0076	.70978	13
48	.70463	.99304	1.0070	.70957	12
49	.70484	.99362	1.0064	.70937	11
50	.70505	.99420	1.0058	.70916	10
51	.70525	.99478	1.0052	.70896	9
52	.70546	.99536	1.0047	.70875	8
53	.70567	.99594	1.0041	.70855	7
54	.70587	.99652	1.0035	.70834	6
55	.70608	.99710	1.0029	.70813	5
56	.70628	.99768	1.0023	.70793	4
57	.70649	.99826	1.0017	.70772	3
58	.70670	.99884	1.0012	.70752	2
59	.70690	.99942	1.0006	.70731	1
60	.70711	1.0000	1.0000	.70711	0
′	Cos	Cot	Tan	Sin	′

134° (314°) (225°) **45°**

TABLE VI SOME IMPORTANT CONSTANTS

$$\pi = 3.14159\ 26536$$
$$e = 2.71828\ 18285$$
$$\log_{10} e = 0.43429\ 44819$$
$$\log_e 10 = 2.30258\ 50930$$
$$\pi \text{ radians} = 180°$$
$$1 \text{ radian} = 57.29578° = 57°17.74677'$$
$$1° = 0.01745\ 32925 \text{ radians}$$
$$1' = 0.00029\ 08882 \text{ radians}$$

TABLE VII GREEK ALPHABET

Letters		*Names*
Α	α	Alpha
Β	β	Beta
Γ	γ	Gamma
Δ	δ	Delta
Ε	ϵ	Epsilon
Ζ	ζ	Zeta
Η	η	Eta
Θ	θ	Theta
Ι	ι	Iota
Κ	κ	Kappa
Λ	λ	Lambda
Μ	μ	Mu
Ν	ν	Nu
Ξ	ξ	Xi
Ο	o	Omicron
Π	π	Pi
Ρ	ρ	Rho
Σ	σ	Sigma
Τ	τ	Tau
Υ	υ	Upsilon
Φ	ϕ	Phi
Χ	χ	Chi
Ψ	ψ	Psi
Ω	ω	Omega

TABLE VIII **SQUARES, CUBES, ROOTS**

n	n^2	$\sqrt{n}$	n^3	$\sqrt[3]{n}$	n	n^2	$\sqrt{n}$	n^3	$\sqrt[3]{n}$
1	1	1.000	1	1.000	51	2,601	7.141	132,651	3.708
2	4	1.414	8	1.260	52	2,704	7.211	140,608	3.733
3	9	1.732	27	1.442	53	2,809	7.280	148,877	3.756
4	16	2.000	64	1.587	54	2,916	7.348	157,464	3.780
5	25	2.236	125	1.710	55	3,025	7.416	166,375	3.803
6	36	2.449	216	1.817	56	3,136	7.483	175,616	3.826
7	49	2.646	343	1.913	57	3,249	7.550	185,193	3.849
8	64	2.828	512	2.000	58	3,364	7.616	195,112	3.871
9	81	3.000	729	2.080	59	3,481	7.681	205,379	3.893
10	100	3.162	1,000	2.154	60	3,600	7.746	216,000	3.915
11	121	3.317	1,331	2.224	61	3,721	7.810	226,981	3.936
12	144	3.464	1,728	2.289	62	3,844	7.874	238,328	3.958
13	169	3.606	2,197	2.351	63	3,969	7.937	250,047	3.979
14	196	3.742	2,744	2.410	64	4,096	8.000	262,144	4.000
15	225	3.873	3,375	2.466	65	4,225	8.062	274,625	4.021
16	256	4.000	4,096	2.520	66	4,356	8.124	287,496	4.041
17	289	4.123	4,913	2.571	67	4,489	8.185	300,763	4.062
18	324	4.243	5,832	2.621	68	4,624	8.246	314,432	4.082
19	361	4.359	6,859	2.668	69	4,761	8.307	328,509	4.102
20	400	4.472	8,000	2.714	70	4,900	8.367	343,000	4.121
21	441	4.583	9,261	2.759	71	5,041	8.426	357,911	4.141
22	484	4.690	10,648	2.802	72	5,184	8.485	373,248	4.160
23	529	4.796	12,167	2.844	73	5,329	8.544	389,017	4.179
24	576	4.899	13,824	2.884	74	5,476	8.602	405,224	4.198
25	625	5.000	15,625	2.924	75	5,625	8.660	421,875	4.217
26	676	5.099	17,576	2.962	76	5,776	8.718	438,976	4.236
27	729	5.196	19,683	3.000	77	5,929	8.775	456,533	4.254
28	784	5.292	21,952	3.037	78	6,084	8.832	474,552	4.273
29	841	5.385	24,389	3.072	79	6,241	8.888	493,039	4.291
30	900	5.477	27,000	3.107	80	6,400	8.944	512,000	4.309
31	961	5.568	29,791	3.141	81	6,561	9.000	531,441	4.327
32	1,024	5.657	32,768	3.175	82	6,724	9.055	551,368	4.344
33	1,089	5.745	35,937	3.208	83	6,889	9.110	571,787	4.362
34	1,156	5.831	39,304	3.240	84	7,056	9.165	592,704	4.380
35	1,225	5.916	42,875	3.271	85	7,225	9.220	614,125	4.397
36	1,296	6.000	46,656	3.302	86	7,396	9.274	636,056	4.414
37	1,369	6.083	50,653	3.332	87	7,569	9.327	658,503	4.431
38	1,444	6.164	54,872	3.362	88	7,744	9.381	681,472	4.448
39	1,521	6.245	59,319	3.391	89	7,921	9.434	704,969	4.465
40	1,600	6.325	64,000	3.420	90	8,100	9.487	729,000	4.481
41	1,681	6.403	68,921	3.448	91	8,281	9.539	753,571	4.498
42	1,764	6.481	74,088	3.476	92	8,464	9.592	778,688	4.514
43	1,849	6.557	79,507	3.503	93	8,649	9.644	804,357	4.531
44	1,936	6.633	85,184	3.530	94	8,836	9.695	830,584	4.547
45	2,025	6.708	91,125	3.557	95	9,025	9.747	857,375	4.563
46	2,116	6.782	97,336	3.583	96	9,216	9.798	884,736	4.579
47	2,209	6.856	103,823	3.609	97	9,409	9.849	912,673	4.595
48	2,304	6.928	110,592	3.634	98	9,604	9.899	941,192	4.610
49	2,401	7.000	117,649	3.659	99	9,801	9.950	970,299	4.626
50	2,500	7.071	125,000	3.684	100	10,000	10.000	1,000,000	4.642
n	n^2	$\sqrt{n}$	n^3	$\sqrt[3]{n}$	n	n^2	$\sqrt{n}$	n^3	$\sqrt[3]{n}$

ANSWERS TO ODD-NUMBERED PROBLEMS

PROBLEMS 1.2

1 (*a*) and (*e*); (*c*) and (*d*).
3 $\varnothing$, $\{2\}$, $\{9\}$, $\{2, 9\}$; all but $\{2, 9\}$
5 $\varnothing$, $\{1\}$, $\{2\}$, $\{3\}$, $\{1, 2\}$, $\{1, 3\}$, $\{2, 3\}$, $\{1, 2, 3\}$; all but $\{1, 2, 3\}$
7 $\varnothing$, $\{1\}$, $\{2\}$, $\{3\}$, $\{4\}$, $\{1, 2\}$, $\{1, 3\}$, $\{1, 4\}$, $\{2, 3\}$, $\{2, 4\}$, $\{3, 4\}$, $\{1, 2, 3\}$, $\{1, 2, 4\}$, $\{1, 3, 4\}$, $\{2, 3, 4\}$, $\{1, 2, 3, 4\}$; all but $\{1, 2, 3, 4\}$
9 The number of subsets of a set with n elements is 2^n. Consider the set $\{a_1, a_2, \ldots, a_n\}$. We can form a subset by deciding at each stage whether or not each a is a member of this subset. Thus there are two choices for each a (in or out), and since there are n elements, the total number of subsets is $2 \times 2 \times \cdots \times 2 = 2^n$.
11 $A = \{1, 1\}, \{2, 2\}, \{3, 3\}, \{4, 4\}$
13 $C = \{1, 1\}, \{1, 3\}, \{3, 1\}, \{3, 3\}$
15 $E = \{1, 3\}, \{3, 1\}, \{2, 2\}$
17 $G = \{1, 1\}, \{1, 2\}, \{1, 3\}, \{1, 4\}, \{2, 1\}, \{2, 2\}, \{2, 3\}, \{2, 4\}, \{3, 1\}, \{3, 2\}, \{3, 3\}, \{3, 4\}, \{4, 1\}, \{4, 2\}, \{4, 3\}, \{4, 4\}$
19 $I = \varnothing$
21 $I = D$, $J = G$, D and $I \subset$ all others, $F \subset H$, all others $\subset J$ and G.
23 $2n \leftrightarrow 4n$ for $n = 1, 2, 3, \ldots$
25 Not possible
27 Not possible
29 First name $\leftrightarrow$ last name, assuming no duplications. When there are duplications, may not be possible.

PROBLEMS 1.3

1 (*a*) Statement, (*b*) statement, (*c*) open sentence, (*d*) open sentence, (*e*) statement
3 (*a*), (*c*), and (*d*)
5 $3x \neq 6$; $\{1, 3, 4, 5\}$
7 $(x - 2)(x - 1)(x - 3) \neq 0$; $\{4, 5\}$
9 $(x^2 \neq 4)$ or $(x + 2 \neq 5)$; $\{1, 2, 3, 4, 5\}$

PROBLEMS 1.4

1 $\{1, 2, 3, 4, 5, 6, 8, 10\}$ 3 $\{1, 2, 3, 4, 5, 7, 9, 10\}$
5 $\{1, 4, 6, 8, 10\}$ 7 $\varnothing$
9 $\varnothing$ 11 $\varnothing$
13 $\{x \mid x$ is a student and is over 30$\}$
15 $\{x \mid x$ is a Boeing 747 and is owned by United Air Lines$\}$
17 A, U 19 $\varnothing, U$ 21 $A = B$ 23 $B \subset A$
25 $A = B$ 27 $A \subset B$ 29 $B \subset A$

PROBLEMS 1.5

1 $P \subseteq Q$ 3 It is false. 5 F
7 T 9 T. Note that $P = \varnothing$

PROBLEMS 1.7

1 If a positive integer is divisible by 9, then the sum of its digits is divisible by 3.
3 If $x^2 > 0$, then $x \neq 0$.
5 [If x^2 is even, then x is even] and [If x is even, then x^2 is even].
7 Converse: If x^2 is even, then x is even.
Inverse: If x is odd, then x^2 is odd.
Contrapositive: If x^2 is odd, then x is odd.
9 Converse: If T is isosceles, then T is equiangular.
Inverse: If T is not equiangular, then T is not isosceles.
Contrapositive: If T is not isosceles, then T is not equiangular.
11 Converse: If L and M do not intersect, then they are parallel.
Inverse: If L and M are not parallel, then they intersect.
Contrapositive: If L and M intersect, then they are not parallel.
13 $\forall_x(q_x \to p_x)$
15 $\forall_x(\sim q_x \to \sim p_x)$
17 If $x \neq -4$, then x is positive or $x^2 \neq 16$.
If $x \neq -4$ and x is negative, then $x^2 \neq 16$.
If $x \neq -4$ and $x^2 = 16$, then x is positive.
19 If Q is not a parallelogram, then $AB \neq CD$ or $AB \nparallel CD$.
If Q is not a parallelogram and $AB = CD$, then $AB \nparallel CD$.
If Q is not a parallelogram and $AB \parallel CD$, then $AB \neq CD$.
21 (Fig. 1.7.21)

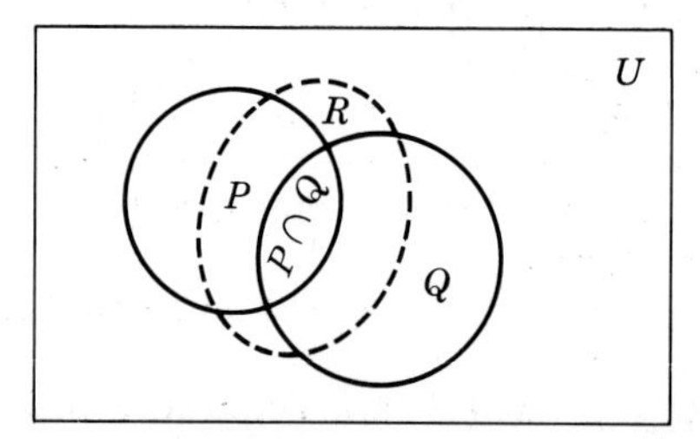

FIGURE 1.7.21

23 (Fig. 1.7.23) (*a*) To show: If $(P \cap Q) \subset R$, then $(R' \cap P) \subset Q'$. $R' \cap P$ is shaded and lies outside Q. Hence it is contained in Q'. (*b*) To show: If $(R' \cap P) \subset Q'$, then $(P \cap Q) \subset R$. $P \cap Q$ is shaded and lies outside R'. Hence it is contained in R.

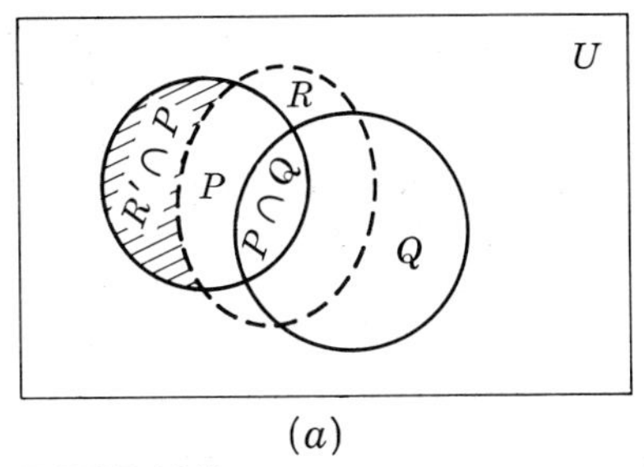

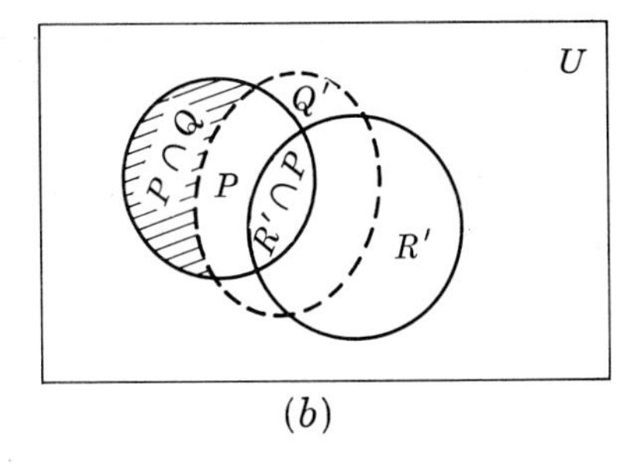

FIGURE 1.7.23

25 A sufficient condition that a triangle be isosceles is that its base angles be equal.
27 A sufficient condition that two lines be parallel is that they be perpendicular to the same line.
29 A sufficient condition that $x = 2$ is that $5x - 3 = x + 5$.
31 A necessary condition that two triangles are congruent is that their pairs of corresponding angles are equal.
33 A necessary condition that x and y are both even is that $x + y$ is even.
35 A necessary condition that $a + b$ is divisible by 3 is that $10a + b$ is divisible by 3.
37 Two triangles are congruent only if their pairs of corresponding angles are equal.
39 x and y are both even only if $x + y$ is even.
41 $a + b$ is divisible by 3 only if $10a + b$ is divisible by 3.
43 A necessary condition that a triangle is isosceles is that its base angles are equal.
A sufficient condition that the base angles of a triangle are equal is that it be isosceles.
45 A necessary condition that two lines are parallel is that they are perpendicular to the same line.
A sufficient condition that two lines are perpendicular to the same line is that they are parallel.
47 A necessary condition that $x = 2$ is that $5x - 3 = x + 5$.
A sufficient condition that $5x - 3 = x + 5$ is that $x = 2$.
49 The pairs of corresponding angles of two triangles are equal only if the triangles are congruent.
51 $x + y$ is even only if x and y are both even.
53 $10a + b$ is divisible by 3 only if $a + b$ is divisible by 3.
55 A necessary and sufficient condition that two lines are parallel is that they do not intersect.
Or: A necessary and sufficient condition that two lines do not intersect is that they are parallel.
57 A necessary and sufficient condition that x^2 is odd is that x is odd.
Or: A necessary and sufficient condition that x is odd is that x^2 is odd.
59 No. Promise is converse of that required for her to win.

PROBLEMS 1.8

1 Step 5 is the converse of the given conditional.
3 Step 3 follows from (1) and the converse of (2), but not from (1) and (2).

PROBLEMS 1.9

1 In triangle ABC, assume $a \neq b$ and $\angle A = \angle B$. This contradicts the theorem that if the base angles of a triangle are equal, it is isosceles.
3 In quadrilateral $ABCD$ assume that the diagonals are unequal and that it is a rectangle. This contradicts a known property of a rectangle.
5 Assume there are two distinct coplanar lines perpendicular to L at P. Then the sum of the angles on one side of L at P is greater than 180—impossible.
7 Assume that two distinct circles have three distinct points in common. Then they are the same circle, for three noncollinear points determine a circle.
9 Let the tangent intersect the circle at $Q \neq P$. Then triangle OPQ is isosceles, one base angle is a right angle, and so it has two right angles—impossible.

PROBLEMS 1.11

1 Counterexample: $3 + 5 = 8$, and 8 is even.
3 Counterexample: $x = 1$, for example
5 Counterexample: $x = -2$
7 Counterexample: $x = -1$, then $(-1)^2 = 1$ and $\sqrt{1} = 1 \neq -1$. Any negative x will do.

PROBLEMS 2.2

1 Two
3 $5 + 7 = 12$
5 All integers, rationals, or reals
7 Set of nonzero reals
9 Commutative
11 For instance, $5 - 2 \neq 2 - 5$
13 No
15 Yes
17 Addition and multiplication
19 Yes, no
21 No
23 Addition and multiplication

PROBLEMS 2.4

1 True, Commutative
3 False, Associative
5 True, Commutative
7 False, Commutative
9 True, Associative
11 False, Commutative
13 True, Distributive
15 False, Distributive
17 $a + b + c = a + (b + c) = (b + c) + a = b + c + a$
19 Associative Property
21 Either $(a \times b \times c) \times d$ or $a \times (b \times c \times d)$
23 No. Although $a \div 1 = a$, $1 \div a \neq a$.
25 $\frac{1}{2}, \frac{4}{3}, -\frac{5}{6}$, 1, none
27 None, 0, 1, 0, none
29 Distributive Property and Commutative Property of Multiplication
31 Only (b)
33 $(a + b) \times (1/c) = [a \times (1/c)] + [b \times (1/c)] = a \div c + b \div c$
35 False
37 -79
39 -434
41 1
43 2, 3
45 3, -3
47 True. Distributive and commutative properties
49 False. Find counterexample such as $24 \div (2 + 4) \neq (24 \div 2) + (24 \div 4)$.
51 True. $x = -b/a$
53 Yes. It is commutative and associative. 0 is the identity. $a(\neq -1)$ has the addiplicative inverse $-a/(1 + a)$.
55 (1) Given; (2) Theorem 1; (3) Associative Property; (4) Multiplicative inverse; (5) Multiplicative identity

57 (1) Given; (2) Theorem 1; (3) Associative Property; (4) Multiplicative inverse; (5) Multiplicative identity

PROBLEMS 2.5

1 Addition and multiplication only
3 Yes, one **5** Yes **7** No
9 2, 3, 5, 7, 11, 13, 17, 19, 23, 29 **11** $24 = 2^3 \times 3$

PROBLEMS 2.6

In Probs. 1 to 30 only the key step is indicated. The complete solution should be in the form shown in Illustrations 3 to 6.

1 $k^2 + (2k + 1) = (k + 1)^2$

3 $5k + 2 + \frac{k}{2}(5k - 1) = \frac{k + 1}{2}(5k + 4)$

5 $(k + 1)^3 + 2(k + 1) = (k^3 + 2k) + 3k^2 + 3k + 3$
7 $(k + 1)(k + 2)(k + 3) = k(k + 1)(k + 2) + 3(k + 1)(k + 2)$
9 $2^k + 2^k - 1 = 2^{k+1} - 1$

11 $\frac{k}{k + 1} + \frac{1}{(k + 1)(k + 2)} = \frac{k + 1}{k + 2}$

13 $(k + 1)^3 + \frac{k^2(k + 1)^2}{4} = \frac{(k + 1)^2(k + 2)^2}{4}$

15 $a + kd + \frac{k}{2}[2a + (k - 1)d] = \left(\frac{k + 1}{2}\right)(2a + kd)$

17 Let $4^k - 1 = 3x$. Then $4^{k+1} - 1 = 4 \cdot 3x + 3$.
19 Let $3^{2k} - 1 = 4x$. Then $3^{2k+2} - 1 = 3^2 \cdot 4x + 8$.
21 Use the hint. **23** Use the hint. **25** $1^{k+1} = 1 \cdot 1^k$
27 $a_1 \times \cdots \times a_{2k+2} = (a_1 \times \cdots \times a_{2k}) \times a_{2k+1} \times a_{2k}$
29 $a_1 + \cdots + a_{k+1} = (a_1 + \cdots + a_k) + a_{k+1}$
31 Fails for $n = 1$, but II can be proved.
33 True for $n = 2$, but II is false.
37 Any integer except ± 1 will do.
39 No

PROBLEMS 2.9

3 Every natural number n can be written as $n/1$.
5 $x = b - a$ is a solution. If $a + x_1 = b$ and $a + x_2 = b$, $x_1 - x_2 = 0$ and so $x_1 = x_2$.

7 $\frac{a/b}{c/d} = \left(\frac{a}{b}\right) \cdot \frac{1}{(c/d)} = \frac{a}{b} \cdot \frac{d}{c} = \frac{ad}{bc}$

9 Suppose the rational number r is nearer to 0 than every other rational number. Then what about $r/2$?
11 $0.\overline{428571}$ **13** $0.\overline{5}$ **15** $2.\overline{09}$
17 $\frac{5}{9}$ **19** $11\frac{34}{99}$ **21** $3\frac{7421}{9990}$
23 There are only a finite number of remainders less than the divisor. Eventually one will repeat.

27 It is false that if a^2 is divisible by 4, then a is divisible by 4. Counterexample: $a = 2$.
29 Put $2 - \sqrt{3} = a/b$. Then $\sqrt{3} = 2 - (a/b) = (2b - a)/b$ which is rational.

PROBLEMS 2.12

1 $4 > 1, 6 > -2, 3 > -3, -2 > -5, 0 > -4$
3 $a < c < b$ **5** 9, 14, 12, 2, 23 **7** (Fig. 2.12.7)

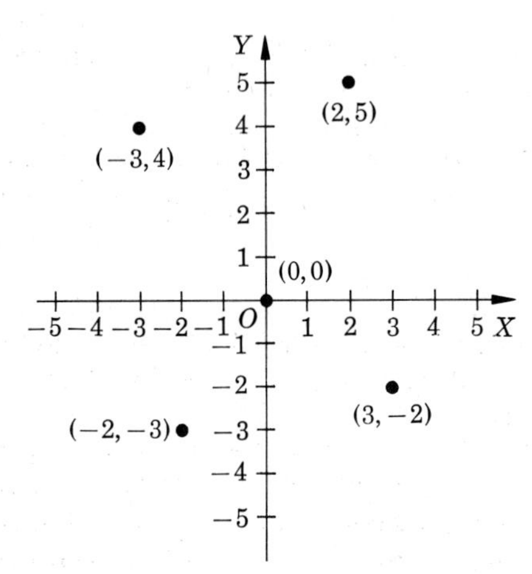

FIGURE 2.12.7

9 $x+, y+; x-, y-$ **11** IV, I, III, II, II **13** 3, 3, 10, 12, 4
15 $|x_2 - x_1| = |x_1 - x_2|$
17 $|x_2 - x_1|^2 = (x_2 - x_1)^2 = (x_1 - x_2)^2$
19 $|x_2' - x_1'| = |(a + x_2) - (a + x_1)| = |x_2 - x_1|$
21 $K = C + 273°$
23 $|x_2' - x_1'| = |ax_2 - ax_1| = |a(x_2 - x_1)| = |a| \cdot |x_2 - x_1|$
25 $I = 12F$
27 $|x_2' - x_1'| = |(ax_2 + b) - (ax_1 + b)| = |a(x_2 - x_1)|$
$= |a| \cdot |x_2 - x_1|$
29 $F = \frac{9}{5}C + 32$; $-40°$

PROBLEMS 2.13

1 $11 + 5i$ **3** $-12 - 10i$ **5** $2 - 18i$
7 $1 + 8i$ **9** $-1 - 10i$ **11** $5 + 2i$
13 $14 - 2i$ **15** $-11 - 13i$ **17** $35 + 62i$
19 8 **21** 20 **23** $36 - 54i$
25 $-42 + 21i$ **27** -36 **29** -1
31 $(18 - i)/25$ **33** $5 + 4i$ **35** $(22 + 6i)/13$
37 $(12 + 15i)/41$ **39** $(2 - 4i)/7$ **41** $x = \frac{10}{13}, y = \frac{11}{13}$
43 $x = \frac{17}{13}, y = \frac{20}{13}$ **45** $x = 0, y = -2$ **55** (7,6)
57 (16,15) **59** $(\frac{5}{2}, -\frac{1}{2})$

PROBLEMS 3.2

1 $-a^2 + ab + 4c$ **3** $p^2 + 6q^3 - 4pq$
5 $10x^2y + 5xy^2 + x^2 - 2y^2 + xy + 2$ **7** $2x^3 - 6y^3 + xy - 6y$

9 $x^2 - 2y^2 + 3x + 5a^2 + 4b^2 + 3y$
11 $4x^3 + 10x^2y - 3xy^2 + 10xy + 2$
13 $5x^2 - 8y^2 - 3r^2 + 5s^2$ **15** $7x^4 + 8xy - 6a^2$
17 $-5xy - 2x^2 - 4y^4 + 4y^2$ **19** $-a^2 - 14b^2 - 13ab$

PROBLEMS 3.3

1 $-6a^9b^5$ **3** $24x^7y^5wz^5$
5 $3a^2 - 9ab - 30b^2$ **7** $9x^2 + 12xy + 4y^2$
9 $8p^3 + 12p^2q + 6pq^2 + q^3$ **11** $4x^2 - 25y^2$
13 $4x^4 - 19x^3 + 29x^2 - 29x + 15$
15 $-4x^6 - 13x^5 + 16x^4 + 13x^3 - 10x^2 + 6x - 8$
17 $12a^7 + 8a^4 - 22a^4b^2 + 3a^3b^3 + 6ab^4 - 12ab^2 + 2b^3 - b^5$
19 $10x^6 - x^5y + 4x^4y^2 + 11x^3y^3 - 7x^2y^4 - 2xy^5 + y^6$

PROBLEMS 3.4

1 10, 20, 35, 1, 3 **3** $10 + 10 = 20$ **5** $1 + 3 + 3 + 1 = 8$
7 $x^5 + 5x^4(2y) + 10x^3(2y)^2 + 10x^2(2y)^3 + 5x(2y)^4 + (2y)^5$
$= x^5 + 10x^4y + 40x^3y^2 + 80x^2y^3 + 80xy^4 + 32y^5$
9 $r^6 - 6r^5s + 15r^4s^2 - 20r^3s^3 + 15r^2s^4 - 6rs^5 + s^6$
11 $x^6 + 6x^5(\frac{1}{3}y) + 15x^4(\frac{1}{3}y)^2 + 20x^3(\frac{1}{3}y)^3 + 15x^2(\frac{1}{3}y)^4 + 6x(\frac{1}{3}y)^5 + (\frac{1}{3}y)^6$
$= x^6 + 2x^5y + \frac{5}{3}x^4y^2 + \frac{20}{27}x^3y^3 + \frac{5}{27}x^2y^4 + \frac{2}{81}xy^5 + \frac{1}{1{,}029}y^6$
13 $(3/x)^3 + 3(3/x)^2x^2 + 3(3/x)(x^2)^2 + (x^2)^3 = (27/x^3) + 27 + 9x^3 + x^6$
15 $1^5 + 5(1)^4(0.01) + 10(1)^3(0.01)^2 + 10(1)^2(0.01)^3 + 5(1)(0.01)^4$
$+ (0.01)^5 = 1 + 0.05 + 0.001 + 0.00001 + 0.00000005 + 0.0000000001$
$= 1.0510100501$
17 $\binom{13}{7} = 1{,}716$
19 Coefficient in $\binom{7}{5}(2x)^2(-3y)^5$, or $-20{,}412$
21 Coefficient in $\binom{4}{3}(x^2)^1(y^3)^3$, or 4
23 $x^3 + y^3 + z^3 + 3(x^2y + x^2z + xy^2 + xz^2 + y^2z + yz^2) + 6xyz$

PROBLEMS 3.5

	Quotient	*Remainder*
1	$x^2 + 4x - 5$	0
3	$3x^2 + 5x + 2$	5
5	$2x^2 + 3x + 5$	$2x - 4$
7	$4x^3 + x - 3$	$-x + 7$
9	$x^2 + 3$	$4x^2 + x + 2$
11	$5x^3 + 2x^2 - x - 6$	$6x + 2$
13	$x^2 + 1$	0
15	$x^4 - x^2 + 1$	-2
17	$x^5 + x^4y + x^3y^2 + x^2y^3 + xy^4 + y^5$	0
19	$\frac{1}{3}x^2 + 2x + 4$	2

PROBLEMS 3.6

1 $(x + 3)(y + 4)$ **3** $(x^2 + 1)(x - 1)$ **5** $(5 - v)(x + y)$
7 $(x + 4)(x - 2)$ **9** $(x - 9)(x - 4)$ **11** $(8 + x)(4 + x)$
13 $(y - 9x)(y - 5x)$ **15** No integral factors **17** $(3x + 2)(x + 1)$
19 $(3x - 2)(x - 2)$ **21** $(7x - 3)(2x + 4)$ **23** $(5x + 2y)(2x - y)$
25 $(10x + 12)(x - 1)$ **27** $(6x - 9)(2x + 3) = 3(2x - 3)(2x + 3)$
29 $(4x + 5)(3x - 1)$ **31** No integral factors **33** $(2x - 5)(2x + 5)$

35 $(x + \sqrt{5})(x - \sqrt{5})$ **37** $(2x + 3i)(2x - 3i)$
39 $(x + 5 + x + 2)(x + 5 - x - 2) = (2x + 7)(3)$
41 $(x^2 + y)(x^2 - y)$
43 $(x + 2 + y - 3)(x + 2 - y + 3) = (x + y - 1)(x - y + 5)$
45 $(2x - y)(4x^2 + 2xy + y^2)$
47 $(x + 2)(x^2 - 2x + 4)$
49 $(a + 2 - 3b)(a + 2 + 3b)$
51 $(x^2 - 3)(x^2 - 4) = (x + \sqrt{3})(x - \sqrt{3})(x + 2)(x - 2)$
53 $x(x + 4y)(2x - y)$
55 $(3x - y)(x - 2y) + 4(x - 2y) + 5(3x - y) + 20$
$= (x - 2y + 5)(3x - y + 4)$

PROBLEMS 4.2

1 $\dfrac{x - 4}{x + 4}$ **3** $\dfrac{x - 5}{x - 4}$ **5** Does not simplify.
7 $\dfrac{2(x + 2)}{x + 5}$ **9** $\dfrac{2x - 3}{x - 4}$ **11** $\dfrac{c + d}{4 - y}$
13 Does not simplify. **15** $a^2 - b^2$ **17** $\dfrac{x^2(a - 1)}{a^2 - 5}$
19 Does not simplify. **21** Any x except 0, $\frac{9}{32}$ **23** Any x except 0, 1
25 Any x except $x = 0$

PROBLEMS 4.3

1 $\dfrac{5x - 2}{x^2 - 4}$ **3** $\dfrac{6a + 33b}{-4a^2 + 9b^2}$
5 $\dfrac{3a - 5b}{a^2 - ab}$ **7** $\dfrac{7x + 4y}{x^2y + xy^2}$
9 $\dfrac{x^2 - 5x}{(x^2 - 4)(x + 1)}$ **11** $\dfrac{-x^4 + x^3 + 11x^2 + 12x}{(x + 1)^2(x + 2)^2}$
13 $\dfrac{2x^2 - 13}{x^2 - 4}$ **15** $\dfrac{3x^3 - 3x^2 + 2x + 2}{x(x^2 - 1)}$
17 $\dfrac{6x + 10}{(x + 1)(x + 2)(x + 3)}$ **19** $\dfrac{7x - (10 + 15i)}{x^2 - 4x + 13}$

PROBLEMS 4.4

1 $\dfrac{x + 3}{4}$ **3** $\dfrac{x + 4}{x + 3}$
5 $\dfrac{(x + 2)^3(x - 2)}{(x + 1)^4}$ **7** $\dfrac{(x + 2)(x - 3)(x - 4)}{(x - 2)(x - 1)}$
9 $\frac{15}{4}$ **11** $\dfrac{x(x - 1)}{2(x + 1)}$
13 $\dfrac{(2x - 5)(4x + 5)}{(2x + 7)(2x - 3)}$ **15** 1
17 $\dfrac{y^2}{x(x + y)}$ **19** $\dfrac{p}{r}$

PROBLEMS 4.5

1 $\dfrac{-3x}{(x-3)(x-2)}$

3 $\dfrac{-x^2}{(x-3)(x+2)}$

5 $\dfrac{x^2-2xy-y^2}{x^2+2xy-y^2}$

7 $\dfrac{70x^3-33x}{35x^2-6}$

9 $\dfrac{2x^2+7x+17}{x(x+1)(x+4)}$

11 $\dfrac{9x^2-64x-31}{(x+1)(x+2)(x-3)}$

13 $\dfrac{3x^2+5x-12}{2x^2+3x-2}$

15 $\dfrac{(3x^2+4x+1)(17x^2-24x+9)}{x^2(2x+1)(4x-3)}$

17 $\dfrac{(x^3+3x^2+2x-3)(x-3)(x-2)}{(x^3-2x^2-2x+6)(x+3)(x+2)}$

19 $\dfrac{-3x^2+4x+6xi-2}{x^2-2x+5}$

PROBLEMS 5.2

1 5^9 **3** 4^2 **5** $\dfrac{6^1}{3^3}$ **7** 9^9

9 $\dfrac{6^5+4^4 6^4+(4^4)(6^2)}{4}$

11 $\dfrac{x^8y^2+2x^6y^5+y^7}{x^2}$

13 $\dfrac{a^3b^3-c^6}{a^4c^7}$

15 $\dfrac{2y^8+5y^4-3}{y^5}$

17 $\dfrac{5x+2x^3-4x^4}{4+2x^3}$

19 $\dfrac{4x^2+5x+3}{6x^6+x^4}$

PROBLEMS 5.3

1 $3, -3, 3, 2$

3 $\frac{1}{5}, \frac{1}{4}, -\frac{1}{2}, \frac{1}{2}$

5 $5, 7, 6^{1/6}, 1/(3^{1/3})$

7 $3x^{1/3}-2x^{-2}$

9 $a+2a^{1/2}b^{1/2}+b$

11 $x^{-2}-x^{-5/2}y^{5/2}+x^{1/2}y^{3/2}-y^4$

13 $x^{3/2}+4x^{-9/2}-x^{-4}$

15 $\dfrac{3y^{4/5}-4+5y^{7/5}}{4y^{13/5}-1}$

17 $\dfrac{4s^{23/5}-3r^{4/5}s^{17/5}}{2r^{7/5}s^5+7r^{12/5}}$

PROBLEMS 5.6

1 $|x+2|$

3 $|x+1|+|x-1|$; $x=0$ is one counterexample.

5 $\dfrac{|3x-2|}{3x-2}$ **7** $\dfrac{x\sqrt{x^2+1}}{|x|}$ **9** $x+4$ or $|x+4|$

11 $-x-4$ or $|x+4|$ **13** $2x+5$ **15** $-2x-5$

17 1 **19** $-\sqrt{30}$ **21** $12i$ **23** -12

25 -4 **27** $-21i$ **29** $3\sqrt{2}$ **31** 0

33 $(-2+16i)\sqrt{2}$

37 $(-a)^{p/q}=[(-a)^{1/q}]^p$

39 Yes

41 No. $(a^b)^c = a^{bc} \neq a^{(b^c)}$

43 No

PROBLEMS 5.7

1 $\frac{2\sqrt{5}}{5}$

3 $+4(\sqrt{2}-\sqrt{3})$

5 $\frac{4\sqrt{x-2}}{x-2}$

7 $\frac{2(\sqrt{x-2}+\sqrt{x+1})}{-3}$

9 $\frac{12\sqrt{2}+3\sqrt{3}}{29}$

11 $\frac{1}{4}$

13 $\frac{7\sqrt{3}+3\sqrt{7}}{21}$

15 $\frac{4x+x\sqrt{x}-8}{4x-x^2}$

17 $\frac{4x^2}{1-x}$

19 2.89

21 1.09

PROBLEMS 6.4

1 $\frac{-5\pm\sqrt{5}}{2}$

3 $\frac{1}{4}$

5 $\frac{1}{3}, -\frac{5}{2}$

7 $-1 \pm i$

9 $2+3i, 4+3i$

11 $x^2+5x+\frac{25}{4}=\frac{13}{4}$

13 $x^2-8x+16=13$

15 $4x^2-10x+\frac{25}{4}=-\frac{3}{4}$

17 Sum $=-5$, product $=9$

19 Sum $=-\frac{5}{3}$, product $=\frac{4}{3}$

21 Sum $=\frac{-5+i}{2}$, product $=1$

23 $\frac{49}{4}$

25 $\frac{25}{8}$

27 $\pm\frac{2\sqrt{2}}{5}$

29 $x^2-3x+5=0$

31 $2x^2-x+5=0$

33 $x^2-(2+i)x+3+4i=0$

35 $\left(x-\frac{5+\sqrt{13}}{2}\right)\left(x-\frac{5-\sqrt{13}}{2}\right)$

37 $3\left(x+\frac{1+i\sqrt{14}}{3}\right)\left(x+\frac{1-i\sqrt{14}}{3}\right)$

39 $(6x-7)(9x+5)$

41 $-\frac{5}{2}$

43 $x^2-39x+25=0$

45 $x^2+11x+15=0$

PROBLEMS 6.5

1 -2. (Fig. 6.5.1)

3 4. (Fig. 6.5.3)

5 -4. (Fig. 6.5.5)

7 $-3, -4$. (Fig. 6.5.7)

9 No real solution. (Fig. 6.5.9)

11 $\frac{1}{2}$ (double). (Fig. 6.5.11)

13 ± 2. (Fig. 6.5.13)

15 -3 (double). (Fig. 6.5.15)

17 y increases as x approaches $\pm\infty$.

19 $y=0$ when $x=r_1$ or r_2.

21 Since there are no real roots, y is never zero.

23 The highest or lowest point occurs when $x+(b/2a)=0$. Hence $x=-b/2a$; $y=(-b+4ac)/4a$.

PROBLEMS 6.7

1 $U=R$ except $0, -3$; $(2, -2)$

3 $U=R$ except ± 1; (3)

5 $U=R$ except $0, 1, -3$; $(-1, \frac{15}{13})$

7 $U=R$ except -1; (4,1)

9 $U=R$ except ± 3; $(\frac{7}{2})$

11 3

13 No solution

15 1

17 0, -4

19 4

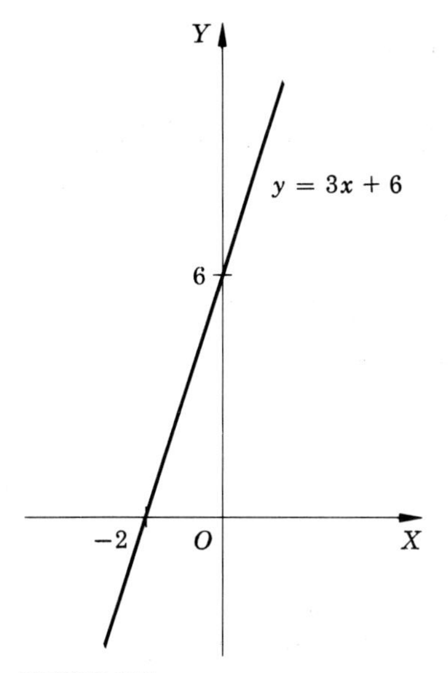

FIGURE 6.5.1

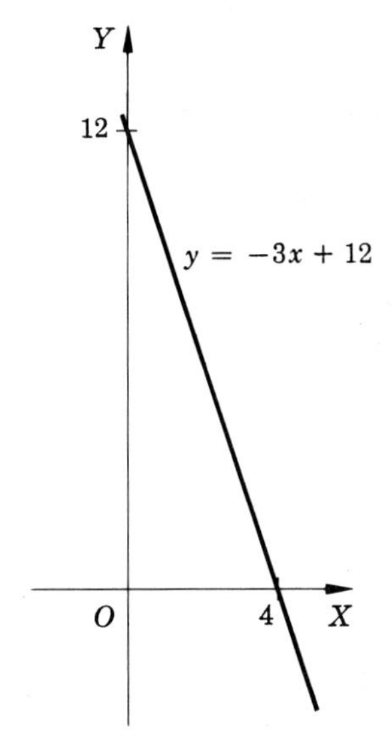

FIGURE 6.5.3

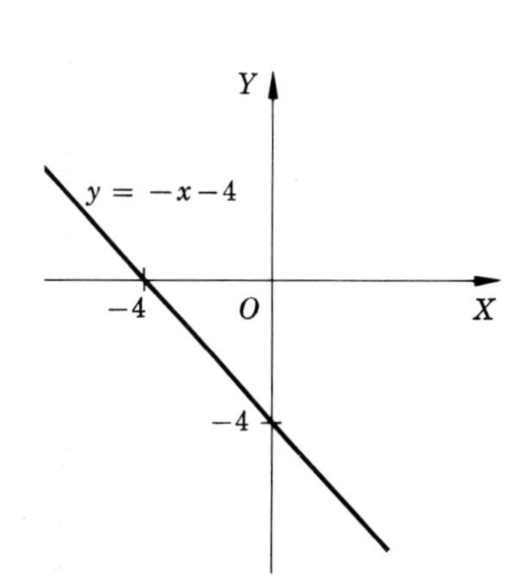

FIGURE 6.5.5

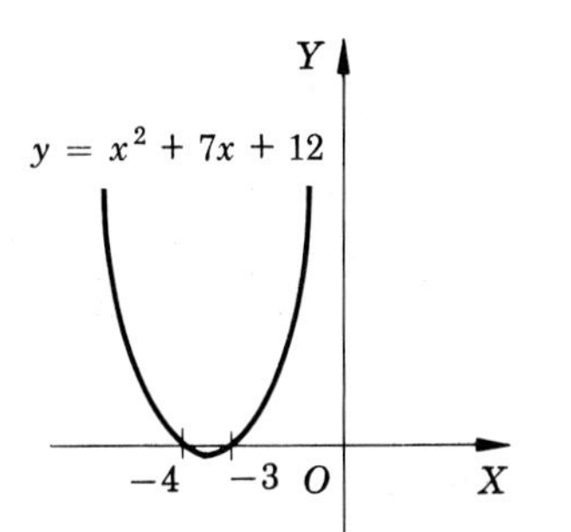

FIGURE 6.5.7

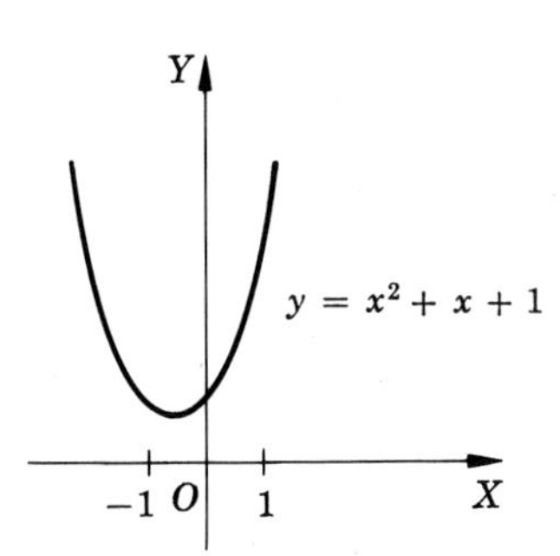

FIGURE 6.5.9

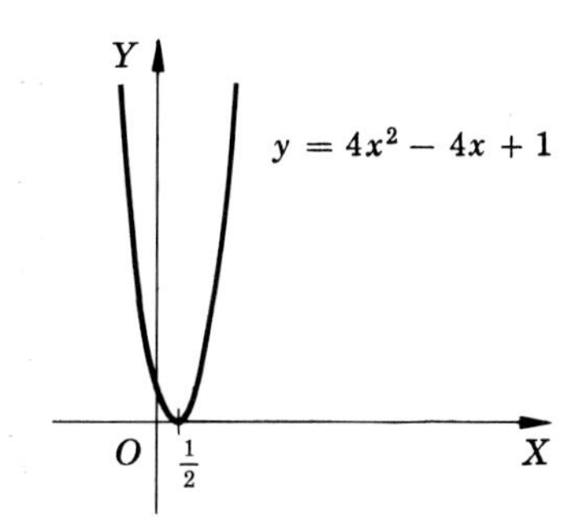

FIGURE 6.5.11

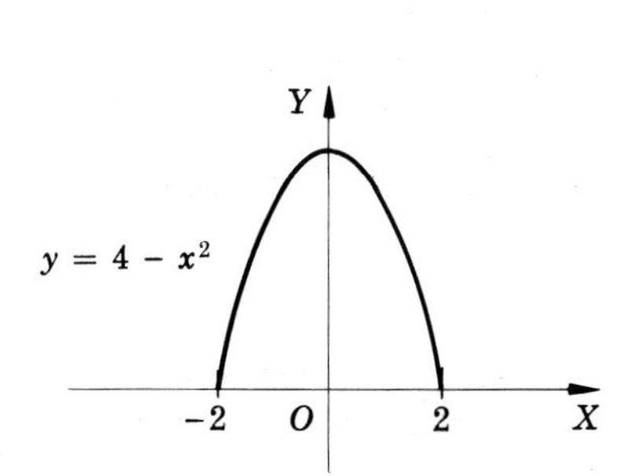

FIGURE 6.5.13

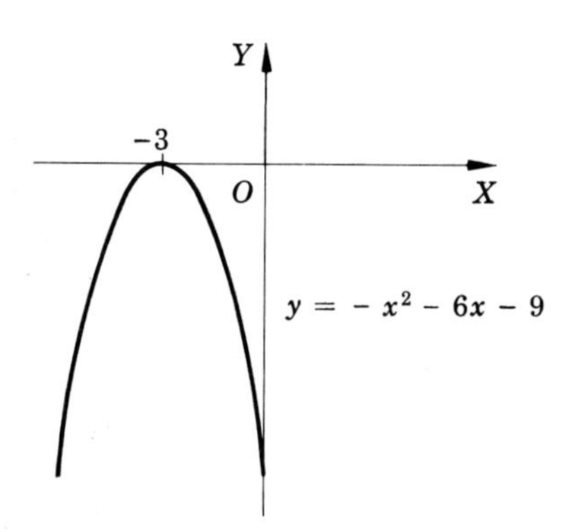

FIGURE 6.5.15

PROBLEMS 6.8

1 (1,2) (Fig. 6.8.1)
3 (2,−3) (Fig. 6.8.3)
5 (0,5) (Fig. 6.8.5)
7 No solution (Fig. 6.8.7)
9 Line of solutions (Fig. 6.8.9)

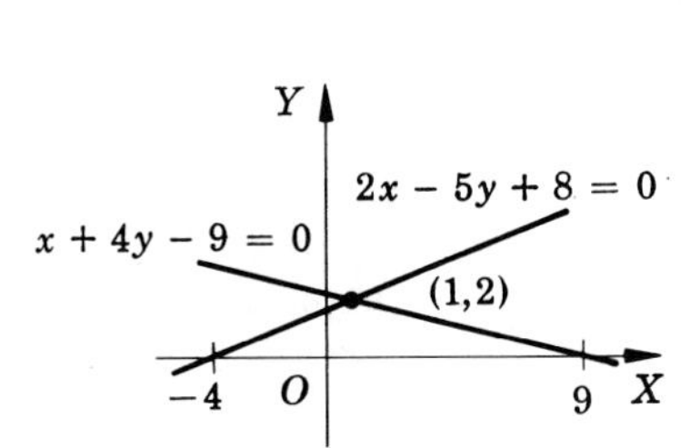

FIGURE 6.8.1

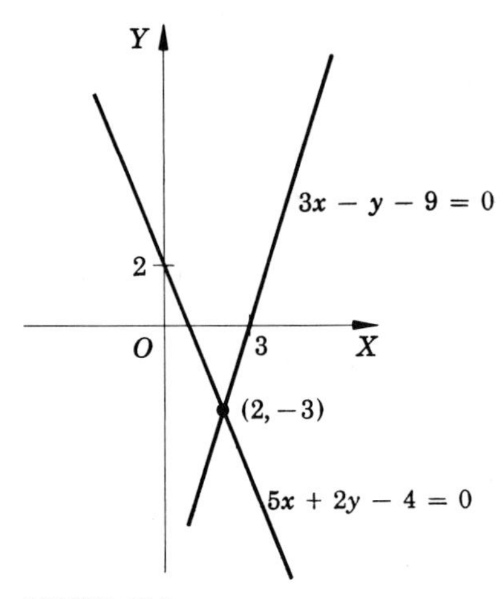

FIGURE 6.8.3

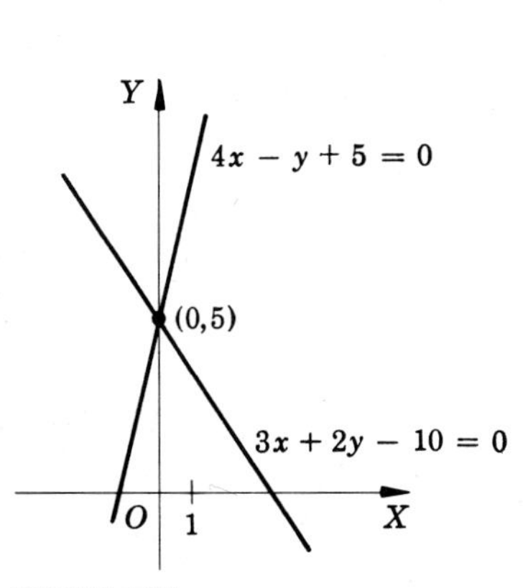

FIGURE 6.8.5

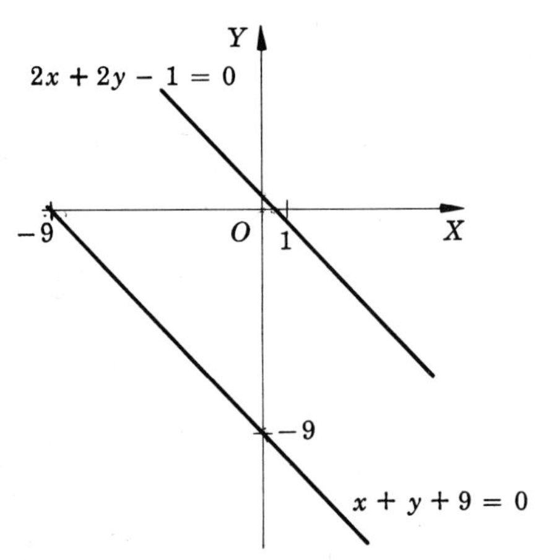

FIGURE 6.8.7

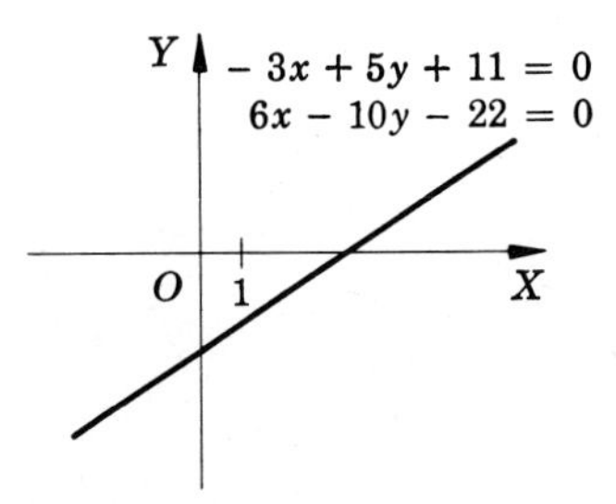

FIGURE 6.8.9

PROBLEMS 6.9

1 $x - y + 1 = 0$

3 $x - 2y + 1 = 0$

5 $x + 3y - 13 = 0$

7 $2x - 3y + 11 = 0$

9 A line parallel to both the given lines

PROBLEMS 6.10

1 (1,1,1)

3 (1,0,−4)

5 Line of solutions

7 Plane of solutions

9 No solution

PROBLEMS 6.11

1 (1,0); (−4, 5) (Fig. 6.11.1)

3 (3,−1); (−4,−8) (Fig. 6.11.3)

5 (2,3) (Fig. 6.11.5)

7 No solution (Fig. 6.11.7)

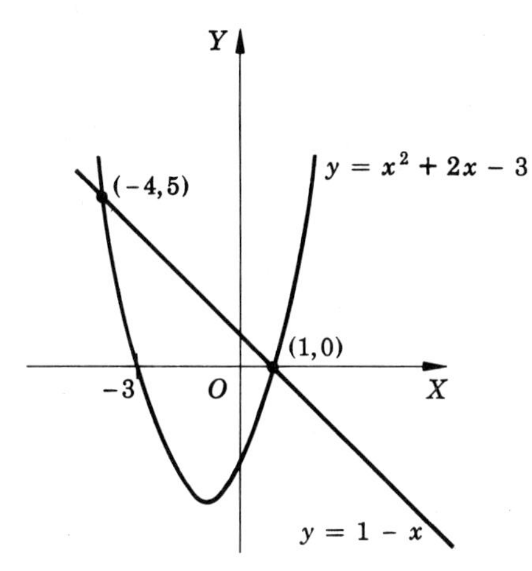

FIGURE 6.11.1

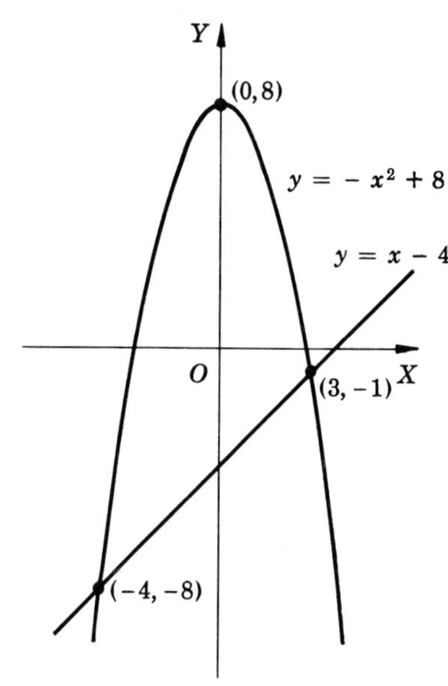

FIGURE 6.11.3

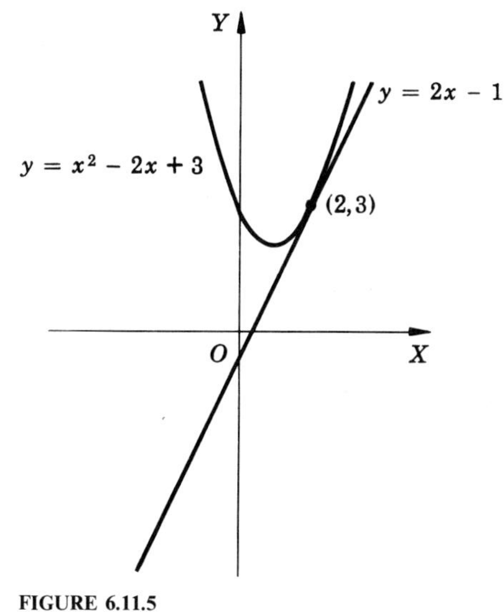

FIGURE 6.11.5

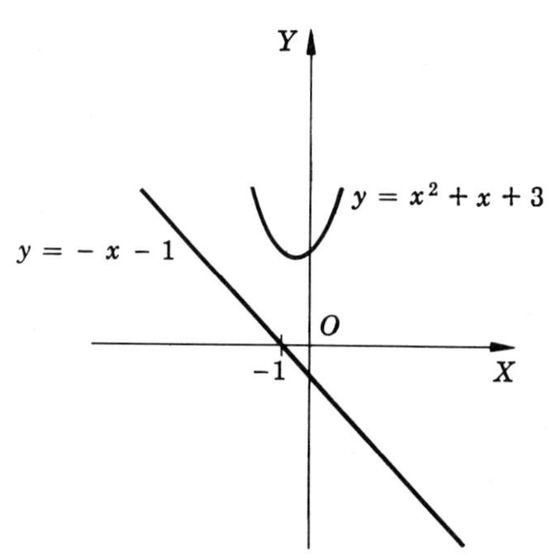

FIGURE 6.11.7

9 (2,2); (0,2) (Fig. 6.11.9)
11 (3,1); (−5,−1) (Fig. 6.11.11)
13 (5,0)
15 (−3,4); (4,−3)
17 $(\sqrt{5},\sqrt{5}), (-\sqrt{5},-\sqrt{5})$
19 (2,1,3)

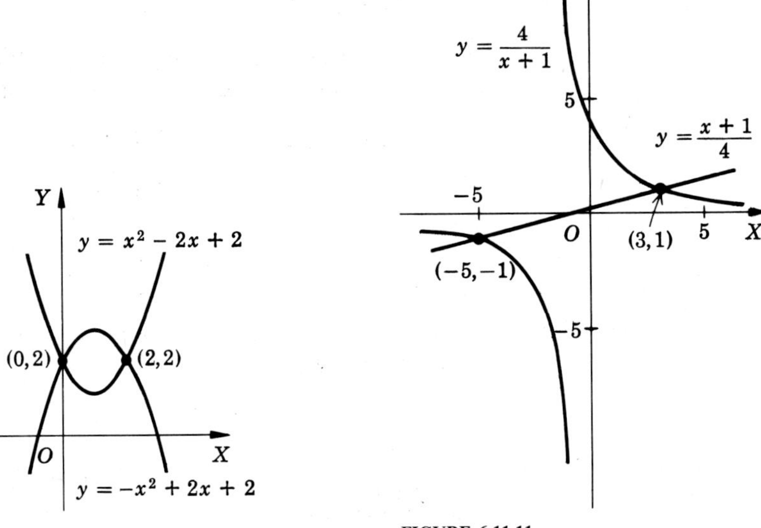

FIGURE 6.11.9

FIGURE 6.11.11

PROBLEMS 6.12

1 27
3 6 and 9
5 70 yd by 120 yd
7 12,236
9 $A = 12, B = 8$
11 60¢ and 15¢
13 Notar 40¢, Coffin Nail 50¢
15 Newspaper 1,000; TV 150
17 5 francs, 7 shillings, 21 marks
19 Neverstart 40 miles, Everknock 51 miles

PROBLEMS 6.13

1 $x' = 2x - 3$
3 $x' = (4x - 20)/3$; 20
5 From Eq. (2),

$$\begin{aligned} d'^2 &= (x_2' - x_1')^2 + (y_2' - y_1')^2 \\ &= [(x_2 + a) - (x_1 + a)]^2 + [(y_2 + b) - (y_1 + b)]^2 \\ &= (x_2 - x_1)^2 + (y_2 - y_1)^2 = d^2 \end{aligned}$$

7 Given (x,y), there is a unique (x',y'). Given (x',y'), there is a unique (x,y), namely: $x = x' - a$, $y = y' - b$.
9 From Eq. (2),

$$\begin{aligned} d'^2 &= (ax_1 - ax_2)^2 + (ay_1 - ay_2)^2 \\ &= a^2[(x_1 - x_2)^2 + (y_1 - y_2)^2] = a^2d^2 \end{aligned}$$

Hence $d' = |a|d$. If a rectangle has sides p and q, then $A = pq$. So

$$A' = (|a|p)(|a|q) = a^2pq = a^2A$$

Two triangles whose corresponding sides have proportional lengths are similar.

11 Let $x' = x \neq 0$; $y' = y \neq 0$. Then $x = ax$, $y = ay$. So $a = 1$.

13 Equation (2) is unchanged.

15 Given (x,y), there is a unique (x',y'). Given (x',y') there is a unique (x,y), namely: $x = -x'$, $y = y'$.

17 From Eq. (2),

$$\begin{aligned} d'^2 &= [(ax_2 + by_2) - (ax_1 + by_1)]^2 + [(-bx_2 + ay_2) - (-bx_1 + ay_1)]^2 \\ &= (a^2 + b^2)[(x_2 - x_1)^2 + (y^2 - y_1)^2] = d^2 \end{aligned}$$

19 Let $x' = x \neq 0$, $y' = y \neq 0$. Then

$$\begin{cases} x = ax + by \\ y = -bx + ay \end{cases}$$

or

$$\begin{cases} x = ax + by \\ y = ay - bx \end{cases} \quad \text{or} \quad \begin{cases} x^2 = ax^2 + bxy \\ y^2 = ay^2 - bxy \end{cases}$$

So $x^2 + y^2 = a(x^2 + y^2)$ or $a = 1$, and then $b = 0$.

21 We must solve

$$\begin{cases} x = 2x + y \\ y = x + 2y \end{cases} \quad \text{or} \quad \begin{cases} 0 = x + y \\ 0 = x + y \end{cases}$$

Any point on the line $x + y = 0$ is a solution.

23 We must solve

$$\begin{cases} x = a_1x + b_1y \\ y = a_2x + b_2y \end{cases} \quad \text{or} \quad \begin{cases} (a_1 - 1)x + b_1y = 0 \\ a_2x + (b_2 - 1)y = 0 \end{cases}$$

From Theorem 9 there is the unique solution (0,0) unless $(a_1 - 1)(b_2 - 1) - a_2b_1 = 0$ or $a_1b_2 - a_2b_1 - b_2 - a_1 + 1 = 0$.

25 The equations can be solved uniquely for (x,y), in terms of (x',y') by virtue of Theorem 9.

PROBLEMS 7.3

1 (1,7) **3** (1,−2) **5** (4,−3)

7 (−5,8,6) **9** (12,−2,24)

11 $20\mathbf{i} - 6\mathbf{j} + 17\mathbf{k} = (20,-6,17)$

13 $(a,b,c) + (d,e,f) = (a + d, b + e, c + f)$
$(d,e,f) + (a,b,c) = (d + a, e + b, f + c)$
Since the right-hand sides are equal (Chap. 2), so are the left-hand sides.

15 $(a,b,c) + (0,0,0) = (a + 0, b + 0, c + 0) = (a,b,c)$

17 7 **19** 8 **21** 0 **23** $\sqrt{21}$

25 $\sqrt{10}$ **27** 1 **29** $\sqrt{21}$

31 $k[(a_1,b_1,c_1) + (a_2,b_2,c_2)] = k[a_1 + a_2, b_1 + b_2, c_1 + c_2]$
$= [k(a_1 + a_2), k(b_1 + b_2), k(c_1 + c_2)]$
$k(a_1,b_1,c_1) + k(a_2,b_2,c_2) = (ka_1,kb_1,kc_1) + (ka_2, kb_2, kc_2)$
$= [(ka_1 + ka_2), (kb_1 + kb_2), (kc_1 + kc_2)]$
$= [k(a_1 + a_2), k(b_1 + b_2), k(c_1 + c_2)]$
The second part has a similar proof.

PROBLEMS 7.5

1 $\begin{pmatrix} 5 & 3 \\ -2 & 10 \end{pmatrix}$ **3** $\begin{pmatrix} -1 & -10 \\ 1 & 5 \end{pmatrix}$ **5** $\begin{pmatrix} 14 & 7 \\ 6 & 11 \end{pmatrix}$

7 Proof similar to that of Probs. 7.3, example 13

11 (15), a one-by-one matrix

13 $\begin{pmatrix} -3 \\ -7 \end{pmatrix}$ **15** $\begin{pmatrix} 31 & 17 \\ -16 & 14 \end{pmatrix}$ **17** $\begin{pmatrix} 7 & 10 \\ 9 & 38 \end{pmatrix}$

19 $\begin{pmatrix} 9 & 13 \\ -1 & -17 \end{pmatrix}$ **21** $\begin{pmatrix} 22 & -5 \\ 20 & -2 \\ 30 & -1 \end{pmatrix}$ **23** $\begin{pmatrix} 2 & 1 & 4 \\ -10 & -5 & -20 \\ 4 & 2 & 8 \end{pmatrix}$

25 $\begin{pmatrix} 19 & -9 & 23 \\ 5 & 14 & -4 \\ 6 & 1 & 3 \end{pmatrix}$ **27** $\begin{pmatrix} 38 & 0 & 0 \\ 0 & 38 & 0 \\ 0 & 0 & 38 \end{pmatrix}$ **29** $\begin{pmatrix} 2 & 2 \\ -1 & 3 \\ 4 & 1 \end{pmatrix}$

31 $\begin{pmatrix} 0 & 0 \\ 0 & 0 \end{pmatrix}$ No **33** $\begin{pmatrix} -5 & 4 \\ 25 & 4 \end{pmatrix}$ and $\begin{pmatrix} 7 & 8 & -2 \\ 4 & 0 & 8 \\ 0 & 4 & -8 \end{pmatrix}$

35 AB is a 2×2 matrix, BA is a 3×3 matrix.

37 0 **39** 0

41 $\begin{pmatrix} 2 & -5 \\ 1 & 3 \end{pmatrix}\begin{pmatrix} x \\ y \end{pmatrix} = \begin{pmatrix} -8 \\ 7 \end{pmatrix}$ **43** $\begin{pmatrix} 2 & -3 & 4 \\ 1 & -2 & 1 \\ 4 & 5 & -5 \end{pmatrix}\begin{pmatrix} x \\ y \\ z \end{pmatrix} = \begin{pmatrix} 3 \\ 0 \\ 4 \end{pmatrix}$

45 $X' = AX$ where $A = \begin{pmatrix} a & b \\ c & d \end{pmatrix}$

$X'' = BX'$ where $B = \begin{pmatrix} p & q \\ r & s \end{pmatrix}$

$X'' = B(AX) = (BA)X$

or $x'' = (ap + cq)x + (bp + dq)y$

$y'' = (ar + cs)x + (br + ds)y$

47 $4x^2 + 4xy + 5y^2$

PROBLEMS 7.6

1 $\triangle = 11, A^{-1} = \frac{1}{11}\begin{pmatrix} 4 & -3 \\ 1 & 2 \end{pmatrix}$ **3** $\triangle = -8, A^{-1} = -\frac{1}{8}\begin{pmatrix} 0 & -4 \\ -2 & -3 \end{pmatrix}$

5 $\triangle = 0$, no inverse **7** $\triangle = 1, A^{-1} = \begin{pmatrix} 1 & 0 \\ 0 & 1 \end{pmatrix}$

9 $\triangle = 1, A^{-1} = \begin{pmatrix} \frac{1}{\sqrt{2}} & \frac{1}{\sqrt{2}} \\ -\frac{1}{\sqrt{2}} & \frac{1}{\sqrt{2}} \end{pmatrix}$

PROBLEMS 7.7

1 -2 **3** -41 **5** 19 **7** -76

9 $x + 2y - 5$

13 $\frac{1}{5}\begin{pmatrix} 2 & 4 & 1 \\ 3 & 1 & -1 \\ 3 & 6 & -1 \end{pmatrix}$ **15** $\frac{1}{7}\begin{pmatrix} 1 & 4 & -3 \\ -3 & -5 & 9 \\ 4 & 2 & -5 \end{pmatrix}$

17 $\triangle = 0$, no inverse **19** $\frac{1}{43}\begin{pmatrix} 3 & -7 & 3 \\ 4 & 5 & 4 \\ -26 & -11 & 17 \end{pmatrix}$

PROBLEMS 7.8

1 (1,2)
3 (2,1)
5 No solution
7 (1,2,1)
9 (2,−1,2)
11 Line of solutions
13 (−3,2,1)

15
$$x = \frac{\begin{vmatrix} d_1 & b_1 & c_1 \\ d_2 & b_2 & c_2 \\ d_3 & b_3 & c_3 \end{vmatrix}}{\begin{vmatrix} a_1 & b_1 & c_1 \\ a_2 & b_2 & c_2 \\ a_3 & b_3 & c_3 \end{vmatrix}} \qquad y = \frac{\begin{vmatrix} a_1 & d_1 & c_1 \\ a_2 & d_2 & c_2 \\ a_3 & d_3 & c_3 \end{vmatrix}}{\begin{vmatrix} a_1 & b_1 & c_1 \\ a_2 & b_2 & c_2 \\ a_3 & b_3 & c_3 \end{vmatrix}} \qquad z = \frac{\begin{vmatrix} a_1 & b_1 & d_1 \\ a_2 & b_2 & d_2 \\ a_3 & b_3 & d_3 \end{vmatrix}}{\begin{vmatrix} a_1 & b_1 & c_1 \\ a_2 & b_2 & c_2 \\ a_3 & b_3 & c_3 \end{vmatrix}}$$

17 $x = k, y = -10k, z = -7k$
19 $x = -15k, y = -5k, z = 13k$
21 (6,−2,−5)
23 (11,−14,9)
25 $(1,-2,5) \wedge (2,-4,10) = (0,0,0)$
27 (7,−4,−5)
29 (0,0,−24) **31** **k** **33** **j**
35 (0,0,0)

PROBLEMS 8.3

1 $x > -5$ (Fig. 8.3.1)
3 $x \leq -3$ (Fig. 8.3.3)
5 $x > 1$ (Fig. 8.3.5)
7 $x > 3$ or $x < -3$ (Fig. 8.3.7)
9 $-4 < x < 6$ (Fig. 8.3.9)
11 $-2 \leq x \leq 7$ (Fig. 8.3.11)
13 No x
15 $-7 < x < 3$ (Fig. 8.3.15)
17 $x \geq 1$ or $x \leq -8$ (Fig. 8.3.17)
19 $2.9 < x < 3.1$ (Fig. 8.3.19)
21 $|x - 3| < 4$
23 $|2x - 4| < 6$
25 $|x - 3| > 2$
27 $|3x + 4| < 5$

−6 −4 −2 0 2

FIGURE 8.3.1

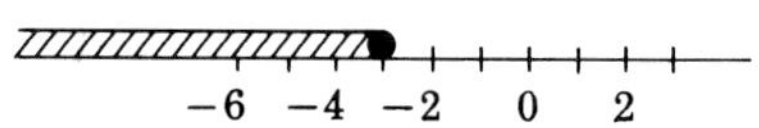

FIGURE 8.3.3

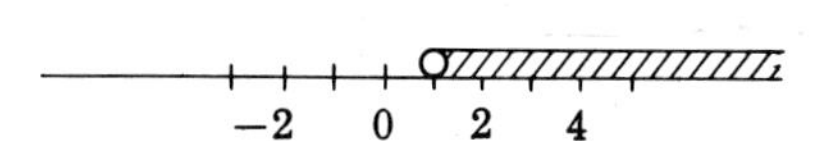

FIGURE 8.3.5

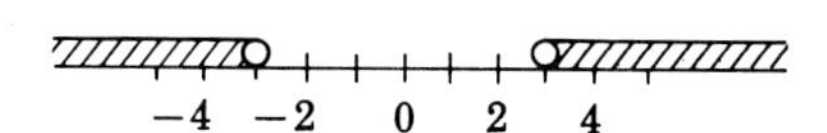

FIGURE 8.3.7

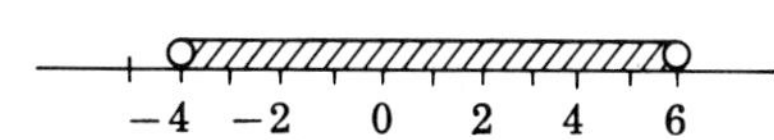

FIGURE 8.3.9

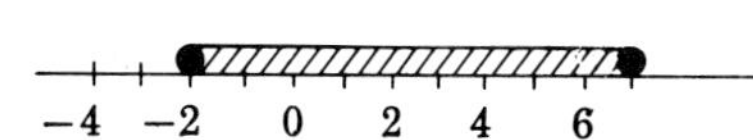

FIGURE 8.3.11

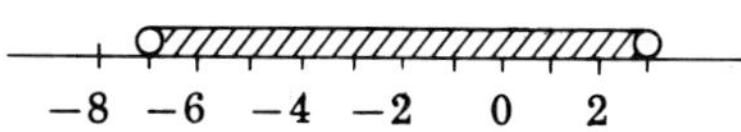

FIGURE 8.3.15

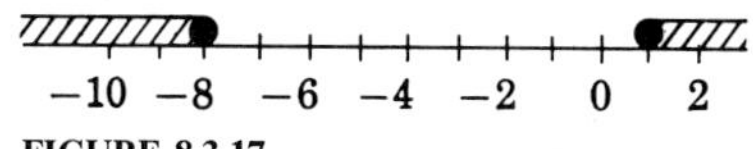

FIGURE 8.3.17

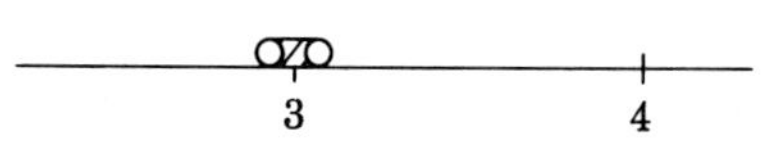

FIGURE 8.3.19

29 $x_1' - x_2' = a(x_1 - x_2)$. Since $a > 0$ and $x_1 - x_2 > 0$, it follows that $x_1' - x_2' > 0$.

31 II is: for all $k \geq 1$; if $5^k \geq 1 + 4k$, then $5^{k+1} \geq 1 + 4(k + 1) = 5 + 4k$. From the hypothesis $5^{k+1} \geq 5 + 20k > 5 + 4k$.

33 II is: for all $k \geq 1$ and $h \geq -1$; if $(1 + h)^k \geq 1 + kh$, then $(1 + h)^{k+1} \geq 1 + (k + 1)h = 1 + kh + h$. From the hypothesis (since $1 + h \geq 0$), $(1 + h)^{k+1} \geq 1 + kh + h + kh^2 \geq 1 + kh + h$.

35 II is: for all $k \geq -1$; if $2k^3 - 9k^2 + 13k + 25 > 0$, then $2(k + 1)^3 - 9(k + 1)^2 + 13(k + 1) + 25 > 0$. By expansion, $2(k + 1)^3 - 9(k + 1)^2 + 13(k + 1) + 25 = (2k^3 - 9k^2 + 13k + 25) + 6(k - 1)^2 > 0$.

37 $(1 - h)^n(1 + h)^n = (1 - h^2)^n$. If n is odd, $(1 - h^2)^n \leq 1$ for all h. If n is even, $(1 - h^2)^n \leq 1$ for $|h| \leq \sqrt{2}$. Also $(1 + h)^n > 0$ for all n if $h > -1$. Hence for $-1 < h \leq \sqrt{2}$, $(1 - h)^n \leq 1/(1 + h)^n$ by dividing by $(1 + h)^n$.

39 **(1)** $(1 - h)^n \leq 1/(1 + h)^n$ if $-1 < h \leq \sqrt{2}$ by Prob. 37
(2) $(1 + h)^n \geq 1 + nh$ if $h \geq -1$ by Prob. 33, and $(1 + nh) > 0$ if $h \geq 0$
(3) $1/(1 + h)^n \leq 1/(1 + nh)$ if $h \geq 0$ by (2) and Prob. 38
(4) From (1) and (3), $(1 - h)^2 \leq 1/(1 + nh)$ if $0 \leq h \leq \sqrt{2}$

41 Since $|x| < 3$, $2x^3 - 3x^2 - 5x + 3 \leq 2(27) + 3(9) + 5(3) + 3 = 99$

45 $-1 < x - 3 < 1, 3 < x + 1 < 5, 7 < x + 5 < 9$. So $|x + 1| > 3$,

$$|x + 5| < 9, \left|\frac{x + 5}{x + 1}\right| < \frac{9}{3} = 3.$$

47 $-1 < x - 2 < 1, 3 < x + 2 < 5, |x + 2| < 5$,
$|x^2 - 4| = |x - 2| \times |x + 2| < 1 \times 5 = 5$

49 $-0.01 < x - 2 < 0.01, 3.99 < x + 2 < 4.01, |x + 2| < 4.01$.
$|x^2 - 4| < 0.01 \times 4.01 = 0.0401$

PROBLEMS 8.4

1 $\{x \mid x < -3 \text{ or } x > 1\}$ (Fig. 8.4.1)
3 ~~$\{x \mid x \geq 1\}$ (Fig. 8.4.3)~~ all reals
5 $\{x \mid x \leq -\frac{1}{2} \text{ or } x \geq 3\}$ (Fig. 8.4.5)
7 $\{x \mid -3 < x < -2\}$ (Fig. 8.4.7)

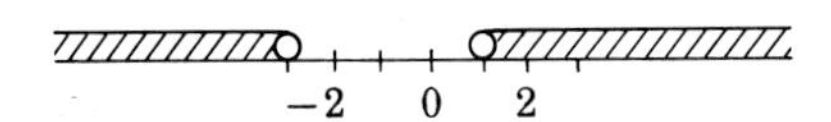

FIGURE 8.4.1

−2 0 2 4 10
FIGURE 8.4.3

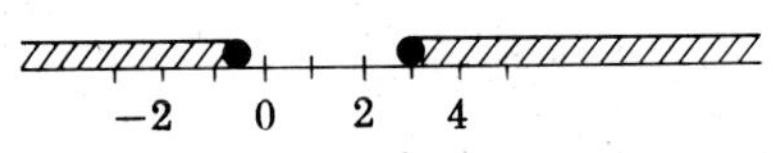

FIGURE 8.4.5

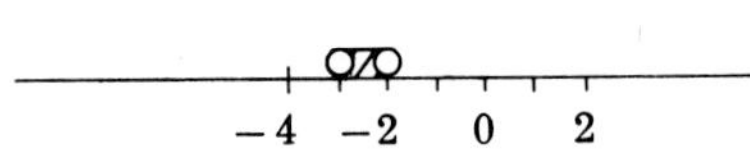

FIGURE 8.4.7

9 All x
11 No x
13 $\{x \mid -3 < x < -1 \text{ or } x > 4\}$ (Fig. 8.4.13)
15 $\{x \mid x \leq 5\}$ (Fig. 8.4.15)

17 Equivalent to $(x+3)(x+2)(x+1)(x-1) < 0$
$\{x \mid -3 < x < -2 \text{ or } -1 < x < 1\}$ (Fig. 8.4.17)

19 Equivalent to $2(x+7)(x+6)(x+5)(x+2) > 0$
$\{x \mid x < -7 \text{ or } -6 < x < -5 \text{ or } x > -2\}$ (Fig. 8.4.19)

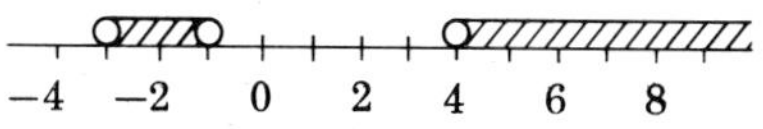

−4 −2 0 2 4 6 8

FIGURE 8.4.13

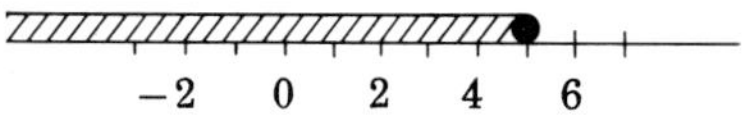

−2 0 2 4 6

FIGURE 8.4.15

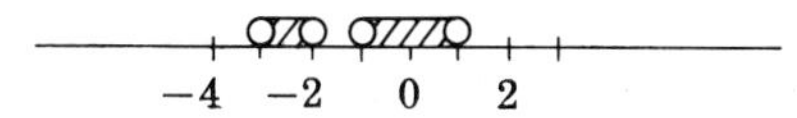

FIGURE 8.4.17

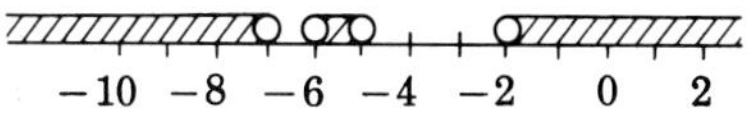

−10 −8 −6 −4 −2 0 2

FIGURE 8.4.19

21 II is: for all $k \geq 5$; if $2^k > k^2$, then $2^{k+1} > k^2 + 2k + 1$. From the hypothesis $2^{k+1} > 2k^2$. Prove that $2k^2 > k^2 + 2k + 1$, or $k^2 - 2k - 1 > 0$. Now $k^2 - 2k - 1 = (k-4)(k+2) + 7$ which is > 0 since $k \geq 5$.

23 I is true. II is false

25 I is false.

27 I is false.

29 I is false.

31 We are given $(a+b)(a-b) > 0$ and $a > 0$, $b > 0$. Hence $a + b > 0$, and consequently $a - b > 0$.

33 $a > c, b > d$. Hence $a + b > c + d$.

35 Equality when $a \geq 0$ and $b \geq 0$.

37 $|(a-b) + b| \leq |a-b| + |b|$; $|a| \leq |a-b| + |b|$; $|a| - |b| \leq |a-b|$.

39 Equality when $a_1 = kb_1$; $a_2 = kb_2$.

41 From Cauchy's Inequality (Prob. 40)

$$\frac{(a_1b_1 + \cdots + a_nb_n)^2}{(a_1^2 + \cdots + a_n^2)(b_1^2 + \cdots + b_n^2)} \leq 1$$

Now rewrite in vector notation.

PROBLEMS 8.6

1 (Fig. 8.6.1)

3 (Fig. 8.6.3)

5 Entire plane

7 (Fig. 8.6.7)

9 (Fig. 8.6.9)

11 (Fig. 8.6.11)

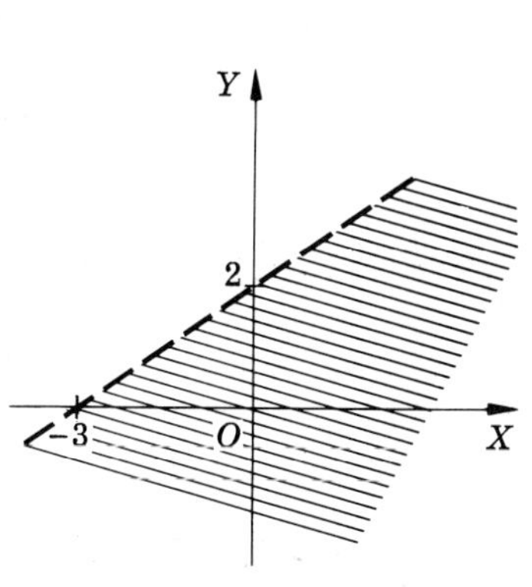

FIGURE 8.6.1

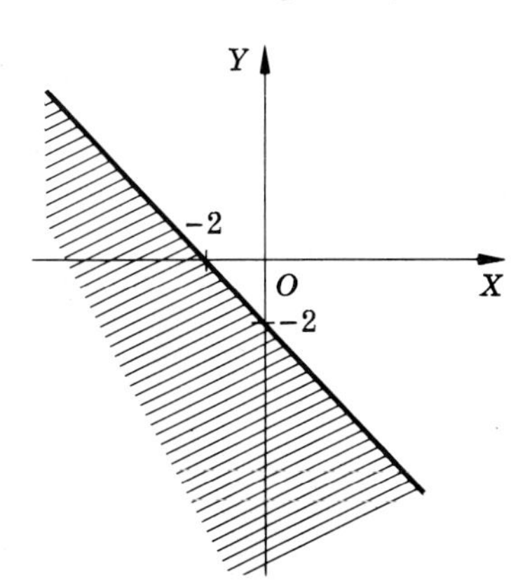

FIGURE 8.6.3

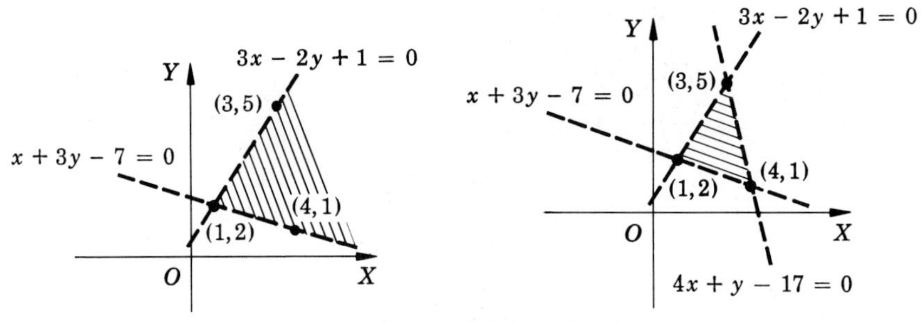

FIGURE 8.6.7

FIGURE 8.6.9

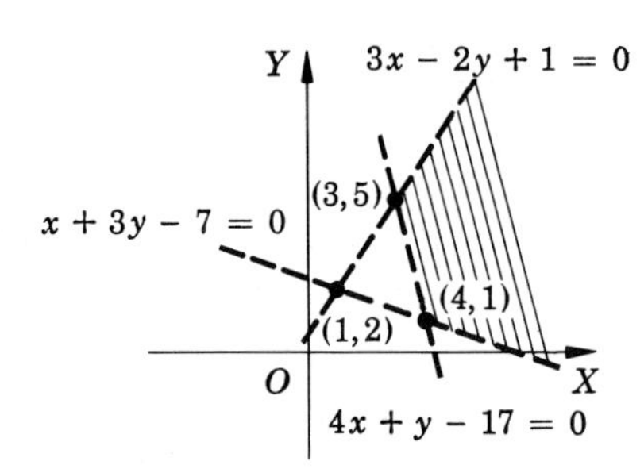

FIGURE 8.6.11

PROBLEMS 8.8

1 (Fig. 8.8.1)
3 (Fig. 8.8.3)
5 (Fig. 8.8.5)
7 (Fig. 8.8.7)
9 Entire plane
11 (Fig. 8.8.11)
13 Empty set

15

r	0···0	1···1	2···2	3
p	10···18	10···15	9···12	9

When 180 sec left to play; impossible if 90 sec left to play.

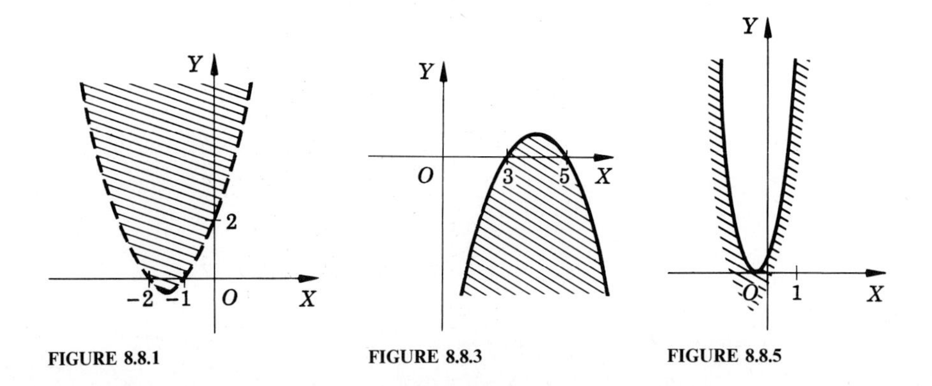

FIGURE 8.8.1

FIGURE 8.8.3

FIGURE 8.8.5

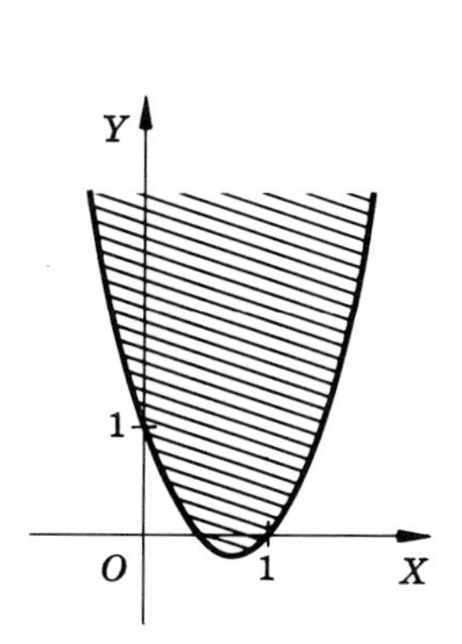

FIGURE 8.8.7

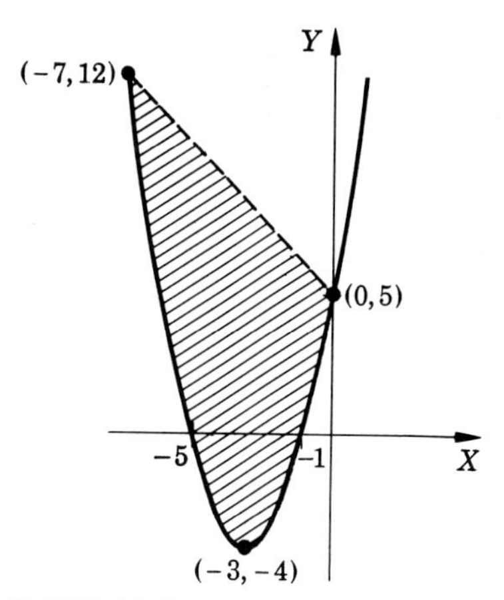

FIGURE 8.8.11

PROBLEMS 8.9

1 Cost $= 10a + 20b$. The minimum is attained at all points of segment QR (Fig. 8.20).

3 $I = (r/5) + (p/10)$; $r = 0$, $p = 10$

5 H: $100x + 200y \geq 2{,}500$ (a) $C = 10x + 30y$
Min. = \$250 at (25,0)

P: $50x + 10y \geq 350$ (b) $T = kx + 2ky$
Min. = $25k$ along segment PQ

F: $30x \geq 150$ (c) $W = x + y$
Min. = 15 at (5,10)

$x \geq 0$; $y \geq 0$

(Fig. 8.9.5)

Y
35
30
25
20
15
10
5
O 5 10 15 20 25 X
(Minimum weight = 15)
P(5, 10)
(Minimum cost = \$250)
Q(25, 0)
H
F P

FIGURE 8.9.5

7 $12x + 25y \geq 435$ (*a*) Cost = $5{,}000x + 12{,}000y$
$10x + 5y \geq 125$ Min. = \$186,000 at (30,3)
$7x + 25y \geq 285$ (*b*) Time = $40x + 50y$
$x \geq 0;\ y \geq 0$ Min. = 950 hr at (5,15)
(Fig. 8.9.7)

9 Cost: $20x + 5y \leq 625$ Errors = $0.18x + 0.22y$
Tests: $x + y \geq 35$ Min. = 6.50 at (30,5)
Time: $10x + 25y \leq 875$
$x \geq 0;\ y \geq 0$
(Fig. 8.9.9)

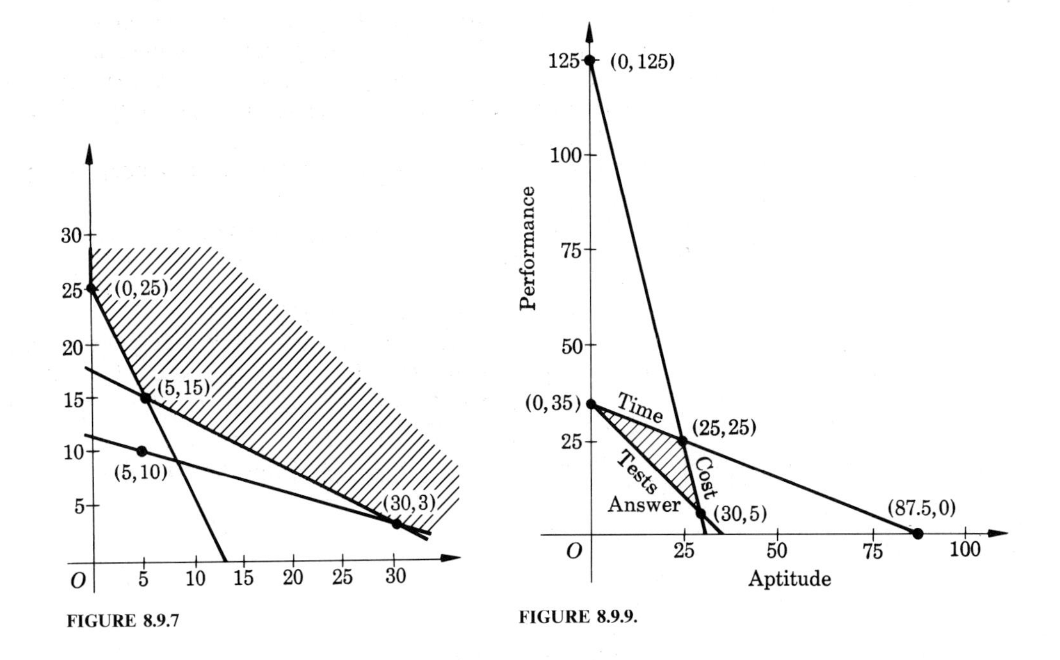

FIGURE 8.9.7

FIGURE 8.9.9.

11 No
13 Max. 15 at (3,5); min. 8 at (1,2)
15 Max. = 20; min. = 6
17 $f(x,y)$ is constant; every point is both a max. and a min.
19 $f(x,y)$ is a constant on R.

PROBLEMS 9.1

1 Domain: all reals; image: all reals; single-valued, one-to-one
3 Domain: all reals; image: all reals; single-valued, one-to-one
5 Domain: all reals; image: $\{y \mid y \geq -1\}$; single-valued, not one-to-one
7 Domain: $\{x \mid -9 \leq x \leq 9\}$; image: $\{y \mid -9 \leq y \leq 9\}$; multivalued
9 Domain: all reals; image: all reals; multivalued
11 Domain: all reals; image: all reals; multivalued
13 Domain: $\{x \mid x \leq 2\}$; image: $\{y \mid y \leq 1\}$; single-valued, one-to-one
15 Domain: $x = 1$; image: $y = 2$; single-valued, one-to-one

17 Domain: $\{3, 7\}$; image: $\{7, 8\}$; single-valued, one-to-one
19 Domain: $\{1, 2\}$; image: $\{2, 3\}$; single-valued, one-to-one

PROBLEMS 9.4

1 Function
3 Not a function
5 Domain: all reals; image: all reals; one-to-one, onto
7 Domain: all reals; image: all reals; one-to-one, onto
9 Domain: $\{1, 2\}$; image: $\{7\}$; not one-to-one, into
11 Domain: $\{5, 8, 7\}$; image: $\{1, 2\}$; not one-to-one, into
13 Domain: $\{2, 3\}$; image: $\{7\}$; not one-to-one, into
15 Domain: all reals; image: $\{0, 1\}$; not one-to-one, into
17 Domain: all reals; image: $\{0, 1, 2\}$; not one-to-one, into
19 Domain: all reals; image: $\{y | 0 < y \leq \frac{1}{4}\}$; not one-to-one, into
21 Domain: all reals except $0, -3$; image: $\{y | y > 0\} \cup \{y | y \leq -\frac{4}{9}\}$; not one-to-one, into
23 Domain: all reals except $-2, 2$; image: all reals except $0, -\frac{1}{4}$; one-to-one, into
25 (*a*) Image: $\{y | -\frac{81}{4} \leq y \leq 0\}$; (*b*) image: $\{0\}$
27 $f(0) = 1$, $f(1) = 3$, $f(2) = 9$, $f(a + h) = 2a^2 + 4ah + 2h^2 + 1$, $f(h) = 2h^2 + 1$
29 Some of the answers are (Fig. 9.4.29):
(*a*) $y = +\sqrt{x^2 - 81}$; domain: $\{x | x \leq -9 \text{ or } x \geq 9\}$; image: $\{y | y \geq 0\}$
(*b*) $y = -\sqrt{x^2 - 81}$; domain: $\{x | x \leq -9 \text{ or } x \geq 9\}$; image: $\{y | y \leq 0\}$
(*c*) $y = \begin{cases} +\sqrt{x^2 - 81}, & x = -9 \\ -\sqrt{x^2 - 81}, & x < -9 \text{ or } x \geq 9 \end{cases}$
Domain: $\{x | x \leq -9 \text{ or } x \geq 9\}$; image: $\{y | y \leq 0\}$

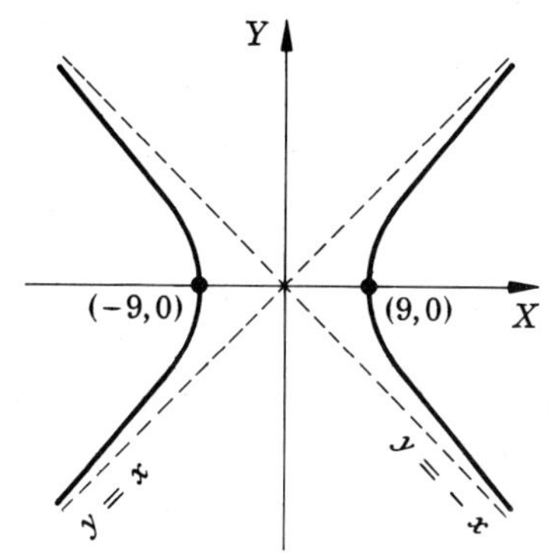

FIGURE 9.4.29

PROBLEMS 9.5

1 $(f + g)(x) = \dfrac{2x + 3}{(x + 1)(x + 2)}$ Domain: all reals except $-1, -2$

$(f - g)(x) = \dfrac{1}{(x + 1)(x + 2)}$ Domain: all reals except $-1, -2$

$(fg)(x) = \dfrac{1}{(x+1)(x+2)}$ Domain: all reals except $-1, -2$

$\left(\dfrac{f}{g}\right)(x) = \dfrac{x+2}{x+1}$ Domain: all reals except $-1, -2$

3 $(f+g)(x) = \dfrac{x^2+3x+3}{x+2}$ Domain: all reals except -2

$(f-g)(x) = \dfrac{-(x^2+3x+1)}{x+2}$ Domain: all reals except -2

$(fg)(x) = \dfrac{x+1}{x+2}$ Domain: all reals except -2

$\left(\dfrac{f}{g}\right)(x) = \dfrac{1}{(x+2)(x+1)}$ Domain: all reals except $-1, -2$

5 $(f+g)(x) = x^2+5x+5$ Domain: all reals
$(f-g)(x) = -(x^2+3x+3)$ Domain: all reals
$(fg)(x) = (x+1)(x+2)^2$ Domain: all reals

$\left(\dfrac{f}{g}\right)(x) = \dfrac{x+1}{(x+2)^2}$ Domain: all reals except -2

7 $(f+g)(x) = \dfrac{x+1}{x^2}$ Domain: all reals except 0

$(f-g)(x) = \dfrac{2x^2+x-1}{x^2}$ Domain: all reals except 0

$(fg)(x) = \dfrac{-(x^3+x^2-x-1)}{x^3}$ Domain: all reals except 0

$\left(\dfrac{f}{g}\right)(x) = \dfrac{x}{1-x}$ Domain: all reals except $-1, 0, 1$

9 $(g \circ f)(x) = \dfrac{3}{x^2} - \dfrac{2}{x} = \dfrac{3-2x}{x^2}$ Domain: all reals except 0

11 $(g \circ f)(x) = 1/x$ Domain: all reals except 0
$(f \circ g)(x) = 1/x$ Domain: all reals except 0
13 $(g \circ f)(x) = (x-4)^2$ Domain: all reals
$(f \circ g)(x) = x^2 - 4$ Domain: all reals
15 $(g \circ f)(x) = 3x^2$ Domain: all reals
$(f \circ g)(x) = 3x^2 + 6x + 2$ Domain: all reals
17 (*a*) $(x-3)\sqrt{x-3}$; (*b*) $1/x^2$

PROBLEMS 9.6

1 (Fig. 9.6.1) **3** (Fig. 9.6.3) **5** (Fig. 9.6.5)
7 (Fig. 9.6.7) **9** (Fig. 9.6.9) **11** (Fig. 9.6.11)
13 (Fig. 9.6.13) **15** (Fig. 9.6.15) **17** (Fig. 9.6.17)
19 (Fig. 9.6.19) **21** (Fig. 9.6.21) **23** (Fig. 9.6.23)
25 (Fig. 9.6.25) **27** (Fig. 9.6.27) **29** (Fig. 9.6.29)
31 (Fig. 9.6.31) **33** (Fig. 9.6.33) **35** (Fig. 9.6.35)
37 (Fig. 9.6.37) **39** (Fig. 9.6.39) **41** (Fig. 9.6.41)
43 (Fig. 9.6.43)

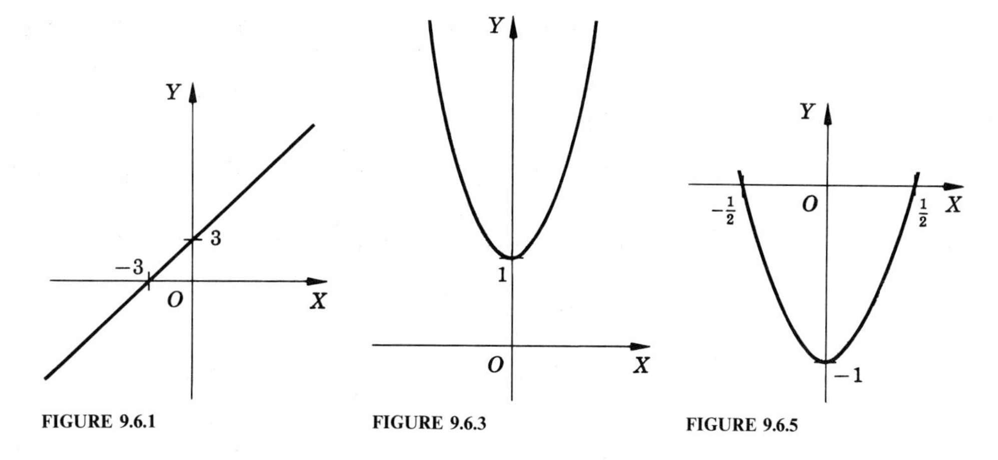

FIGURE 9.6.1 FIGURE 9.6.3 FIGURE 9.6.5

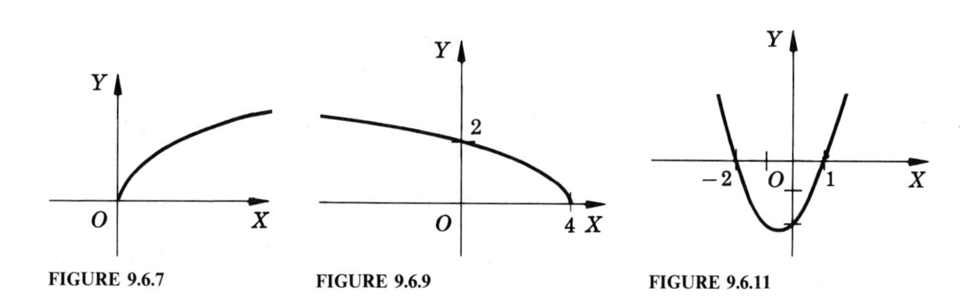

FIGURE 9.6.7 FIGURE 9.6.9 FIGURE 9.6.11

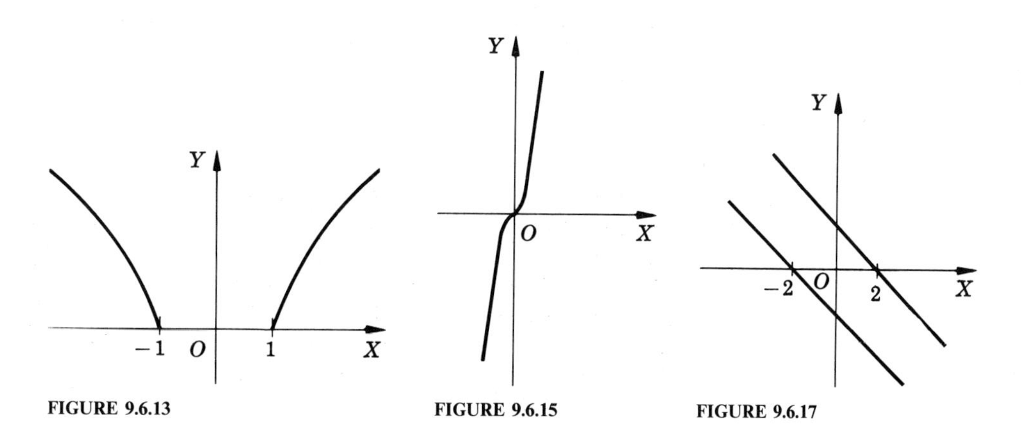

FIGURE 9.6.13 FIGURE 9.6.15 FIGURE 9.6.17

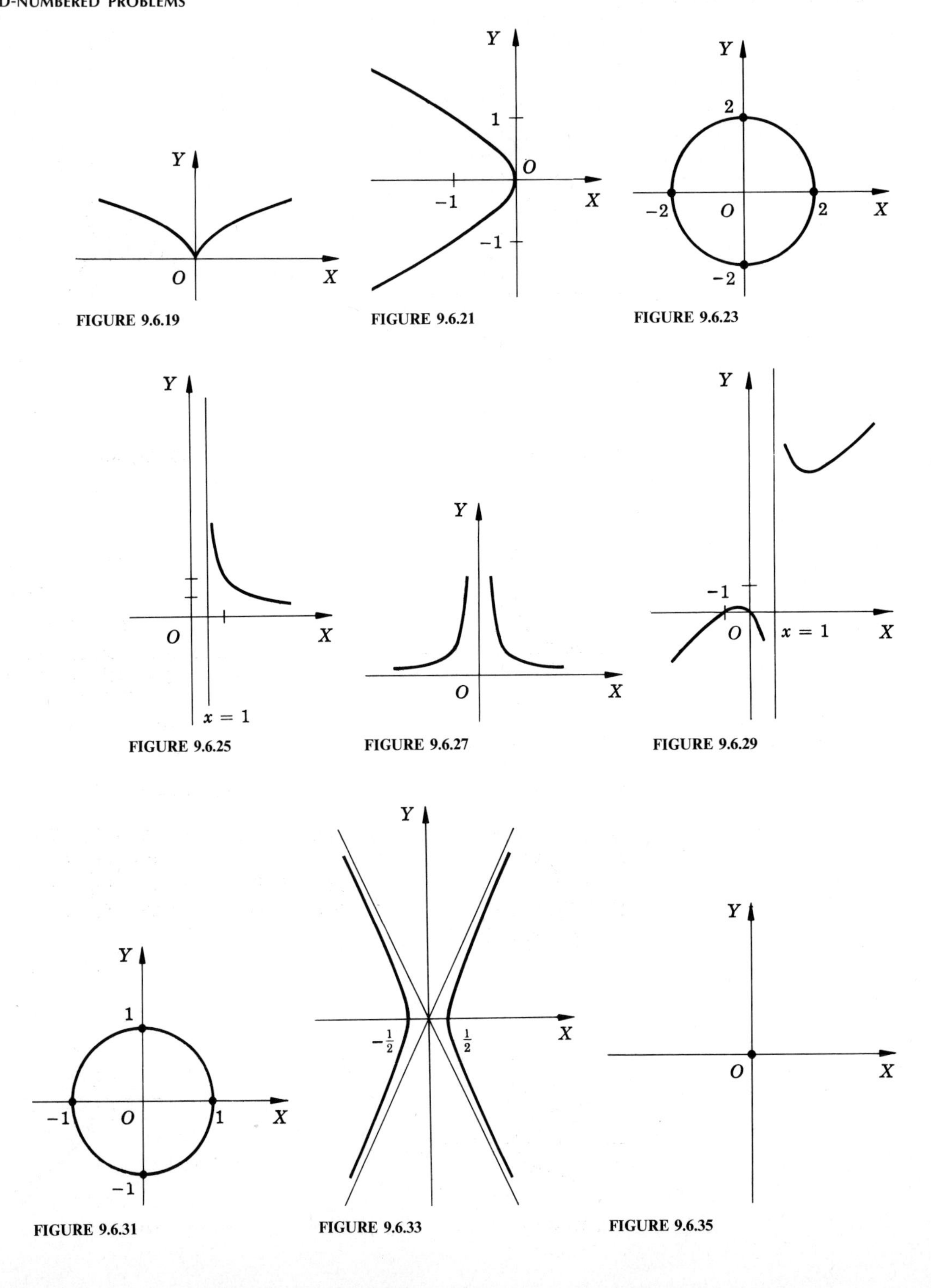

FIGURE 9.6.19

FIGURE 9.6.21

FIGURE 9.6.23

FIGURE 9.6.25

FIGURE 9.6.27

FIGURE 9.6.29

FIGURE 9.6.31

FIGURE 9.6.33

FIGURE 9.6.35

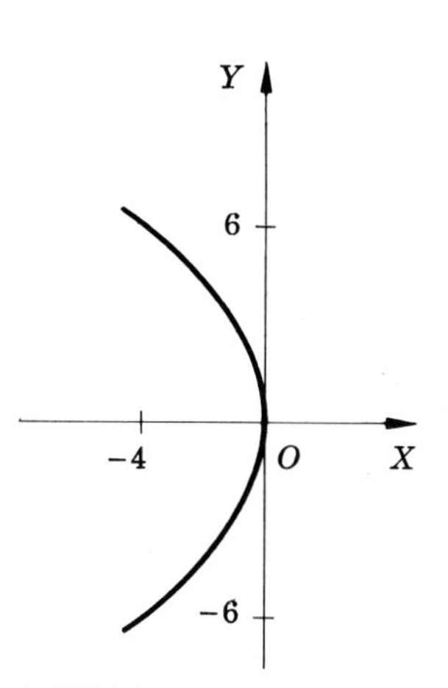

FIGURE 9.6.37

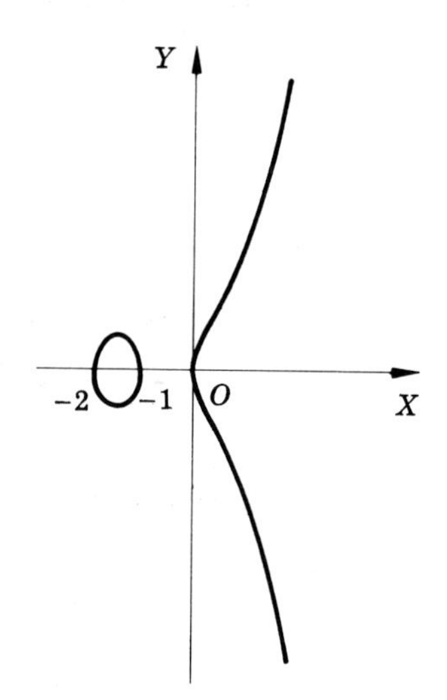

FIGURE 9.6.39

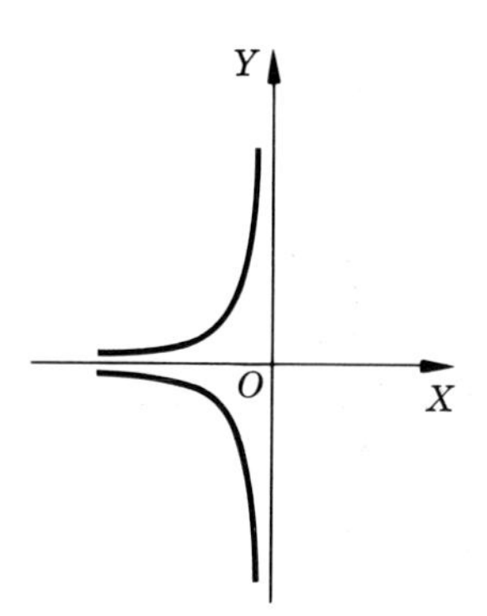

FIGURE 9.6.41

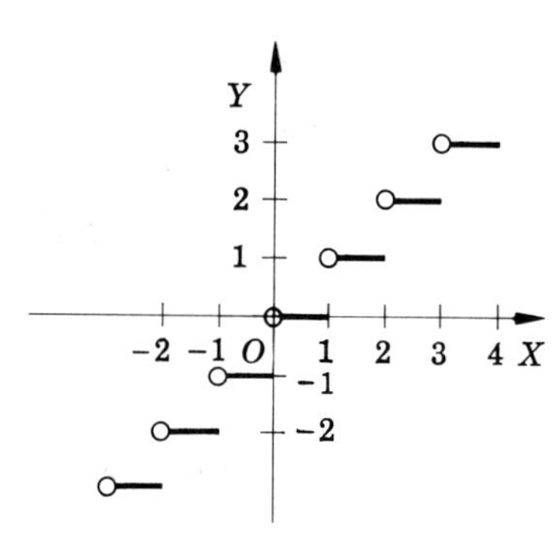

FIGURE 9.6.43

PROBLEMS 9.7

1 (Fig. 9.7.1) **3** (Fig. 9.7.3) **5** (Fig. 9.7.5)
7 (Fig. 9.7.7) **9** (Fig. 9.7.9) **11** (Fig. 9.7.11)
13 (Fig. 9.7.13) **15** (Fig. 9.7.15) **17** (Fig. 9.7.17)
19 (Fig. 9.7.19) **21** (Fig. 9.7.21) **23** (Fig. 9.7.23)
25 (Fig. 9.7.25)

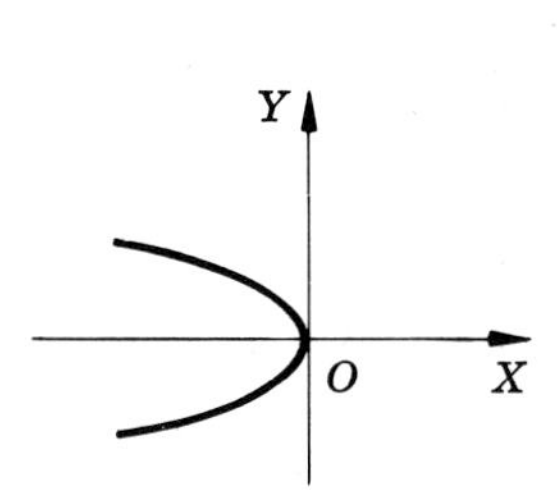

FIGURE 9.7.1

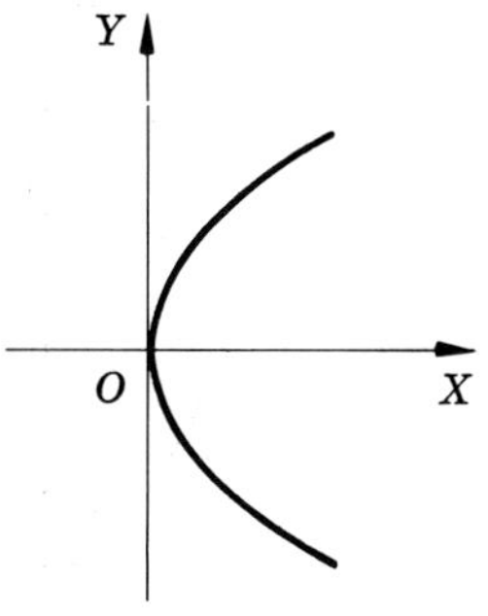

FIGURE 9.7.3

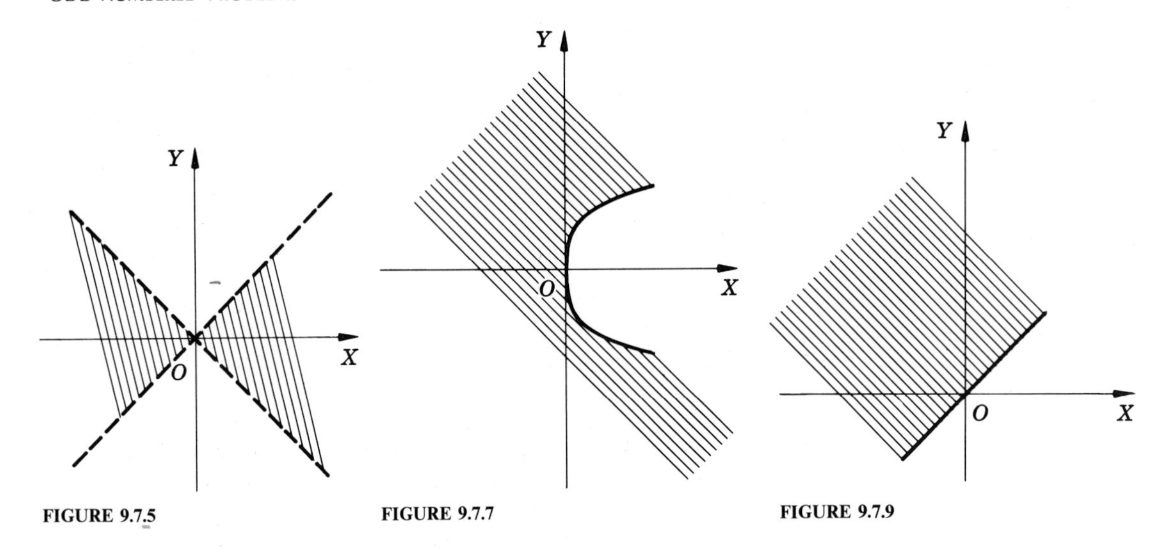

FIGURE 9.7.5

FIGURE 9.7.7

FIGURE 9.7.9

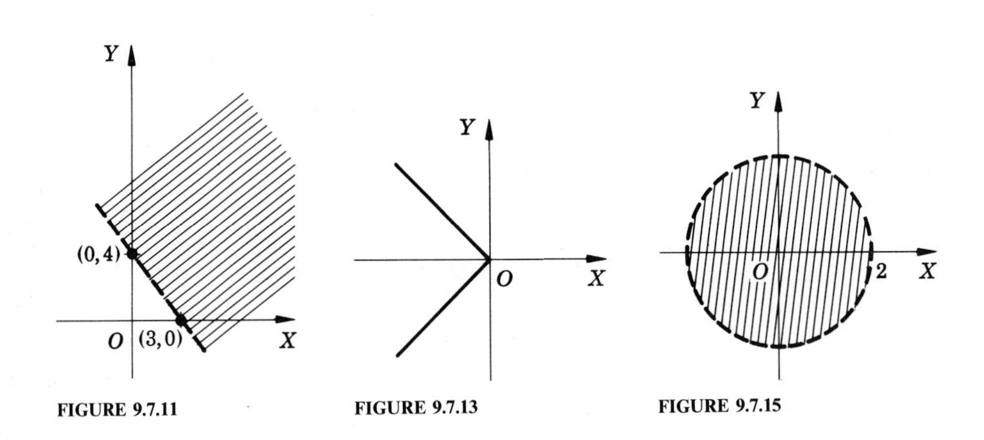

FIGURE 9.7.11

FIGURE 9.7.13

FIGURE 9.7.15

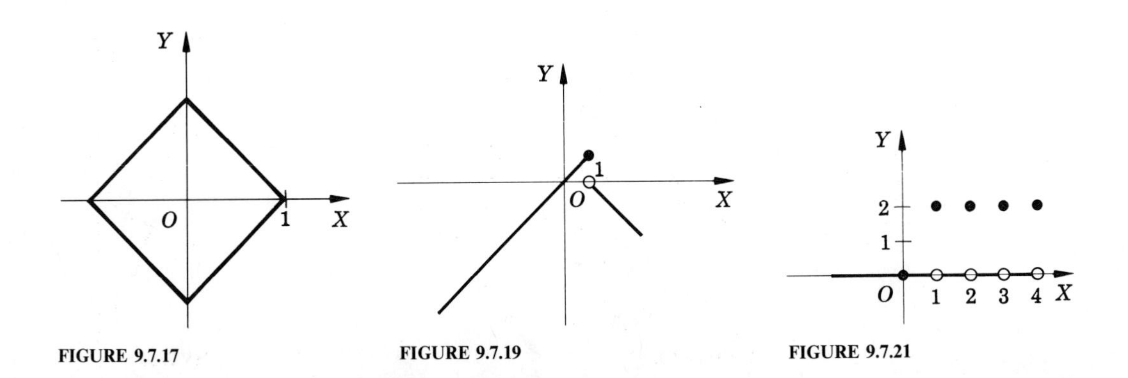

FIGURE 9.7.17

FIGURE 9.7.19

FIGURE 9.7.21

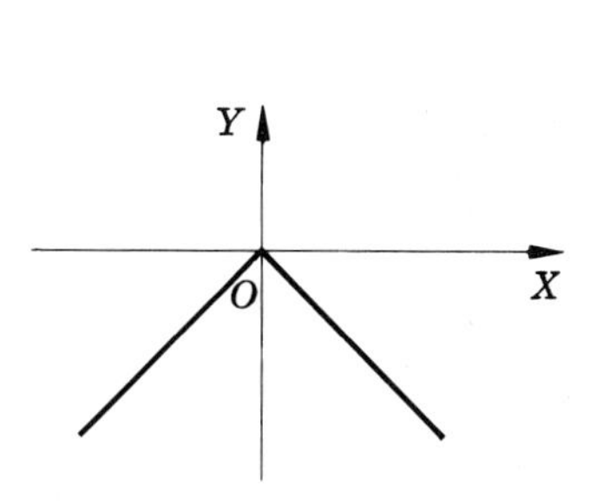

FIGURE 9.7.23

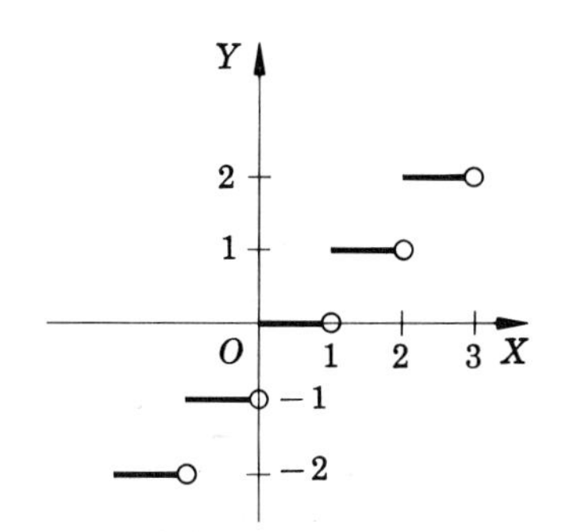

FIGURE 9.7.25

PROBLEMS 9.8

1 $f^{-1}(x) = x$; domain: all reals; image: all reals (Fig. 9.8.1)
3 $f^{-1}(x) = (x/3) - 2$; domain: all reals; image: all reals (Fig. 9.8.3)
5 $f^{-1}(x) = x/a + b/a$; domain: all reals; image: all reals (Fig. 9.8.5)
7 Does not exist. (Fig. 9.8.7)
9 $f^{-1}(x) = (x + 8)^{1/3}$; Domain: all reals; image: all reals (Fig. 9.8.9)
11 Does not exist. (Fig. 9.8.11)
13 $f(x) = x^2$, $g(x) = x + 1$; no

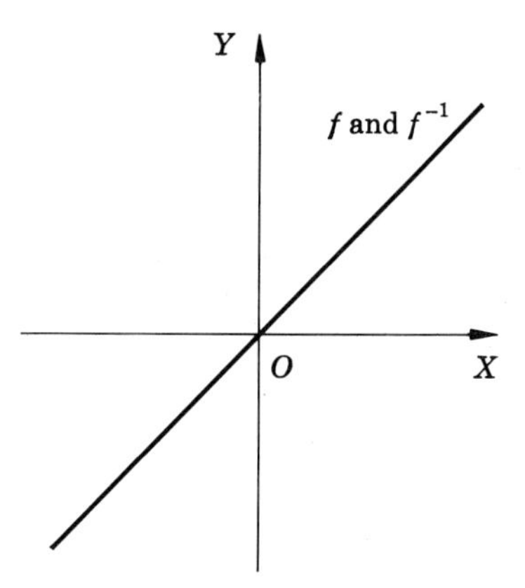

FIGURE 9.8.1

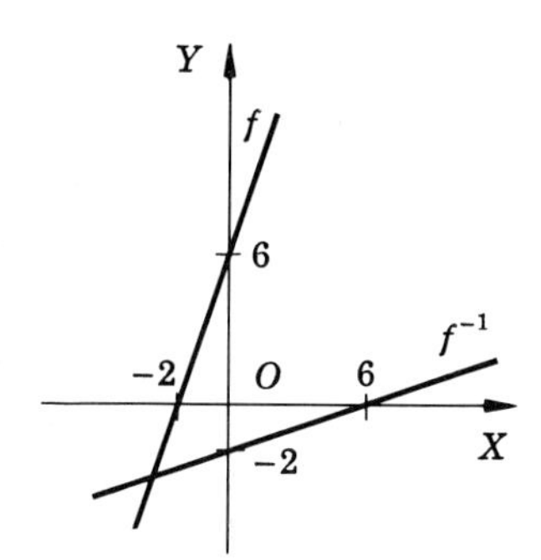

FIGURE 9.8.3

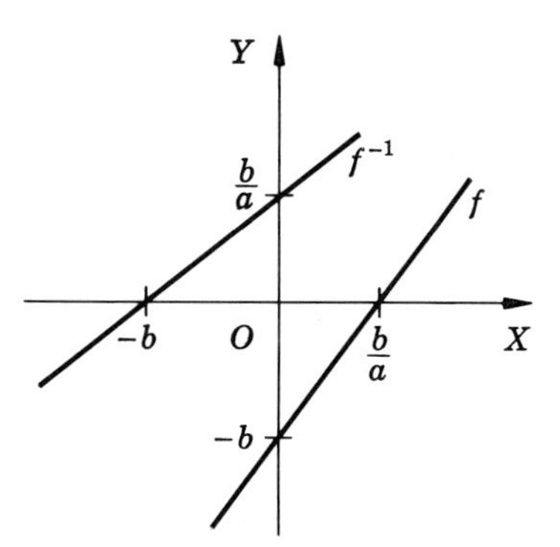

FIGURE 9.8.5

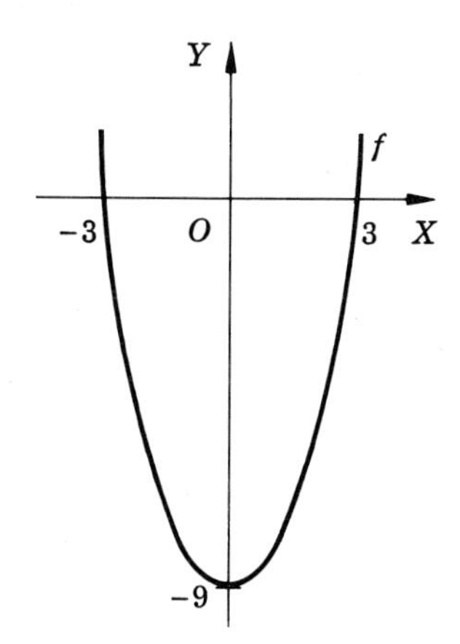

FIGURE 9.8.7

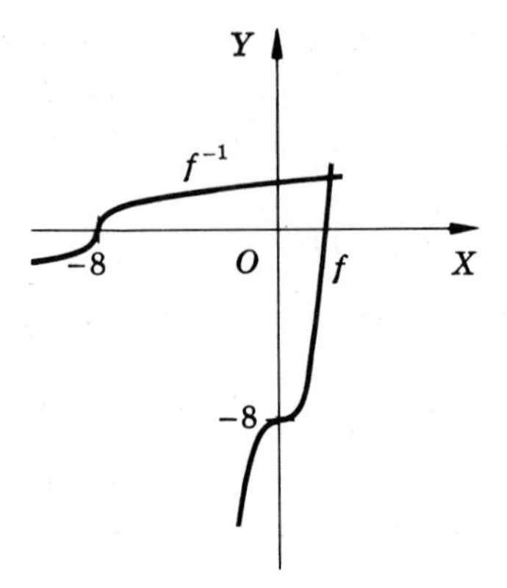

FIGURE 9.8.9

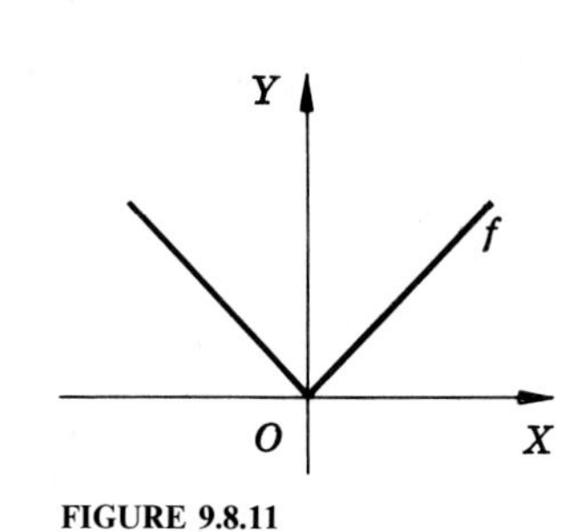

FIGURE 9.8.11

PROBLEMS 9.9

1 $f(x) = (8 - 3x)/4$; d_f: all reals; i_f: all reals
3 $f(x) = \sqrt{2 - 2x^2}$; d_f: $|x| \leq 1$; i_f: $0 \leq y \leq \sqrt{2}$
5 $f(x) = (1 - x)/x^2$; d_f: $x < 0$, $x > 0$; i_f: $y \geq -\frac{1}{4}$
7 $f(x) = -|x| + 1$: d_f: all reals; i_f: $y \leq 1$
9 $f(x) = x$; d_f: all reals; i_f: all reals
11 $f(x) = |x| + 1$; d_f: all reals; i_f: $y \geq 1$, and other answers
13 $f(v) = v^2/2g$; d_f: all reals; i_f: $s \geq 0$

PROBLEMS 10.4

1 (*a*)   3 (*a*), (*b*), (*c*)   5 (*a*)   7 (*a*), (*c*)
9 (*a*), (*c*)   11 (*c*)
13 (*a*) Polynomial, all reals; (*b*) explicit, $x \geq 0$; (*c*) explicit, $x \geq 0$
15 (*a*) Rational, all reals except $-1, 6$; (*b*) polynomial, all reals; (*c*) explicit, $-2 \leq x \leq 2$
17 (*a*) Explicit, $x \geq 2$; (*b*) rational, all reals except 0; (*c*) polynomial, all reals

PROBLEMS 10.5

1 Polynomial; domain: all reals; image: all reals; zeros: $\{0\}$ (Fig. 10.5.1)
3 Explicit algebraic; domain: $x \geq 0$; image: $y \geq 0$; zeros: $\{0\}$ (Fig. 10.5.3)
5 Polynomial; domain: all reals; image: $y \leq 1$; zeros: $\{0(\text{double}), 1, -1\}$ (Fig. 10.5.5)
7 Explicit algebraic; domain: $|x| \leq 1$; image: $0 \leq y \leq 1$; zeros: $\{0(\text{double}), 1, -1\}$; endpoints of the domain: $\{-1, 1\}$ (Fig. 10.5.7)
9 Polynomial; domain: all reals; image: all reals; zeros: $\{0(\text{double}), -2\}$ (Fig. 10.5.9)
11 Explicit algebraic; domain: $x \geq -2$; image: $y \geq 0$; zeros: $\{0(\text{double}), -2\}$; left endpoint of the domain: $\{-2\}$ (Fig. 10.5.11)
13 Rational, domain: all reals; image: $0 < y \leq 1$ (Fig. 10.5.13)
15 Polynomial; domain: all reals; image: (by advanced methods) $y \geq \frac{511}{256}$ (Fig. 10.5.15)
17 Rational; domain: all reals; image: $0 < y \leq 3$ (Fig. 10.5.17)
19 Explicit algebraic; domain: $\{x \mid -1 \leq x \leq 1\}$; image: $0 \leq y \leq 1$; zeros: $\{-1, 1\}$; endpoints: $\{-1, 1\}$ (Fig. 10.5.19)

21 Explicit algebraic; domain: $\{x \mid -2 \le x \le 1 \text{ or } 3 \le x\}$; image: $y \le 0$; zeros: $\{1, -2, 3\}$; endpoints: $\{-2, 1, 3\}$ (Fig. 10.5.21)

23 Implicit algebraic; domain: $-\frac{1}{2\sqrt{2}} \le x \le \frac{1}{2\sqrt{2}}$; image: $-\frac{1}{\sqrt{2}} \le y \le \frac{1}{\sqrt{2}}$; zeros: $\left\{-\frac{1}{2\sqrt{2}}, \frac{1}{2\sqrt{2}}\right\}$ (Fig. 10.5.23)

25 Implicit algebraic; domain: all reals; image: all reals; zeros: $\{0\}$ (Fig. 10.5.25)

27 The lines $y = x$, $x = 0$, $y = 0$; domain: all reals; image: all reals (Fig. 10.5.27)

29 n even: domain: all reals; image: $y \ge 0$; zero: $\{0\}$
n odd: domain: all reals; image: all reals; zero: $\{0\}$

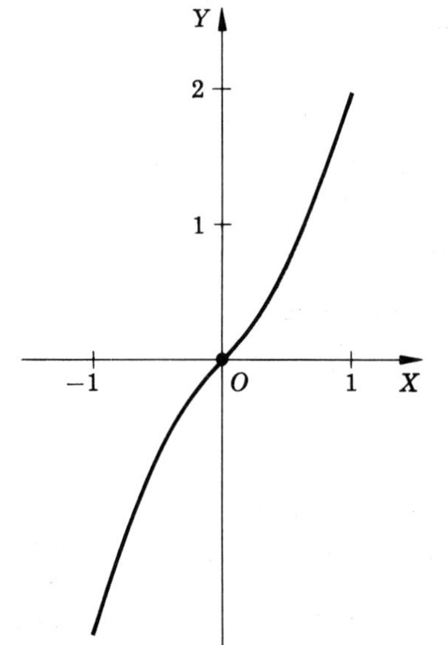

FIGURE 10.5.1

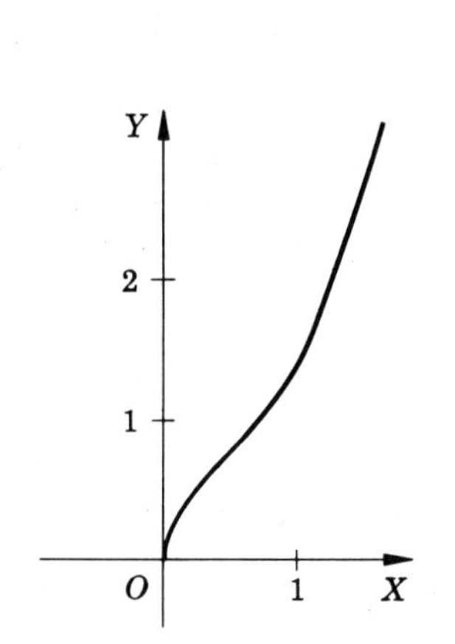

FIGURE 10.5.3

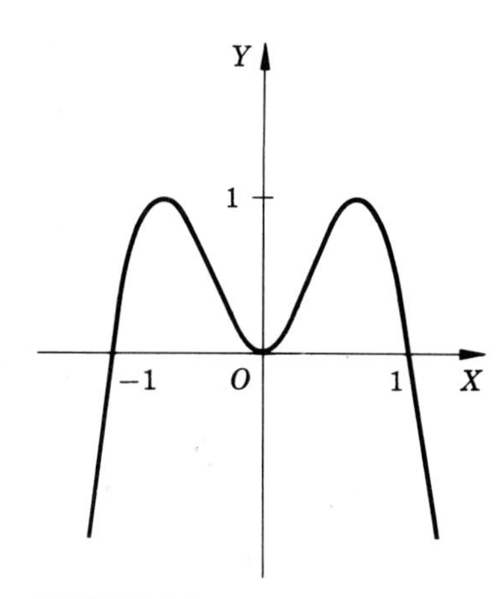

FIGURE 10.5.5

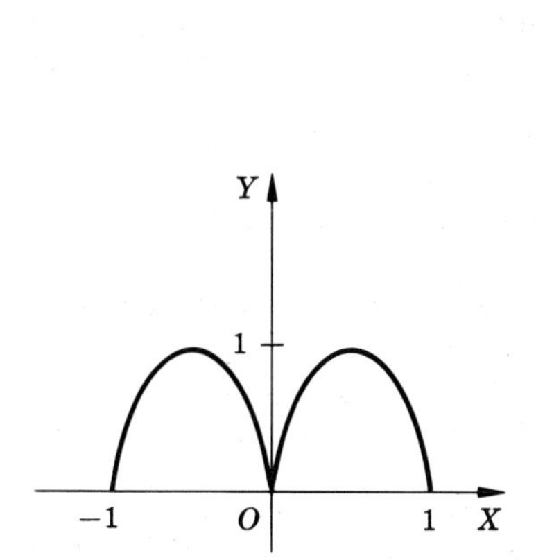

FIGURE 10.5.7

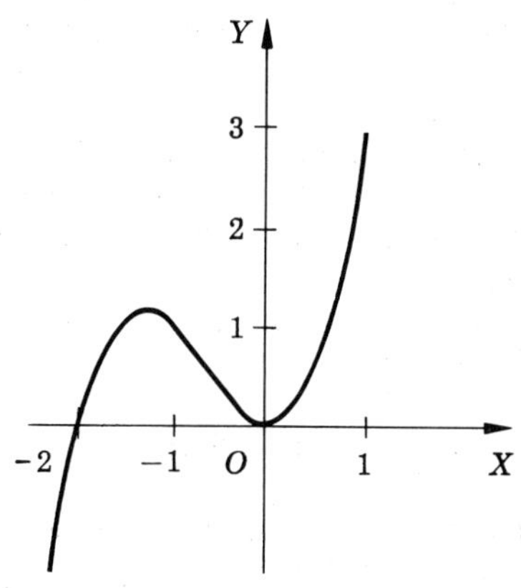

FIGURE 10.5.9

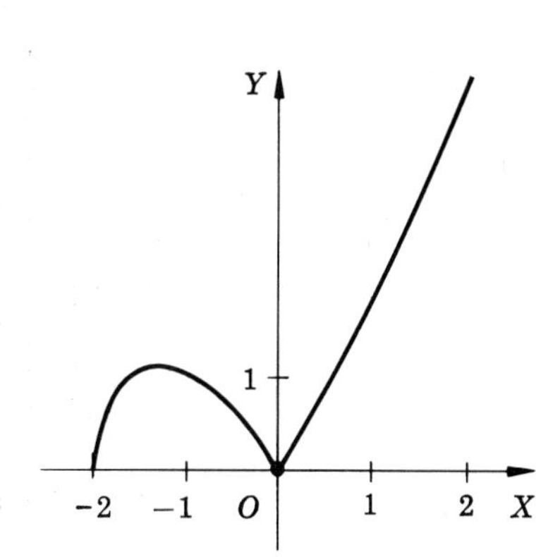

FIGURE 10.5.11

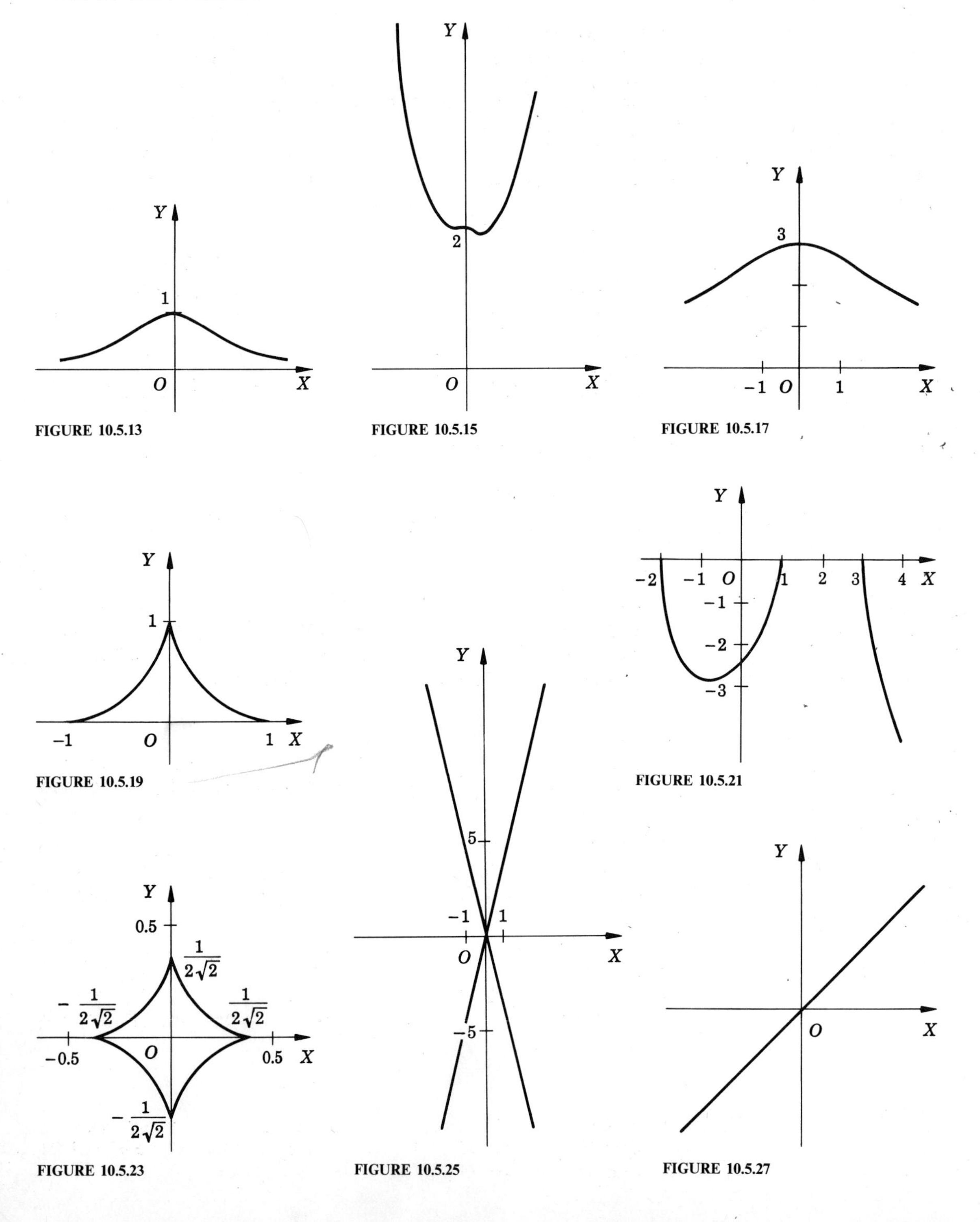

FIGURE 10.5.13

FIGURE 10.5.15

FIGURE 10.5.17

FIGURE 10.5.19

FIGURE 10.5.21

FIGURE 10.5.23

FIGURE 10.5.25

FIGURE 10.5.27

PROBLEMS 10.6

1 $x^2 - 3x = 0$ **3** $x + i = 0$
5 $x^2 + (1 - i)x - i = 0$ **7** $x^3 + x^2 + x + 1 = 0$
9 $x^2 - 2i = 0$ **11** $x^3 - x^2$
13 $x^3 - (1 - \sqrt{3})x^2 - (4 + 2\sqrt{3})x - 2\sqrt{3} - 2$
15 $x^3 - (a + b)x^2 + abx$
17 x^3 **19** 1 **21** 0 **23** 0
25 $\pm 2, \pm 2i$ **27** $\pm i$, 1 (double) **29** 4 **31** 1

PROBLEMS 10.7

1 0, 1 **3** 4, 0 **5** $\frac{11}{4}, \frac{23}{4}$
7 $3x^2 + 5x + 6, 7$ **9** $x^2 - 5x + 21, -71$
11 $6x^3 + 6x^2 + 5x + 5, 6$ **13** $(x/2) - \frac{11}{4}, \frac{5}{8}$
15 $2x^2 - \frac{2}{3}x - \frac{25}{9}, \frac{268}{27}$

17
$$\begin{array}{rrrr|l} 1 & 0 & -18 & 27 & 3 \\ & 3 & 9 & -27 & \\ \hline 1 & 3 & -9 & 0 & \end{array}$$

19
$$\begin{array}{rrrrr|l} 1 & 2 & 2 & 2 & 1 & -1 \\ & -1 & -1 & -1 & -1 & \\ \hline 1 & 1 & 1 & 1 & 0 & \end{array}$$

21
$$\begin{array}{rrrrrr|l} 1 & -\frac{7}{2} & -\frac{3}{2} & 1 & -\frac{1}{2} & \frac{3}{8} & \frac{1}{2} \\ & \frac{1}{2} & -\frac{3}{2} & -\frac{3}{2} & -\frac{1}{4} & -\frac{3}{8} & \\ \hline 1 & -3 & -3 & -\frac{1}{2} & -\frac{3}{4} & 0 & \end{array}$$

25 None **27** $\frac{2}{3}$ **29** 2

PROBLEMS 10.8

1 $-\frac{1}{2}, 2$ **3** $-k, 3$ **5** $-1, 2, \frac{1}{2}$
7 1 (double) **9** $\pm\frac{1}{2}, \pm\frac{1}{2}i$ **11** $-\frac{1}{2}, 2, -3$
13 0 (double), 1, -1 **15** 9, -6 **17** 0, 5, 3
19 0 (triple), $(-1 \pm i\sqrt{3})/2$ **21** $\frac{1}{2} \pm \frac{1}{2}\sqrt{17}, \frac{1}{2} \pm \frac{1}{2}\sqrt{5}$

PROBLEMS 10.9

1 $2, -3, -\frac{1}{2}$ **3** $\frac{2}{5}, \frac{4}{3}, -3$ **5** $\frac{1}{2}, (-3 \pm \sqrt{5})/2$
7 $-1, \frac{1}{2}, -i, i$ **9** $\frac{3}{2}, -2, (1 \pm i\sqrt{15})/4$ **11** $\frac{1}{3}, \frac{2}{3}, (-3 \pm i\sqrt{7})/2$

PROBLEMS 10.10

1 -1.7 **3** 0.6 **5** -1.0 **7** -0.5
9 0.4 **11** ± 0.7 **13** 0.2 **15** ± 2.0

PROBLEMS 11.1

3 $\frac{1}{72}$ **5** 1 **7** 4 **9** $\frac{8}{243}$
11 $\frac{9}{4}$ **13** 13 **15** a^x **17** a^{4m-2}
19 $6^{25/12} \times 10^{23/12}$ **21** (Fig. 11.1.21) **23** (Fig. 11.1.23)
25 (Fig. 11.1.25)

PROBLEMS 11.2

1 1.6487 **3** 0.94176 **5** 40.172 **7** 0.36788
9 0.00958 **11** (Fig. 11.2.11) **13** (Fig. 11.2.13) **15** (Fig. 11.2.15)
17 (Fig. 11.2.17) **19** (Fig. 11.2.19)

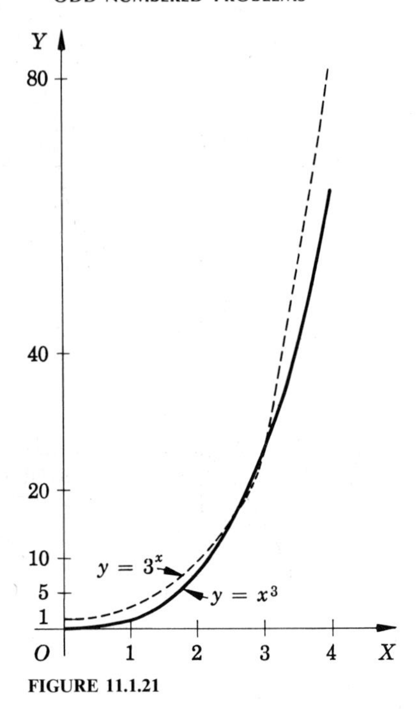

FIGURE 11.1.21

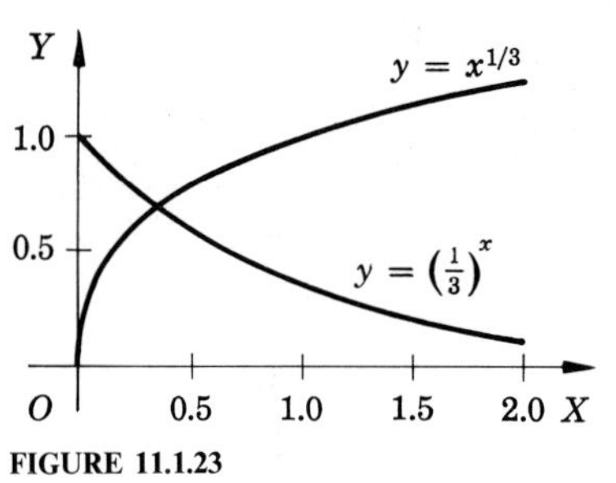

FIGURE 11.1.23

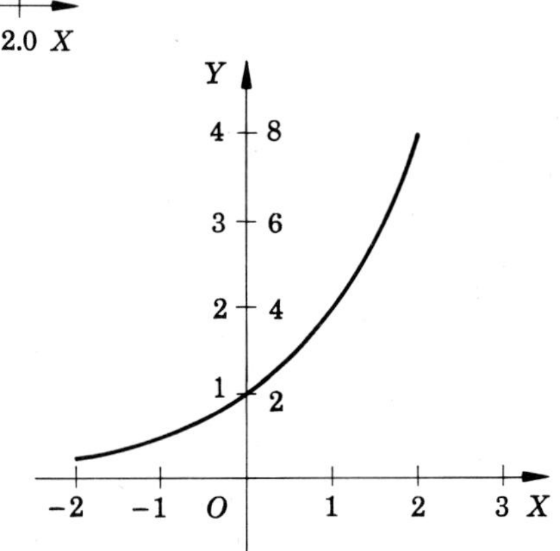

FIGURE 11.1.25

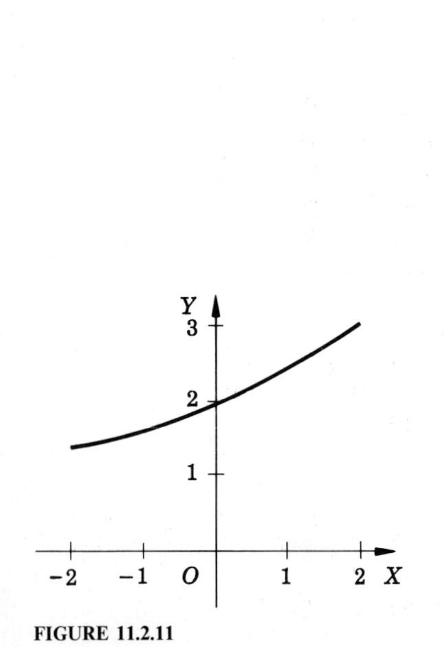

FIGURE 11.2.11

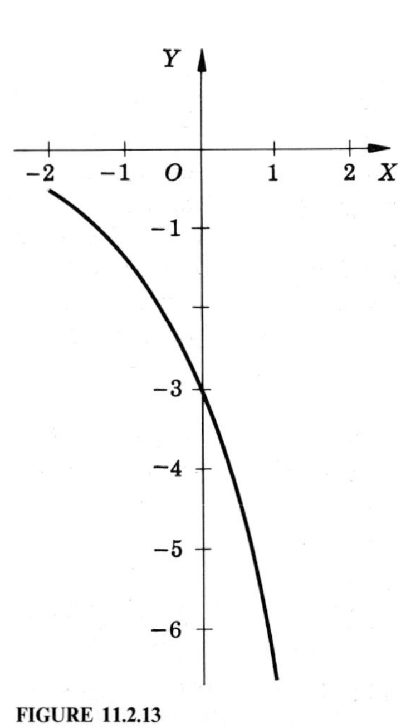

FIGURE 11.2.13

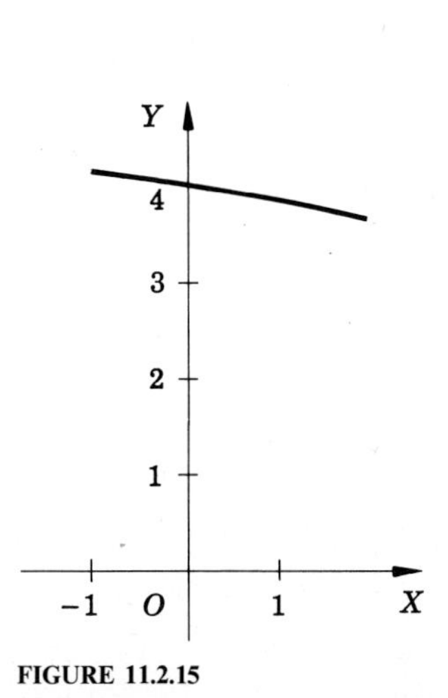

FIGURE 11.2.15

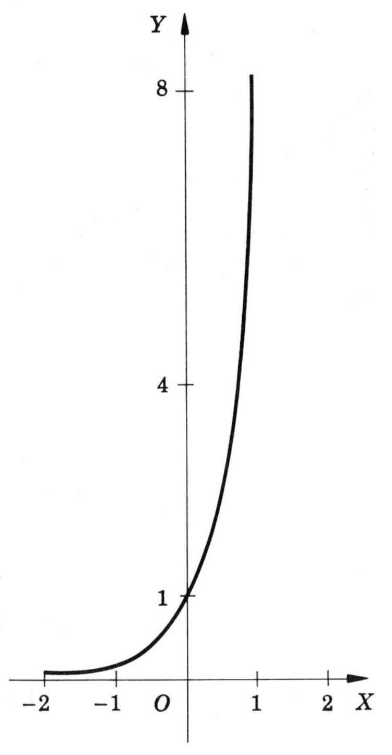

FIGURE 11.2.17

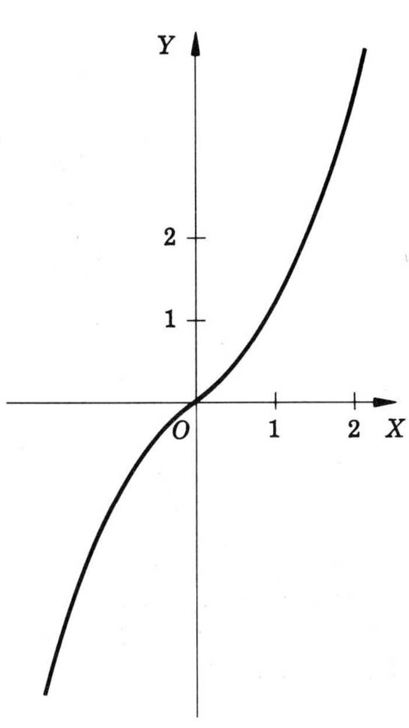

FIGURE 11.2.19

PROBLEMS 11.3

1 3.737 **3** 8.162 **5** 199.5 **7** 3.737
9 8.166 **11** 196.5 **13** 6 **15** 6
17 $2\sqrt[3]{2}$ **21** $a^{x(1+\log_a b)}$

23

$y = 5^{6x}$	Domain: all reals; image: $y > 0$
$y = \frac{1}{6}\log_5 x$	Domain: $x > 0$; image: all reals
$y = 6^{5x}$	Domain: all reals; image: $y > 0$
$y = \frac{1}{5}\log_6 x$	Domain: $x > 0$, image: all reals
$y = 2^{-x}$	Domain: all reals; image: $y > 0$
$y = -\log_2 x$	Domain: $x > 0$; image: all reals

PROBLEMS 11.4

1 (Fig. 11.4.1) **3** (Fig. 11.4.3) **5** (Fig. 11.4.5) **7** (Fig. 11.4.7)
9 (Fig. 11.4.9) **11** (Fig. 11.4.11) **13** (Fig. 11.4.13)
15 3.322 **17** −2 **19** 0.5

PROBLEMS 11.5

1 $\dfrac{\log_e \frac{2}{3}}{1.2}$ **3** $\dfrac{\log_e 0.2}{-0.5}$ **5** $\dfrac{17}{7 \times 3^{0.1}}$ **7** $\dfrac{\log_e 2}{-0.25}$

9 $10 \ln 4$ **11** 1.4 years **13** 3×10^{-16}

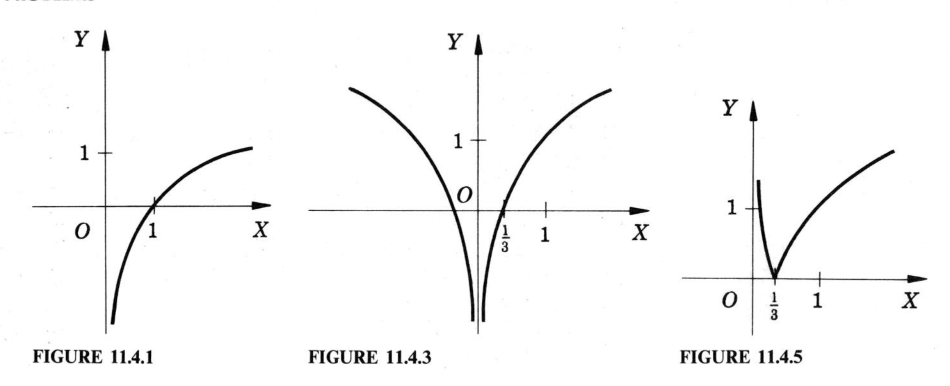

FIGURE 11.4.1 FIGURE 11.4.3 FIGURE 11.4.5

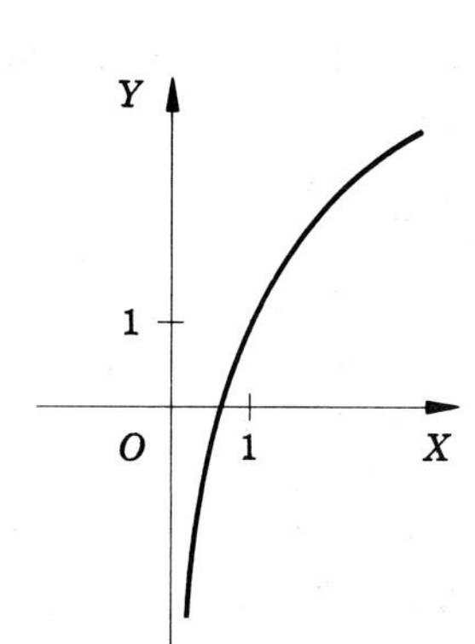

FIGURE 11.4.7

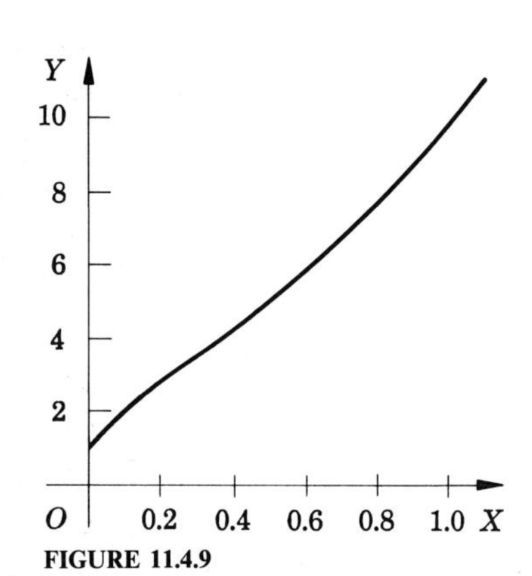

FIGURE 11.4.9

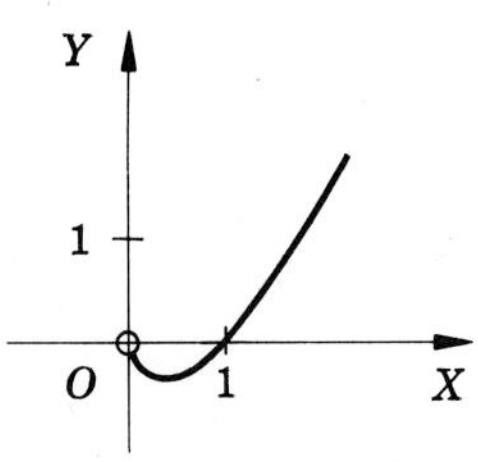

FIGURE 11.4.11

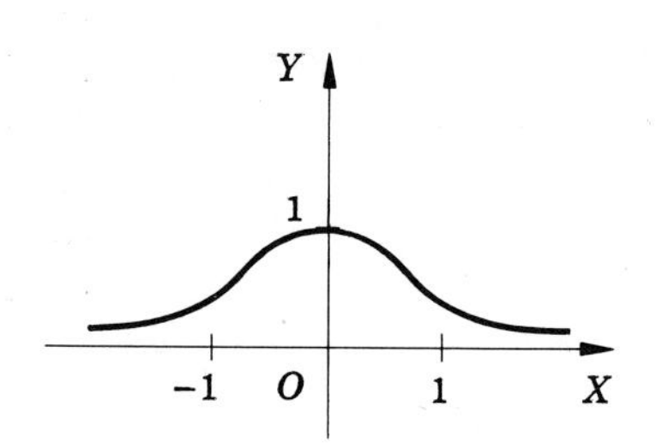

FIGURE 11.4.13

PROBLEMS 11.6

1 1,000 **3** $y = 4x + 4$ **5** $s = e^{bT}$ **7** semilog
9 log–log **11** log–log **13** log–log **15** semilog

PROBLEMS 12.2

1 $2\sqrt{13}$ **3** 15 **5** $2\sqrt{10}$ **17** $|x_1 - x_2|$

PROBLEMS 12.6

1 $\sqrt{2}/2, -\sqrt{2}/2$ **3** $0, -1$ **5** $-\sqrt{2}/2, -\sqrt{2}/2$
7 $-1, 0$ **9** $-\sqrt{2}/2, \sqrt{2}/2$ **11** $0, 1$

13	$-\sqrt{3}/2, -\frac{1}{2}$	**15**	$-\frac{1}{2}, \sqrt{3}/2$	**17**	$-\sqrt{3}/2, \frac{1}{2}$
19	$-\sqrt{3}/2, -\frac{1}{2}$	**21**	$-\frac{1}{2}, -\sqrt{3}/2$	**23**	$\sqrt{2}/2, -\sqrt{2}/2$
25	$\frac{1}{2}, \sqrt{3}/2$	**27**	$-\frac{1}{2}, \sqrt{3}/2$	**29**	$-\sqrt{2}/2, -\sqrt{2}/2$
31	$4\sqrt{3}, 4$	**33**	$-3, 0$	**35**	$-5, -5\sqrt{3}$
37	$5, -5\sqrt{3}$	**39**	$-\frac{1}{2}, \frac{1}{2}$	**41**	$-3\sqrt{3}/2, -\frac{3}{2}$
43	$1, -1$	**45**	$150\sqrt{3}, 150$	**47**	$r = 4, \theta = 270°$
49	$r = 2, \theta = 300°$	**51**	$r = 100, \theta = 330°$	**53**	$r = 3, \theta = 210°$
55	$r = \sqrt{6}/2, \theta = 225°$	**57**	$r = 16, \theta = 210°$	**59**	$r = 1, \theta = 150°$
61	$r = 12, \theta = 120°$	**63**	$r = 16, \theta = 0°$	**65**	$r = \sqrt{2}, \theta = 135°$
67	$r = 7, \theta = -90°$	**69**	$r = 2, \theta = -120°$	**71**	$\sqrt{13}$
73	$\sqrt{101 - 5(\sqrt{6} - \sqrt{2})}$			**75**	$\sqrt{10}$

PROBLEMS 12.7

Answers are given from left to right as the sine, cosine, tangent, cotangent, secant, and cosecant of each angle.

1 0, 1, 0, undefined, 1, undefined
3 $\frac{1}{2}\sqrt{3}, \frac{1}{2}, \sqrt{3}, \frac{1}{3}\sqrt{3}, 2, \frac{2}{3}\sqrt{3}$
5 $\frac{1}{2}\sqrt{3}, -\frac{1}{2}, -\sqrt{3}, -\frac{1}{3}\sqrt{3}, -2, \frac{2}{3}\sqrt{3}$
7 $\frac{1}{2}, -\frac{1}{2}\sqrt{3}, -\frac{1}{3}\sqrt{3}, -\sqrt{3}, -\frac{2}{3}\sqrt{3}, 2$
9 $-\frac{1}{2}, -\frac{1}{2}\sqrt{3}, \frac{1}{3}\sqrt{3}, \sqrt{3}, -\frac{2}{3}\sqrt{3}, -2$
11 $-\frac{1}{2}\sqrt{3}, -\frac{1}{2}, \sqrt{3}, \frac{1}{3}\sqrt{3}, -2, -\frac{2}{3}\sqrt{3}$
13 $-\frac{1}{2}\sqrt{3}, \frac{1}{2}, -\sqrt{3}, -\frac{1}{3}\sqrt{3}, 2, -\frac{2}{3}\sqrt{3}$
15 $-\frac{1}{2}, \frac{1}{2}\sqrt{3}, -\frac{1}{3}\sqrt{3}, -\sqrt{3}, \frac{2}{3}\sqrt{3}, -2$
17 $-\frac{1}{2}\sqrt{2}, -\frac{1}{2}\sqrt{2}, -1, -1, \sqrt{2}, -\sqrt{2}$
19 $-1, 0, -, 0, -, -1$
21 $-\frac{1}{2}\sqrt{2}, -\frac{1}{2}\sqrt{2}, 1, 1, -\sqrt{2}, -\sqrt{2}$
23 $0, -1, 0, -, -1, -$
25 $\frac{1}{2}\sqrt{2}, -\frac{1}{2}\sqrt{2}, -1, -1, -\sqrt{2}, \sqrt{2}$
27 $1, 0, -, 0, -, 1$
29 $\frac{1}{2}\sqrt{2}, \frac{1}{2}\sqrt{2}, 1, 1, \sqrt{2}, \sqrt{2}$
31 $0, 1, 0, -, 1, -$

PROBLEMS 12.8

1	$-\sin\theta$	**3**	$\cos\theta$	**5**	$\cot\theta$	**7**	$-\sin\theta$
9	$\tan\theta$	**11**	$-\cos\theta$	**13**	$\cot\theta$	**15**	$\sin\theta$
17	$-\cos\theta$	**19**	$-\cot\theta$	**21**	$\sin\theta$	**23**	$-\sin\theta$
25	$-\tan\theta$	**27**	$\sin\theta$	**29**	$-\cos\theta$		
31	$(-1)^k\cos\theta$	**33**	$(-1)^k\sec\theta$	**35**	$\cot\theta$		
37	$(-1)^k\cos\theta$	**39**	$(-1)^{k+1}\sin\theta$	**41**	$\cot\theta$		
43	$-\tan\theta$						

PROBLEMS 12.9

1	0.95389	**3**	0.61889	**5**	0.30763	**7**	−0.27871
9	0.86603	**11**	0.43340	**13**	−0.44268	**15**	−0.98557
17	0.67816	**19**	−1	**21**	0.50806	**23**	−0.77447

25 -0.33023 **27** -15.096 **29** 0.28867 **31** -0.34954
33 -0.12506 **35** -3.3611 **37** -4.6448 **39** 2.0918
41 $40°19.7'$ **43** $200°10.8'$ **45** $45°45'$ **47** $250°50.1'$
49 $20°46.6'$ **51** $201°$ **53** $46°50'$ **55** $184°48.5'$

PROBLEMS 12.10

1 $A = 22°37.2', B = 67°22.8', a = 5$
3 $A = 11°18.6', b = 5, c = 5.0990$
5 $A = 50°, a = 76.604, b = 64.279$
7 $A = 24°13.1', a = 1.3494, c = 3.2895$
9 $A = 53°11.4', B = 36°48.6', a = 18.975$
11 30 in. **13** 30.614 in. **15** 30.902 in.
17 $53°7.4'$ **19** $35°15.9'$ **23** 17,824 ft^2

PROBLEMS 12.11

1 $L = 11.66, \theta = 59°2.1'$ **3** $L = 19.78, \theta = 28°20.6'$
5 Speed = 200.9 miles/hr, direction = north by $5°42.6'$ west
7 $x = 10\sqrt{3}, y = 10$ **9** 9.430 lb **11** $120°$
13 $t = 5\sqrt{2}$ lb, $c = 5\sqrt{2}$ lb
15 HINT: in line **17** 175 lb

PROBLEMS 12.12

1 $C = 80°, b = 10.642, c = 20.960$
3 $B = 17°14.9', C = 31°19', b = 75.3$
5 $C = 139°12.3', a = 4.7, c = 18.234$
7 No solution
9 $A_1 = 48°23.9', B_1 = 97°20.8', b_1 = 20.704$
$A_2 = 131°36.1', B_2 = 14°8.6', b_2 = 5.1008$
13 Length of resultant = 6.306, angle between vectors = $52°1.8'$

PROBLEMS 12.13

1 $C = 111°48.2'$ **3** $A = 90°$ **5** $C = 88°5.6'$
7 $c = 18.754$ **9** $b = 19.769$ **11** $a = 1.8635$
13 $a = 3.687, b = 11.954$ **17** $A = 48°11.4', B = 58°24.7'$
19 About 2.47 miles

PROBLEMS 13.3

1 $90°, 30°, 60°, 45°$ **3** $30°, 120°, 210°, 300°, 330°$
5 $-45°, -90°, -135°, -180°$ **7** $2{,}880°, 3{,}060°, 28.6487°, 19.0991°$
9 $0, \pi/6, \pi/4, \pi/3, \pi/2$ **11** $5\pi/9, 1.205\pi, 0.0944\pi, 2\pi/3, \pi/15$
13 $\cos 0 = \sec 0 = 1, \sin 0 = \tan 0 = 0$

15 $$\sin\frac{\pi}{3} = \frac{\sqrt{3}}{2}, \cos\frac{\pi}{3} = \frac{1}{2}, \tan\frac{\pi}{3} = \sqrt{3}$$

$$\cot\frac{\pi}{3} = \frac{1}{\sqrt{3}}, \sec\frac{\pi}{3} = 2, \csc\frac{\pi}{3} = \frac{2}{\sqrt{3}}$$

17 $$\sin\frac{4\pi}{3} = -\frac{\sqrt{3}}{2}, \cos\frac{4\pi}{3} = -\frac{1}{2}, \tan\frac{4\pi}{3} = \sqrt{3}$$

$\cot\frac{4\pi}{3} = \frac{1}{\sqrt{3}}$, $\sec\frac{4\pi}{3} = -2$, $\csc\frac{4\pi}{3} = -\frac{2}{\sqrt{3}}$

19 $\sin 5\pi = \tan 5\pi = 0$, $\cos 5\pi = \sec 5\pi = -1$
21 $\sin 2n\pi = \tan 2n\pi = 0$, $\cos 2n\pi = \sec 2n\pi = 1$
23 (Fig. 13.3.23)

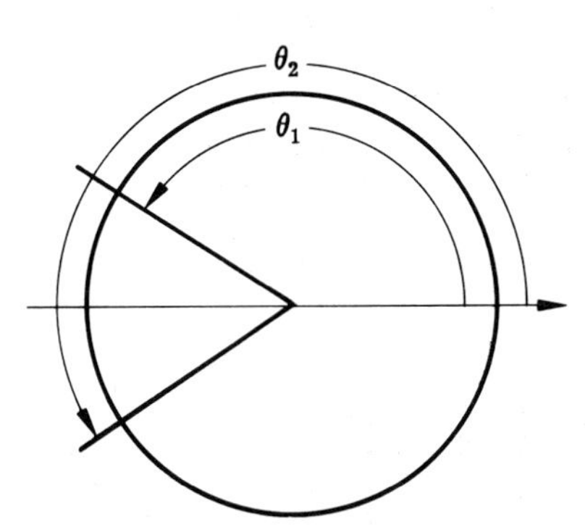

FIGURE 13.3.23

25 HINT: If $\cos\theta = x$, then $\cos(-\theta) = x$.
29 HINT: If $\sin(\theta + 2\pi) = y$, then $\sin\theta = y$.
If $\sin(\theta + \pi) = y$, then $\sin\theta = -y$.
31 $(\pi/4) + 2n\pi$, $(5\pi/4) + 2n\pi$
33 0.99923 **35** 0.62099 **37** 0.18964 **39** 0.72484
41 34.233 **43** −34.233

PROBLEMS 13.4

1 $\sec\theta = OT$, $\csc\theta = OS$, $\cot\theta = BS$ (Fig. 13.4.1)

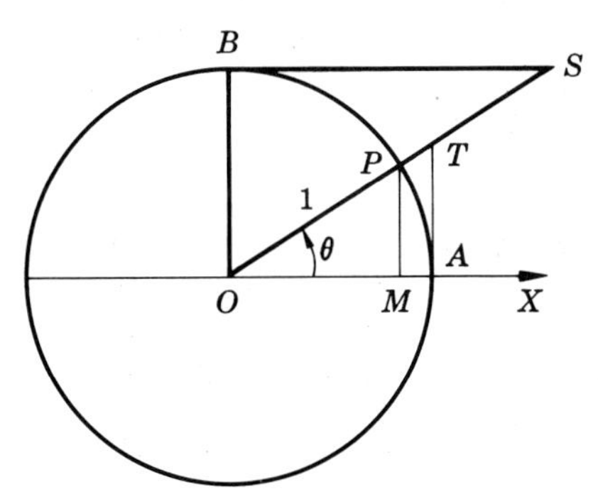

FIGURE 13.4.1

3 (Fig. 13.4.3) **5** (Fig. 13.4.5) **7** (Fig. 13.4.7)
9 (Fig. 13.4.9) **11** (Fig. 13.4.11) **13** (Fig. 13.4.13)

15 (Fig. 13.4.15) **17** (Fig. 13.4.17) **19** (Fig. 13.4.19)
21 (Fig. 13.4.21)

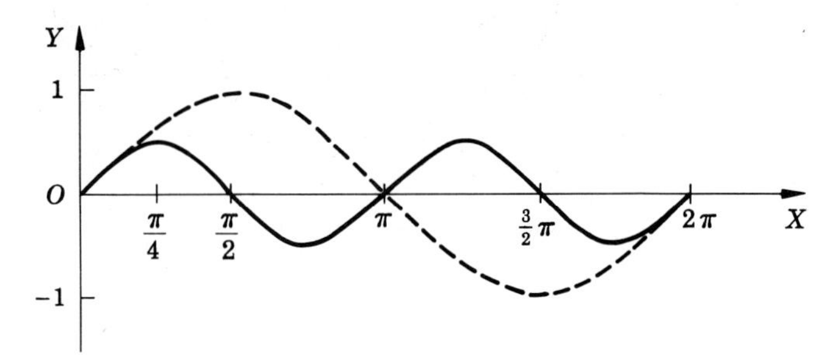

FIGURE 13.4.3

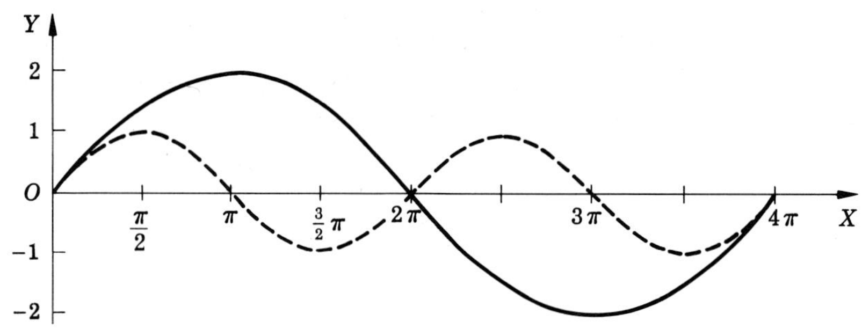

FIGURE 13.4.5

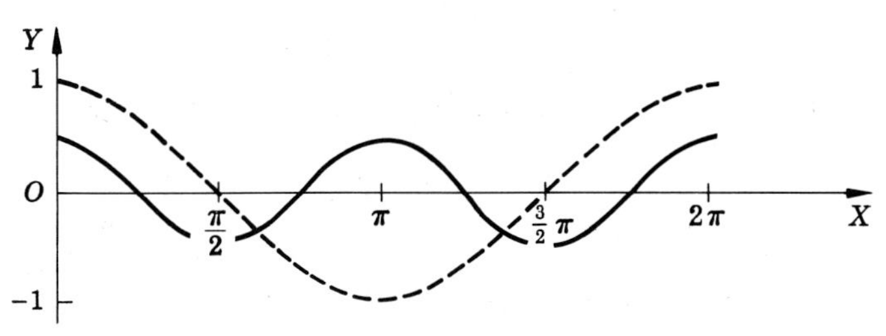

FIGURE 13.4.7

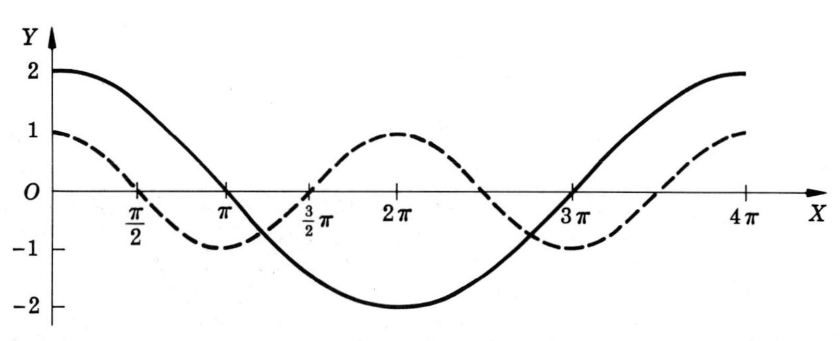

FIGURE 13.4.9

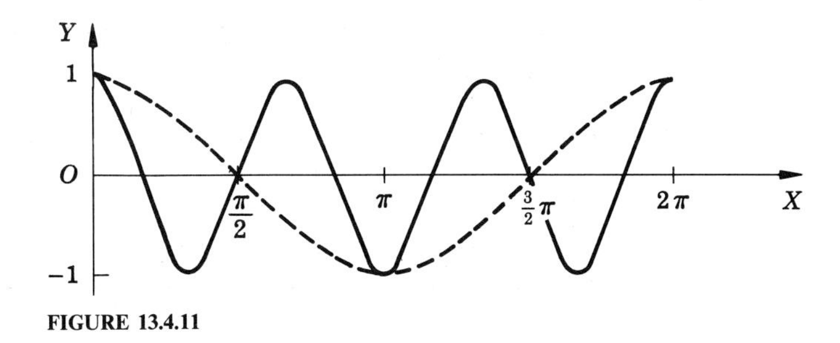

FIGURE 13.4.11

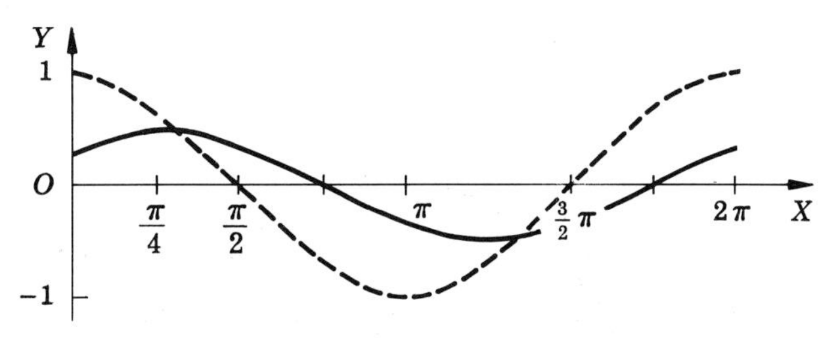

FIGURE 13.4.13

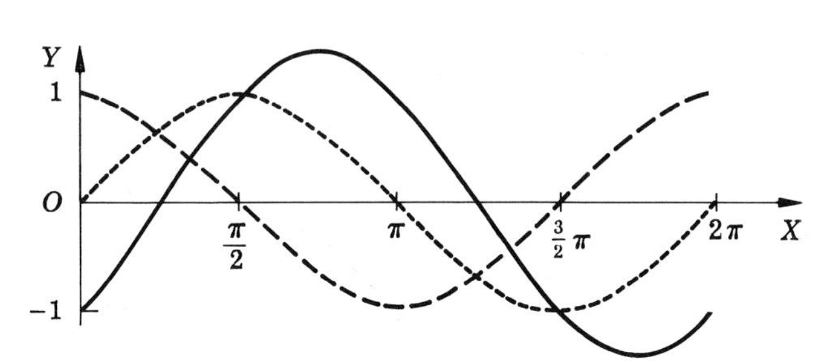

FIGURE 13.4.15

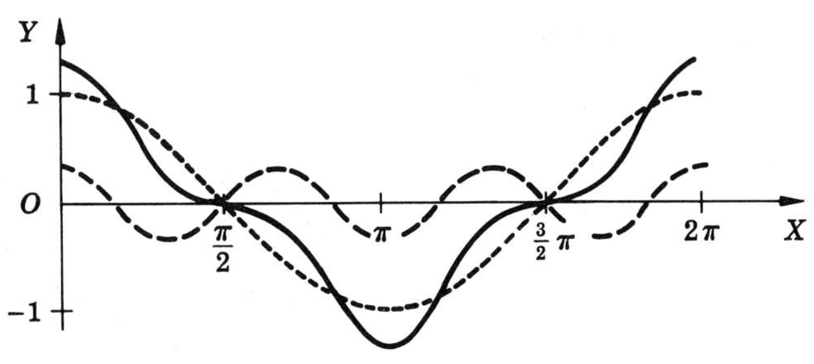

FIGURE 13.4.17

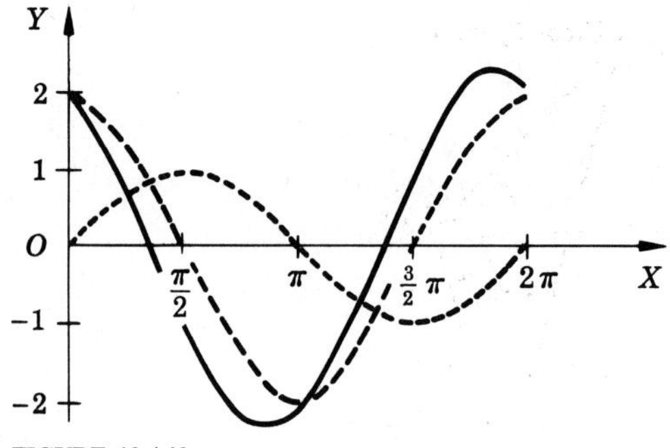

FIGURE 13.4.19

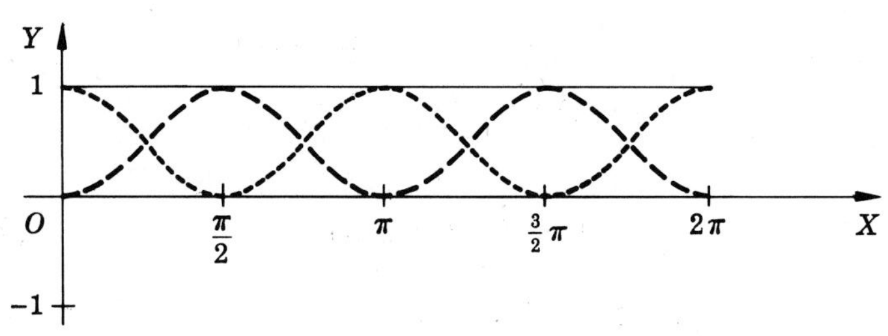

FIGURE 13.4.21

23 and **25**:

Quadrant	as θ varies from	sec θ varies from	cot θ varies from
I	0 to $\pi/2$	1 to ∞	∞ to 0
II	$\pi/2$ to π	$-\infty$ to -1	0 to $-\infty$
III	π to $3\pi/2$	-1 to $-\infty$	∞ to 0
IV	$3\pi/2$ to 2π	∞ to 1	0 to $-\infty$

PROBLEMS 13.5

1 $2, 2\pi, -\pi/4$ **3** $\frac{1}{2}, \pi, -\pi/6$ **5** $3, 2\pi, \pi/4$
7 $2, 4\pi, \pi$ **9** $\frac{1}{3}, 2\pi/3, 0$ **11** $\pi, -\pi/2$
13 $\pi, -\pi/4$

PROBLEMS 13.6

1 $\dfrac{\sqrt{3}}{2}$ **3** $\dfrac{\sqrt{3}}{2}$ **5** $\dfrac{\sqrt{2}(1-\sqrt{3})}{4}$ **7** $\dfrac{\sqrt{2}(\sqrt{3}+1)}{4}$

9 -1 **11** $-\dfrac{\sqrt{3}+3}{\sqrt{3}-3}$ **13** $\dfrac{2}{\sqrt{3}}$ **15** $\dfrac{4}{\sqrt{2}(1+\sqrt{3})}$

17 $\mp\sin\theta$ **19** $-\cos\theta$ **21** $-\cos\theta$ **23** $\mp\cot\theta$
25 $\pm\csc\theta$ **29** $\pm\frac{33}{65}, \pm\frac{63}{65}$ **31** $\sin 4$ **33** $\tan 29.6$
35 (Fig. 13.6.35)

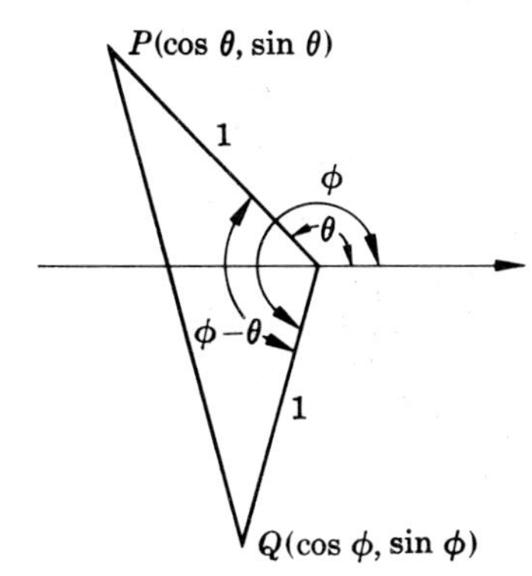

FIGURE 13.6.35

PROBLEMS 13.7

15 $-\frac{21}{221}$ 17 $\frac{140}{221}$ 19 $-\frac{21}{220}$ 21 $\sqrt{3}/2$
23 $\frac{1}{2}\sqrt{2+\sqrt{3}}$

PROBLEMS 13.9

1 $\{\pi n\}, \left\{\pm\frac{\pi}{3}+2\pi n\right\}$
3 $\left\{\frac{\pi}{4}\pm\frac{\pi}{12}+\pi n\right\}$
5 $\left\{\pm\frac{\pi}{9}+\frac{2n\pi}{3}\right\}$
7 $\{\pi n\}, \left\{\frac{\pi}{2}\pm\frac{\pi}{3}+2\pi n\right\}$
9 $\left\{\pi\pm\frac{\pi}{2}+2\pi n\right\}, \left\{\pi\pm\frac{\pi}{3}+2\pi n\right\}$
11 $\{0.20861+(2n+1)\pi\}, \{-0.20861+2n\pi\}$
13 $\left\{\frac{\pi}{2}+\frac{\pi n}{2}\right\}, \left\{\frac{\pi}{2}\pm\frac{\pi}{6}+\pi n\right\}$
15 No solution
17 $\left\{\frac{\pi}{3}+2n\pi\right\}$
19 $\left\{\pi\pm\frac{\pi}{6}+2\pi n\right\}, \left\{\pm\frac{\pi}{3}+2\pi n\right\}$
21 $\{\pm 0.89565+2\pi n\}, \{(2n+1)\pi\}$
23 $\{\pi n\}$
25 $\{\pi n\}, \left\{\pm\frac{\pi}{12}+\frac{\pi n}{2}\right\}$

PROBLEMS 13.10

1 $\pi/6$ 3 $-\pi/2$ 5 $\pi/3$ 7 $5\pi/6$
9 $-\pi/6$ 11 π 13 Does not exist. 15 $-\frac{1}{8}$
17 $\pi/3$ 19 $\pi/2$

PROBLEMS 13.11

1 $2 \text{ cis } 45°$
3 $30 \text{ cis } 270°$
5 $3 \text{ cis } 0°$
7 $1 \text{ cis } 150°$
9 $5 \text{ cis } [\text{Tan}^{-1}(-\frac{4}{3})+\pi]$
11 $2\sqrt{3}+2i$
13 $-5-5\sqrt{3}i$
15 $\frac{1}{2}\sqrt{6}+\frac{1}{2}\sqrt{6}i$
17 $-2-2i$
19 -6
21 $-4+4\sqrt{3}i$, $8 \text{ cis } 120°$
23 $1 \text{ cis } 80°$
25 $12 \text{ cis } (3\pi/4)$
27 $4 \text{ cis } (-\pi/8)$
29 $\frac{11}{6} \text{ cis } 45°$
31 $\frac{1}{16} \text{ cis } 60°$

33 $16 \operatorname{cis}(-\pi/4)$

35 $1, i, -1, -i$

37 $\operatorname{cis}\left[\dfrac{(4k+1)\pi}{8}\right], k = 0, 1, 2, 3$

PROBLEMS 14.4

1 $x - 4y + 11 = 0;\ m = \frac{1}{4}$

3 $2x - 5y - 26 = 0;\ m = \frac{2}{5}$

5 $y - 6 = 0;\ m = 0$

7 $2x - y = 0;\ m = 2$

9 $4x - 3y - 12 = 0;\ m = \frac{4}{3}$

11 $m = -\frac{1}{3};\ a = 6,\ b = 2;\ (x/6) + (y/2) = 1;\ y = -(x/3) + 2$

13 $m = \frac{3}{4};\ a = 4,\ b = -3;\ (x/4) + [y/(-3)] = 1;\ y = (3x/4) - 3$

15 $m = -\frac{1}{7};\ a = 5,\ b = \frac{5}{7};\ (x/5) + (7y/5) = 1;\ y = -(x/7) + \frac{5}{7}$

17 No slope; $a = 2$, no b; no other forms.

19 $y + 3 = (x - 4)/\sqrt{3}$

21 $x + 2y - 4 = 0$ or $3x + 2y = 0$

23 $bx + ay = 0$

PROBLEMS 14.5

1 $2x - 4y - 14 = 0$

3 $-3x + 2y - 5 = 0$

5 $3x - y - 7 = 0$

7 $3x + y - 14 = 0$

9 $2x + 3y - 22 = 0$

11 $2x + y - 6 = 0$

13 $(\frac{3}{2}, 1)$

15 $(2, \frac{9}{2})$

17 $(7,8)$

19 $(3,11)$

23 $\overline{AB} = \sqrt{5}, \overline{AC} = 4\sqrt{2}, \overline{BC} = \sqrt{13}$

25 $m_{AB} = \frac{4}{7}, m_{AC} = (-8)/(-14) = \frac{4}{7}$

27 Median from A: $m = \frac{5}{3}$
from B: $m = -\frac{2}{3}$
from C: $m = \frac{1}{9}$

29 $(x + 3)^2 + (y - 2)^2 = 4$

31 $(-1,1)$

33 $\bar{x} = \dfrac{r_2x_1 + r_1x_2}{r_1 + r_2}, \bar{y} = \dfrac{r_2y_1 + r_1y_2}{r_1 + r_2}$

PROBLEMS 14.6

1 $\dfrac{1}{\sqrt{2}}, \dfrac{1}{\sqrt{2}}$

3 $\dfrac{5}{\sqrt{41}}, \dfrac{4}{\sqrt{41}}$

5 $-\dfrac{1}{\sqrt{37}}, \dfrac{6}{\sqrt{37}}$

7 $-1, 0$

9 $-\dfrac{2}{\sqrt{5}}$

11 $\dfrac{-32}{\sqrt{29}\sqrt{37}}$

13 $\pm\left(\dfrac{5}{\sqrt{29}}, \dfrac{-2}{\sqrt{29}}\right)$

15 $\pm\left(\dfrac{5}{\sqrt{41}}, \dfrac{4}{\sqrt{41}}\right)$

17 $-\dfrac{\sqrt{5}}{2}$

19 1

21 $2x - y - 5 = 0$

23 $3x - y + 7 = 0$

25 0

27 $\dfrac{3}{\sqrt{10}}$

29 $\dfrac{7}{\sqrt{85}}$

31 $\dfrac{3}{\sqrt{53}\sqrt{26}}$

33 1

35 $\frac{2}{5}$

37 $\dfrac{14}{\sqrt{45}}$

39 $\frac{7}{5}$

41 $\dfrac{7}{\sqrt{10}}$

43 $3x - y = \pm\sqrt{2}(x - 2y)$

45 $d = 2$ in each case.

49 A unit vector perpendicular to the line

PROBLEMS 14.9

1 $(x + 2)^2 + (y + 1)^2 = 4$ (Fig. 14.9.1)
3 $(x + 1)^2 + (y - 4)^2 = 16$ (Fig. 14.9.3)
5 $(x - \frac{1}{2})^2 + y^2 = \frac{25}{4}$ (Fig. 14.9.5)
7 $(x - 3)^2 + (y - 4)^2 = 16$ (Fig. 14.9.7)
9 $(x - \frac{3}{2})^2 + (y - 2)^2 = \frac{25}{4}$ (Fig. 14.9.9)
11 $(x - 3)^2 + (y - 3)^2 = 9$, $(x + 3)^2 + (y - 3)^2 = 9$
$(x + 3)^2 + (y + 3)^2 = 9$, $(x - 3)^2 + (y + 3)^2 = 9$ (Fig. 14.9.11)
13 $(3,-4), r = 5$ (Fig. 14.9.13)
15 $(3,-2), r = \sqrt{13}$) (Fig. 14.9.15)
17 $(1,-2), r = \sqrt{3}$ (Fig. 14.9.17)
19 $(3,-4)$, r (not real). Equation is satisfied by no point.
21 $(x + \frac{1}{2})^2 + (y - 3)^2 = \frac{25}{4}$
23 $(x - h)^2 + (y - k)^2 = 1$
25 $(x - h)^2 + (y - k)^2 = 1$ where $h^2 + k^2 = 4$

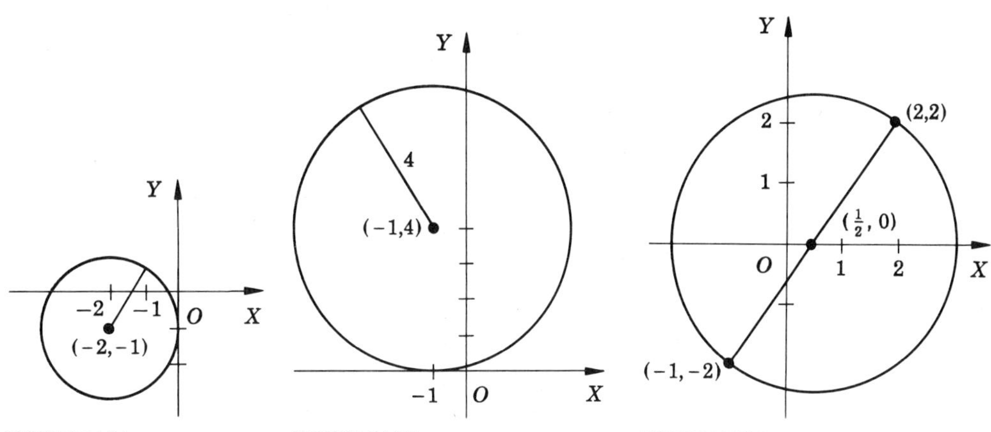

FIGURE 14.9.1

FIGURE 14.9.3

FIGURE 14.9.5

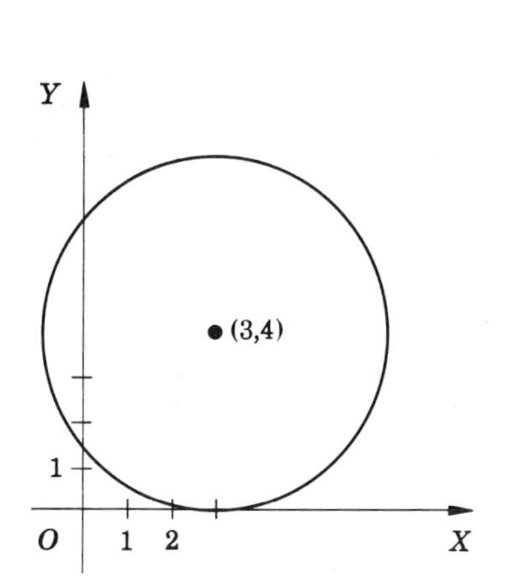

FIGURE 14.9.7

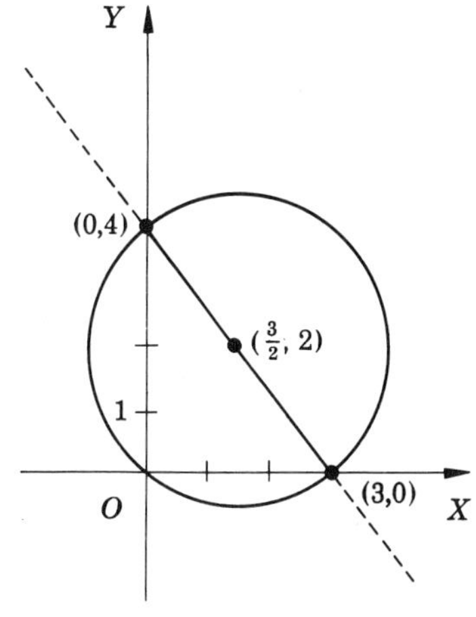

FIGURE 14.9.9

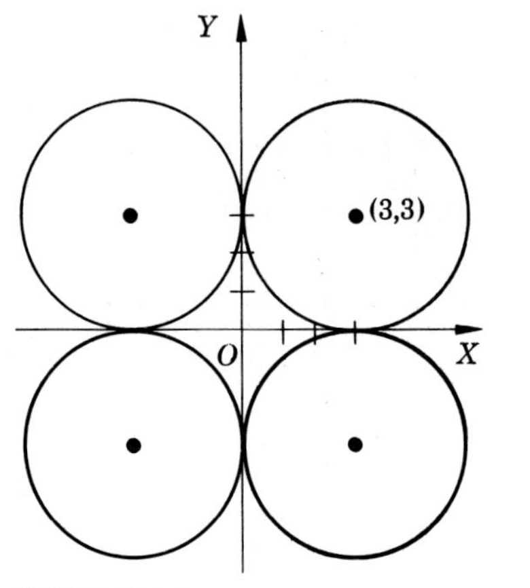

FIGURE 14.9.11

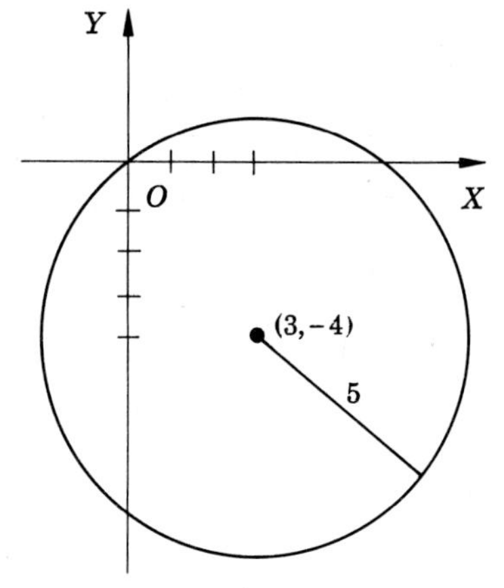

FIGURE 14.9.13

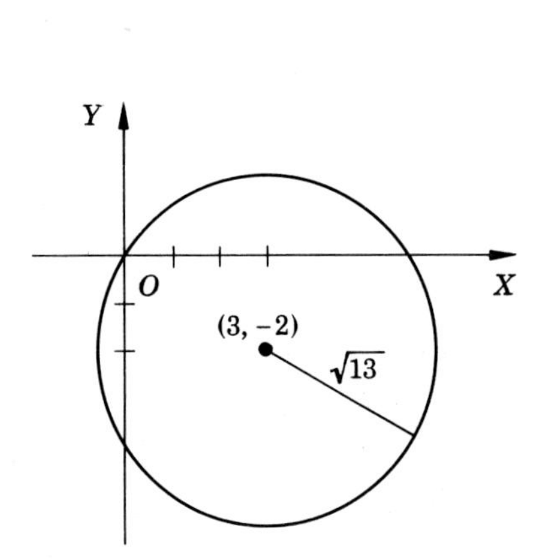

FIGURE 14.9.15

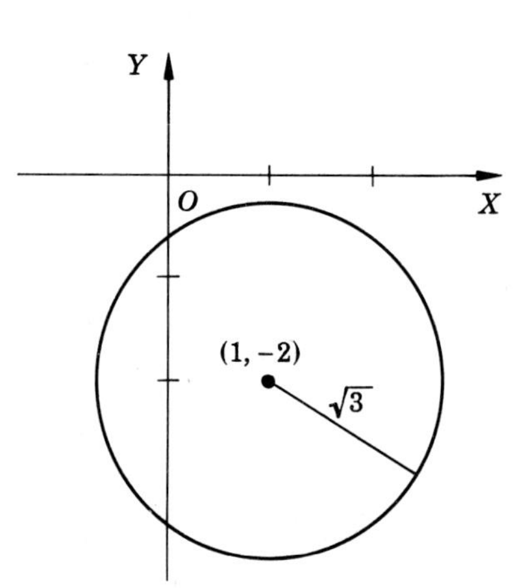

FIGURE 14.9.17

PROBLEMS 14.10

1 $F(4,0), x = -4$ (Fig. 14.10.1)

3 $F(0,\frac{7}{2}), y = -\frac{7}{2}$ (Fig. 14.10.3)

5 $F(\frac{1}{4},0), x = -\frac{1}{4}$ (Fig. 14.10.5)

7 $F(-1,0), x = 1$ (Fig. 14.10.7)

9 $F(2,-2), x = 4$ (Fig. 14.10.9)

11 $x^2 = -4(y + 3)$ (Fig. 14.10.11)

13 $x^2 = -4y$ (Fig. 14.10.13)

15 $x^2 = 3y$ (Fig. 14.10.15)

17 $y^2 = 16x$ (Fig. 14.10.17)

19 $y^2 = -2x$ (Fig. 14.10.19)

21 $(x + 1)^2 = 4(y + 2)$ (Fig. 14.10.21)

23 $(0,0), (-4,-4)$

25 $\dfrac{x^2}{100} + \dfrac{(y + 1)^2}{64} = 1$

27 $y^2 = 12(x + 2)$

29 $x = m + \sqrt{m^2 - 1}$; only one intersection if $m = \pm 1$. The lines are tangents to the parabola at (1,1) and (−1,1) respectively (Fig. 14.10.29).

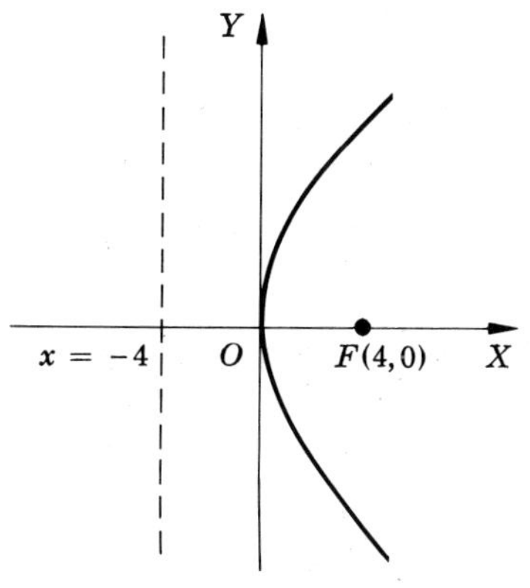

FIGURE 14.10.1

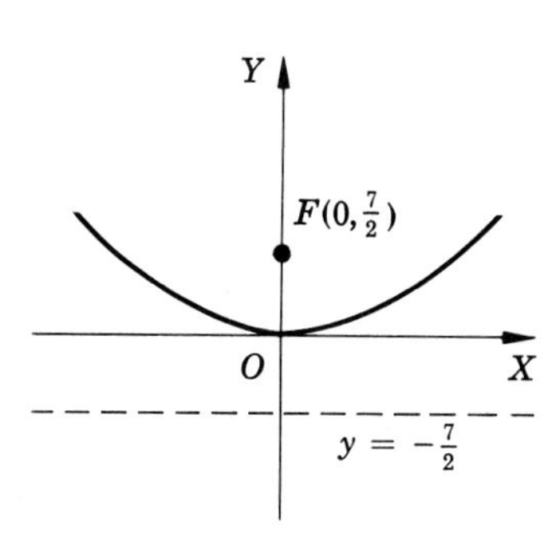

FIGURE 14.10.3

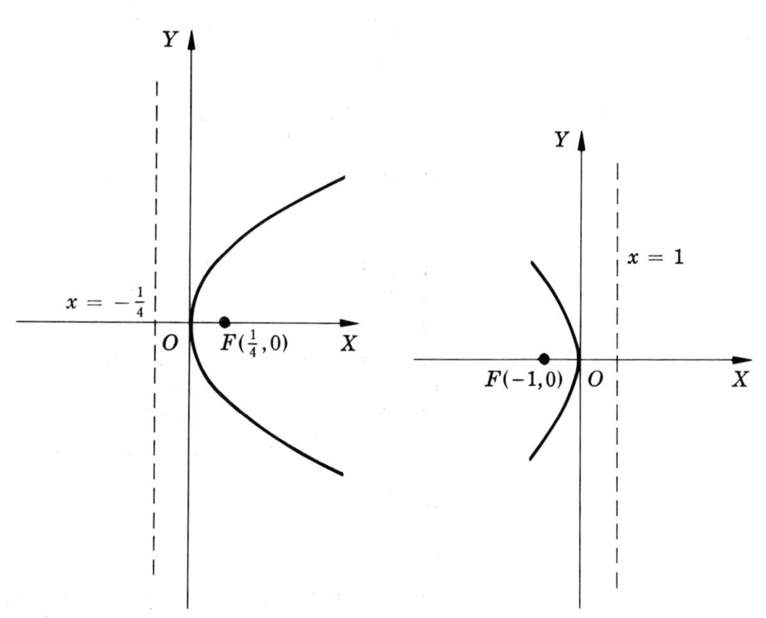

FIGURE 14.10.5

FIGURE 14.10.7

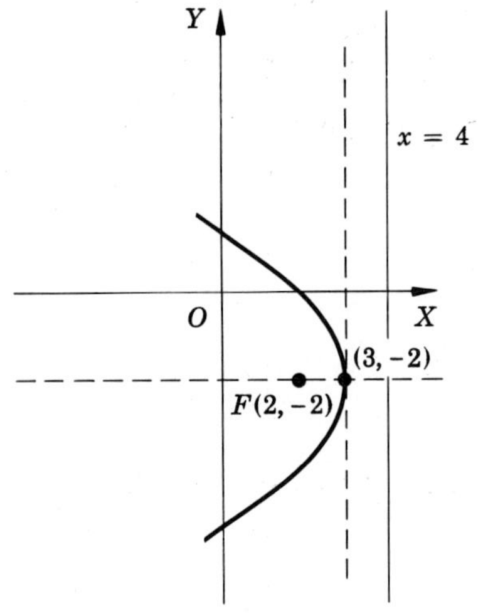

FIGURE 14.10.9

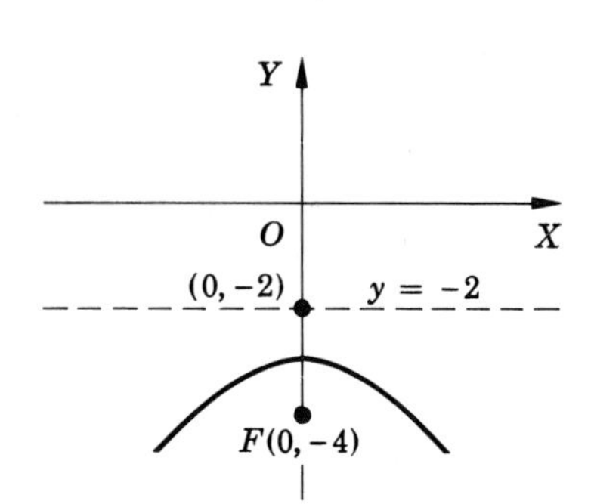

FIGURE 14.10.11

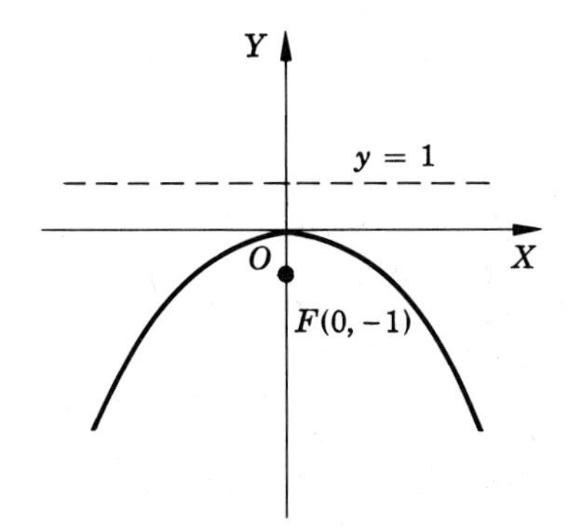

FIGURE 14.10.13

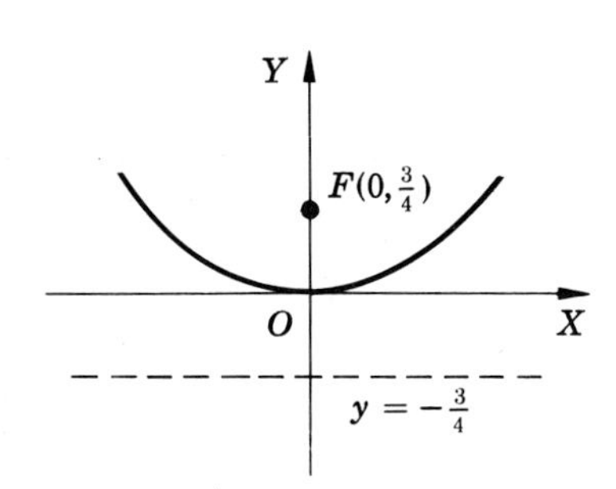

FIGURE 14.10.15

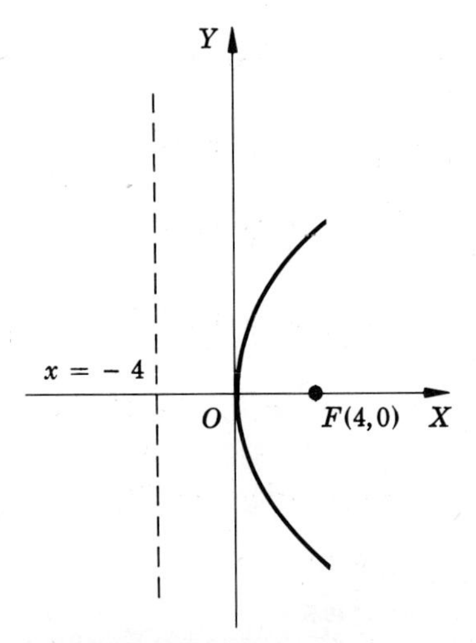

FIGURE 14.10.17

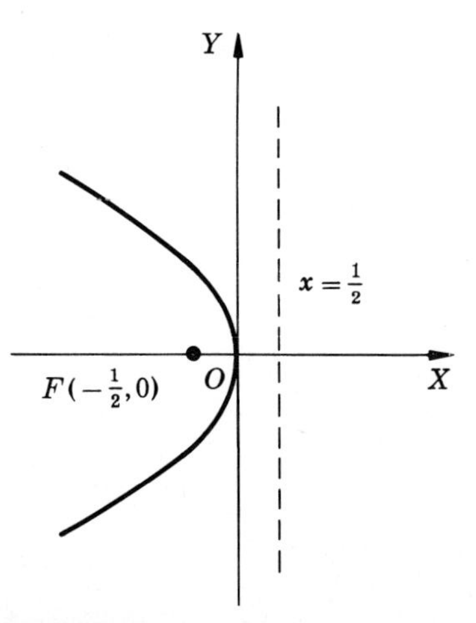

FIGURE 14.10.19

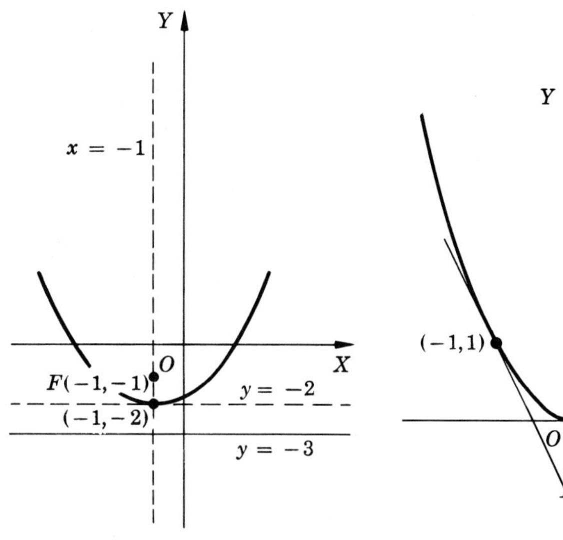

FIGURE 14.10.21

FIGURE 14.10.29

PROBLEMS 14.11

1 V: $(2\sqrt{3},0), (-2\sqrt{3},0)$
F: $(\sqrt{3},0), (-\sqrt{3},0)$ (Fig. 14.11.1)

3 V: $(5,0), (-5,0)$
F: $(3,0), (-3,0)$ (Fig. 14.11.3)

5 V: $(12,0), (-12,0)$
F: $(6\sqrt{3},0), (-6\sqrt{3},0)$ (Fig. 14.11.5)

7 V: $(0,10), (0,-10)$
F: $(0,8), (0,-8)$ (Fig. 14.11.7)

9 V: $(5,-3), (1,-3)$
F: $(3+\sqrt{3}, -3), (3-\sqrt{3}, -3)$ (Fig. 14.11.9)

11 $\dfrac{x^2}{16} + \dfrac{y^2}{12} = 1$ (Fig. 14.11.11)

13 $\dfrac{x^2}{109} + \dfrac{y^2}{100} = 1$ (Fig. 14.11.13)

15 $\dfrac{x^2}{9} + \dfrac{y^2}{4} = 1$ (Fig. 14.11.15)

17 $\dfrac{x^2}{9} + \dfrac{y^2}{8} = 1$ (Fig. 14.11.17)

19 $\dfrac{x^2}{16} + \dfrac{(y-2)^2}{7} = 1$ (Fig. 14.11.19)

21 $\dfrac{x^2}{8} + \dfrac{y^2}{9} = 1$ (Fig. 14.11.21)

23 $\dfrac{x^2}{4} - \dfrac{(y+1)^2}{5} = 1$

27 Ellipse. Center (h,k), major axis $2a$, minor axis $2b$ (Fig. 14.11.27)

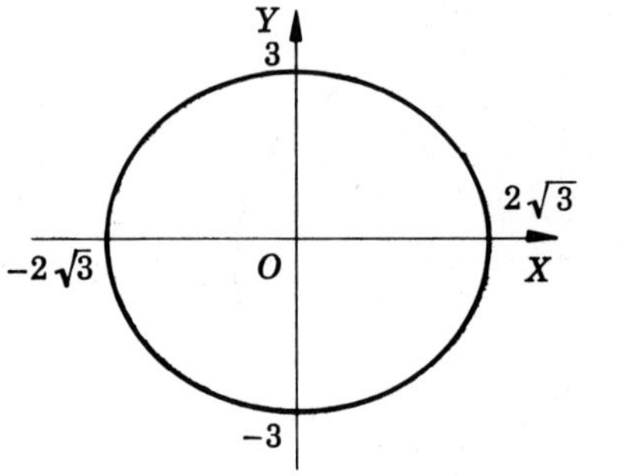

FIGURE 14.11.1

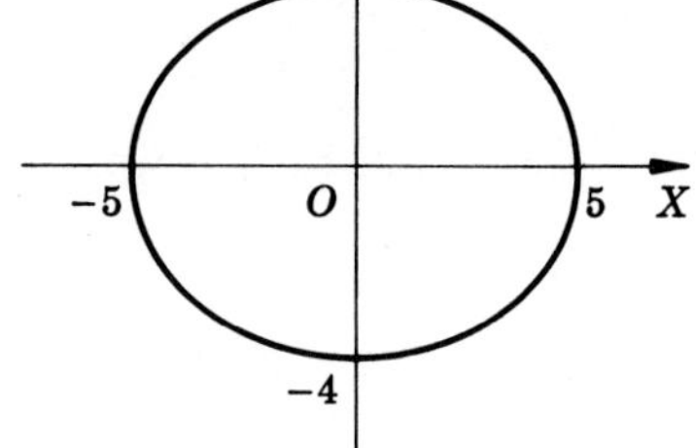

FIGURE 14.11.3

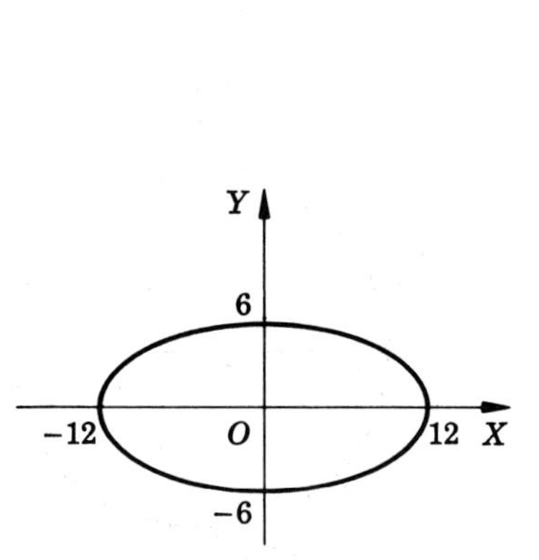

FIGURE 14.11.5

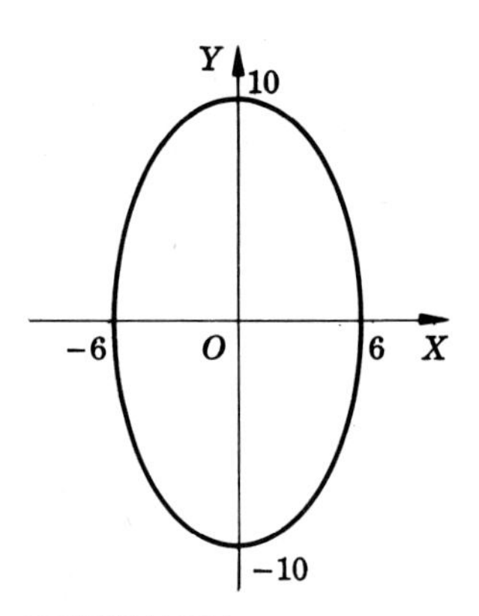

FIGURE 14.11.7

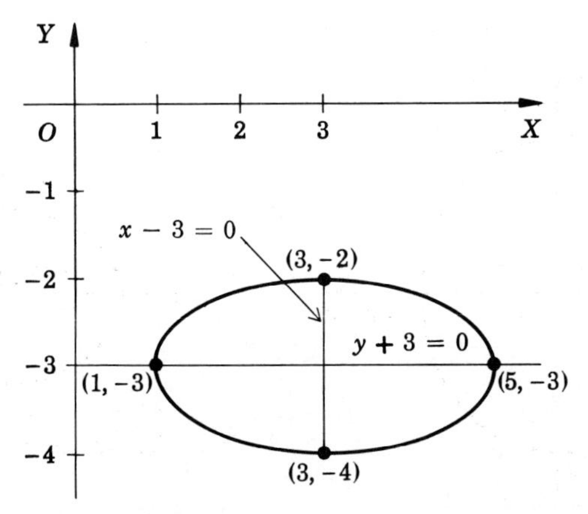

FIGURE 14.11.9

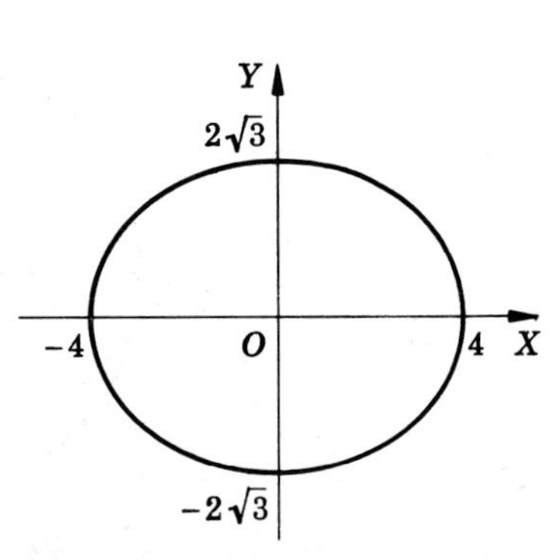

FIGURE 14.11.11

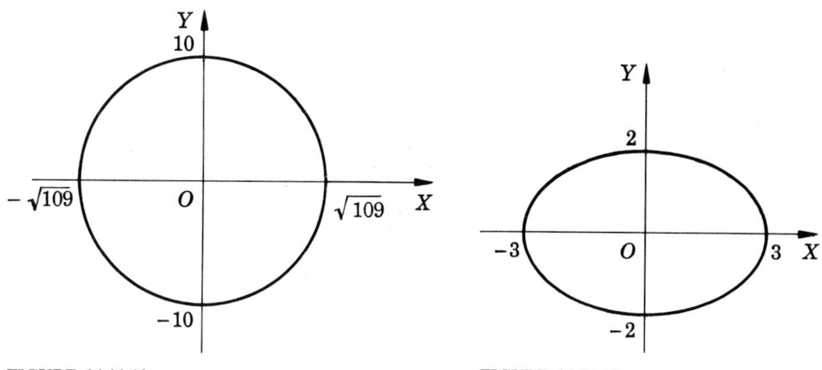

FIGURE 14.11.13

FIGURE 14.11.15

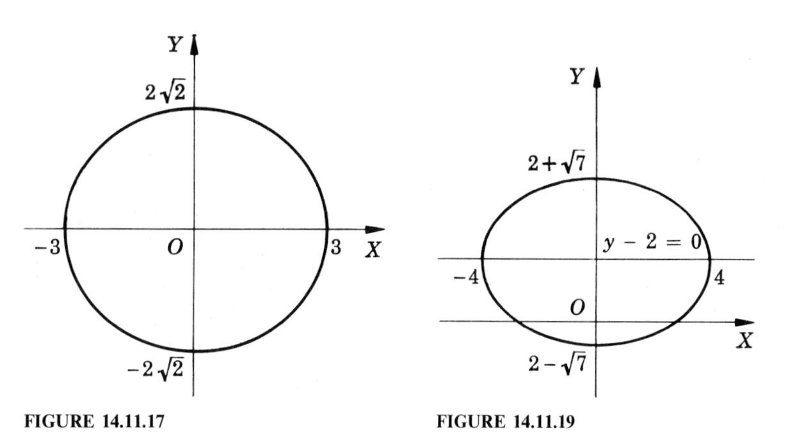

FIGURE 14.11.17

FIGURE 14.11.19

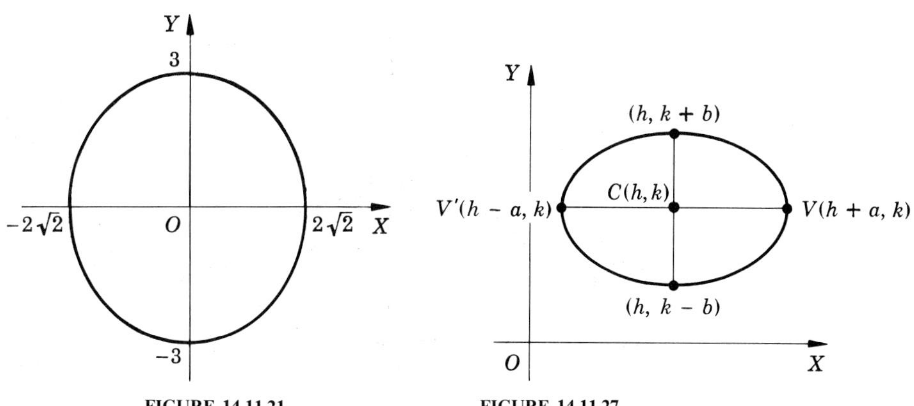

FIGURE 14.11.21

FIGURE 14.11.27

PROBLEMS 14.12

1 $F(\pm\sqrt{34},0)$, $V(\pm 5,0)$ (Fig. 14.12.1)
3 $F(0,\pm\sqrt{160})$, $V(0,\pm 4)$ (Fig. 14.12.3)
5 $F(0,\pm\sqrt{5})$, $V(0,\pm 1)$ (Fig. 14.12.5)
7 $F(0,\pm\sqrt{34})$, $V(0,\pm 5)$ (Fig. 14.12.7)
9 $F(-3\pm\sqrt{65},2)$, $V(-3\pm 7,2)$ (Fig. 14.12.9)

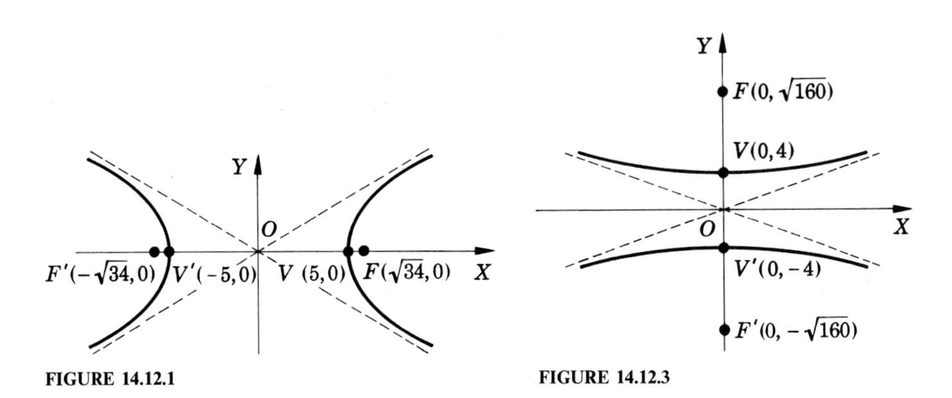

FIGURE 14.12.1

FIGURE 14.12.3

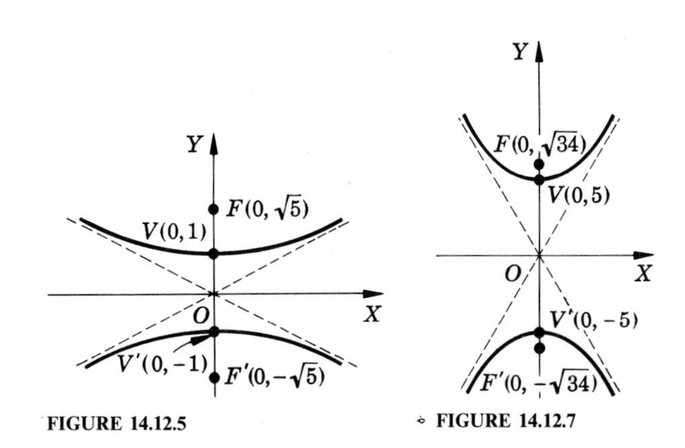

FIGURE 14.12.5

FIGURE 14.12.7

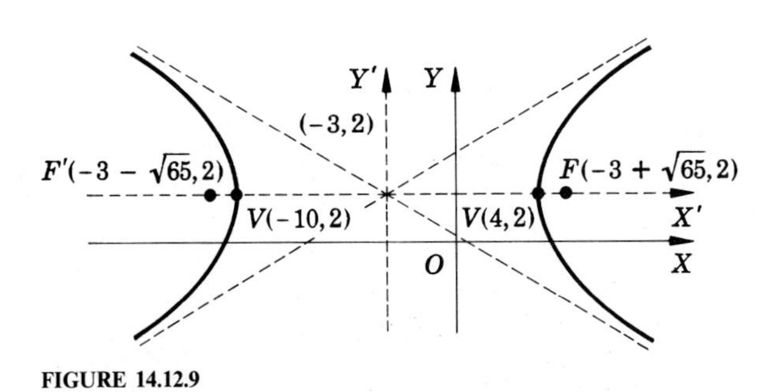

FIGURE 14.12.9

11 $\frac{x^2}{4} - \frac{y^2}{12} = 1, \frac{x}{2} = \pm\frac{y}{\sqrt{12}}$

13 $\frac{x^2}{15} - y^2 = 1, \frac{x}{\sqrt{15}} = \pm y$

15 $-\frac{x^2}{9} + \frac{y^2}{16} = 1, \frac{x}{3} = \pm\frac{y}{4}$

17 $\frac{(x-3)^2}{16} - \frac{(y-2)^2}{9} = 1, \frac{x-3}{4} = \pm\frac{y-2}{3}$

19 $-\frac{(x+3)^2}{64} + \frac{(y-5)^2}{36} = 1, \frac{x+3}{8} = \pm\frac{y-5}{6}$

21 $\frac{x^2}{25} - \frac{y^2}{9} = 1, \frac{x}{5} = \pm\frac{y}{3}$

23 $\frac{x^2}{36} - \frac{y^2}{45} = 1, \frac{x}{6} = \pm\frac{y}{\sqrt{45}}$

25 $\frac{x^2}{36} - \frac{(y+1)^2}{45} = 1, \frac{x}{6} = \pm\frac{y+1}{\sqrt{45}}$

27 $-\frac{(x-3)^2}{96} + \frac{(y+3)^2}{100} = 1, \frac{x-3}{\sqrt{96}} = \pm\frac{y+3}{10}$

29 Hyperbola. Center (h,k), transverse axis $2a$, conjugate axis $2b$, $V(h \pm a, k)$, $F(h + c, k)$, asymptotes $\frac{x-h}{a} = \pm\frac{y-k}{b}$ (Fig. 14.12.29)

31 $(x + y - 2)(x + y - 1) = 0$. Two parallel lines (Fig. 14.12.31)

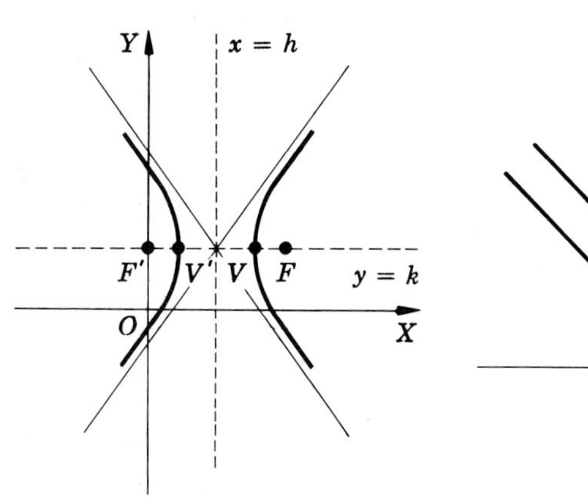

FIGURE 14.12.29

FIGURE 14.12.31

PROBLEMS 14.14

1 $(x + 3)^2 + (y - 2)^2 = 16$
3 $(x^2/25) + (y^2/9) = 1$
7 $x^2(1 - r^2) + y^2(1 - r^2) + 2ax(1 + r^2) + a^2(1 - r^2) = 0$, a circle
9 $y^2 = -2ax + a^2$, a parabola

11 $x^2 + y^2 - |x| - |y| = 0$, a circle
13 $2a^2x^2 + 2y^2 = (1 + a^2)r$, an ellipse
15 $rx^2 - y^2 + ay = 0$, a hyperbola; $rx^2 + y^2 - ay = 0$, an ellipse

PROBLEMS 14.15

1 $3r\cos\theta + r\sin\theta = 4$
3 $r = 4$
5 $r\cos^2\theta = \sin\theta$
7 $\dfrac{r^2\cos^2\theta}{16} + \dfrac{r^2\sin^2\theta}{25} = 1$
9 $\dfrac{r^2\cos^2\theta}{16} - \dfrac{r^2\sin^2\theta}{9} = 1$
11 $r^2 - 2r\cos\theta + \cos^2\theta = 1$
13 $x^2 + y^2 = 4y$
15 $x^2 + y^2 = -4x$
17 $(x^2 + y^2 - x)^2 = 9(x^2 + y^2)$
19 $(x^2 + y^2 + y)^2 = x^2 + y^2$
21 $x^2 = 1 - 2y$
23 $x^2 + y^2 = 4$
25 $\sqrt{3}y - x = 0$
27 $y = x\tan\dfrac{2}{x^2 + y^2}$

PROBLEMS 14.16

1 $\dfrac{(x-1)^2}{9} + \dfrac{y^2}{8} = 1$ (Fig. 14.16.1)
3 $\dfrac{(x-3)^2}{1} - \dfrac{y^2}{8} = 1$ (Fig. 14.16.3)
5 $x^2 = 2(y + \frac{1}{2})$ (Fig. 14.16.5)
7 $2x^2 + 2y^2 - x + y\sqrt{3} = 0$ (Fig. 14.16.7)
9 $(x^2 + y^2)^3 = 4x^2y^2$ (Fig. 14.16.9)
11 $(x^2 + y^2)^5 = 16x^2y^2(x^2 - y^2)^2$ (Fig. 14.16.11)
13 $(x^2 + y^2)^3 = (x^2 - y^2)^2$ (Fig. 14.16.13)
15 $(x^2 + y^2)^5 = (x^4 - 6x^2y^2 + y^4)^2$ (Fig. 14.16.15)
17 $x^2 + y^2 = (x^2 + y^2 - 2y)^2$ (Fig. 14.16.17)
19 $x^2 + y^2 = (x^2 + y^2 - y)^2$ (Fig. 14.16.19)
21 $4(x^2 + y^2) = (x^2 + y^2 - y)^2$ (Fig. 14.16.21)
23 $\dfrac{y}{x} = \tan\dfrac{\pi}{\sqrt{x^2 + y^2}}$ (Fig. 14.16.23)
25 $\dfrac{y}{x} = -\tan\dfrac{\pi}{x^2 + y^2}$ (Fig. 14.16.25)
27 $(x^2 + y^2)^2 = 2xy$ (Fig. 14.16.27)
29 $(\frac{1}{2}, 60°)$, $(\frac{1}{2}, 300°)$ and geometrically, the origin (Fig. 14.16.29)
31 Let the circle be $r = 2a\cos\theta$ and let the fixed point be the origin. Then $r = a\cos\theta$.
33 $r = k\theta^2$

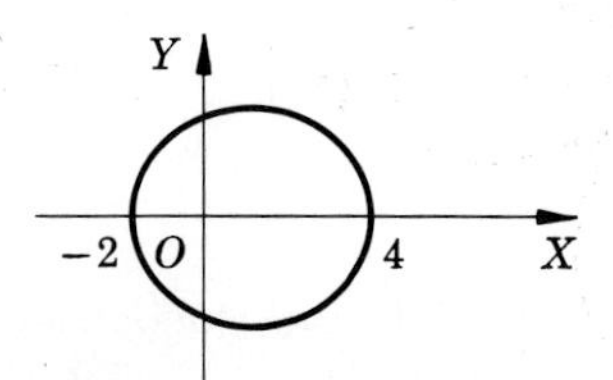

FIGURE 14.16.1

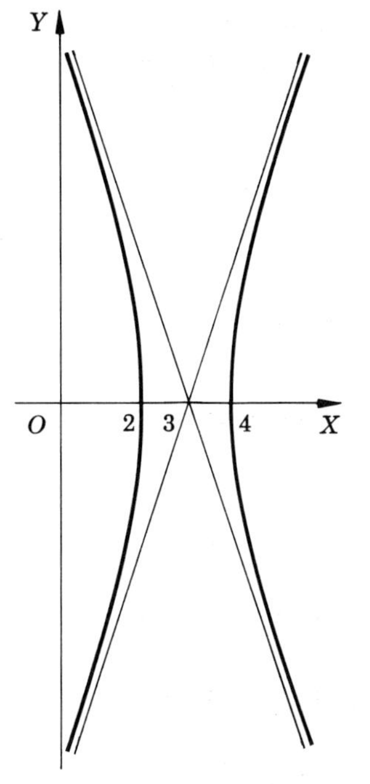

FIGURE 14.16.3

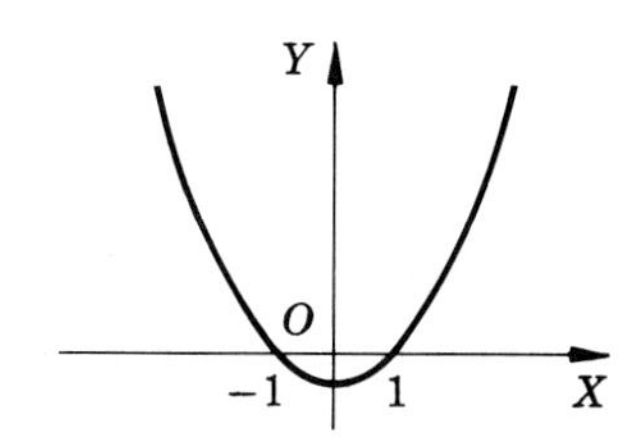

FIGURE 14.16.5

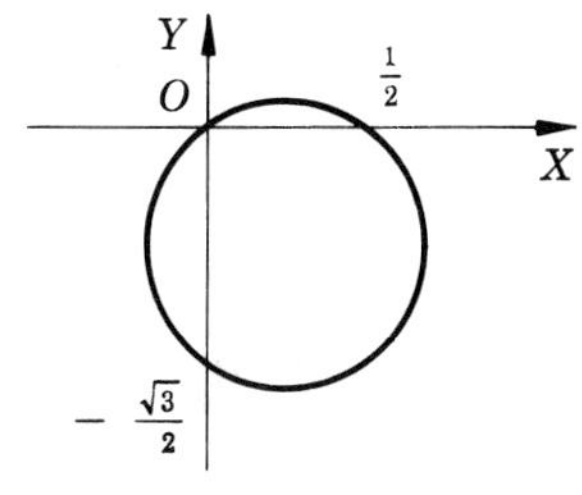

FIGURE 14.16.7

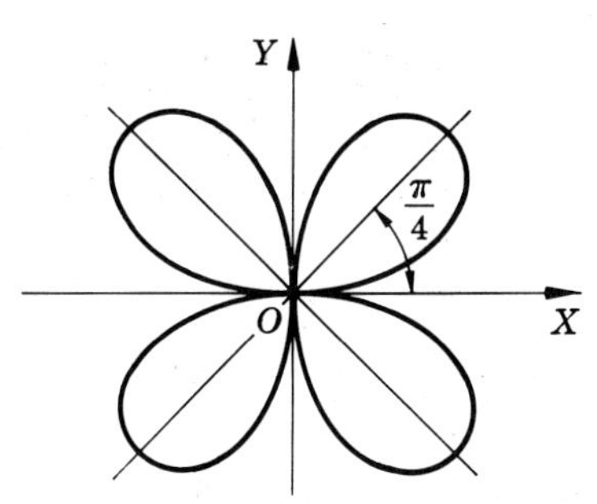

FIGURE 14.16.9

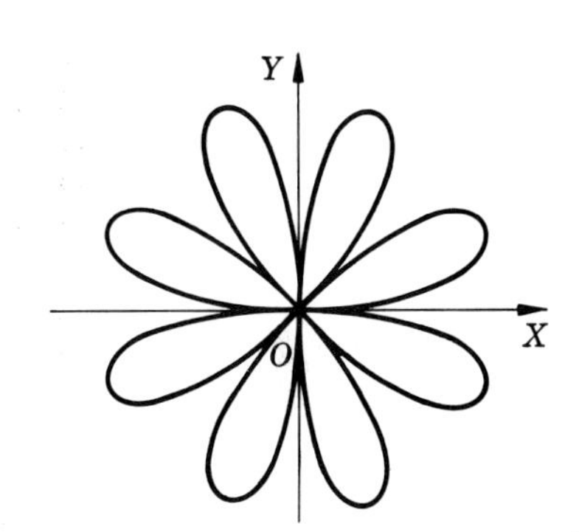

FIGURE 14.16.11

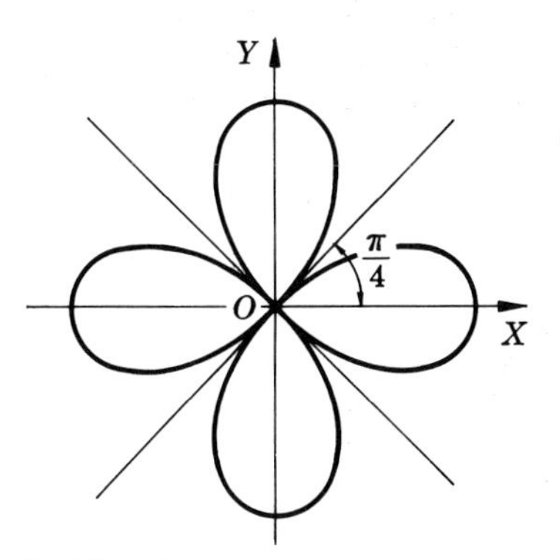

FIGURE 14.16.13

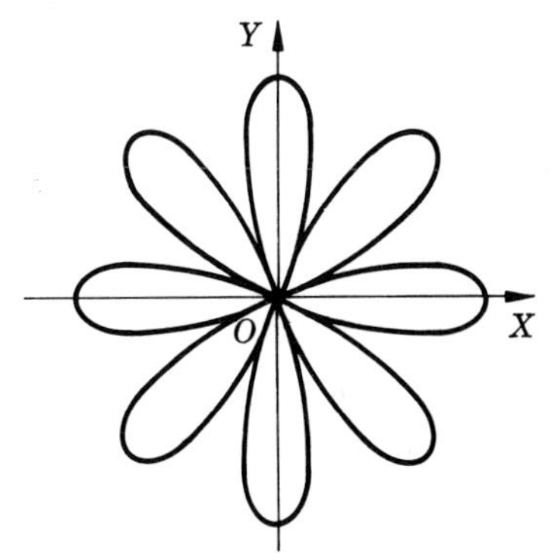

FIGURE 14.16.15

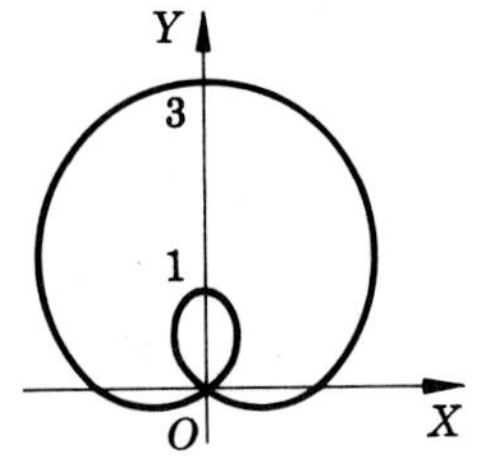

FIGURE 14.16.17

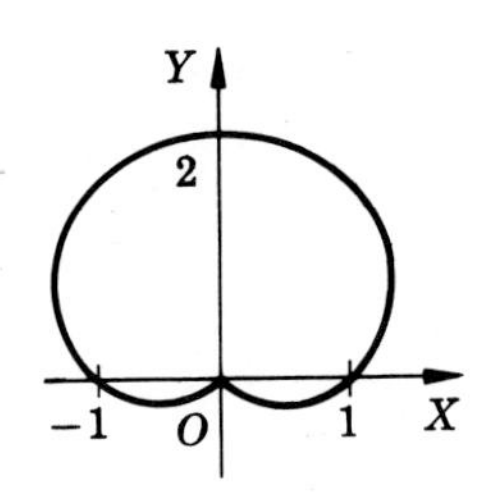

FIGURE 14.16.19

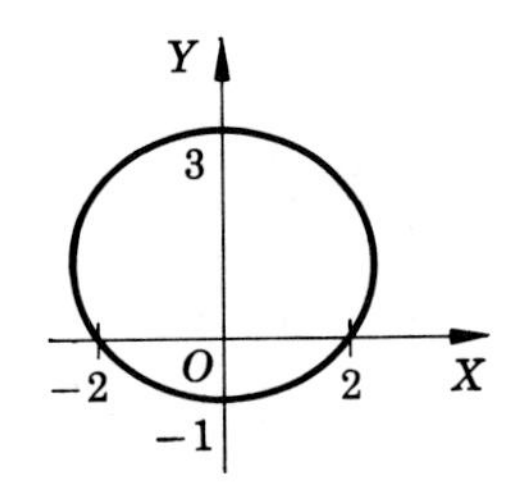

FIGURE 14.16.21

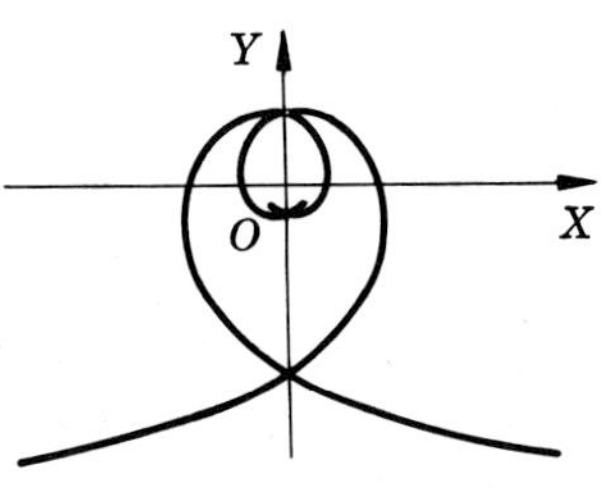

FIGURE 14.16.23

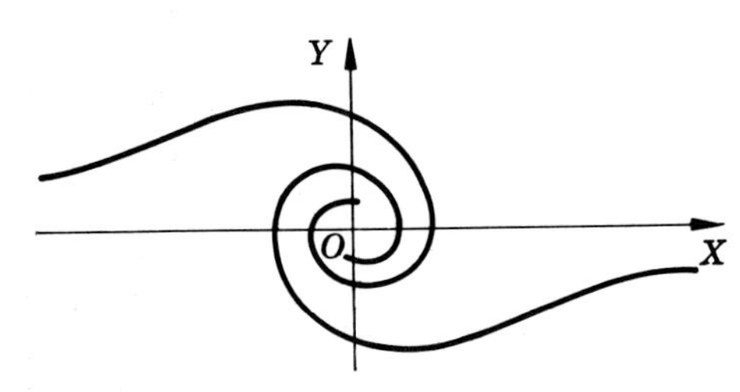

FIGURE 14.16.25

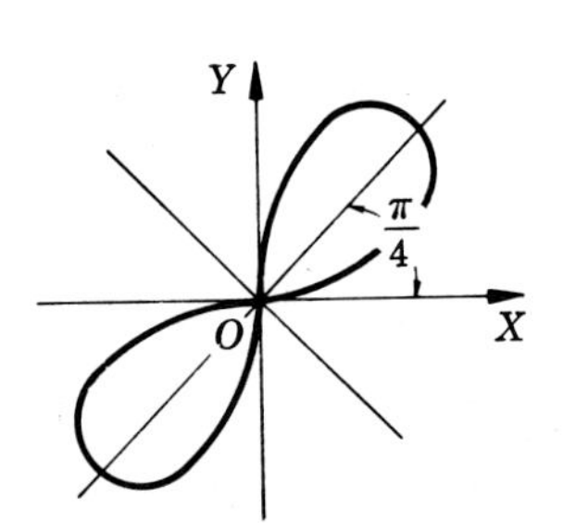

FIGURE 14.16.27

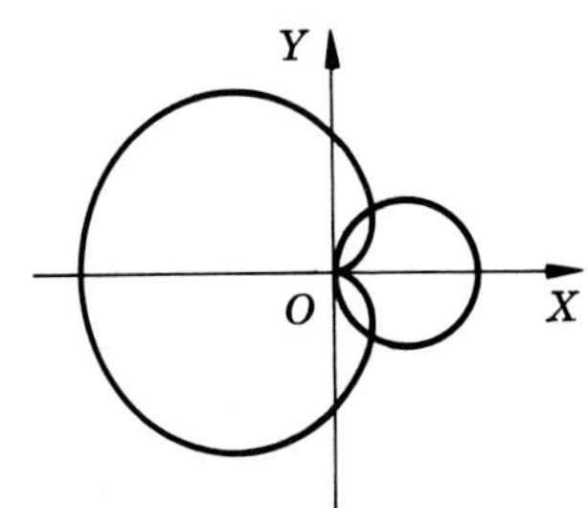

FIGURE 14.16.29

PROBLEMS 14.17

1 $3x - 2y = 29$, straight line
3 $(y + 2)^2 = x$, parabola
5 $x = y^2 + 3y + 1$, parabola
7 $x^2 = -(y - 1)$, parabola
9 $9x = -y^2 + 9$, parabola
11 $x^2 + y^2 = a^2 + b^2$, circle
13 $y = x$, straight line
15 $(x - 2)^2 = y + 1$, parabola
17 $(x^2/4) + (y^2/1) = 1$, ellipse
19 $y^2 = x - 4$, parabola
21 $x^3 + y^3 = xy$
23 $4y^2(1 - y^2) = x^2$
25 $x^2 + y^2 = a^2$, circle
27 $x = 4p/t^2, y = 4p/t$

PROBLEMS 15.3

1 -1 **3** 1 **5** 4 **7** 4
9 -1 **11** 1 **13** 2 **15** 1
17 $\dfrac{\frac{3}{2}\sin 1}{1 + \sin 1}$ **19** 1 **21** 2 **23** -1
25 (Fig. 15.3.25)

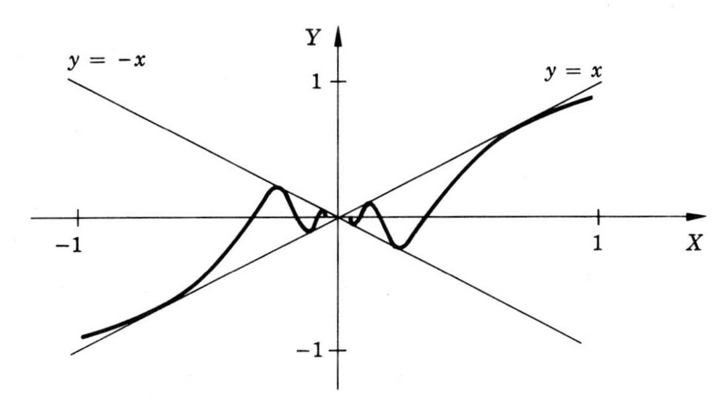

FIGURE 15.3.25

PROBLEMS 15.4

1 $\frac{2}{3}$ 3 $\frac{1}{3}$ 5 $k/3$ 7 $\frac{4}{3}$

9 Sum of zigzag lengths is always 2. Hence the limit is 2. But each individual zigzag sum is $\sqrt{2}$ times the length of its diagonal. Hence the limit is not the length of the diagonal.

PROBLEMS 15.5

1 $(4 - \sqrt{2})/3$ 3 $\frac{5}{12}$ 5 $\frac{1}{4}$ 7 $\frac{1}{4}$
9 $\frac{8}{5}$ 11 $\frac{1}{2}$ 13 $(3^7 + 2^7)/7$ 15 1
17 $b^{3n} - a^{3n}$

PROBLEMS 15.6

1 $\frac{32}{3}$ 3 $\frac{27}{6}$ 5 2 7 $\frac{1}{12}$
9 $\frac{5}{12}$ 11 $64\frac{4}{5}$ 13 $\frac{16}{15}$ 15 $\frac{41}{6}$
17 0 19 −4 21 0
23 Symmetry with respect to the Y-axis

PROBLEMS 15.8

1 12 3 $\frac{7}{6}$ 5 $-\frac{2}{15}$
7 $[27(3)^{1/3} - 18(2)^{1/3}]/4$ 9 Not possible 11 $(a/3) + (b/2) + c$
13 $(7a^2/3) + 3ab + b^2$ 15 $24p$ 17 $xy^2/2$
19 $x^4z/5$ 21 $x^4y^2/5$ 23 $\frac{4}{5}qa^{5/4}$
25 $(3b^3/8) - b^2$ 27 $(-3b^2 - 12b + 36)/2$ 29 $\frac{1}{6}$

PROBLEMS 15.9

1 2,904 ft-lb 3 16 in.-lb 5 125 (62.4 lb)
7 $1{,}575w$ 9 $\pi p^3/32$ 11 $4\pi b^2a^3/3$
13 $(\pi x_0^3/30)(6A^2x_0^2 + 15ABx_0 + 10B^2)$
15 $(\pi w/4)(50^4)$
17 $\int_0^{\pi/4} \pi \cos^2\theta\, d\theta = \frac{\pi^2}{8} + \frac{\pi}{4} \approx \pi(0.6427)$

PROBLEMS 16.2

1 $y = 3x - 5$ 3 $y = 8x - 7$ 5 $y = 6x - 4$
7 $y = 4x + 5$ 9 $y = 12x - 17$ 11 $y = 8x - 3$
13 $y = mx + b$ 15 $y = 4x + 2$

PROBLEMS 16.3

1 (*a*) 8, (*b*) 4, (*c*) 56 3 (*a*) $-15\frac{1}{2}$, (*b*) $-\frac{191}{2}$, (*c*) $-191\frac{1}{2}$
5 (*a*) 20, (*b*) 16, (*c*) 28 7 (*a*) 2, (*b*) 14, (*c*) 6
9 (*a*) 1, (*b*) 1 11 (*a*) 56, (*b*) 24
13 (*a*) 0, (*b*) 2 15 (*a*) 20, (*b*) 20
17 (*a*) 11, (*b*) 42 19 (*a*) $2t_1$, (*b*) 2

PROBLEMS 16.4

1 $24x^5 + 1/(2\sqrt{x}) + 2$ 3 $(9/x^4) - x^2$ 5 $2x + 3$
7 $2x + 2$ 9 $-3(\frac{1}{2} - x)^2$ 11 $3x^2 + 2$
13 $y = 0, m = 0$ 15 $y = x + 3, m = 1$ 17 $y = bx + c, m = b$
19 $y = x - 3, m = 1$ 21 $y = 7x - \frac{19}{3}$
23 (*a*) $y = 0$; (*b*) $x = 0$ 25 $Ax_0^2 + Bx_0 + C$
27 $108t_0^5$ 29 14 31 $2x_0 + 1$
33 0 35 $-nm$ 37 $-3x^{-4}$
39 $D_x(x + 1)(x - 1) = D_x(x^2 - 1) = 2x \neq D_x(x + 1)D_x(x - 1)$
41 $D_x\left(\frac{x + 1}{x + 1}\right) = D_x 1 = 0 \qquad \frac{D_x(x + 1)}{D_x(x + 1)} = \frac{1}{1} = 1$

PROBLEMS 16.6

1 $v = 3t_0^2 + \frac{1}{2}t_0^{-1/2} - 1$
$a = 6t_0 - \frac{1}{4}t_0^{-3/2}$
3 $v = 3t_0^2 + 2t_0 + 2$
$a = 6t_0 + 2$
5 $v = 4t_0$
$a = 4$
7 $v = \frac{1}{2}(t_0 - 3)^{-1/2} + 4t_0$
$a = -\frac{1}{4}(t_0 - 3)^{-3/2} + 4$
9 $\frac{1}{2}(t^{-1/2}) + 2t$
$-\frac{1}{4}(t^{-3/2}) + 2$
$\frac{3}{8}(t^{-5/2})$
11 $6t(t^2 + 1)^2$
$30t^4 + 36t^2 + 6$
$120t^3 + 72t$
13 1, 0, 0
15 $2(t + t^2)(1 + 2t)$
$12t^2 + 12t + 2$
$24t + 12$
17 $\frac{1}{2}(2t + 3)^{-3/4}$
$-\frac{3}{4}(2t + 3)^{-7/4}$
$\frac{21}{8}(2t + 3)^{-11/4}$
19 $3(x^2 + 5x + 25)^2(2x + 5)$
21 $6(6x^2 - 3x)^2(12x - 3)$
23 $\frac{2}{5}(x^2 + x^3)^{-3/5}(2x + 3x^2)$
25 $-(x/2)(b - x^2)^{-3/4}$
27 26, 0
29 $-34, -18$

PROBLEMS 16.7

1 Min. $(-2,0)$
3 Max. $(\frac{4}{5},\frac{6}{5})$
5 Min. $(1,-1)$
7 Infl. $(0,0)$
9 Min. $(0,0)$;
max. $(\frac{4}{9}, \frac{32}{243})$;
infl. $(\frac{2}{9}, \frac{16}{243})$
11 Max. $(0,0)$;
min. $(\pm 1, -1)$;
infl. $(\pm\sqrt{\frac{1}{3}}, -\frac{5}{9})$
13 Infl. $(0,0)$
15 Min. $(0, \sqrt{2})$
17 No max. or min.;
infl. $(-3,0)$
19 Max. $(0,1)$;
min. $(\pm 4,0)$

21 Min. $(2 + \sqrt{\frac{1}{3}}, -\frac{2}{3}\sqrt{\frac{1}{3}})$; max. $(2 - \sqrt{\frac{1}{3}}, \frac{2}{3}\sqrt{\frac{1}{3}})$; infl. (2,0)
23 Infl. (1,0)
25 Infl. $(2,-30)$
27 Infl. (0,1)
29 Infl. $(0,0)(1,-2)$
31 No min.
33 No min.
35 4
37 1
41 $\frac{5}{2}$ weeks
43 $\sqrt{3}$ mile from P; same for 50 miles; one mile
45 Abs. max. 0
Abs. min. $-4(\frac{10}{6})^3$.

PROBLEMS 16.8

1 (*a*) 36 in.^{3}/sec; (*b*) 72 in.^{2}/sec
3 $(25/\sqrt{2})$ ft/sec
5 $250{,}025/\sqrt{250{,}026}$
7 $(25/108\pi)$ in./sec
9 -9 in.^{2}/sec

PROBLEMS 16.9

1 $6x^2 + 6x - 4$
3 $40x^3 - 6x^2 + 18x$
5 $-12x^{-4} - 4x^{-3}$
7 $18x^2 + 8x^{-5}$
9 $-3ax^{-4} - 2bx^{-3}$
11 $(n + 4)Ax^{n+3} + (n + 3)Bx^{n+2} + C$
13 $x^3 + x^2 - 4x + C$
15 $\dfrac{x^5}{5} + x^3 - 2x + C$
17 $x^{-1} + (x^3/3) + C$
19 $\dfrac{-x^{-4}}{4} + \dfrac{x^6}{6} + x^4 - x^2 + C$
21 $\dfrac{ax^5}{5} + \dfrac{bx^3}{3} + \dfrac{cx^{-2}}{2} + D$
23 $-\dfrac{ax^{-5}}{5} - \dfrac{bx^{-6}}{6} + C$
25 $x^{p+1} + Bx^{q+1} + C$
27 2
29 0
31 $9a + (9b/2) + 3c$
33 $a - \frac{1}{2}$
35 $(7a^3/3) - 3a^2$
37 320
39 $\frac{1075}{648}$
41 $-\frac{21}{32}$
43 $\frac{1}{6}$
45 1
47 2
49 $\frac{1}{2}$
51 (*a*) $6t_0 + 1$; (*b*) 1
53 (*a*) $3t_0$; (*b*) 0; (*c*) $\frac{5}{2}$

PROBLEMS 16.10

1 (*a*) $V = 160 - 16t, a = -16$; (*b*) 160; (*c*) 800; (*d*) 20
3 (*a*) $V = 64 - 2t, a = -2$; (*b*) 64; (*c*) 32^2; (*d*) 64
5 (*a*) $V = 160 - 16t, a = -16$; (*b*) 160; (*c*) 1,200; (*d*) $10 + 5\sqrt{6}$
7 (*a*) $V = 16 - 8t, a = -8$; (*b*) 16; (*c*) 48; (*d*) $2 + 2\sqrt{3}$
9 $y = 200 + 20t - 16t^2$
11 $y = (5 \times 5{,}280) + \dfrac{400 \times 5{,}280t}{60 \times 60} - 16t^2$

PROBLEMS 17.3

13 $\cosh(-x) = \cosh x$
15 $\operatorname{sech}(-x) = \operatorname{sech} x$
17 $\coth(-x) = -\coth x$
19 $\coth^{-1} x = \frac{1}{2}\ln\dfrac{x+1}{x-1}, |x| > 1$
21 $\operatorname{Cosh}^{-1} x = \ln(x + \sqrt{x^2 - 1}), x \geq 1$

INDEX

Absolute value, 48
  function, 244
Acceleration, 497
Addition:
  of fractions, 105
  of functions, 246
  of matrices, 171
  of vectors, 167
Addition theorems, 375
Additive identity, 45
Additive inverse, 45
Affine transformation, 75, 160
Algebra of functions, 246
AM radio, 372
Amplitude, 370
Analytic geometry, 402
Angle:
  directed, 322
  generalized, 323
  interior of, 320
  measure of, 321
  signed, 322
  special, 327
  standard position of, 320
  between two directed lines, 417
  vertex of, 320
Antidifferentiation, 523
Arc length, 356
Area, 457
  of a circle, 458
  under graph of a polynomial, 477
  of a triangle, 187
  under $y = x^2$, 469
  under $y = x^n$, 473
  under $y = f(x)$, 480
Associative property of a binary operation, 41
Asymptotes, 254
Axiom, 2
Axis:
  conjugate, 436
  major, 431
  minor, 431
  transverse, 435

Binary operation, 39
Binomial coefficients, 89
Binomial theorem, 88

Cancellation, 109
Cartesian product, 4
Chain rule, 509
Circle:
  area of, 458
  circumference of, 356
  equation of, 425
Cofactor, 183
Commutative property of a binary operation, 40
  generalized, 42
Complement, 12
Completion of the square, 130
Complex numbers:
  conjugate, 76
  graphical representation of, 81
  imaginary part of, 76
  polar form, 395
  real part, 76
  roots of, 397
Composite of functions, 248
Compound interest, 307
Conditionals, 16
  alternative expressions for, 23
  derived, 20
  true, 16
Conic sections, 424
Constant, 9
Constants, values of, 571
Continuity, 276
Contrapositive, 20
Converse, 20
Convex polygonal set, 223
Coordinates, 70
  on a line, 70
  in a plane, 72
  transformation of, 74, 160
Correspondence, one-to-one, 6
Cosecant, 330
Cosine:
  of an angle, 325
  of a real number, 360
Cosines, Law of, 351
Cotangent, 330
Counterexample, 34
Cramer's rule, 189
Critical point, 514
Cross product, 191
Cube roots, tables of, 572
Cubes, tables of, 572
Curve fitting, 313
Cycloid, 454

Decimal expansions:
  infinite, 38
  nonrepeating, 66
  repeating, 65
De Moivre's theorem, 397
Derivative:
  chain rule for, 509
  of a product, 508
  of a quotient, 508
  second, 508
  of a sum, 504
  of x^n, 506

Determinant:
of a general matrix, 185
theorems about, 184–186
of a 3×3 matrix, 184
of a 2×2 matrix, 183
Diet problem, 222
Differentiation, 492
Dilitation, 75, 162
Direct proof, 27
Direction cosines, 415
Directrix, 428
Disproof, methods of, 34
Distance:
between parallel lines, 418
between point and line, 419
between two points, 318
in polar coordinates, 327
in rectangular coordinates, 318
Distributive property, 47
Divisibility of natural numbers, 55
Division, 46
algorithm, 94
Domain, 237, 242, 251
Double-angle formulas, 381

e, 300
e^x, 300
tables of, 541
Eccentricity, 440
Ellipse:
definition, 430, 440
eccentricity, 440
focus of, 430
major axis of, 431
minor axis of, 431
parametric equations of, 453
in polar coordinates, 445
vertices of, 431
Equality of symbols, 39
Equations, solution of, 128, 134
containing fractions, 139
containing radicals, 140
linear, 130
quadratic, 130
simultaneous linear, 143, 149
simultaneous linear and quadratic, 152
trigonometric, 387
Euler's formula, 538
exp x, 300
Explicit algebraic functions, 274
Exponential functions, 297
Exponents, 113
fractional, 119
negative, 116
zero, 116

Factorial, 89
Factoring, 95
difference of two squares, 98
removal of common factor, 96
trinomials, 96
Factorization, unique of natural numbers, 56
Falling bodies, 528
Feasible solution, 225
Field, 48
ordered, 49
FM radio, 373
Focus, 428, 430, 434
Football problem, 216
Fractions, algebraic, 101
addition of, 105
compound, 111
division of, 108
equality of, 102
multiplication of, 108
simplification of, 101
Frequency, 370
Functions, 241
algebra of, 246
composite of two, 248
constant, 243
derived from equations, 267
domain of, 242
restriction of, 263
explicit algebraic, 274
exponential, 297
graphs of, 250
hyperbolic, 301, 531
identity, 266
image of, 242
implicit, 269
into, 242
inverse, 260
logarithmic, 302
monotone, 265
onto, 242
polynomial, 271, 276
power, 313
range of, 242
rational, 272
real-valued, 242
trigonometric, 313
Fundamental Theorem of Algebra, 82, 282
Fundamental Theorem of Calculus, 522, 524

Generalized commutative property, 42
Graph:
of linear equation, 135
of pair of linear equations, 146
of quadratic equation, 136
of relations, 250
Greek alphabet, 571

Half-angle formulas, 380
Homogeneous linear equations, systems of, 190
Hyperbola, 434, 440
conjugate axis, 436
eccentricity, 440
in polar coordinates, 445
transverse axis, 435
vertices, 435
Hyperbolic functions, 301, 531
Hyperbolic trigonometry, 534

Identity:
of sets, 4
trigonometric, 383
Identity elements, 45
Identity function, 266
Identity matrix, 175
If and only if, 24
Image, 237, 242, 251
Imaginary number, pure, 78
Inclination, 403
Indirect proof, 29
Induction, mathematical, 56
axiom of, 59
principle of, 59
Inequalities, 69, 195
addition of, 196
involving absolute values, 200
linear, 199
graph of, 210
simultaneous, 213
multiplication of: by a negative number, 198
by a positive number, 196
quadratic, 205
graph of, 219

Inequalities:
  sense of, 198
  transitive property of, 196
Inflection point, 516
Inner product, 168
  interpretation of, 352
Integral, 484
Integration, 457
  applications, 486
  definition, 482
  theorems about, 483
Intercept, 251
  form of equation of a line, 408
Intersection of sets, 14
Into function, 242
Inverse, 20
  elements, 45
  function, 260
    trigonometric, 391
  of a matrix, 180
Irrational numbers, 66

Law of Cosines, 351
Law of Sines, 347
Law of Tangents, 355
Law of Trichotomy, 48
Least common denominator, 105
Length of a vector, 169
Like terms, 85
Limits, 461
  definition of, 462
  theorems about, 463
Line, 403
  directed, 414
  direction cosines of, 415
  equation of: general, 408
    normal form, 422
    point-slope form, 406
    slope-intercept form, 408
    two-intercept form, 408
    two-point form, 406
  inclination of, 403
  parallel, 410, 418
  perpendicular, 411
  slope of, 405
Linear programming, 222
Linear transformation, 163
Locus, 402, 427, 428, 434, 439
Log-log paper, 312
Logarithmic function, 302
  properties of, 302
Logarithmic scale, 310
Logarithms:
  common, 302
    tables of, 542
  natural, 302
    tables of, 544
Lumber problem, 217

Mapping, 233
Mathematical model, 35
Matrices, 165, 170
  addition of, 171
  equality of, 171
  identity, 175
  inverse of, 180
  nonsingular, 182
  product by a scalar, 171
  product of two, 172, 174
  singular, 182
  zero, 171
Maxima and minima, 512
  relative, 514
  tests for, 515
Method of equivalence, 129
Midpoint of a segment, 412
Minor, 183
Modulation amplitude, 373
Modulation frequency, 373
Monotone function, 265
Multiple-angle formulas, 380
Multiplicative identity, 45
Multiplicative inverse, 46
Multivalued, 238

Natural numbers, 54
Necessary condition, 23
Negation, 10
Negative real numbers, 48
Nonsingular matrix, 182
Notations for functions, 242
Number-of-roots theorem, 283
Numbers:
  classification of, 83
  complex, 75, 395
  irrational, 38, 66
  natural, 54
  prime, 55
  rational, 63
    decimal expansions of, 65
  real, 38

Odd roots, 123
One-to-one correspondence, 6
One-to-one functions, 235, 238, 242
Only if, 23
Onto functions, 242
Open sentences, 8
  equivalent, 10
  negation of, 10
Operation, binary, 39
Opposite, 46
Optimal solution, 225
Ordered field, 49
Ordered pair, 40
Outer product, 191
  interpretation of, 353

Parabola, 137, 427
  directrix, 428
  eccentricity, 445
  focus, 428
  in polar coordinates, 445
Parameter, 451
Parametric equations, 450
Pascal's triangle, 88
Period, 370
Periodic function, 366
Phase, 373
Pi (π), 358
Polar coordinates, 324, 441
Polynomial equations:
  complex roots of, 290
  number-of-roots theorem, 283
  rational roots of, 291
  real roots of, 293
  roots of, 289
Polynomial functions, 271, 276
  complex, 281
  rational, 281
  real, 281
Polynomials, 84
  addition of, 85
  division of, 93
  multiplication of, 86
Positive real numbers, 48
Power function, 313
Prime number, 55
Primitive, 522
Product of three real numbers, 42
Product formulas, 381
Proof, methods of, 27, 29, 32

Quadrants, 71
Quadratic equations, 130
Quadratic formula, 131

Radian, 361
Radicals, 113
Radioactive decay, 309
Radius vector, 324
Range, 242
Rates, related, 520
Rational function, 272
Rational numbers, 63
  decimal expansions of, 65
Rational root theorem, 291
Rationalization of denominators, 125
Real number system, properties of, 44
Real numbers, 38
  geometric interpretation of, 69
  use in the plane, 70
Reduction formulas, 332
Relation, 323
  domain of, 237
  image of, 237
  multivalued, 238
  one-to-one, 238
  single-valued, 238
Remainder theorem, 282
Right triangles, solution of, 338
Roots of real numbers:
  even, 121
  odd, 123
Rotation, 163

Scalar, 168
Secant, 330
Segment, length of, 70, 72
Semilog paper, 312
Sets, 3
  complement of, 12
  element of, 4
  empty, 4
  intersection of, 14
  subset, 5
    proper, 5
  truth, 9
  union of, 13
  universal, 9
Simultaneous equations, linear, 143, 149
    homogeneous, 190
    and quadratic, 152
    solution by Cramer's rule, 189
Simultaneous equations, linear:
    solution by matrices, 188
Sine:
  of an angle, 325
  of a real number, 360
Sine wave:
  amplitude of, 370
  frequency of, 370
  period of, 370
  phase of, 373
  wave length of, 371
Sines, Law of, 347
Single-valued, 238
Singular matrix, 182
Slide rule, 310
Slope, 405
Solution set, 128, 134
Square root:
  of negative numbers, 122
  tables of, 572
  of 2, 67
Squares, tables of, 572
Stationary point, 514
Subset, 5
  proper, 5
Subtraction, 46
Sufficient condition, 23
Sum of three real numbers, 42
Symmetry, 252
Synthetic division, 285

Tangent:
  of an angle, 330
  to a curve, 492
Tensor product, 177
Theorem, 3
Transformation of coordinates, 74, 160
  affine, 75, 160
  dilitation, 75, 162
  linear, 163
  rotation, 163
  translation, 75, 162
Translation, 75, 162
Triangles:
  area of, 187
  solution of, 347, 351
    ambiguous case, 348
Triangles:
  solution of: right, 338
Trichotomy, Law of, 195
Trigonometric equations, 387
Trigonometric functions:
  of angles, 317
    tables of, 549
  graphs of, 364
  half-angle formulas for, 380
  inverse, 391
  multiple-angle formulas for, 380
  product formulas, 381
  of real numbers, 359
    tables of, 547
  variation of, 364
Trigonometric identities, 331, 383
Trigonometric tables, use of, 335
Truth set, 9

Undefined words, 1
Union of sets, 13
Universal set, 9

Variable, 9
  dependent, 244
  independent, 244
Vector moment, 192
Vector product, 191
Vectors, 165
  addition of, 167
  application of statics, 341
  components of, 166
  length of, 169
  perpendicular, 170
  products of: inner, 168
    outer (cross, vector), 191
    tensor, 177
Velocity, 469
Volume of solid of revolution, 490

Wave length, 371
Word problems, 157
Work, 488

Zeno's paradoxes, 473
Zeros of polynomial, 276

THE SYMBOLS USED IN THIS BOOK AND THE PAGES ON WHICH THEY ARE DEFINED

page	*symbol*	*definition*
4	$\{\ \}$	notation for a set
4	$\emptyset$	the empty set
4	$x \in A$	x is an element of the set A
4	$\{3\}$	the set of which 3 is the only element
4	$\{2, 4, 7, 9\}$	the set whose elements are 2, 4, 7, and 9
4	$R = \{x \mid x \text{ is a real number}\}$	the set of real numbers
4	$R \times R$	the set of ordered pairs of real numbers
5	$\subseteq$	inclusion for sets
5	$\subset$	proper inclusion for sets
9	U	universal set
9	p_x, q_x	open sentences with variable x
10	P, Q	truth sets of p_x and q_x, respectively
10	$p_x \leftrightarrow q_x$	equivalence of p_x and q_x
10	$\sim p_x$	negation of p_x
12	A'	complement of the set A
13	$\cup$	union of sets
14	$\cap$	intersection of sets
16	$p_x \rightarrow q_x$	the conditional: if p_x, then q_x
16	$\forall_x$	for all x
39	$(a,b) \rightarrow (a * b)$	binary operation
45	a^{-1}	inverse of a
48	$a\vert$	absolute value
65	$1.\overline{142857}$	repeating decimal
69	$<$	less than
69	$>$	greater than
69	$\leq$	less than or equal to
69	$\geq$	greater than or equal to
72	(x,y)	coordinates of a point in a plane
75	$a + bi$	complex number
76	i	unit imaginary number, $i^2 = -1$
89	$n!$	n factorial
89	$0!$	zero factorial
89	$\binom{n}{r}$	binomial coefficient
119	$\sqrt[n]{a} = a^{1/n}$	positive nth root of a